# Advanced Mobile Robotics

# Advanced Mobile Robotics

## Volume 1

Special Issue Editor

**DaeEun Kim**

MDPI • Basel • Beijing • Wuhan • Barcelona • Belgrade

*Special Issue Editor*
DaeEun Kim
Yonsei University
Korea

*Editorial Office*
MDPI
St. Alban-Anlage 66
4052 Basel, Switzerland

This is a reprint of articles from the Special Issue published online in the open access journal *Applied Sciences* (ISSN 2076-3417) from 2018 to 2019 (available at: https://www.mdpi.com/journal/applsci/special_issues/Advanced_Mobile_Robotics).

For citation purposes, cite each article independently as indicated on the article page online and as indicated below:

LastName, A.A.; LastName, B.B.; LastName, C.C. Article Title. *Journal Name* **Year**, *Article Number*, Page Range.

**Volume 1**
**ISBN 978-3-03921-916-2 (Pbk)**
**ISBN 978-3-03921-917-9 (PDF)**

**Volume 1-3**
**ISBN 978-3-03921-942-1 (Pbk)**
**ISBN 978-3-03921-943-8 (PDF)**

Cover image illustrated by Danho Kim.

# Contents

# About the Special Issue Editor

**DaeEun Kim** received his BE and MS from the Department of Computer Science and Engineering of Seoul National University, South Korea, and the University of Michigan at Ann Arbor, USA, respectively. He received his Ph.D. from the University of Edinburgh, UK, in 2002. From 2002 to 2006 he was a research scientist at the Max Planck Institute for Human Cognitive and Brain Sciences in Munich, Germany. Currently he is a professor in Yonsei University in Seoul, Korea. His research interests are in the areas of biorobotics, autonomous robots, artificial intelligence, artificial life, neural networks and neuroethology.

*applied sciences*

MDPI

*Editorial*

# Special Feature on Advanced Mobile Robotics

**DaeEun Kim**

School of Electrical and Electronic Engineering, Yonsei University, Shinchon, Seoul 03722, Korea; daeeun@yonsei.ac.kr

Received: 29 October 2019; Accepted: 31 October 2019; Published: 4 November 2019

## 1. Introduction

Mobile robots and their applications are involved with many research fields including electrical engineering, mechanical engineering, computer science, artificial intelligence and cognitive science. Mobile robots are widely used for transportation, surveillance, inspection, interaction with human, medical system and entertainment. This Special Issue handles recent development of mobile robots and their research, and it will help find or enhance the principle of robotics and practical applications in real world.

The Special Issue is intended to be a collection of multidisciplinary work in the field of mobile robotics. Various approaches and integrative contributions are introduced through this Special Issue. Motion control of mobile robots, aerial robots/vehicles, robot navigation, localization and mapping, robot vision and 3D sensing, networked robots, swarm robotics, biologically-inspired robotics, learning and adaptation in robotics, human-robot interaction and control systems for industrial robots are covered.

## 2. Advanced Mobile Robotics

This Special Issue includes a variety of research fields related to mobile robotics. Initially, multi-agent robots or multi-robots are introduced. It covers cooperation of multi-agent robots or formation control. Trajectory planning methods and applications are listed. Robot navigations have been studied as classical robot application. Autonomous navigation examples are demonstrated. Then services robots are introduced as human-robot interaction. Furthermore, unmanned aerial vehicles (UAVs) or autonomous underwater vehicles (AUVs) are shown for autonomous navigation or map building. Path planning problem has been a well-established field but new intelligent approaches are introduced. Quadruped robots and biped robots are presented. Also, robot manipulators are handled with their accuracy control. Control methods with snake robots or exoskeleton are shown. Further, various experiments and tests of wheeled robots are demonstrated. Learning and adaptation is a key issue in robotics. Many researchers have developed new algorithms or applications based on learning and adaptation. Finally, a variety of applications with a new style of actuators are introduced.

A transportation problem with multi-robots is a demanding work, where a team of robots is supposed to collect a set of samples scattered in an environment and transport them to a storage facility. Burlacu at al. [1] showed that the task can be transformed to an optimal assignment problem with mathematical modeling and suboptimal relaxations are available for the solution. Multiple mobile robots often experience path planning problems and a real-time navigation algorithm with obstacle avoidance has been suggested by Huang et al. [2] based on a fish swarm algorithm to guide two phases of global and local path planning. In some environments, multiple mobile robots together manipulate an object in a cooperative way. Fan et al. [3] considered possible modes and mode transitions for cooperative planning. Mobile robots also handle autonomous transportation of pallets in smart factory logistics. Li et al. [4] used an intelligent mobile robot platform consisting of master–slave parallel robots.

Multi-agent robots have been applied to formation control by Kowalczyk [5] and its control is integrated with distributed goal assignment. Biologically-inspired learning and adaptation can even be applied to control of networked mobile robots. Self-evolving formation control with mobile robots was demonstrated by Xu et al. [6]. Cooperation among multi-agent robots is needed in search and rescue tasks. The navigation of the robot swarm and the consensus of the robots have been tested for a victim detection task by Cardona and Calderon [7]. Another application of multi-robot systems can be found in the welding process. Trajectory planning for the position/force cooperative control in the multi-robot manipulators has been addressed by Gan et al. [8].

Trajectory planning has been a challenging issue in industrial robots. Robotic operations in logistics are involved with pick and place operations of a manipulator. Planning a trajectory has been handled with deep reinforcement learning by Iriondo et al. [9]. Chen and Li [10] suggested geodesic trajectory planning with constraints on the end-effector and joint which is involved with the trajectory properties of the end-effector. Optical polishing also needs trajectory planning, and the accuracy of trajectory and runtime of trajectory can influence the polishing quality. Thus, Zhao and Guo [11] proposed that applying a B-spline curve method improves the performance.

In robot navigation, loop closure detection is important to reduce the cumulative errors of pose estimation for a mobile robot. A new loop closure detection algorithm with multi-scale deep feature fusion, that is, CNN (convolutional neural network) has been introduced by Chen at al. [12]. In low-texture environments, data association and closed-loop detection are challenging problems in the SLAM (simultaneous localization and mapping) method. Wang et al. [13] showed that the data association process and the back-end optimization stage with sensors, the IMU (Inertial Measurement Unit) sensor and a 2D LiDAR (Light Detection and Ranging), can be improved to enhance navigation performance. The SLAM algorithm is applied to a non-flat road with a 3D LiDAR sensor by Wang et al. [14]. The data association problem for map consistency is solved with iterative matching algorithm, reducing the computation cost. Alonso-Ramirez et al. [15] showed that mobile robots can detect and recognize household furniture, using the analysis and integration of geometric features over 3D points with a color-depth camera. A spatial model of the environment is demanding work for the navigation map. Villaseñor et al. [16] introduced a new object-mapping algorithm, approximating point clouds with multiple ellipsoids.

Mobile service robot needs to handle the human–machine interactive scene, and Wang et al. [17] proposed a topological map construction pipeline with regional dynamic growth algorithm; the map has a representation of topological information as well as occupied information. Autonomous service robots in an indoor complex environment need to find paths with obstacles and also interact with patients. Chien et al. [18] showed an adaptive neuro-fuzzy system with 3D depth camera, infrared sensors and sonar sensors for path planning of a service robot.The service robot also used facial features for personal recognition.

Recently, control of UAVs (unmanned aerial vehicles) has been a challenging problem and an example of a task is that a robot needs to avoid collision with an enemy UAV in its flying path to the goal. Cheng et al. [19] formulated this as a Markov decision process and applied temporal-difference reinforcement learning to the robot control. The learned policy can achieve a good performance to reach the goal without colliding with the enemy. For a UAV controller, fast and iterative real-time auto-tuning of parameters has been tested for altitude control by Giernacki [20]. It considered environmental disturbances as well as change of environmental conditions. UAV (Unmanned Aerial Vehicle) application can be available for wheat crops. Wang et al. [21] investigated the working efficiency of a UAV with sprayers in the field. Spraying technology using UAVs (Unmanned Aerial Vehicles) can be applied to agricultural production for protecting plants against pesticides. Wen et al. [22] showed a PWM (Pulse Width Modulation) spray system with UAVs, based on the plant diseases and insect pests map in the target area. It also considers actual droplet deposition and deposition density in the operation unit, using PID (Proportional Integrative Derivative) control. Nguyen et al. [23] proposed a robust fault diagnosis method for quadcopter UAVs (Unmanned Aerial Vehicles). It uses a sliding mode observer to estimate the fault magnitude and location. Inspired by the flight mechanism of the insect dragonfly, a wing root control mechanism was introduced to the MAV (micro air vehicle) stabilizing hovering, by Jang and Yang [24]. It was shown that the mechanism can control the flight mode easily.

A nonlinear robust adaptive control scheme was proposed by Fan et al. [25] to handle the path following control problem, for example, steering a USV (unmanned surface vessel) to follow the desired path with disturbances. They used radial basis function neural networks for the controller. An attitude-tracking control was applied to an AUV (autonomous underwater vehicle) by Wang et al. [26]. It was involved with a disturbance-rejection control for hover and transition mode of the vehicle. AUVs (autonomous underwater vehicles) need spot hover and high-speed capabilities to explore an ocean. For this application, Wang et al. [27] presented an adaptive nonlinear control to an AUV with tri-tiltrotor. Li et al. [28] investigated a particular model of remotely operated vehicles (ROVs) as autonomous underwater vehicles (AUVs), involved with an ocean current model and a cable disturbing force.

Many robotic tasks are involved with pathplanning. Path planning is a challenging issue in robotic tasks. Jung et al. [29] proposed a new path planning method to handle curvilinear obstacles. Zeng et al. [30] presented reinforcement learning with subgoal graphs, leading to near-optimal subgoal sequences as motion-planning policies. A Tetris-inspired reconfigurable cleaning robot was demonstrated with efficient tiling path planning by Kouzehgar et al. [31]. Multi-criteria decision making wasused to handle two objectives, energy and area coverage. An interesting problem of mobile robots is path planning to a specific target position in a cluttered environment. Xue [32] presented a multi-objective evolutionary algorithm for the path planning problem. Another approach, authored by Gawron and Michałek [33], is available for the path planning problem of mobile robots. They showed that their path planning, which is collision-free, satisfies curvature constraints, and preserves continuity of the curvature arc-length derivative.

The structure design and recovery for a damaged quadruped robot has been tackled by Chattunyakit et al. [34]. They showed a caterpillar-inspired quadruped robot whichimitatesthe prolegs of caterpillars, and a mudskipper-inspired crawling algorithm based on reinforcement learning, which improves the adaptation of locomotion. Hayat et al. [35] designed a quadruped wheeled robot and showed its kinematic formulation. Jia et al. [36] tackled the motion stability of quadruped robots with dynamic gait. A dynamic stability criterion and measurement is proposed in the approach.

There have also been many studies with biped robots. Reinforcement learning can be applied to efficient gait control of a biped robot. Gil et al. [37] showed a reinforcement learning mechanism to handle stability and efficiency of movement, thus improving speed and precision of the trajectory. Yang et al. [38] showed an interesting work to transform the complex motion of robot turning into a simple translational motion. The inertial forces can be analyzed for the turning walk of humanoid robots. Bai et al. [39] showed a miniaturized continuous hopping robot consisting of a servo motor and the clockwork spring so that it has a good energy storage speed. Biped climbing robots need to move in a complex truss environment. Gu et al. [40] proposed a grip planning method for biped robots to produce optimal collision-free grip sequences under kinematic constraints. Glass façade is a challenging problem, since frames between glass panels become barriers for a cleaning robot, degrading the performance of area coverage. Nansai et al. [41] presented a new style of façade cleaning robot with a biped mechanism with active suction system.

Handling robot manipulators is one of non-trivial problems. Kinematics and their system model are important factors to solve the problem. Vo et al. [42] proposed an adaptive sliding-model control to industrial robotic manipulators. It uses a system model with radial-basis function neural network. The control system provides high tracking accuracy as well as fast response time. Kelemen et al. [43] introduced a new approach for the inverse kinematics solution of a redundant manipulator. The method considers weight matrices to prioritize tasks, thus controlling the robot behavior efficiently. A quasi-analytic inverse kinematics approach has been suggested for an active slave manipulator in the surgical robot by Bai et al. [44]. The approach can meet the real-time and high-accuracy requirements of control for the robot.

Sanfilippo et al. [45] demonstrated snake robot locomotion in a cluttered environment. The system uses a perception-driven locomotion and handles compliant motion and fine torque control for elastic joints. A snake robot needs to raise its head to obtain visual space, for example, to track a flying object. Zhang et al. [46] analyzed head-raising motion of the snake robot, which is related to the angle sequences of roll, pitch and yaw. Another snake-like robot and its analysis was presented by Nansai et al. [47]. Singular configuration analysis was provided in the work.

Recently, exoskeletons draw much attention from researchers. Nomura et al. [48] suggested a novel power assist control for a powered exoskeleton. They used motion sensors on the wearer's body to detect the walking motion quickly, where electromyography is not required. Li et al. [49] showed that an augmentation exoskeleton can be developed for load carriage. The mean activities of muscle increase significantly with exoskeleton assistance.

Wheeled robots still have interesting issues in mobile robotics. To support autonomous vehicle driving on the road, a method to imitate the lane-changing operation of excellent drivers was introduced by Geng et al. [50]. As a result, the ride comfort of the vehicle was improved. Four-wheel steering and four-wheel drive (4WS4WD) vehicles include redundant manipulations in mobile robots. Tan et al. [51] used model predictive control and particle swarm optimization as an optimization process for steering angles and wheel forces. A new application of robot control can be found in the work of climbing robots. Xu et al. [52] presented a model of a three-wheel-drive climbing robot with high-altitude safety recovery mechanism, engaged in automatic inspection of bridge cables. Ikeda et al. [53] proposed step-climbing tactics, such that a mobile robot with manipulators can help a heavy hand cart climb a step. The wheeled robot holds or pushes a hand cart by imitating human motion.

Learning and adaptation are important key issues in robotics. An adaptive system based on reservoir computing and recurrent neural networks has been applied to couple control signals and robotic behaviors by Melidis and Marocco [54]. Yamauchi and Suzuki [55] focused on designing a base action set for a complex task with a wheel robot, and developed an algorithm to search for the base action to change the environment. Kim [56] showed an agent model to chase a high-speed evader. It controls the relative speed of the pursuer with respect to the evader, depending on the distance between them. Kuo et al. [57] showed an obstacle avoidance approach based on velocity potential function. They focused on curvature constraints for a mobile robot.

New actuators have been developed and tested for a variety of applications by researchers. A passive ski robot without an actuator was developed by Saga et al. [58] to understand the turn mechanism, ski deflection and skier posture mechanics during sliding. It can reveal the factors affecting ski turns, for example, the center of gravity (COG) and the ski shape. New applications of mobile robots are available in an underground coal mine for explosion safety. Novák et al. [59] investigated the safety regulations and practice solutions with tele-operated mobile robots. Zhang et al. [60] argued that a robotic drilling task can be handled with a sliding mode control. Controlling the drilling end-effector achieves dynamic stabilization and tracking accuracy. Sun et al. [61] designed a novel robot to assist human astronauts in a space station, and demonstrated its walking, rolling and sliding motion. For pneumatic positioning and force-control systems, Kanno et al. [62] proposed a three-port poppet-type servo valve to reduce air leakage of the spool-type servo valves. It is effective even in experiments with pressure and position control. A new class of actuator was studied especially in micro-robots. Chen et al. [63] proposed a high step-up ratio flyback converter for a piezoelectric bimorph actuator in micro mobile robot. Medical devices and rehabilitation mechanisms are often involved with smart materials such as electro-rheological fluids, magneto-rheological fluids and shape memory alloys. Sohn et al. [64] introduced various systems for those robots and medical devices, depending on design configuration or operating principles.

**Acknowledgments:** We would like to thank all the authors and reviewers who contributed to this Special Issue. There have been many ceaseless efforts from the authors, and the reviewers' feedback and comments greatly helped the authors to improve the papers. Also, we would like to express our special thanks to the editorial team and Daria Shi, the managing editor of Applied Sciences.

**Conflicts of Interest:** The author declares no conflict of interest.

## References


1. Burlacu, A.; Kloetzer, M.; Mahulea, C. Numerical Evaluation of Sample Gathering Solutions for Mobile Robots. *Appl. Sci.* **2019**, *9*, 791. [CrossRef]
2. Huang, Y.; Li, Z.; Jiang, Y.; Cheng, L. Cooperative Path Planning for Multiple Mobile Robots Via HAFSA and an Expansion Logic Strategy. *Appl. Sci.* **2019**, *9*, 672. [CrossRef]
3. Fan, C.; Shirafuji, S.; Ota, J. Modal Planning for Cooperative Non-Prehensile Manipulation by Mobile Robots. *Appl. Sci.* **2019**, *9*, 462. [CrossRef]
4. Li, G.; Lin, R.; Li, M.; Sun, R.; Piao, S. A Master-Slave Separate Parallel Intelligent Mobile Robot Used for Autonomous Pallet Transportation. *Appl. Sci.* **2019**, *9*, 368. [CrossRef]
5. Kowalczyk, W. Formation Control and Distributed Goal Assignment for Multi-Agent Non-Holonomic Systems. *Appl. Sci.* **2019**, *9*, 1311. [CrossRef]
6. Xu, S.S.D.; Huang, H.C.; Chiu, T.C.; Lin, S.K. Biologically-Inspired Learning and Adaptation of Self-Evolving Control for Networked Mobile Robots. *Appl. Sci.* **2019**, *9*, 1034.
7. Cardona, G.A.; Calderon, J.M. Robot Swarm Navigation and Victim Detection Using Rendezvous Consensus in Search and Rescue Operations. *Appl. Sci.* **2019**, *9*, 1702. [CrossRef]
8. Gan, Y.; Duan, J.; Chen, M.; Dai, X. Multi-Robot Trajectory Planning and Position/Force Coordination Control in Complex Welding Tasks. *Appl. Sci.* **2019**, *9*, 924. [CrossRef]
9. Iriondo, A.; Lazkano, E.; Susperregi, L.; Urain, J.; Fernandez, A.; Molina, J. Pick and Place Operations in Logistics Using a Mobile Manipulator Controlled with Deep Reinforcement Learning. *Appl. Sci.* **2019**, *9*, 348. [CrossRef]
10. Chen, Y.; Li, L. Predictable Trajectory Planning of Industrial Robots with Constraints. *Appl. Sci.* **2018**, *8*, 2648. [CrossRef]
11. Zhao, D.; Guo, H. A Trajectory Planning Method for Polishing Optical Elements Based on a Non-Uniform Rational B-Spline Curve. *Appl. Sci.* **2018**, *8*, 1355. [CrossRef]
12. Chen, B.; Yuan, D.; Liu, C.; Wu, Q. Loop Closure Detection Based on Multi-Scale Deep Feature Fusion. *Appl. Sci.* **2019**, *9*, 1120. [CrossRef]
13. Wang, Z.; Chen, Y.; Mei, Y.; Yang, K.; Cai, B. IMU-Assisted 2D SLAM Method for Low-Texture and Dynamic Environments. *Appl. Sci.* **2018**, *8*, 2534. [CrossRef]
14. Wang, J.; Zhao, M.; Chen, W. MIM–SLAM: A Multi-Level ICP Matching Method for Mobile Robot in Large-Scale and Sparse Scenes. *Appl. Sci.* **2018**, *8*, 2432. [CrossRef]
15. Alonso Ramirez, O.; Marin Hernandez, A.; Rios Figueroa, H.V.; Devy, M.; Pomares Hernandez, S.E.; Rechy Ramirez, E.J. A Graph Representation Composed of Geometrical Components for Household Furniture Detection by Autonomous Mobile Robots. *Appl. Sci.* **2018**, *8*, 2234. [CrossRef]
16. Villasenor, C.; Arana Daniel, N.; Alanis, A.Y.; Lopez Franco, C.; Gomez Avila, J. Multiellipsoidal Mapping Algorithm. *Appl. Sci.* **2018**, *8*, 1239. [CrossRef]
17. Wang, F.; Liu, Y.; Xiao, L.; Wu, C.; Chu, H. Topological Map Construction Based on Region Dynamic Growing and Map Representation Method. *Appl. Sci.* **2019**, *9*, 816. [CrossRef]
18. Chien, J.C.; Dang, Z.Y.; Lee, J.D. Navigating a Service Robot for Indoor Complex Environments. *Appl. Sci.* **2019**, *9*, 491. [CrossRef]
19. Cheng, Q.; Wang, X.; Yang, J.; Shen, L. Automated Enemy Avoidance of Unmanned Aerial Vehicles Based on Reinforcement Learning. *Appl. Sci.* **2019**, *9*, 669. [CrossRef]
20. Giernacki, W. Iterative Learning Method for In-Flight Auto-Tuning of UAV Controllers Based on Basic Sensory Information. *Appl. Sci.* **2019**, *9*, 648. [CrossRef]
21. Wang, G.; Lan, Y.; Yuan, H.; Qi, H.; Chen, P.; Ouyang, F.; Han, Y. Comparison of Spray Deposition, Control Efficacy on Wheat Aphids and Working Efficiency in the Wheat Field of the Unmanned Aerial Vehicle with Boom Sprayer and Two Conventional Knapsack Sprayers. *Appl. Sci.* **2019**, *9*, 218. [CrossRef]
22. Wen, S.; Zhang, Q.; Deng, J.; Lan, Y.; Yin, X.; Shan, J. Design and Experiment of a Variable Spray System for Unmanned Aerial Vehicles Based on PID and PWM Control. *Appl. Sci.* **2018**, *8*, 2482. [CrossRef]
23. Nguyen, N.P.; Hong, S.K. Sliding Mode Thau Observer for Actuator Fault Diagnosis of Quadcopter UAVs. *Appl. Sci.* **2018**, *8*, 1893. [CrossRef]

24. Jang, J.H.; Yang, G.H. Design of Wing Root Rotation Mechanism for Dragonfly-Inspired Micro Air Vehicle. *Appl. Sci.* **2018**, *8*, 1868. [CrossRef]
25. Fan, Y.; Huang, H.; Tan, Y. Robust Adaptive Path Following Control of an Unmanned Surface Vessel Subject to Input Saturation and Uncertainties. *Appl. Sci.* **2019**, *9*, 1815. [CrossRef]
26. Wang, T.; Wang, J.; Wu, C.; Zhao, M.; Ge, T. Disturbance-Rejection Control for the Hover and Transition Modes of a Negative-Buoyancy Quad Tilt-Rotor Autonomous Underwater Vehicle. *Appl. Sci.* **2018**, *8*, 2459. [CrossRef]
27. Wang, T.; Wu, C.; Wang, J.; Ge, T. Modeling and Control of Negative-Buoyancy Tri-Tilt-Rotor Autonomous Underwater Vehicles Based on Immersion and Invariance Methodology. *Appl. Sci.* **2018**, *8*, 1150. [CrossRef]
28. Li, X.; Zhao, M.; Ge, T. A Nonlinear Observer for Remotely Operated Vehicles with Cable Effect in Ocean Currents. *Appl. Sci.* **2018**, *8*, 867. [CrossRef]
29. Jung, J.W.; So, B.C.; Kang, J.G.; Lim, D.W.; Son, Y. Expanded Douglas–Peucker Polygonal Approximation and Opposite Angle-Based Exact Cell Decomposition for Path Planning with Curvilinear Obstacles. *Appl. Sci.* **2019**, *9*, 638. [CrossRef]
30. Zeng, J.; Qin, L.; Hu, Y.; Hu, C.; Yin, Q. Combining Subgoal Graphs with Reinforcement Learning to Build a Rational Pathfinder. *Appl. Sci.* **2019**, *9*, 323. [CrossRef]
31. Kouzehgar, M.; Elara, M.R.; Philip, M.A.; Arunmozhi, M.; Prabakaran, V. Multi-Criteria Decision Making for Efficient Tiling Path Planning in a Tetris-Inspired Self-Reconfigurable Cleaning Robot. *Appl. Sci.* **2019**, *9*, 63. [CrossRef]
32. Xue, Y. Mobile Robot Path Planning with a Non-Dominated Sorting Genetic Algorithm. *Appl. Sci.* **2018**, *8*, 2253. [CrossRef]
33. Gawron, T.; Michalek, M.M. A G3-Continuous Extend Procedure for Path Planning of Mobile Robots with Limited Motion Curvature and State Constraints. *Appl. Sci.* **2018**, *8*, 2127. [CrossRef]
34. Chattunyakit, S.; Kobayashi, Y.; Emaru, T.; Ravankar, A.A. Bio-Inspired Structure and Behavior of Self-Recovery Quadruped Robot with a Limited Number of Functional Legs. *Appl. Sci.* **2019**, *9*, 799. [CrossRef]
35. Hayat, A.A.; Elangovan, K.; Elara, M.R.; Teja, M.S. Tarantula: Design, Modeling, and Kinematic Identification of a Quadruped Wheeled Robot. *Appl. Sci.* **2019**, *9*, 94. [CrossRef]
36. Jia, Y.; Luo, X.; Han, B.; Liang, G.; Zhao, J.; Zhao, Y. Stability Criterion for Dynamic Gaits of Quadruped Robot. *Appl. Sci.* **2018**, *8*, 2381. [CrossRef]
37. Gil, C.R.; Calvo, H.; Sossa, H. Learning an Efficient Gait Cycle of a Biped Robot Based on Reinforcement Learning and Artificial Neural Networks. *Appl. Sci.* **2019**, *9*, 502. [CrossRef]
38. Yang, T.; Zhang, W.; Chen, X.; Yu, Z.; Meng, L.; Huang, Q. Turning Gait Planning Method for Humanoid Robots. *Appl. Sci.* **2018**, *8*, 1257. [CrossRef]
39. Bai, L.; Zheng, F.; Chen, X.; Sun, Y.; Hou, J. Design and Experimental Evaluation of a Single-Actuator Continuous Hopping Robot Using the Geared Symmetric Multi-Bar Mechanism. *Appl. Sci.* **2019**, *9*, 13. [CrossRef]
40. Gu, S.; Zhu, H.; Li, H.; Guan, Y.; Zhang, H. Optimal Collision-Free Grip Planning for Biped Climbing Robots in Complex Truss Environment. *Appl. Sci.* **2018**, *8*, 2533. [CrossRef]
41. Nansai, S.; Onodera, K.; Veerajagadheswar, P.; Elara, M.R.; Iwase, M. Design and Experiment of a Novel Façade Cleaning Robot with a Biped Mechanism. *Appl. Sci.* **2018**, *8*, 2398. [CrossRef]
42. Vo, A.T.; Kang, H.J. An Adaptive Neural Non-Singular Fast-Terminal Sliding-Mode Control for Industrial Robotic Manipulators. *Appl. Sci.* **2018**, *8*, 2562. [CrossRef]
43. Kelemen, M.; Virgala, I.; Liptak, T.; Mikova, L.; Filakovsky, F.; Bulej, V. A Novel Approach for a Inverse Kinematics Solution of a Redundant Manipulator. *Appl. Sci.* **2018**, *8*, 2229. [CrossRef]
44. Bai, L.; Yang, J.; Chen, X.; Jiang, P.; Liu, F.; Zheng, F.; Sun, Y. Solving the Time-Varying Inverse Kinematics Problem for the Da Vinci Surgical Robot. *Appl. Sci.* **2019**, *9*, 546. [CrossRef]
45. Sanfilippo, F.; Helgerud, E.; Stadheim, P.A.; Aronsen, S.L. Serpens: A Highly Compliant Low-Cost ROS-Based Snake Robot with Series Elastic Actuators, Stereoscopic Vision and a Screw-Less Assembly Mechanism. *Appl. Sci.* **2019**, *9*, 396. [CrossRef]
46. Zhang, X.; Liu, J.; Ju, Z.; Yang, C. Head-Raising of Snake Robots Based on a Predefined Spiral Curve Method. *Appl. Sci.* **2018**, *8*, 2011. [CrossRef]

47. Nansai, S.; Iwase, M.; Itoh, H. Generalized Singularity Analysis of Snake-Like Robot. *Appl. Sci.* **2018**, *8*, 1873. [CrossRef]
48. Nomura, S.; Takahashi, Y.; Sahashi, K.; Murai, S.; Kawai, M.; Taniai, Y.; Naniwa, T. Power Assist Control Based on Human Motion Estimation Using Motion Sensors for Powered Exoskeleton without Binding Legs. *Appl. Sci.* **2019**, *9*, 164. [CrossRef]
49. Li, H.; Cheng, W.; Liu, F.; Zhang, M.; Wang, K. The Effects on Muscle Activity and Discomfort of Varying Load Carriage With and Without an Augmentation Exoskeleton. *Appl. Sci.* **2018**, *8*, 2638. [CrossRef]
50. Geng, G.; Wu, Z.; Jiang, H.; Sun, L.; Duan, C. Study on Path Planning Method for Imitating the Lane-Changing Operation of Excellent Drivers. *Appl. Sci.* **2018**, *8*, 814. [CrossRef]
51. Tan, Q.; Dai, P.; Zhang, Z.; Katupitiya, J. MPC and PSO Based Control Methodology for Path Tracking of 4WS4WD Vehicles. *Appl. Sci.* **2018**, *8*, 1000. [CrossRef]
52. Xu, F.; Jiang, Q.; Lv, F.; Wu, M.; Zhang, L. The Dynamic Coupling Analysis for All-Wheel-Drive Climbing Robot Based on Safety Recovery Mechanism Model. *Appl. Sci.* **2018**, *8*, 2123. [CrossRef]
53. Ikeda, H.; Kawabe, T.; Wada, R.; Sato, K. Step-Climbing Tactics Using a Mobile Robot Pushing a Hand Cart. *Appl. Sci.* **2018**, *8*, 2114. [CrossRef]
54. Melidis, C.; Marocco, D. Effective Behavioural Dynamic Coupling Through Echo State Networks. *Appl. Sci.* **2019**, *9*, 1300. [CrossRef]
55. Yamauchi, S.; Suzuki, K. Algorithm for Base Action Set Generation Focusing on Undiscovered Sensor Values. *Appl. Sci.* **2019**, *9*, 161. [CrossRef]
56. Kim, J. Controllers to Chase a High-Speed Evader Using a Pursuer with Variable Speed. *Appl. Sci.* **2018**, *8*, 1976. [CrossRef]
57. Kuo, P.L.; Wang, C.H.; Chou, H.J.; Liu, J.S. A Real-Time Hydrodynamic-Based Obstacle Avoidance System for Non-Holonomic Mobile Robots with Curvature Constraints. *Appl. Sci.* **2018**, *8*, 2144. [CrossRef]
58. Saga, T.; Saga, N. Alpine Skiing Robot Using a Passive Turn with Variable Mechanism. *Appl. Sci.* **2018**, *8*, 2643. [CrossRef]
59. Novak, P.; Kot, T.; Babjak, J.; Konecny, Z.; Moczulski, W.; Lopez, A.R. Implementation of Explosion Safety Regulations in Design of a Mobile Robot for Coal Mines. *Appl. Sci.* **2018**, *8*, 2300. [CrossRef]
60. Zhang, L.; Dhupia, J.S.; Wu, M.; Huang, H. A Robotic Drilling End-Effector and Its Sliding Mode Control for the Normal Adjustment. *Appl. Sci.* **2018**, *8*, 1892. [CrossRef]
61. Sun, Z.; Li, H.; Jiang, Z.; Song, Z.; Mo, Y.; Ceccarelli, M. Prototype Design and Performance Tests of Beijing's Astronaut Robot. *Appl. Sci.* **2018**, *8*, 1342. [CrossRef]
62. Kanno, T.; Hasegawa, T.; Miyazaki, T.; Yamamoto, N.; Haraguchi, D.; Kawashima, K. Development of a Poppet-Type Pneumatic Servo Valve. *Appl. Sci.* **2018**, *8*, 2094. [CrossRef]
63. Chen, C.; Liu, M.; Wang, Y. A Dual Stage Low Power Converter Driving for Piezoelectric Actuator Applied in Micro Mobile Robot. *Appl. Sci.* **2018**, *8*, 1666. [CrossRef]
64. Sohn, J.W.; Kim, G.W.; Choi, S.B. A State-Of-The-Art Review on Robots and Medical Devices Using Smart Fluids and Shape Memory Alloys. *Appl. Sci.* **2018**, *8*, 1928. [CrossRef]

*applied sciences*

MDPI

*Article*

# Numerical Evaluation of Sample Gathering Solutions for Mobile Robots

**Adrian Burlacu [1,†], Marius Kloetzer [1,*,†] and Cristian Mahulea [2,†]**

1 Department of Automatic Control and Applied Informatics, "Gheorghe Asachi" Technical University of Iasi, 700050 Iasi, Romania; aburlacu@ac.tuiasi.ro
2 Department of Computer Science and Systems Engineering, University of Zaragoza, 50018 Zaragoza, Spain; cmahulea@unizar.es
* Correspondence: kmarius@ac.tuiasi.ro
† The authors contributed equally to this work.

Received: 20 January 2019; Accepted: 19 February 2019; Published: 23 February 2019

**Abstract:** This paper applies mathematical modeling and solution numerical evaluation to the problem of collecting a set of samples scattered throughout a graph environment and transporting them to a storage facility. A team of identical robots is available, where each robot has a limited amount of energy and it can carry one sample at a time. The graph weights are related to energy and time consumed for moving between adjacent nodes, and thus, the task is transformed to a specific optimal assignment problem. The design of the mathematical model starts from a mixed-integer linear programming problem whose solution yields an optimal movement plan that minimizes the total time for gathering all samples. For reducing the computational complexity of the optimal solution, we develop two sub-optimal relaxations and then we quantitatively compare all the approaches based on extensive numerical simulations. The numerical evaluation yields a decision diagram that can help a user to choose the appropriate method for a given problem instance.

**Keywords:** sample gathering problem; mobile robots; mathematical modeling; numerical evaluation; centralized architecture; optimization

## 1. Introduction

Much robotics research develops automatic planning procedures for autonomous agents such that a given mission is accomplished under an optimality criterion. The missions are usually related to standard problems such as navigation, coverage, localization, and mapping [1,2]. Some works provide strategies directly implementable on particular robots with complicated dynamics and multiple sensors [3]. Other research aims to increase the task expressiveness [4], e.g., starting from Boolean-inspired specifications [5,6] up to temporal logic ones [7–10], even if the obtained plans may be applied only to simple robots.

It is often common to construct discrete models for the environment and robot movement capabilities, by using results from multiple areas such as systems theory, computational geometry [11,12], and discrete event systems [13,14].

The current research is focused on solving a sample gathering problem. The considered task belongs to the general class of optimal assignment problems [15], since the sample can correspond to jobs and the robots to machines. The minimization of the overall time for gathering all samples (yielded by the "slowest" agent) thus translates to so-called min-max problems [16] or bottleneck assignment problems [15] (Chapter 6.2). However, these standard frameworks do not consider different numbers of jobs and machines, nor machines (agents) with limited energy amounts.

A broad taxonomy of allocations in multi-robot teams is presented in [17,18], according to which our problem belongs to the class of assignments of single-robot tasks (one task requiring one robot)

in multi-task robot systems (a robot can move, pick up, and deposit samples). Again, the general solutions assume utility estimates for different job–machine pairs.

For such problems, various Mixed-Integer Linear Programming Problem (MILP) formulations are given as in [17,19,20]. Some resemble our problem, but they are not an exact fit because of the specificities that all samples should be eventually gathered into the same node, a robot can carry one sample at a time, there are limited amounts of energy, and we do not know a priori a relationship between number of samples and number of robots. Furthermore, we provide a second MILP for the case of problems infeasible due to energy requirements. We further relax the complex MILP solutions into sub-optimal solutions as non-convex Quadratic Programming (QP) and iterative heuristics. Our goal is to draw rules of choosing the appropriate method for a given problem, based on extensive tests.

Some preliminary mathematical formulations that generalize traveling salesman problems are included in [21], with targeted application to exploring robots that must collect and analyze multiple heterogeneous samples from a planetary surface. Other works focus on specific applications as task allocation accomplished by agents with different dynamics [22], or allocation in scenarios with heterogenous robots that can perform different tasks [23]. Various works propose auction-based mechanisms for various assignments problems or develop and apply distributed algorithms for specific cases with equal number of agents and tasks [24]. Research [25] assumes precedence constraints on available tasks and builds solutions based on integer programming forms and auction mechanisms. However, we do not include auction-based methods here. The closest solution we propose may be our iterative heuristic algorithm from Section 4.2, which can be viewed as a specific greedy allocation method [17].

Our problem can be seen as a particular case of Capacity and Distance constrained Vehicle Routing Problem (CDVRP) [26] with capacity equal to one and the distance constrains related to the limited energy. For this problem, many algorithms have been provided, for the exact methods [27,28] and for heuristic methods [29–32]. However, our problem is different in multiple aspects. First, in the CDVRP problem, the capacity of each vehicle (robot in our case) should be greater than the demand of each vertex [26]. In our case this cannot hold since the robot capacity is one (we assume that each robot can transport maximum one good) and the number of goods at the vertices is, in many cases, greater than one. Second, up to our knowledge, the heuristics considered in this work, which include relaxing the optimal solution of the MILP to a QP problem, have not been considered for the CDVRP. Finally, most of the works on CDVRP try to characterize worst-case scenarios through cost differences between heuristic and optimal methods.

In this paper, we are also interested in the computational complexity and we evaluate the proposed solutions using numerical simulations. To the best of our knowledge, none of the mentioned works contains a directly applicable formulation that yields a solution for our specific problem. This work builds on solutions reported in [33–35]. In [33] we constructed a MILP problem that solves the minimum time sample gathering problem. Different than in [33], we also design a MILP solution that can be used when the initial problem is infeasible due to scarce energy limits. Besides the MILP formulations, one of the main goals of this research are to provide computationally efficient sub-optimal solutions for the targeted problems. The MILP solution was relaxed to a QP formulation in [34]. Here we also construct an iterative sub-optimal solution inspired by [35] as a QP alternative. Based on extensive simulations that involve the three proposed solutions, we conclude with a decision scheme that helps a user to choose the appropriate method for a specific problem instance.

The purpose of this work is to plan a team of mobile agents such that they gather the samples scattered throughout the environment into a storage facility. The problem's hypothesis consists in a team of mobile robots which must bring to a deposit region a set of samples that exist in an environment at known locations. As main contributions we claim the new mathematical models for the considered problems and the numerical evaluation of the obtained solutions.

The environment is modeled by a graph where an arc weight corresponds to consumed energy and time for moving between the linked nodes. Each robot has limited energy, and the goal is to collect all samples in minimum time. Because the robots are initially deployed in the storage (deposit) node and given the static nature of the environment, it is customary to build movement plans before the agents start to move. The problem reduces to a specific case of optimal assignment or task allocation [15,17,18]. The paper combines results from our previous research reported in [34,36,37]. We first build an optimal solution involving a MILP formulation. Then, we design two solutions with lower complexities, one as a Quadratic Programming (QP) relaxation of the initial MILP, and the other as an Iterative Heuristic (IH) algorithm. A numerical evaluation between the formulated solutions yields criteria as time complexities for computing robotic plans and the difference of costs between these plans. Based on these criteria, a decision diagram is provided such that a user can easily choose the appropriate method to embed.

The remainder of the paper is structured as follows. Section 2 formulates the targeted problem, outlines the involved assumptions and introduces an example that will be solved throughout the subsequent sections. Section 3 details two optimal solutions, from which the first will be relaxed to a QP formulation, while the second can be used when robots have low energy supplies. The sub-optimal methods based on QP optimization or IH algorithm are presented in Section 4. The developed methods are numerically evaluated and compared in Section 5 and rules are given for choosing the proper method for a specific problem instance.

## 2. Problem Formulation

Consider a team of $N_R$ identical robots that are labeled with elements of set $R = \{r_1, r_2, \ldots, r_{N_R}\}$. The robots "move" on a weighted graph $G = (V, E, c)$, where $V = \{v_1, v_2, \ldots, v_{|V|}\}$ is the finite number of nodes (also called locations or vertices), $E \subseteq V \times V$ is the adjacency relationship corresponding to graph edges, and $c : E \to \mathbb{R}_+$ is a cost (weight) function.

We mention that there are multiple approaches for creating such finite-state abstractions of robot control capabilities in a given environment [1,2]. A widely used idea is to partition the free space into a set of regions via cell decomposition methods, each of these regions corresponding to a node from $V$ [6,11,38]. The graph edges correspond to possible robot movements between adjacent partition regions, i.e., $\forall v, v' \in V$, if an agent can move from location $v$ to $v'$ without visiting any other node from graph, then $(v, v') \in E$. Each edge corresponds to a continuous feedback control law for the robot such that the desired movement is produced, and various methods exist for designing such control laws based on agent dynamics and partition types [39,40]. Alternatives to cell decomposition methods, as visibility graphs or generalized Voronoi diagrams, can also produce discrete abstractions in form of graphs or transition systems [1].

We assume that the graph $G$ is connected, and the adjacency relationship $E$ is symmetric, i.e., if a robot from $R$ can travel from location $v$ to $v'$, then it can also move from $v'$ to $v$. For any $(v, v') \in E$, we consider that the cost $c(v, v')$ represents the amount of *energy* spent by the robot for performing the movement from $v$ to $v'$ and that $c(v', v) = c(v, v')$. By assuming identical agents with constant velocity and a homogenous environment (i.e., the energy for following an arc is proportional with the distance between linked nodes), we denote the *time* necessary for performing the movement from location $v$ to $v'$ by $\gamma \cdot c(v, v')$, where $\gamma \in \mathbb{R}_+$ is a fixed value.

Initially, all agents are deployed in a storage (deposit) node $v_{|V|}$ (labeled for simplicity as the last node in graph $G$), and each robot $r \in R$ has a limited amount of energy, given by map $\mathcal{E} : R \to \mathbb{R}_+$, for performing movements on abstraction $G$. For homogenous environments and constant moving speeds, energy $\mathcal{E}(r)$ can be easily linked with battery level of robot $r$, with the distance it can travel, or the sum of costs of followed edges.

There are $N_S$ *samples* or valuable items scattered throughout the environment graph $G$. The samples are indexed (labeled) with elements of set $S = \{1, 2, \ldots, N_S\}$, while a map $\pi : S \to V$ shows to which node each sample belongs.

**Problem 1.** *For every robot $r \in R$ find a moving strategy on $G$ such that:*

- *the team of robots gathers (collects) all samples from graph G in the storage node* $v_{|V|}$ *within minimum time;*
- *each robot can carry at most one sample at any moment;*
- *the total amount of energy spent by each robot is at most its initially available energy.*

**Remark 1** (NP-hardness). *Our problem is related to the so-called Set Partitioning Problems (SPPs) [17], which are employed in various task allocation problems for mobile agents. A SPP formulation aims to find a partition of a given set such that a utility function defined over the set of acceptable partitions with real values is maximized. Various SPPs are solved by using Operations Research formulations that employ different standard optimization problems. In our case, the given set is S (samples to be collected), while the desired partition should have* $N_R$ *disjoint subsets of S, each subset corresponding to the samples a robot should collect. The utility relates to the necessary time required for collecting all samples (being a maximum value over utilities of elements of obtained partition), while the partition is acceptable if each robot has enough energy to collects its samples. The maximization over individual utilities show that our problem is more complicated than standard SPPs. Since an SPP is NP-hard [41], we conclude that Problem 1 is also NP-hard. Therefore, we expect computationally intensive solutions for optimally solving Problem 1, while sub-optimal relaxations may be used when an optimal solution does not seem tractable.*

Since the sample deployment and robot energy limits are known, the searched solution is basically an off-line computed plan (sequence of nodes) for each robot such that the mission requirements are fulfilled. The first requirement from Problem 1 can be regarded as a global target for the whole team (properly assign robots to collect samples such that the overall time for accomplishing the task is minimized), while the last two requirements are related to robot capabilities. As in many robot planning approaches where global tasks are accomplished, we do not account for inter-robot collisions when developing movement plans. In real applications, such collisions can be avoided by using local rules during the actual movement, and the time (or energy) offset induced by such rules is negligible with respect to the total movement time (or required energy). Clearly, in some cases Problem 1 may not have a solution due to insufficient available energy for robots, these situations will be discussed during solution description.

**Example 1.** *For supporting the problem formulation and solution development, we introduce an example that will be discussed throughout the next sections. Thus, we assume an environment abstracted to the graph from Figure 1, composed by 10 nodes* $(v_1, v_2, \ldots, v_{10})$*, with the deposit* $v_{10}$*. The costs for moving between adjacent nodes are marked on the arcs from Figure 1, e.g.,* $c(v_{10}, v_8) = 2$*. The team consists of 3 robots labeled with elements of set* $R = \{r_1, r_2, r_3\}$*. For simplicity of exposition, we consider* $\gamma = 1$ *(the constant that links the moving energy with necessary time) and equal amounts of energy for robots,* $\mathcal{E}(r) = 100, \forall r \in R$*. There are 14 samples scattered in this graph (labeled with numbers from 1 to 14), with locations given by map* $\pi$*, e.g.,* $\pi(9) = \pi(10) = v_3$*.*

*The problem requires a sequence of movements for each robot such that all the 14 samples are gathered into storage* $v_{10}$*, each robot being able to carry one sample at any time.*

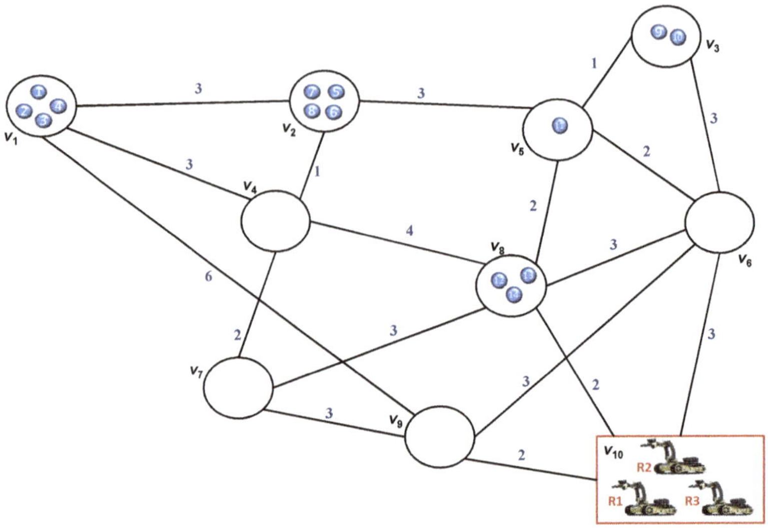


**Figure 1.** Example: environment graph with 10 nodes, 3 robots, and 14 samples. The robot moving costs are marked on graph edges, and the samples are represented by the blue discs.

## 3. Mathematical Model and Optimal Solution

To solve Problem 1, defined in the previous section, our approach consists from the following main steps:

(i) Determine optimal paths in graph $G$ from storage node to all nodes containing samples.
(ii) Formulate linear constraints for correctly picking samples and for not exceeding robot's available energy based on a given allocation of each robot to pick specific samples.
(iii) Create a cost function based on robot-to-sample allocation and on necessary time for gathering all samples.
(iv) Cast the above steps in a form suitable for applying existing optimization algorithms and thus find the desired robot-to-sample allocations.

Step (i) is accomplished by running a Dijkstra algorithm [42] on the weighted graph $G$, with source node $v_{|V|}$ and with multiple destination nodes: $v \in V$ for which $\exists s \in S$ such that $\pi(s) = v$. Please note that a single run of Dijkstra algorithm returns minimum cost paths to multiple destinations.

Let us denote with $path(v)$ the obtained path (sequence of nodes) from deposit $v_{|V|}$ to node $v \in V$ and with $\omega(v)/2$ its cost. Due to symmetrical adjacency relationship of $G$, the retour path from $v$ to storage is immediately constructed by following $path(v)$ in inverse order. The retour path is denoted by $path^{-1}(v)$ and it has the same cost $\omega(v)/2$. (Because are interested in the round-trip cost between nodes $v_{|V|}$ and $v$, we denote the one-way cost by $\omega(v)/2$ and thus the cost of the full path is simply denoted by $\omega(v)$.). Therefore, for collecting and bringing to storage location the sample $s \in S$, a robot spends $\omega(\pi(s))$ for the round-trip given by $path(\pi(s))$, $path^{-1}(\pi(s))$.

For solving steps (ii)–(iv), let us first define a decision function as $x : N_R \times N_S \to \{0, 1\}$ by:

$$x(r,s) = \begin{cases} 1, & \text{if robot } r \text{ picks sample } s \\ 0, & \text{otherwise} \end{cases}, \forall r \in R, s \in S. \tag{1}$$

The actual values returned by map $x$ for any pair $(r,s) \in R \times S$ give the most important part of solution to Problem 1, and these values are unknown, yet. The outcomes of $x$ will constitute the decision variables in an optimization problem that is described next.

Step (ii) is formulated as the following set of linear constraints:

$$\begin{aligned} &\sum_{r\in R} x(r,s) = 1, && \forall s \in S \\ &\sum_{s\in S} \Big(\omega\big(\pi(s)\big)\cdot x(r,s)\Big) \leq \mathcal{E}(r), && \forall r \in R \end{aligned}, \tag{2}$$

where the first set of equalities impose that exactly one robot is sent to collect each sample, while the subsequent inequalities guarantee that the energy spent by robot $r$ for collecting all its assigned samples does not exceed its available energy.

The cost function from step (iii) corresponds to the first requirement of Problem 1. It means to minimize the maximum time among all robots required for collecting the assigned samples, and it formally translates to finding outcomes of $x$ that minimize the objective function $J(x)$ from

$$J(x) = \min_{x} \max_{r\in R} \left( \gamma \cdot \sum_{s\in S} \Big(\omega\big(\pi(s)\big)\cdot x(r,s)\Big)\right). \tag{3}$$

The objective function from (3) and the constraints from (2) form a minimax optimization problem [43,44]. However, decision function $x$ should take binary values. To use available software tools when solving for values of $x$, step (iv) transforms the minimax optimization into a MILP problem [44,45], by adding an auxiliary variable $z \in \mathbb{R}_+$ and additional constraints that replace the max term from (3). This results in the following MILP optimization:

$$\begin{aligned} \min_{x,z} \quad & z \\ \text{s.t.:} \quad & \sum_{r\in R} x(r,s) = 1,\ \forall s \in S \\ & \sum_{s\in S} \Big(\omega\big(\pi(s)\big)\cdot x(r,s)\Big) \leq \mathcal{E}(r),\ \forall r \in R \\ & \gamma \cdot \sum_{s\in S} \Big(\omega\big(\pi(s)\big)\cdot x(r,s)\Big) \leq z,\ \forall r \in R \\ & x(r,s) \in \{0,1\},\ \forall (r,s) \in R \times S \\ & z \geq 0 \end{aligned}. \tag{4}$$

The MILP (4) can be solved by using existing software tools [46–48]. The solution is guaranteed to be globally optimal because both the feasible set defined by the linear constraints from (4) and the objective function (z) are convex [45,49]. Thus, solution of (4) gives the optimal outcomes for $x$ (unknown decision variables $x(r,s)$), as well as the time $z$ in which the team solves Problem 1.

Please note that the obtained map $x$ indicates the samples that must be collected by each robot, as in (1). However, it does not impose any order for collecting these samples. For imposing a specific sequencing, each robot is planned to collect its allocated samples in the ascending order of the necessary costs. This means that robot $r$ first picks the sample $s$ for which $x(r,s) = 1$ and $\omega\big(\pi(s)\big) \leq \omega\big(\pi(s')\big)$, $\forall s' \in S$ with $x(r,s') = 1$, and so on for the other samples. The optimum path for collecting sample $s$ from node $\pi(s)$ was already determined in step (i). Under the above explanations, Algorithm 1 includes the steps for obtaining an optimal solution for Problem 1.

**Algorithm 1:** Optimal solution

**Input:** $G, R, S, \mathcal{E}, \pi$

**Output:** Robot movement plans

1 Find on graph $G$ paths $path(v)$ and costs $\omega(v)$ for every $v = \pi(s), s \in S$
2 Solve MILP optimization (4)
3 **if** *solutions z and x are obtained* **then**
4     **for** $r \in R$ **do**
5         $plan_r = \emptyset$
6         $S_{r:collects} = \{s \in S \mid x(r,s) = 1\}$
7         Sort set $S_{r:collects}$ in ascending order based on costs $\omega(\pi(s)), s \in S_{r:collects}$
8         **for** $s \in S_{r:collects}$ **do**
9             Append $path(\pi(s))$ to plan of robot $r$, $plan_r$
10             Insert command to collect sample $s$ in $plan_r$
11             Append $path^{-1}(\pi(s))$ to $plan_r$
12             Insert command to deposit sample $s$ in $plan_r$
13     *Return* plans $plan_r, \forall r \in R$
14 **else**
15     Problem 1 is infeasible
16     *Return*

**Example 2.** *We apply the optimal solution from this section on the example introduced in Section 2. The Dijkstra algorithm (line 1 from Algorithm 1) returns paths and corresponding energy for collecting each sample, e.g.,* $path(\pi(1)) = path(v_1) = v_{10}, v_9, v_1$ *and* $\omega(\pi(1)) = 16$. *The MILP (4) was solved in about 0.7 s and it returned an optimal solution with* $z = 54$ *(time for fulfilling Problem 1) and allocation map x. Based on robot-to-sample allocations x, lines 3–13 from Algorithm 1 yield the robotic plans for collecting samples from the following nodes (the sequences of nodes and the collect/deposit commands are omitted due to their length):*

$$\begin{array}{ll} \textit{Robot } r_1 \textit{ collects samples from:} & \\ (v_8),\ (v_8),\ (v_2),\ (v_1),\ (v_1) & (\text{time}: 54) \\ \textit{Robot } r_2 \textit{ collects samples from:} & \\ (v_3),\ (v_2),\ (v_2),\ (v_1) & (\text{time}: 54) \\ \textit{Robot } r_3 \textit{ collects samples from:} & \\ (v_8),\ (v_5),\ (v_3),\ (v_2),\ (v_1) & (\text{time}: 52) \end{array} \tag{5}$$

In the remainder of this section we focus on the situation in which the mobile robots cannot accomplish Problem 1 due to energy constraints.

**Remark 2. Relaxing infeasible problems:** *If MILP (4) is infeasible, this means that Problem 1 cannot be solved due to insufficient available energy of robots for collecting all samples.*

*Intuitive argument.* Whenever (4) has a non-empty feasible set (the set defined by the linear constraints), it returns an optimal solution from this set [45,49]. The feasible set can become empty only when the first two sets of constraints and the fourth ones from (4) are too stringent. The third set of constraints cannot imply the emptiness of feasible set, because there is no upper bound on $z$. It results that (4) has no solution whenever the first, second and fourth sets of its constraints cannot simultaneously hold. The fourth constraints cannot be relaxed, and therefore only the first two sets may imply the infeasibility of (4). This proves the remark, since the first constraints require all samples to be collected, while the second ones impose upper bounds on consumed energy.

MILP (4) has a non-empty feasible when the required energy for collecting all samples is small enough, or when the available energy limits $\mathcal{E}(r)$ are large enough. This is because the connectedness of $G$ implies that the coefficients $\omega(\pi(s))$ are finite, $\forall s \in S$. Therefore, the first two sets of constraints from (4) could be satisfied even by an initial solution of form $x(r,s) = 1$ for a given $r \in R$, $\forall s \in S$, and $x(r',s) = 0$ for any $r' \in R \setminus \{r\}$.

This aspect yields the idea that one can relax the first constraints from (4) whenever there is no solution, i.e., collect as many samples as possible with the available robot energy. Such a formulation is given by the MILP problem (6), which allows that some samples are not collected (inequalities in first constraints) and imposes a penalty in the cost function for each uncollected sample. Basically, for a big enough value of $W > 0$ from (6), any uncollected sample would increase the value of the objective function more than the decrease resulted from saved energy. The constant $W$ can be lower-bounded by:

$$W > \gamma \cdot \sum_{s\in S} \omega(\pi(s)).$$

For this lower bound, the cost function increases whenever a sample $s$ is not collected, because the term $z$ decreases with $\gamma \cdot \omega(\pi(s))$ and the second term increases with more than this value. Since MILP optimization returns a global optimum, minimizing the cost function under constraints from (6) guarantees that the largest possible number of samples are collected while minimizing the necessary time.

Observe that when (6) is employed, the returned value of the minimized function does not represent the time for collecting all samples, but this time is given by the returned $z$.

$$\begin{aligned}
\min_{x,z} \quad & z - W \cdot \sum_{s\in S} \sum_{r\in R} x(r,s) \\
\text{s.t.:} \quad & \sum_{r\in R} x(r,s) \leq 1\,, \ \forall s \in S \\
& \sum_{s\in S} \Big(\omega(\pi(s)) \cdot x(r,s)\Big) \leq \mathcal{E}(r)\,, \ \forall r \in R \\
& \gamma \cdot \sum_{s\in S} \Big(\omega(\pi(s)) \cdot x(r,s)\Big) \leq z\,, \ \forall r \in R \\
& x(r,s) \in \{0,1\}\,, \ \forall (r,s) \in R \times S \\
& z \geq 0
\end{aligned} \tag{6}$$

The optimization problem from Equation (6).

## 4. Sub-Optimal Planning Methods

In the general case, a MILP optimization is NP-hard [50]. The computational complexity increases with the number of integer variables and with the number of constraints, but exact complexity orders or upper bounds on computational time cannot be formulated [51]. These notes imply that for some cases the complexity of MILP (4) or (6) may render the solution from Section 3 as being computationally intractable, although the optimization is run off-line, i.e., before robot movement.

This section includes two approaches for overcoming this issue. Section 4.1 reformulates the MILP problem (4) as in [34] and obtains a QP formulation. Section 4.2 proposes an IH algorithm, inspired by allocation ideas from [35].

### 4.1. Quadratic Programming Relaxation

We aim to relax the binary constraints from MILP (4), and for accomplishing this we embed them into a new objective function. The idea starts from various penalty formulations defined in [52], some being related to the so-called big M method [49].

Let us replace the binary constraints $x(r,s) \in \{0,1\}$ from (4) with lower and upper bounds of 0 and respectively 1 for outcomes of map $x$. At the same time, add to the cost function of (4) a penalty term depending on $M > 0$, as shown in the objective

$$\min_{x,z} z + M \cdot \sum_{r\in R}\sum_{s\in S} \Big(x(r,s) \cdot (1 - x(r,s))\Big). \tag{7}$$

For a large enough value of the penalty parameter $M$, the minimization of the new cost function from (7) tends to yield a binary value for each variable $x(r,s)$. This is because only binary outcomes of $x$ imply that the second term from sum (7) vanishes, while otherwise this term has a big value due to the large $M$. The quadratic objective from (7) can be re-written in a standard form of an objective function of a QP problem, and together with the remaining constraints from (4) we obtain the QP formulation:

$$\begin{aligned}
\min_{x,z}\quad & z + M \cdot \sum_{r\in R}\sum_{s\in S} x(r,s) \;-\; M \cdot \sum_{r\in R}\sum_{s\in S} \big(x(r,s)\big)^2 \\
\text{s.t.:}\quad & \sum_{r\in R} x(r,s) = 1\,,\ \forall s \in S \\
& \sum_{s\in S} \Big(\omega\big(\pi(s)\big) \cdot x(r,s)\Big) \le \mathcal{E}(r)\,,\ \forall r \in R \\
& \gamma \cdot \sum_{s\in S} \Big(\omega\big(\pi(s)\big) \cdot x(r,s)\Big) \le z\,,\ \forall r \in R \\
& 0 \le x(r,s) \le 1\,,\ \forall (r,s) \in R \times S \\
& z \ge 0
\end{aligned} \tag{8}$$

Under the above informal explanations and based on formal proofs from [52], the MILP (4) and QP (8) have the same global minimum for a sufficiently large value of parameter $M$ (The actual value of $M$ is usually chosen based on numerical ranges of other data from the optimization problem, as values returned by maps $\omega$ and $\mathcal{E}$.).

**Remark 3** (Sub-optimality or failure)**.** *Please note that the cost function from* (8) *is non-convex, because of the negative term in* $x(r,s)^2$. *Therefore, optimization* (8) *could return local minima, while in some cases the obtained values of* $x$ *may even be non-binary. If the obtained outcomes of* $x$ *are binary, the value of completion time* $z$ *for collecting all samples is directly returned as the cost of QP* (8)*, while otherwise a large cost is obtained due to the non-zero term in* $M$ *from* (7).

The QP optimization (8) can be solved with existing software tools [46,48]. As noted, it may return a sub-optimal solution. Nevertheless, such a sub-optimal solution is preferable when the optimal solution from Section 3 is computationally intractable. If the QP returns a (local minimum) solution with non-integer values for outcomes of map $x$, then this result cannot be used for solving Problem 1.

**Example 3.** *Consider again the example from the end of Section 2. The paths in* $G$ *and outcome values of map* $\omega$ *were already computed as in Section 3, where the optimal cost from MILP* (4) *was 54. QP* (8) *was solved in less than 0.4 s and it led to a sub-optimal total time of 60 for bringing all samples in* $v_{10}$. *The robots were allocated to collect samples as follows:*

$$\begin{array}{ll}
\textit{Robot } r_1 \textit{ collects samples from:} & \\
(v_8),\ (v_3),\ (v_2),\ (v_1),\ (v_1) & (\text{time}: 60) \\
\textit{Robot } r_2 \textit{ collects samples from:} & \\
(v_8),\ (v_8),\ (v_3),\ (v_2),\ (v_1) & (\text{time}: 48) \\
\textit{Robot } r_3 \textit{ collects samples from:} & \\
(v_5),\ (v_2),\ (v_2),\ (v_1) & (\text{time}: 52)
\end{array} \tag{9}$$

A similar QP relaxation may be constructed for MILP (6) by considering $M >> W$.

### 4.2. Iterative Solution

This subsection proposes an alternative sub-optimal allocation method, described in Algorithm 2. The method iteratively picks an uncollected sample whose transport to deposit requires minimum energy (line 7) and assigns it to a robot that has spent less energy (time) than other agents (lines 8–15). If a robot does not have enough energy to pick the current sample $s$, it is removed from further assignments (lines 16–17), because the remaining samples would require more energy than $\omega(\pi(s))$. If the current sample $s$ cannot be allocated to any robot, the procedure is stopped (lines 18–19), and in this case some samples remain uncollected. When Algorithm 2 reaches line 20, the robot assignments constitute a solution to Problem 1 for collecting all samples from $S$. The robot-to-sample allocations returned by Algorithm 2 can be easily transformed to robot plans, as in lines 4–13 from Algorithm 1. The total time for completing the mission can be easily computed by maximizing over the times spent by each robot.

**Algorithm 2:** Iterative heuristic solution

**Input:** $R, S, w, \mathcal{E}$
**Output:** Robot-to-sample assignments
1 $R_{assign} = R$
2 $S_{uncollected} = S$
3 Set $x(r,s) = 0, \forall (r,s) \in R \times S$
4 Let $\mathcal{E}_{consumed}(r) = 0, \forall r \in R$
5 **while** $S_{uncollected} \neq \emptyset$ **do**
6 $S_0 = S_{uncollected}$
7 Pick $s \in S_{uncollected}$ s.t. $\omega(\pi(s)) = \min_{s \in S_{uncollected}} \omega(\pi(s))$
8 Sort $R_{assign}$ based on ascending order of consumed robot energy ($\mathcal{E}_{consumed}$)
9 **for** $r \in R_{assign}$ **do**
10 **if** $w(s) \leq \mathcal{E}(r)$ **then**
11 $x(r,s) = 1$ (assign sample $s$ to robot $r$)
12 $\mathcal{E}(r) := \mathcal{E}(r) - w(s)$
13 $\mathcal{E}_{consumed}(r) := \mathcal{E}_{consumed}(r) + w(s)$
14 $S_{uncollected} := S_{uncollected} \setminus \{s\}$
15 Break "for" loop
16 **else**
17 $R_{assign} := R_{assign} \setminus \{r\}$
18 **if** $S_{uncollected} = S_0$ **then**
19 *Return* current robot-to-sample allocations $x$
20 *Return* robot-to-sample allocations $x$

Under these ideas, Algorithm 2 can be seen as a greedy approach (first collect samples that require less energy/time), while the allocations to robots with less spent energy tries to reduce the overall time until the samples are collected.

Observe that this IH solution always returns a solution for collecting some (if not all) samples, whereas MILP from Section 3 may become computationally impracticable, while QP (8) may fail in providing a solution (Remark 3). Moreover, the software implementation of Algorithm 2 does not require additional tools, in contrast with specific optimization packages needed by MILP and QP solutions. A detailed analysis of the three methods is the goal of Section 5.

**Example 4.** *For illustrating Algorithm 2 on the example considered in the previous sections, we give here the sample picking costs:* $\omega(\pi(s)) = 16$, $s = 1, \ldots, 4$, $\omega(\pi(s)) = 14$, $s = 5, \ldots, 8$, $\omega(\pi(s)) = 10$, $s = 9, 10$, $\omega(\pi(11)) = 8$, $\omega(\pi(s)) = 4$, $s = 12, \ldots, 14$. *First, IH solution allocates sample 12 to* $r_1$, *then 13 and 14 to* $r_2$, $r_3$, *sample 11 to* $r_1$ *and so on. Algorithm 1 was run in 0.015 s and it yielded the following robotic plans:*

$$\begin{array}{ll} \textit{Robot } r_1 \textit{ collects samples from:} & \\ (v_8),\ (v_5),\ (v_2),\ (v_2),\ (v_1) & (\text{time}: 56) \\ \textit{Robot } r_2 \textit{ collects samples from:} & \\ (v_8),\ (v_3),\ (v_2),\ (v_1),\ (v_1) & (\text{time}: 60) \\ \textit{Robot } r_3 \textit{ collects samples from:} & \\ (v_8),\ (v_3),\ (v_2),\ (v_1) & (\text{time}: 44) \end{array} \tag{10}$$

## 5. Numerical Evaluation and Comparative Analysis

### 5.1. Additional Examples

Besides the remarks and examples from Sections 3 and 4, we present some slight modifications of the Example from Section 2 with the purpose of emphasizing the need for a comparative analysis of the three proposed solutions.

**Example A:** Let us add one more sample in node $v_1$ of the environment from Figure 1, leading to a total number of 15 samples. By running the MILP optimization (4), a solution was obtained in almost 40 s and it leads to an optimum time of 60. The QP relaxation (8) was run in almost 0.4 s (negligible increase from example from Section 4.1), and it implied a time cost of 62 for collecting all samples. The IH solution from Algorithm 2 yielded a cost of 60 in 0.016 s (practically no different to in Section 4.2). The actual robotic plans are omitted for this case.

**Example B:** If we assume a team of 4 robots for Example A, the MILP running time exhibits a significant decrease, being solved in less than 0.1 s. The QP optimization was solved in 0.5 s, and the IH in less than 0.02 s. The resulted time costs were 44 for MILP (optimum) and 50 for QP and IH (sub-optimum).

**Example C:** By adding one more robot to the team from Example B, the MILP optimization did not finish in 1 h, so it can be declared computationally unfeasible for this situation. The QP optimization finished in slightly more than 0.5 s, while the IH in around 0.02 s. Both QP and IH solutions returned a cost of 40.

Similar modifications of the above examples suggested the following empirical ideas:

- The running time of the MILP optimization may exhibit unpredictable behaviors with respect to the team size and to the number and position of samples, leading to impossibility of obtaining a solution in some cases;
- In contrast to MILP, the running times of the QP and IH solutions have insignificant variations when small changes are made in the environment;
- When MILP optimization finishes, the sub-optimal costs obtained by solutions from Section 4 are generally acceptable when compared to the optimal cost;
- In some cases the QP cost was better than the one obtained by IH, while in other cases the vice versa, but again the observed differences were fairly small.

The above ideas were formulated only based on a few variations of the same example. However, they motivate the more extensive comparison performed in the next subsections between the computation feasibility and outcomes of MILP, QP, and IH solutions.

**Remark 4.** *As mentioned, complexity orders of MILP and QP solutions cannot be formally given. However, as is customary in some studies, we here recall the number of unknowns and constraints of these optimizations. MILP (4) and QP (8) have each* $N_R \times N_S + 1$ *unknowns (from which* $N_R \times N_S$ *are binary in case of MILP (4))* *and $2N_R + N_S$ linear constraints. IH solution has complexity order $\mathcal{O}(N_R \times N_S)$, based on iterative loops from Algorithm 2 but, in all our studies the execution time of the algorithm is very small.*

**Real-time example:** Sample collecting experiments were implemented on a test-bed platform by using two Khepera robots equipped with plows for collecting items [53]. For exemplification, a movie is available at https://www.youtube.com/watch?v=2BQiWvquP7w. In the mentioned scenario, the graph environment is obtained from a cell decomposition [1,11] and a greedy method is employed for planning the robots. Although the collision avoidance problem is not treated in this paper, in the mentioned experiment the possible collisions are avoided by pausing the motion of one robot. In future work we intend to embed formal tools inspired by resource allocation techniques for collision and deadlock avoidance [54,55].

### 5.2. Numerical Experiments

All the simulations to be presented were implemented in MATLAB [48] and were performed on a computer with Intel i7 quad-core processor and 8 GB RAM.

The numerical experiments were run for almost 15 days, and they were organized by considering the following aspects:

(i) Time complexity orders cannot be a priori given for MILP or for non-convex QP optimization problems. Thus, the time for obtaining a planning solution solving Problem 1 is to be recorded as an important comparison criterion.

(ii) The complexity of MILP, QP, and IH solutions does not directly depend on the size of environment graph $G$, except for the initial computation of map $\omega$ that further embeds the necessary information from the environment structure. Therefore, the running time of either solution is influenced by two parameters: the number of robots ($N_R$) and the number of samples ($N_S$).

(iii) Based on item (ii), we consider variation ranges $(N_R, N_S) \in [2, \ldots, 10] \times [2, \ldots, 50]$, with unit increment steps for $N_R$ and $N_S$. Please note that the cases of 1 robot and/or 1 sample are trivial, and therefore are not included in the above parameter intervals.

(iv) To obtain reliable results for item (i), for each pair $(N_R, N_S)$ we have run a set of 50 trials. For each trial we generated a random distribution of samples in a 50-node graph. To maintain focus on time complexity, we assumed sufficiently large amounts of robot energy $\mathcal{E}$, such that Problem 1 is not infeasible due to these limitations.

(v) For each trial, the MILP optimization was deemed *failed* whenever (4) did not return a solution in less than 1 min. This is because in multiple situations we observed that if no solution is obtained in less than 30–40 s, then the MILP (4) does not finish even after 2–3 h.

(vi) For each trial from item (iv), the QP solution was deemed *failed* whenever it yielded non-binary outcomes of map $x$ (see Remark 3). The IH solution is always *successful*.

(vii) For each proposed solution, for each pair $(N_R, N_S)$ from (iii) and based on trials from (iv), we computed the following comparison criteria:

- *success rate*, showing the percentage of trials when the solution succeeded in outputting feasible plans;
- *computation time*, averaged over the successful trials of a given $(N_R, N_S)$ instance;
- *solution cost*, i.e., time for gathering all samples, for each successful situation.

### 5.3. Results

The results from item (vii) allow us to draw some rules that guide a user to choose MILP, QP, or IH solution when solving a specific instance of Problem 1. The following figures present and comment these results, leading to the decision diagram from the end of this section.

Figure 2 illustrates the success rates of the optimization problems from Sections 3 and 4.1, respectively. For a clearer understanding, Figure 3 presents pairs $(N_R, N_S)$ when MILP (4) fails in more than 50% from each set of 50 trials (see item (v) from Section 5.2). It is noted that MILP generally returns optimal solutions for small values of $N_R$ and $N_S$ and fails (because of optimization time limit) for larger values. QP returns usually returns feasible solutions, excepting some cases with small values of $N_R$ and $N_S$.

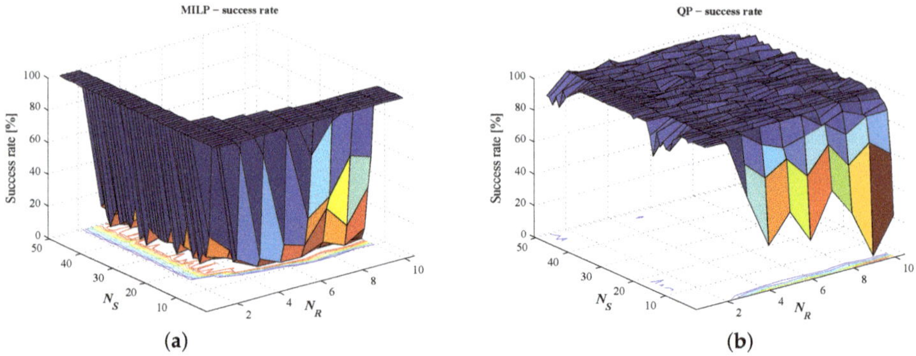


**Figure 2.** Success rates of MILP (**a**) and QP (**b**) vs. number of robots $N_R$ and number of samples $N_S$.

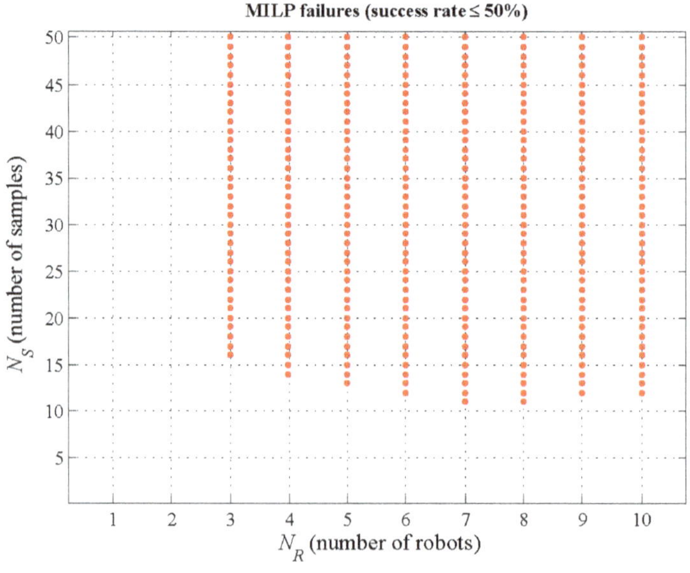


**Figure 3.** MILP (4): 2D projection for failures, defined as success rate of less than 50%.

Figure 4 presents the average computation time over the successful trials, for each solution we proposed. The representation is omitted for pairs $(N_R, N_S)$ when there are less than 5 (from 50) successful trials—as it is often the case for MILP solver, when $N_R \geq 3$ and $N_S \geq 13$. MILP time may sudden variations, whereas the times for QP and IH indicates predictable behaviors. The IH time is very small (note axis limits in Figure 4) and exhibits negligible variations versus $N_R$ and almost linear increases versus $N_S$, due to the main iteration loop from Algorithm 2.

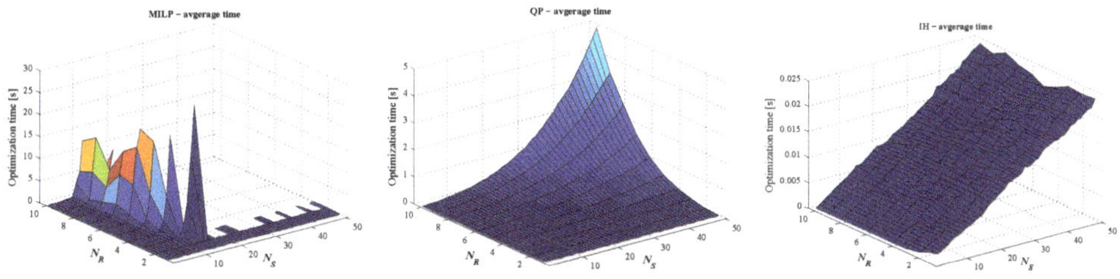


(**a**) (**b**) (**c**)

**Figure 4.** Average optimization times, for $(N_R, N_S)$ pairs for which at least 5 feasible solutions from the 50 tests were obtained: (**a**) MILP (4), (**b**) QP (8), (**c**) IH from Algorithm 2.

To suggest the confidence intervals of values plotted in Figure 4, we mention that:

- For $N_R = 3$ and $N_S = 15$, when the success rate of each optimization exceeds 98%, the standard deviations of optimization times are: 15 for MILP, 0.01 for QP, 0.002 for IH;
- For $N_R = 10$ and $N_S = 50$, when QP has 96% success rate, the standard deviations of optimization times are: 0.23 for QP and 0.003 for IH.

Figure 5 presents the averaged relative differences of costs yielded by the three proposed solutions for solving Problem 1. As visible in Figure 5a, the QP (sub-optimal) cost is usually less than 120% . . . 130% of optimal MILP cost. More specifically, from all the 25,000 trials, in 8443 cases (about 33%) both optimizations succeeded. After averaging cost differences versus $(N_R, N_S)$, we obtained 348 points for representing Figure 5a, and in 330 cases the cost difference was less than 20%. Figure 5b compares the costs yielded by the sub-optimal solutions from Section 4 by representing variations of the IH cost related to the QP one. It follows that usually IH yields a higher cost than QP, but the difference decreases below 5% . . . 10% with the increase in problem complexity. Further studies can be conducted towards formulating a conjecture that gives a formal tendency for the variation of cost difference based on problem size. However, one issue for such a study is mainly given by the necessity of obtaining the optimal cost even for large problems, i.e., solving large MILP optimizations.

For more complex problems ($N_R > 10$, $N_S > 50$), the time tendencies from Figure 4b,c and the cost differences from Figure 5b suggest that the IH solution is preferable as a good trade-off between planning complexity and resulted cost.

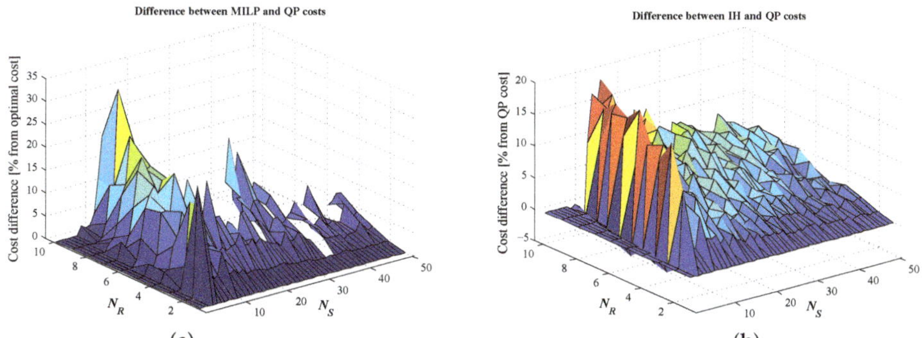


(**a**) (**b**)

**Figure 5.** Differences between costs induced by the three solutions: (**a**) difference between QP and MILP costs, related to the optimal MILP cost; (**b**) difference between IH and QP sub-optimal costs, related to QP cost.

Extensive simulations yielded quantitative comparison criteria. Based on this information a decision scheme Figure 6 is given for indicating the proper method to be used in a specific problem instance when a fast computation scenario is considered.

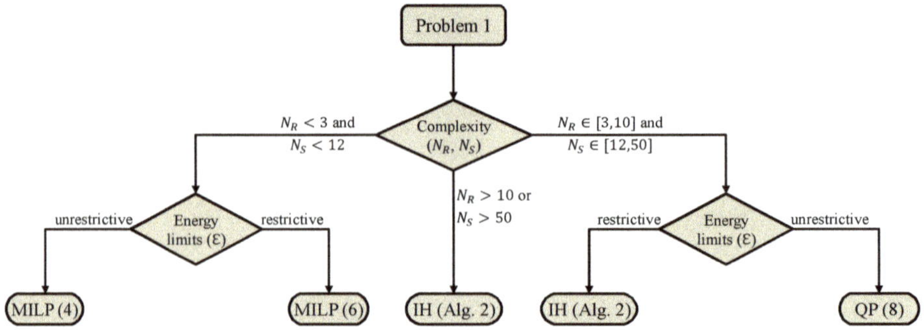


**Figure 6.** Decision diagram for choosing an appropriate solution for Problem 1 when a fast computational scenario is considered.

## 6. Conclusions

This paper details three methods for planning a team of robots such that multiple samples (items) from the environment are collected and deposited into a storage location. The environment is represented by a weighted graph, and the robots have limited amounts of energy for performing movements on this graph. The goal is to plan the agents such that the samples are collected in minimum time, under the assumption that each robot can carry at most one item at a time. The first solution is given by a MILP formulation that can be too complex to solve when there are many samples or many robots in the team. The second solution provides a QP relaxation that represents in some cases a good trade-off between the time for finding movement plans and the cost difference from the optimal one. The third solution is an IH algorithm that yields plans even when the QP fails due to low amounts of available energy for robots.

Based on the results reported in the previous subsection, the recommendations for a user that solves Problem 1 are the following: MILP (6) or IH algorithm are to be used when the robots have small amounts of energy in comparison with the energy required for moving to sample locations. Otherwise, MILP (4) and QP (8) may fail in providing any solution for such restrictive scenarios.

The usage of QP or IH solutions generally yields an acceptable loss in total time for accomplishing the mission whenever the MILP optimization becomes computationally intractable on a decently powerful computer. We emphasize that the above recommendations resulted from an intensive campaign of numerical simulations and they could not be drawn by only inspecting the formal solutions. Extensive simulations yielded quantitative comparison criteria. Based on this information a decision scheme is given for indicating the proper method to be used in a specific problem instance. Besides suggestions from Figure 6, we recall that MILP and QP solutions involve existing optimization tools.

A real-time experiment was performed for illustrating a sample gathering solution. Although the collision avoidance problem is not treated in this paper, in the mentioned experiment the possible collisions are avoided by pausing the motion of one robot. In future work we intend to embed formal tools inspired by resource allocation techniques for collision and deadlock avoidance, while considering the effects of acceleration and deceleration of the mobile robots and restricted energy.

**Author Contributions:** The authors contributed equally to this work, each of them being involved in all research aspects.

**Funding:** This research was partially supported by a grant of Romanian Ministry of Research and Innovation: CNCS-UEFISCDI project PN-III-P1-1.1-TE-2016-0737.

**Conflicts of Interest:** The author declares no conflict of interest.

## References


1. Choset, H.; Lynch, K.; Hutchinson, S.; Kantor, G.; Burgard, W.; Kavraki, L.; Thrun, S. *Principles of Robot Motion: Theory, Algorithms and Implementation*; MIT Press: Cambridge, MA, USA, 2005.
2. LaValle, S.M. *Planning Algorithms*; Cambridge University Press: Cambridge, UK, 2006.
3. Siciliano, B.; Khatib, O. *Springer Handbook of Robotics*; Springer: Berlin, Germany, 2008.
4. Belta, C.; Bicchi, A.; Egerstedt, M.; Frazzoli, E.; Klavins, E.; Pappas, G.J. Symbolic Planning and Control of Robot Motion. *IEEE Robot. Autom. Mag.* **2007**, *14*, 61–71. [CrossRef]
5. Imeson, F.; Smith, S.L. A Language For Robot Path Planning in Discrete Environments: The TSP with Boolean Satisfiability Constraints. In Proceedings of the IEEE Conference on Robotics and Automation, Hong Kong, China, 31 May–7 June 2014; pp. 5772–5777.
6. Mahulea, C.; Kloetzer, M. Robot Planning based on Boolean Specifications using Petri Net Models. *IEEE Trans. Autom. Control* **2018**, *63*, 2218–2225. [CrossRef]
7. Fainekos, G.; Girard, A.; Kress-Gazit, H.; Pappas, G. Temporal logic motion planning for dynamic robot. *Automatica* **2009**, *45*, 343–352. [CrossRef]
8. Ding, X.; Lazar, M.; Belta, C. {LTL} receding horizon control for finite deterministic systems. *Automatica* **2014**, *50*, 399–408. [CrossRef]
9. Schillinger, P.; Bürger, M.; Dimarogonas, D. Simultaneous task allocation and planning for temporal logic goals in heterogeneous multi-robot systems. *Int. J. Robot. Res.* **2018**, *37*, 818–838. [CrossRef]
10. Kloetzer, M.; Mahulea, C. LTL-Based Planning in Environments With Probabilistic Observations. *IEEE Trans. Autom. Sci. Eng.* **2015**, *12*, 1407–1420. [CrossRef]
11. Berg, M.D.; Cheong, O.; van Kreveld, M. *Computational Geometry: Algorithms and Applications*, 3rd ed.; Springer: Berlin, Germany, 2008.
12. Kloetzer, M.; Mahulea, C. A Petri net based approach for multi-robot path planning. *Discret. Event Dyn. Syst.* **2014**, *24*, 417–445. [CrossRef]
13. Cassandras, C.; Lafortune, S. *Introduction to Discrete Event Systems*; Springer: Berlin, Germany, 2008.
14. Silva, M. Introducing Petri nets. In *Practice of Petri Nets in Manufacturing*; Springer: Berlin, Germany, 1993; pp. 1–62.
15. Burkard, R.; Dell'Amico, M.; Martello, S. *Assignment Problems*; SIAM e-Books; Society for Industrial and Applied Mathematics (SIAM): Philadelphia, PA, USA, 2009.
16. Mosteo, A.; Montano, L. *A Survey of Multi-Robot Task Allocation*; Technical Report AMI-009-10-TEC; Instituto de Investigaci6n en Ingenieria de Arag6n, University of Zaragoza: Zaragoza, Spain, 2010.
17. Gerkey, B.; Matarić, M. A formal analysis and taxonomy of task allocation in multi-robot systems. *Int. J. Robot. Res.* **2004**, *23*, 939–954. [CrossRef]
18. Korsah, G.; Stentz, A.; Dias, M. A comprehensive taxonomy for multi-robot task allocation. *Int. J. Robot. Res.* **2013**, *32*, 1495–1512. [CrossRef]
19. Shmoys, D.; Tardos, É. An approximation algorithm for the generalized assignment problem. *Math. Program.* **1993**, *62*, 461–474. [CrossRef]
20. Atay, N.; Bayazit, B. Emergent task allocation for mobile robots. In Proceedings of the Robotics: Science and Systems, Atlanta, GA, USA, 27–30 June 2007.
21. Cardema, J.; Wang, P. Optimal Path Planning of Multiple Mobile Robots for Sample Collection on a Planetary Surface. In *Mobile Robots: Perception & Navigation*; Kolski, S., Ed.; IntechOpen: London, UK, 2007; pp. 605–636.
22. Chen, J.; Sun, D. Coalition-Based Approach to Task Allocation of Multiple Robots with Resource Constraints. *IEEE Trans. Autom. Sci. Eng.* **2012**, *9*, 516–528. [CrossRef]
23. Das, G.; McGinnity, T.; Coleman, S.; Behera, L. A Distributed Task Allocation Algorithm for a Multi-Robot System in Healthcare Facilities. *J. Intell. Robot. Syst.* **2015**, *80*, 33–58. [CrossRef]
24. Burger, M.; Notarstefano, G.; Allgower, F.; Bullo, F. A distributed simplex algorithm and the multi-agent assignment problem. In Proceedings of the American Control Conference (ACC), San Francisco, CA, USA, 29 June–1 July 2011; pp. 2639–2644.

25. Luo, L.; Chakraborty, N.; Sycara, K. Multi-robot assignment algorithm for tasks with set precedence constraints. In Proceedings of the IEEE International Conference on Robotics and Automation (ICRA), Shanghai, China, 9–13 May 2011; pp. 2526–2533.
26. Toth, P.; Vigo, D. (Eds.) *The Vehicle Routing Problem*; Society for Industrial and Applied Mathematics: Philadelphia, PA, USA, 2001.
27. Laporte, G.; Nobert, Y. Exact Algorithms for the Vehicle Routing Problem. In *Surveys in Combinatorial Optimization*; Martello, S., Minoux, M., Ribeiro, C., Laporte, G., Eds.; North-Holland Mathematics Studies; North-Holland: Amsterdam, The Netherlands, 1987; Volume 132, pp. 147–184.
28. Toth, P.; Vigo, D. Models, relaxations and exact approaches for the capacitated vehicle routing problem. *Discret. Appl. Math.* **2002**, *123*, 487–512. [CrossRef]
29. Raff, S. Routing and scheduling of vehicles and crews: The state of the art. *Comput. Oper. Res.* **1983**, *10*, 63–211. [CrossRef]
30. Christofides, N. Vehicle routing. In *The Traveling Salesman Problem: A guided Tour of Combinatorial Optimization*; Wiley: New York, NY, USA, 1985; pp. 410–431.
31. Laporte, G. The vehicle routing problem: An overview of exact and approximate algorithms. *Eur. J. Oper. Res.* **1992**, *59*, 345–358. [CrossRef]
32. Chen, J.F.; Wu, T.H. Vehicle routing problem with simultaneous deliveries and pickups. *J. Oper. Res. Soc.* **2006**, *57*, 579–587. [CrossRef]
33. Kloetzer, M.; Burlacu, A.; Panescu, D. Optimal multi-agent planning solution for a sample gathering problem. In Proceedings of the IEEE International Conference on Automation, Quality and Testing, Robotics (AQTR), Cluj-Napoca, Romania, 22–24 May 2014.
34. Kloetzer, M.; Ostafi, F.; Burlacu, A. Trading optimality for computational feasibility in a sample gathering problem. In Proceedings of the International Conference on System Theory, Control and Computing, Sinaia, Romania, 17–19 October 2014; pp. 151–156.
35. Kloetzer, M.; Mahulea, C. An Assembly Problem with Mobile Robots. In Proceedings of the IEEE 19th Conference on Emerging Technologies Factory Automation, Barcelona, Spain, 16–19 September 2014.
36. Panescu, D.; Kloetzer, M.; Burlacu, A.; Pascal, C. Artificial Intelligence based Solutions for Cooperative Mobile Robots. *J. Control Eng. Appl. Inform.* **2012**, *14*, 74–82.
37. Kloetzer, M.; Mahulea, C.; Burlacu, A. Sample gathering problem for different robots with limited capacity. In Proceedings of the International Conference on System Theory, Control and Computing, Sinaia, Romania, 13–15 October 2016; pp. 490–495.
38. Tumova, J.; Dimarogonas, D. Multi-agent planning under local LTL specifications and event-based synchronization. *Automatica* **2016**, *70*, 239–248. [CrossRef]
39. Belta, C.; Habets, L. Constructing decidable hybrid systems with velocity bounds. In Proceedings of the 43rd IEEE Conference on Decision and Control, Nassau, Bahamas, 14–17 December 2004; pp. 467–472.
40. Habets, L.C.G.J.M.; Collins, P.J.; van Schuppen, J.H. Reachability and control synthesis for piecewise-affine hybrid systems on simplices. *IEEE Trans. Autom. Control* **2006**, *51*, 938–948. [CrossRef]
41. Garey, M.; Johnson, D. "Strong" NP-Completeness Results: Motivation, Examples, and Implications. *J. ACM* **1978**, *25*, 499–508. [CrossRef]
42. Cormen, T.; Leiserson, C.; Rivest, R.; Stein, C. *Introduction to Algorithms*, 2nd ed.; MIT Press: Cambridge, MA, USA, 2001.
43. Ding-Zhu, D.; Pardolos, P. *Nonconvex Optimization and Applications: Minimax and Applications*; Spinger: New York, NY, USA, 1995.
44. Polak, E. *Optimization: Algorithms and Consistent Approximations*; Spinger: New York, NY, USA, 1997.
45. Floudas, C.; Pardolos, P. *Encyclopedia of Optimization*, 2nd ed.; Spinger: New York, NY, USA, 2009; Volume 2.
46. Makhorin, A. GNU Linear Programming Kit. 2012. Available online: http://www.gnu.org/software/glpk/ (accessed on 4 January 2019).
47. SAS Institute. The Mixed-Integer Linear Programming Solver. 2014. Available online: http://support.sas.com/rnd/app/or/mp/MILPsolver.html (accessed on 4 January 2019).
48. The MathWorks. *MATLAB®R2014a (v. 8.3)*; The MathWorks: Natick, MA, USA, 2006.
49. Wolsey, L.; Nemhauser, G. *Integer and Combinatorial Optimization*; Wiley: New York, NY, USA, 1999.
50. Vazirani, V.V. *Approximation Algorithms*; Springer: New York, NY, USA, 2001.

51. Earl, M.; D'Andrea, R. Iterative MILP Methods for Vehicle-Control Problems. *IEEE Trans. Robot.* **2005**, *21*, 1158–1167. [CrossRef]
52. Murray, W.; Ng, K. An Algorithm for Nonlinear Optimization Problems with Binary Variables. *Comput. Optim. Appl.* **2010**, *47*, 257–288. [CrossRef]
53. Tiganas, V.; Kloetzer, M.; Burlacu, A. Multi-Robot based Implementation for a Sample Gathering Problem. In Proceedings of the International Conference on System Theory, Control and Computing, Sinaia, Romania, 11–13 October 2013; pp. 545–550.
54. Kloetzer, M.; Mahulea, C.; Colom, J.M. Petri net approach for deadlock prevention in robot planning. In Proceedings of the IEEE Conference on Emerging Technologies Factory Automation (ETFA), Cagliari, Italy, 10–13 September 2013; pp. 1–4.
55. Roszkowska, E.; Reveliotis, S. A Distributed Protocol for Motion Coordination in Free-Range Vehicular Systems. *Automatica* **2013**, *49*, 1639–1653. [CrossRef]

*applied sciences*

MDPI

*Article*

# Cooperative Path Planning for Multiple Mobile Robots via HAFSA and an Expansion Logic Strategy

**Yiqing Huang *, Zhikun Li, Yan Jiang and Lu Cheng**

College of Electrical Engineering, Anhui Polytechnic University, Wuhu 241000, China; lzk52170@163.com (Z.L.); Jiangyan117@163.com (Y.J.); chenglu073@163.com (L.C.)

* Correspondence: yiqhuang@ahpu.edu.cn

Received: 31 January 2019; Accepted: 14 February 2019; Published: 16 February 2019

**Featured Application: In various unknown environments, cooperative path planning problem of multiple mobile robots is becoming more and more important. The efficiency and reliability can be greatly improved by the cooperation of multiple mobile robots. The novel obstacle avoidance and real-time navigation algorithm presented in this article may be useful for marine exploration, military, aerospace and mining detection. Also, for the developed real-time navigation algorithm, the presented hybrid artificial fish swarm algorithm and expansion logic strategy are helpful not only for accelerating the convergence rate, but also for improving decision-making ability.**

**Abstract:** The cooperative path planning problem of multiple mobile robots in an unknown indoor environment is considered in this article. We presented a novel obstacle avoidance and real-time navigation algorithm. The proposed approach consisted of global path planning and local path planning via HAFSA (hybrid artificial fish swarm algorithm) and an expansion logic strategy. Meanwhile, a kind of scoring function was developed, which shortened the time of local path planning and improved the decision-making ability of the path planning algorithm. Finally, using STDR (simple two dimensional robot simulator) and RVIZ (robot operating system visualizer), a multiple mobile robot simulation platform was designed to verify the presented real-time navigation algorithm. Simulation experiments were performed to validate the effectiveness of the proposed path planning method for multiple mobile robots.

**Keywords:** path planning; multiple mobile robots; artificial fish swarm algorithm; expansion logic strategy

## 1. Introduction

Mobile robots can be equipped with different sensors and tools to afford a variety of services such as home care, mining detection and object handling [1,2]. One of the fundamental issues with mobile robots performing tasks is ensuring that they can navigate safely in an unknown indoor environment. Therefore, path planning is crucial for the successful application of mobile robots. The goal of mobile robot path planning is to find a motion path from a starting position to a target position in an environment with obstacles [3–5]. For the past two decades, there has been a great deal of research on the problem of mobile robot path planning. For example, a novel motion map was constructed for mobile robots based on the BIE (boundary integral equation) method, and then, a point-to point path planning problem was addressed in a known environment with static obstacles [6]. Furthermore, an improved three-dimensional-like grid map was developed to represent the environment model [7], and then, a simple but efficient path planning algorithm was presented to solve robot navigation problems in a static environment. The authors designed an autonomous multi-goal navigation system for picking up or delivering tasks in mobile robotics and a multi-goal path planning method based on the Lin–Kernighan heuristics (LKH) algorithm for intelligent service mobile robots in Reference [8].

There are also some intelligent methods that can be applied to mobile robot path planning. The cross probability and the mutation probability for GA (genetic algorithm) were improved and the improved algorithm was applied to the path planning problem of mobile robots in Reference [9], whereas the authors proposed an intelligent motion planning and navigation method for omnidirectional mobile robots via a fuzzy logic algorithm in Reference [10]. Moreover, the developed navigation system is especially suitable for real-time path planning applications. A novel optimal hierarchical global path planning method for mobile robots in a cluttered environment was presented in Reference [11]. In this method, a combination of the triangular decomposition approach, constrained multi-objective PSO (particle swarm optimization) and Dijkstra's algorithm is presented in order to obtain an optimal path planning trajectory. In addition, due to good feedback information and better distributed computing, the authors proposed a path planning method for mobile robots via an improved ant colony algorithm in grid maps in Reference [12].

Using multiple mobile robots rather than a single mobile robot can improve working capability and performance. Therefore, recently, research on multiple mobile robots has become a hot topic. In previous studies, only static obstacles in the unknown environment were considered. For holonomic wheeled mobile robots in static environments, an optimal multiple mobile robot path planning method based on adaptive charged system search (CSS) algorithms was addressed [13]. However, path planning methods in an unknown environment with dynamic obstacles are even more acute in multiple mobile robot areas. In Reference [14], a new path planning approach for coordinating multiple mobile robots was presented and the authors developed an online strategy to adjust path planning for avoidance of dynamic obstacles. Furthermore, a biologically inspired neural-network-based intelligent method was proposed for a multiple robot system with moving obstacles [15]. The proposed method could plan the paths of multiple robots to avoid collision with dynamic obstacles.

Although these previously-developed navigation algorithms have shown good performance for solving robot path planning, they have also shown some limitations such as slow convergence and a local optimum. The local optimal problem is the most common problem in solving path planning. Motivated by the aforementioned reasons, we reconstructed an analytical real-time cooperative navigation algorithm to accommodate multiple mobile robot systems. The proposed EAFSA (empirical artificial fish swarm algorithm) was used to avoid falling into local optimal problems and to realize global path planning for a single mobile robot. Then, an expansion logic strategy was introduced to avoid collisions between multiple mobile robots and an environment with obstacles. A multiple mobile robot simulation system was developed using STDR (simple two dimensional robot simulator) and RVIZ (robot operating system visualizer) software. Finally, the presented method was proven to be effective by experiments conducted in a simulated environment.

The main contributions of the article are summarized as follows. (1) EAFSA is presented to solve the global path planning problem for a single mobile robot; (2) an expansion logic strategy and a kind of scoring function are proposed for a multiple mobile robot real-time navigation algorithm. The presented real-time navigation algorithm is helpful not only for accelerating the convergence rate, but also for improving decision-making ability.

The remainder of this paper is organized as follows: Section 2 describes the process of the presented HAFSA. Section 3 presents the developed expansion logic strategy and scoring function. Section 4 shows the results of a simulation to demonstrate the performance of the proposed algorithms. The concluding remarks are given in Section 5.

## 2. Hybrid Artificial Fish Swarm Algorithm

The artificial fish swarm algorithm (AFSA) is a novel swarm intelligent optimization method inspired by natural fish swarm behavior. It has been successfully used in the field of wireless telemedicine systems [16], fault diagnosis [17], indoor visible light positioning [18], floating wind turbines [19], etc.

The basic idea of AFSA can be described as follows: If the position of each artificial fish is $X = (x_1, x_2, \cdots, x_n)$ and the size of the fish population is *Num*, $Y$ denotes the food concentration of the artificial fish in the current position and $Y = f(X)$ is fitness or the objective function at position $X$. Each artificial fish tries to find an optimal position to satisfy their food needs using preying behavior, swarming behavior, following behavior and random behavior [20].

**(1) Preying Behavior**

If the current state of an artificial fish is $X_i(t)$, $X_j(t)$ is the random state of its visual distance, and $X_{i+1}(t)$ is the next position of $X_i(t)$. If food concentration is $Y_i < Y_j$, the artificial fish swims a *step* in the direction of $X_j(t)$. Otherwise, it randomly selects a state again and judges whether it satisfies the aforementioned condition. In other words, preying behavior can be expressed by the following equation:

$$X_{i+1}(t) = \begin{cases} X_i(t) + \frac{X_j(t) - X_i(t)}{\|X_j(t) - X_i(t)\|} \times step \times rand(0,1) & if\ Y_i < Y_j \\ X_i(t) + Visual \times rand(0,1) & if\ Y_i \geq Y \end{cases}. \quad (1)$$

**(2) Swarming Behavior**

When $N_F$ is the number of artificial fishes in the current position $X_i(t)$, $X_c(t)$ is the center position of the artificial fishes in their current neighborhood. $if\ Y_c/N_F > \delta Y_i$ is satisfied, the artificial fish moves to a center position, according to Equation (2), due to high food concentration and to avoid crowding each other. Otherwise, the artificial fish executes preying behavior.

$$X_i(t) + \frac{X_c(t) - X_i(t)}{\|X_c(t) - X_i(t)\|} \times step \times rand(0,1) \quad if\ Y_c/N_F > \delta Y_i \quad (2)$$

**(3) Following Behavior**

Let $X_{\max}(t)$ be the local best companion with food concentration $Y_{\max}$ in the current neighborhood of $X_i(t)$. $if\ Y_{\max}/N_F > \delta Y_i$ is satisfied, the artificial fish moves to a position according to Equation (3). Otherwise, the next position of the artificial fish can be obtained by preying behavior.

$$X_i(t) + \frac{X_{\max}(t) - X_i(t)}{\|X_{\max}(t) - X_i(t)\|} \times step \times rand(0,1) \quad if\ Y_{\max}/N_F > \delta Y_i \quad (3)$$

**(4) Random Behavior**

The artificial fish chooses an arbitrary state or position randomly in its *Visual* field, and then it swims towards the selected state. Random behavior is a default behavior and it can be described as

$$X_i(t) + Visual \times rand(0,1), \quad (4)$$

where $X_i(t)$ is the current state of the artificial fish and $X_{i+1}(t)$ is the next position of $X_i(t)$.

Given the above consideration, an effective hybrid fish swarm algorithm (HFSA) with experiential learning and a detection operator was presented to solve the local optimal problem and realize global path planning for a single mobile robot.

### 2.1. Experiential Learning

In this section, an experiential learning strategy is presented to improve the performance of AFSA. Experiential learning strategies include adjustment of the step size and food concentration for the artificial fish. The step size of an artificial fish is fixed in traditional AFSA. However, as is known, the step size determines the convergence rate. If the step size is too small, the artificial fish will reach the optimal solution slowly and the global search ability will be decreased. Therefore, the artificial fish is easy to fall into a local optimum. Conversely, if the step size is too large, the convergence speed will be increased and oscillation will occur later in the algorithm iteration. Therefore, it is necessary to select an appropriate step size to ensure the global convergence speed and improve the accuracy of the optimal solution. In this article, a logarithmic function was used to update the step size by Equation (5);

$$b = \log_N p, p = 1,2,3 \cdots N, \tag{5}$$

where $p$ indicates the current iteration number, and $N$ is the maximum number of iterations.

Then, the step size in the population update formula could be obtained by

$$step = \log_N p \cdot (X_\varepsilon(t) - X_i(t))p = 1,2,3 \cdots N, i = 1,2,3 \cdots NUM, \tag{6}$$

where $X_i(t)$ is the current state of the artificial fish, $X_{i+1}(t)$ is the next position of $X_i(t)$ and $X_\varepsilon(t)$ is the state that needs to be searched.

Finally, the position of the updated solution could be expressed as follows:

$$X_{i+1}(t) = X_i(t) + \log_{NC} p \cdot (X_\varepsilon(t) - X_i(t))p = 1,2,3 \cdots N, i = 1,2,3 \cdots NUM. \tag{7}$$

As shown in Equation (7), with the increase in the number of iterations, the moving step gradually adapts to the change of the iteration numbers. The local search ability is increased by the improved moving factor. As a result, the artificial fish can locate the search direction quickly, move to the target area, maintain the global search ability of the optimal solution and accelerate the convergence speed.

On the other hand, the food concentration of the artificial fish was represented as the ability of the solution to solve an optimization problem. The current position of the artificial fish with the highest food concentration was the optimal solution of the optimization problem. In this article, a weight coefficient function was used to design the food concentration, which could decrease the food concentration of the artificial fish and avoid the problem that the suboptimal solution of the fish swarm algorithm would interfere with the global solution. The food concentration equation was designed as follows:

$$H_{\exp} = w\sqrt{(x_i - x_g)^2 + (y_i - y_g)^2}, \tag{8}$$

where the weight coefficient is $w \in [1, 1.5)$.

### 2.2. Detection Operator

A detection operator was developed to optimize the resulting path trajectory. If $p(kx, ky)$ is the position coordinate of the optimal fish group solved at the $k$th time, $n$ is the dimension of the grid-based map and $k_t$ is the number of the solutions that have been optimized. $goal(x, y)$ is the position coordinate of the target point. The detection operator $R(t)$ can be described as follows:

$$R(t) = \begin{cases} \lceil D(k)/(2k_t) \rceil \times \sqrt{2} & 0 < t < count \\ 2\sqrt{2} & t \geq count \end{cases} \tag{9}$$

$$D(k) = \frac{\|p(kx, ky) - goal(x, y)\|}{\sqrt{2}} \tag{10}$$

$$count = \lfloor [0.27n + 0.5] \times 75\% \rfloor, \tag{11}$$

where $D(k) = \frac{\|p(kx,ky) - goal(x,y)\|}{\sqrt{2}}$, $count = \lfloor [0.27n + 0.5] * 75\% \rfloor$, $\lfloor \cdot \rfloor$ and $\lceil \cdot \rceil$ round toward negative or positive infinity.

The outline of the presented algorithm is described in the following steps:

Step 1. Initialize the population size *NUM*, the parameters *step* and *visual*, and the maximum number of iterations *N*.
Step 2. Update the step size and position of the artificial fish using Equation (7).
Step 3. Calculate the food concentration for each artificial fish using Equation (8) and record the optimal value in the bulletin board.
Step 4. Perform preying behavior, swarming behavior, following behavior and random behavior.
Step 5. Check the termination condition. If the stopping condition is satisfied, terminate the iteration process and output optimal solution. Otherwise, return to Step 2.

## 3. Local Path Planning Based on an Expansion Logic Strategy

In this section, an expansion logic strategy is presented to avoid collisions between multiple mobile robots and an environment with obstacles. It plays a decisive role in real-time navigation of mobile robots. Figure 1 shows the obstacle information in an unknown environment. For any polygonal obstacle, the minimum circumscribed circle (MCCI) method and wire envelopes method can be used to perform obstacle expansion operations. In this article, as we sought a rapid expansion method, we adopted the endpoint connection method to generate a circular equation instead of the MCCI method. The expansion logic strategy is described as follows:

If the vertices of n-sided polygonal obstacles are denoted as $p_i(x_i, y_i)$, the Euclidean distance $d_{ij}$ between two vertices $p_i(x_i, y_i)$ and $p_j(x_j, y_j)$ can be expressed by the following equation:

$$d_{ij} = \sqrt{(x_i - x_j)^2 + (y_i - y_j)^2}. \tag{12}$$

Therefore, the diameter of a range circle is given by

$$d_{\max} = \max\{d_{ij}\}, i, j = 1, 2, \cdots, n \tag{13}$$

and the circle equation is given by Equation (10).

$$(x - \frac{x_i + x_j}{2})^2 + (y - \frac{y_i + y_j}{2})^2 = (\frac{d_{\max}}{2})^2 \tag{14}$$

In this paper, an environmental map with n-sided polygonal obstacles is shown in Figure 1a. Following the circle Equation (14), two concentric circles with diameters $1.2d_{\max}$ and $1.4d_{\max}$ could be obtained (Figure 1b). As the figure shows, the collision probability was 0.85 for the complementary set area of the intersection between the first circle and the polygonal obstacles. Furthermore, the collision probabilities were 0.65 and 0.45 when the robots moved in the other two grid areas.

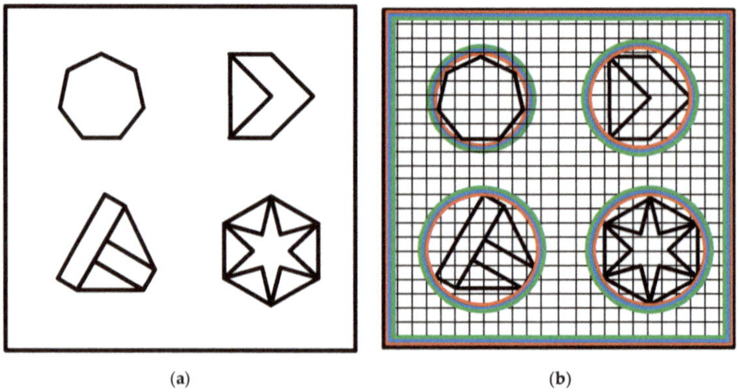

**Figure 1.** (**a**) Environmental map information; (**b**) expansion logic operation for obstacles.

To evaluate the grid-based environment map, a kind of scoring function is given by Equation (15), which shortens the time of local path planning and improves decision-making ability.

$$score = 100 - \lceil \frac{dist}{l} \rceil, \tag{15}$$

where *dist* denotes the Euclidean distance between the starting point and the target point, and $l$ represents the Euclidean distance between the starting point and the current position of the robot.

From the vertical line of the motion direction, we scored the surrounding grids by Equation (15) and the obtained grid scores are shown in Figure 2a. Then, in accordance with the current position information and Equation (12), the mobile robot selected the grid that was the closest to the target point as the position of the next moment. If the distance between the grids was equal, preference was given to the high score grid. Finally, the local path planning trajectory, which is indicated by a dotted line, could be obtained using the expansion logic strategy (Figure 2b).

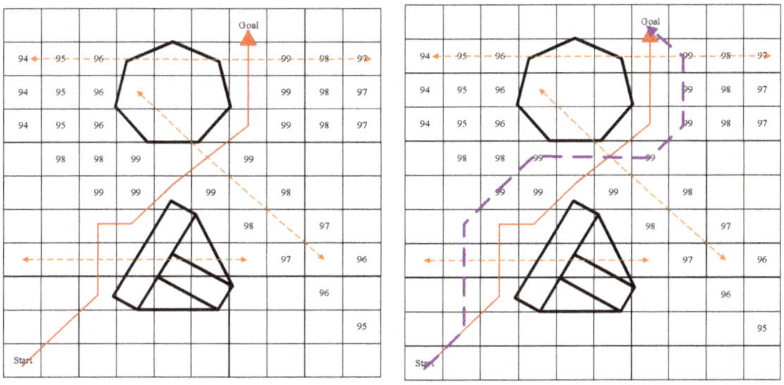


(**a**) (**b**)

**Figure 2.** (**a**) Local path planning and grid scores; (**b**) local path planning using the expansion logic strategy.

## 4. Simulation Experiments

In this section, to verify the superiority of the presented algorithm for a single mobile robot and $20 \times 20$ grid-based environment maps, the path planning results under the presented hybrid artificial fish swarm algorithm and the traditional fish swarm algorithm are shown in Figure 3.

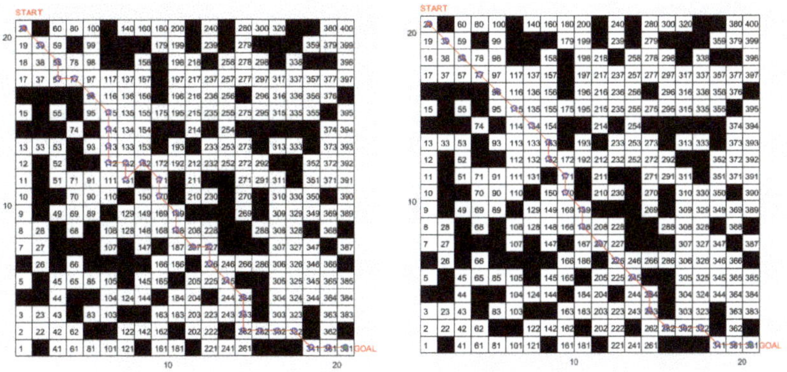


(**a**) (**b**)

**Figure 3.** (**a**) Artificial fish swarm algorithm; (**b**) the presented hybrid artificial fish swarm algorithm.

For 20 × 20 and 40 × 40 grid-based environment maps, the path planning results based on the adaptive artificial fish swarm algorithm (AAFA) [21], fuzzy logic (FL) [22], the improved genetic algorithm (IGA) [23] and our method are shown in Table 1. As can be seen in Figure 3 and Table 1, optimization performance and iteration time can be improved by using the presented method.

**Table 1.** Performance comparison of four algorithms.

| Environment Map | Algorithms | The Longest Path Length | The Optimal Path Length | The Average Path Length | Iteration Time/s |
|---|---|---|---|---|---|
| 20 × 20 grids | AAFA | 35.1283 | 30.0348 | 33.6231 | 16.5182 |
| | FL | 34.3848 | 29.7990 | 32.0919 | 12.2304 |
| | IGA | 32.3254 | 29.6325 | 30.8652 | 16.1826 |
| | Our method | 30.3848 | 29.2132 | 29.7990 | 9.3102 |
| 40 × 40 grids | AAFA | 80.2372 | 73.7103 | 79.2293 | 98.5621 |
| | FL | 75.1838 | 69.4975 | 72.3407 | 90.4073 |
| | IGA | 66.4723 | 62.9002 | 65.7213 | 97.7652 |
| | Our method | 64.0833 | 61.4264 | 62.7549 | 74.5801 |

Furthermore, simulation results were performed on a group of mobile robots. A multiple mobile robot navigation system, which included three mobile robots labeled as Robot 0, Robot 1 and Robot 2, was designed using STDR (simple two dimensional robot simulator) and RVIZ (robot operating system visualizer) software. Each mobile robot was equipped with four ultrasonic sensors and a radar detector with laser detection capabilities. The environment map and obstacle information of the simulation experiments are shown in Figure 4a. If the size of the empirical fish swarm was $N = 50$, and the visual distance of an artificial fish was $v = 10$, the crowd factor was $\delta = 0.618$.

As shown in Figure 4, the initial poses of the three mobile robots were (1, 1, 0), (1, 5, 0) and (1, 9, 0). The goal positions were (10, 14), (17, 1) and (18, 13).

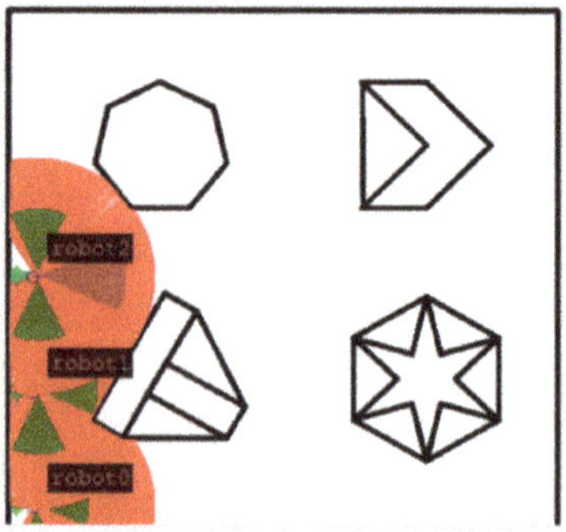


(**a**)

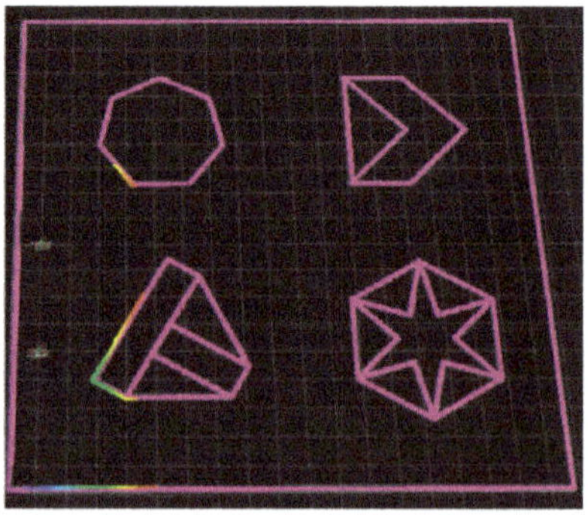

(**b**)

**Figure 4.** (**a**) The initial poses of the three mobile robots; (**b**) simulation visual interface of multiple mobile robots.

The global path planning trajectories of the three mobile robots are shown in Figure 5. As shown in Figure 5a, Robot 0 (initial pose was (1, 1, 0)), Robot 1 (initial pose was (1, 5, 0)) and Robot 2 (initial pose was (1, 9, 0)) started to move, and then they updated their motion paths using the expansion logic strategy to avoid mutual collision. Meanwhile, many feasible paths could be obtained by the presented EFSA. Therefore, we can see that the global path planning results for mobile robots were not unique (Figure 5b).

Figure 6a illustrates the position information (the poses of the three mobile robots were (3, 10, 0), (9, 1, 0) and (10, 10, 0)) at a specific time in the simulation experiment. Local and global paths of multiple mobile robots are shown in Figure 6b.

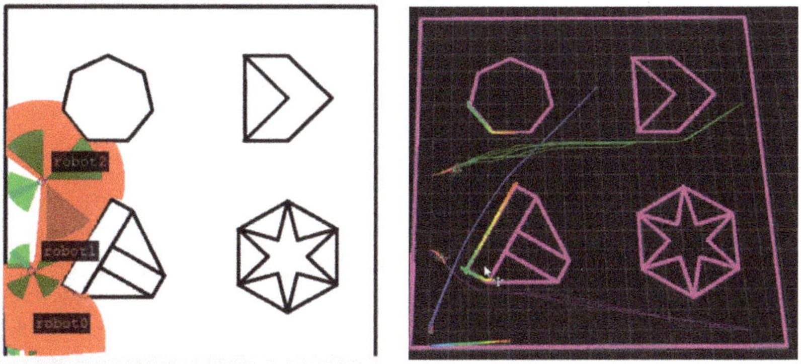


(**a**) (**b**)

**Figure 5.** (**a**) The initial poses of the three mobile robots; (**b**) global path planning trajectories.

(**a**) (**b**)

**Figure 6.** (**a**) Position information at a specific time; (**b**) local and global paths of multiple mobile robots.

The positions of Robot 0, Robot 1 and Robot 2 are shown in Figure 7a when the robots reached the target points. Figure 7b illustrates the final paths of the three mobile robots after reaching the target points.

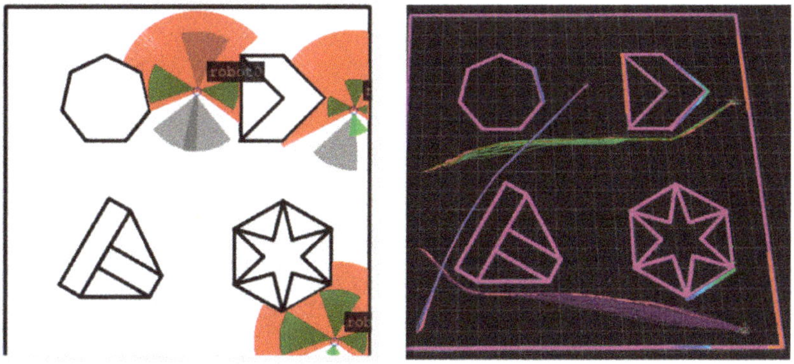


(**a**) (**b**)

**Figure 7.** (**a**) The robots reached the target points; (**b**) the final paths of the three mobile robots.

## 5. Conclusions

This article focused on the cooperative path planning problem of multiple mobile robots in an unknown environment with obstacles. HAFSA (hybrid artificial fish swarm algorithm) was proposed to solve the local optimal problem and realize cooperative path planning for multiple mobile robots. An experiential learning strategy was presented to improve the performance of AFSA and a detection operator was developed to optimize the resulting path trajectory. In particular, an expansion logic strategy was used to avoid collision between multiple mobile robots and an environment with obstacles. In order to evaluate a grid-based environment map, a kind of scoring function was designed. Finally, a multiple mobile robot simulation system was developed by utilizing STDR and RVIZ software; simulation experimental results validated the effectiveness of the proposed real-time navigation algorithm.

**Author Contributions:** Validation, Z.L. and Y.J.; formal analysis, L.C.; writing—original draft preparation, Y.H.; supervision, Y.H.; project administration, Y.H.

**Funding:** This work was supported in part by Natural Science Research Key Projects in Universities of Anhui Province (KJ2018A0110), Natural Science Foundation of Anhui Province (1608085MF146) and the Foundation for talented young people of Anhui Polytechnic University (2016BJRC004, 2016BJRC008).

**Conflicts of Interest:** The authors declare no conflict of interest.

## References

1. Cheng, H.; Chen, H.; Liu, Y. Topological Indoor Localization and Navigation for Autonomous Mobile Robot. *IEEE Trans. Autom. Sci. Eng.* **2015**, *12*, 729–738. [CrossRef]
2. Jhang, J.; Lin, C.; Lin, C.; Young, K. Navigation Control of Mobile Robots Using an Interval Type-2 Fuzzy Controller Based on Dynamic-group Particle Swarm Optimization. *Int. J. Control Autom. Syst.* **2018**, *16*, 2446–2457. [CrossRef]
3. Raja, P.; Pugazhenthi, S. Optimal path planning of mobile robots: A review. *Int. J. Phys. Sci.* **2012**, *7*, 1314–1320. [CrossRef]
4. Han, J.; Seo, Y. Mobile robot path planning with surrounding point set and path improvement. *Appl. Soft Comput.* **2017**, *57*, 35–47. [CrossRef]
5. Cherni, F.; Boujelben, M. Autonomous mobile robot navigation based on an integrated environment representation designed in dynamic environments. *Int. J. Autom. Control* **2017**, *11*, 35–53. [CrossRef]
6. Mantegh, I.; Jenkin, M.R.M.; Goldenberg, A.A. Path Planning for Autonomous Mobile Robots Using the Boundary Integral Equation Method. *J. Intell. Robot. Syst.* **2010**, *59*, 191–220. [CrossRef]
7. Wang, Y.; Cao, W. A global path planning method for mobile robot based on a three-dimensional-like map. *Robotica* **2014**, *32*, 611–624. [CrossRef]
8. Hernández, K.; Bacca, B.; Posso, B. Multi-goal Path Planning Autonomous System for Picking up and Delivery Tasks in Mobile Robotics. *IEEE Lat. Am. Trans.* **2017**, *15*, 232–238. [CrossRef]
9. Wu, M.; Chen, E.; Shi, Q.; Zhou, L.; Chen, Z.; Li, M. Path planning of mobile robot based on improved genetic algorithm. In Proceedings of the Chinese Automation Congress (CAC), Jinan, China, 20–22 October 2017; pp. 6696–6700.
10. Zavlangas, P.G.; Tzafestas, S.G. Motion control for mobile robot obstacle avoidance and navigation: A fuzzy logic-based approach. *Syst. Anal. Model. Simul.* **2003**, *43*, 1625–1637. [CrossRef]
11. Mac, T.T.; Copot, C.; Tran, D.T.; de Keyser, R. A hierarchical global path planning approach for mobile robots based on multi-objective particle swarm optimization. *Appl. Soft Comput.* **2017**, *59*, 68–76. [CrossRef]
12. Akka, K.; Khaber, F. Mobile robot path planning using an improved ant colony optimization. *Int. J. Adv. Robot. Syst.* **2018**, *15*. [CrossRef]
13. Precup, R.E.; Petriu, E.M.; Radae, M.B.; Voisan, E.I.; Dragan, F. Adaptive Charged System Search Approach to Path Planning for Multiple Mobile Robots. In Proceedings of the 2nd IFAC Conference on Embedded Systems, Computer Intelligence and Telematics, Maribor, Slovenia, 22–24 June 2015; Volume 48, pp. 294–299.
14. Liu, S.; Sun, D.; Zhu, C. Coordinated Motion Planning for Multiple Mobile Robots along Designed Paths with Formation Requirement. *IEEE/ASME Trans. Mechatron.* **2011**, *16*, 1021–1031. [CrossRef]

15. Li, H.; Yang, S.X.; Seto, M.L. Neural-network-based path planning for a multi-robot system with moving obstacles. *IEEE Trans. Syst. Man Cybern. Part C Appl. Rev.* **2009**, *39*, 410–419. [CrossRef]
16. Umarani, P.; Thangaraj, P. Improving Group Communication in Wireless Telemedicine System Using Fish Swarm Algorithm. *J. Med. Imaging Health Inform.* **2016**, *6*, 1576–1580. [CrossRef]
17. Zhu, J.; Wang, C.; Hu, Z.; Kong, F.; Liu, X. Adaptive variational mode decomposition based on artificial fish swarm algorithm for fault diagnosis of rolling bearings. *Proc. Inst. Mech. Eng. Part C J. Mech. Eng. Sci.* **2016**, *231*, 635–654. [CrossRef]
18. Wen, S.; Cai, X.; Guan, W.; Jiang, J.; Chen, B.; Huang, M. High-precision indoor three-dimensional positioning system based on visible light communication using modified artificial fish swarm algorithm. *Opt. Eng.* **2018**, *57*, 106102. [CrossRef]
19. Jin, X.; Xie, S.; He, J.; Lin, Y.; Wang, Y.; Wang, N. Optimization of tuned mass damper parameters for floating wind turbines by using the artificial fish swarm algorithm. *Ocean Eng.* **2018**, *167*, 130–141. [CrossRef]
20. He, Q.; Hu, X.; Ren, H.; Zhang, H. A novelartificial fish swarm algorithm for solving large-scale reliability-redundancy application problem. *ISA Trans.* **2015**, *59*, 105–113. [CrossRef] [PubMed]
21. Yiqing, H.; Panpan, W.; Mengru, Y. Path planning of mobile robots based on logarithmic function adaptive artificial fish swarm algorithm. In Proceedings of the 36th Chinese Control Conference, Dalian, China, 26–28 July 2017; pp. 4819–4823.
22. Singh, N.H.; Thongam, K. Mobile Robot Navigation Using Fuzzy Logic in Static Environments. *Procedia Comput. Sci.* **2018**, *125*, 11–17. [CrossRef]
23. Tuncer, A.; Yildirim, M. Dynamic path planning of mobile robots with improved genetic algorithm. *Comput. Electr. Eng.* **2012**, *38*, 1564–1572. [CrossRef]

*applied sciences*

MDPI

*Article*

# Modal Planning for Cooperative Non-Prehensile Manipulation by Mobile Robots

**Changxiang Fan [1,*], Shouhei Shirafuji [2] and Jun Ota [2]**

[1] Department of Precision Engineering, Graduate School of Engineering, The University of Tokyo, 7-3-1 Hongo, Bunkyo-ku, Tokyo 113-8656, Japan
[2] Research into Artifacts, Center for Engineering, The University of Tokyo, 5-1-5 Kashiwanoha, Kashiwa-shi, Chiba 277-8568, Japan; shirafuji@race.u-tokyo.ac.jp (S.S.); ota@race.u-tokyo.ac.jp (J.O.)
* Correspondence: fan@race.u-tokyo.ac.jp; Tel.: +81-04-7136-4276

Received: 28 December 2018; Accepted: 24 January 2019; Published: 29 January 2019

**Abstract:** If we define a mode as a set of specific configurations that hold the same constraint, and if we investigate their transitions beforehand, we can efficiently probe the configuration space by using a manipulation planner. However, when multiple mobile robots together manipulate an object by using the non-prehensile method, the candidates for the modes and their transitions become enormous because of the numerous contacts among the object, the environment, and the robots. In some cases, the constraints on the object, which include a combination of robot contacts and environmental contacts, are incapable of guaranteeing the object's stability. Furthermore, some transitions cannot appear because of geometrical and functional restrictions of the robots. Therefore, in this paper, we propose a method to narrow down the possible modes and transitions between modes by excluding the impossible modes and transitions from the viewpoint of statics, kinematics, and geometry. We first generated modes that described an object's contact set from the robots and the environment while ignoring their exact configurations. Each multi-contact set exerted by the robots and the environment satisfied the condition necessary for the force closure on the object along with gravity. Second, we listed every possible transition between the modes by determining whether or not the given robot could actively change the contacts with geometrical feasibility. Finally, we performed two simulations to validate our method on specific manipulation tasks. Our method can be used in various cases of non-prehensile manipulations by using mobile robots. The mode transition graph generated by our method was used to efficiently sequence the manipulation actions before deciding the detailed configuration planning.

**Keywords:** non-prehensile manipulation; manipulation planning; contact planning; manipulation action sequences

## 1. Introduction

Spatial restrictions make it almost impossible to manipulate a big object in a narrow space by using big-scaled manipulators. For instance, it is impractical to carry an industrial manipulator into our house to move furniture by grasping and lifting. Owing to their small size and flexibility in motion, multiple mobile robots can be adopted to perform tasks in a narrow space [1]. These robots can move in a narrow environment to approach and manipulate objects, but these robots cannot grasp big objects as large-scale industrial manipulators. Therefore, non-prehensile methods [2,3], which involves manipulation without grasping, is practical for such cases. For instance, a preferred way to manipulate a big object is to push it along the floor, or to pivot it with a vertex that makes contact with the floor. In certain cases, the object keeps contacting the environment in such manipulations and the restrictions on the object motion caused by the contacts complicates the kinematics in the manipulation. Furthermore, when multiple mobile robots perform a manipulation task, they themselves form a complex coordinated system [4]. Consequently, non-prehensile manipulations that use multiple mobile robots (as shown in Figure 1) require convoluted manipulation planning than standard manipulation, which comprehensively takes into account each robot's kinematics, the surrounding constraints, and their changes.

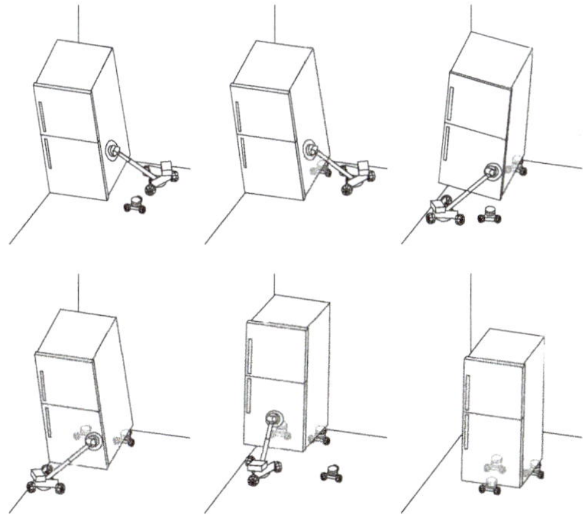

**Figure 1.** Example of non-prehensile transportation adopting multiple mobile robot: Two types of mobile robots move the refrigerator placed at the corner of a room.

Multiple robot motion planning is often faced with the high dimensional configuration space [5]. Typical planners for such problems are sample-based, such as RRT [6] and PRM [7]. However, manipulation planning problems often encounter particular *multi-modal* structures [8], if the contacts among the robots, the manipulated objects, and the environment change during the process of manipulation. Here, a *mode* refers to a certain set of configurations that hold the same constraints in motion (e.g., the object motion keeping a set of contacts with the environment). Possible configurations under the same constraints form sub-spaces in the configuration space. This requires the planner to be capable not only to probe the sub-spaces of each mode but also to cross among the different sub-spaces. For the application of typical sample-based methods on their original spaces, the expansiveness among the configurations of the different modes is more difficult to achieve than those of the same mode [9]. For instance, we can generate a transitable configuration of the system for a current configuration by sampling the configuration the belongs to the sub-space, while keeping the contact set among the robots, the objects, and the environment unchanged. However, in case the contact set changes, it would be necessary to check the connectivity between the different sub-spaces corresponding to the contact states, which would complicate the problem.

Therefore, typical sample-based methods are usually applied to the modes' sub-spaces after splitting the configuration space into sub-spaces according to the modes. The sample-based methods become realizable by deciding the sequence of modes where connectivity is guaranteed beforehand. Maeda et al. [10] and Miyazawa et al. [11] sampled the manipulation states in a configuration space where a sequence of modes existed and the modes' transitions were prior defined. Some planners have been proposed for creating the modes' roadmap to guide the manipulation sequencing [8,12–14]. In particular, Lee [13] adopted a PRM-based planner [15,16] to split the modes by comprehensively considering the multi-contacts between the object, robots, and the environment to obtain the necessary modes to pass through for a manipulation task. Mode transitions are mainly derived from the compliant transformations of contacts between the object, the environment, and the contacts of the given robots; two robot contacts in the examples [13] were adopted to realize the subsequent mode transitions.

However, when multiple mobile robots manipulated objects by using non-prehensile methods, mode transitions became more complex, because besides considering the mode transitions caused by the changes in the environmental contact, robot contacts also had to be considered. One way of addressing this complexity is by eliminating unfeasible modes in statics from the enormous combinations of contacts among them before considering the transitions among modes. An object should be under sufficient constraints in each mode; otherwise, the robots will fail to manipulate the object (e.g., the object drops down in an unexpected direction because of the lack of constraints). This means that the contacts from the environment and the robots should be able to form full constraints on the objects for non-prehensile manipulations. If we consider the environmental contact alone, the resultant constraints would vary in different contact states. For example, when an object–floor contact state changes from face–face contact into vertex–face contact, the constraints exerted by the floor would reduce. This results in various least requirements of robot constraints under different environmental constraints. Sometimes, we require the object to keep stationary contact with the environment, so that the robot can use friction to move the object (e.g., inclining a box); sometimes, we require the robots to manipulate the object to slide along the contacting part (e.g., sliding a box on a floor). To distinguish these manipulations, the environmental contact should be identified into the fixed contact and the sliding contact. The restraint placed on the object's degrees of freedom by a contact is different between the cases when it is fixed or sliding. Thus, we consider the necessary amount of robot for manipulations, both when relative sliding happened and did not happen. Therefore, a proper consideration of the individual robot's kinematics and the changes in the environmental constraints is essential when splitting the modes.

Furthermore, a robot contact changes when the robot makes or breaks contact with the object either actively or passively (by the object's motion). For example, in Figure 1, a large robot with a manipulator actively makes contact with the object and pushes it over the small robots, and the small robots passively make contact with the object by the action of the large robot. The distinction between the active and passive action appears in many manipulation tasks, and the possible sequence of actions depends on this distinction of actions. Thus, the planner should be able to reason about all such possible mode transitions.

By addressing the problems peculiar to non-prehensile manipulations using the mobile robots described above, we propose a method to generate the modes and transitions between the modes for contact planning. In our method, for a given set of contacts, we identified the modes by analyzing its constraints and least requirements to constrain the object's motion. We determine the mode transitions based on how the robot contact influenced the contact state of a targeted object. The mobile robots were divided into active robots and passive robots according to how their contacts changed in the state transition. Finally, the manipulation actions were determined by sequencing a series of modes. We applied the same concept to the limited cases also [17]. In this paper, we propose a more generalized framework to determine the action sequence including the distinction between the fixed and the sliding contact states.

In the second section, we describe the problem statement, and in the third section, we introduce the generation of the contact state. In the fourth section, the state transition is investigated to sequence the action series. In the fifth section, we describe the applications of our methodology by conducting two simulations of the specified manipulation cases.

## 2. Problem Statement

In this paper, for simplicity, we have only manipulated objects having convex polyhedrons. Furthermore, we consider that the objects, the environment, and the robots are all rigid. A set of contacts on an object having environmental contact and robotic contact were represented as a contact state and defined as a mode. A contact state is described without the exact position of contact and configuration of the object and the robots.

For a description of the contact state without the exact configurations, we adopt the concept of principal contact (PC) [18] to express the contact states. A PC is a contacting pair between the geometrical primitives (a vertex, an edge, or a face). A PC is denoted by $c = (a, b)$, where $a$ is a geometrical primitive on the object and $b$ is a geometrical primitive on a polyhedron of the environment or a robot. Furthermore, we denote the object–environment PC and object–robot PC by $c^{\mathrm{e}}$ and $c^{\mathrm{r}}$, respectively, for clarity.

In our method, we need to distinguish whether relative sliding happening on a contact for determining the constraints required to realize the full constraints of the object. For example, Figure 2 shows the difference of the required constraints in the quasi-static manipulation of an object that lies on a floor with an edge contact when a robot contacts with it as a point–face contact (in the planar case). Manipulation without relative sliding on the point with the help of floor's static friction, as shown in Figure 2c, requires a robot to tilt the object. However, for manipulation with relative sliding between the object and the floor, it is difficult to predict the resultant motion of the object because of dynamic friction, as shown in Figure 2b. To guarantee that the target motion will be realized, an additional robot is required (see Figure 2c), and the number of robots required is different between the two cases.

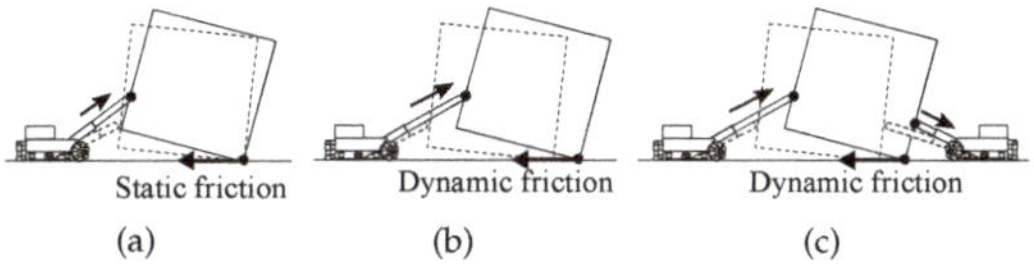


**Figure 2.** Frictional constraint on object's motion in fixed and sliding cases: (**a**) A robot tilts an object with fixed contact on the floor; (**b**) a robot tilts an object with sliding contact; (**c**) two robots tilts an object with sliding contact.

Therefore, we distinguish whether relative sliding happens or not on a contact. A non-sliding PC is defined as *static* PC, denoted by $c^{\mathrm{e,st}} = (a, b)^{\mathrm{e,st}}$. Correspondingly, a sliding PC is defined as *dynamic* PC, denoted by $c^{\mathrm{e,dn}} = (a, b)^{\mathrm{e,dn}}$.

When multiple robots manipulate objects, not all of them actively make contact with an object. A contact occurs either when the robot actively touches the object or passively touches it when the object is moved. Furthermore, some robots move an object actively and change its state, whereas some robots operate as auxiliaries in the manipulation task. The distinction between the active and passive functions of a contact is important for considering the possible transitions of the states. Accordingly, the contacts are divided into active and passive and are denoted as $c^{\mathrm{r,st,ac}}$ and $c^{\mathrm{r,st,ps}}$, respectively. A robot with active joints can also act as a passive robot. Whether the robot always acts passively or actively or is switchable between passive and active depends on the function of the robot. A contact between an object and a robot usually does not slide. Therefore, for simplicity, we omit the subscript for the static and the dynamic contacts if the contact is between an object and a robot and write as $c^{\mathrm{r,ac}}$ and $c^{\mathrm{r,ps}}$. However, a contact between an object and the environment is passive. Therefore, we omit the subscript for the passive and the active contacts if the contact is between an object and the environment and write as $c^{\mathrm{e,st}}$ and $c^{\mathrm{e,dn}}$.

The set of PCs on an object is called contact formation (CF) [18] and denoted by $C$. To describe the original CF proposed by Xiao and Zhang [18], which does not concern the sliding of the contact point, we call a CF without distinguishing the static and dynamic as the *primitive* contact states, and denote it as $\widehat{C}$. For example, $\{(a_1, b_1)^{\mathrm{e}}, (a_2, b_2)^{\mathrm{e}}\}$ is the primitive state of $\{(a_1, b_1)^{\mathrm{e,st}}, (a_2, b_2)^{\mathrm{e,st}}\}$ or $\{(a_1, b_1)^{\mathrm{e,st}}, (a_2, b_2)^{\mathrm{e,dn}}\}$. Furthermore, when CFs are the same from the viewpoint of the primitive contact states, we describe them as *isogenous*. For example, $\{(a, b)^{\mathrm{e,st}}\}$ and $\{(a, b)^{\mathrm{e,dn}}\}$ are isogenous, and $\{(a_1, b_1)^{\mathrm{e,st}}, (a_2, b_2)^{\mathrm{e,st}}\}$ and $\{(a_1, b_1)^{\mathrm{e,st}}, (a_2, b_2)^{\mathrm{e,dn}}\}$ are also isogenous.

In this paper, we distinguish the CFs consisting of the contact states between an object and the environment, which we call environmental contact formation (ECF). We also distinguish the CFs consisting of the contact states between the robots and the environment, which we call robot contact formation (RCF); see Figure 3.

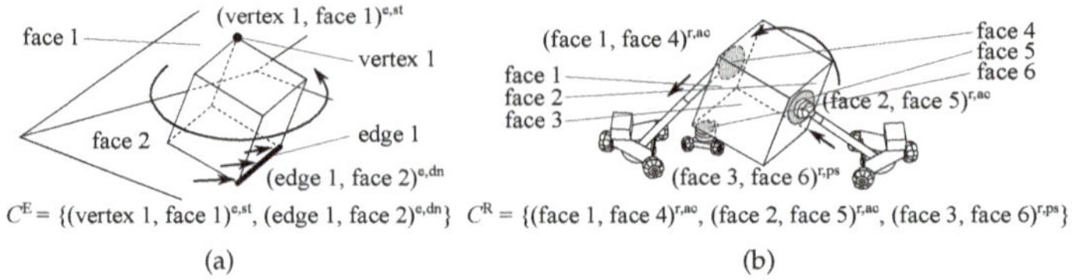


**Figure 3.** Examples of contact formation (CF): (**a**) environmental contact formation (ECF) when a cuboid contacts with two surfaces of the environment by its vertex and edge; (**b**) robot contact formation (RCF) when a cuboid contacts with two active robots and a passive robot by its faces.

Furthermore, we describe them as $C^{\mathrm{E}}$ and $C^{\mathrm{R}}$, respectively. As the result, we denote the mode $s$ in the manipulation as the set of ECF and RCF:

$$s = \{C^{\mathrm{E}}, C^{\mathrm{R}}\}.$$

A multi-contact set on an object was supposed to form the full constraints, so that the robots manipulated the object quasi-statically. Gravity closure [19,20], which is force closure that includes the gravitational force, is introduced later in this paper as the requirement for a mode. Our first goal is to identify all possible modes from the viewpoint of gravity closure.

If one mode can directly transform into another mode without any intermediate ones, it is a possible mode transition. Mode transitions can be described by a graph that comprises nodes and arcs, where the nodes represent individual modes, and the arcs between them represent the transitions; the value of an individual arc represented the cost of the state transition [21]. Our second goal is to generate this mode graph by taking into account some restrictions on robots and the geometrical relationships between an object and the environment. Using the resultant graph, the manipulation action sequences are determined by searching for paths from a given initial mode to a targeted mode before applying a sample-based method to probe the configuration spaces determined by modes.

In the following sections, we have made the following assumptions. The shape and the gravity center of the object and the shape of the environment are given. Furthermore, we have assumed that the type of contact that a robot generates with the object is given, and it does not change in the manipulation. The number of robots is also given. We have considered only the possible RCF on the targeted object, and the contacts between the robots and the environment were ignored; the robots generally make contact with a face (a floor) in the environment.

## 3. Generation of Modes

In this section, we identified all possible modes from the viewpoint of gravity closure. Given a set of mobile robots and an object lying in a certain environment, we obtained the possible modes using the following steps: (i) The possible ECFs were specially identified. We identified the ECFs before identifying the RCFs because the ECF restrains the robot's accessible area. (ii) For the identified ECF, we identified the possible RCF. (iii) Finally, by combining the ECFs and the RCFs, full constraints could be achieved.

### 3.1. Generation of ECFs

The geometrical relationship between an object's shape and the shape of its environment determines the possible ECFs. A PRM-based sampling strategy has been proposed to investigate the possible CFs between objects [15,16,22]. In their method, the target object's orientation was incrementally changed with collisions checked to obtain the possible CFs. If a new CF appeared, it was recorded on the list, along with the object's current orientation. In this way, all the possible CFs between the objects could be probed; each CF was recorded with an available object orientation. We adopted this method to investigate the possible ECFs in our planner.

As mentioned in the previous section, we distinguished the contacts into static and dynamic contact based on the resultant CFs obtained by their method. For the given CFs obtained by their method, every PC was split into static and dynamic PCs, and ECFs are given as all combinations of the split static and dynamic PCs. For example, given a CF $C = \{c_1, c_2\}$, the ECFs are $C_1^E = \{c_1^{e,st}, c_2^{e,st}\}$, $C_2^E = \{c_1^{e,st}, c_2^{e,dn}\}$, $C_3^E = \{c_1^{e,dn}, c_2^{e,st}\}$, and $C_4^E = \{c_1^{e,dn}, c_2^{e,dn}\}$.

Using the above method, we created all the possible environmental contact states in a non-prehensile manipulation. However, one thing that needs to be considered in non-prehensile manipulation is that the object does not make contact with the environment when it is loaded onto the mobile robot, as shown in Figure 4a. Thus, this individual state is added with the notation $C^E = \varnothing$.

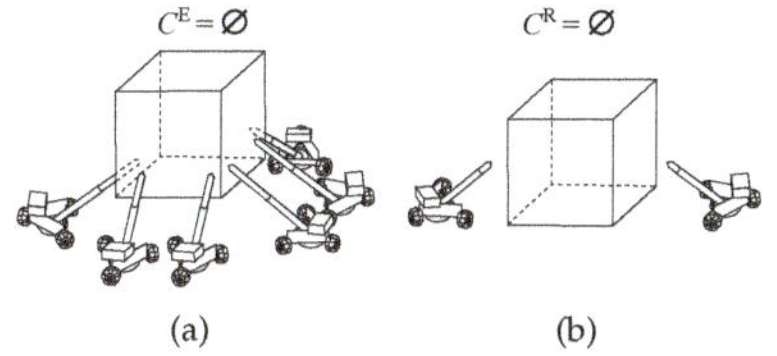


**Figure 4.** Examples of cases where (**a**) ECF is empty and (**b**) RCF is empy.

### 3.2. Generation of RCFs

RCF was easier to generate than ECF. The geometrical relationships between an object and the robots and the penetrations among them were not checked in this phase of planning because robots could locate flexibly around the targeted object, different from the case of generating the ECFs where geometrical relationships and penetrations should be checked because the geometry of the environment did not change.

When creating the possible RCF, only the robot contact on the object is considered; therefore, all the object's geometrical primitives are assumed to be accessible for robots. PCs between an object and the robots is given a definite label for each of the robots, such as $c_1^{r,ac}, c_2^{r,ac}, c_2^{r,ps}, \ldots c_n^{r,ac}$, where $n$ is the number of robots. As mentioned earlier, whether the robot always acts in passive, in active, or as switchable between passive and active depends on the functions of the robot. Then, all possible PCs between an object and the robots are combined to generate RCFs. For example, when all possible PCs between an accessible object's geometrical primitive and robots are given by $c_1^{r,ac}$, $c_2^{r,ps}$, the possible RCFs are obtained as $C_1^R = \{c_1^{r,ac}\}$, $C_2^R = \{c_2^{r,ps}\}$ and $C_3^R = \{c_1^{r,ac}, c_2^{r,ps}\}$ by combining the PCs.

In some states, the object lay in a stable pose, and could itself keep the contact state with the environment without any support from the robot, as shown in Figure 4b. In such cases, the RCF is an empty set given as $C^R = \varnothing$, and it is added to possible RCFs.

### 3.3. ECF-RCF Combination

With ECFs and RCFs created, the contact states of the manipulated object are generated by matching the RCFs with ECFs. In robotic manipulations, generally, force closure is required to achieve full constraints on the targeted object, which means the non-negative combination of primitive wrenches on an object can balance any external load [23]. In the case of non-prehensile manipulation, Maeda et al. [19] and Aiyama et al. [20] proposed gravity closure, which is the force closure formed by robot contacts, environmental contacts, and gravity. External loads applied on an object can be described by wrenches, and they expand the wrench vector space [24]. If sufficient wrench vector bases are provided by the gravitational force and the contacts placed on an object, gravity closure can be achieved. However, the wrench vector bases given by contacts depend on the location and direction of the contacts determined by the configurations of the object and the robots, whereas these configurations are not concerned in the phase to generate the modes and their transitions.

Therefore, we combine an ECF and an RCF if the possible dimension of the wrench space spanned by the ECF and the RCF and the gravity satisfies the condition necessary to realize gravity closure, through which impossible modes are omitted from the viewpoint of statics.

Before calculating the dimensions of the wrench spaces spanned by the ECF and by the RCF for determining their combinations, we eliminated the infeasible combinations for a given ECF by checking whether the object's geometrical primitives were accessible to robots under any of the object's configuration. In this contact planner, we considered only the case when an object's geometrical primitives make absolute contact with the geometrical primitive of the environment; this situation would obviously disable a robot from accessing them. Therefore, for the PCs in ECF, a surface under the PC $(\text{face}, \text{face})^e$, an edge under the PC $(\text{edge}, \text{face})^e$, and a vertex under the PC $(\text{vertex}, \text{face})^e$ are not accessible to robots, as shown in Figure 5. The ECF that contains a certain geometrical primitive of the object that contacts the surface of the environment will not be considered to combine with an RCF that contains the same geometrical primitive of the object that contacts with robots.

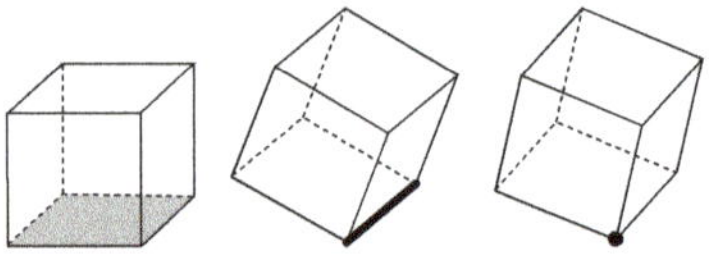


**Figure 5.** Three types of principal contacts (PCs) that a robot cannot access regardless of the object's configuration: (**a**) $(\text{face}, \text{face})^e$; (**b**) $(\text{edge}, \text{face})^e$; and (**c**) $(\text{vertex}, \text{face})^e$.

With the geometrical feasibility validated, the ECF and RCF individuals are combined according to the condition for gravity closure. Here, we considered the possible dimensions of two kind of wrench vector spaces correspondingly spanned by two kinds of forces: the passive and active forces. The passive force is applied by contacts with the environment and passive robots, and the active force is applied by contacts with the active robots and gravity. The active and passive forces were derived for the given ECFs and RCFs, respectively. Based on this, we combined the ECF and RCF, by taking into account only the derived maximum dimension of the wrench vector bases of the passive and active forces to satisfy the necessary condition of gravity closure. Configurations of the objects and the robots that meet the gravity closure will be decided in our future studies after we decide the mode transitions.

When force is applied on a contact point, the resultant wrench on the object is expressed as

$$w = \begin{bmatrix} f \\ p \times f \end{bmatrix}, \tag{1}$$

where $\boldsymbol{f}$ is the force applied to the contact, and $\boldsymbol{p}$ is the position of contact with respect to the object coordinate frame, as shown in Figure 6a. Certain manipulation studies have taken into account the case when a point contact also can resist a torque, which is called as a soft finger contact. However, we have not considered this type of contact in this paper.

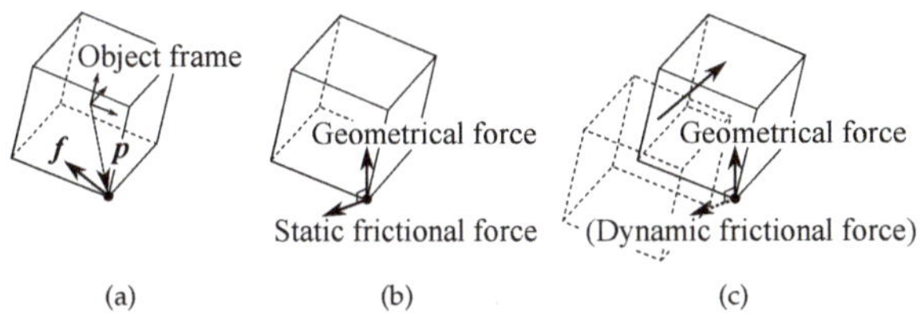


**Figure 6.** Force applied on a point of contact. (**a**) Definition of force. (**b**) Type of forces applied on a contact when the contact is fixed. (**c**) Types of forces applied on a contact when the contact is sliding.

There are two cases when a contact resists an external force: the geometrical case and the frictional case (Figure 6a,b). The former is the force caused on the object not to penetrate into the environment or the robot. The latter is the frictional force, and there are also two types of frictional forces: static and dynamic. We ignore dynamic friction when we consider the wrench bases to span gravity closure to ensure manipulation, as explained in the section of the problem statement.

For static friction on a point, the frictional cone, which is determined by the linear relationship between the geometrical force and the static frictional force, is usually considered when checking the conditions necessary for force closure. We dealt with static friction as decomposed basis independent of the geometrical force without considering the frictional cone (see Figure 7a) because our aim was to omit the modes that could not realize gravity closure regardless of the object's configuration. Therefore, the wrench space for the $j$th static or dynamic contact is expressed as

$$W_j = \langle \boldsymbol{w}_{j1}, \boldsymbol{w}_{j2}, \boldsymbol{w}_{j3} \rangle, \quad W_j = \langle \boldsymbol{w}_{j1} \rangle, \tag{2}$$

respectively, where $\boldsymbol{w}_{j1}$ is the basis for the geometrical force, and $\boldsymbol{w}_{j2}$ and $\boldsymbol{w}_{j3}$ are the bases for the static frictional force. The geometric and static frictional forces applied on edge–edge–cross contact can also be represented in the same manner (see Figure 7b). To deal with edge–face contacts and face–face contacts, we consider them as two point contacts on the edge and three non-collinear point contacts on the face (such as the boundary points of the contacting area), as shown in Figure 7c,d. The wrench space spanned by the bases is calculated by summing up the bases for all PCs in the given CF as

$$W = W_1 \cup W_2 \cup \ldots, \cup W_n, \tag{3}$$

where $n$ is the number of bases sets on the object defined in the above manner.

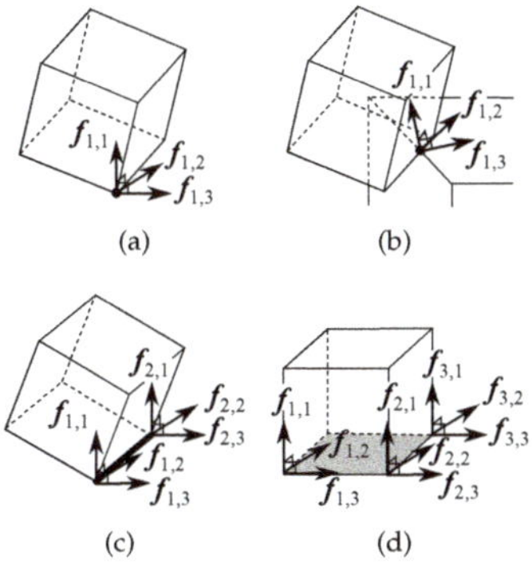


**Figure 7.** Wrench bases of the force applied on the static contact in the case of (**a**) point–face contact, (**b**) edge–face contact, (**c**) face–face contact, and (**d**) edge–edge–cross contact.

Then, the wrench spaces were separately determined for (a) the passive force applied by the contacts from the environment and the passive robots (denoted by $W^{\mathrm{ps}}$) and (b) the active force applied by gravity and the contacts from the active robots (denoted by $W^{\mathrm{ac}}$); see Figure 8. To achieve gravity closure, the wrench vectors consisting of the above forces must positively span $\mathbb{R}^6$ as follows:

$$\mathrm{pos}(W^{\mathrm{ps}} \cup W^{\mathrm{ac}}) = \mathbb{R}^6. \tag{4}$$

The reason why we separate the space into $W^{\mathrm{ps}}$ and $W^{\mathrm{ac}}$ is that the force applied by the environmental contacts and the passive robots cannot realize the force closure by themselves, and the active robots or the gravitational force must act to cause internal force on the object. According to the Carathéodory Theorem, if vectors in a wrench set positively span $\mathbb{R}^6$, the wrench set should contain at least seven vector frames in the six-dimensional space [24]. Thus, at least seven vector frames must exist in the wrench vector space consisting of forces applied by ECFs, RCFs, and gravity to achieve gravity closure. If the total dimension of the three kinds of wrench bases is less than seven, gravity closure would not be realized regardless of the configuration. Therefore, we define the necessary condition for gravity closure when combining ECFs and RCFs to omit the impossible modes as follows. Let $\dim(W^{\mathrm{ps}})$ and $\dim(W^{\mathrm{ac}})$ be the dimensions of $W^{\mathrm{ps}}$ and $W^{\mathrm{ac}}$, respectively, then an ECF and an RCF can be combined for any configuration of the object and the robots satisfying

$$\dim(W^{\mathrm{ps}}) + \dim(W^{\mathrm{ac}}) \geq 7. \tag{5}$$

To determine the modes satisfying Equation (5), we calculate the maximum dimensions of $W^{\mathrm{ps}}$ and $W^{\mathrm{ac}}$ as follows:

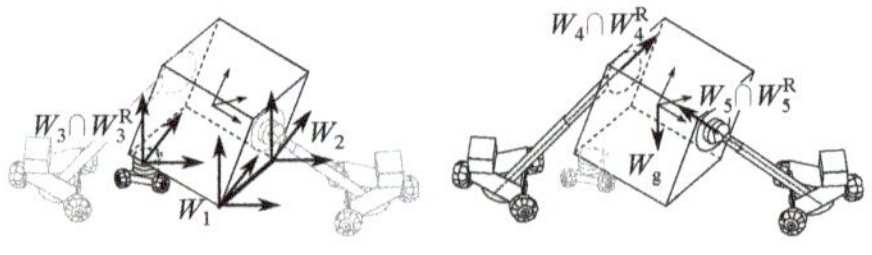


**Figure 8.** Wrench spaces spanned by (**a**) forces applied by the environment–object and the passive robot–object contacts and (**b**) forces applied by the active robot–object contacts and the gravitational force.

Let there be $k$ sets of bases between an object and environment, $l$ sets of bases between an object and passive robots, and $m$ sets of bases between an object and active robots. Let $W_1, \ldots, W_{k+l+m}$ be the wrench space spanned by the contacts, which is derived from Equation (2), and the subscripts correspond with the above order. The wrench spaces simply spanned by the passive and active contacts and gravity are given by

$$W_{\mathrm{c}}^{\mathrm{ps}} = W_1 \cup \ldots \cup W_k \cup W_{k+1} \cup \ldots \cup W_{k+l} \tag{6}$$

and

$$W_{\mathrm{c}}^{\mathrm{ac}} = W_{k+l+1} \cup \ldots \cup W_{k+l+m} \cup W_{\mathrm{g}}, \tag{7}$$

respectively, where $W_{\mathrm{g}}$ is the wrench space spanned by gravity.

The wrench space spanned by RCFs requires more consideration because not only the contacts between the object and the robots, but also the kinematics of the robots affect the spanned wrench space on the object. The limitation on the available actuators of the robot, the passive joints in the robot, or other kinematics restrictions determine the possible wrench. The possible wrench space for a given robot kinematics and its configuration is also represented by the wrench bases aside from the bases of contact. For example, Figure 9 shows the wrench bases of a robot to push an object and the force bases of a robot to carry an object; these robots have passive joints and can be used in the latter section. The robot to push an object has a basis represented as a force along with the axis of the linear actuator, as shown in Figure 9a. The robot to carry an object has three bases represented as forces passing through the axes of the passive universal joint, as shown in Figure 9b. Let $W_j^{\mathrm{R}}$ represent the wrench space spanned by the robot based on its kinematics. By substituting the wrench spaces spanned by the forces applied by robot–object contacts in Equations (6) and (7) with the wrench spaces spanned by the robots based on their kinematics, the followings equations are obtained

$$W_{\mathrm{r}}^{\mathrm{ps}} = W_1 \cup \ldots \cup W_k \cup W_{k+1}^{\mathrm{R}} \cup \ldots \cup W_{k+l}^{\mathrm{R}} \tag{8}$$

and

$$W_{\mathrm{r}}^{\mathrm{ac}} = W_{k+l+1}^{\mathrm{R}} \cup \ldots \cup W_{k+l+m}^{\mathrm{R}} \cup W_{\mathrm{g}}. \tag{9}$$

Taking into account the kinematics of the robots, $W^{\mathrm{ps}}$ and $W^{\mathrm{ac}}$ are obtained by

$$W^{\mathrm{ps}} = W_{\mathrm{c}}^{\mathrm{ps}} \cap W_{\mathrm{r}}^{\mathrm{ps}} = W_1 \cup \ldots \cup W_k \cup [(W_{k+1} \cup \ldots \cup W_{k+l}) \cap (W_{k+1}^{\mathrm{R}} \cup \ldots \cup W_{k+l}^{\mathrm{R}})] \tag{10}$$

and

$$W^{\mathrm{ac}} = W_{\mathrm{c}}^{\mathrm{ac}} \cap W_{\mathrm{r}}^{\mathrm{ac}} = W_{\mathrm{g}} \cup [(W_{k+l+1} \cup \ldots \cup W_{k+l+m}) \cap (W_{k+l+1}^{\mathrm{R}} \cup \ldots \cup W_{k+l+m}^{\mathrm{R}})], \tag{11}$$

respectively. By the dimension theorem for union, $\dim(W^{\mathrm{ps}})$ and $\dim(W^{\mathrm{ac}})$ are given by

$$\dim(W^{\mathrm{ps}}) = \dim(W_{\mathrm{c}}^{\mathrm{ps}}) + \dim(W_{\mathrm{r}}^{\mathrm{ps}}) - \dim(W_{\mathrm{c}}^{\mathrm{ps}} \cup W_{\mathrm{r}}^{\mathrm{ps}}) \tag{12}$$

and

$$\dim(W^{\mathrm{ac}}) = \dim(W_{\mathrm{c}}^{\mathrm{ac}}) + \dim(W_{\mathrm{r}}^{\mathrm{ac}}) - \dim(W_{\mathrm{c}}^{\mathrm{ac}} \cup W_{\mathrm{r}}^{\mathrm{ac}}), \tag{13}$$

respectively.

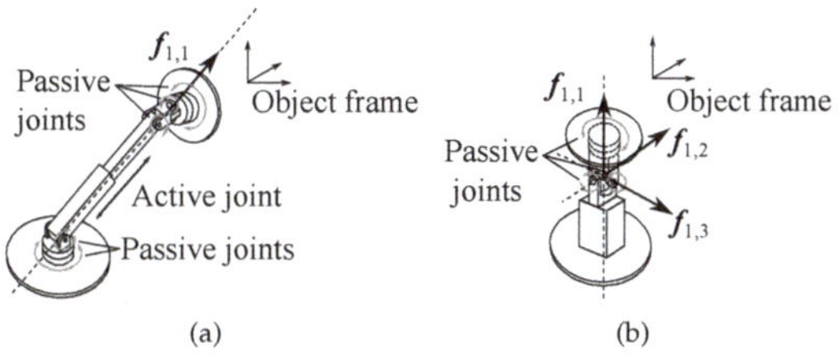


**Figure 9.** Examples having passive joints and their wrench bases. (**a**) Robot with five passive joints and a linear actuator. (**b**) Robots with three passive joints.

The maximum of $\dim(W^{\mathrm{ps}})$ and $\dim(W^{\mathrm{ac}})$ are difficult to derive analytically because the exact configuration of each contact state is not concerned in the current planning level. Fortunately, the maximum dimension of a contact state is a constant, except when the object lies in a singular configuration that causes dimensional degeneration. However, such kinds of singular configurations are caused only in specified configurations, such as a point or a line in the wrench space. Therefore, we sampled several configurations of objects and robots to calculate the maximum of $\dim(W^{\mathrm{ps}})$ and $\dim(W^{\mathrm{ac}})$. If one resultant dimension was smaller than the others, the corresponding configuration was designated as a singularity in the given ECF and RCF. The sampled $W^{\mathrm{ps}}$ and $W^{\mathrm{ac}}$ that provides the maximum dimension for a given ECF and RCF is considered to be the dimension of the wrench space spanned by $W^{\mathrm{ps}}$ and $W^{\mathrm{ac}}$.

Based on the obtained $\dim(W^{\mathrm{ps}})$ and $\dim(W^{\mathrm{ac}})$, ECFs and RCFs are then combined as modes if the necessary condition (i.e., Equation (5)) could be met. In this way, the modes that would appear when the robots manipulate an object are created. However, this procedure does not guarantee that the generated modes will always satisfy gravity closure, but it reduces the number of modes, as shown in latter section, by omitting the impossible modes from the viewpoint of statics.

## 4. Generation of Transitions Among Modes

In non-prehensile manipulation planning, the transition among modes is complicated because of the multiple types of contacts and their relationships. Furthermore, to obtain feasible transitions among modes, the possibility of transitions must be well reasoned by taking into account the difference between ECF and RCF in our method.

For the ECF, we consider the transition between two ECFs is determined by geometrical restrictions; therefore, we adopted the goal-contact-relaxation (GCR) graph [15,16,22] to analyze the transition among ECFs. For the RCF, because of the difference between the active robot and the passive robot, further consideration for transition is required. In this section, we propose a method to decide the transitions among modes by taking into account the following requirements for mode transition. By comprehensively considering the transition of ECF and RCF, we can obtain the mode graph, which describes the transition between the modes and represents the manipulation action sequences.

### Either ECF or RCF Changes

In non-prehensile manipulations, robots change the contacts between an object and its environment to realize the target motion. Certain manipulation planning requires that ECF and RCF changes do not occur simultaneously. Actually, simultaneous change is highly improbable in the actual manipulation. Consequently, we assume that in a mode transition process, either the RCF changes while maintaining the ECF or the RCF keeps contact to manipulate the object and to change the ECF, as shown in Figure 10a.

### Connection of ECFs in GCR Graph

The possible change of contacts between an object and the environment depends on their geometry. The object's CF can transit to its neighboring CF, which connects with it in a GCR graph [15,22]. A GCR graph is a topological graph that represents the transition among the contact states of objects. Nodes that represent the contact states of the object are connected by arcs if the corresponding states can transit to each other by compliant motions. Let $\mathcal{G}$, $V(\mathcal{G})$, and $E(\mathcal{G})$ be a GCR graph, nodes, edges in the graph, respectively. Given two primitive CFs, $\widehat{C}_a \in V(\mathcal{G})$ and $\widehat{C}_b \in V(\mathcal{G})$, $\{\widehat{C}_a, \widehat{C}_b\} \in E(\mathcal{G})$ if it is possible for the two nodes to transit to each other. Though an ECF is further split into the static and dynamic in our method, we adopted a GCR graph to judge the transition between ECFs by checking whether the set of their primitive CFs is an arc in the GCR, as shown in Figure 10b. If the corresponding primitive CFs are the same, they can transit to each other.

### Action of Active Robot

The active robot actively exerts contact forces to an object or releases contact from an object. Furthermore, the active robots changed the contact state in a system. In comparison, the passive robots could not exert contact to the object actively. They got contact with an object or released contact from an object when the object moved. Therefore, transitions that caused by the passive contact changes do not occur without the existence of active robots. Similarly, without the existence of active robots, ECFs cannot change except for the transition between isogenous ECFs, as shown in Figure 10c.

### Object Motion

As explained above, if a contact is exerted by the environment or a passive robot, the contact changes only under the object's motion. Therefore, if the object is fully constrained by the environment and the passive robots alone, the active robots cannot move the object and the transition is not caused, except for the transition between the isogenous CFs, as shown in Figure 10d.

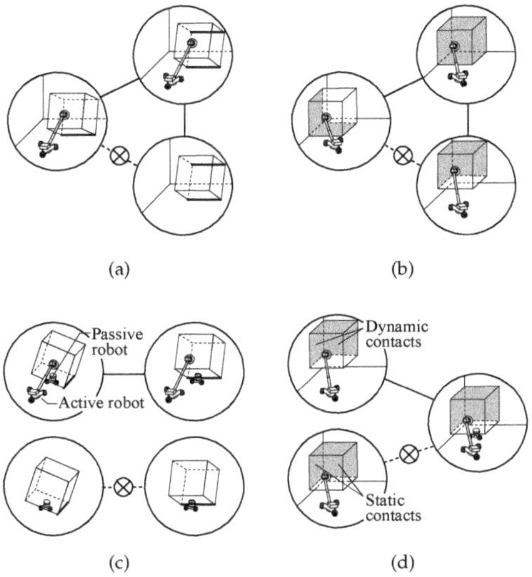


**Figure 10.** Examples of possible and impossible mode transitions. (**a**) Either ECF or RCF changes. (**b**) ECFs must be the same or connected in a goal-contact-relaxation (GCR) Graph. (**c**) Action of active robots causes a mode transition. (**d**) Object's motion caused change of ECFs and contacts of passive robots.

### 4.1. Mode Graph

With the above-mentioned requirements for transition, a mode graph, which is a graph representing the possible manipulation sequences of multiple robots, is obtained as follows. Let $\mathcal{M}$, $V(\mathcal{M})$, and $E(\mathcal{M})$ be a mode graph, nodes, and the graph edges, respectively. Let $\mathcal{G}$ be a corresponding GCR graph of ECFs with $V(\mathcal{G})$ and $E(\mathcal{G})$ as its nodes and edges, respectively. Given the two modes, $s_a = \{C_a^E, C_a^R\}, s_b = \{C_b^E, C_b^R\}$, where $s_a \in V(\mathcal{M}), s_b \in V(\mathcal{M}), \{s_a, s_b\} \in E(\mathcal{M})$, the following conditions are satisfied:

1. ECF is changed: $C_a^R = C_b^R$ and $C_a^E \neq C_b^E$:
   (a) ECFs with their primitive CFs connected in $\mathcal{G}$ are transitable under the motion of the non-fully-constrained object caused by the active robots:
   $\{\widehat{C}_a^E, \widehat{C}_b^E\} \in E(\mathcal{G})$, $C_a^R$ and $C_b^R$ include active PC, besides gravity, and $\dim(W^{ps}) < 6$ in either or both $s_a$ and $s_b$.
   (b) Change in isogenous contacts:
   $\widehat{C}_a^E = \widehat{C}_b^E$
2. RCF is changed: $C_a^R \neq C_b^R$ and $C_a^E = C_b^E$.
   (a) The active robot is added or removed:
   $(C_a^R \backslash C_b^R) \cup (C_b^R \backslash C_a^R)$ comprises only an active PC.
   (b) The passive robot is added or removed under the motion of a non-fully-constrained object caused by the active robots:
   $(C_a^R \backslash C_b^R) \cup (C_b^R \backslash C_a^R)$ that comprise only a passive PC; $C_a^R$ and $C_b^R$ include an active PC, and $\dim(W^{ps}) < 6$ in either or both $s_a$ and $s_b$.
   (c) Change in isogenous contacts:
   $\widehat{C}_a^R = \widehat{C}_b^R$

### 4.2. Cost for Transition between Modes

In the generated state transition graph, the nodes were connected by arcs if the corresponding states were able to transit to each other. When we plan the manipulation based on a mode graph, we take into account the cost for transition between the nodes, which is a value defined for each arc according to the targeted manipulation task. For example, as seen in previous section, there are five kinds of connections, including transitions between the isogenous states. They may have different costs. Transitions between the isogenous states have different costs because they are just internal transitions without changes of the contact set on the object.

Given two states, we define the cost function for the transition between them as

$$l(s_{\mathrm{a}}, s_{\mathrm{b}}) = k, \tag{14}$$

where $k$ is the cost for $s_{\mathrm{a}}$ to transform into $s_{\mathrm{b}}$. In general, the manipulation path is determined by searching for path from the initial to the final states in the graph to lower the total cost. In that case, the objective function is defined as

$$\min \sum_{i=1}^{i=n-1} l(s_i, s_{i+1}), \tag{15}$$

where $s_1$ is the initial mode, and $s_n$ is the final mode.

## 5. Simulations

In this section, we showed two examples of mode generation based on the proposed method for the non-prehensile manipulation planning. We showed how the method narrows down the possible modes and their transitions and what manipulation sequences can be chosen from them based on the costs defined on the transitions.

In both examples, a certain number of robots were given, and here the important thing was that we needed to use the given robot to generate valid modes and to search for feasible manipulation paths, where the number of robots was enough to achieve the gravity closure. By searching for the paths with lowest total costs in the state transition graph, we obtained the least amount of manipulation action sequences. Since we only concerned about the transformation of contact sets in the mode transitions, we viewed the cost for each transition as the same. Therefore, we set the cost of transition between the two given modes as $k = 1$ if they contained different primitive states, either when the environmental contact or when any kind of robot contact changed. Otherwise, we set the cost as $k = 0$. Dijkstra's Algorithm was used to search for the shortest path from the initial state to the final state. We inhibited paths where the modes transited back and forth.

In the following examples, we show only the examples for generating modes and determining transitions, and we do not deal with the configuration space of an object–robot system. Therefore, although we show some figures in which the object and the robots are placed in specified configurations, they are only examples of configurations to help understated the resultant modes and transitions.

### 5.1. Example 1 and Result Discussion

In the first example, we used two types of mobile robots. The first type of robot was a pusher robot that we developed [25–27], as shown in Figure 11b. In this robot, we realized the safety manipulation that avoided the robot from falling; we did this by restricting the force that the robot could apply to the environment by using passive points. The pusher robot has a linear actuator. A face contact between the manipulator and an object acted as an active contact to change the mode of the robot–object system. The pusher robot could move to the target position by using wheels, but those wheels were lifted up and not used while the robot manipulated an object. Because of the special mechanism with passive joints, the kinematics of the robot (shown in Figure 9a) is given as explained in the previous section. See the details of the mechanism in our previous studies [25–27].

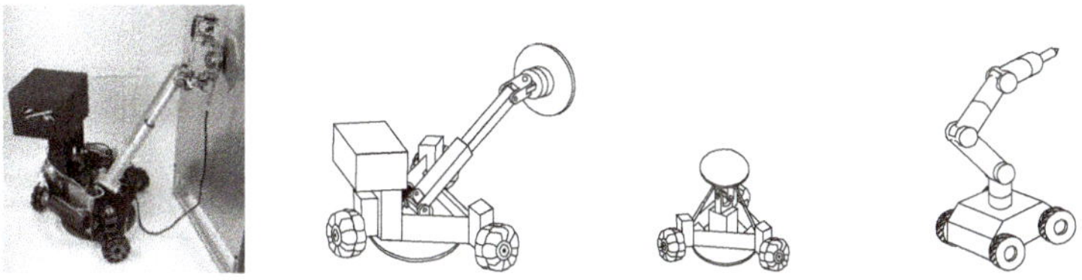

(a) (b) (c) (d)

**Figure 11.** Mobile robots used in examples. (**a**) Pusher robot. (**b**) Schematic of a pusher robot. (**c**) Schematic of a transporter robot. (**d**) Mobile robot with a six-axis manipulator.

Another type of robot is a transporter robot, as shown in Figure 11c. A transporter robot (see Figure 9b) has a structure similar to the pusher robot, but it does not have a linear actuator and has fewer passive joints (as explained in the previous section). The transporter robot can also move to the target position by using wheels but those wheels are not used until it starts transporting the object. As a result, a transporter robot supports an object passively during manipulation, and a face contact between the robot and an object acted as a passive contact. We assume that a pusher robot and a transporter robot act as an active contact and a passive contact, respectively, and these functions are not changed during manipulation for simplicity.

In this manipulation, the task is to load a cuboid up to the transporter robots by using the pusher robots so that the object can be transferred away by the transporter robots. In the initial mode, the cuboid lies against a corner of the two adjacent walls, as shown in Figure 11a. The target final mode is the object held by the three transporter robots, as shown in Figure 11b. We consider the interactions between the cuboid and the environment when loading the cuboid up to the transporter robots by using the proposed method.

As shown in Figure 12, the faces, the edges, and the vertices of the cuboid are denoted as $f_1, f_2, \ldots, f_6$, $e_1, e_2, \ldots, e_{12}$, and $v_1, v_2, \ldots, v_8$, respectively. The walls are denoted as $F_1$ and $F_2$, and the floor is denoted as $F_3$. For simplicity, we assume that the frictional coefficient of the walls is small enough to deal with the contacts on them as friction-less contacts. Therefore, the ECFs including $F_1$ and $F_2$ are always dynamic contacts. We used six pusher robots whose geometrical primitives were denoted as $r_1, r_2, \ldots, r_6$ and three transporter robots whose geometrical primitives were denoted as $r_7, r_8, r_9$. For simplicity, the transporter robots and the object always made contact at the bottom of the object. Therefore, $r_7, r_8, r_9$ consisted of RCFs only with $f_2$. The initial state is given as $s_1 = \{C_1^E, C_1^R\}$, where $C_1^E = \{(f_6, F_1)^{e,dn}, (f_2, F_3)^{e,st}, (f_1, F_2)^{e,dn}\}$, and $C_1^R = \varnothing$. The final state is given as $s_G = \{C_G^E, C_G^R\}$, where $C_G^E = \varnothing$, and $C_G^R = \{(f_2, r_7)^{r,ps}, (f_2, r_8)^{r,ps}, (f_2, r_9)^{r,ps}\}$.

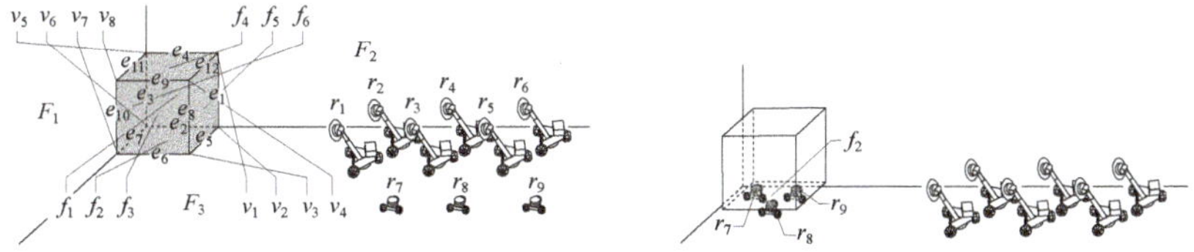


**Figure 12.** (**a**) Initial mode and (**b**) Target final mode. The targeted object for manipulation is a cuboid. There are two walls and the floor as the environment. We used six pusher robots and three transporter robots.

By applying the proposed method to generate the mode from the viewpoint of statics, 166,553,714 modes were generated, with the type and the corresponding number of robots determined according to the necessary condition to achieve gravity closure. As shown in Table 1, the total number of combinations between the possible ECFs and RCFs was 167,180,587, and we can see that proposed protocol reduced the number of modes. The reduction in the total number of modes was not dramatical because by considering all the possible contact between the robots and the object's geometrical primitives, there existed a huge amount of RCFs. For each ECF, our method eliminated the unvailable robot combinations. However, for most of the ECFs, the environmental constraints enabled most of the robot combinations to provide sufficient constraints to achieve the gravity closure, so the reduction of the candidates of robot combination was not really much. Further, after considering the possible geometrical primitives of the object that made contact with the robots, this reduction became less significant.

The mode transition graph was generated from the above modes by using the rules defined in the previous section. Also, as shown in Table 1, the resultant number of transitions was $8.6 \times 10^9$, where we can see that the proposed rules significantly reduced the candidates of transition between modes, which originally would be $1.4 \times 10^{16}$. We can see a dramatic reduction in the number of mode transitions here because one contact states can only change to another one under the constraints of geometrical boundary relationships and under the rules that the robot contact changes one by one. With such geometrical boundary relationships and the rules to change the robot contacts considered, a mode has only very few or even no transitable modes, whereas all the other modes will be its transitable modes and a large number of meaningless transitions will be caused if our method is not adopted.

Finally, by searching the mode graph, a total of 27,216 paths which minimized the cost were obtained in the graph.

**Table 1.** Comparison among the various parameters obtained by Example 1.

| | Before Adopting Our Method | After Adopting Our Method |
|---|---|---|
| Generated modes | 167,180,587 | 166,553,714 |
| Number of mode transitions | $1.4 \times 10^{16}$ | $8.6 \times 10^9$ |

Table 2 shows one of the obtained paths with the lowest cost in the mode graph. As mentioned above, the paths in the modes were decided without planning the exact configuration of robots. Therefore, the configurations of the object and the robots shown in Table 2 are examples of the modes. In the path shown in Table 2, the object was pushed by a pusher robot and rotated about the object's edge $e_5$ alongside the wall $F_1$. A transporter robot was then inserted underneath it. The pusher robot then moved the object so that only the object's vertex $v_3$ was in contact with the floor $F_3$, and another transporter robot was inserted underneath the object. Finally, with two transporter robots at the bottom, the object lost contact with the floor $F_3$ through contact with the pusher robot, which allowed the third transporter robot to enter underneath the object. Thus, the object was finally loaded onto three transporter robots.

We can see that the obtained path is reasonable. From the viewpoint of statics and geometry, it was also reasonable to include the paths in which only the combinations of geometrical primitives were different. The resultant paths also contained those paths that were impossible to realize from the viewpoint of the force balance determined by the configurations of the object and the robots. Those paths will be omitted in the later planning phase when deciding the object–robot system's exact configurations based on the transitions of the modes; they will not be dealt with in the current phase where only the transitions of the modes are decided. The important result is that the candidates of the mode transposition were narrowed down, as shown Table 1, and this facilitated the planning to determine the configurations.

**Table 2.** One of the manipulation paths that minimized the cost.

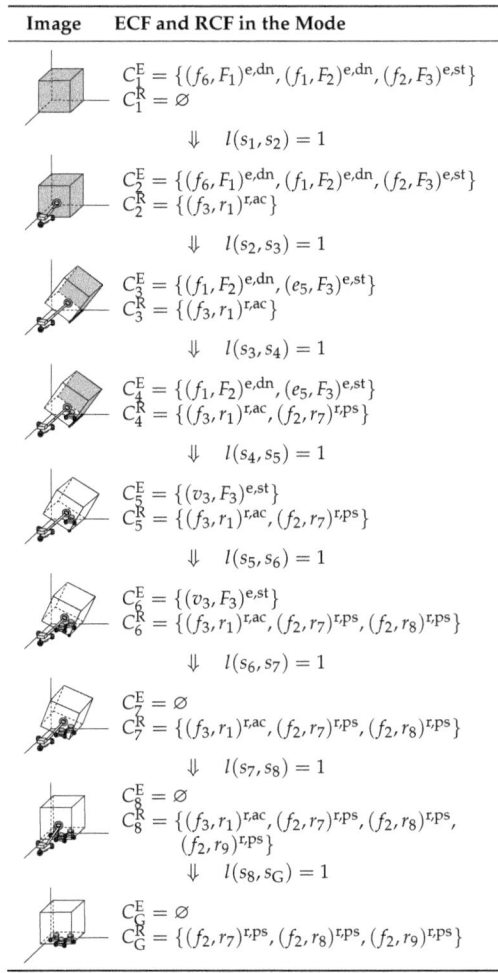

| Image | ECF and RCF in the Mode |
|---|---|
| 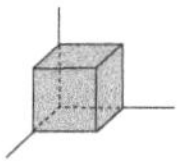 | $C_1^E = \{(f_6, F_1)^{e,dn}, (f_1, F_2)^{e,dn}, (f_2, F_3)^{e,st}\}$ <br> $C_1^R = \varnothing$ |
| | $\Downarrow \quad l(s_1, s_2) = 1$ |
| 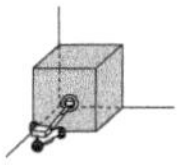 | $C_2^E = \{(f_6, F_1)^{e,dn}, (f_1, F_2)^{e,dn}, (f_2, F_3)^{e,st}\}$ <br> $C_2^R = \{(f_3, r_1)^{r,ac}\}$ |
| | $\Downarrow \quad l(s_2, s_3) = 1$ |
| 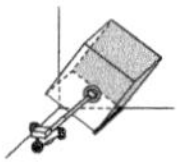 | $C_3^E = \{(f_1, F_2)^{e,dn}, (e_5, F_3)^{e,st}\}$ <br> $C_3^R = \{(f_3, r_1)^{r,ac}\}$ |
| | $\Downarrow \quad l(s_3, s_4) = 1$ |
| 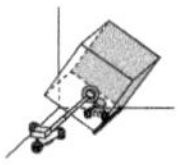 | $C_4^E = \{(f_1, F_2)^{e,dn}, (e_5, F_3)^{e,st}\}$ <br> $C_4^R = \{(f_3, r_1)^{r,ac}, (f_2, r_7)^{r,ps}\}$ |
| | $\Downarrow \quad l(s_4, s_5) = 1$ |
| 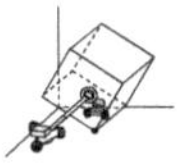 | $C_5^E = \{(v_3, F_3)^{e,st}\}$ <br> $C_5^R = \{(f_3, r_1)^{r,ac}, (f_2, r_7)^{r,ps}\}$ |
| | $\Downarrow \quad l(s_5, s_6) = 1$ |
| 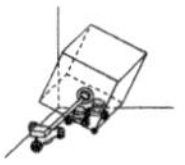 | $C_6^E = \{(v_3, F_3)^{e,st}\}$ <br> $C_6^R = \{(f_3, r_1)^{r,ac}, (f_2, r_7)^{r,ps}, (f_2, r_8)^{r,ps}\}$ |
| | $\Downarrow \quad l(s_6, s_7) = 1$ |
| 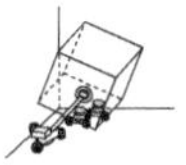 | $C_7^E = \varnothing$ <br> $C_7^R = \{(f_3, r_1)^{r,ac}, (f_2, r_7)^{r,ps}, (f_2, r_8)^{r,ps}\}$ |
| | $\Downarrow \quad l(s_7, s_8) = 1$ |
| 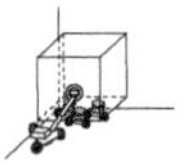 | $C_8^E = \varnothing$ <br> $C_8^R = \{(f_3, r_1)^{r,ac}, (f_2, r_7)^{r,ps}, (f_2, r_8)^{r,ps},$ <br> $(f_2, r_9)^{r,ps}\}$ |
| | $\Downarrow \quad l(s_8, s_G) = 1$ |
| 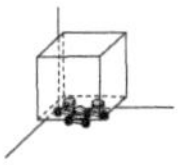 | $C_G^E = \varnothing$ <br> $C_G^R = \{(f_2, r_7)^{r,ps}, (f_2, r_8)^{r,ps}, (f_2, r_9)^{r,ps}\}$ |

### 5.2. Example 2 and Result Discussion

In the second example, a cuboid is loaded onto a step by using mobile robots, as shown in Figure 13. The faces, the edges, and the vertices of the cuboid are denoted as $f_1, f_2, \ldots, f_6, e_1, e_2, \ldots, e_{12}$, and $v_1, v_2, \ldots, v_8$, respectively. The floor is denoted as $F_1$ and the faces consisting of the step are denoted as $F_2$ and $F_3$.

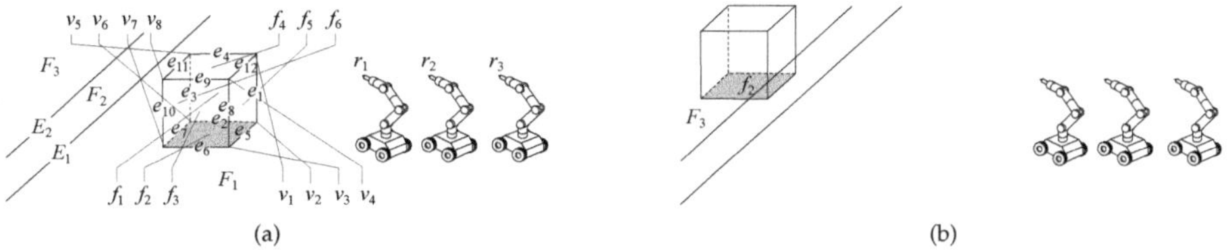


**Figure 13.** (**a**) Initial mode and (**b**) target final mode. Targeted object for manipulation is cuboid. There is a step in the environment. We used six pusher robots to bring the cuboid on the step.

A mobile robot with a standard six-axis manipulator is shown in Figure 11d. The robot generates a frictional point of contact with the object. We used three robots and denoted their geometrical primitives as $r_1, r_2$, and $r_3$. There were no obstacles around the step; therefore, the manipulation could be performed by either using or not using the environment. By using the proposed method, we generated the modes with the necessary number of robots to achieve gravity closure. Then, in the created mode graph, we took the initial state as $s_1 = \{\{(f_2, F_1)^{\mathrm{e,st}}\}, \varnothing\}$ and the final state as $s_G = \{\{(f_2, F_3)^{\mathrm{e,st}}\}, \varnothing\}$ and searched for the manipulation paths with the minimum total cost.

Using the procedure given in the previous example, we generated 292,415 modes and $2.9 \times 10^6$ transitions. The total number of combinations of the ECFs and RCFs was 299,967, and the total number of combinations of the generated modes was $4.5 \times 10^{10}$, as shown in Table 3. For the same reason we analyzed in Example 1, our method also narrowed down the number of possible modes, and especially, dramatically decreased the transitions between them.

**Table 3.** Comparison among the various parameters obtained by Example 2.

| | Before Adopting Our Method | After Adopting Our Method |
|---|---|---|
| Generated modes | 299,967 | 292,415 |
| Number of mode transitions | $4.5 \times 10^{10}$ | $2.9 \times 10^6$ |

Our search produced 672,352 paths with a minimum total cost. In these obtained paths, the number of robots involved varied. In the two paths shown in Table 4, we found that the number of robots engaged in the two paths were different.

In Path 1, only two robots were needed to perform the manipulation task whereas three robots were needed in Path 2 because the floor was used to exert constraints on the object. Therefore, fewer robots were needed in Path 1. We could choose the path according to some other criteria based on the obtained paths. For example, the manipulation action sequences obtained from Path 1 would be more preferable than those obtained from Path 2 because fewer robots were adopted. Besides, as compared with the previous example, relative sliding occured on the face contact principals in the action sequences of Path 1 between the mode $s_5 = \{\{(e_5, F_1)^{\mathrm{e,dn}}\}, \{(f_3, r_1)^{\mathrm{r,ac}}, (f_5, r_2)^{\mathrm{r,ac}}\}\}$ and the mode $s_6 = \{\{(e_5, F_1)^{\mathrm{e,dn}}, (f_2, E_2)^{\mathrm{e,dn}}\}, \{(f_3, r_1)^{\mathrm{r,ac}}, (f_5, r_2)^{\mathrm{r,ac}}\}\}$; therefore, the transition from static contact to dynamic contact appeared between the mode $s_4 = \{\{(e_5, F_1)^{\mathrm{e,st}}\}, \{(f_3, r_1)^{\mathrm{r,ac}}, (f_5, r_2)^{\mathrm{r,ac}}\}\}$ and $s_5$. A similar transition can also be seen in Path 2.

In the non-prehensile manipulation by multiple mobile robots, there were many possible manipulation sequences each with a difference number of robots. As shown in above example, our method can list up possible patterns of manipulation taking into account their possibility from the viewpoint of statics.

**Table 4.** Two paths of the manipulation paths which minimized the cost.

| **Path 1** | | **Path 2** | |
|---|---|---|---|
| **Image** | **ECF and RCF in the Mode** | **Image** | **ECF and RCF in the Mode** |
| 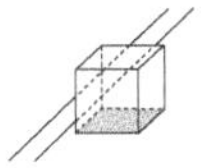 | $C^E_1 = \{(f_2, F_1)^{e,st}\}$ <br> $C^R_1 = \varnothing$ | 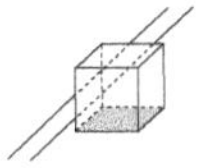 | $C^E_1 = \{(f_2, F_1)^{e,st}\}$ <br> $C^R_1 = \varnothing$ |
| | $\Downarrow \; l(s_1, s_2) = 1$ | | $\Downarrow \; l(s_1, s_2) = 1$ |
| 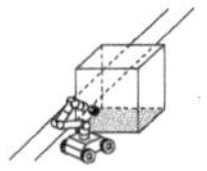 | $C^E_2 = \{(f_2, F_1)^{e,st}\}$ <br> $C^R_2 = \{(f_3, r_1)^{r,ac}\}$ | 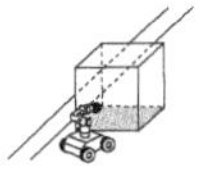 | $C^E_2 = \{(f_2, F_1)^{e,st}\}$ <br> $C^R_2 = \{(f_3, r_1)^{r,ac}\}$ |
| | $\Downarrow \; l(s_2, s_3) = 1$ | | $\Downarrow \; l(s_2, s_3) = 1$ |
| 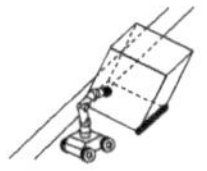 | $C^E_3 = \{(e_5, F_1)^{e,st}\}$ <br> $C^R_3 = \{(f_3, r_1)^{r,ac}\}$ | 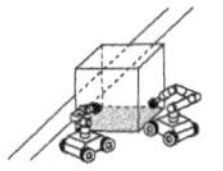 | $C^E_3 = \{(f_2, F_1)^{e,st}\}$ <br> $C^R_3 = \{(f_3, r_1)^{r,ac}, (f_5, r_2)^{r,ac}\}$ |
| | $\Downarrow \; l(s_3, s_4) = 1$ | | $\Downarrow \; l(s_3, s_4) = 1$ |
| 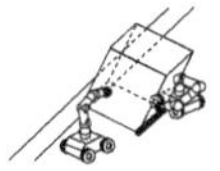 | $C^E_4 = \{(e_5, F_1)^{e,st}\}$ <br> $C^R_4 = \{(f_3, r_1)^{r,ac}, (f_5, r_2)^{r,ac}\}$ | 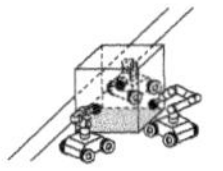 | $C^E_4 = \{(f_2, F_1)^{e,st}\}$ <br> $C^R_4 = \{(f_3, r_1)^{r,ac}, (f_5, r_2)^{r,ac}, (f_1, r_3)^{r,ac}\}$ |
| | $\Downarrow \; l(s_4, s_5) = 0$ | | $\Downarrow \; l(s_4, s_5) = 1$ |
| 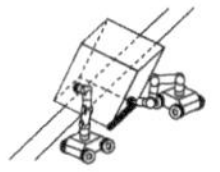 | $C^E_5 = \{(e_5, F_1)^{e,dn}\}$ <br> $C^R_5 = \{(f_3, r_1)^{r,ac}, (f_5, r_2)^{r,ac}\}$ | 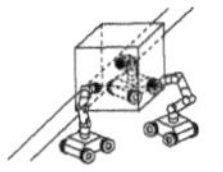 | $C^E_5 = \varnothing$ <br> $C^R_5 = \{(f_3, r_1)^{r,ac}, (f_5, r_2)^{r,ac}, (f_1, r_3)^{r,ac}\}$ |
| | $\Downarrow \; l(s_5, s_6) = 1$ | | $\Downarrow \; l(s_5, s_6) = 1$ |
| 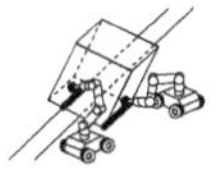 | $C^E_6 = \{(e_5, F_1)^{e,dn}, (f_2, E_2)^{e,dn}\}$ <br> $C^R_6 = \{(f_3, r_1)^{r,ac}, (f_5, r_2)^{r,ac}\}$ | 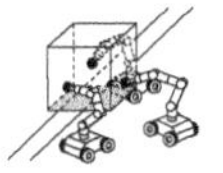 | $C^E_6 = \{(f_2, F_3)^{e,dn}\}$ <br> $C^R_6 = \{(f_3, r_1)^{r,ac}, (f_5, r_2)^{r,ac}, (f_1, r_3)^{r,ac}\}$ |
| | $\Downarrow \; l(s_6, s_7) = 1$ | | $\Downarrow \; l(s_6, s_7) = 0$ |
| 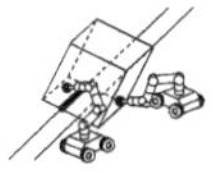 | $C^E_7 = \{(f_2, E_2)^{e,dn}\}$ <br> $C^R_7 = \{(f_3, r_1)^{r,ac}, (f_5, r_2)^{r,ac}\}$ | 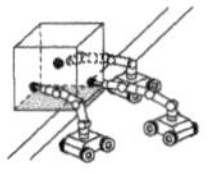 | $C^E_7 = \{(f_2, F_3)^{e,st}\}$ <br> $C^R_7 = \{(f_3, r_1)^{r,ac}, (f_5, r_2)^{r,ac}, (f_1, r_3)^{r,ac}\}$ |
| | $\Downarrow \; l(s_7, s_8) = 1$ | | $\Downarrow \; l(s_7, s_8) = 1$ |
| 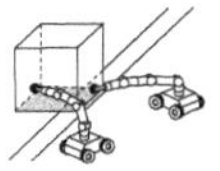 | $C^E_8 = \{(f_2, F_3)^{e,st}\}$ <br> $C^R_8 = \{(f_3, r_1)^{r,ac}, (f_5, r_2)^{r,ac}\}$ | 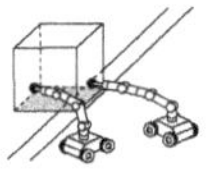 | $C^E_8 = \{(f_2, F_3)^{e,st}\}$ <br> $C^R_8 = \{(f_3, r_1)^{r,ac}, (f_5, r_2)^{r,ac}\}$ |
| | $\Downarrow \; l(s_8, s_9) = 1$ | | $\Downarrow \; l(s_8, s_9) = 1$ |
| 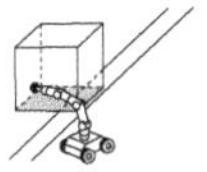 | $C^E_9 = \{(f_2, F_3)^{e,st}\}$ <br> $C^R_9 = \{(f_3, r_1)^{r,ac}\}$ | 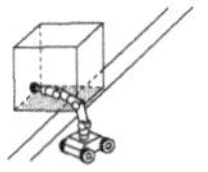 | $C^E_9 = \{(f_2, F_3)^{e,st}\}$ <br> $C^R_9 = \{(f_3, r_1)^{r,ac}\}$ |
| | $\Downarrow \; l(s_9, s_G) = 1$ | | $\Downarrow \; l(s_9, s_G) = 1$ |
| 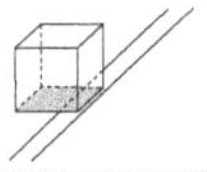 | $C^E_G = \{(f_2, F_3)^{e,st}\}$ <br> $C^R_G = \varnothing$ | 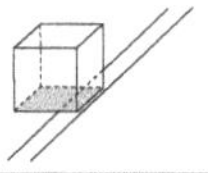 | $C^E_G = \{(f_2, F_3)^{e,st}\}$ <br> $C^R_G = \varnothing$ |

## 6. Conclusions

In this paper, we proposed the modal planning method for multi-contact non-prehensile manipulation using multiple mobile robots. After defining a mode as a set of configurations that hold the same contact state and investigating the transition between modes, a manipulation planner can efficiently probe the configuration space even if the states under varying constraints are difficult to sample in the configuration space. When multiple mobile robots manipulate the object using non-prehensile methods, the modes and their consequent transitions become enormous because of the numerous contacts made by the environment and the robots on the object. For such situations, we proposed a method to narrow down the possible modes and their transitions beforehand by excluding the invalid modes and transitions. In our proposed method, we generated modes that described an object's contact states with the robots and the environment while ignoring their exact configurations, provided each multi-contact set satisfied the necessary condition for gravity closure on the object (along with gravity). Secondly, we investigated the valid transition between the modes by taking into account whether the given robot could actively change an object's contact state under feasible geometrical relationships. Finally, we conducted two simulations on specific manipulation tasks to validate our method and confirmed that the number of modes and transitions had significantly reduced. Also, it was feasible to obtain the sequence of modes obtained by searching the shortest path.

Our method can be adopted to probe the modal spaces in the variant cases of non-prehensile manipulation by mobile robots. If prior sequencing manipulation actions by adopting the generated mode transition graph, the manipulation planner can avoid the heavy computation of searching in a whole large configuration space. Thus, determining the sequence of modes is usually the first hierarchy in manipulation planning, where we do not consider the exact configuration of a system that comprises objects and robots. However, these configurations must be determined to complete manipulation planning by considering certain factors, such as the achievement force closure, the movability of the object, and accessibility for mobile robots. We can obtain those configurations based on the prior determined modal space, by applying sample-based methods to each modal space. This will be significantly more efficient than directly applying sample-based methods to probe a whole configuration space. In our future studies, we propose to further investigate this topic.

**Author Contributions:** Conceptualization, J.O., S.S. and C.F.; Methodology, C.F., S.S. and J.O.; Software, C.F. and S.S.; Validation, C.F. and S.S.; Data curation, C.F. and S.S.; Formal analysis, C.F., S.S. and J.O.; Investigation, C.F., S.S. and J.O.; Resources, J.O.; Writing–original draft preparation, C.F. and S.S.; Writing–review and editing, C.F., S.S. and J.O.; Visualization, S.S. and C.F.; Supervision, S.S. and J.O.; Project administration, J.O.; Funding acquisition, J.O.

**Funding:** This research received no external funding.

**Acknowledgments:** Changxiang Fan is financially supported by China Scholarship Council (CSC).

**Conflicts of Interest:** The authors declare no conflict of interest.

## Abbreviations

The following abbreviations are used in this manuscript:

| | |
|---|---|
| PC | principal contact |
| CF | contact formation |
| ECF | environmental contact formation |
| RCF | robot contact formation |
| GCR | goal-contact-relaxation |

## References

1. Stilwell, D.J.; Bay, J.S. Toward the development of a material transport system using swarms of ant-like robots. In Proceedings of the IEEE International Conference on Robotics and Automation, Atlanta, GA, USA, 2–6 May 1993; pp. 766–771.
2. Mason, M.T. *Mechanics of Robotic Manipulation*; MIT Press: Cambridge, MA, USA, 2001.
3. Ruggiero, F.; Lippiello, V.; Siciliano, B. Nonprehensile Dynamic Manipulation: A Survey. *IEEE Robot. Autom. Lett.* **2018**, *3*, 1711–1718. [CrossRef]
4. Arai, T.; Ota, J. Let us work together-Task planning of multiple mobile robots. In Proceedings of the 1996 IEEE/RSJ International Conference on Intelligent Robots and Systems' 96, Osaka, Japan, 8 November 1996; Volume 1, pp. 298–303.
5. Latombe, J.C. *Robot Motion Planning*; Springer US: Boston, MA, USA, 1991; Volume 124.
6. LaValle, S.M. *Rapidly-Exploring Random Trees: A New Tool for Path Planning*; Technical Report TR98-11; Computer Science Department, Iowa State University, IA, USA, 1998.
7. Kavraki, L.; Svestka, P.; Latombe, J.; Overmars, M.H. Probabilistic Roadmaps for Path Planning in High-Dimensional Configuration Spaces. *IEEE Trans. Robot. Autom.* **1996**, *12*, 566–588. [CrossRef]

8. Hauser, K.; Latombe, J.C. Multi-modal motion planning in non-expansive spaces. *Int. J. Robot. Res.* **2010**, *29*, 897–915. [CrossRef]
9. Hsu, D.; Kavraki, L.E.; Latombe, J.C.; Motwani, R.; Sorkin, S. On finding narrow passages with probabilistic roadmap planners. In *Proceedings of the Workshop on the Algorithmic Foundations of Robotics, Robotics: The Algorithmic Perspective*; A. K. Peters, Ltd.: Natick, MA, USA, 1998; pp. 141–154.
10. Maeda, Y.; Arai, T. Planning of graspless manipulation by a multifingered robot hand. *Adv. Robot.* **2005**, *19*, 501–521. [CrossRef]
11. Miyazawa, K.; Maeda, Y.; Arai, T. Planning of graspless manipulation based on rapidly-exploring random trees. In Proceedings of the 6th IEEE International Symposium on Assembly and Task Planning: From Nano to Macro Assembly and Manufacturing, Montreal, QC, Canada, 19–21 July 2005; pp. 7–12.
12. Barry, J.; Kaelbling, L.P.; Lozano-Pérez, T. A hierarchical approach to manipulation with diverse actions. In Proceedings of the 2013 IEEE International Conference on Robotics and Automation, Karlsruhe, Germany, 6–10 May 2013; pp. 1799–1806.
13. Lee, G.; Lozano-Pérez, T.; Kaelbling, L.P. Hierarchical planning for multi-contact non-prehensile manipulation. In Proceedings of the 2015 IEEE/RSJ International Conference on Intelligent Robots and Systems (IROS), Hamburg, Germany, 28 September–2 October 2015; pp. 264–271.
14. Schmitt, P.S.; Neubauer, W.; Feiten, W.; Wurm, K.M.; Wichert, G.V.; Burgard, W. Optimal, sampling-based manipulation planning. In Proceedings of the 2017 IEEE International Conference on Robotics and Automation (ICRA), Singapore, 29 May–3 June 2017; pp. 3426–3432.
15. Xiao, J. Goal-contact relaxation graphs for contact-based fine motion planning. In Proceedings of the 1997 IEEE International Symposium on Assembly and Task Planning, Marina del Rey, CA, USA, 7–9 August 1997; pp. 25–30.
16. Xiao, J.; Ji, X. Automatic generation of high-level contact state space. *Int. J. Robot. Res.* **2001**, *20*, 584–606. [CrossRef]
17. Fan, C.; Shirafuji, S.; Ota, J. Least Action Sequence Determination in the Planning of Non-Prehensile Manipulation with Multiple Mobile Robots. In *the 15th International Conference on Intelligent Autonomous Systems (IAS-15)*; Springer: Cham, Switzerland, 2018; pp. 174–185.
18. Xiao, J.; Zhang, L. A General Strategy to Determine Geometrically Valid Contact Formations from Possible Contact Primitives. In Proceedings of the IEEE International Conference on Robotics And Automation, Atlanta, GA, USA, 2–6 May 1993; Volume 1, p. 2728.
19. Maeda, Y.; Aiyama, Y.; Arai, T.; Ozawa, T. Analysis of object-stability and internal force in robotic contact tasks. In Proceedings of the IEEE/RSJ International Conference on Intelligent Robots and Systems, Osaka, Japan, 8 November 1996; Volume 2, pp. 751–756.
20. Aiyama, Y.; Arai, T.; Ota, J. Dexterous assembly manipulation of a compact array of objects. *CIRP Ann.* **1998**, *47*, 13–16. [CrossRef]
21. LaValle, S.M. *Planning Algorithms*; Cambridge University Press: Cambridge, UK, 2006.
22. Ji, X.; Xiao, J. Planning motions compliant to complex contact states. *Int. J. Robot. Res.* **2001**, *20*, 446–465. [CrossRef]
23. Ponce, J.; Faverjon, B. On computing three-finger force-closure grasps of polygonal objects. *IEEE Trans. Robot. Autom.* **1995**, *11*, 868–881. [CrossRef]
24. Murray, R.M.; Li, Z.; Sastry, S.S. *A Mathematical Introduction to Robotic Manipulation*; CRC Press: Boca Raton, FL, USA, 1994.
25. Shirafuji, S.; Terada, Y.; Ota, J. Mechanism Allowing a Mobile Robot to Apply a Large Force to the Environment. In *The 14th International Conference on Intelligent Autonomous Systems (IAS-14)*; Springer: Cham, Switzerland, 2016; pp. 795–808.
26. Shirafuji, S.; Terada, Y.; Ito, T.; Ota, J. Mechanism allowing large-force application by a mobile robot, and development of ARODA. *Robot. Auton. Syst.* **2018**, *110*, 92–101. [CrossRef]
27. Ito, T.; Shirafuji, S.; Ota, J. Development of a Mobile Robot Capable of Tilting Heavy Objects and its Safe Placement with Respect to Target Objects. In Proceedings of the 2018 IEEE International Conference on Roboics and Biomimetics (ROBIO2018), Kuala Lumpur, Malaysia, 12–15 December 2018; pp. 716–722.

*applied sciences*

MDPI

*Article*

# A Master-Slave Separate Parallel Intelligent Mobile Robot Used for Autonomous Pallet Transportation

**Guo Li [1], Rui Lin [2,*], Maohai Li [2], Rongchuan Sun [2] and Songhao Piao [1,*]**

1 Department of Computer Science and Technology, Harbin Institute of Technology, Harbin 150001, China; 14b903020@hit.edu.cn

2 School of Mechanical and Electrical Engineering, Soochow University, Suzhou 215000, China; limaohai@suda.edu.cn (M.L.); sunrongchuan@suda.edu.cn (R.S.)

* Correspondence: linrui@suda.edu.cn (R.L.); piaosh@hit.edu.cn (S.P.)

Received: 27 December 2018; Accepted: 21 January 2019; Published: 22 January 2019

**Abstract:** This work reports a master-slave separate parallel intelligent mobile robot for the fully autonomous transportation of pallets in the smart factory logistics. This separate parallel intelligent mobile robot consists of two independent sub robots, one master robot and one slave robot. It is similar to two forks of the forklift, but the slave robot does not have any physical or mechanical connection with the master robot. A compact driving unit was designed and used to ensure access to the narrow free entry under the pallets. It was also possible for the mobile robot to perform a synchronous pallet lifting action. In order to ensure the consistency and synchronization of the motions of the two sub robots, high-gain observer was used to synchronize the moving speed, the lifting speed and the relative position. Compared with the traditional forklift AGV (Automated Guided Vehicle), the mobile robot has the advantages of more compact structure, higher expandability and safety. It can move flexibly and take zero-radius turn. Therefore, the intelligent mobile robot is quite suitable for the standardized logistics factory with small working space.

**Keywords:** intelligent mobile robot; pallet transportation; master-slave; compact driving unit; high-gain observer

## 1. Introduction

Logistics is an important part of the manufacturing enterprise [1,2], and the pallet is one of the logistics carriers in the factory [3]. The automation of pallet transportation is an important factor which affects the smart factory logistics [4]. Pallet handling equipment mainly refers to manually operated hydraulic trucks and electric or fuel forklifts, both of which require a large working space for operation. When the goods are delivered, it is necessary for the operator to handle the loading and unloading of the goods. A large amount of labor is required, which does not meet the requirements of factory automation.

In recent years, with the development of factory intelligence, flexible manufacturing systems and automated warehousing, the autonomous navigation forklift or forklift AGV (Automated Guided Vehicle) has gradually become an important device for solving the internal logistics of pallet transportation in the factory [5,6]. Compared with the traditional forklift, the autonomous navigation forklift AGV is improved with intelligent components such as external sensors, navigation and positioning modules, software scheduling, and power management system. Forklift AGV can transport the pallets placed on the ground to meet the unmanned logistics requirements [7–9]. As early as 2015, the forklift company, for example, Dematic or Jungheinrich, launched a forklift AGV based on the traditional forklift. For AGV products based on forklift, a single steering wheel or a differential fixed double steering wheel with a front integrated steering motor was generally used as the driving wheel, and the two-forked casters with two auxiliary wheels. Forklift AGV could move freely by controlling a single steering wheel or a differential double steering wheel. Forklift AGV used the traditional forklift structure and had the function of stacking high pallets. However, due to its own structural characteristics, the overall size and its self-weight are both large. And the arrangement of the driving wheels and the attached wheels determined that this type of AGV cannot achieve zero-radius rotation. It requires a rather large working space [10].

In order to adapt to small working space of factory logistics, another type of pallet transportation robot was designed and used. The pallet transportation robot moved underneath the pallet rack to lift and delivered the pallet to a destination, which was placed on a pallet rack with a certain height in advance. In 2007, Professor Manuel Weber of the University of Stuttgart, Germany, proposed and designed the double-pronged pallet handling robot, the Doppelkufen system [11], to minimize the required transportation space. The Doppelkufen system used a visual recognition ribbon to guide between different production lines. In order to realize autonomous navigation, the two separate double forks connected with each other physically, and peripheral sensors were assembled to develop a compact forklift robot capable of zero-radius rotation and omnidirectional motion, for example, Agumos G130 [12] of Melkus Mechatronic GmbH and Nipper [13] of Dutch F3 Design. Combining with the advantages of both the traditional forklift AGV and the compact forklift AGV, INTREST Services GmbH of Germany introduced the Agilox product [14]. Agilox could perform palletizing automatically. It had a heavy self-weight and high height and cannot move underneath the pallet directly like a forklift. However the pallet needed to be placed on a custom-height pallet rack in advance.

In order to solve the problems, this paper mainly introduces a master-slave separate parallel intelligent mobile robot for autonomous pallet transportation in the smart factory logistics. The separate parallel intelligent mobile robot consists of two independent sub robots, which is similar to two forks of the forklift, but these sub robots do not have any physical connection with each other to achieve the pallet transportation. It can move underneath the narrow free entry of the pallet and has a load capacity of up to 1 ton. It was designed with a compact driving unit, in addition to actual motions, such as linear motion, oblique linear motion, and zero-radius turn. It could also perform synchronous lifting and laying of the pallet. The slave robot follows the master robot synchronously and consistently. Each robot had two driving units. In order to synchronize the motion and trajectory of these two sub robots, a nonlinear control system was used [15]. The application of nonlinear control systems in robotics control was very common, from the simplest two-wheel driving self-balancing robot dynamics [16,17], the robotic arm joint PD control [18], to the cluster control of UAV formations [19–21]. In the design we used the synchronic control strategy of slave robot dynamic adaptive following master robot. By analyzing the practical effects of sliding mode control [22] and high-gain observer [23] of the nonlinear control tools, we selected high-gain observer to obtain the following speed and state control of slave robot which was to ensure that the following errors converged and achieved synchronization of the master-slave robot motion.

Compared with the manual and forklift truck, the separate parallel intelligent mobile robot can fulfill an autonomous pallet transportation. Compared with the traditional intelligent forklift AGV, it has the advantages of compact structure, small self-weight and large payload, high expandability and safety. It can implement zero-radius turn and requires small channel width. It can also be combined with automation equipment or storage system seamlessly. So it is very suitable for small and standardized logistics warehouses.

The rest of this paper is organized as follows. Section 2 describes the mechanical structure of master-slave separate parallel intelligent mobile robot, such as compact driving units, communication modules, and safety protection modules. High-gain observer is adopted to ensure the consistency and synchronization of actual motions of two sub robots of intelligent mobile robot in Section 3. Experimental results and related analyses are given in Section 4, followed by conclusions in Section 5.

## 2. Separate Parallel Intelligent Mobile Robot

### 2.1. A Compact Driving Unit

As the EPAL Euro pallet is widely used in the factory logistics (PALETTE EUR-EPAL, dimension: L1200 mm × W800 mm and L1200 mm × W1000 mm), a compact driving unit was designed ensure that the two sub robots could move into the free entry of the standard pallet and lift the pallet up. Each sub robot has front and rear driving units. Separate parallel intelligent mobile robot implemented synchronous motions such as linear motion, oblique linear motion, and zero-radius turn. It could also perform synchronous lifting and laying of the pallet.

#### 2.1.1. Motor Performance Calculation

In the actual motions of the master-slave separate parallel robot during the pallet transportation, it is necessary to overcome some resistance, such as the rolling resistance from the ground and the air resistance from the air. The resistance expressions $\sum F$ are:

$$\sum F = F_f + F_w + F_i + F_j \tag{1}$$

where, $F_f$ is tolling resistance; $F_w$ is air resistance; $F_i$ is climbing resistance; $F_j$ is acceleration resistance.

Tolling resistance for a sub robot is

$$F_f = \mu mg \tag{2}$$

where, $\mu$ is coefficient of tolling resistance. $m$ represents the its own weight and maximum payload.

The resistance during acceleration of the sub robot is:

$$F_j = ma \tag{3}$$

Due to surface roughness, climbing resistance of the sub robot is

$$F_i = mg \sin \alpha \tag{4}$$

The rolling diameter of sub robot's driving wheel is set to $D$. The total motion-resistance force is $\sum F$. Then total resistance moment $\sum M$ is:

$$\sum M = \sum F \cdot \frac{D}{2} \tag{5}$$

As each unit has two driving wheels, the driving moment of every wheel $M_d$ is

$$M_d = \frac{\sum M}{4} \tag{6}$$

According to the design requirements, a sub-robot's self-weight and maximum payload $m$ = 600 kg. The normal velocity is 0.7 m/s. The acceleration is $a = 0.3$ m/s$^2$. The surface slope of working ground is $\alpha = 2°$. The ground friction coefficient is $\mu = 0.02$.

#### 2.1.2. Mechanism Design

Each sub robot uses front and rear driving units and one driving unit consists of two DC motors, reducers and driving wheels. The power transmission sequence is: DC motor, planetary reducer, small timing pulley, timing belt, large timing pulley, and driving wheel. The slip ring is assembled to simplify wiring during the zero-radius turn of the driving units. The slip ring is divided into upper and lower parts: the upper part is the stator, and the lower part is the rotor. The stator is fixed with the integral support plate, and the rotor rotates together with the driving mechanism. Slip ring not only solves the problem of wiring of power and cable during the lifting and laying of the pallet, but also provides direction angle value for the front and rear driving units through the built-in multi-turn absolute encoder, which provides input for the nonlinear synchronization control in Section 3. It is known that the pitch, length and other parameters of the screw can lift the pallet up precisely to a certain height by controlling the number of wheel rotations of the driving units. Since the screw has a self-holding characteristic, the pallet does not fall down by itself even though power outage may occur unexpectedly. The compact driving unit is shown in Figure 1.

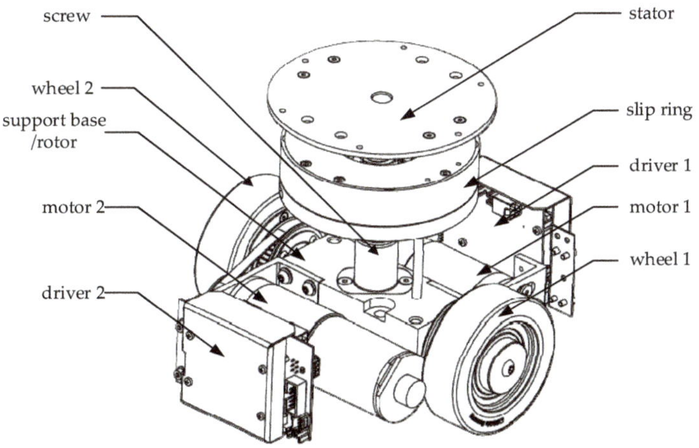


**Figure 1.** Three dimensional sketches of the compact driving unit.

### 2.1.3. Analysis of Motion Models

Motion models of separate parallel intelligent mobile robot are mainly based on the master robot's motion, such as visual navigation, and laser-based navigation. Slave robot follows master robot to dynamically adjust the relative speed and position. The two sub robots implement synchronous actual motions, such as linear motion, oblique linear motion, zero-radius turn, and lifting and laying. The motion model is shown in Figure 2.

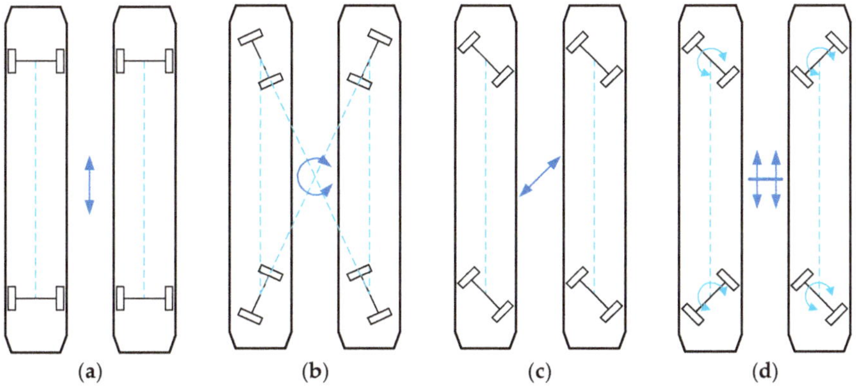


**Figure 2.** Analysis of motion model. (**a**) Forwards/Backwards; (**b**) Rotation; (**c**) Sideways/crab motion; (**d**) Lifting and laying.

### 2.2. Schematic Diagram of Separate Parallel Intelligent Mobile Robot

The main components of master-slave separate parallel intelligent mobile robot are: driving units, control and power supply modules, sensor modules, shell cover and safety protection modules, as shown in Figure 3. The individual modules of the sub robot are mechanically independent but electrically connected to each other. Each sub robot consists of two driving units, two safety protection modules, two sensor modules, and control and power supply modules with the cover. Safety sensors and infrared data transmission module are installed around the sides of each sub robot. The shell cover is the mounting base, and each module is mounted on the cover by a cross recessed countersunk head screw.

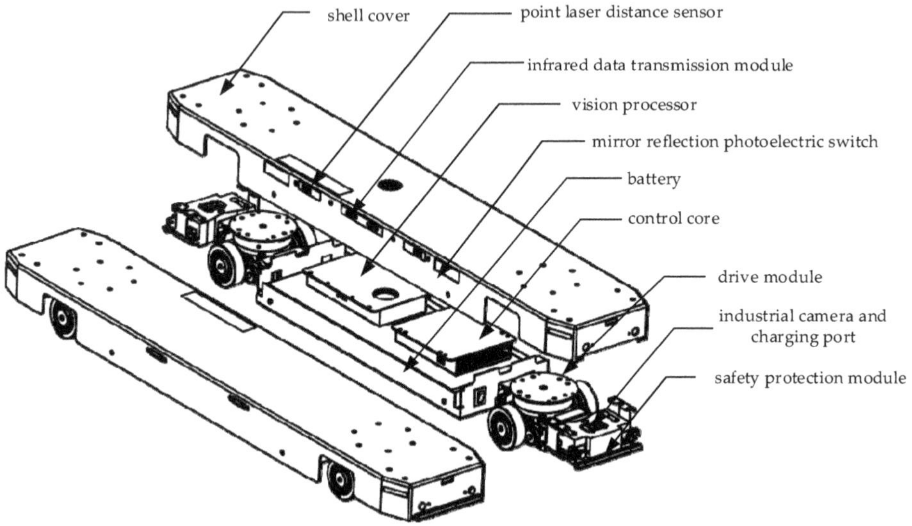


**Figure 3.** Schematic diagram of separate parallel intelligent mobile robot.

## 3. Synchronization Nonlinear Control

The most significant and complex problem in the real-time control system of the master-slave separate parallel intelligent mobile robot is the synchronization control of actual motions of the two sub robots, which is also a very difficult nonlinear control system. To get good performance of the synchronization control, complex nonlinear control strategy should be used. Therefore, this section mainly introduces some nonlinear control strategies, such as the input of nonlinear control system and high-gain observer control tool.

### 3.1. Control Strategies of Actual Motions

Different control strategies should be used in master robot and slave robot, while separate parallel intelligent mobile robot is moving. As the master robot, the only control principle is ensuring that the front and rear driving units work precisely and move as ordered. But for the slave robot, its control strategy is much more complex. (1) Self-motion conforms to self-physical model. (2) Following the master robot on front-rear axis and left-right axis, which are two different axes (X-axis and Y-axis). The repetitive position accuracy of master robot is normally distributed and related to the control accuracy of the driving units, the smoothness of the ground and the operation cycle time of the feedback control system. While the precision of driving wheel is higher and the smoothness of ground is better or the operation cycle time of control system is shorter, the variance of repetitive position accuracy will get smaller. As to the slave robot, the factors affecting repetitive position accuracy are much more complex, which include master robot's position error, slave robot's following error, inherent errors, etc.

### 3.1.1. Control Strategy of Master Robot Motions

As mentioned above, master robot only needs to move according to the driving task demands. Therefore, PID control strategy is suitable for master robot to move to the destination or lift up and lay down the pallet by controlling its driving motors. What is more, in simulation diagram, module "Constant" means setting trajectory, and "Integrator" means that the odometer is the integral result of velocity and time. In "Control section", the speeds of the two driving units are adjusted dynamically based on the position feedback. Because the system is an open loop including integral part, a proportional part in feedback control can make master robot motion system gradually stable. Control strategy of master robot motion is shown in Figure 4.

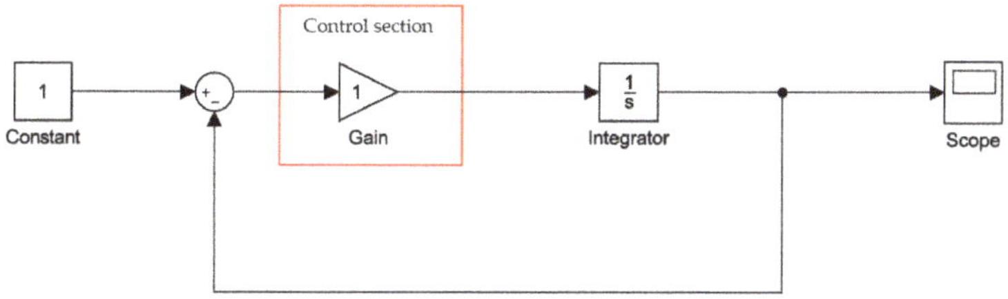


**Figure 4.** Control strategy of master robot motion.

### 3.1.2. Control Strategy of Slave Robot Motions

In fact, slave robot control system is a very complex nonlinear control strategy. Here, the main control ideas of slave robot are described in this section. The controlled object for slave robot are driving units. Control goal is to eliminate position errors on X-axis and Y-axis. The feedback signal includes the distance data between master robot and slave robot, and the SIGN signal indicates if master robot position leads or lags on Y-axis.

In order to keep the system in a steady stable state, the system's energy derivative should be non-positive definite according to Lyapunov's second stability theorem. Therefore, in this control system the driving motors' speed should tend to be stable in the process of system. Because slave robot cannot know the accurate error in real time between master robot and slave robot on Y-axis, which means it is impossible for slave robot to follow master robot without position error on Y-axis. Therefore, high-gain observer is used for keep slave robot stable within a certain error value while following master robot on Y-axis. Actually, when high-gain observer is used, it will affect the motions on both X-axis and Y-axis. Because when driving units move in the oblique linear direction, slave robot motion on X-axis and Y-axis are coupling. The slave robot moving on Y-axis will also result in moving on X-axis. Therefore, high-gain observer should be used when slave robot motion on X-axis and-Y axis are in the positive correlation.

In order to assure the sub robot motion on X-axis and Y-axis are in positive correlation, the moving direction of driving units should be adjusted accordingly. Because the control system energy will not be changed if only directions of driving units are adjusted. In fact, by adjusting driving motors' direction, the motions on X-axis and Y-axis are able to be changed as ordered. By using the above control strategy, the following errors between master robot and slave robot will be converged, and slave robot is able to keep stable within a certain error range. The control strategy of slave robot is shown in Figure 5.

Given the initial position offset and angle offset, the errors on X-axis and Y-axis will approach zero but not reaches an asymptotically stability using high-gain observer control strategy. In the actual motions, the following errors between master robot and slave robot can be observed by industrial camera. And jitters may occur during slave robot motion, which also verifies the accuracy of high-gain observer control tool.

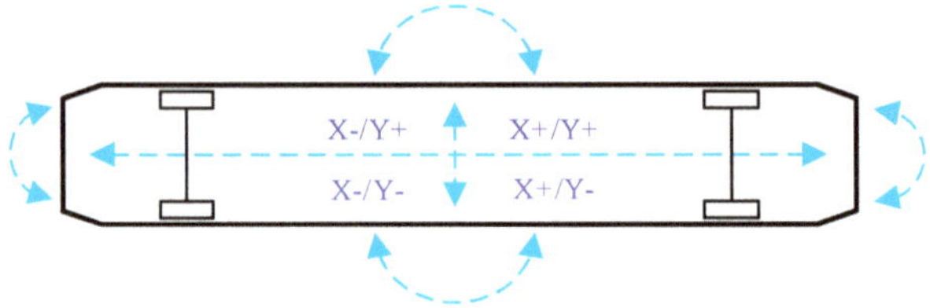


**Figure 5.** Control strategy of slave robot.

### *3.2. Input of Nonlinear Control System*

Separate parallel intelligent mobile robot uses master-slave robot structure. In order to achieve high performance of real-time control and response, the chip with over 600 MHZ main frequency should be selected as main control chip. What is more, RTOS system should be embedded for using time slices efficiently. And infrared data transmission module is used for real-time communication between master robot and slave robot. In this way, the operation cycle time of the control system can be within 5 milliseconds, which is an important factor of affecting master-slave robot real-time synchronization control.

In order to reduce the following errors and achieve high repetitive position accuracy, high-precision sensors should be used, for example distance sensors. In this way, sensors will get relative distance data, which are also the input parameters of slave robot nonlinear control system. Here are the input parameters of slave robot nonlinear control system:

(1) Distance between master robot and slave robot on X-axis: the precise distance between slave robot and master robot can be obtained by two sets of high-precision point laser distance sensors which are installed on the relative side of the front and rear of slave robot.

(2) Distance between master robot and slave robot on Y-axis: position error on Y-axis can be roughly measured by tree groups of mirror reflection photoelectric switches, which return a set of SIGN function signal to the control system. That is to say, if the difference between slave robot and master robot is greater than a preset value, the Y-axis distance sensor returns +1 signal to the control system. If the difference is within a preset range, the Y-axis distance sensor returns 0 signal to the control system. Otherwise, the Y-axis distance sensor returns -1 signal.

Obviously, the procedure of slave robot following master robot is a nonlinear control process. The input parameters of the nonlinear control system include the distance on X-axis and position error SIGN signal on Y-axis. And the output is the moving speed of each driving motor. The goal of the slave robot control system is to make the following process stable and keep the following errors between master robot on X-axis and Y-axis within a preset value.

### *3.3. High-Gain Observer*

High-gain observer is a good tool for the nonlinear control process in which some state variables cannot be measured accurately due to technical or economic reasons. In some practical cases, the tool can be modified to produce an output feedback controller [14].

Suppose that a nonlinear system can be transformed into the form:

$$\begin{cases} \dot{x} = Ax + g(y, u) \\ \hat{y} = Cx \end{cases} \tag{7}$$

where $(A, C)$ is observable. This form is special because the nonlinear function $g$ depends only on the output $y$ and the control input $u$. Taking the observer as:

$$\dot{\hat{x}} = A\hat{x} + g(y, u) + H(y - C\hat{x}) \tag{8}$$

it can be easily seen that the estimation error $\tilde{x} = x - \hat{x}$ satisfies the linear equation:

$$\dot{\tilde{x}} = (A - HC)\tilde{x} \tag{9}$$

Hence, designing $H$ such that $A - HC$ is Hurwitz guarantees asymptotic error convergence, that is $\lim_{x\to\infty} \tilde{x}(t) = 0$. Aside from the fact that the observer works only for a special class of nonlinear systems, its main drawback is the assumption that the nonlinear function $g$ is perfectly known. Any error in modeling $g$ will be reflected in the estimation error equation.

The upper nonlinear observer is suitable for most system. But for the sub-style parallel intelligent mobile robot, slave robot should follow master robot synchronously and timely. There are also many measurement deviations. We should give a special design of the observer gain that makes the observer robust to uncertainties in modeling the nonlinear functions. High-gain observer can guarantee that the output feedback controller recovers the performance of the state feedback controller when the observer gain is sufficiently high.

We use the observer in output feedback stabilization. The main results in Section 4 are separation principles that allow us to separate the design into two tasks. First, we design a state feedback controller that stabilizes the system and meets other specifications. Then, an output feedback controller is obtained by replacing the state $x$ by its estimate $\hat{x}$ provided by high-gain observer. A key property that makes this separation possible is the design of the state feedback controller to be globally bounded in $x$. High-gain observer can be used in a wide range of control problems.

Consider the second-order nonlinear system:

$$\begin{cases} \dot{x}_1 = x_2 \\ \dot{x}_2 = \phi(x, u) \\ y = x_1 \end{cases} \tag{10}$$

where $x = [x_1, x_2]^T$. Suppose $u = y(x)$ is a locally Lipschitz state feedback control law that stabilizes the origin $x = 0$ of the closed-loop system.

To implement this feedback control using only measurements of the output $y$, we use the observer:

$$\begin{cases} \dot{\hat{x}}_1 = \hat{x}_2 + h_1(y - \hat{x}_1) \\ \dot{\hat{x}}_2 = \phi_0(\hat{x}, u) + h_2(y - \hat{x}_1) \end{cases} \tag{11}$$

We set $\tilde{x}_1(0) = 0$ and $\tilde{x}_2(0) = 0$. Where $\phi_0(x, u)$ is a nominal model of the nonlinear function $\phi(x, u)$. The estimation error:

$$\tilde{x} = \begin{bmatrix} \tilde{x}_1 \\ \tilde{x}_2 \end{bmatrix} = \begin{bmatrix} x_1 - \hat{x}_1 \\ x_2 - \hat{x}_2 \end{bmatrix} \tag{12}$$

Satisfies the equation:

$$\begin{cases} \dot{\tilde{x}}_1 = -h_1\tilde{x}_1 + \tilde{x}_2 \\ \dot{\tilde{x}}_2 = -h_2\tilde{x}_1 + \delta(x, \tilde{x}) \end{cases} \tag{13}$$

where $\delta(x, \tilde{x} = \phi(x, \gamma(\hat{x})) - \phi_0(\hat{x}, \gamma(\hat{x})))$. We want to design the observer gain $H = [h_1, h_2]^T$ such that $\lim_{t\to\infty}\tilde{x}(t) = 0$. In the absence of the disturbance term $\delta$, asymptotic error convergence is achieved by designing $H$ such that:

$$A_o = \begin{bmatrix} -h_1 & 1 \\ -h_2 & 0 \end{bmatrix} \tag{14}$$

is Hurwitz. For this second-order system. $A_o$ is Hurwitz for any positive constants $h_1$ and $h_2$. In the presence of $\delta$, we need to design $H$ with the additional goal of rejecting the effect of $\delta$ on $x$. This is ideally achieved, for any $\delta$, if the transfer function:

$$G_o(s) = \frac{1}{s^2 + h_1 s + h_2} \begin{bmatrix} 1 \\ s + h_1 \end{bmatrix} \tag{15}$$

From $\delta$ to $x$ is identically zero. While this is not possible, we can make $\sup_{\omega \in R}|G_o(j\omega)$ arbitrarily small by choosing $h_2 >> h_1 >> 1$. In particular, taking

$$h_1 = \frac{\alpha_1}{\varepsilon}, h_2 = \frac{\alpha_2}{\varepsilon^2} \tag{16}$$

for some positive constant $\alpha_1$, $\alpha_2$ and $\varepsilon$, with $\varepsilon << 1$, it can be shown that:

$$G_o(s) = \frac{\varepsilon}{(\varepsilon s)^2 + \alpha_1 \varepsilon s + \alpha_2} \begin{bmatrix} \varepsilon \\ \varepsilon s + \alpha_1 \end{bmatrix} \tag{17}$$

Hence, $\lim_{\varepsilon\to 0} G_o(s) = 0$. The disturbance rejection property of high-gain observer can be also seen in the time domain by representing the error equation in the singularly perturbed form.

The initial conditions are $\varepsilon = 0.01$, $h_1 = 100$, and $h_2 = 1000$. We can obtain the best high-gain observer for synchronization control of these two sub robots of separate parallel intelligent mobile robot.

## 4. Experiment and Analysis

The separate parallel intelligent mobile robot consists of two independent sub robots, a master robot and a slave robot, which are similar to two forks of the forklift, but the sub robot does not have any physical connection with master robot. Each robot has front and rear driving units. These two sub robots collaborate to deliver the pallet synchronously. This section will analyze the steady state error in order to synchronize the actual motions of these two sub robots. The center of gravity distribution and offset error during flexible docking to the free entry of the pallet will be also analyzed.

### *4.1. Steady State Error Analysis of Actual Motions*

This section will summarize the steady state error of actual motions based on both simulation in Simulink and convergence of results.

#### 4.1.1. Linear Motion

When the master-slave mobile robot is moving linearly frontwards or backwards, slave robot should follow master robot synchronously. In order to get the value of front or rear deviation, three non-uniformly distributed are used to get relative position, which is set to be SIGN (+1, −1, 0). Slave robot should align its velocity of every driving motor by high-gain observer in real time. Otherwise, jitter of slave robot will be intolerable. Then slave robot cannot follow master robot synchronously. The procedure of adjusting relative velocity and position will be convergent and easy to reach a steady state after a period of time. The jitter of slave robot following master robot will also be reduced. The simulation in Simulink and convergence of results are shown in Figure 6. As shown in Figure 6b, there are following errors at the beginning of linear motion. In Figure 6b, the abscissa unit is second and ordinate unit is millimeter. Self-adjusting process during the following procedure will then eliminate.

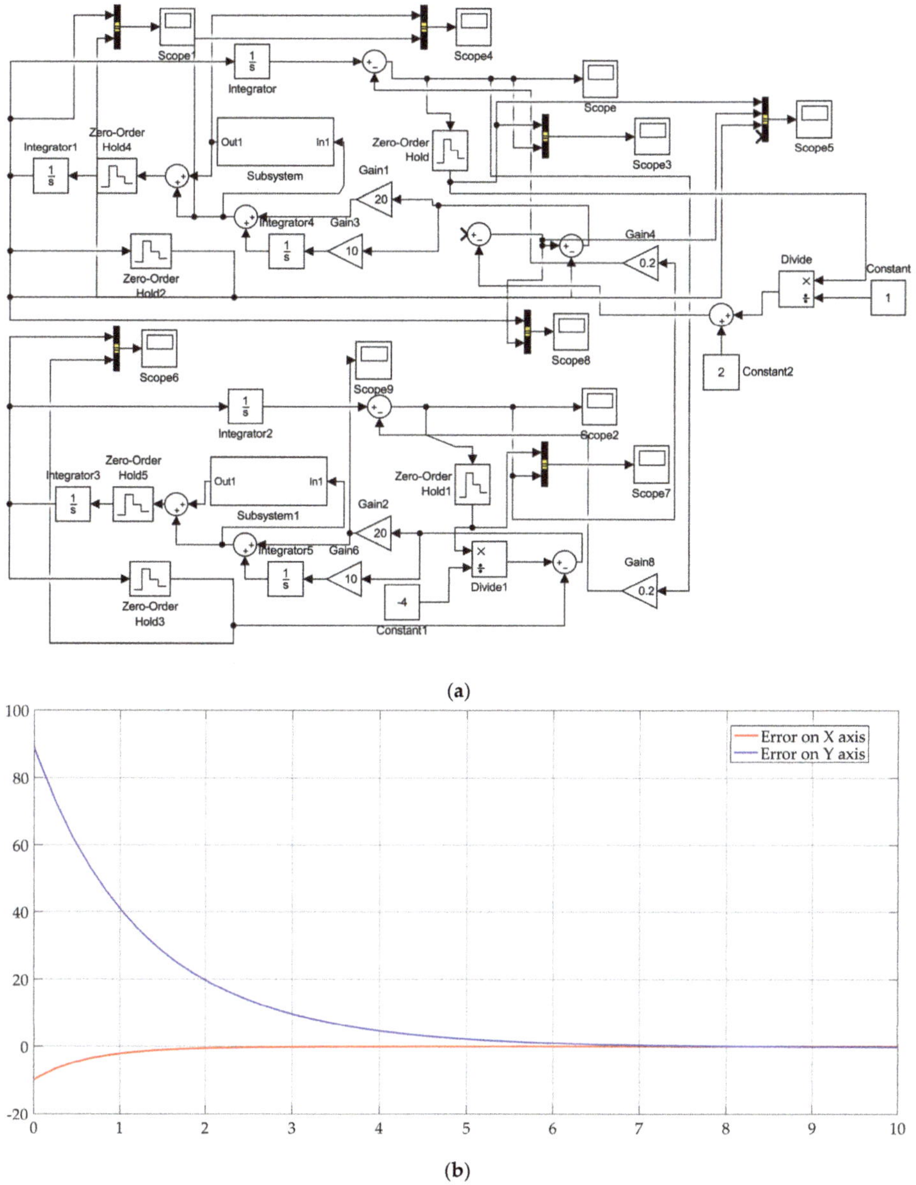


**Figure 6.** The simulation in Simulink and convergence of results for linear motion. (**a**) Simulink simulative model graph; (**b**) The model reaches a steady convergent state.

### 4.1.2. Oblique Linear Motion

*Integrator* and *integrator1* are two integrations of the model which describe the driving motor of the sub robot. Also this represents that the speed difference of driving motors would not directly cause the offset of robot motion. Simple Zero-Order-Hold is sample hold. Due to the integral element could lead to time-delay and vibration, the system would not use integral element in system offset reducing. So only *integrator4* is added to *I* in outer-loop in Simulink to reduce steady state error in the actual following motions. Theoretically, when the initial offset of following motion is 15 mm and angle is $-5°$, the following errors on X-axis and Y-axis will approximate 0 after reaching the steady state. As shown in Figure 7, the model reaches a steady convergent state. In Figure 7b, the abscissa unit is second and ordinate unit is millimeter.

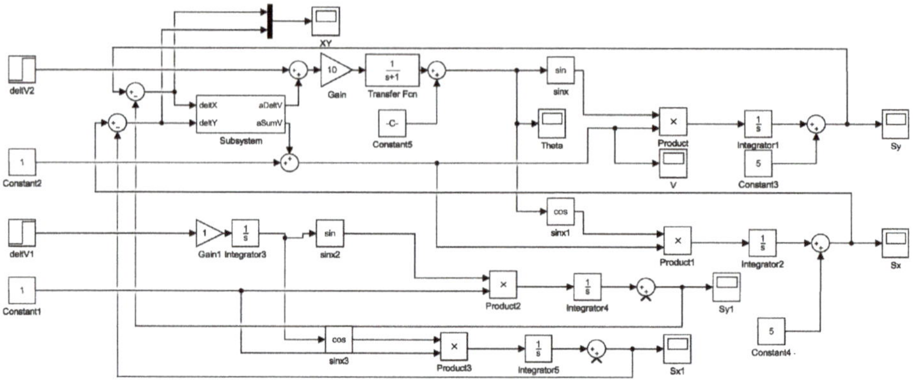


(**a**)

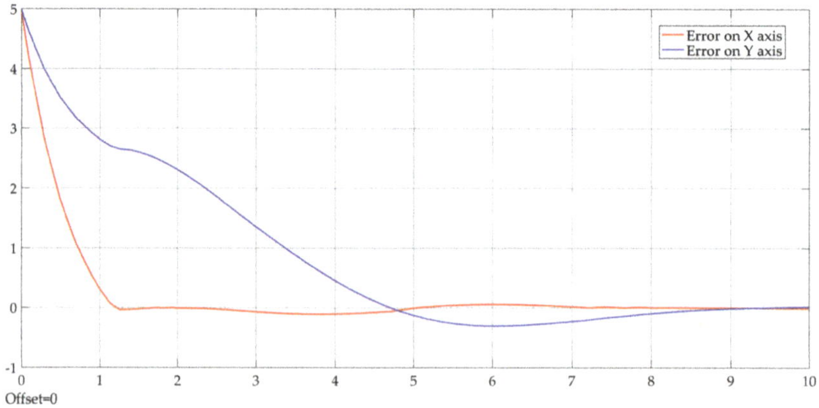


(**b**)

**Figure 7.** The simulation in Simulink and convergence of results for oblique linear motion. (**a**) Simulink simulative model graph; (**b**) The model reaches a steady convergent state.

### 4.1.3. Zero-radius Turn

When the master-slave separate parallel intelligent mobile robot takes a zero-radius turn, slave robot should follow master robot's position synchronously and timely on both X-axis and Y-axis. While rotating, the motion analysis of driving units should have a unique solution in the robot motion model. Then the velocities of slave robot's front and rear driving motors will cancel out each other on Y-axis. And the velocity on X-axis of the sub robot's driving motors is taken as the linear velocity while rotating. When the position following errors on X-axis and Y-axis occur on slave robot, the velocity and direction of driving units should be adjusted in real time, by using Lyapunov second stability theorem. It should be noted that when slave robot is adjusted, it is not guaranteed that both the position following errors on X-axis and Y-axis converge simultaneously. However the energy derivative of control system is always non-position define. The simulation in Simulink and convergence of results for zero-radius turn are shown in Figure 8. In Figure 8b, the abscissa unit is second and ordinate unit is millimeter.

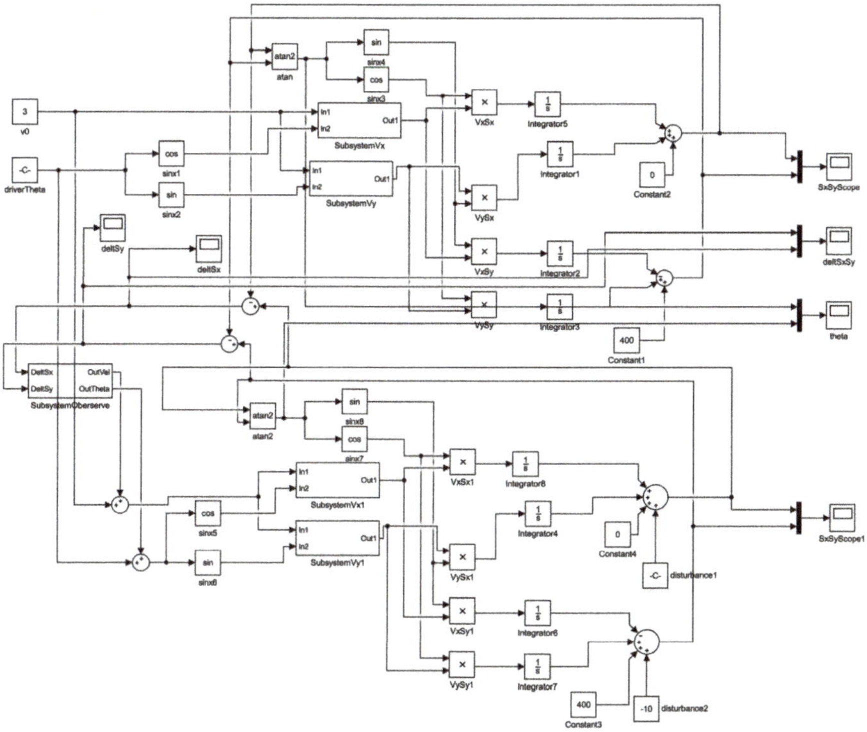


(**a**)

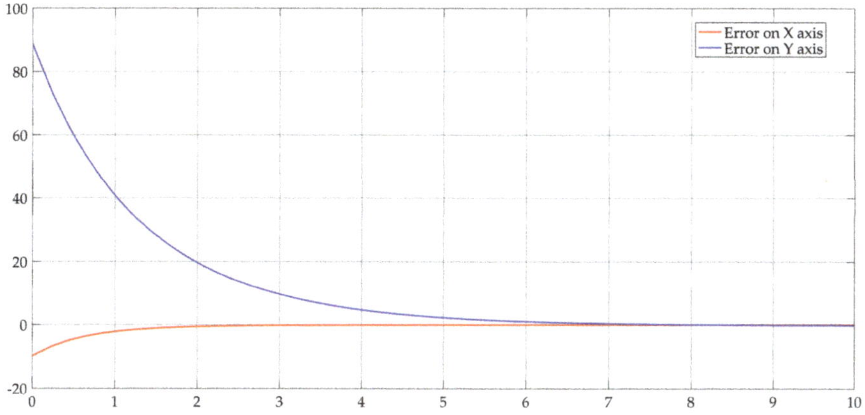


(**b**)

**Figure 8.** The simulation in Simulink and convergence of results for zero-radius turn. (**a**) Simulink simulative model graph; (**b**) The model reaches a steady convergent state.

#### 4.1.4. Steady State Error Factors

As we know, in the ideal case, the zero offset error means that slave robot follows master robot absolutely according to calibrated data, and there is no error. So the most significant factor that affects the stability of the system is the accuracy of calibrated data. (1) Camera installation error: The installation angle deviation of camera and driving unit is unavoidable. The camera should be calibrated first to reduce the initial angle deviation. (2) Encoder calibration offset: The error of encoder deviation would be transferred from inner-loop to the outer-loop. Thus the encoder calibration offset would directly lead to the error of system stability. (3) Accuracy of driving control: In actual motions, the precision of each driving motor is up to 1 rpm. As the result, the precision of the following velocity and position control could not be well guaranteed. (4) Mechanical installation error: The mechanical installation error represents installation tolerance error such as driving unit and other mechanical parts. All these errors would result in the difference between calibrated zero position and actual zero position.

As we known, the errors of actual motions of these two sub robots cannot reach zero. This section analyzes most of the factors that may affect the steady state errors of the following motions. The two driving units of slave robot may jitter during the adjusting control. Surface roughness of the road and the curve degree of vision-based color bar on the road can both have significant effect on the stability of the master-slave parallel robot' motion. As discussed above, various factors affect the steady state error. The steady state error for the master-slave mobile robot measured in the experiment is about 3mm, which is not easy to eliminate.

### *4.2. Accurate Flexible Docking to the Free Entry of the Pallet*

To ensure that the two sub robots can move underneath the pallet accurately and lift up and lay down the pallet synchronously, an evaluation of docking to the free entry is required. After arriving at the pre-docking position, the mobile robot should recognize and compute the accurate position the free entry of pallet by laser sensor or vision sensor, and then move into the free entry smoothly. In this section, the maximum deviation angle error $\beta_{\max}$ during docking to the free entry is analyzed through synthetical consideration of the following error and positioning error.

#### 4.2.1. The Ideal Case (No Direct Collision)

In order to lift the pallet up, the sub robot should move underneath the pallets accurately and safely. In the ideal case, the shell cover of the sub robot won't collide with the edge of the free entry of the pallets, as shown in Figure 9a. $\beta_1 = (90^\circ - \theta_1)$ represents the intersection angle of the sub robot and the free entry of the pallet.

$$\begin{cases} AC = l_r - \frac{w_r - w_h}{2} \tan \delta \\ BC = \frac{w_p - w_r}{2} \pm \varepsilon \\ \cos \theta_1 = \frac{BC}{AC} \end{cases} \tag{18}$$

where, $l_r = 1180$ mm is the length of the basis of the sub robot. $w_r = 212$ mm is the width of the basis of the sub robot. $w_h = 170$ mm is the width of the head of the sub robot. $\delta = 70^\circ$ is the conical chamfer of the head of the sub robot. $w_p = 382.5$ mm represents the width of free entry of the pallet. $\varepsilon = 10$ mm is the maximum following error of the sub robot.

Relative deviation angle $\beta_1$:

$$3.84^\circ < \beta_1 = (90^\circ - \theta_1) < 4.87^\circ \tag{19}$$

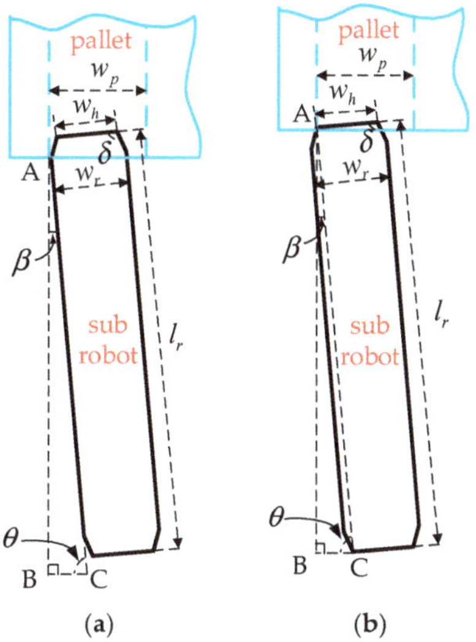


**Figure 9.** The maximum deviation angle error of a sub robot docking to the free entry of the pallet. (**a**) The ideal case; (**b**) The extreme case.

#### 4.2.2. The Extreme Case (Contact Guidance)

Although the shell cover of the sub robot will collide with the free entry slightly in some conditions, the sub robot can move underneath the pallets yieldingly, as shown in Figure 9b. $\beta_2 = (90^\circ - \theta_2)$ represents the included angle of the sub robot and the free entry of the pallet.

$$\begin{cases} AC = l_r \\ BC \approx \frac{l_r - w_r}{2} + \frac{w_r - w_h}{2} \pm \varepsilon \\ \cos\theta_2 = \frac{BC}{AC} \end{cases} \quad (20)$$

Relative deviation angle $\beta_2$:

$$4.70^\circ < \beta_2 = (90^\circ - \theta_2) < 5.65^\circ \quad (21)$$

By combing Equation (19) with (21), we have the the maximum allowable angle error $\beta_{\max}$:

$$-\max(\beta_1, \beta_2) < \beta_{\max} < \max(\beta_1, \beta_2) \quad (22)$$

Therefore, $\beta_{\max} \in \left( -5.65^\circ \quad 5.65^\circ \right)$.

### 4.3. Analysis of Lifting and Laying

The main procedures of pallet transportation for intelligent mobile robot are that it firstly moves underneath the pallet, then lifts the pallet up and delivers the pallet to the destination, and finally lays it down. In order to control motions of the two sub robots synchronously, this section analyzes the pallet offset error, including longitudinal and transverse offset error during lifting and laying.

#### 4.3.1. Analysis of Center of Gravity

As we know, the sub robot may roll over while the four driving motors are at two parallel lines at a certain moment. Relative velocity and position control loops are used during the lifting and laying of the pallet. The front and rear driving wheels are always vertical through position feedback. The initial relative angle of front and rear driving units is shown in Figure 10. Thus the center of gravity of robot will be within the base of the sub robot during the procedure of lifting and laying.

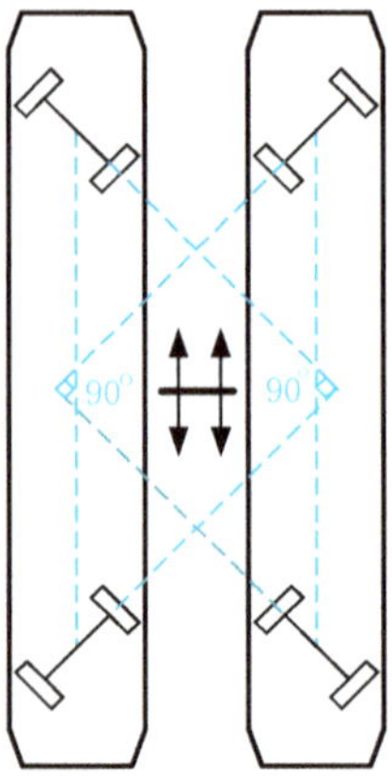


**Figure 10.** The initial relative angle of front and rear driving units before lifting and laying.

4.3.2. Analysis of Pallet Offset Error

During the lifting process, the offset errors of the position of the pallet include longitudinal offset error and transverse offset error. They are both caused by uneven ground and jittering of sub-robot's motion. It is difficult to guarantee the level of the ground, so changing or optimizing the robot motion is the most effective way.

- Analysis of pallet longitudinal offset errors

The longitudinal offset errors of the pallet cannot be completely eliminated. It is unable to measure the absolute accurate deviation between pallet position and robot position, so no correction can be made. There must be cumulative error between them. What we should do is to ensure the absolute stop position between lifting and laying as consistent as possible. The two sub robots should be controlled to stop accurately and synchronously while height position of the pallet changes. The control strategy of accurate stop is shown in Figure 11.

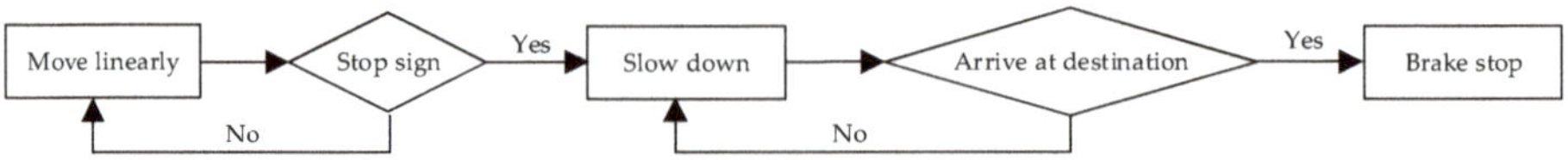


**Figure 11.** Control strategy of accurate stop.

- Analysis of pallet transverse offset errors

The transverse offset of the pallet refer to errors in the left and right direction. In general, the mobile robot moves underneath of the pallet and then lift the pallet up. But once there are some transverse offset errors, the mobile robot may be stuck under the free entry of the pallet. It may be caused by uneven ground or jitters of the sub robot's motions. In order to reduce the transverse offset errors, dynamic position adjustment is required during lifting and laying. The dynamic adjustment strategy is shown in Figure 12. The rotation angle of each driving unit is divided into 2 areas. (1) Angle adjustment areas. If the angle of rear driving unit is in the range of angle that needs to be adjusted, the rear driving unit will follow the front one, and they will be perpendicular. Within the angle adjustment areas, the center of gravity of the sub robot can always maintain a relatively large area, which will prevent the sub robot to roll over. (2) Position adjustment areas. If the angle of driving unit is in the range of position that needs to be adjusted, the speed of the driving motor will be changed according to the position error collected by the camera or navigation sensor. It ensures that the sub robot performs left and right position feedback during the procedure of lifting and laying. According to the position adjustment areas, it is possible to optimize the transverse offset errors between the sub robot and the pallet.

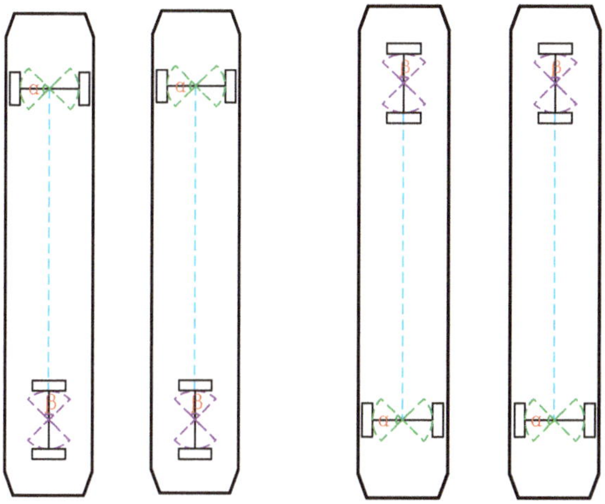

(a) (b)

**Figure 12.** Dynamic adjustment strategy of master-slave robot during lifting and laying. (**a**) Angle adjustment areas; (**b**) Position adjustment areas.

When the front or rear driving unit lies in the angle adjustment area, we adjust the speeds of the two motors of this driving unit synchronously. The two motors may accelerate or decelerate at the same speeds. But when the front or rear driving unit lies in the position adjustment area, we adjust the speeds of the two motors of this driving unit asynchronously. The two motors may accelerate or decelerate at the different speeds.

### *4.4. Actual Running Performance*

The master-slave separate parallel intelligent mobile robot was designed and assembled. The performance was tested by actual delivering the pallets in the factory. The two sub robots could also lift up and lay down the pallet synchronously. The mobile robot moved flexibly at the crossroads. And detailed specifications of the master-slave separate parallel intelligent mobile robot are listed in Table 1.

**Table 1.** The specifications of the master-slave separate parallel intelligent mobile robot.

| Feature | Specification | Feature | Specification |
| --- | --- | --- | --- |
| max payload | 1000 kg | swing radius | >730 mm |
| max speed | 0.7 m/s | navigation | color bar/vision |
| outline dimension | L1180 × W212 × H98 mm | running time | 10 h |
| weight | 40 kg | charging time | 2 h |
| uplift height | 28 mm | turning radius | 0 mm |
| uplift time | 16 s | pallets | PALETTE EUR-EPAL |

## 5. Conclusions

To pick up and transport pallets autonomously in the factory, a master-slave separate parallel intelligent mobile robot was designed. The mobile robot consists of two independent sub robots, which are similar to two forks of the forklift, but master robot does not have any physical connection with slave robot. In order to adapt to the compact space of the entry of pallet, four compact motion driving units are designed for these two sub robots. High-gain observer is used to control the following speed timely. The state control of slave robot is to ensure that the following error converges and achieve synchronization of the two sub robots' motions. The experiment results demonstrate that the separate parallel intelligent mobile robot can transport the pallet autonomously. The two sub robots can fulfill synchronous motions, such as linear motion, oblique linear motion, and zero-radius turn. They can also lift up and lay down the pallet synchronously. The mobile robot moves flexibly and is quite suitable for the standardized logistics factory with small working space. In our future research, the autonomous navigation module based on the 2D laser will be assembled with the intelligent mobile robot. Then the mobile robot can move freely for pallet transportation in the factory.

**Author Contributions:** Data curation, G.L. and R.L.; Formal analysis, R.S.; Investigation, M.L. and R.S.; Methodology, R.L., M.L. and S.P.; Project administration, S.P.; Software, G.L.; Writing—original draft, R.L.; Writing—review & editing, R.L. and S.P.

**Funding:** This research was funded by the Projects of the National Natural Science Foundation (61673288).

**Acknowledgments:** Thanks associate Professor Hongmiao Zhang for giving suggestions to improve this manuscript. We would like to take the opportunity to thank the reviewers for their thoughtful and meaningful comments.

**Conflicts of Interest:** The authors declare no conflict of interest.

## References

1. Behzad, E.; Sara, B.; Ben, W. The evolution and future of manufacturing: A review. *J. Manuf. Syst.* **2016**, *39*, 79–100.
2. Jeremie, C. *A Look into Logistics Automation. ROBO Global Defining the Universe of Robotics and Automation*; ETF Securities: London, UK, 2017; pp. 1–18.
3. Paweł, Z. Model of Forklift Truck Work Efficiency in Logistic Warehouse System. In *Logistics Operations, Supply Chain Management and Sustainability*; Springer: Cham, Switzerland, 2014; pp. 467–479.
4. Armesto, L.; Tornero, J.; Torres, J. Transport process automation with industrial forklifts. *IFAC Proc.* **2003**, *36*, 33453–33455. [CrossRef]
5. Kelen, C.T.; Marcelo, B.; Glauco, A.P. Automatic routing of forklift robots in warehouse applications. *ABCM Symp. S. Mechatron.* **2010**, *4*, 335–344.
6. Wang, J.; Zhang, N.; He, Q. Application of Automated Warehouse Logistics in Manufacturing Industry. In Proceedings of the IEEE 2009 ISECS International Colloquium on Computing, Communication, Control, and Management, Sanya, China, 8–9 August 2009.
7. Li, L.Y.; Liu, Y.H.; Fang, M.; Zheng, Z.-Z.; Tang, H.-B. Vision-Based Intelligent Forklift Automatic Guided Vehicle (AGV). In Proceedings of the 2015 IEEE International Conference on Automation Science and Engineering (CASE), Fort Worth, TX, USA, 21–24 August 2015; pp. 264–265.
8. Seelinger, M.; Yoder, J.D. Automatic Pallet Engagment by a Vision Guided Forklift. In Proceedings of the 2005 IEEE International Conference on Robotics and Automation, Barcelona, Spain, 18–22 April 2005.
9. Qian, J.; Zi, B.; Wang, D.; Ma, Y.; Zhang, D. The Design and Development of an Omni-Directional Mobile Robot Oriented to an Intelligent Manufacturing System. *Sensors* **2017**, *17*, 2073. [CrossRef] [PubMed]
10. Patel, P.; Parekh, R.; Panchal, R.; Solanki, V. Material Handling System Based on Automated Guided Vehicle: Scope, Limitation and Application. In Proceedings of the International Conference on Emerging Trends in Mechanical Engineering, Gujarat, India, 6–7 July 2017.
11. Weber, M.; Vorwerk, C. Neuartiges antriebskonzept zum fahren, lenken und heben. *Log. J. Proc.* **2010**, *6*. [CrossRef]
12. Melkus, A. Fahreloses transportfarzeug: EP, No. EP3216747A1. 7 March 2016.
13. F3-Design. Available online: https://www.dinostretchhood.com/nipper/ (accessed on 21 November 2018).
14. AGILOX SYSTEM. Available online: http://agilox.net/ (accessed on 19 July 2018).
15. Hassan, K.K. *Nonlinear Systems*, 3rd ed.; Prentice Hall: Englewood, IL, USA, 2002; pp. 551–625.

16. Kim, S.; Kwon, S. Nonlinear Control Design for a Two-Wheeled Balancing Robot. In Proceedings of the 2013 10th International Conference on Ubiquitous Robots and Ambient Intelligence, Jeju, Korea, 30 October–2 November 2013; pp. 486–487.
17. Osama, J.; Mohsin, J.; Yasar, A.; Khubab, A. Modeling, Control of a two-Wheeled Self-Balancing Robot. In Proceedings of the IEEE International Conference on Robotics and Emerging Allied Technologies in Engineering (iCREATE), Islamabad, Pakistan, 22–24 April 2014.
18. Antonio, H.J.; Yu, W. High-Gain Observer-Based PD COntrol for Robot Manipulator. In Proceedings of the IEEE Xplore Conference on American Control Conference, Chicago, IL, USA, 28–30 June 2000; Volume 4, pp. 2518–2522.
19. Voos, H. Nonlinear Control of a Quadrotor micro-UAV Using Feedback-Linearization. In Proceedings of the IEEE International Conference on Mechatronics, Malaga, Spain, 14–17 April 2009; pp. 1–6.
20. Goodarzi, F.; Lee, D.; Lee, T. Geometric Nonlinear PID Control of a Quadrotor UAV on SE(3). In Proceedings of the 2013 European Control Conference (ECC), Zürich, Switzerland, 17–19 July 2013; pp. 3845–3850.
21. Hu, Q.; Fei, Q.; Wu, Q. Research and application of nonlinear control techniques for quad rotor UAV. *J. Univ. Sci. Technol. China* **2012**, *42*, 706–710.
22. Young, K.D.; Utkin, V.I.; Ozguner, U. A control engineer's guide to sliding mode control. *IEEE Trans. Control Syst. Technol.* **1999**, *7*, 328–342. [CrossRef]
23. Boizot, N.; Busvelle, E.; Gauthier, J.P. An adaptive high-gain observer for nonlinear systems. *Automatica* **2010**, *46*, 1483–1488. [CrossRef]

*applied sciences*

MDPI

*Article*

# Formation Control and Distributed Goal Assignment for Multi-Agent Non-Holonomic Systems

**Wojciech Kowalczyk**

Institute of Automation and Robotics, Poznań University of Technology (PUT), Piotrowo 3A, 60-965 Poznań, Poland; wojciech.kowalczyk@put.poznan.pl; Tel.: +48-61-665-2043

Received: 5 March 2019; Accepted: 25 March 2019; Published: 29 March 2019

**Abstract:** This paper presents control algorithms for multiple non-holonomic mobile robots moving in formation. Trajectory tracking based on linear feedback control is combined with inter-agent collision avoidance. Artificial potential functions (APF) are used to generate a repulsive component of the control. Stability analysis is based on a Lyapunov-like function. Then the presented method is extended to include a goal exchange algorithm that makes the convergence of the formation much more rapid and, in addition, reduces the number of collision avoidance interactions. The extended method is theoretically justified using a Lyapunov-like function. The controller is discontinuous but the set of discontinuity points is of zero measure. The novelty of the proposed method lies in integration of the closed-loop control for non-holonomic mobile robots with the distributed goal assignment, which is usually regarded in the literature as part of trajectory planning problem. A Lyapunov-like function joins both trajectory tracking and goal assignment analyses. It is shown that distributed goal exchange supports stability of the closed-loop control system. Moreover, robots are equipped with a reactive collision avoidance mechanism, which often does not exist in the known algorithms. The effectiveness of the presented method is illustrated by numerical simulations carried out on the large formation of robots.

**Keywords:** formation of robots; non-holonomic robot; stability analysis; Lyapunov-like function; target assignment; goal exchange; path following; switching control

## 1. Introduction

The idea to use artificial potential fields to control manipulators and mobile robots was introduced by Khatib [1] in 1986. In this approach both attraction to the goal and repulsion from the obstacles are negated gradients of the artificial potential functions (APF). His paper not only presents the theory, but also a solution of the practical problem, implemented in the Puma 560 robot simulator. It is worth noting that much earlier, in 1977, Laitmann and Skowronski [2] investigated control of two agents avoiding collision with each other. This work was purely theoretical. The authors continued their work in the following years [3].

Since the 1990s, intensive research on the trajectory tracking control for non-holonomic mobile robots has been conducted [4–6]. The algorithm presented further in this paper is based on the method from [4]. This method considers a single, differentially-driven mobile robot moving in a free space. Its goal is to track a desired trajectory. The paper includes a stability analysis.

The last decade has seen a lot of publications on multiple mobile robot control. In [7], the goal of multiple mobile robots is to track desired trajectories avoiding inter-agent collisions. The same type of task is considered further in this paper. The tracking controller is different, as is the formula of APF, but the method of combining trajectory tracking and collision avoidance is the same. In [8], the same type of task is considered, but the dynamics of mobile platforms and uncertainties of its parameters are taken into account. Kowalczyk et al. [9] propose a vector-field-orientation algorithm for multiple mobile robots moving in the environment with circle shaped static obstacles. The dynamic properties of mobile platforms are also taken into account. The paper [10] presents a kinematic controller for the formation of robots that move in a queue. The goal is to keep desired displacements between robots and avoid collisions in the transient states. Hatanaka et al. [11] investigate a cooperative estimation problem for visual sensor networks based on multi-agent optimization techniques. The paper [12] addresses the formation control problem for fleets of autonomous underwater vehicles. Yoshioka et al. [13] deal with formation control strategies based on virtual structure for multiple non-holonomic systems.

Recent years have also seen a number of publications on barrier functions. In [14], coordination control for a group of mobile robots is combined with collision avoidance provided by a safety barrier. If the coordination control command leads to collision, the safety barrier dominates the controller and computes a safe control closest to coordination control law. In the method proposed in [15] control barrier functions are unified with performance objectives expressed as control Lyapunov functions. The authors of the paper [16] provide a theoretical framework to synthesize controllers using finite time convergence control barrier functions guided by linear temporal logic specifications for continuous time multi-agent systems.

The paper [17] addresses the problem of optimal goal assignment for the formation of holonomic robots moving on a plane. A linear bottleneck assignment problem solution is used to minimize the maximum completion time or maximum distance for any robot in the formation. The authors of [18] consider formation of non-holonomic mobile vehicles that has to change the geometrical shape of the formation. Goal assignment minimizes the total distance travelled by agents. The exemplary application indicated by the authors is reconfiguration of the formation when it approaches a narrow passage. In [19,20], goal assignment based on distances squared is proposed and tested for large formations, but the collision avoidance is resolved at the trajectory planning level. This makes this approach less robust in the case of unpredictable disturbances (which are natural in real applications). The second of these papers proposes a solution to the problem of collision avoidance with static and dynamic obstacles present in the environment. In [21], collision avoidance is obtained by applying safety constraints in optimal trajectory generation. The robots do not have non-holonomic constraints (they are quadrotors) and they have to change the shape of the formation using goal assignment that minimizes the total distance travelled by the robots. Turpin et al. [22] present concurrent assignment and planning of trajectories (CAPT) algorithms. The authors propose two variations of the algorithm: centralized and decentralized, and test them on a group of holonomic mobile robots moving in a three-dimensional space. The solution is based on the Hungarian assignment algorithm. The above-mentioned works do not deal with the problem of closed-loop control. For this reason stability issues were not considered there.

In comparison to the above two approaches, here the user or the higher level controller determines the locations of desired trajectories according to the needs (this can be considered as an expected feature in many applications). In addition, trajectory generation is decoupled from the closed-loop control (the system is modular). Such a solution is considered as a design pattern in robotics. The algorithm is responsible for tracking these trajectories and reacting to the risk of collisions between agents at the same time. Even if initial states of individual robots are far from the desired ones, the collision avoidance module works correctly. Furthermore, the algorithm characterizes conceptual and computational simplicity. The paper considers preserving data integrity during the goal exchange as it requires a simultaneous change of states in remote systems. This subject is omitted in the literature on goal assignment in multi-robot systems. To the best of the author's knowledge, no work has been published so far that proposes closed-loop control for multiple non-holonomic mobile robots combined with target assignment with analysis based on a Lyapunov-like function for both the tracking algorithm and target assignment. Panagou et al. [23] propose a similar method, but it assumes that agents are fully-actuated (modeled using integrators), and the analysis is based on multiple Lyapunov-like functions. This algorithm also uses a different criterion for goal exchange.

The algorithm proposed in this paper is applicable mainly to the homogeneous formations of non-holonomic mobile robots but also in the scenarios when two or more robots of the same type are involved in task execution. For a high number of applications of multiple mobile robots this situation occurs, e.g., exploration, mapping, safety, and surveillance.

In Section 2 a control algorithm for the formation of non-holonomic mobile robots is described. Section 3 analysis stability. The simulation results are presented in Section 4. The goal exchange algorithm is introduced in Section 5. Section 6 provides stability analysis of extended algorithm. Section 7 details distributed implementation. Some generalization of the proposed goal exchange rule is given in Section 8. Section 9 discusses the problem of maintaining data integrity in the process of the goal exchange. Section 10 offers simulation results for the goal exchange algorithm. Section 11 presents simulation results for limited wheel velocity controls. In the last Section concluding remarks are provided.

## 2. Control Algorithm

The kinematic model of the $i$-th differentially-driven mobile robot $R_i$ ($i = 1 \dots N$, $N$—number of robots) is given by the following equation:

$$\dot{q}_i = \begin{bmatrix} \cos\theta_i & 0 \\ \sin\theta_i & 0 \\ 0 & 1 \end{bmatrix} u_i \tag{1}$$

where vector $q_i = [x_i \;\; y_i \;\; \theta_i]^\top$ denotes the pose and $x_i, y_i, \theta_i$ are the position coordinates and orientation of the robot with respect to a global, fixed coordinate frame. Vector $u_i = \begin{bmatrix} v_i & \omega_i \end{bmatrix}^\top$ is the control vector with $v_i$ denoting the linear velocity and $\omega_i$ denoting the angular velocity of the platform.

The task of the formation is to follow the virtual leader that moves with desired linear and angular velocities $[v_l \;\; \omega_l]^T$. The robots are expected to imitate the motion of the virtual leader. They should have the same velocities as the virtual leader. The position coordinates $[x_l \;\; y_l]^T$ of the virtual leader are used as a reference position for the individual robots but each of them has different displacement with respect to the leader:

$$x_{id} = x_l + d_{ix} \qquad y_{id} = y_l + d_{iy}, \tag{2}$$

where $[d_{ix} \;\; d_{iy}]^T$ is desired displacement of the $i$-th robot. As the robots position converge to the desired values their orientations $\theta_i$ converge to the orientation of the virtual leader $\theta_l$.

The collision avoidance behaviour is based on the APF. This concept was originally proposed in [1]. All robots are surrounded by APFs that raise to infinity near objects border $r_j$ ($j$—number of the robots/obstacles) and decreases to zero at some distance $R_j$, $R_j > r_j$.

One can introduce the following function [6]:

$$B_{aij}(l_{ij}) = \begin{cases} 0 & \text{for} \quad l_{ij} < r_j \\ e^{\frac{l_{ij}-r_j}{l_{ij}-R_j}} & \text{for} \quad r_j \le l_{ij} < R_j \\ 0 & \text{for} \quad l_{ij} \ge R_j \end{cases}, \tag{3}$$

that gives output $B_{aij}(l_{ij}) \in \langle 0, 1)$. The distance between the $i$-th robot and the $j$-th robot is defined as the Euclidean length $l_{ij} = \|[x_j \; y_j]^\top - [x_i \; y_i]^\top\|$.

Scaling the function given by Equation (3) within the range $\langle 0, \infty)$ can be given as follows:

$$V_{aij}(l_{ij}) = \frac{B_{aij}(l_{ij})}{1 - B_{aij}(l_{ij})}, \tag{4}$$

that is used later to avoid collisions.

In further description, the terms 'collision area' or 'collision region' are used for locations fulfilling the condition $l_{ij} < r_j$. The range $r_j < l_{ij} < R_j$ is called 'collision avoidance area' or 'collision avoidance region' (Figure 1).

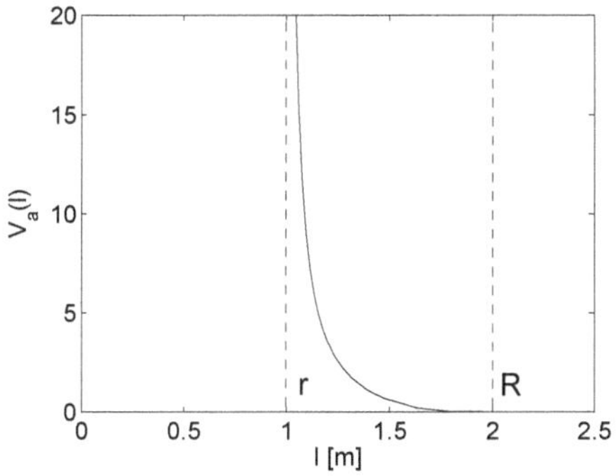


**Figure 1.** Artificial potential functions (APF) as a function of distance to the centre of the robot (indexes omitted for simplicity).

The goal of the control is to drive the formation along the desired trajectory avoiding collisions between agents. It is equivalent to bringing the following quantities to zero:

$$\begin{aligned} p_{ix} &= x_{id} - x_i \\ p_{iy} &= y_{id} - y_i \\ p_{i\theta} &= \theta_l - \theta_i. \end{aligned} \tag{5}$$

**Assumption 1.** $\forall\{i,j\}, i \neq j, ||[x_{id}\ y_{id}]^T - [x_{jd}\ y_{jd}]^T|| > R_j$.

**Assumption 2.** *If robot i gets into the collision avoidance region of any other robot j, $j \neq i$ its desired trajectory is temporarily frozen ($\dot{x}_{id} = 0$, $\dot{y}_{id} = 0$). If the robot leaves the avoidance area its desired coordinates are immediately updated. As long as the robot remains in the avoidance region, its desired coordinates are periodically updated at certain discrete instants of time. The time period $t_u$ of this update process is relatively large in comparison to the main control loop sample time.*

Assumption 1 comes down to the statement that desired paths of individual robots are planned in such a way that in steady state all robots are out of the collision avoidance regions of other robots.

Assumption 2 means that the tracking process is temporarily suspended because collision avoidance has a higher priority. Once the robot is outside the collision detection region, it updates the reference to the new values. In addition, when the robot is in the collision avoidance region its reference is periodically updated. This low-frequency process supports leaving the unstable equilibrium points that occur e.g., when one robot is located exactly between the other robots and its goal.

The system error expressed with respect to the coordinate frame fixed to the robot is described below:

$$\begin{bmatrix} e_{ix} \\ e_{iy} \\ e_{i\theta} \end{bmatrix} = \begin{bmatrix} \cos(\theta_i) & \sin(\theta_i) & 0 \\ -\sin(\theta_i) & \cos(\theta_i) & 0 \\ 0 & 0 & 1 \end{bmatrix} \begin{bmatrix} p_{ix} \\ p_{iy} \\ p_{i\theta} \end{bmatrix}. \tag{6}$$

Using the above equations and non-holonomic constraint $\dot{y}_i \cos(\theta_i) - \dot{x}_i \sin(\theta_i) = 0$ the error dynamics between the leader and the follower are as follows:

$$\begin{aligned}\dot{e}_{ix} &= e_{iy}\omega_i - v_i + v_l \cos e_{i\theta}\\ \dot{e}_{iy} &= -e_{ix}\omega_i + v_l \sin e_{i\theta}\\ \dot{e}_{i\theta} &= \omega_l - \omega_i.\end{aligned} \tag{7}$$

One can introduce the position correction variables that consist of position error and collision avoidance terms:

$$\begin{aligned}P_{ix} &= p_{ix} - \sum_{j=1, j\neq i}^{N} \frac{\partial V_{aij}}{\partial x_i}\\ P_{iy} &= p_{iy} - \sum_{j=1, j\neq i}^{N} \frac{\partial V_{aij}}{\partial y_i}.\end{aligned} \tag{8}$$

$V_{aij}$ depends on $x_i$ and $y_i$ according to Equation (5). It is assumed that the robots avoid collisions with each other and there are no other obstacles in the taskspace. The correction variables can be transformed to the local coordinate frame fixed in the mass centre of the robot:

$$\begin{bmatrix} E_{ix}\\ E_{iy}\\ e_{i\theta}\end{bmatrix} = \begin{bmatrix} \cos(\theta_i) & \sin(\theta_i) & 0\\ -\sin(\theta_i) & \cos(\theta_i) & 0\\ 0 & 0 & 1\end{bmatrix} \begin{bmatrix} P_{ix}\\ P_{iy}\\ p_{i\theta}\end{bmatrix}. \tag{9}$$

Differentiating the first two equations of (5) with respect to the $p_{ix}$ and $p_{iy}$ respectively one obtains:

$$\frac{\partial x_i}{\partial p_{ix}} = -1 \qquad \frac{\partial y_i}{\partial p_{iy}} = -1. \tag{10}$$

Using (10) one can write:

$$\begin{aligned}\frac{\partial V_{aij}}{\partial p_{ix}} &= \frac{\partial V_{aij}}{\partial x_i}\frac{\partial x_i}{\partial p_{ix}} = -\frac{\partial V_{aij}}{\partial x_i}\\ \frac{\partial V_{aij}}{\partial p_{iy}} &= \frac{\partial V_{aij}}{\partial y_i}\frac{\partial y_i}{\partial p_{iy}} = -\frac{\partial V_{aij}}{\partial y_i}.\end{aligned} \tag{11}$$

Taking into account Equations (8) and (9) the gradient of the APF can be expressed with respect to the local coordinate frame fixed to the $i$-th robot:

$$\begin{bmatrix} \frac{\partial V_{aij}}{\partial e_{ix}}\\ \frac{\partial V_{aij}}{\partial e_{iy}}\end{bmatrix} = \begin{bmatrix} \cos\theta_i & \sin\theta_i\\ -\sin\theta_i & \cos\theta_i\end{bmatrix} \begin{bmatrix} \frac{\partial V_{aij}}{\partial p_{ix}}\\ \frac{\partial V_{aij}}{\partial p_{iy}}\end{bmatrix}. \tag{12}$$

Equation (12) can be verified easily by taking partial derivatives of $V_{aij}(d_{ix} - p_{ix}, d_{iy} - p_{iy}) = V_{aij}(d_{ix} - p_{ix}(e_{ix}, e_{iy}), d_{iy} - p_{iy}(e_{ix}, e_{iy}))$ with respect to $e_{ix}$, $e_{iy}$ and taking into account the inverse transformation of the first two equations of Equation (6).

Using Equation (11), the above equation can be written as follows:

$$\begin{bmatrix} \frac{\partial V_{aij}}{\partial e_{ix}}\\ \frac{\partial V_{aij}}{\partial e_{iy}}\end{bmatrix} = \begin{bmatrix} -\cos\theta_i & -\sin\theta_i\\ \sin\theta_i & -\cos\theta_i\end{bmatrix} \begin{bmatrix} \frac{\partial V_{aij}}{\partial x_i}\\ \frac{\partial V_{aij}}{\partial y_i}\end{bmatrix}. \tag{13}$$

Equations (9) and (12) can be transformed to the following form:

$$\begin{aligned} E_{ix} &= p_{ix}\cos(\theta_i) + p_{iy}\sin(\theta_i) + \sum_{j=1,j\neq i}^{N} \frac{\partial V_{aij}}{\partial e_{ix}} \\ E_{iy} &= -p_{ix}\sin(\theta_i) + p_{iy}\cos(\theta_i) + \sum_{j=1,j\neq i}^{N} \frac{\partial V_{aij}}{\partial e_{iy}} \\ e_{i\theta} &= p_{i\theta}, \end{aligned} \quad (14)$$

where each derivative of the APF is transformed from the global coordinate frame to the local coordinate frame fixed to the robot. Finally, the correction variables expressed with respect to the local coordinate frame (Figure 2) are as follows:

$$\begin{aligned} E_{ix} &= e_{ix} + \sum_{j=1,j\neq i}^{N} \frac{\partial V_{aij}}{\partial e_{ix}} \\ E_{iy} &= e_{iy} + \sum_{j=1,j\neq i}^{N} \frac{\partial V_{aij}}{\partial e_{iy}}. \end{aligned} \quad (15)$$

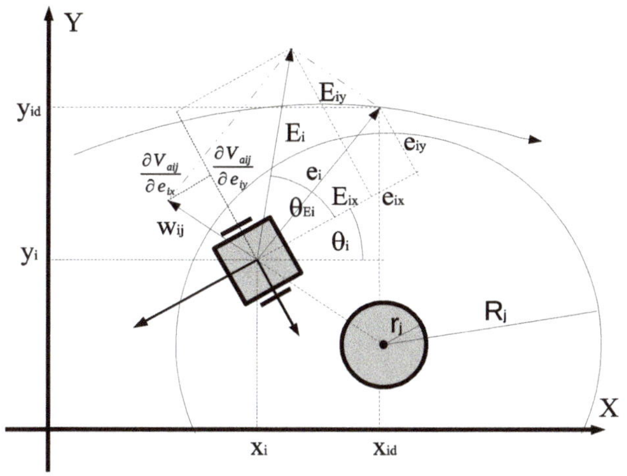


**Figure 2.** Robot in the environment with an obstacle.

Note the similarity of the structure of Equations (8) (updated by Equations (11)) and (15).

The trajectory tracking algorithm from [4] was chosen based on the author's experience. It is simple, easy to implement, and, above all, it is effective. Tracking control with persistent excitation [24] and the vector-field-orientation method [25] were also taken into account. The former gives much worse convergence time, the latter gives even better convergence but it is more difficult to implement on a real robot.

The control for $N$ robots extended by the collision avoidance is as follows:

$$\begin{aligned} v_i &= v_l \cos e_{i\theta} + k_1 E_{ix} \\ \omega_i &= \omega_l + k_2 \mathrm{sgn}(v_l) E_{iy} + k_3 e_{i\theta}, \end{aligned} \quad (16)$$

where $k_1$, $k_2$ and $k_3$ are positive constant design parameters.

**Assumption 3.** *If the value of the linear control signal is less than considered threshold value $v_t$, i.e., $|v| < v_t$ ($v_t$-positive constant), it is replaced with a new value $\tilde{v} = S(v)v_t$, where*

$$S(v) = \begin{cases} -1 & \text{for} \quad v < 0 \\ 1 & \text{for} \quad v \geq 0 \end{cases}, \tag{17}$$

*(indexes omitted for simplicity).*

Substituting Equation (16) for (7) error dynamics is given by the following equations:

$$\begin{aligned} \dot{e}_{ix} &= e_{iy}\omega_i - k_1 E_{ix} \\ \dot{e}_{iy} &= -e_{ix}\omega_i + v_l \sin e_{i\theta} \\ \dot{e}_{i\theta} &= -k_2 \mathrm{sgn}(v_l) E_{iy} - k_3 e_{i\theta} \end{aligned} . \tag{18}$$

Transforming (18) using (16) and taking into account Assumption 2 (when the robot gets into the collision avoidance region, in the collision avoidance state, velocities $v_l$ and $\omega_l$ are replaced with 0 value) error dynamics can be expressed in the following form:

$$\begin{aligned} \dot{e}_{ix} &= k_3 e_{iy} e_{i\theta} - k_1 E_{ix} \\ \dot{e}_{iy} &= -k_3 e_{i\theta} e_{ix} \\ \dot{e}_{i\theta} &= -k_3 e_{i\theta} \end{aligned} . \tag{19}$$

Orientation error $e_{i\theta}$ decreases exponentially to zero (refer to the last equation in (19)).

In Figure 3 a schematic diagram of the control system is presented. The following signal vectors are marked: $[x\ y]^T = [x_1\ \dots\ x_N\ y_1\ \dots\ y_N]^T, \theta = [\theta_1\ \dots\ \theta_N]^T$, $[x_d\ y_d]^T = [x_{1d}\ \dots\ x_{Nd}\ y_{1d}\ \dots\ y_{Nd}]^T$, $[v\ \omega]^T = [v_1\ \dots\ v_N\ \omega_1\ \dots\ \omega_N]^T$, $[p_x\ p_y]^T = [p_{1x}\ \dots\ p_{Nx}\ p_{1y}\ \dots\ p_{Ny}]^T$, $[e_x\ e_y]^T = [e_{1x}\ \dots\ e_{Nx}\ e_{1y}\ \dots\ e_{Ny}]^T$, $[E_x\ E_y]^T = [E_{1x}\ \dots\ E_{Nx}\ E_{1y}\ \dots\ E_{Ny}]^T$.

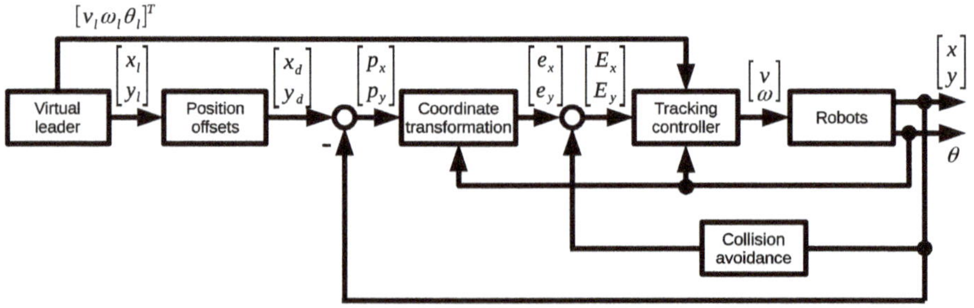


**Figure 3.** Control system.

## 3. Stability of the System

In this section stability analysis of the closed-loop system is presented. When the $i$-th robot is out of the collision regions of the other robots (APF takes the value zero) the analysis given in [4] is actual and will not be repeated here. Further the analysis for the situation in which the $i$-th robot is in the collision region of other robot is presented.

For further analysis a new variable is introduced: $\theta_{iE} = Atan2(-E_{iy}, -E_{ix})$ ($Atan2(\bullet, \bullet)$ is a version of the $Atan(\bullet)$ function covering all four quarters of the Euclidean plane)—auxiliary orientation variable.

**Proposition 1.** *The system (1) with controls (16) is stable if the desired trajectories fulfil the condition* $\theta_{iE} \notin \left\langle \frac{\pi}{2} \pm \theta_{E\Delta} \pm \pi d \right\rangle$ *($d = 0, \pm 1, \pm 2, \ldots$), where* $\theta_{E\Delta}$ *is a small constant.*

As stated in [7], if $\theta_{iE} \in \left\langle \frac{\pi}{2} \pm \theta_{E\Delta} \pm \pi d \right\rangle$ (the combination of obstacle position and reference trajectory drive the robot into the neighbourhood of a singular configuration where the condition in Proposition 1 does not hold) one solution is to add perturbation to the desired signal. The system can also leave the neighbourhood of the singularity easily since the robot can reorient itself in place if the condition is not satisfied. This requires a special procedure to be implemented, which will not be discussed here.

**Proof.** Consider the following Lyapunov-like function:

$$V = \sum_{i=1}^{N} \left[ \frac{1}{2}(e_{ix}^2 + e_{iy}^2 + e_{i\theta}^2) + \sum_{j=1, j\neq i}^{N} V_{aij} \right]. \tag{20}$$

When the robot is outside of the collision avoidance region, i.e., $l_{ij} \geq R_j$, the system is equivalent to the one presented in [4] (robot moving in a free space) and stability analysis presented in this paper still holds.

If the robot is in the collision avoidance region of the other robot time derivative of the Lyapunov-like function is calculated as follows:

$$\frac{dV}{dt} = \sum_{i=1}^{N} \left[ e_{ix}\dot{e}_{ix} + e_{iy}\dot{e}_{iy} + e_{i\theta}\dot{e}_{i\theta} + \sum_{j=1, j\neq i}^{N} \left( \frac{\partial V_{aij}}{\partial e_{ix}}\dot{e}_{ix} + \frac{\partial V_{aij}}{\partial e_{iy}}\dot{e}_{iy} \right) \right]. \tag{21}$$

Taking into account Equation (15) the above formula can be transformed to the following form:

$$\frac{dV}{dt} = \sum_{i=1}^{N} \left[ E_{ix}\dot{e}_{ix} + E_{iy}\dot{e}_{iy} + e_{i\theta}\dot{e}_{i\theta} \right]. \tag{22}$$

Next, using Equation (19) one obtains:

$$\dot{V} = \sum_{i=1}^{N} \left[ k_3 E_{ix} e_{iy} e_{i\theta} - k_3 E_{iy} e_{ix} e_{i\theta} - k_3 e_{i\theta}^2 - k_1 E_{ix}^2 \right]. \tag{23}$$

Substituting $E_{ix} = D_i \cos\theta_{iE}$ and $E_{iy} = D_i \sin\theta_{iE}$, $D_i = \sqrt{E_{ix}^2 + E_{iy}^2}$ in the above equation one obtains:

$$\begin{aligned}
\dot{V} &= \sum_{i=1}^{N} \left[ k_3 D_i \cos\theta_{iE} e_{iy} e_{i\theta} - k_3 D \sin\theta_{iE} e_{ix} e_{i\theta} - k_3 e_{i\theta}^2 - k_1 D_i^2 \cos^2\theta_{iE} \right] \\
&= \sum_{i=1}^{N} \left[ -\frac{1}{2} k_3 e_{i\theta}^2 + k_3 D_i \cos\theta_{iE} e_{iy} e_{i\theta} - \frac{1}{2} k_3 e_{i\theta}^2 + k_3 D_i \sin\theta_{iE} e_{ix} e_{i\theta} - k_1 D_i^2 \cos^2\theta_{iE} \right] \\
&= \sum_{i=1}^{N} \left\{ -k_3 \left[ \left(\frac{e_{i\theta}}{\sqrt{2}} - \frac{1}{\sqrt{2}} D_i \cos\theta_{iE} e_{iy}\right)^2 - \frac{1}{2} D_i^2 \cos^2\theta_{iE} e_{iy}^2 \right] \right. \\
&\quad \left. -k_3 \left[ \left(\frac{e_{i\theta}}{\sqrt{2}} + \frac{1}{\sqrt{2}} D_i \sin\theta_{iE} e_{ix}\right)^2 - \frac{1}{2} D_i^2 \sin^2\theta_{iE} e_{ix}^2 \right] - k_1 D_i^2 \cos^2\theta_{iE} \right\}.
\end{aligned}$$

To simplify further calculations, new scalar functions are introduced:

$$a_i = \frac{e_{i\theta}}{\sqrt{2}} - \frac{1}{\sqrt{2}} D_i \cos\theta_{iE} e_{iy}, \qquad b_i = \frac{e_{i\theta}}{\sqrt{2}} + \frac{1}{\sqrt{2}} D_i \sin\theta_{iE} e_{ix}. \tag{24}$$

Taking into account (24) $\dot{V}$ can be written as follows:

$$\dot{V} = \sum_{i=1}^{N} \left[ -k_3 a_i^2 - k_3 b_i^2 - k_1 D_i^2 \cos^2\theta_{iE} + \frac{1}{2} k_3 D_i^2 \cos^2\theta_{iE} e_{iy}^2 + \frac{1}{2} k_3 D_i^2 \sin^2\theta_{iE} e_{ix}^2 \right] \tag{25}$$

$$\dot{V} \leq \sum_{i=1}^{N} \left[ -k_3 a_i^2 - k_3 b_i^2 - k_1 D_i^2 \cos^2\theta_{iE} + \frac{1}{2} k_3 D_i^2 e_{iy}^2 + \frac{1}{2} k_3 D_i^2 e_{ix}^2 \right]. \tag{26}$$

The closed-loop system is stable ($\dot{V} \leq 0$) if the following condition is fulfilled:

$$\sum_{i=1}^{N} \left[ k_1 \cos^2\theta_{iE} - \frac{1}{2} k_3 (e_{ix}^2 + e_{iy}^2) \right] > 0. \tag{27}$$

As $\cos^2\theta_{iE} > 0$ because of the assumption in Proposition 1 it cannot be arbitrarily small, the condition (27) can be met by setting a sufficiently high value of $k_1$ or by reducing $k_3$.

Note that the error dynamics (19) with frozen reference velocities can be decomposed into two subsystems (Figure 4). The origin of the system $\Sigma_2$ is exponentially stable if $k_3 > 0$. Each of the subsystems is input to state stable (ISS). Stability of the origin may be concluded invoking the small-gain theorem for ISS systems [26].

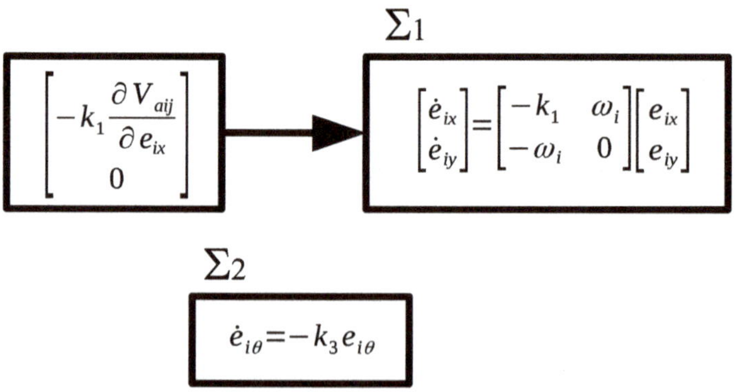

**Figure 4.** Diagram of the control system in the collision avoidance mode.

The boundedness of the output of the collision avoidance subsystem is necessary to prove stability. Taking the first equation in (19), one can state that if $\frac{\partial V_{aij}}{\partial e_{ix}}$ is sufficiently high (that happens if the robot is very close to the obstacle; there is no problem of boundedness in the other cases (refer to the properties of the APF, Figure 1)), i.e., $\frac{\partial V_{aij}}{\partial e_{ix}} \gg e_{ix}$, $\frac{\partial V_{aij}}{\partial e_{ix}} \gg e_{iy}$, and $\frac{\partial V_{aij}}{\partial e_{ix}} \gg e_{i\theta}$ the error dynamics can be approximated as follows:

$$\dot{e}_{ix} \cong -k_1 \frac{\partial V_{aij}}{\partial e_{ix}}. \tag{28}$$

From the Equation (28) it is clear that $\dot{e}_{ix}$ and $\frac{\partial V_{aij}}{\partial e_{ix}}$ have different signs and as a result $\frac{\partial V_{aij}}{\partial e_{ix}}\dot{e}_{ix} < 0$. To fulfil the condition that $\dot{V}_{aij} = \frac{\partial V_{aij}}{\partial e_{ix}}\dot{e}_{ix} + \frac{\partial V_{aij}}{\partial e_{iy}}\dot{e}_{iy}$ is less than zero the second term on the right hand side must be less than the first one taking their absolute values. This can be obtained by reducing $k_3$ parameter (refer to Equation (16)). The property $\dot{V}_{aij} \leq 0$ guarantees boundedness of both $V_{aij}$ and $\frac{\partial V_{aij}}{\partial e_{ix}}$. Finally one can state that the collision avoidance block that is input to the system shown in Figure 4 also has bounded output and both error components $e_{ix}$ and $e_{iy}$ in $\Sigma_1$ are bounded.

The above is true if the robot is not located close to the boundary of more than one robot at a time. This situation is unlikely because it leads to high controls that increase the distance between the robots quickly and therefore this will not be considered further. □

As shown in [7] collision avoidance is guaranteed if $\dot{V}_{aij} \leq 0$ and $\lim_{||[x_i\ y_i]^\top - [x_j\ y_j]^\top|| \to r^+} V_{aij} = +\infty, i \neq j$.

Each robot needs information about positions of other robots in its neighbourhood to avoid collision (their orientations are not needed). It can be obtained using on-board sensors with the range equal to or greater than $R$. In addition, robots need to know their position and orientation errors to calculate the tracking component of the control. This requirement imposes the use of a system allowing localization with respect to the global coordinate frame, because usually, the motion task is defined with respect to it. The author plans to conduct experiments on real robots in the near future. The OptiTrack motion capture system will be used to obtain coordinates of robots (positions and orientations) which is enough for control purposes.

## 4. Simulation Results

In this section a numerical simulation for a group of $N = 48$ mobile robots is presented. The initial coordinates (both positions and orientations) were random. The goal of the formation was to follow a circular reference trajectory at the same time avoiding collisions between agents. The formation had a shape of a circle. The assignments of robots to particular goal points were also random.

The following settings of the algorithm were used: $k_1 = 0.5$, $k_2 = 0.5$, $k_3 = 1.0$, $t_u = 1$ s, $r = 0.3$ m, $R = 1.2$ m.

Figure 5a shows paths of robots on the $(x, y)$ plane. To make the presentation clearer in Figure 5a–g signals of 45 robots are grey while the 3 selected ones are highlighted in black. In Figure 5h the three selected inter-agent distances are highlighted in black. Figure 5b,c present graphs of $x$ and $y$ coordinates as a function of time. The robots converged to the desired values in 115 s. Figure 5d shows a time graph of the orientations. In Figure 5e,f linear and angular controls, respectively, are shown. Initially and in the transient state, their values were high, exceeding the maximum values of a typical mobile platform. In practical implementation, they should be scaled down to realizable values. Figure 5g presents a time graph of the freeze procedure (refer to Assumption 2) of all robots. Although the drawings are not easily readable (because they include the 'freeze' signal of all robots), one can interpret that the last collision avoidance interaction ends in 108 s. In Figure 5h relative distances between robots are shown. Important information that can be read from this drawing is that no pair of robots reaches the inter-agent distance lower than or equal to $r = 0.3$ m (dashed line). This means that no collision has occurred.

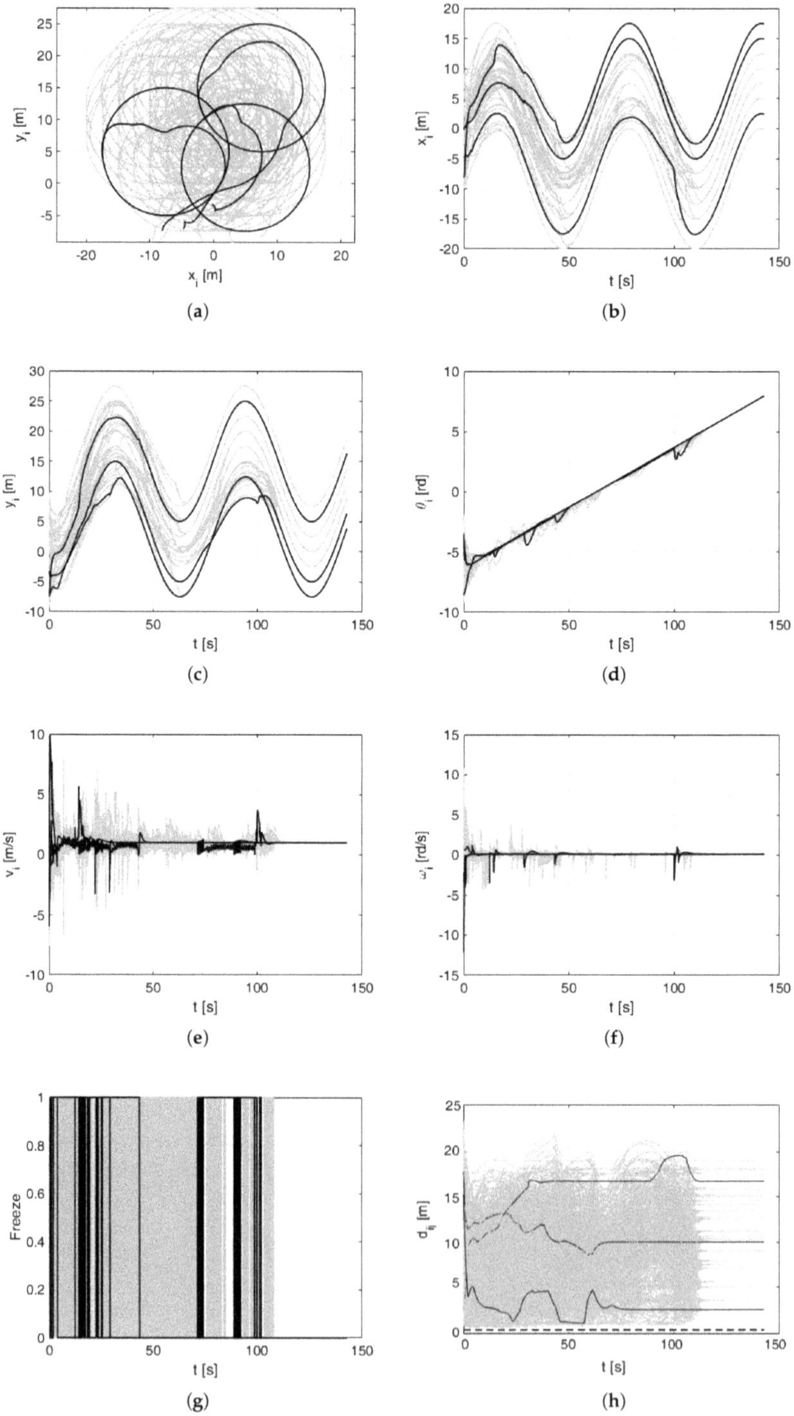


**Figure 5.** Numerical simulation 1: trajectory tracking for $N = 48$ robots. (**a**) locations of robots in $xy$-plane, (**b**) $x$ coordinates as a function of time, (**c**) $y$ coordinates as a function of time, (**d**) robot orientations as a function of time, (**e**) linear velocity controls, (**f**) angular velocity controls, (**g**) 'freeze' signals, (**h**) distances between robots.

## 5. Goal Exchange

This section presents a new control that introduces the ability to exchange goal between agents. The block diagram of the new control is shown in Figure 6. Two new blocks are included: goal switching and permutation block.

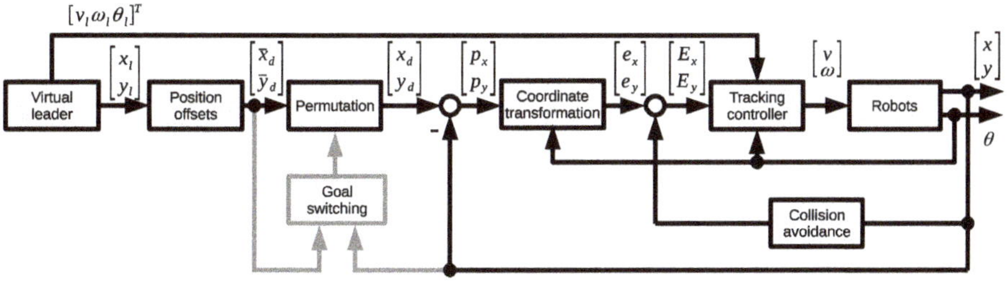


**Figure 6.** Control system with goal switching.

In the method proposed here Equation (2) is replaced as follows:

$$\bar{x}_{id} = x_l + d_{ix} \qquad \bar{y}_{id} = y_l + d_{iy}. \tag{29}$$

The new variables $\bar{x}_{id}$ and $\bar{y}_{id}$ are not representing the goal position of robot $i$, but the goal that can be assigned to any robot in the formation.

One can introduce the following aggregated goal coordinate vectors: $\bar{x}_d = [\bar{x}_{1d} \ \ldots \ \bar{x}_{Nd}]^T$ and $\bar{y}_d = [\bar{y}_{1d} \ \ldots \ \bar{y}_{Nd}]^T$ (numbers in lower index represent the numbers of the goals). The assignment of goals to particular robots is computed using $N \times N$ permutation matrix $P(t)$:

$$x_d = P(t)\bar{x}_d \qquad y_d = P(t)\bar{y}_d. \tag{30}$$

Resulting vectors contained goal coordinates assigned to particular robots $x_d = [x_{1d} \ \ldots \ x_{Nd}]^T$ and $y_d = [y_{1d} \ \ldots \ y_{Nd}]^T$ (number in lower index represents the number of the robot).

An additional control loop is introduced that acts asynchronously to the main control loop (Figure 6).

Let's assume that at some instant of time $t_1$ an arbitrary goal $m$ is assigned to the robot $k$ and another goal $n$ is assigned to the robot $l$. This can be written as:

$$\begin{array}{l} [x_{kd} \ y_{kd}]^T = [\bar{x}_{md} \ \bar{y}_{md}]^T \\ [x_{ld} \ y_{ld}]^T = [\bar{x}_{nd} \ \bar{y}_{nd}]^T. \end{array} \tag{31}$$

There are ones in permutation matrix $P(t_1)$ at element $(m,k)$ and $(n,l)$ and all other elements in rows $m$, $n$ and columns $k$, $l$ were zero.

At some discrete instant in time $t_s >= t_1$ for the pair of robots $k$ and $l$ and their goals $m$ and $n$ the following switching function was computed:

$$\sigma = \begin{cases} 1 & \text{if } ||p_{mk}||^2 + ||p_{nl}||^2 > ||p_{nk}||^2 + ||p_{ml}||^2 \\ 0 & \text{otherwise} \end{cases}, \tag{32}$$

where $p_{ij} = [\bar{x}_{id} - x_j \ \ \bar{y}_{id} - y_j]^T$.

If the switching function $\sigma$ takes the value of 1 matrix $P$ is changed as follows:

$$P(t) = S_{mn}P(t_s^-), \tag{33}$$

where $P(t_s^-)$ is the permutation matrix before switching.

The elementary matrix $S_{mn}$ is a row-switching transformation. It swaps row $m$ with row $n$ and it takes the following form:

$$S_{mn} = \begin{array}{c@{\quad}l} \left[\begin{array}{ccccccccc} 1 & 0 & \dots & 0 & \dots & 0 & \dots & 0 & 0 \\ 0 & 1 & \dots & 0 & \dots & 0 & \dots & 0 & 0 \\ \vdots & \vdots & & \vdots & & \vdots & & \vdots & \vdots \\ 0 & 0 & \dots & 0 & \dots & 1 & \dots & 0 & 0 \\ \vdots & \vdots & & \vdots & & \vdots & & \vdots & \vdots \\ 0 & 0 & \dots & 1 & \dots & 0 & \dots & 0 & 0 \\ \vdots & \vdots & & \vdots & & \vdots & & \vdots & \vdots \\ 0 & 0 & \dots & 0 & \dots & 0 & \dots & 1 & 0 \\ 0 & 0 & \dots & 0 & \dots & 0 & \dots & 0 & 1 \end{array}\right] & \begin{array}{l} \\ \\ \\ \leftarrow m\text{-th row} \\ \\ \leftarrow n\text{-th row} \\ \\ \\ \\ \end{array} \\ \begin{array}{ccccccccc} \phantom{0} & \phantom{0} & \phantom{\dots} & \uparrow & \phantom{\dots} & \uparrow & \phantom{\dots} & \phantom{0} & \phantom{0} \\ & & & m\text{-th} & & n\text{-th} & & & \\ & & & \text{col.} & & \text{col.} & & & \end{array} & \end{array}. \tag{34}$$

Transformation (33) describes a process of the goal exchange between agents $k$ and $l$ at time $t_s$. After that goal $m$, was assigned to robot $l$ and goal $n$ was assigned to robot $k$ that is equivalent to the following equalities:

$$\begin{array}{c} [x_{kd}\ y_{kd}]^T = [\bar{x}_{nd}\ \bar{y}_{nd}]^T \\ [x_{ld}\ y_{ld}]^T = [\bar{x}_{md}\ \bar{y}_{md}]^T. \end{array} \tag{35}$$

Note that the process of goal exchange is asynchronous with the main control loop. It operated at lower frequency because it required communication between remote agents, which is time consuming. The low frequency subsystem is highlighted in grey in Figure 6.

## 6. Stability of the System with Target Assignment

The goal exchange procedure significantly improves system convergence and reduces the number of collision avoidance interactions between agents. On the other hand, its execution time was not critical for the control of the system.

Stability analysis of the control system with goal switching was conducted using the same Lyapunov-like function (20) as in Section 3.

**Proposition 2.** *The procedure given by Equation (33) results in a decrease of the Lyapunov-like function Equation (20).*

**Proof.** A hypothetical position error $p_{ij}$ can be expressed in a local coordinate frame by the following transformation

$$e_{ij} = \begin{bmatrix} \cos(\theta_j) & \sin(\theta_j) \\ -\sin(\theta_j) & \cos(\theta_j) \end{bmatrix} p_{ij}. \tag{36}$$

that is invariant under scaling, and thus, the following equality holds true:

$$||e_{ij}|| = ||p_{ij}|| \tag{37}$$

(notice that index $i$ is the number of the goal and index $j$ is the number of the robot).

Using Equation (37) the switching function (32) can be rewritten as follows:

$$\sigma = \begin{cases} 1 & \text{if } ||e_{mk}||^2 + ||e_{nl}||^2 > ||e_{nk}||^2 + ||e_{ml}||^2 \\ 0 & \text{otherwise} \end{cases}. \tag{38}$$

To carry out further analysis the Lyapunov-like function (20) will be rewritten as follows:

$$V = \overbrace{\sum_{i=1}^{N} \frac{1}{2} e_{i\theta}^2}^{V_\theta} + \overbrace{\sum_{i=1}^{N} \sum_{j=1, j \neq i}^{N} V_{aij}}^{V_a} + \frac{1}{2}(e_{1x}^2 + e_{1y}^2) + \ldots$$
$$+ \underbrace{\frac{1}{2}(e_{kx}^2 + e_{ky}^2)}_{V_{pk}} + \ldots + \underbrace{\frac{1}{2}(e_{lx}^2 + e_{ly}^2)}_{V_{pl}} + \ldots + \frac{1}{2}(e_{Nx}^2 + e_{Ny}^2). \quad (39)$$

Terms $V_{pk}$ and $V_{pl}$ are related to the position errors of robot $k$ and $l$, respectively. Notice that other terms of the Lyapunov-like function $V$ are invariant under goal assignment as $V_a$ depends only on the distances between agents and $V_\theta$ remains constant because all agents share the same reference orientation $\theta_l$. The position error terms related to robots that were not involved in goal exchange were also invariant under goal exchange.

Two cases will be considered further: case 1 at $t_s^-$, and case 2 at $t_s$.

In case 1 the sum of position terms of robots $k$ and $l$ can be transformed using (36) and (31) as follows:

$$\begin{aligned}
V_{p1} &= V_{pk} + V_{pl} = \frac{1}{2}(e_{kx}^2 + e_{ky}^2) + \frac{1}{2}(e_{lx}^2 + e_{ly}^2) \\
&= \frac{1}{2}||[e_{kx}\ e_{ky}]^T||^2 + \frac{1}{2}||[e_{lx}\ e_{ly}]^T||^2 \\
&= \frac{1}{2}||[p_{kx}\ p_{ky}]^T||^2 + \frac{1}{2}||[p_{lx}\ p_{ly}]^T||^2 \\
&= \frac{1}{2}(p_{kx}^2 + p_{ky}^2) + \frac{1}{2}(p_{lx}^2 + p_{ly}^2) \\
&= \frac{1}{2}([x_{kd} - x_k\ \ y_{kd} - y_k])^2 + \frac{1}{2}([x_{ld} - x_l\ \ y_{ld} - y_l])^2 \\
&= \frac{1}{2}([\bar{x}_{md} - x_k\ \ \bar{y}_{md} - y_k])^2 + \frac{1}{2}([\bar{x}_{nd} - x_l\ \ \bar{y}_{nd} - y_l])^2 \\
&= \frac{1}{2}(||p_{mk}||^2 + ||p_{nl}||^2).
\end{aligned}$$

In case 2 the sum of position terms of robot $k$ and $l$, repeating the initial steps above and taking into account (35) is given by:

$$\begin{aligned}
V_{p2} &= V_{pk} + V_{pl} = \\
&= \frac{1}{2}([x_{kd} - x_k\ \ y_{kd} - y_k])^2 + \frac{1}{2}([x_{ld} - x_l\ \ y_{ld} - y_l])^2 \\
&= \frac{1}{2}([\bar{x}_{nd} - x_k\ \ \bar{y}_{nd} - y_k])^2 + \frac{1}{2}([\bar{x}_{md} - x_l\ \ \bar{y}_{md} - y_l])^2 \\
&= \frac{1}{2}(||p_{nk}||^2 + ||p_{ml}||^2).
\end{aligned}$$

Note that $V_{p1}$ (omitting the constant multiplier $\frac{1}{2}$) is the left hand side of the inequality in the first condition of the switching function (32) while $V_{p2}$ is the right hand side of this condition (also omitting the multiplier). This leads to the conclusion that as $V_{p1} > V_{p2}$, goal exchange results in a rapid decrease (discontinuous) of the Lyapunov-like function. □

All other properties of the $V$ still hold including $\dot{V} \leq 0$ if the condition given by Equation (27) is fulfilled.

**Proposition 3.** *The procedure given by Equation (33) results in a decrease of the sum of the position errors squared.*

**Proof.** As $V_a$ and $V_\theta$ in Equation (39) are invariant under the goal assignment and the sum of all other terms represent the sum of the position errors squared (omitting the constant multiplier $\frac{1}{2}$), the proof of Proposition 3 comes directly from Proposition 2. □

Notice that the sum of the position errors squared can be easily expressed in the global coordinate frame using equality $e_{ix}^2 + e_{iy}^2 = ||[e_{ix} \;\; e_{iy}]^T||^2 = ||[p_{ix} \;\; p_{iy}]^T||^2 = p_{ix}^2 + p_{iy}^2$ (refer to Equation (6)) to transform the position error of each robot in (39).

The tracking algorithm presented in Section 2 together with the goal exchange procedure resulted in time intervals continuous algorithm with discrete optimization that uses the Lyapunov-like function as a criterion to be minimized. The discontinuities occur in two situations: when the robot is in the collision avoidance region (the reference trajectory is temporarily frozen and then unfrozen) and when a pair of agents exchange the goals. The set of these discontinuity points was of zero measure. Note that including goal exchange in the control reinforces fulfilling condition (27) because it supports reduction of the component $\sum_{i=1}^{N} \frac{1}{2} k_3 (e_{ix}^2 + e_{iy}^2)$.

The presented algorithm does not guarantee optimal solutions but each goal exchange improves the quality of the resulting motion. The total improvement depends significantly on the initial state of the system. In extreme cases, there is a situation in which the initial coordinates are close to optimal. The procedure may lead to no goal exchange, and thus no improvement, even though communication costs have been incurred. On the other hand, if the initial coordinates are not special, benefits of using the procedure are usually considerable.

## 7. Distributed Goal Exchange

The procedure described in Section 5 can be implemented in a distributed manner. Both key components of the goal switching algorithm; computation of the switching function (32) and permutation matrix transformation (33), involve only two agents. Reliable connection between them was needed as the process was conducted in a sequence of steps. After the agents establish the connection one of them ($i$-th) sends its position coordinates $(x_i, y_i)$ and goal location to the other ($j$-th). The second robot computes switching function (32) which is the verdict on the goal exchange. If it is negative the robots disconnect and continue motion to their goals, otherwise robot $j$ sends its position coordinates $(x_j, y_j)$ and goal location to robot $i$. This part is critical in the process and must be designed carefully to ensure correct task execution.

To make distributed goal exchange possible, the robots must be equipped with the on-board radio modules allowing inter-agent communication. Even if not all pairs of agents were capable of communicating, the goal exchange algorithm improved the result. On the other hand, without communication between all pairs of agents the algorithm did not fail. In the vast majority of environments there was no problem with the communication range (current technology allows communication through many routers, base stations and peers). The exceptions may be space and oceanic applications. They will not be considered here. The author plans to conduct laboratory experiments where there are no restrictions on communication. From the practical point of view, robots need to know the network addresses of other robots and sequentially attempt to establish connection and, if successful, exchange the goals.

## 8. On Some Generalization

The condition (32) can be rewritten in more general form as follows:

$$\sigma = \begin{cases} 1 & \text{if } ||p_{mk}||^n + ||p_{nl}||^n > ||p_{nk}||^n + ||p_{ml}||^n \\ 0 & \text{otherwise} \end{cases} \tag{40}$$

Taking $n = 1$ leads to the shortest path criterion that seems to be natural in many cases because the shortest path induces lower motion cost. Unfortunately this observation may not be true in a cluttered case. This will be shown later in this section. Note that in the cases for $n \neq 2$ the Lyapunov analysis is much more complex. In [23] the similar method that results in the shortest total distance to the goals is presented.

Several specific scenarios for the simple case of two robots are analysed further. These scenarios should be treated as an approximation of a real case because typically paths of real robots are not straight lines in the case of the platforms that are not fully actuated (like the differentially driven mobile platform considered here).

In Figure 7 two robots are the same distance away from their goals. Initially the goals were assigned as marked with dashed arrows. The goal exchange procedure led to the assignment marked with continuous arrows. The resulting paths were less collisional or even non-collisional (as they are parallel). The new assignment was optimal using both the shortest path criterion ($n = 1$) and quadratic criterion ($n = 2$).

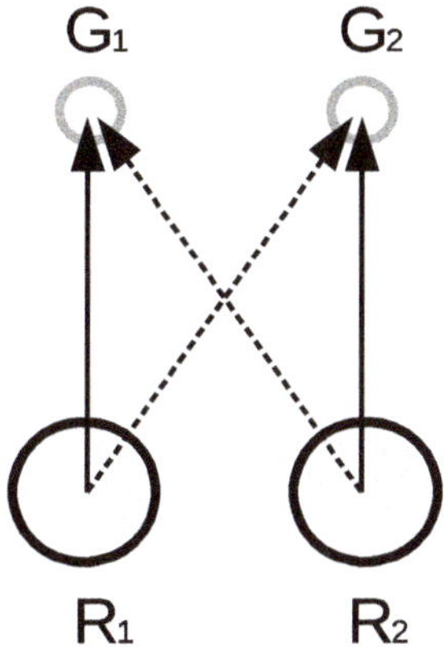


**Figure 7.** Two robot-goal assignment—case 1.

Figure 8 shows a case in which two robots and their goals lie on a straight line. Initially goal 2 is assigned to robot 1 and goal 1 is assigned to robot 2. This situation caused a saddle point because during the motion to the goal $R_1$ stays on the path of $R_2$. One can observe that the shortest path criterion ($n = 1$) produces exactly the same result for both possible goal assignments, while quadratic one ($n = 2$) produces the result marked by continuous arrows. By the assignment $R_1 - G_1$ and $R_2 - G_2$ the goals can be reached by the robots without bypass manoeuvre. This is one of the examples showing the significant advantage of the quadratic criterion over the shortest path criterion proposed in [23].

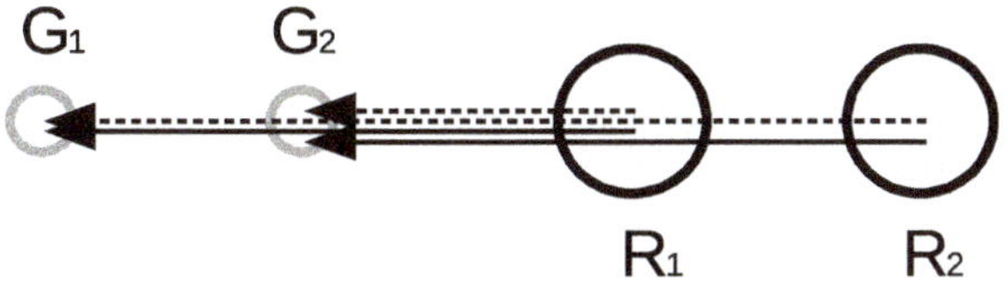


**Figure 8.** Two robot-goal assignment—case 2.

In Figure 9 $R_1$ is exactly at the goal $G_2$ assigned to it. $G_1$ was assigned to $R_2$. The quadratic criterion resulted in the opposite assignment (continuous arrows). Notice that for the shortest path criterion the collision avoidance interaction between $R_1$ and $R_2$ was possible. For this type of situation the quadratic criterion resulted in goal exchange for all locations in the hatched circle. The opposite situation is presented in Figure 10. Fields of squares shown in the figures represent values of cost functions for two possible goal assignments (compare two left-hand side squares with right-hand square). The quadratic criterion resulted (in comparison to the shortest path) in a higher cost function for assigning far goals to the robots. It favoured a larger number of short assignments instead of the lower number of farther ones. This promoted the reduction of collision interaction situations.

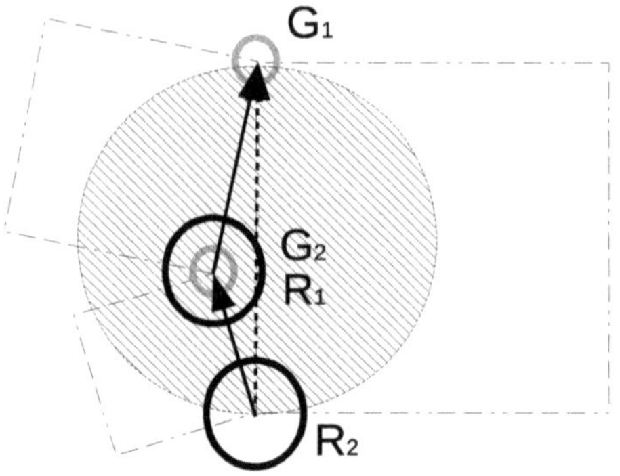


**Figure 9.** Two robot-goal assignment—case 3.

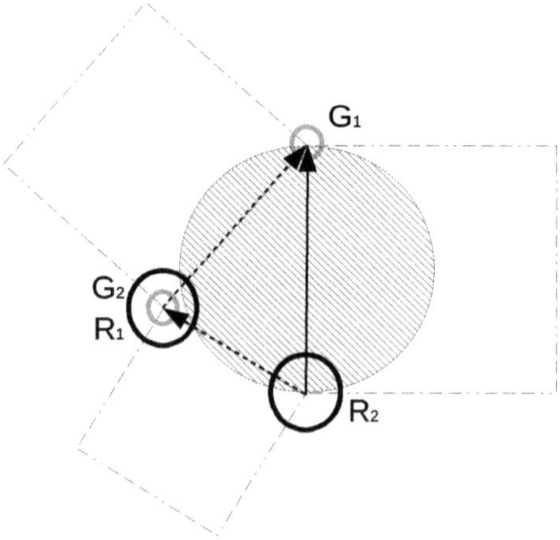


**Figure 10.** Two robot-goal assignment—case 4.

In Figure 11 certain positions of the robot $R_1$ and goals $G_1$ and $G_2$ have been assumed. If robot $R_2$ is located in the hatched area the quadratic criterion assigns it to $G_2$, otherwise it is assigned to $G_1$. Some initial configuration may have led to temporary collision interaction states (i.e., when $R_2$ initially is located to the left of the $R_1$) but the saddle point occurrence was not possible. Dashed circles on the sides represent examples of boundary locations (for goal exchange) of the robot $R_2$.

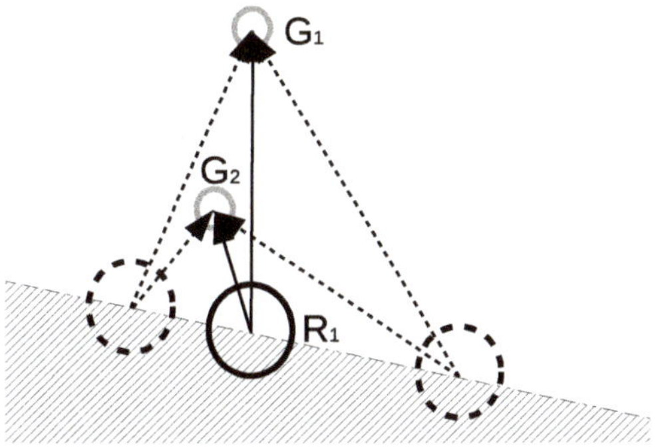


**Figure 11.** Two robot-goal assignment—case 5.

The considered cases do not cover all possible scenarios, but since there is no formal guarantee that the number of collision avoidance interactions between agents is reduced, they illustrate, together with simulation results, that in typical situations goal exchange procedure leads to the simplification of the control task.

## 9. Ensuring Integrity

As the goal exchange process involves two agents that are physically separated machines and they communicate through a wireless link the goal exchange process is at the risk of failure. Using a reliable communication protocol like Transmission Control Protocol (TCP) and dividing the process into a sequence of stages acknowledged by the remote host the fault-tolerance of the system can be increased. Assuming that the transmitted packets are encrypted (which is standard nowadays) and the implementation is relatively simple (the author believes that software bugs can be corrected) byzantine fault tolerance (BFT) [27] is not considered here.

The first stage of the goal exchange process is establishing the connection between agents. It is proposed to use the TCP connection because this protocol uses sequence numbers, acknowledgements, checksums and it is the most reliable, widely used communication protocol. Agents know network addresses of the other agents. They can be given in advance, provided by the higher-level system or obtained using a dedicated network broadcasting service. The attempt to connect to the agent that is already involved in the goal exchange process should be rejected. This can be easily and effectively implemented using TCP.

The second stage is transmission of the robot location coordinates and the goal from agent one to another. The receiver computes $\sigma$ (Equation (32)) and sends back the obtained value. If $\sigma = 0$ agent closes the connection, otherwise they go to the third step.

The third step is a goal exchange that is the most critical part of the process. It must be guaranteed that no goal stays unassigned and no goal can be assigned to more than one robot in the case of agent/communication failure. To fulfil this condition the goal exchange must have all properties of database transaction: it must be atomic, consistent, isolated and durable. In practice this idealistic solution can be approximated by applying one of the widely used algorithms: two-phase commit protocol [28], three-phase commit protocol [29], or Paxos [30]. All of them introduce a coordinator block that is the central point of the algorithm. It can be run (for example as a separate process) on the one of the machines involved in the goal exchange procedure. Notice that in the case of failure (communication error, agent failure, etc.) the operation of goal exchange is aborted. This leads to slower convergence of robots to their desired values but is not critical for the task execution.

## 10. Simulation Results for Goal Exchange

This section presents numerical simulation of the algorithm extended with goal exchange procedure. The initial conditions are exactly the same as in Section 4 (results for the algorithm without goal exchange). The parameters of the controllers were also the same. The initial value of the permutation matrix was the identity matrix $P(0) = I$.

Figure 12a shows paths of robots on the $(x, y)$ plane. As in the previous experiment signals of 3 robots (out of 48) are highlighted in black. Figure 12b,c present graphs of $x$ and $y$ coordinates as a function of time. The robots converge to the desired values in less than 20 s, a significantly better result than 115 s (without goal exchange). This experiment was completed at 28 s due to faster convergence. Figure 12d shows a time graph of the orientations. In Figure 12e,f linear and angular controls are plotted. They reach constant values in less than 20 s. Figure 12g presents a time graph of the freeze procedure (refer to Assumption 2). It can be seen that the last collision avoidance interaction ends at 13 s. In Figure 12h relative distances between robots are shown. No pair of robots reaches inter-agent distance lower than or equal to $r = 0.3$ m (dashed line). It means that no collision has occurred.

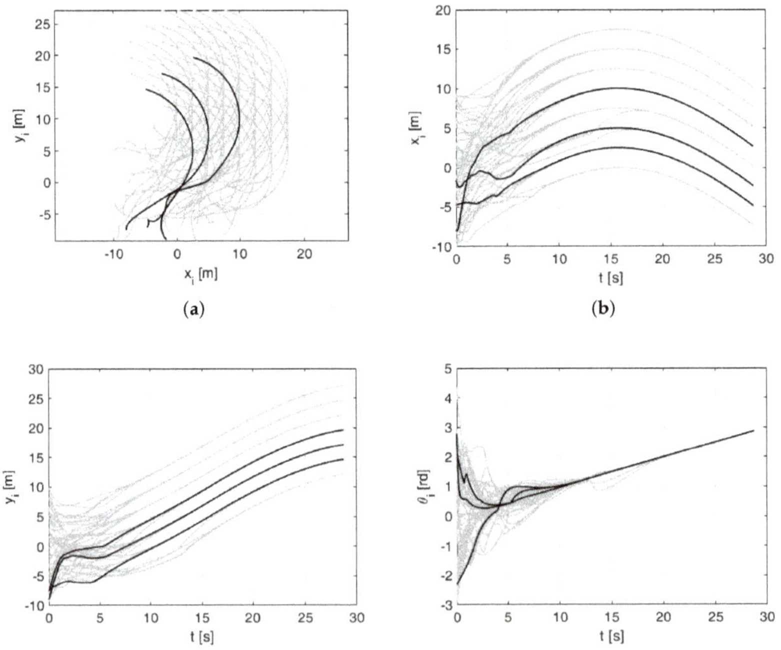


(a) (b) (c) (d)

**Figure 12.** *Cont.*

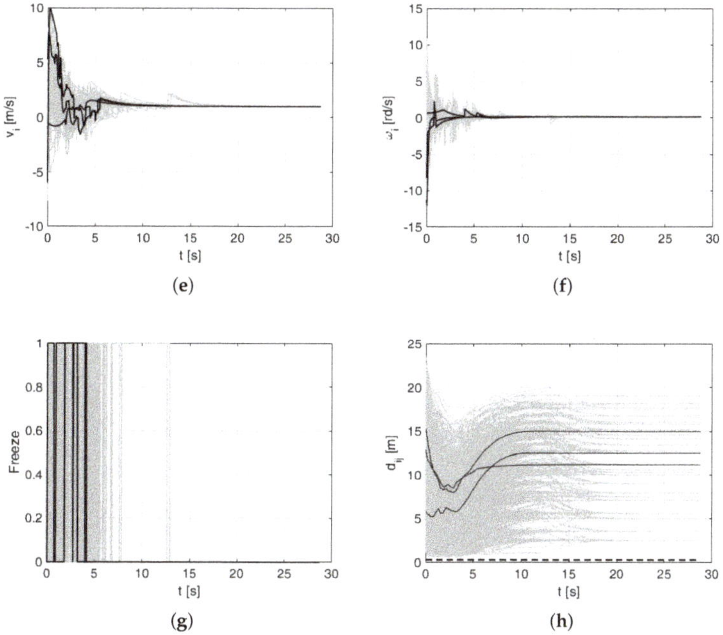


**Figure 12.** Numerical simulation 2: trajectory tracking for $N = 48$ robots with goal exchange. (**a**) locations of robots in $xy$-plane, (**b**) $x$ coordinates as a function of time, (**c**) $y$ coordinates as a function of time, (**d**) robot orientations as a function of time, (**e**) linear velocity controls, (**f**) angular velocity controls, (**g**) 'freeze' signals, (**h**) distances between robots.

Notice that in the presented simulation no communication delay of the goal exchange procedure was taken into account. Depending on the quality of the network single goal exchange may take even hundreds of milliseconds. On the other hand, as the procedure involves only a pair of agents, many of such pairs can execute goal exchange at the same time. Of course another limitation was the bandwidth of the communication network. These issues will be investigated by the author in the near future. In the presented numerical simulations the number of goal switchings was 238, which is a significant number.

Visualizations of the exemplary experiments are available on the website http://wojciech.kowalczyk.pracownik.put.poznan.pl/research/target-assignment/ts.html.

## 11. Simulation Results for Saturated Wheel Controls

As the APFs used to avoid collisions are unbounded, the algorithm should be tested for the limited wheel velocities (resulting from the motor velocity limits). This section presents a numerical simulation for the mobile robots with the wheel diameter of 0.0245 m, the distance between wheels amounting to 0.148 m and the maximum angular velocity of 48.6 rd/s.

A special scaling procedure was applied to the wheel controls. The desired wheel velocities were scaled down when at least one of the wheels exceeds the assumed limitation. The scaled control signal $u_{iws}$ is calculated as follows:

$$u_{iws} = s_i u_{iw}, \tag{41}$$

where

$$s_i = \begin{cases} \frac{\omega_{max}}{\omega_{io}} & \text{if} \quad \omega_{io} > \omega_{max} \\ 1 & \text{otherwise} \end{cases}, \tag{42}$$

and

$$\omega_{io} = \max\{|\omega_{iR}|, |\omega_{iL}|\}, \tag{43}$$

where $\omega_{iR}$, $\omega_{iL}$ denote right and left wheel angular velocity, $\omega_{max}$ is the predefined maximum allowed angular velocity for each wheel.

Figure 13a,b show time graphs of the right and left wheels of the platforms. As in the previous experiments, signals of three robots (out of 48) are highlighted in black. It can be clearly observed that both of them were limited to $\pm 48.6$ rd/s. Linear and angular velocities of the platform are shown in Figure 13c,d. Their velocities are lower in comparison to the non-limited case (refer to Figure 12e,f). Figure 13e presents relative distances between the robots. The area below dashed line represents the collision region. It can be seen that no pair of robots has reached it—no collision occurred during this experiment.

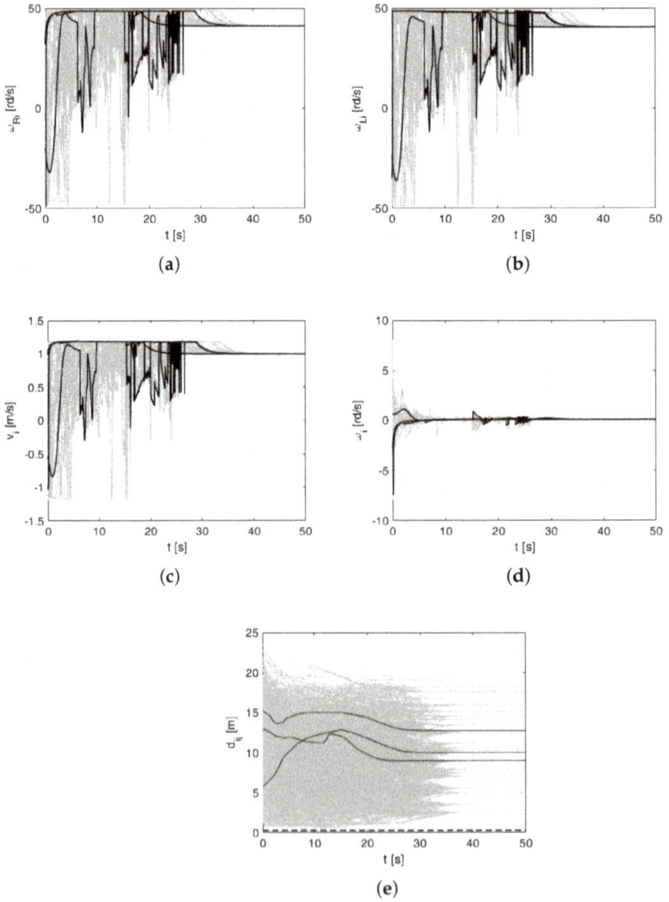


**Figure 13.** Numerical simulation 3: trajectory tracking for $N = 48$ robots with goal exchange and saturated wheel controls. (**a**) right wheel velocities, (**b**) left wheel velocities, (**c**) linear velocity controls, (**d**) angular velocity controls, (**e**) distances between robots.

## 12. Conclusions

This paper presents a new control algorithm for the formation of non-holonomic mobile robots. Inter-agent communication is used to check if exchange of goals between robots reduces the system's overall Lyapunov-like function and the sum of position errors squared. The procedure is verified by numerical simulations for large group of non-holonomic mobile robots moving in the formation. The simulations also include the case in which wheel velocity controls are limited. A significant improvement of system convergence is shown. The author plans to conduct extensive tests of the presented algorithm on real two-wheeled mobile robots in the near future.

**Funding:** This work is supported by statutory grant 09/93/DSPB/0811.

**Conflicts of Interest:** The authors declare no conflict of interest.

## References

1. Khatib, O. Real-time Obstacle Avoidance for Manipulators and Mobile Robots. *Int. J. Robot. Res.* **1986**, *5*, 90–98. [CrossRef]
2. Leitmann, G.; Skowronski, J. Avoidance Control. *J. Optim. Theory Appl.* **1977**, *23*, 581–591. [CrossRef]
3. Leitmann, G. Guaranteed Avoidance Strategies. *J. Optim. Theory Appl.* **1980**, *32*, 569–576. [CrossRef]
4. De Wit, C.C.; Khennouf, H.; Samson, C.; Sordalen, O.J. Nonlinear Control Design for Mobile Robots. *Recent Trends Mob. Rob.* **1994**, 121–156._0005. [CrossRef]
5. Morin, P.; Samson, C. Practical stabilization of driftless systems on Lie groups: The transverse function approach. *IEEE Trans. Autom. Control* **2003**, *48*, 1496–1508. [CrossRef]
6. Kowalczyk, W.; Michałek, M.; Kozłowski, K. Trajectory Tracking Control with Obstacle Avoidance Capability for Unicycle-like Mobile Robot. *Bull. Pol. Acad. Sci. Tech. Sci.* **2012**, *60*, 537–546. [CrossRef]
7. Mastellone, S.; Stipanovic, D.; Spong, M. Formation control and collision avoidance for multi-agent non-holonomic systems: Theory and experiments. *Int. J. Robot. Res.* **2008**, 107–126. [CrossRef]
8. Do, K.D. Formation Tracking Control of Unicycle-Type Mobile Robots With Limited Sensing Ranges. *IEEE Trans. Control Syst. Technol.* **2008**, *16*, 527–538. [CrossRef]
9. Kowalczyk, W.; Kozłowski, K.; Tar, J.K. Trajectory Tracking for Multiple Unicycles in the Environment with Obstacles. In Proceedings of the 19th International Workshop on Robotics in Alpe-Adria-Danube Region (RAAD 2010), Budapest, Hungary, 23–27 June 2010; pp. 451–456. [CrossRef]
10. Kowalczyk, W.; Kozłowski, K. Leader-Follower Control and Collision Avoidance for the Formation of Differentially-Driven Mobile Robots. In Proceedings of the MMAR 2018—23rd International Conference on Methods and Models in Automation and Robotics, Międzyzdroje, Poland, 27–30 August 2018.
11. Hatanaka, T.; Fujita, M.; Bullo, F. Vision-based cooperative estimation via multi-agent optimization. In Proceedings of the 49th IEEE Conference on Decision and Control (CDC), Atlanta, GA, USA, 15–17 December 2010; pp. 2492–2497. doi:10.1109/CDC.2010.5717384. [CrossRef]
12. Millan, P.; Orihuela, L.; Jurado, I.; Rubio, F. Formation Control of Autonomous Underwater Vehicles Subject to Communication Delays. *IEEE Trans. Control Syst. Technol.* **2014**, *22*, 770–777. [CrossRef]
13. Yoshioka, C.; Namerikawa, T. Formation Control of Nonholonomic Multi-Vehicle Systems based on Virtual Structure. *IFAC Proc. Vol.* **2008**, *41*, 5149–5154. [CrossRef]
14. Borrmann, U.; Wang, L.; Ames, A.D.; Egerstedt, M. Control Barrier Certificates for Safe Swarm Behavior. *IFAC-PapersOnLine* **2015**, *48*, 68–73. [CrossRef]
15. Ames, A.D.; Xu, X.; Grizzle, J.W.; Tabuada, P. Control Barrier Function Based Quadratic Programs for Safety Critical Systems. *IEEE Trans. Autom. Control* **2017**, *62*, 3861–3876. [CrossRef]
16. Srinivasan, M.; Coogan, S.; Egerstedt, M. Control of Multi-Agent Systems with Finite Time Control Barrier Certificates and Temporal Logic. In Proceedings of the 2018 IEEE Conference on Decision and Control (CDC), Miami Beach, FL, USA, 17–19 December 2018; pp. 1991–1996. doi:10.1109/CDC.2018.8619113. [CrossRef]
17. Akella, S. Assignment Algorithms for Variable Robot Formations. In Proceedings of the 12th International Workshop on the Algorithmic Foundations of Robotics, San Francisco, CA, USA, 18–20 December 2016.

18. Caldeira, A.; Paiva, L.; Fontes, D.; Fontes, F. Optimal Reorganization of a Formation of Nonholonomic Agents Using Shortest Paths. In Proceedings of the 2018 13th APCA International Conference on Automatic Control and Soft Computing (CONTROLO), Ponta Delgada, Azores, Portugal, 4–6 June 2018.
19. Alonso-Mora, J.; Baker, S.; Rus, D. Multi-robot formation control and object transport in dynamic environments via constrained optimization. *Int. J. Robot. Res.* **2017**, *36*, 1000–10217. [CrossRef]
20. Alonso-Mora, J.; Breitenmoser, A.; Rufli, M.; Siegwart, R.; Beardsley, P. Image and Animation Display with Multiple Mobile Robots. *Int. J. Robot. Res.* **2012**, *31*, 753–773. [CrossRef]
21. Desai, A.; Cappo, E.; Michael, N. Dynamically feasible and safe shape transitions for teams of aerial robots. In Proceedings of the 2016 IEEE/RSJ International Conference on Intelligent Robots and Systems (IROS), Daejeon Convention Center, Daejeon, Korea, 9–14 October 2016; pp. 5489–5494.
22. Turpin, M.; Michael, N.; Kumar, V. Capt: Concurrent Assignment and Planning of Trajectories for Multiple Robots. *Int. J. Robot. Res.* **2014**, *33*, 98–112. [CrossRef]
23. Panagou, D.; Turpin, M.; Kumar, V. Decentralized goal assignment and trajectory generation in multi-robot networks: A multiple Lyapunov functions approach. In Proceedings of the 2014 IEEE International Conference on Robotics and Automation (ICRA), Hong Kong, China, 31 May–5 June 2014; pp. 6757–6762. [CrossRef]
24. Loria, A.; Dasdemir, J.; Alvarez Jarquin, N. Leader—Follower Formation and Tracking Control of Mobile Robots Along Straight Paths. *IEEE Trans. Control Syst. Technol.* **2016**, *24*, 727–732. [CrossRef]
25. Michałek, M.; Kozłowski, K. Vector-Field-Orientation Feedback Control Method for a Differentially Driven Vehicle. *IEEE Trans. Control Syst. Technol.* **2010**, *18*, 45–65. [CrossRef]
26. Khalil, H.K. *Nonlinear Systems*, 3rd ed.; Prentice-Hall: New York, NY, USA, 2002.
27. Castro, M.; Liskov, B. Practical Byzantine Fault Tolerance. In Proceedings of the Third Symposium on Operating Systems Design and Implementation, New Orleans, LA, USA, 22–25 February 1999.
28. Bernstein, P.; Hadzilacos, V.; Goodman, N. *Concurrency Control and Recovery in Database Systems*; Addison Wesley Publishing Company: Boston, MA, USA, 1987; ISBN 0-201-10715-5.
29. Skeen, D.; Stonebraker, M. A Formal Model of Crash Recovery in a Distributed System. *IEEE Trans. Softw. Eng.* **1983**, *9*, 219–228. [CrossRef]
30. Lamport, L. The Part-Time Parliament. *ACM Trans. Comput. Syst.* **1998**, *16*, 133–169. [CrossRef]

*applied sciences*

MDPI

*Article*

# Biologically-Inspired Learning and Adaptation of Self-Evolving Control for Networked Mobile Robots

**Sendren Sheng-Dong Xu [1], Hsu-Chih Huang [2,*], Tai-Chun Chiu [1] and Shao-Kang Lin [2]**

1 Graduate Institute of Automation and Control, National Taiwan University of Science and Technology, Taipei City 10607, Taiwan; sdxu@mail.ntust.edu.tw (S.S.-D.X.); ifuzzy2015@gmail.com (T.-C.C.)
2 Department of Electrical Engineering, National Ilan University, Yilan City 26047, Taiwan; cacs2015@gmail.com
* Correspondence: hchuang@niu.edu.tw

Received: 16 January 2019; Accepted: 5 March 2019; Published: 12 March 2019

**Abstract:** This paper presents a biologically-inspired learning and adaptation method for self-evolving control of networked mobile robots. A Kalman filter (KF) algorithm is employed to develop a self-learning RBFNN (Radial Basis Function Neural Network), called the KF-RBFNN. The structure of the KF-RBFNN is optimally initialized by means of a modified genetic algorithm (GA) in which a Lévy flight strategy is applied. By using the derived mathematical kinematic model of the mobile robots, the proposed GA-KF-RBFNN is utilized to design a self-evolving motion control law. The control parameters of the mobile robots are self-learned and adapted via the proposed GA-KF-RBFNN. This approach is extended to address the formation control problem of networked mobile robots by using a broadcast leader-follower control strategy. The proposed pragmatic approach circumvents the communication delay problem found in traditional networked mobile robot systems where consensus graph theory and directed topology are applied. The simulation results and numerical analysis are provided to demonstrate the merits and effectiveness of the developed GA-KF-RBFNN to achieve self-evolving formation control of networked mobile robots.

**Keywords:** biologically-inspired; self-learning; formation control; mobile robots

## 1. Introduction

Networked mobile robots that are capable of self-learning have received growing attention in the mobile robotics research community [1–3]. This emerging technology has surpassed the conventional robotic system by taking advantage of robot collaboration, system robustness, scalability, and greater flexibility. This modern robotic system has been commonly applied in manufacturing, military applications, surveillance, etc. to perform complex tasks [4–6]. Some self-learning strategies have been proposed to develop motion controllers for networked mobile robots [6,7]. Among them, an RBFNN incorporating the gradient descent method is regarded as a powerful tool for designing the self-learning controllers of networked mobile robots [7,8].

The RBFNN introduced by Broomhead and Lowe is a three-layer feedforward artificial neural network in which radial basis functions are used as activation functions [9,10]. This methodology takes advantage of fast learning capability and universal approximation. To date, it is a useful neural network architecture for addressing many engineering problems [11,12]. However, traditional RBFNNs adopt a gradient descent approach for training the neural network that is not capable of noise reduction [10–12]. In other words, these studies did not consider the uncertainty and noise induced in the process and measurement phases. This paper presents a Kalman filter based RBFNN and its application to self-learning control of networked mobile robots.

The Kalman filter is a state estimation technique introduced by R.E. Kalman [13]. It is a classic state estimation technique used widely in engineering applications, including spacecraft navigation, motion planning in robotics, signal processing and wireless sensor networks [14–16] because of its ability to extract useful information from noisy data. It is an optimal estimator for evaluating the internal state of a dynamic system under certain process patterns and/or measurement disturbances in the physical environment [13,17,18]. The objective of the Kalman filter is to minimize the mean squared error between the actual and estimated data. Although the proposed KF-RBFNN is useful for designing learning control schemes with noise reduction, the initial network parameters influences the system performance, namely that the selection of centers, widths and output weights for the Gaussian functions is an important consideration.

Parameter-tuning of an RBFNN is challenging when using this neural network to solve multidimensional optimization problems. Over the years, several methods have been developed for addressing this RBFNN optimization problem [19–21]. However, these traditional RBFNN methods may cause the output to converge to local optimum when the dimensionality of the problem increases [22]. Since RBFNN optimization can be formulated as a search problem, biologically, algorithms are new paths for optimizing RBFNNs. This is a successful hybridization of RBFNNs and evolutionary algorithms. Although there are some metaheuristic algorithms used to develop evolutionary RBFNNs [23–26], there has been no attempt to present an evolutionary KF-RBFNN using a GA to achieve learning control of networked mobile robotic systems.

The GA is one of the most popular evolutionary algorithms for solving optimization problems [27–29]. Although GAs have been widely applied to various optimization problems, these biologically inspired algorithms suffer from premature convergence. In other words, these traditional computing paradigms may converge to local optimum. This paper contributes to the development of a modified GA to improve the search diversity by including the Lévy flight approach. An adaptive determination of crossover and mutation probabilities in the GA is proposed via the Lévy flights. This random walk is very efficient in exploring the search space of the optimization problem. The proposed modified GA metaheuristics is then applied to the design of an optimal GA-KF-RBFNN for self-tuning motion control of networked mobile robots.

Of the increasing demands on networked mobile robot systems, formation using a leader-follower control strategy is one of the most important and is becoming increasing crucial [30–32]. It is a coordinated control in which the leader robot follows a desired trajectory while the follower robots maintain a specified geometrical pattern [32]. Although some studies have addressed this control problem by considering graph theory and consensus control approaches [33–35], these networked mobile robot systems suffer from communication delay problem. This paper presents a pragmatic self-learning optimal GA-KF-RBFNN formation control method for networked mobile robot systems that avoids the communication delay problem.

This paper is structured as follows: a biologically-inspired Kalman filter based RBFNN control technique, called GA-FA-RBFNN control is introduced in Section 2. Section 3 employs the proposed GA-FA-RBFNN to develop a networked mobile robot system to achieve self-evolving formation control. In Section 4, several simulation results are reported to demonstrate the effectiveness of the proposed methods. Finally, Section 5 concludes this paper.

## 2. Biologically-Inspired Kalman Filter Based RBFNN Control

### 2.1. Kalman Filter Algorithm

This section aims to describe the Kalman filter algorithm by which measurements are taken, and the state is estimated at discrete time points. The Kalman filter deals with the general problems encountered in estimating the state of a discrete-time controlled process, which is governed by the following state-space Equation (1) at time index $k$:

$$\begin{aligned} x(k) &= Ax(k-1) + BU(k-1) + w(k) \\ y(k) &= Cx(k) \\ z(k) &= Hy(k) + v(k) \end{aligned} \tag{1}$$

where $A$, $B$, and $C$ are matrices in the state-space Equation (2). $w(k)$ is the process noise and $v(k)$ is the measurement noise. $z(k)$ is the measured signal and $H$ is the sensor matrix. The probability of the process noise $w(k)$ is $p(w)$ and the probability of measurement noise $v(k)$ is $p(v)$. The process noise covariance of $p(w)$ is $Q$ and the measured noise covariance of $p(v)$ is $R$. In Kalman filtering, $p(w)$ and $p(v)$ are independent white noises with normal probability distributions, expressed by:

$$\begin{aligned} p(w) &\sim N(0, Q) \\ p(v) &\sim N(0, R) \end{aligned} \tag{2}$$

Figure 1 presents the structure of Kalman filter algorithm in which the estimated state $\hat{x}(k-1)$ and the error covariance $P(k-1)$ are included [13]. As shown in Figure 1, The Kalman filter algorithm consists of two phases: time update phase (predictor) and measurement update phase (corrector). The following summarizes the two important phases in classical Kalman filter algorithm.

1. Time update phase:
    a. At time step $k-1$, calculate $\hat{x}(k-1)$ and $P(k-1)$.
    b. Update the estimation of state vector $\hat{x}^-(k)$ and the estimation of error covariance matrix $\hat{P}^-(k)$.
2. Measurement update phase:
    a. Update the optimal gain $K(k)$ of Kalman filter.
    b. Update the estimation of state vector $\hat{x}(k)$ using $z(k)$, $\hat{x}^-(k)$ and $K(k)$.
    c. Update the estimation of error covariance matrix $p(k)$ by utilizing $K(k)$ and $P^-(k)$ for next iteration in the Kalman filter algorithm process.

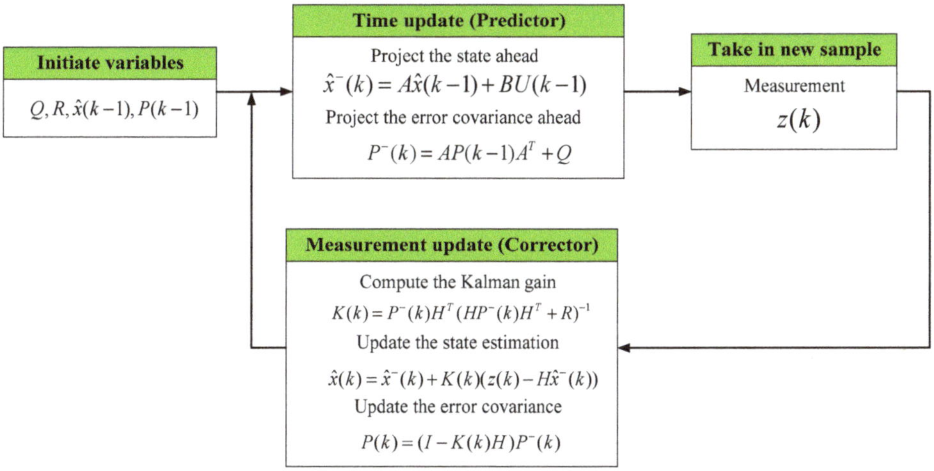


**Figure 1.** Structure of the classical Kalman filter.

### 2.2. Classical RBFNN

Figure 2 presents the structure of a classical RBFNN. This feed forward multilayer neural network has three layers, comprising the input layer, hidden layer and output layer. As shown in Figure 2, the inputs of the hidden layer in the RBFNN structure are the linear combinations of the weights and the input vector $[x_1\ x_2,\ x_3, \ldots \ldots,\ x_n]^T$. These vectors are then mapped by means of a radial basis functions in each node. Finally, the output layer of RBFNN generates a vector $y_p$ for $m$ outputs by linear combination of the outputs of the hidden nodes. This kind of artificial neural network has been regarded as a powerful tool that can approximate any continuous function with satisfactory accuracy [11]. In Figure 2, the output of the RBFNN is expressed by:

$$y_m = \sum_{j=1}^{m} w_j h_j \tag{3}$$

where $h_j$ is the radial basis vector. This vector is described by using the following Gaussian function:

$$h_j = \exp\left(\frac{-\|X - C_j\|^2}{2b_j^2}\right), j = 1, 2, \ldots, m \tag{4}$$

where $\|\bullet\|$ is the Euclidean norm operation, $C_j = [c_{j1}, c_{j2}, \ldots, c_{jm}]^T$ is the center vector of the $j$th node, $B = [b_1, b_2, \ldots, b_m]^T$ is the basis width vector. $W = [w_1, w_2, \ldots, w_m]^T$ is the weight vector in the RBFNN. Typically, this neural network is initialized with a randomly determined of RBFNN parameters, including $C_j$, $B$, and $W$.

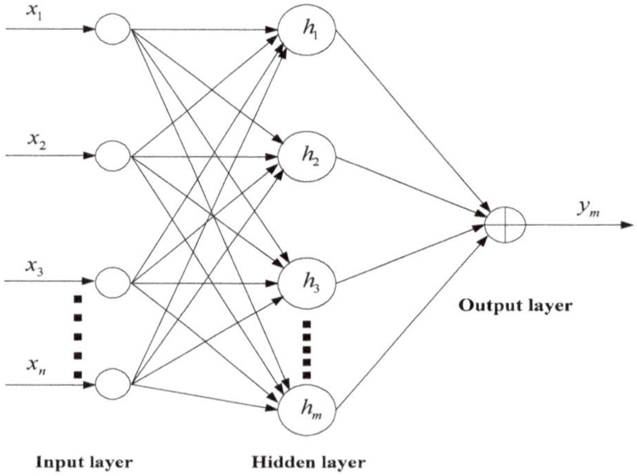


**Figure 2.** Structure of the classical RBFNN (Radial Basis Function Neural Network).

Gradient descent is an effective method for training RBFNN networks compared with other conventional training approaches [9,10]. In gradient descent training with one neuron in the output layer, the weights are updated at each time step by using the following rules:

$$w_j(k+1) = w_j(k) + \eta e(k) h_j(k), \tag{5}$$

$$C_{ji}(k+1) = C_{ji}(k) + \eta e(k) w_j h_j \frac{x_i(k) - C_{ji}(k)}{b_j^2(k)}, \tag{6}$$

$$b_j(k+1) = b_j(k) + \eta e(k) w_j h_j \frac{\|X(k) - C_j(k)\|^2}{b_j^3(k)}, \tag{7}$$

where $e(k)$ represents the error at the $k$th time step and $\eta$ denotes the learning rate. Since the initial parameters for an RBFNN using the gradient descent method are determined either by a trial-and-error approach or randomly set, convergence to a local optimum solution is inevitable [10]. To improve the learning performance of the RBFNN, this paper has developed a Kalman filter to train the RBFNN network structure based on a gradient descent approach. The proposed KF-RBFNN is applied to the self-learning control of networked mobile robots.

### 2.3. Kalman Filter Based RBFNN Control

Figure 3 depicts the block diagram of the RBFNN-based control. In Figure 3, the error between the real output $y(k)$ and the estimated output of the neural network $y_m(k)$ are considered to develop a self-learning RBFNN. The cost function or performance index is defined by the squared estimation error:

$$J(k) = \frac{1}{2}[y(k) - y_m(k)]^2. \tag{8}$$

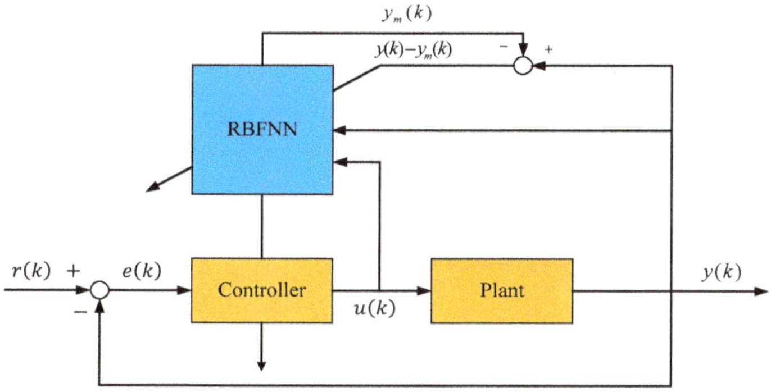


**Figure 3.** Block diagram of the RBFNN control scheme.

Considering the gradient descent method in Equations (5)–(7), the structure parameters: $w_j$, $C_{ji}$, and $b_j$ are updated online at every sampling point, expressed by:

$$\begin{aligned}
&\Delta w_j = [y(k) - y_m(k)]h_j \\
&w_j(k) = w_j(k-1) + \eta\Delta w_j + \varsigma\left[w_j(k-1) - w_j(k-2)\right] \\
&\Delta b_j = [y(k) - y_m(k)]w_j h_j \tfrac{\|X - C_j\|^2}{b_j^3} \\
&b_j(k) = b_j(k-1) + \eta\Delta b_j + \varsigma\left[b_j(k-1) - b_j(k-2)\right] \\
&\Delta c_{ji} = [y(k) - y_m(k)]w_j \tfrac{x_j - c_{ji}}{b_j^2} \\
&c_{ji}(k) = c_{ji}(k-1) + \eta\Delta c_{ji} + \varsigma\left[c_{ji}(k-1) - c_{ji}(k-2)\right]
\end{aligned}, \tag{9}$$

where $\varsigma$ denotes the momentum factor and $\eta$ is the learning rate of the neural network.

Based on Figure 3, the proposed KF-RBFNN control scheme depicted in Figure 4 considers the process noise $w(k)$ and measurement noise $v(k)$. In the proposed intelligent KF-RBFNN control scheme, the KF-RBFNN serves as an on-line learning and adapting mechanism in the intelligent controller. As shown in Figure 4, the effects of the process uncertainty $w(k)$ and the measurement noise $v(k)$ in the control scheme can be reduced via the implementation of the Kalman filter. To retrain the uncertainty and noise, the measured output $z(k)$ is employed to derive an estimation $\hat{y}(k)$ of the output $y(k)$ in the proposed KF-RBFNN. Both the control signal $u(k)$ and output $y_m(k)$ of the RBFNN are fed into the neural network for on-line learning. Moreover, the estimated output value $y_m(k)$ of the RBFNN is then utilized to update the control parameters to achieve self-learning control using the Kalman filter.

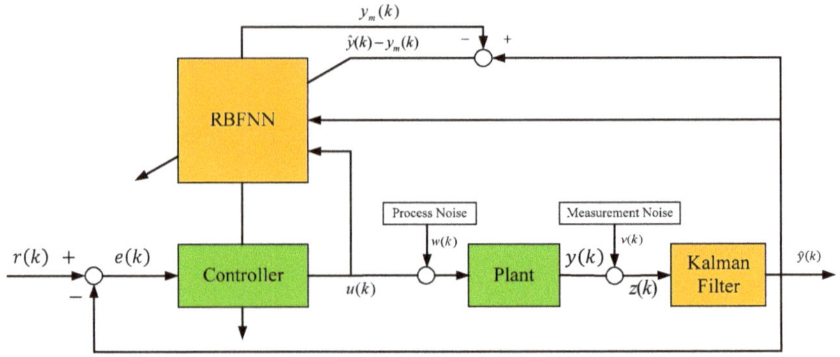


**Figure 4.** Block diagram of the KF-RBFNN control scheme.

## 2.4. Evolutionary KF-RBFNN Control

### 2.4.1. Modified GA with Lévy Flight

The GA developed by John Holland is a search algorithm that is inspired by Charles Darwin's theory of natural evolution. This evolutionary algorithm reflects the process of natural selection where the fittest individuals are selected for reproduction, thereby producing offspring of the next generation. This stochastic optimization technique starts with a set of solutions (chromosomes), called a population. This paradigm employs probabilistic rules to evolve a population from one generation to the next via the genetic operators: reproduction, crossover, and mutation. This paradigm is widely used to solve multidimensional optimization problems. When applying a GA to deal with optimization problems, an initial population of feasible solutions is generated. Each feasible solution is encoded as a chromosome string. These chromosomes are evaluated using a predefined fitness function or objective function based on the optimization problems. Figure 5 presents the flowchart of a GA. The initial population is randomly generated and the fitness function is defined before the GA evolutionary process begins. This study employs tournament selection, single-point crossover and single-point mutation strategies to develop a modified GA paradigm.

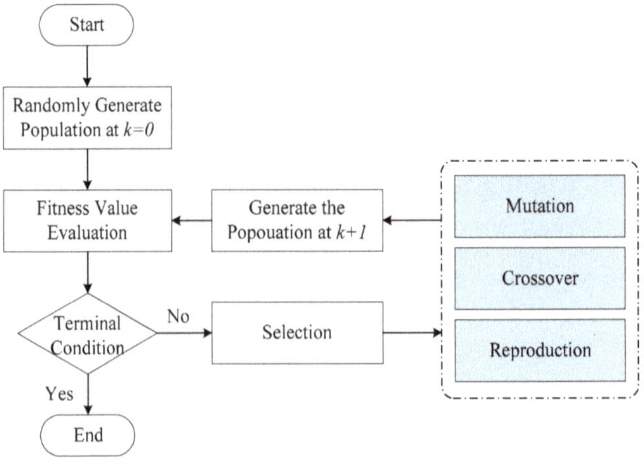


**Figure 5.** Flowchart of evolutionary GA.

Crossover and mutation are important operations for generating new individuals in the GAs. The performance of a GA is sensitive to the control parameter setting. The probabilities of crossover and mutation are significant parameters that influence the convergence performance of a GA. The use of unsuitable probability for crossover and mutation can result in poor convergence performance. More precisely, choosing suitable parameter values is a problem dependent task and requires previous experiences. Most studies adopt fixed crossover and mutation probabilities; this paper employs the Lévy flight which is a specialized random walk to increase the search diversity, expressed by:

$$Levy(\beta) = \left| \frac{\Gamma(\beta+1) \times \sin\left(\frac{\pi\beta}{2}\right)}{\Gamma\left(\frac{1+\beta}{2}\right) \times \beta \times 2^{\left(\frac{\beta-1}{2}\right)}} \right|^{\frac{1}{\beta}}, \tag{10}$$

where $\Gamma$ denotes the gamma function and $\beta$ is a constant $(1 < \beta \leq 3)$. The proposed modified GA is then applied to optimally set the initial parameters of the KF-RBFNN.

### 2.4.2. GA-Based KF-RBFNN

In the proposed KF-RBFNN, the accuracy and performance are influenced by the selections of the radial functions that are defined by a center vector $B$, width vector $C_j$, and weight vector $W$. In other words, proper tuning of these parameters is an important part of the optimal KF-RBFNN designs. This study employs the modified GA to develop a parameter tuning process that optimizes the KF-RBFNN. When applying the GA to address this issue, a chromosome is defined by the RBFNN parameter sequence $Chromosome = \{C_j, B, W\}$ and the optimal RBFNN structure can be evolved via the GA process with Lévy flight. The following fitness function (root mean square error, RMSE) for the $N_s$ sample is used to evaluate the GA chromosomes.

$$Fitness = \sqrt{\frac{1}{N_S}\sum_{k=1}^{N_s}\left(y_p{}^*(k) - y_p(k)\right)^2} \tag{11}$$

where $y_p(k)$ is the output and $y_p{}^*(k)$ is the predicted output at the $k^{\text{th}}$ sampling time. The following describes the GA process for KF-RBFNN optimization.

Step 1: Initialize the GA computing with Lévy flight.
Step 2: Each GA chromosome in the population contains genes to represent the KF-RBFNN parameters, meaning that $Chromosome = \{C_j, B, W\}$.
Step 3: Construct the KF-RBFNN using $Chromosome = \{C_j, B, W\}$ and evaluate the performance using the fitness function (11).
Step 4: Perform GA crossover and mutation with the probabilities set by Lévy flight.
Step 5: Update the GA population.
Step 6: Check the termination criterion. Go to Step 3 or output the optimized GA individual $Chromosome^* = \left\{C_j^*, B^*, W^*\right\}$ for the proposed GA-KF-RBFNN.

## 3. Application to Self-Evolving Control of Networked Mobile Robots

### 3.1. Modeling and Lyapunov-Based Control

Figure 6 depicts the geometrical structure of a mobile robot with four Swedish wheels for the proposed networked mobile robot system. Compared to the conventional differential-drive (non-holonomic) mobile robots, this kind of mobile robot with omnidirectional capability has superior mobility. The kinematic model of the four-wheeled Swedish mobile robot is expressed by:

$$\begin{bmatrix} v_1(t) \\ v_2(t) \\ v_3(t) \\ v_4(t) \end{bmatrix} = \begin{bmatrix} r\omega_1(t) \\ r\omega_2(t) \\ r\omega_3(t) \\ r\omega_4(t) \end{bmatrix} = T(\theta(t)) \begin{bmatrix} \dot{x}(t) \\ \dot{y}(t) \\ \dot{\theta}(t) \end{bmatrix}, \tag{12}$$

where:

$$T(\theta(t)) = \begin{bmatrix} -\sin(\delta+\theta) & \cos(\delta+\theta) & L \\ -\cos(\delta+\theta) & -\sin(\delta+\theta) & L \\ \sin(\delta+\theta) & -\cos(\delta+\theta) & L \\ \cos(\delta+\theta) & \sin(\delta+\theta) & L \end{bmatrix},$$

$\delta$ is $\pi/4$; $r$ represents the radius of the Swedish wheel; $L$ denotes the distance from the Swedish wheel's center to the geometric center of the mobile robot; $v_i(t)$ and $\omega_i(t), i = 1,2,3,4$ respectively denote the linear and angular velocities of each omnidirectional wheel. $[x(t)\ y(t)\ \theta(t)]^T$ is the pose vector that includes the position and orientation of the mobile robot measured at time $t$.

In mobile robotic research, robots with over three degrees-of-freedom (DOFs) are classified as redundant robots because they provide redundancy. Note that $T(\theta(t))$ in Equation (12) is singular for any $\theta$ in this redundant mobile robot system. This study adopts the pseudo inverse matrix approach to address the redundant control problem of mobile robots. Considering the left pseudo-inverse matrix $T^{\#}(\theta(t))$ of $T(\theta(t))$ by using $T^{\#}(\theta(t))P(\theta(t)) = I$, the matrix $T^{\#}(\theta(t))$ is expressed by:

$$T^{\#}(\theta(t)) = \begin{bmatrix} \frac{-\sin(\delta+\theta)}{2} & \frac{-\cos(\delta+\theta)}{2} & \frac{\sin(\delta+\theta)}{2} & \frac{\cos(\delta+\theta)}{2} \\ \frac{\cos(\delta+\theta)}{2} & \frac{-\sin(\delta+\theta)}{2} & \frac{-\cos(\delta+\theta)}{2} & \frac{\sin(\delta+\theta)}{2} \\ \frac{1}{4L} & \frac{1}{4L} & \frac{1}{4L} & \frac{1}{4L} \end{bmatrix}. \tag{13}$$

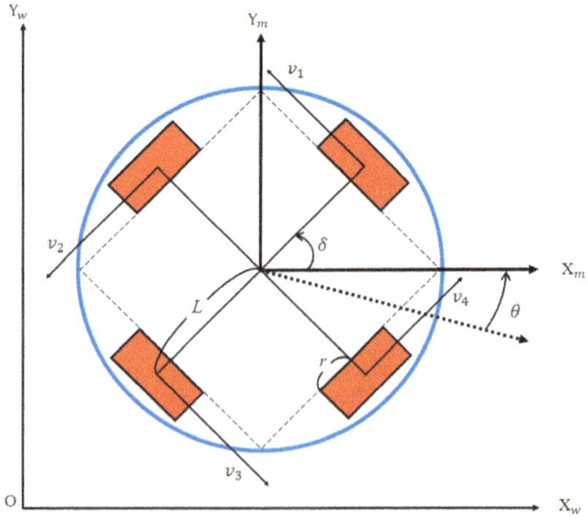


**Figure 6.** Geometry of the omnidirectional mobile robot with four Swedish wheels.

Combining Equations (12) and (13), the kinematics of the four-wheeled omnidirectional mobile robot is derived as follows:

$$\begin{bmatrix} \dot{x}(t) \\ \dot{y}(t) \\ \dot{\theta}(t) \end{bmatrix} = T^{\#}(\theta(t)) \begin{bmatrix} v_1(t) \\ v_2(t) \\ v_3(t) \\ v_4(t) \end{bmatrix}. \tag{14}$$

After the kinematics analysis of the Swedish mobile robots, the next step is to design a motion control law and prove its stability using Lyapunov theory. The current pose of the omnidirectional Swedish mobile robot at time $t$ is defined as $S = \begin{bmatrix} x(t) & y(t) & \theta(t) \end{bmatrix}^T$, and the desired reference trajectory of the Swedish mobile robot is expressed as $S_r = \begin{bmatrix} x_r(t) & y_r(t) & \theta_r(t) \end{bmatrix}^T$. With the two pre-defined vectors, the tracking error of the mobile robot is given by:

$$S_e = \begin{bmatrix} x_e(t) \\ y_e(t) \\ \theta_e(t) \end{bmatrix} = \begin{bmatrix} x(t) \\ y(t) \\ \theta(t) \end{bmatrix} - \begin{bmatrix} x_r(t) \\ y_r(t) \\ \theta_r(t) \end{bmatrix} = S - S_r, \tag{15}$$

which gives:

$$\dot{S}_e = \begin{bmatrix} \dot{x}_e(t) \\ \dot{y}_e(t) \\ \dot{\theta}_e(t) \end{bmatrix} = \begin{bmatrix} \dot{x}(t) \\ \dot{y}(t) \\ \dot{\theta}(t) \end{bmatrix} - \begin{bmatrix} \dot{x}_r(t) \\ \dot{y}_r(t) \\ \dot{\theta}_r(t) \end{bmatrix} = T^{\#}(\theta(t)) \begin{bmatrix} r\omega_1(t) \\ r\omega_2(t) \\ r\omega_3(t) \\ r\omega_4(t) \end{bmatrix} - \begin{bmatrix} \dot{x}_r(t) \\ \dot{y}_r(t) \\ \dot{\theta}_r(t) \end{bmatrix}. \tag{16}$$

The goal of control law design is to derive the angular velocity vector $\begin{bmatrix} \omega_1(t) & \omega_2(t) & \omega_3(t) & \omega_4(t) \end{bmatrix}^T$ for tracking the desired differentiable trajectory $\begin{bmatrix} x_r(t) & y_r(t) & \theta_r(t) \end{bmatrix}^T$ with asymptotical stability. Based on the PID (Proportional-Integral-Derivative) control strategy, the following redundant control law is proposed:

$$\begin{bmatrix} v_1(t) \\ v_2(t) \\ v_3(t) \\ v_4(t) \end{bmatrix} = T(\theta(t)) \left( -K_P \begin{bmatrix} x_e(t) \\ y_e(t) \\ \theta_e(t) \end{bmatrix} - K_I \begin{bmatrix} \int_0^t x_e(\tau)d\tau \\ \int_0^t y_e(\tau)d\tau \\ \int_0^t \theta_e(\tau)d\tau \end{bmatrix} - K_D \begin{bmatrix} \dot{x}_e(t) \\ \dot{y}_e(t) \\ \dot{\theta}_e(t) \end{bmatrix} + \begin{bmatrix} \dot{x}_r(t) \\ \dot{y}_r(t) \\ \dot{\theta}_r(t) \end{bmatrix} \right), \tag{17}$$

where $K_P, K_I$ and $K_D$ are the control matrices. They are diagonal and positive, thus $K_P = diag[k_{xp}\ k_{yp}\ k_{\theta p}]$, $K_I = diag[k_{xi}\ k_{yi}\ k_{\theta i}]$, and $K_D = diag[k_{xd}\ k_{yd}\ k_{\theta d}]$. By substituting Equations (17) into (16), the closed-loop error system is obtained:

$$\dot{S}_e = \begin{bmatrix} \dot{x}_e(t) \\ \dot{y}_e(t) \\ \dot{\theta}_e(t) \end{bmatrix} = \left( -K_P \begin{bmatrix} x_e(t) \\ y_e(t) \\ \theta_e(t) \end{bmatrix} - K_I \begin{bmatrix} \int_0^t x_e(\tau)d\tau \\ \int_0^t y_e(\tau)d\tau \\ \int_0^t \theta_e(\tau)d\tau \end{bmatrix} - K_D \begin{bmatrix} \dot{x}_e(t) \\ \dot{y}_e(t) \\ \dot{\theta}_e(t) \end{bmatrix} \right). \tag{18}$$

To prove the asymptotical stability of the closed-loop error system in (18) via Lyapunov theory, the following Lyapunov function is selected:

$$V(t) = \frac{1}{2} \begin{bmatrix} x_e(t) & y_e(t) & \theta_e(t) \end{bmatrix} \begin{bmatrix} x_e(t) \\ y_e(t) \\ \theta_e(t) \end{bmatrix} + \frac{1}{2} \begin{bmatrix} \int_0^t x_e(\tau)d\tau & \int_0^t y_e(\tau)d\tau & \int_0^t \theta_e(\tau)d\tau \end{bmatrix} K_I \begin{bmatrix} \int_0^t x_e(\tau)d\tau \\ \int_0^t y_e(\tau)d\tau \\ \int_0^t \theta_e(\tau)d\tau \end{bmatrix} + \frac{1}{2} \begin{bmatrix} x_e(t) & y_e(t) & \theta_e(t) \end{bmatrix} K_D \begin{bmatrix} x_e(t) \\ y_e(t) \\ \theta_e(t) \end{bmatrix} > 0$$

one obtains:

$$\dot{V}(t) = -\begin{bmatrix} x_e(t) & y_e(t) & \theta_e(t) \end{bmatrix} K_P \begin{bmatrix} x_e(t) \\ y_e(t) \\ \theta_e(t) \end{bmatrix} < 0.$$

Since $\dot{V}$ is negative definite, the asymptotical stability is therefore proven. The proposed motion control law can steer the mobile robot to achieve $S \rightarrow S_r$ as $t \rightarrow \infty$.

### 3.2. GA-KF-RBFNN Self-Learning Control

Figure 7 depicts the GA-KF-RBFNN self-learning control scheme, in which the noise from the process and measurement phases are included. As shown in Figure 7, the proposed GA-KF-RBFNN is employed to online adjust the parameters $K_P = diag[k_{xp}\ k_{yp}\ k_{\theta p}]$, $K_I = diag[k_{xi}\ k_{yi}\ k_{\theta i}]$, and $K_D = diag[k_{xd}\ k_{yd}\ k_{\theta d}]$ of the mobile robot. It is worthy to mention that the control matrices in Equation (17) are online adjusted at every sampling point to achieve tracking control. This GA-KF-RBFNN evolutionary online tuning method with noise reduction outperforms the traditional off-line and hand-tuning approaches.

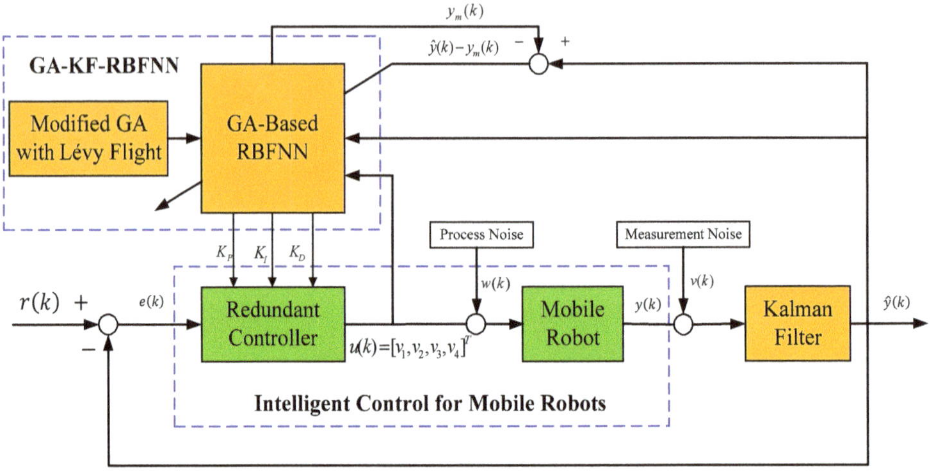


**Figure 7.** Block diagram of the GA-KF-RBFNN redundant control scheme for mobile robot.

In Figure 7, the control law $u(k) = [v_1, v_2, v_3, v_4]^T$ and the output $y_m(k)$ of the RBFNN are fed into the network for on-line learning. Moreover, the estimated output value $y_m(k)$ of the RBFNN is then employed to update the control matrices $K_P = diag[k_{xp}\ k_{yp}\ k_{\theta p}]$, $K_I = diag[k_{xi}\ k_{yi}\ k_{\theta i}]$, and $K_D = diag[k_{xd}\ k_{yd}\ k_{\theta d}]$ of the four-wheeled omnidirectional mobile robot to achieve the auto-tuning control with a Kalman filter.

### 3.3. Leader-Follower Formation Control of Networked Mobile Robots

The leader-follower model is the main trend of networked mobile robotics, where a leader robot and several follower robots are included in a multi-robot system [32]. For networked mobile robots, formation control is an important topic that the leader robot tracks the desired trajectory while the follower robots maintain the formation shape. Figure 8 depicts a leader-follower networked mobile robotic system to achieve triangular formation control with three robots.

In this paper, all mobile robots are independently controlled by using Equation (17) to accomplish leader-follower formation control, and the control parameters are self-tuned via the GA-KF-RBFNN paradigm. Compared to traditional consensus multiple robot systems with directed graph topology, the proposed broadcast leader-follower networked mobile robot system circumvents the delay problem. The position and orientation of the robots are broadcasted via the network. To maintain the desired formation shape, the geometrical relationship of the leader robot and follower robots in Figure 7 is calculated. Since the data flow is broadcasted online to every robot, the communication delay issues that occur in the consensus multiple robot system are therefore avoided. The proposed GA-KF-RBFNN self-evolving formation control for networked mobile robots not only reduces the system noises, but also avoids the communication delay.

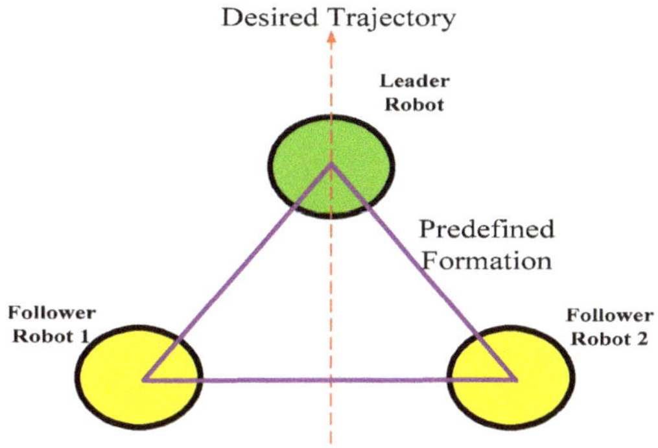


**Figure 8.** Leader-follower formation control with three mobile robots.

## 4. Simulations, Comparative Analysis, and Discussion

This section aims to conduct several simulations to examine the effectiveness of the proposed methods. The proposed networked mobile robot system consists of three four-wheeled Swedish omnidirectional mobile robots, including one leader and two follower mobile robots in a broadcast communication environment. The desired formation shape is triangular as shown in Figure 8. The system parameters in the proposed networked mobile robot system are $L$ = 0.25 m and $r$ = 5.08 cm.

The first simulation is conducted to demonstrate the performance of the circular formation control using the proposed GA-KF-RBFNN control approach. The number of iterations for the modified GA is 150, and the probabilities of crossover and mutation are determined by the Lévy flight. The circular trajectory for the leader robot is expressed as $\begin{bmatrix} x_r(t) & y_r(t) & \theta_r(t) \end{bmatrix}^T = \begin{bmatrix} 1.75\cos(\omega_i t)\text{ m} & 1.75\sin(\omega_i t)\text{ m} & \pi/4\text{ rad} \end{bmatrix}^T, \omega_i = 0.35$ rad/ sec. Figure 9 depicts the simulation result of the circular formation control. The three omnidirectional mobile robots are initially placed at different poses in the workspace. The desired trajectory for leader robot is a circular trajectory and the two follower robots aim to maintain a triangular formation. The tracking error of the leader mobile robot is presented in Figure 10. As shown in Figure 10, the leader mobile robot successfully tracks the desired circular trajectory in 4 s. Figure 11 depicts the formation error of follower robot #1 and Figure 12 depicts the formation error of follower robot #2. These simulation results demonstrate that the proposed GA-KF-RBFNN optimization is capable of accomplishing the self-learning formation control of networked mobile robotic systems.

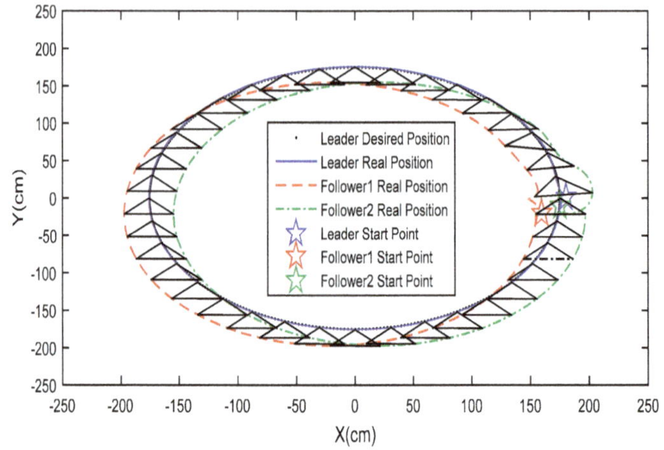


**Figure 9.** Simulation result of circular formation control.

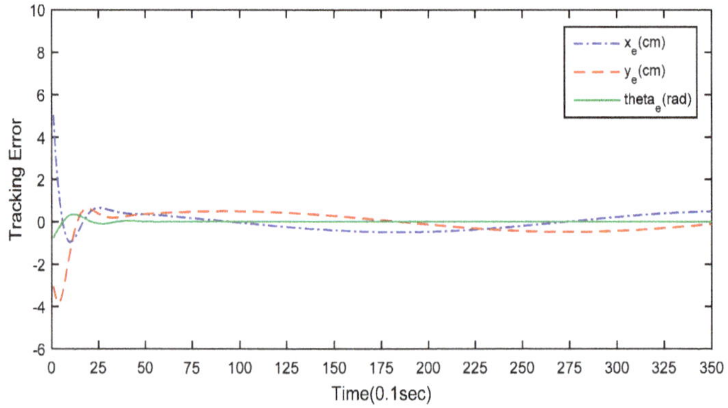


**Figure 10.** Tracking error of the leader mobile robot for circular formation control.

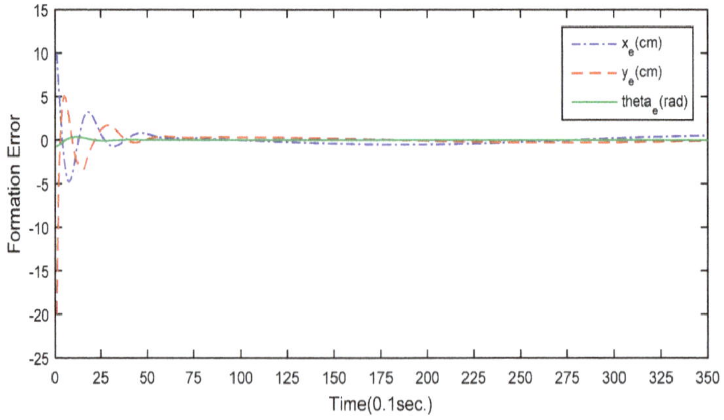


**Figure 11.** Formation error of follower robot #1 for circular formation control.

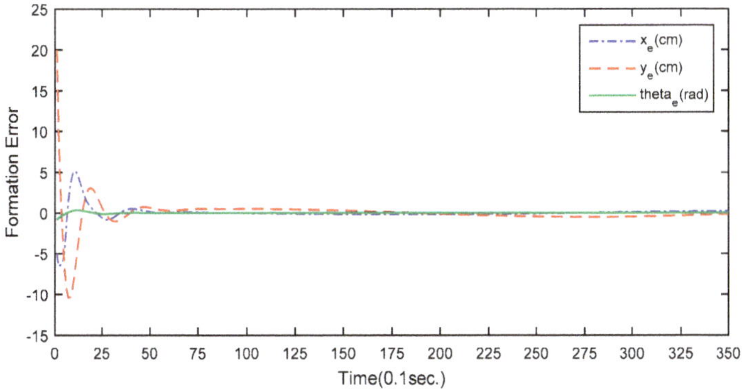


**Figure 12.** Formation error of follower mobile robot #2 for circular formation control.

The second simulation is provided to show the effectiveness of the proposed GA-KF-RBFNN formation control for daisy curve tracking. The desired trajectory of the leader robot is a special daisy curve with three petals, expressed by $\begin{bmatrix} x_r(t) & y_r(t) & \theta_r(t) \end{bmatrix}^T = \begin{bmatrix} \frac{1}{2} + b(a + r'\cos(p\omega_i t))\cos(\omega_i t)\ \text{m} & \frac{1}{2} + b(a + r'\cos(p\omega_i t))\sin(\omega_i t)\ \text{m} & \pi/4\ \text{rad} \end{bmatrix}^T$, $b = 0.2$ m, $a = 0.6$ m, $r' = 0.42$ m, $p = 3$, and $\omega_i = 0.35$ rad/sec. To illustrate the noise reduction capability of the proposed GA-KF-RBFNN self-learning controller, a Gaussian noise is added into the process and measurement phases. Figure 13 presents the simulation result of daisy curve formation control using the proposed GA-KF-RBFNN and Figure 14 depicts the tracking error of the leader robot. Moreover, Figures 15 and 16 present the formation error of follower robot #1 and follower robot #2, respectively, for daisy curve formation control. Both the tracking performance and formation behavior are guaranteed. These simulation results clearly indicate that the proposed metaheuristic GA-KF-RBFNN self-evolving control scheme with noise reduction achieves the formation control task for networked mobile robots. This approach outperforms the traditional consensus control methods where the uncertainty and self-adaptation are not considered.

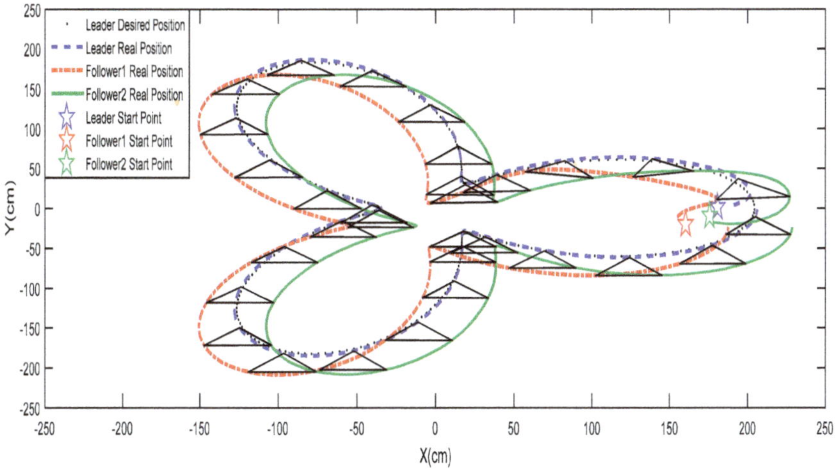


**Figure 13.** Simulation result for the daisy curve formation control.

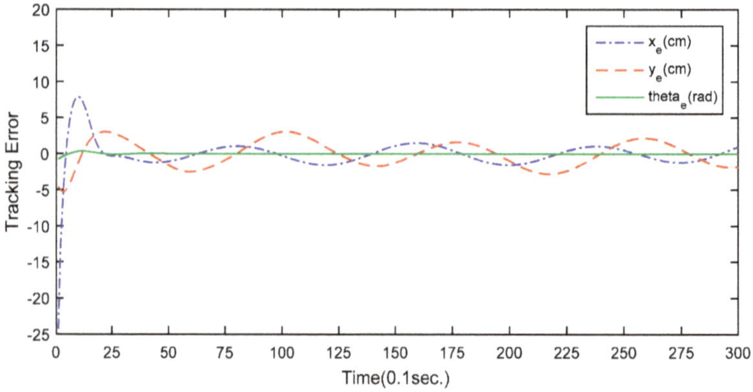


**Figure 14.** Tracking error of the leader mobile robot for daisy curve formation control.

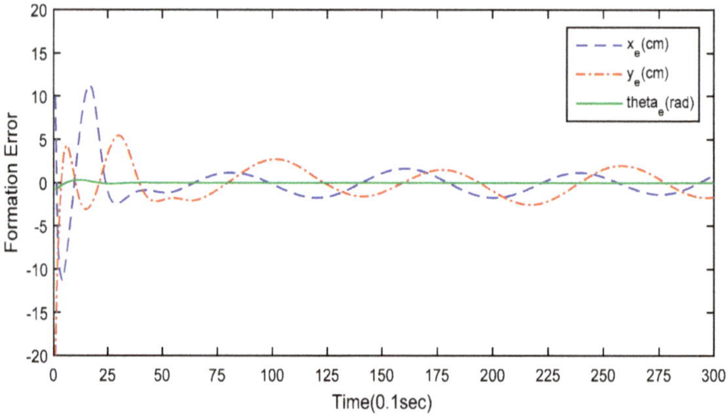


**Figure 15.** Formation error of follower robot #1 for daisy curve formation control.

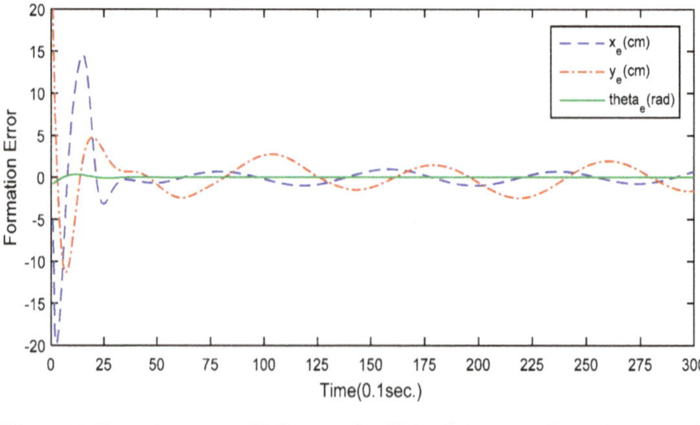


**Figure 16.** Formation error of follower robot #2 for daisy curve formation control.

In order to provide a comparative analysis to illustrate the merits of the proposed methods over other existing approaches, Table 1 lists the comparison of the proposed GA-KF-RBFNN formation control and traditional controllers. As shown in Table 1, the proposed biologically-inspired

GA-KF-RBFNN learning and adaptation for self-evolving of networked mobile robots is superior to the traditional formation control methods. With the modified GA metaheuristics and Kalman filter, both the noise and delay problem are avoided. The proposed GA-KF-RBFNN formation control strategy is applicable to all kinds of mobile robots, including different-drive [36] and three-wheeled mobile robotic systems.

**Table 1.** A comparative analysis of the formation controllers for networked robots.

| | Evolutionary Strategy | Noise Reduction | Distributed Formation | Leader-Follower Approach | Consensus Delayed Formation | Omnidirectional Capability |
|---|---|---|---|---|---|---|
| [1,2] | No | No | Yes | Yes | Yes | No |
| [3–6] | No | No | Yes | Yes | Yes | No |
| [7–9] | No | No | Yes | Yes | Yes | No |
| [10–13] | No | No | Yes | Yes | Yes | No |
| [31–34] | No | No | Yes | Yes | Yes | No |
| This Study | Yes | Yes | Yes | Yes | No | Yes |

## 5. Conclusions

This paper has presented a biologically-inspired GA-KF-RBFNN learning and adaptation method for the self-evolving control of networked mobile robots. The Kalman filter algorithm is employed to develop a self-learning RBFNN by considering uncertainty and noises. Moreover, the structure of the proposed KF-RBFNN is optimally initialized by means of the modified GA in which a Lévy flight strategy is applied. With the derived kinematic model of a four-wheeled omnidirectional mobile robot and broadcast leader-follower model, the GA-KF-RBFNN is utilized to design a self-evolving motion control law for a networked mobile robotic system. This approach overcomes the problem of communication delay found in conventional consensus networked robotic systems. Simulation results illustrate the merits of the proposed intelligent networked mobile robot system which uses a GA-KF-RBFNN to achieve self-learning formation control and consider the uncertainty and noise.

**Author Contributions:** Conceptualization, methodology, analysis, writing—review and editing, S.S.-D.X. and H.-C.H.; Experiments, T.-C.C. and S.-K.L.

**Funding:** This research was funded by the Ministry of Science and Technology, Taiwan, under grant number MOST 107-2221-E-011-145, MOST 106-2221-E-197-002 and MOST 107-2221-E-197-028.

**Conflicts of Interest:** The authors declare no conflict of interest.

## References

1. Liu, Y.; Gao, J.; Shi, X.; Jiang, C. Decentralization of virtual linkage in formation control of multi-agents via consensus strategies. *Appl. Sci.* **2018**, *8*, 2020. [CrossRef]
2. Keviczky, T.; Borrelli, F.; Fregene, K.; Godbole, D.; Balas, G. Decentralized receding horizon control and coordination of autonomous vehicle formations. *IEEE Trans. Control Syst. Technol.* **2008**, *16*, 19–33. [CrossRef]
3. Qian, D.; Xi, Y. Leader-follower formation maneuvers for multi-robot systems via derivative and integral terminal sliding mode. *Appl. Sci.* **2018**, *8*, 1045. [CrossRef]
4. Qu, Z.; Wang, J.; Hull, R.A.; Martin, J. Cooperative control design and stability analysis for multi-agent systems with communication delays. In Proceedings of the 2006 IEEE International Conference of Robotics and Automation, Orlando, FL, USA, 15–19 May 2006; pp. 970–975.
5. Kantaros, Y.; Zavlanos, M.M. Global planning for multi-robot communication networks in complex environments. *IEEE Trans. Robot.* **2016**, *32*, 1045–1061. [CrossRef]
6. Pan, L.; Lu, Q.; Yin, K.; Zhang, B. Signal source localization of multiple robots using an event-triggered communication scheme. *Appl. Sci.* **2018**, *8*, 977. [CrossRef]
7. Ferrari-Trecate, G.; Galbusera, L.; Marciandi, M.P.E.; Scattolini, R. Model predictive control schemes for consensus in multi-agent systems with single and double integrator dynamics. *IEEE Trans. Autom. Control* **2009**, *54*, 2560–2572. [CrossRef]

8. Dong, F.; Lei, X.; Chou, W. The adaptive radial basis function neural network for small rotary-wing unmanned aircraft. *IEEE Trans. Ind. Electron.* **2014**, *61*, 4808–4815.
9. Yang, C.; Li, Z.; Cui, R.; Xu, B. Neural network-based motion control of an underactuated wheeled inverted pendulum model. *IEEE Trans. Neural Netw. Learn. Syst.* **2014**, *25*, 2004–2016. [CrossRef] [PubMed]
10. Khosravi, H. A novel structure for radial basis function networks-WRBF. *Neural Process. Lett.* **2012**, *35*, 177–186. [CrossRef]
11. Dash, C.S.K.; Dash, A.P.; Cho, S.B.; Wang, G.N. DE+RBFNs based classification: A special attention to removal of inconsistency and irrelevant features. *Eng. Appl. Artif. Intell.* **2013**, *26*, 2315–2326. [CrossRef]
12. Chang, G.W.; Chen, C.I.; Teng, Y.F. Radial-basis-function-based neural network for harmonic detection. *IEEE Trans. Ind. Electron.* **2010**, *57*, 2171–2179. [CrossRef]
13. Tong, C.C.; Ooi, E.T.; Liu, J.C. Design a RBF neural network auto-tuning controller for magnetic levitation system with Kalman filter. In Proceedings of the 2015 IEEE/SICE International Symposium on System Integration (SII), Nagoya, Japan, 11–13 December 2015; pp. 528–533.
14. Hamid, K.R.; Talukder, A.; Islam, A.K. Implementation of fuzzy aided Kalman filter for tracking a moving object in two-dimensional space. *Int. J. Fuzzy Log. Intell. Syst.* **2018**, *18*, 85–96. [CrossRef]
15. Ko, N.Y.; Jeong, S.; Bae, Y. Sine rotation vector method for attitude estimation of an underwater robot. *Sensors* **2013**, *16*, 1213. [CrossRef]
16. Lu, X.; Wang, L.; Wang, H.; Wang, X. Kalman filtering for delayed singular systems with multiplicative noise. *IEEE/CAA J. Autom. Sin.* **2016**, *3*, 51–58.
17. Wang, S.; Feng, J.; Tse, C. Analysis of the characteristic of the Kalman gain for 1-D chaotic maps in cubature Kalman filter. *IEEE Signal Process. Lett.* **2013**, *20*, 229–232. [CrossRef]
18. Foley, C.; Quinn, A. Fully probabilistic design for knowledge transfer in a pair of Kalman filters. *IEEE Signal Process. Lett.* **2018**, *25*, 487–490. [CrossRef]
19. Xie, S.; Xie, Y.; Huang, T.; Gui, W.; Yang, C. Generalized predictive control for industrial processes based on neuron adaptive splitting and merging RBF neural network. *IEEE Trans. Ind. Electron.* **2019**, *66*, 1192–1202. [CrossRef]
20. Gutiérrez, P.A.; Hervas-Martinez, C.; Martínez-Estudillo, F.J. Logistic regression by means of evolutionary radial basis function neural networks. *IEEE Trans. Neural Netw.* **2011**, *22*, 246–263. [CrossRef] [PubMed]
21. Huang, F.; Zhang, W.; Chen, Z.; Tang, J.; Song, W.; Zhu, S. RBFNN-based adaptive sliding mode control design for nonlinear bilateral teleoperation system under time-varying delays. *IEEE Access* **2019**, *7*, 11905–11912. [CrossRef]
22. Zao, Y.; Zheng, Z. A robust adaptive RBFNN augmenting backstepping control approach for a model-scaled helicopter. *IEEE Trans. Control Syst. Technol.* **2015**, *23*, 2344–2352.
23. Tian, J.; Li, M.; Chen, F.; Feng, N. Learning subspace-based RBFNN using coevolutionary algorithm for complex classification tasks. *IEEE Trans. Neural Netw. Learn. Syst.* **2016**, *27*, 47–61. [CrossRef] [PubMed]
24. Chang, G.W.; Shih, M.F.; Chen, Y.Y.; Liang, Y.J. A hybrid wavelet transform and neural-network-based approach for modelling dynamic voltage-current characteristics of electric arc furnace. *IEEE Trans. Power Deliv.* **2014**, *29*, 815–824. [CrossRef]
25. Zhao, Y.; Ye, H.; Kang, Z.S.; Shi, S.S.; Zhou, L. The recognition of train wheel tread damages based on PSO-RBFNN algorithm. In Proceedings of the 2012 8th International Conference on Natural Computation, Chongqing, China, 29–31 May 2012; pp. 1093–1095.
26. Wei, Z.Q.; Hai, X.Z.; Jian, W. Prediction of electricity consumption based on genetic algorithm-RBF neural network. In Proceedings of the 2010 2nd International Conference on Advanced Computer Control, Shenyang, China, 27–29 March 2010; Volume 5, pp. 339–342.
27. Wei, H.; Tang, X.S. A genetic-algorithm-based explicit description of object contour and its ability to facilitate recognition. *IEEE Trans. Cybern.* **2015**, *45*, 2558–2571. [CrossRef] [PubMed]
28. Huang, H.C. FPGA-based hybrid GA-PSO algorithm and its application to global path planning for mobile robots. *Przeglad Elektrotechniczny* **2012**, *7*, 281–284.
29. Ding, S.F.; Xu, L.; Su, C.Y.; Jin, F.X. An optimizing method of RBF neural network based on genetic algorithm. *Neural Comput. Appl.* **2012**, *21*, 333–336. [CrossRef]
30. Chou, C.J.; Chen, L.F. Combining neural networks and genetic algorithms for optimising the parameter design of the inter-metal dielectric process. *Int. J. Prod. Res.* **2012**, *50*, 1905–1916. [CrossRef]

31. Marshall, J.A.; Broucke, M.E.; Francis, B.A. Formations of vehicles in cyclic pursuit. *IEEE Trans. Autom. Control* **2004**, *49*, 1963–1974. [CrossRef]
32. Oh, K.K.; Park, M.C.; Ahn, H.S. A survey of multi-agent formation control. *Automatica* **2015**, *53*, 424–440. [CrossRef]
33. Alonso-Mora, J.; Baker, S.; Siegwart, R. Multi-robot navigation in formation via sequential convex programming. In Proceedings of the IEEE/RSJ International Conference on Intelligent Robots and Systems, Hamburg, Germany, 28 September–2 October 2015.
34. Lee, H.C.; Roh, B.S.; Lee, B.H. Multi-hypothesis map merging with sinogram-based PSO for multi-robot systems. *Electron. Lett.* **2016**, *52*, 1213–1214. [CrossRef]
35. Tsai, C.C.; Chen, Y.S.; Tai, F.C. Intelligent adaptive distributed consensus formation control for uncertain networked heterogeneous Swedish-wheeled omnidirectional multi-robots. In Proceedings of the Annual Conference of the Society of Instrument and Control Engineers of Japan (SICE), Tsukuba, Japan, 20–23 September 2016; pp. 154–159.
36. Tiep, D.K.; Lee, K.; Im, D.Y.; Kwak, B.; Ryoo, Y.J. Design of fuzzy-PID controller for path tracking of mobile robot with differential drive. *Int. J. Fuzzy Log. Intell. Syst.* **2018**, *18*, 220–228. [CrossRef]

applied sciences

MDPI

Article

# Robot Swarm Navigation and Victim Detection Using Rendezvous Consensus in Search and Rescue Operations

**Gustavo A. Cardona [1] and Juan M. Calderon [2,3,*]**

[1] Department of Electrical and Electronics Engineering, Universidad Nacional de Colombia, Bogota 110131, Colombia; gacardonac@unal.edu.co
[2] Department of Electronic Engineering, Universidad Santo Tomás, Bogota 110131, Colombia
[3] Department of Computer Science & Engineering , Bethune-Cookman University, Daytona Beach, FL 32114, USA
* Correspondence: juancalderon@usantotomas.edu.co or calderonj@cookman.edu

Received: 31 December 2018; Accepted: 21 March 2019; Published: 25 April 2019

**Abstract:** Cooperative behaviors in multi-robot systems emerge as an excellent alternative for collaboration in search and rescue tasks to accelerate the finding survivors process and avoid risking additional lives. Although there are still several challenges to be solved, such as communication between agents, power autonomy, navigation strategies, and detection and classification of survivors, among others. The research work presented by this paper focuses on the navigation of the robot swarm and the consensus of the agents applied to the victims detection. The navigation strategy is based on the application of particle swarm theory, where the robots are the agents of the swarm. The attraction and repulsion forces that are typical in swarm particle systems are used by the multi-robot system to avoid obstacles, keep group compact and navigate to a target location. The victims are detected by each agent separately, however, once the agents agree on the existence of a possible victim, these agents separate from the general swarm by creating a sub-swarm. The sub-swarm agents use a modified rendezvous consensus algorithm to perform a formation control around the possible victims and then carry out a consensus of the information acquired by the sensors with the aim to determine the victim existence. Several experiments were conducted to test navigation, obstacle avoidance, and search for victims. Additionally, different situations were simulated with the consensus algorithm. The results show how swarm theory allows the multi-robot system navigates avoiding obstacles, finding possible victims, and settling down their possible use in search and rescue operations.

**Keywords:** swarm-robotics; rendezvous consensus; robot navigation; victim-detection

---

## 1. Introduction

Natural disasters and wars are some of the worst events that humanity has had and will have to face, since in these kinds of situations it is almost impossible to evacuate people in the affected area, causing many more deaths and having a devasting impact. However, the catastrophe is not the only problem, considering that after a catastrophe, the area is still dangerous due to issues such as landslides, debris, damage to the electrical system, and gases, among others. This is a reason rescue teams should be cautious because their lives are still at risk. It causes the survivor search to take longer losing valuable time, which is the most important resource in order to keep survivors alive. It opens the door to some researchers that have taken action in the matter. They involve robotics to help paramedics and rescuers in some tasks, improving their performance. For instance, in [1] they show some advantages, disadvantages, and difficulties that were observed in the use of different robotic platforms that performed SAR (Search And Rescue) tasks on the World Trade Center terrorist attack. This is a good benchmark for research related to this matter that is being developed in recent years and it is present in the robotics road-map.

Robotics is presented as a viable alternative thanks to the use of sensors and actuators which allow to minimize human failures. Taking this into account, some works [2–4] explore the idea of robotics platforms with a hybrid mechanics design. The proposed mechanism allows the robot to navigate on land and air avoiding get stuck in complex environments. Additionally, it is worth highlighting the amount of information that is required to be processed by a single robot in order to satisfying all tasks. Consequently, some researchers focus on the design of semi-autonomous robots as presented in [5,6]. On the other hand, multi-robot research has shown a different alternative to solve the aforementioned task, for instance, the advantages of using robotic swarms in exploration tasks is mentioned in [7,8]. Advantages such as coverage of a larger area in less time in contrast to the use of a single platform. In the same way, the robustness and flexibility of the robotic swarm is higher than a single robot, considering that if a robot from the swarm fails or get stuck, the operation and the success of a mission will not be compromised.

As was previously mentioned, robotics has become important in SAR operations. Taking into account that by using multi-robot systems some tasks such as exploration, mapping, search and location of victims and classification, among others can be optimized. Tasks related to the exploration of an unknown area become the first important step in wanting to save lives as well as mapping the area to locate victims. The work shown in [9,10] uses robot network to accomplish navigation and exploration tasks, making use of Voronoi tessellation, and graph theory. On the other hand, it is necessary to take into account the control law that allows a swarm to create attraction towards the victims and repulsion towards obstacles and robots in order to avoid collisions. This approach is achieved using attractive and repulsive potential functions as shown in [11], among other works that have been made to solve the different challenges that entail the exploration task.

Taking into account that it is difficult to have access to a real disaster scenario, it is worth mentioning that the present work is carried out by simulation. In robotics, simulation has become an essential tool for research, due to the fact that in recent years simulators have reached a high realistic level in comparison with the real environment as shown in [12]. On the other hand, simulators allow testing algorithms without putting people or equipment at risk. In addition, without losing performance if simulator keeps the necessary characteristics of the real world as shown in [13]. As previously mentioned, is difficult to be able of performing the algorithm in a real environment, in addition to this, having available a high number of drones with all the necessary sensors require a huge investment. Reasons why the present algorithm is carried out in V-Rep, which has a good performance in this kind of tasks that involve the search an rescue.

The main contributions of the present paper are threefold. First, the development of a navigation swarm system based on the artificial potential approach, which uses attraction and repulsion forces to keep swarm communication between agents and avoiding collisions, respectively. Additionally, an attraction force pointing to the location of a possible victim is added to the navigation system in the same way as proposed in our previous projects [14–16]. Second, the policy of sub-swarm generation when some robots detect a possible victim, through creating a weighted graph based on the distance in order to perform a k-nearest algorithm and then break the agents' links to the main swarm. Finally, with the generated sub-swarm is performed a control formation algorithm to put all of the agents close to the possible survivor and then apply a sensor network estimation consensus. The existence of a victim is determined by the sensor network consensus modeling each sensor with least-squares theory.

The other sections that complete this paper are organized as follows: Section 2, Related Work, in which the benchmark in robotics search and rescue is exposed. Section 3, Navigation Process, explains how the navigation algorithm is carried out using artificial potential functions and how the sub-swarm is generated.

Section 4, Victim Detection, shows how the formation control around the victim is implemented and then the consensus applied in the sensor network estimation. Section 5, Simulation and Results, shows how the theory was applied in a swarm of drones through V-rep simulator and finally, Section 6 corresponds to the document conclusions and some future work alternatives.

## 2. Related Work

Robotic swarms face a variety of challenges such as navigation, communication and collective decision making. These challenges generate in the swarm characteristics robustness and reliability to the system allowing to perform tasks such as exploration and location of targets. In navigation task, there is a great variety of algorithms such as presented in [17] which addresses the characteristics of different algorithms based on swarm intelligence found in nature, such as ant colony, swarm of bees, and wolf pack, among others. The characteristics of the different algorithms generate a great variety of possibilities to choose according to the needs of the challenge. In [18] they use two bio-inspired algorithms; The first algorithm is the particle swarm optimization algorithm, which has good results in exploration work, but sometimes the swarm stagnates in local minima. The second algorithm is the Darwinian particle swarm optimization algorithm, where the swarm is divided into small sub-swarms which share the best solution with the other sub-swarms avoiding being trapped in local minima.

As mentioned before, the communication in robotic swarm systems is a research topic that tries to solve problems such as noise due to interference in the environment, loss of information and limitations in the transmission distance and bandwidth of the system. In [19] a low resource consumption communication protocol called RRTLAN is presented which can be implemented in an 8-bit micro controller. An asynchronous communication protocol is shown in [20]; this protocol was tested on robotic platforms that search for a moving target using a multi-robot cooperation strategy. In [21] they use the electrical networks of a collapsed building as a communication channel for a cooperative robotic system that is performing exploration tasks. In [22] a communication protocol called Chopin is proposed and implemented, this protocol was implemented using Robotics Operating System (ROS) in real time. The communication process in a swarm system can be a problem depending on the algorithm implemented, considering that they are systems that manage a great amount of information.

Given that a robotic swarm system has a large amount of information coming from each agent of the swarm, a method is required to calculate a consensus value of the acquired information and then a decision is made collectively. In [23] they present a security system for the detection of the intruders which makes a consensus among the sensors located in common areas. In image sensors such as thermal cameras, wireless cameras and security cameras present a large amount of noise due to the environment and lighting, in [24] they use different data fusion techniques implemented in the pixels of the image captured from different cameras. This allows giving greater reliability of the information given by the image resulting from the fusion of the data. On the other hand, platforms designed for search and rescue work need portable energy sources such as batteries or renewable systems. Energy management is very important to allow the platform operates as long as possible without interruptions. In [25] an algorithm is proposed to find an optimal path between two points and using a cross-layer algorithm to improve energy performance. In [26] they propose a system where the transmission of information is not carried out without first reaching a consensus among the different agents. The preliminary consensus is made to eliminate noise faults or malfunction of some sensors and thus achieve energy savings since no redundant information is transmitted.

Finally, the victim's detection is the aim of this robotics application. The detection of survivors trapped under debris is a problem in which robotics can help search and rescue teams, unfortunately, there are scenarios that are not allowed to perform frequently tests to evaluate the performance of robotic systems. In [27] they present a model of the respiratory signal resulting from an ultra-wideband radar, which allows the detection of this signal through obstacles. In [28] they perform mathematical methods to separate the signals given by the ultra-wideband radar, allowing the identification of a person's vital signs. In [29] they present an app which uses the mobile phones of the rescue teams, allowing to record the sounds and estimate the victim's location through the use of a central processing system. Some of the applications of robotic swarms and cooperative robotics in search and rescue are presented in [30]. Some of the mentioned applications stand out; the creation of communication networks as in [31] where a swarm of drones equipped with WIFI and GPS modules is used to create a network, allowing the rescue team to coordinate the search of survivors in case the local communication system is out of service. Another application is the recognition of fire zones without compromising the integrity of firefighters. An example of terrain recognition is exposed in [32] where a 3-dimensional map is created by exchanging information between the platforms that perform the exploration of the environment.

In consequence of the nature of disasters, it is difficult to test robotics platforms that provide support to search and rescue tasks. In [33] they use V-REP for the creation of simulation environments, in addition to this, sensors and actuators can be integrated into predefined or user-created platforms. In [7] present an analysis of programs that allow the simulation of multi-robot systems for the development of bio-inspired algorithms. After carrying out the tests in simulation programs, it is necessary to test the prototypes in an environment with characteristics similar to those found in a disaster. In [34], the Department of National Security and the National Institute of Standards and Technology are coordinated to design three tests that will allow the evaluation of requirements identified by rescuers. In [35] they give standards for the elaboration of a scale scenario and thus be able to analyze the performance of the platforms using sensors and performance matrices. In [36] they develop a test scenario that emulates a disaster zone in which they carry out performance tests of a navigation algorithm in multi-robot systems.

As shown previously, Assistance in disaster areas is a fairly large and necessary field of research. The robots for search and rescue of victims have emerged as a part of the solution to several challenges that must be solved. When we look at the large number of researches that have been done on the subject, it seems that everything is done. However, our approach shows that are still details that can be improved such as, other works use the concept of swarm navigation but almost all of them focuses on the connectivity maintenance. In this work we allow the system to disconnect agents from the swarm depending on possible victim detection. In this way, once a possible victim is detected, a group of robots disconnects from the main swarm that keeps navigating. This sub-swarm determines if there is or not a victim based on a consensus algorithm. Additionally, other authors consider the detection of a victim based only on the performance of a sophisticated type of sensor. In this work, after applying a formation control to sub-swarm, we proposed a classification between victim or false positive based on consensus approach over a modeled sensor network using minimum least squares theory.

Finally, as a method to test the proposed swarm navigation system, an environment with obstacles, victims and aerial robotics were simulated using V-Rep [33]. This is virtual robotics environment tool that allows to proximate to the real implementation approach.

## 3. Navigation Process

In nature, there are several species that display swarm behavior, such as bacteria, insects, birds, fish, lions, hyenas, and horses among others. As can be expected, these groups are composed of individuals, who possess specific distinct capabilities and behaviors, but when they all get together as one group and operate as such, they will behave differently in order to work as a group. One of the more notable examples of this group behavior is a swarm of honey bees choosing a place to build a new honeycomb, this phenomenon was presented by Seeley et. al. in [37], whose paper showed that this swarm behavior, which takes place during the spring and summer, it is best exemplified when the colony outgrows its honeycomb and then proceeds to divide itself to find new grounds and then gets back together as a swarm once it finds the correct location to build a bigger hive. To be more specific, The new site selection begins with several hundred scout bees, who leave the colony in search of potential new sites, once the scouts find an appropriate candidate, they return to the hive in order to report with a wiggle dance where the potential new place is located. Once the scouts select the correct location from the pre-selected sites, they will steer the rest of the swarm to the new location by chemical stimulation. After the selection step is completed, the division process of the hive begins, and the old queen with nearly half of the colony leaves the hive to build the new one, while a younger queen stays with the other half of the colony in the old honeycomb.

It is also important to highlight that the ability of the swarm to navigate can be affected by a set of critical factors, such as collisions between themselves, large obstacles in the route, pheromone communication errors, erroneous scout indications and even poor sense of direction.

We can represent the swarm of bees using graph theory with the application of the concepts worked on by Mesbahi in [38]. One can describe the system by defining $N$ agents representing each member of the bee swarm, or in the case of our application a robot, and a connection representing the communication and transmission of information between the agents, either bees or robots. As mentioned by Mesbahi, we first need to define the agents of the system $(V, E)$, where $V$ is the set of nodes in the system and $E$ stands for the topology of the system, we also need to define $N(i)$ as the neighborhood or adjacent nodes to node $i$. Regarding the links between the agents, we can describe them as a dynamic system.

It is advantageous to apply graph theory to control swarm behavior, because it allows us to apply concepts such as the Laplacian matrix, which represents the system and allows the consensus analysis, connectivity grade calculation, and applying a control law for the whole system. In addition to that, we assume that the system has moderated sensor noise, links are bidirectional given the capability for any robot to share and receive information with the agents within own communication range. The design of the swarm navigation algorithm focuses on the generation of simple navigation rules that are applied to each of the agents individually. However, the sum of the individual decisions of the agents generates a collective behavior typical of a swarm. The applied criteria to the simple navigation algorithm of the swarm agents are based on Reynolds rules [39]. These rules settle down the base conduct for the generation of swarm behavior, such as (1) avoid collisions with neighbors, (2) match speed and direction with neighbors, and (3) stay close to neighbors.

Before establishing the interaction policy to fit the Reynolds' rules, it is necessary to define the agent concept and the kinematic model of the robotics agent used in the current work. Each agent is described as a point in the space with position and velocity. We assumed that each agent can detect the position and velocity of the rest of the swarm agents. The swarm agents interaction is represented by graphs $(V, E)$ where $V = \{1, 2, \ldots, N\}$ is a set of nodes and $E = \{(v_i, v_j) : i, j \in V, i \neq j\}$ represents the communication and sensing topology between the $i^{th}$ agent and each of the other $j^{th}$ swarm agents as depicted in Appendix A.

The dynamic model employed for the robot is a simple integrator model. Without loss of generality, since as exposed in [40] it is possible to impose integrator dynamics for multiple robot structures employing kinematic control

$$\dot{x}_i = u_i, \tag{1}$$

where $\dot{x}_i$ represents the linear speed of a robot and $u$ is the control signal applied to the system in order to impose the desired behavior. The navigation approach in this work consists of the displacement of a robot swarm through an environment $S \in \mathbb{R}^2$, in which, on the one hand, we have areas clear to move composed by the set $S_c$ and on the other hand, $S_o$ the areas with presence of obstacles that forbid the robots to move in it (Note that $(S = S_c \cup S_o)$.

*Interaction between swarm agents:* The interaction of the swarm agents is defined by the attraction and repulsion strategy allowing the navigation into the environment. This strategy allows the agents to keep a comfortable distance over the other swarm agents. The attraction parameter tries to maintain in close proximity every element of the swarm and thus gives the group a mechanism to remain grouped together. The attraction parameter is defined by (2)

$$-k_a x_{ij} \tag{2}$$

where $k_a$ is the attraction force and $x_{ij}$ is the distance between the agent $i$ to $j$, this mechanism can be local (with a restriction by the sensing range) or global (agents can move other agents from the group regardless of how far they are).

The repulsion parameter allows the agent to keep a comfortable distance from the other members of the swarm. This avoids collision between agents when the swarm is moving. The repulsion mechanism can respond in two different ways. First one, when a comfortable distance is reached and its parameter is expressed by (3)

$$[-k(\|x_{ij}\| - d)](x_{ij}) \tag{3}$$

where $\|x_{ij}\| = \sqrt[2]{(x_i - x_j)^T(x_i - x_j)}$, $k > 0$ is the repulsion magnitude and $d$ is the comfortable distance between $i^{th}$ agent and the $j^{th}$ agent. On the other hand, if two agents are two close to each other, the parameter is represented by (4).

$$k_r exp(\frac{-\frac{1}{2}\|x_{ij}\|^2}{r_s^2})(x_{ij}) \tag{4}$$

where $k_r > 0$, represents the repulsion magnitude and $r_s > 0$, is the repulsion range.

*Interaction between agents and obstacles:* The obstacles produce a repulsion force over the agents. This repulsion is modeled using (4) where $k_r > 0$, and $r_s$ is related to the size of the obstacle.

*Interaction between agents and victims:* The victims exert an attraction force, where $k_r < 0$ and $r_s$ is proportional to the victims density in the near area.

*Sub-Swarm Generation:* Every agent is exposed to different attraction and repulsion forces, which help to keep the swarm coherent and at the same time pushes the swarm towards the goal. The victims generate an attraction force over close agents and this force can disturb the normal behavior of the swarm and reduce the velocity of the agents in proximity to the victims. The $k_r$ magnitude was put in place so that it can stop the closest agents to the victims. When agents are navigating close to the victims, at least one agent must stop near the victims. Once at least one agent has stopped close to the victim and generated the alarm about victim detection, there is a relevant variation in the process that has to be exposed, which is the generation of a sub-swarm due to the localization of possible victim. The creation of a sub-swarm its generated by K-nearest neighbors approach. The K-nearest algorithm behaves as a classifier by selecting as his name indicates the $k$ closest robots $j$ to robot $i$. When the neighborhood has been selected we create a graph $G_{ss} = (V_{ss}, E_{ss})$ where $V_{ss}$ is composed by the neighbors of robot $i$ and $E_{ss}$ is constructed taking into account a weighted function $W_{ss} : E_{ss} \to \mathbb{R}$ in the following way $W_{ss} = \frac{1}{x_{ij}}$. Allowing in this way, that the closer robots have more relevance to the classification of robot $i$ between take itself to the sub-swarm or remain in the principal.

As can be seen from Figure 1 the sub-swarm process is given when some robots in the network detect a possible victim. In the instant when a victim is detected the main inconvenience is to decide which robots will remain at the swarm and which of them will generate a new one. Consequently with the K-nearest algorithm in Figure 1 is possible to appreciate when a non-filled dot representing the robot that will be classified as a member of the sub-swarm blue dots or not green dots. It is possible to note that he is assigned to the sub-swarm becoming to a blue dot and breaking communication links with the rest of the main swarm.

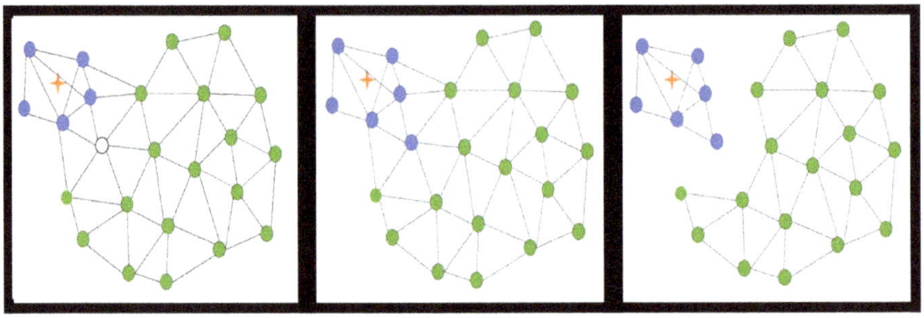

**Figure 1.** Sub-swarm generation.

## 4. Victim Detection

One of the most investigated approaches in robotics applied to the search and rescue missions is non-convex environment navigation including collision avoidance, as was shown in the previous section. Although navigation is really important in search and rescue operations, detection of a victim is the main task in the mission. This is the reason some techniques for the improvement of victims' detection using a mobile robot network have been addressed. Which are formation control and distributed estimation consensus, with the aim of improving the measurements reliability of victims' detection coming from the sensors, based on assigning a better sub-swarm location around the victim and classifying between victims and false positive carrying out a sensor network consensus, respectively. It is worth mentioning that each robot has a sensor capable of sensing if somewhere there is a possible victim, giving place to a sensor network system. In this way, it is not important for the purposes of this project to specify the type of sensor used in the model.

### 4.1. Formation Control

Taking into account, that sub-swarms were generated in the navigation process due to some robots from the robot network detect a possible victim through their sensors and decide to leave the main swarm to create a new sub-swarm. Consequently, an algorithm capable of relocating all the robots in the sub-swarm around to the possible victim location is carried out in order to improve measurements from the sensor network on the detection of a victim. Giving to the sensor network the ability to increase their probability of well-classifying and avoid that a far robot introduces noise in the consensus. The algorithm has the capability to bring robots closer to the victims using the rendezvous algorithm and then take agents to the perimeter of a circle making certain modifications to the rendezvous consensus algorithm exposed in [38]. In which the Laplacian matrix and the state vector are used, giving rise to the consensus equation that models the dynamics of the system, expressed in (5), and explained with more detail in Appendix A. Where the state vector contains the dynamical information of the agents.

$$\dot{x} = -Lx \tag{5}$$

According to [38] the consensus equation shown above will have a space of agreement $A = \{X \in R^n \mid x_i = x_j \forall i, j\}$ in which, when $\dot{x}_i = 0$ there will be a consensus. On the other hand, in the time domain, we have $x(t) = e^{-Lt}x_0$ in which, when the limit when $t$ tends to infinity is evaluated and performing an expansion in Taylor Series is possible to note that except for the first term all the others are zero. The first term turns out to be $x(t) = e^{-\lambda_1 t}((u_1)^T x_0)u_1$ letting that the first eigenvalue is zero because the whole exponential part tends to 1 which causes the following expression $((u_1)^T x_0)u_1$. Extending it to all the agents in the graph, we have that the consensus of a non-directed graph is the average of the initial conditions of the system, this means that the consensus can be written as follows

$$consensus = \sum \frac{x_{0i}}{n}. \tag{6}$$

Taking into account that this rendezvous consensus allows all agents to reach the same position in the space, the next step is to guarantee the connection between the agents while the signal control is modified to reach the goal position for each agent based on the approach shown in [11] which in order to solve the problem of maintenance communication, artificial potential functions was performed used to guarantee connectivity and keeping a distance among agents with the aim to avoid collisions. The artificial potential functions are of the form,

$$\psi_{ij} = \frac{1}{\rho_2^2 - ||x_{ij}||^2} - \frac{1}{||x_{ij}||^2 - \rho_1^2}, \tag{7}$$

where $\psi_{ij}$ is the artificial potential function between the agents $i$ and $j$, $\rho_2$ corresponds to the radius of connectivity of the agents and $x_{ij}$ corresponds to the distance between the agents $i$ and $j$. From these functions, the dynamics of the agents is imposed by $\dot{x}_i = -\sum_{j \in N_i^\sigma} \nabla_{x_i} \psi_{ij}$, note that this function tends to infinite when the distance between the agents tends to $\rho_2$. In fact, the more the distance between agents increases, the more the intensity of the attraction among the robots increments to avoid communication breaks.

The artificial potential functions approach demonstrates connectivity maintenance by evaluating the total energy of the graph connections given by

$$\psi = \frac{1}{2} \sum_{i=1}^{n} \psi_i = \frac{1}{2} \sum_{i=1}^{n} \sum_{j \in N_i^\sigma} \psi_{ij}, \tag{8}$$

where $\psi$ is the total energy of all the links and $\psi_i$ corresponds to the energy of all the links concerning the agent $i$. From this total energy, it is ascertained whether it is growing, maintaining or reducing with time from its derivative with respect to time according to

$$\dot{\psi} = \frac{1}{2} \sum_{i=1}^{n} \sum_{j \in N_i^\sigma} \dot{x}_{ij}^T \nabla_{x_i} \psi_{ij}, \tag{9}$$

demonstrating that the total energy of the links tends to decrease over time and since the energy of the links grows due to the loss of connectivity, it is also demonstrated that the links are maintained. As expected the cooperative behavior of agents in a sub swarm converges to the desired location and then they spread out around the target in a circle formation as shown as follow in Figure 2.

Consequently, as can be seen in Figure 11 agents that have been chosen from the main swarm to generate a sub-swarm do not have an order in space and some of them can be far from the possible victim. The fact that some robots could be far from the victim can cause misdetection due to a possible error in measurements from the sensors produced by noise. As a result of possible errors coming from long distances between robots and the victim, a formation control is proposed to resolve this inconvenient by performing a position consensus in the sub-swarm, setting the possible victim as a target point. Once consensus has been reached, avoiding collisions among robots through a repulsion effect between each robot to the others, the next step is to perform a consensus in a circle shape around the victim, allowing in this way that each robot has approximately the same probability to well classifying. Taking into account that each robot from the sub-swarm has his own sensor to determine if there is or not detecting a victim, the next step is concerning to a distributed consensus in the sensor network that will be shown in the next subsection.

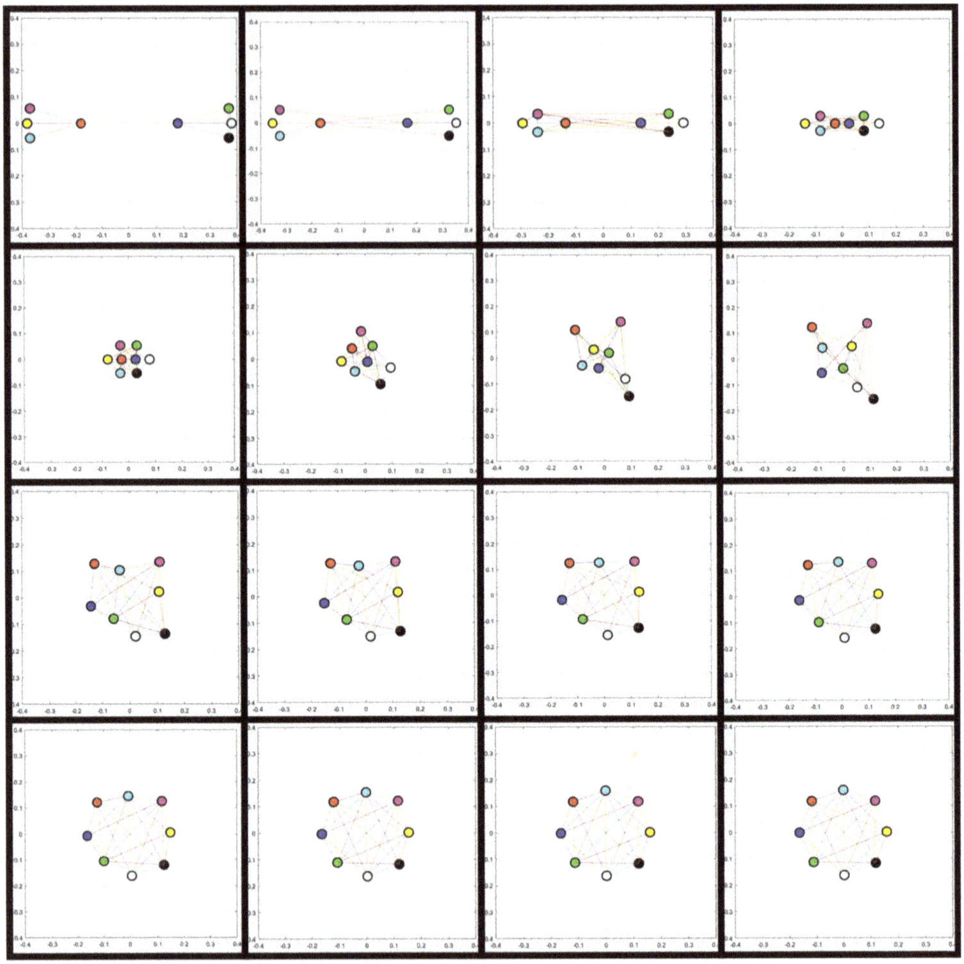

**Figure 2.** Consensus around a circle.

### 4.2. Distributed Estimation Consensus

Sensors are one of the most relevant parts for successful a victim detection process, due to that they are in charge of capturing all information from the environment. It is relevant to note that real sensors also capture noise in the measurements, which can come from electromagnetic noise, poor conditioning or even external factors depending on the type of sensors. This is the reason that in the present work, sensors are modeled using distributed linear least square with the presence of additive zero-mean Gaussian noise as shown in Figure 3 the sensing of a Gaussian signal without noise and in the presence of noise, respectively, based on the work presented in [38]. The underlying model involves the estimation of a linear function of a variable $\beta \in \mathbb{R}^q$ which is additively affected by noise $\varepsilon_i$ in each of the $n$ sensors on the network, how was previously mentioned,

$$\sigma_i = H_i\beta + \varepsilon_i, \tag{10}$$

where $\sigma_i$, $\varepsilon_i \in \mathbb{R}^{p_i \times 1}$ and $H_i \in \mathbb{R}^{p_i x q}$ in which each value of the vector $\sigma_i$ is a measurement channel of *ith* sensor, on the other hand, $H$ is assumed to be of rank $q$ which assures that the measurements are not entirely redundant.

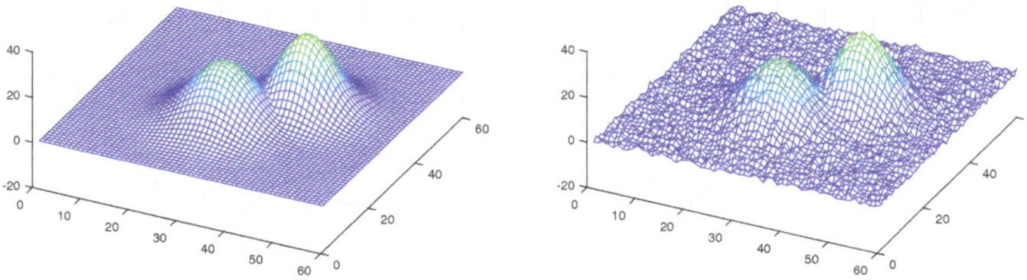


(**a**) Sensing without noise in the signal. (**b**) Sensing with noise in the signal.

**Figure 3.** Additive Gaussian noise applied to a measurements in sensors.

Taking into account that the aim of the Least Squares is to minimize the error, in the centralized case, we have by isolating it $\varepsilon = \sigma - H\beta$ and then applying the square error $\varepsilon^2 = \varepsilon\varepsilon' = (\sigma - H\beta)'(\sigma - H\beta)$ the result of which is $\varepsilon'\varepsilon = \sigma'\sigma - 2\sigma' H\beta + \beta' H' H\beta$which is a function $f(\beta)$ that depends only on $\beta$. In order to find the local or global minimums it is necessary to find the error gradient and the Hessian matrix, which in this case are $\bigtriangledown f(\beta) = \frac{\mathrm{d}f}{\mathrm{d}\beta} = -2H'\sigma + 2H'H\beta$, by isolating $\beta$ it is found a minimum when,

$$\hat{\beta} = (H'H)^{-1} H'\sigma, \tag{11}$$

additionally to verify that is a minimum function, it must satisfy $\frac{\mathrm{d}^2 f}{\mathrm{d}^2\beta} = 2H'H > 0$ taking as $u$ a non-null vector by multiplying $u'H'Hu > 0 \rightarrow (Hu)'(Hu) \rightarrow \|Hu\|^2 > 0$. It is possible to conclude that due to the Hessian matrix it is always positive because the function is convex, and the critical point is not just a minimum is the optimum.

Coming back to the distributed sensor network, it can be written as $\hat{\beta} = \left(\sum_{i=1}^{n} H_i' H_i\right)^{-1} \left(\sum_{i=1}^{n} H_i'\sigma_i\right)$ by solving the follow distributed optimization function $\begin{matrix} min \sum_{i=1}^{n} f_i(\beta) \\ s.t. \beta \in \mathbb{R}^q \end{matrix}$, as was shown previously each $f_i :: \mathbb{R}^q \rightarrow \mathbb{R}$ are convex function as a result the target function is given by the average of the gradient functions of each node as $f^* = \frac{1}{n}\sum_{j=1}^{n}(f_i)$, taking this into account, it can be evidenced that Equation (6) has the same form of the following

$$\hat{\beta} = \frac{1}{n}\sum_{j=1}^{n} \sigma_i, \tag{12}$$

concluding in this way that a consensus among the sensors can be achieved if some condition are satisfied, considering the iteration of a sensor as $\hat{\beta}_i(k+1) = \hat{\beta}_i(k) + \Delta \sum_{j \in N_i} w_{ij}(\hat{\beta}_j(k) - \hat{\beta}_i(k))$ in which $\hat{\beta}_i(k)$ illustrate the estimation of sensor $i$ of variable $\beta$ at time $k$,

$$\lim_{k \rightarrow \infty} \hat{\beta}_i(k) = \left(\frac{1}{n}\sum_{i=1}^{n} \sigma_i\right) \mathbf{1}, \tag{13}$$

if and only if the sensor network is connected and $\rho(L_w(G)) < \frac{2}{\Delta}$ where $\rho(L_w(G))$ correspond to the maximum eigenvalue of the graph in absolute values, then the agents on the network will converge to the average of its initial conditions as shown below in Figure 4.

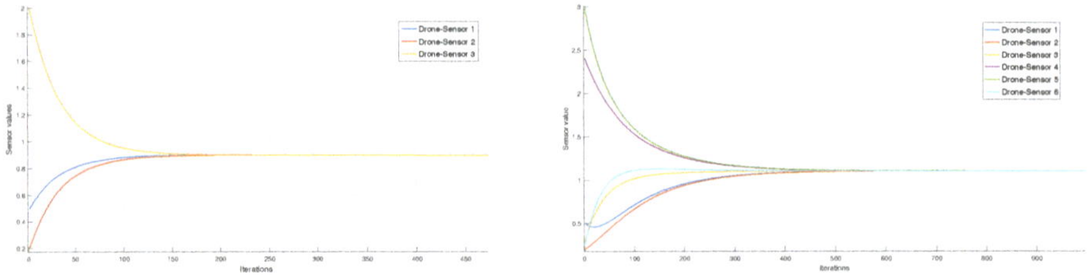


(**a**) Sensor consensus with three agents. (**b**) Sensor consensus with six agents.

**Figure 4.** Sensor network converging due to the consensus applied.

## 5. Simulation Results

In order to perform experiments for swarm navigation, sub-swarm generation, and victim detection, the multi-agent model previously explained was implemented using Matlab and VRep. The Matlab implementation was performed with the aim to evaluate the proposed mathematical model in different scenarios and behavioral cases as previously named. VRep was used to perform simulations of the proposed swarm behavior model over 3D environments with a virtual model of real robots as Drones. The principal objective of the VRep simulations is to validate the proposed system using a more realistic environment as shown in Figure 5.

**Figure 5.** Simulation in a realistic environment and robot platforms using V-Rep.

Six different types of experiments were developed to verify the proposed model in several situations of navigation, sub-swarm generation, and victim detection. Every experiment is shown in both simulation environments as named previously (Matlab and VRep). Both types of simulations were performed using 23 agents or Drones respectively. The trace left by the agent is shown as a solid line behind every agent or

Drone according to the case. The idea of those lines is to show the path followed by every swarm's robot and depict the collective behavior generated by the swarm computational model. In Matlab simulations, the green circles represent the initial locations of the swarm's agents, the red filled circle is the current position of the agent, and the red empty circles are a sample location in the agent path through the simulation time. Next, every experiment is explained in detail and the results are analyzed.

### *5.1. Obstacle Avoidance*

The first version of this experiment is performed using 23 agents and two cases of a single obstacle avoidance are performed. The first case is a small obstacle as shown in Figure 6. This part of the experiment depicts how the swarm goes around the obstacle in order to avoid it. In VRep simulation, the small obstacle is represented by a bunch of trees. This simulation shows how the obstacle is avoided by the swarm's drones flying around the trees. This is possible because the obstacle is relatively small and the agents are able to tolerate the obstacle between the attraction forces. The second version uses a big obstacle as depicted in Figure 7. In the second case, the agents avoid the obstacle by taking a side path. The big obstacle is represented by a building in VRep-simulation, where the swarm of drones navigates describing a side path to the building. This avoiding obstacle strategy occurs because there is not enough space between the attraction forces to allow broad obstacles to stay in between the agents.

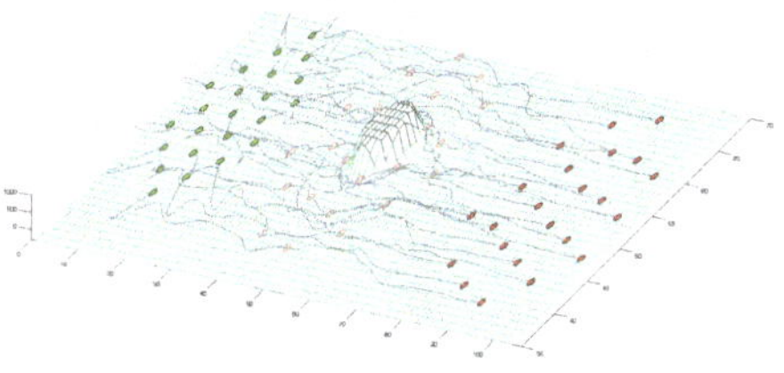

(**a**) Small-obstacle avoidance in Matlab.

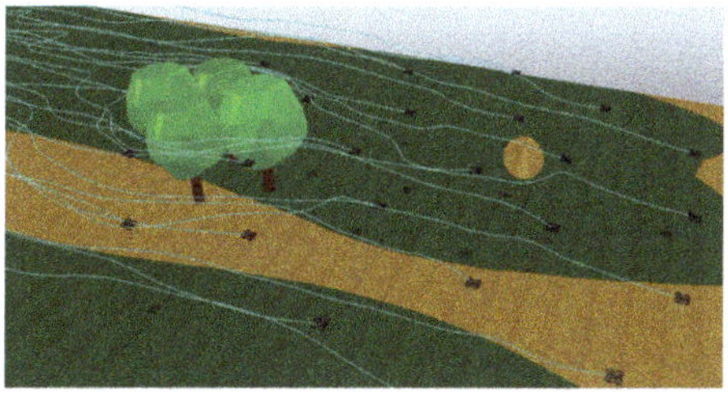

(**b**) Small-obstacle avoidance in a virtual environment.

**Figure 6.** Simulation of robots avoiding small obstacle.

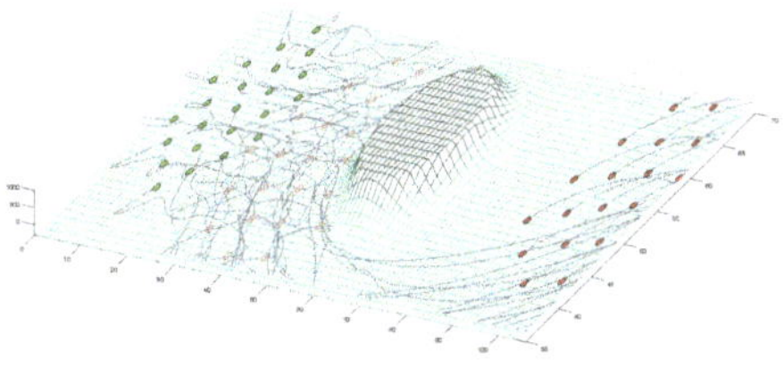

(**a**) Large-obstacle avoidance in Matlab.

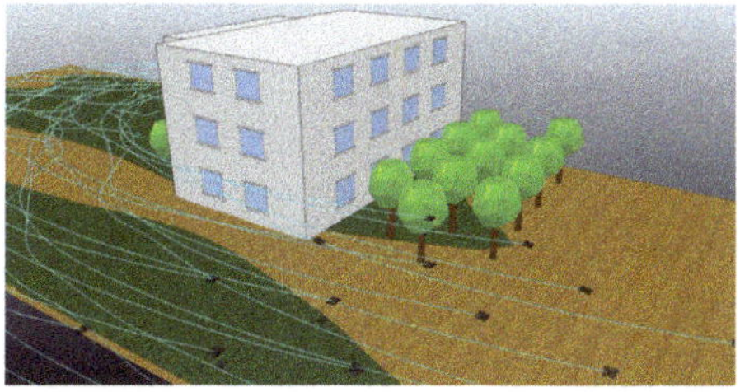

(**b**) Large-obstacle avoidance in a virtual environment.

**Figure 7.** Simulation of robots avoiding large obstacle.

### 5.2. Multiple Obstacles

This part of the experiment depicts how the swarm goes around several obstacles in order to avoid them. This is possible because the obstacles are relatively small and the agents are able to tolerate obstacles between the attraction forces, as depicted by Figure 8. This case uses nine obstacles distributed throughout the area between the start point and the goal point. For the virtual environment, obstacles are represented by nine trees distributed along the search area. It forces the swarm to navigate through obstacles while avoiding them and moving towards to the goal point. Figure 8 shows the agents navigating and exploring the zone. At the same time, the path described by every agent is drawn and it depicts the explored area. VRep Video simulation of the multiple obstacle avoidance in Ref. [41].

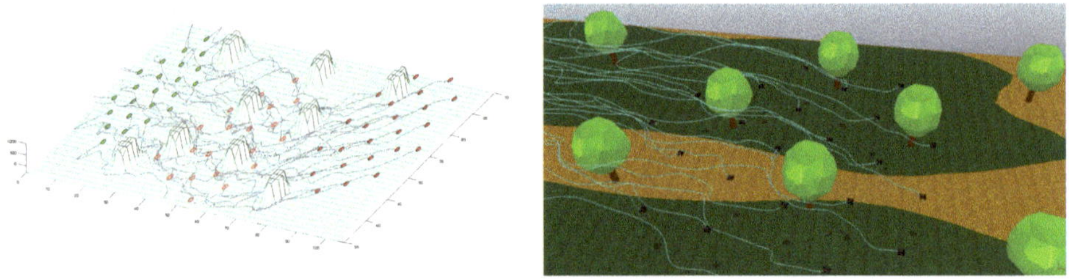

(**a**) Multi-obstacle avoidance in Matlab. (**b**) Multi-obstacle avoidance in a virtual environment.

**Figure 8.** Simulation of robots avoiding multi-obstacle scenario.

### 5.3. Victim Localization

This test uses a flat terrain to show how the swarm localizes victims. This process is accomplished with the use of the sub-swarm generation process. Once an agent has localized a victim, it stops near the potential victim. At the same time, the swarm stops, because of the attraction force between the agents. At that moment the process called "Sub-Swarm Generation" detaches the agents that found victims. The detached agents create a new swarm surrounding the victims. The original swarm restarts its movement towards the target point and leaves behind the swarm agents that are in charge of the victims. Figure 9 shows both simulation styles, where the agents stop near the victims and surround them, while the swarm leaves them behind.

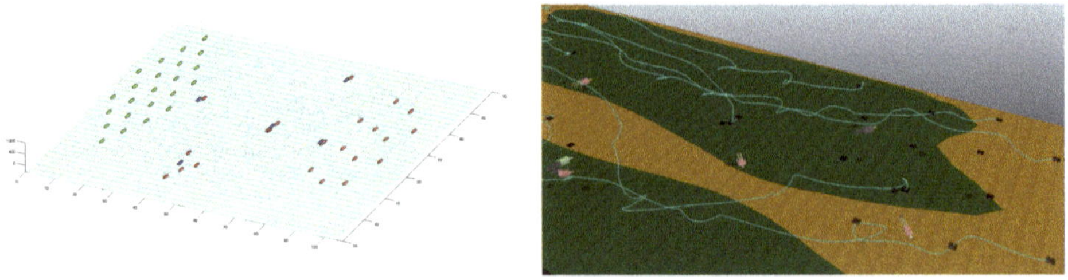

(**a**) Victim localization in Matlab. (**b**) Victim localization in a virtual environment.

**Figure 9.** Victim localization in convex space.

## 5.4. Navigation and Victim Localization

This is a complete case where the swarm navigates through the area full of obstacles and some places with potential victims. The experiment uses 23 agents and 6 victims distributed in 5 victim places, given there are 2 victims in the same place as shown by Figure 10 this case depict how the swarm covered the area navigating through obstacles and localizing potential victims at the same time. The simulation shows how the proposed model generates sub-swarms with more drones in places where there are more victims or where the probability of finding victims is greater. This case is represented in the virtual environment simulation, where there is a place with two victims and the model assigns a sub-swarm with 6 drones. This victims place has more assigned drones that the other places. VRep Video simulation of Navigation and Victim Localization in Ref. [42].

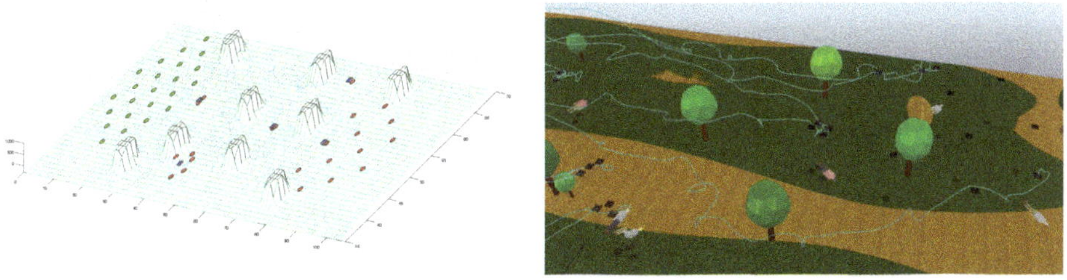

(**a**) Navigation and victim localization in Matlab. (**b**) Navigation and victim localization in a virtual environment.

**Figure 10.** Navigation and victims localization in environment with presence of obstacles.

## 5.5. Sub-Swarm Formation for Distributed Estimation Consensus

This experiment shows two cases of possible victims with different detection probability. The first case presents two victims in an open field with a high probability to be detected by the drone sensors as shown in Figure 11. Given the victim detection is easy for this case, the swarm is made up of 6 agents located in the vicinity of the victims. After sub-swarm creation, the consensus and control formation algorithm is performed as shown in Figure 11 as explained previously in this paper the consensus looks for to approach every drone to the victim, reducing the uncertainty of sensing factor over the possible victim. Additionally, the formation control redistributes the drones in a circle around the victim with the aim to provide better assistance to the victims and balance the sensor measurements. The second case presents a victim with the body partially covered and a sub-swarm of three agents. The number of agents for this case is lower than the first case given the difficulty of the victim detection. Once the victim is detected, the sub-swarm executes the processes of consensus and control formation as shown by Figure 12.

**Figure 11.** Formation control applied to 6 agents sub-swarm generated by localization of a victim.

**Figure 12.** Formation control applied to the 3 agents sub-swarm generated by localization of a victim.

### *5.6. Convergence Analysis of the Distributed Estimation Consensus*

In order to perform a convergence analysis, it is worth mentioning that the estimation algorithm was carried out in a computer with the following specs: 7-th generation core-i7 microprocessor, 32 GB installed memory (RAM) and NVIDIA GEFORCE 940$MX$. Several simulations were performed with the aim to know the algorithm performance under different parameter values. The principal parameter taken into account is the number of agents able to take a measurement, which in this case vary from 3 to 10. Additionally, in the interest to prove the robustness of the estimation algorithm, the measurement values are settled in a random way. In this way, the convergence of the algorithm depending on the number of agents is shown in Figure 13. Where it is appreciable how the convergence is affected by the number of agents. The greater number of agents giving measurements the more time it takes the algorithm to converge.

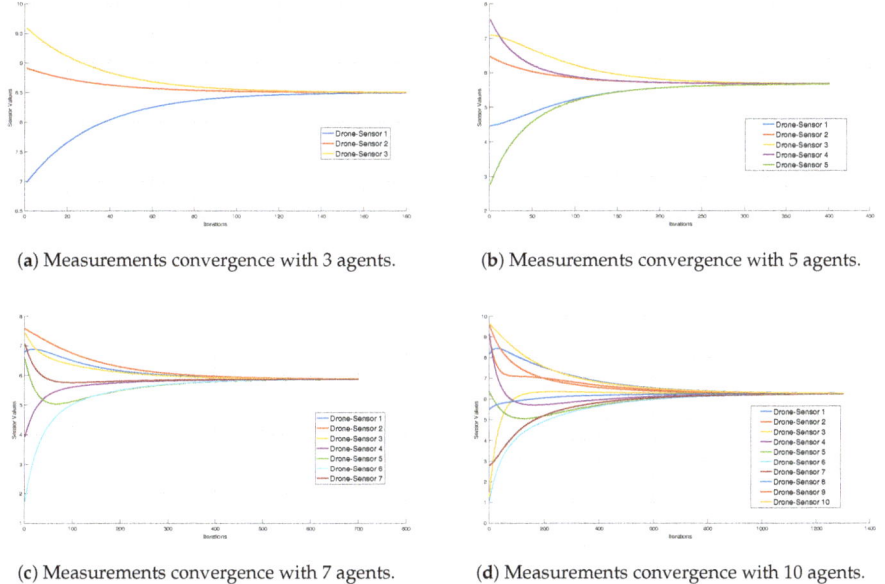

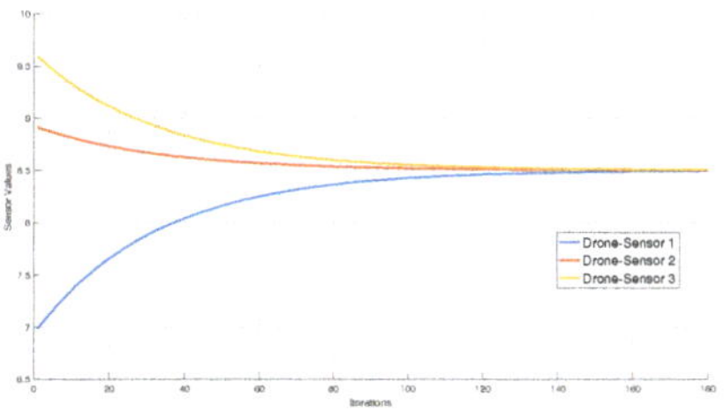


(**a**) Measurements convergence with 3 agents.

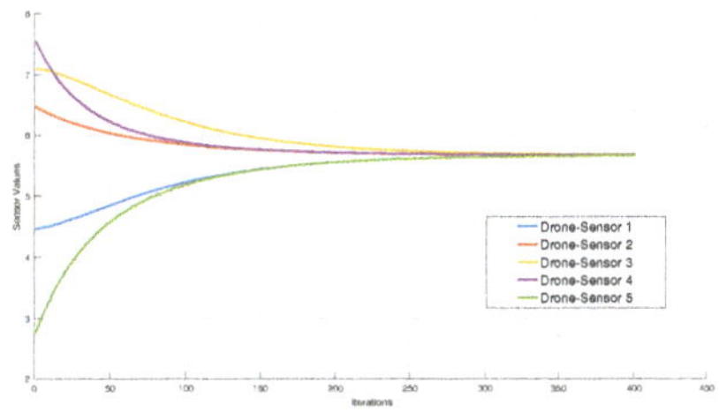


(**b**) Measurements convergence with 5 agents.

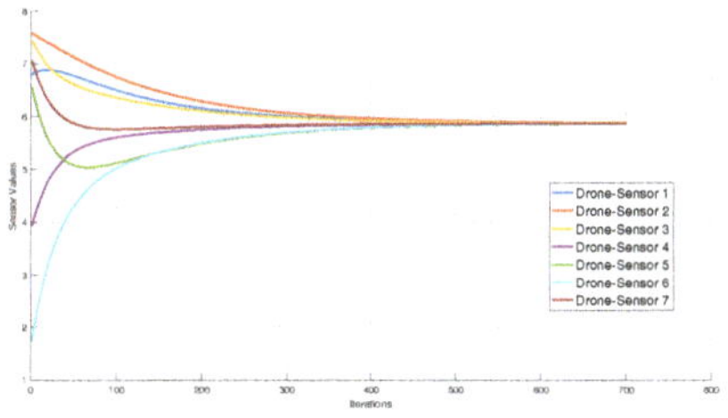


(**c**) Measurements convergence with 7 agents.

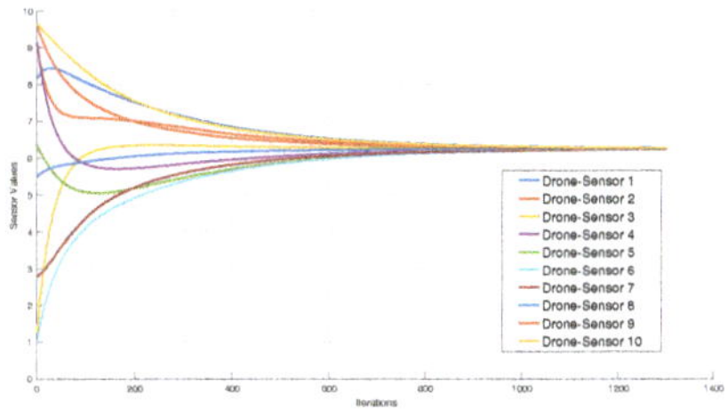


(**d**) Measurements convergence with 10 agents.

**Figure 13.** Convergence analysis with variation in the number of agents and random initial values.

As previously mentioned, several simulation scenarios were performed with the purpose of knowing how the variation of parameters can affect the algorithm convergence. In fact, 30 simulations were carried out for each variation in the number of agents (3, 5, 7 and 10). The results are summarized in Table 1, where is appreciable that by increasing the number of agents also increase the iterations number to converge in the same way the convergence time do. Taking this into account, the mean value $\mu$ and standard deviation $\sigma$ for each case were calculated. In which, the largest number of experiments converge between the mean value and the standard deviation, this corresponds to the approximately the 68 percent of the experiments converge within the interval $[\mu - \sigma, \mu + \sigma]$. The fact that the standard deviation are not large numbers shows that the algorithm is kind of robust under changes in its parameters.

**Table 1.** Convergence data depending on the number of agents.

| Agents Number | Average Convergence Iteration | Standard Deviation | Average Convergence Time |
|---|---|---|---|
| 3 | 176.6 | 8.3 | 90.6 (mS) |
| 5 | 404.9 | 22.4 | 103.9 (mS) |
| 7 | 602.6 | 39.4 | 144.5 (mS) |
| 10 | 1162.4 | 102.3 | 226.6 (mS) |

### *5.7. Effective Coverage Area*

An important parameter that should be taken into account is the effective coverage area. This parameter is extremely important due to developed tasks that are related to the search and localization of human beings in a disaster zone. The idea of swarm drone exploration is supported by the aim to inspect every small space into the disaster zone at least once by at least one drone. The most insignificant failure in the exploration can be translated into the loss of human life, so it is preferable that the same area within the disaster zone be inspected by more than one drone. Generating in this way more robustness to the task of search and localization, this search and rescue approach gives great importance to the use of a drone swarm, which seeks to increase the probability of finding victims. The effective coverage area is defined as a corridor through which the swarm travels. This corridor is as long as the area to explore and the width depends on the number of drones since the width of the area depends on the number of drones that are located across the width of the swarm. The width of the coverage area is directly related to factors such as the number of drones, comfortable distance between drones $(d)$ previously defined in (3), and the radius of sensors coverage $(r)$ used to perform the inspection and search tasks. In order to analyze the exploration degree of the covered zone, several experiments were carried out in which the parameter r was varied in relation to d as shown in Table 2.

The experiment aim is the percentage calculation of the unexplored and explored area by at least one, two, three, or more drones. Figure 14 shows the simulation of the experiment and the representation of different degrees of exploration for the areas covered by the drone in its trajectory towards the endpoint.

The Figure 15 shows different experimental cases where the covered area by the swarm is depicted by several gray color levels through the path to the endpoint. Figure 15 shows how explored area changes according to the relationship between $r$ and $d$. In Figure 15 from (a) to (d) is showed the covered area using a swarm of 23 agents. The covered area increases according to $r$ value approaches to $d$. Figure 15e,f shows how the explored area is affected by the swarm agent number. The covered area percentage in relation to the values of $r$ and $d$ is depicted by Table 2, which describes seven experiments where the rate between $r$ and $d$ is studied from $r = d/15$ to $r = d$. From Table 2 it is concluded that explored area increase significantly through the increase of the r factor. The equilibrium point is reached when $r = d/4$ because is in this point, where 100% and 80% of the entire area is explored by at least two drones.

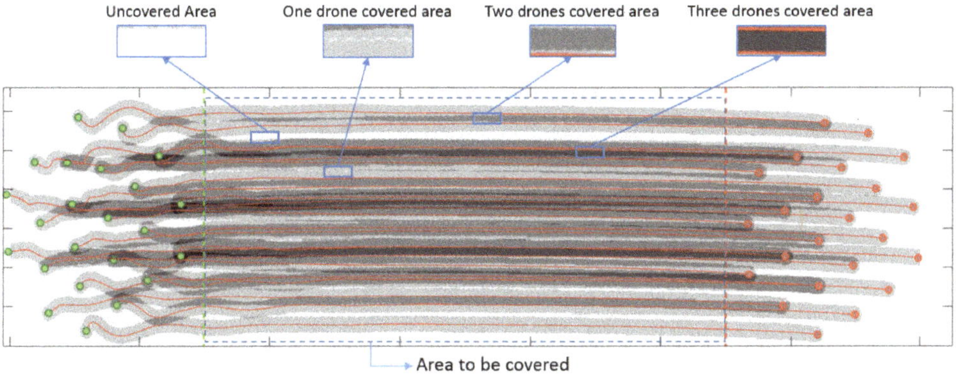


**Figure 14.** Formation control applied to the 3 agents sub-swarm generated by localization of a victim.

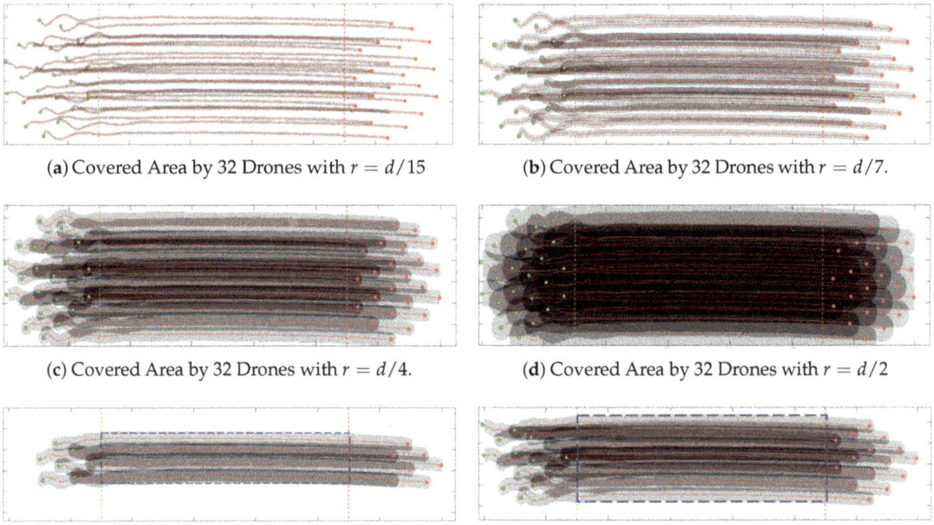

(**a**) Covered Area by 32 Drones with $r = d/15$
(**b**) Covered Area by 32 Drones with $r = d/7$.
(**c**) Covered Area by 32 Drones with $r = d/4$.
(**d**) Covered Area by 32 Drones with $r = d/2$
(**e**) Covered Area by 7 Drones with $r = d/4$.
(**f**) Covered Area by 15 Drones with $r = d/4$.

**Figure 15.** The covered area by a different number of drones with several relation values between $r$ and $d$.

**Table 2.** Percentage of the covered area according to different rate values between $r$ and $d$.

| r | Uncovered | Covered | +2 Drones | +3 Drones | +4 Drones | +5 Drones |
|---|---|---|---|---|---|---|
| d/15 | 43.10% | 56.89% | 9.55% | 0.0014% | 0.0% | 0.0% |
| d/8 | 13.51% | 86.48% | 42.01% | 4.35% | 0.0069% | 0.0% |
| d/5 | 1.73% | 98.26% | 73.28% | 24.57% | 3.06% | 0.0% |
| d/4 | 0.0% | 100% | 87.77% | 54.55% | 20.53% | 0.20% |
| d/3 | 0.0% | 100% | 94.34% | 80.14% | 52.72% | 14.56% |
| d/2 | 0.0% | 100% | 98.17% | 94.04% | 85.42% | 71.26% |
| d | 0.0% | 100% | 100% | 100% | 97.57% | 96.11% |

Table 3 shows how the areas explored by more than one drone increases as the number of drones increases. This result is quite logical, taking into account that increasing the number of drones also increases the number of rows in the swarm. In this case, more drones will follow roads similar to the path taken by the drones of the first rows and they will explore similar disaster zones. The relation between r and d was selected as $r = d/4$ with the aim to ensure the covered area close is to 100% and focuses the results on the percentage of areas explored by more than one drone.

**Table 3.** Percentage of the covered area according to different swarm size and $r = d/4$.

| Drone Number | Uncovered | Covered | +2 Drones | +3 Drones | +4 Drones | +5 Drones |
|---|---|---|---|---|---|---|
| 7 | 0.0% | 100% | 93.14% | 16.63% | 0.01% | 0.0% |
| 12 | 1.02% | 98.87% | 83.88% | 29.8% | 5.03% | 0.0% |
| 17 | 0.0% | 100% | 89.10% | 48.48% | 11.87% | 0.0% |
| 23 | 0.0% | 100% | 87.77% | 54.55% | 20.53% | 0.20% |

## 6. Conclusions and Future Work

An algorithm capable of performing some of the main tasks in search and rescue missions using aerial robotics has been proposed, based on swarm concepts. The algorithm uses the attraction and repulsion forces approach in order to keep the swarm as compact as possible or to be attracted by victims and at the same time avoid collisions or keep a minimum distance between agents, respectively. The algorithm uses the graph theory as the main tool to model and work with the robots' swarm. On this approach, the nodes represent the robots or sensor estimations and links are the distance between them or the weighted distance assigned to their location respect to the victim. Once that behaviors of navigation are achieved in the swarm, the next step is to create the policy that allows the swarm to generate a sub-swarm and disconnect the agents in the sub-swarm from the main one. As explained before, the generation of sub-swarm allows this algorithm to perform two tasks at the same time, which are the classification and assistance to a victim with the sub-swarm and continue the navigation seeking for more victims using the rest of the swarm.

As expected, the navigation process in a non-convex environment using artificial potential functions was successful as well as the implementation of sub-swarm generation in order to classify victims. The use of the k-nearest approach to a weighted graph has a good performance in the generation of sub-swarms close to a target location which could be a possible victim. The sub-swarm allows the system to perform a consensus algorithm at the same time that the other agents from the main swarm can still be navigating looking for more victims. This consensus in order to get closer to the victim and then through formation control surround it, improving in this way the probability of well victim detection. Once the robots get better location around the victim, they accomplish an estimator distributed consensus modeling the sensor by least square approach allowing sub-swarm to evaluate if they are sensing a victim or not.

Several experiments were carried out by using a robot simulator. This system allows using virtual models of real robots and the creation of environments with obstacles and people as victims whit-in. The simulation exposes how the algorithm generates a swarm behavior by allocating each robot behavior. The robots' swarm were capable of navigating without colliding with obstacles in the environment or with themselves and generates sub-swarms when they find a victim. Additionally, when a sub-swarm found a victim is showed how the formation control allows the robots to locate around the victims which is crucial in order to estimate if there is or not a victim in the place.

As future work, we intend to give more attention to the time-varying dynamical graphs in order to model extra behaviors concerning to the sub-swarm generation and the inclusion of heterogeneous agents. On the other hand, it is important to develop an algorithm that allows the system to get information from the detection of a victim, for instance, use of visual information to determine if the robot is in front of a victim. Taking this into account, the next step is to apply this information coming from the sensors in each robot close to the victim in the distributed estimation consensus algorithm exposed in this work. Finally, create an algorithm in charge of the global victim localization that allows the system to inform rescuers the exact localization of a victim.

**Author Contributions:** J.M.C. leads the research project. G.A.C. and J.M.C. developed the technical/theoretical approaches by equal parts.

**Funding:** This work has been funded by "Treceava Convocatoria Interna de Proyectos de Investigacion FODEIN 2019 #1936006" at Universidad Santo Tomas Colombia, entitled "Generación de algoritmos de navegación en enjambres de robots mediante el uso de aprendizaje por refuerzo para el desarrollo de tareas de búsqueda y rescate de víctimas humanas en zonas de desastres". and Department of Computer Science and Engineering of Bethune-Cookman University.

**Acknowledgments:** The authors express their thanks to all who participated in this research for their cooperation, especially thanks to Jose Leon for his help at the beginning of this research. The authors would like to give great thank to the hard work by the peer reviewers and editor.

**Conflicts of Interest:** The authors declare no conflict of interest.

## Appendix A. Basic Concepts of Graph Theory

The graph theory is a tool that allows working in distributed systems. This means it can help to model the multi-agent system dynamics, the reason is important to clarify some concepts used through this work. First of all, let us explain what a graph $G = (V, E)$ could be, It is a representation of agents that is composed of nodes and links, which represent the agents $V = \{v_1, v_2 \dots v_i\}$ and the connection or capability of communication among agents $E = \{v_1v_1, v_1v_2 \dots v_1v_j, v_2v_j \dots v_iv_j\}$respectively. In the same way, it is worth to explain concepts as Laplacian of a graph, adjacency and degree matrix. The Laplacian is composed by the operation between the adjacency matrix and the degree matrix how is shown in Equation (A1)

$$L(G) = D(G) - A(G), \tag{A1}$$

where the adjacency matrix, as its name indicates describes which nodes have a way of connection. for this case bidirectional, in the case of an undirected graph the way to fill this matrix is shown in Equation (A2) where $V_iV_j$ Represents a link among nodes $V_i$ and $V_j$ respectively. On the other hand, the degree matrix expresses in its main diagonal the number of connections that each node has, how is shown in Equation (A3).

$$[A(G)]_{ij} = \begin{cases} 1 & \text{if } V_iV_j \in E \\ 0 & \text{otherwise} \end{cases} \tag{A2}$$

$$D(G) = \begin{bmatrix} d_{in}(1) & & \\ & \ddots & \\ & & d_{in}k \end{bmatrix} \tag{A3}$$

Finally another way to represent the laplacian matrix of a directed weighted graph $L_w(G) = d_{lk}(G)W(G)(d_{lk}(G))'$ is using the concept of incidence matrix defined as Equation (A4) an the weighted matrix $W(G) = diag([w_1, \dots, W_n]')$ where $W_i > 0$ is the weight on the *ith* edge from the graph, indexed consistenly with the column ordering in the corresponding incidence matrix.

$$d_{lk}(G) := \begin{cases} +1 & \text{if } k \in_i^+ \\ -1 & \text{if } k \in_i^- \\ 0 & \text{otherwise} \end{cases} \tag{A4}$$

## References


1. Casper, J.; Murphy, R.R. Human-robot interactions during the robot-assisted urban search and rescue response at the world trade center. *IEEE Trans. Syst. Man Cybern. Part B (Cybern.)* **2003**, *33*, 367–385. [CrossRef] [PubMed]
2. Ventura, R.; Lima, P.U. Search and rescue robots: The civil protection teams of the future. In Proceedings of the 2012 Third International Conference on Emerging Security Technologies, Lisbon, Portugal, 5–7 September 2012; pp. 12–19.
3. Latscha, S.; Kofron, M.; Stroffolino, A.; Davis, L.; Merritt, G.; Piccoli, M.; Yim, M. Design of a Hybrid Exploration Robot for Air and Land Deployment (HERALD) for urban search and rescue applications. In Proceedings of the 2014 IEEE/RSJ International Conference on Intelligent Robots and Systems (IROS 2014), Chicago, IL, USA, 14–18 September 2014; pp. 1868–1873.
4. Seljanko, F. Low-cost electronic equipment architecture proposal for urban search and rescue robot. In Proceedings of the 2013 IEEE International Conference on Mechatronics and Automation (ICMA), Takamatsu, Japan, 4–7 August 2013; pp. 1245–1250.
5. Doroodgar, B.; Ficocelli, M.; Mobedi, B.; Nejat, G. The search for survivors: Cooperative human-robot interaction in search and rescue environments using semi-autonomous robots. In Proceedings of the 2010 IEEE International Conference on Robotics and Automation (ICRA), Anchorage, AK, USA, 3–8 May 2010; pp. 2858–2863.
6. Pfotzer, L.; Ruehl, S.; Heppner, G.; Rönnau, A.; Dillmann, R. KAIRO 3: A modular reconfigurable robot for search and rescue field missions. In Proceedings of the 2014 IEEE International Conference on Robotics and Biomimetics (ROBIO), Bali, Indonesia, 5–10 December 2014; pp. 205–210.
7. Tan, Y.; Zheng, Z.Y. Research advance in swarm robotics. *Def. Technol.* **2013**, *9*, 18–39. [CrossRef]
8. Husni, N.L.; Handayani, A.S.; Nurmaini, S.; Yani, I. Cooperative searching strategy for swarm robot. In Proceedings of the 2017 International Conference on Electrical Engineering and Computer Science (ICECOS), Palembang, Indonesia, 22–23 August 2017; pp. 92–97.
9. Yanguas-Rojas, D.; Cardona, G.A.; Ramirez-Rugeles, J.; Mojica-Nava, E. Victims search, identification, and evacuation with heterogeneous robot networks for search and rescue. In Proceedings of the 2017 IEEE 3rd Colombian Conference on Automatic Control (CCAC), Cartagena, Colombia, 18–20 October 2017; pp. 1–6.
10. Cardona, G.A.; Yanguas-Rojas, D.; Arevalo-Castiblanco, M.F.; Mojica-Nava, E. Ant-Based Multi-Robot Exploration in Non-Convex Space without Global-Connectivity Constraints. In Proceedings of the 2019 European Control Conference (ECC), Naples, Italy, 25–28 June 2019.
11. Zavlanos, M.M.; Egerstedt, M.B.; Pappas, G.J. Graph-theoretic connectivity control of mobile robot networks. *Proc. IEEE* **2011**, *99*, 1525–1540. [CrossRef]
12. Torres-Torriti, M.; Arredondo, T.; Castillo-Pizarro, P. Survey and comparative study of free simulation software for mobile robots. *Robotica* **2016**, *34*, 791–822. [CrossRef]
13. Zhang, F.; Leitner, J.; Milford, M.; Corke, P. Sim-to-real transfer of visuo-motor policies for reaching in clutter: Domain randomization and adaptation with modular networks. *World* **2017**, *7*, 8.
14. León, J.; Cardona, G.A.; Botello, A.; Calderon, J.M. Robot swarms theory applicable to seek and rescue operation. In Proceedings of the International Conference on Intelligent Systems Design and Applications, Porto, Portugal, 16–18 December 2016; Springer: Cham, Switzerland, 2016; pp. 1061–1070.
15. León, J.; Cardona, G.A.; Jaimes, L.G.; Calderon, J.M.; Rodriguez, P.O. Rendezvous Consensus Algorithm Applied to the Location of Possible Victims in Disaster Zones. In Proceedings of the International Conference on Artificial Intelligence and Soft Computing, Zakopane, Poland, 3–7 June 2018; Springer: Cham, Switzerland, 2018; pp. 700–710.
16. Quesada, W.O.; Rodriguez, J.I.; Murillo, J.C.; Cardona, G.A.; Yanguas-Rojas, D.; Jaimes, L.G.; Calderon, J.M. Leader-Follower Formation for UAV Robot Swarm Based on Fuzzy Logic Theory. In Proceedings of the International Conference on Artificial Intelligence and Soft Computing, Zakopane, Poland, 3–7 June 2018; Springer: Cham, Switzerland, 2018; pp. 740–751.

17. de Sousa Paula, P.; de Castro, M.F.; Paillard GA, L.; Sarmento, W.W. A swarm solution for a cooperative and self-organized team of UAVs to search targets. In Proceedings of the 2016 8th Euro American Conference on Telematics and Information Systems (EATIS), Cartagena, Colombia, 28–29 April 2016; pp. 1–8.
18. Couceiro, M.S.; Rocha, R.P.; Ferreira, N.M. A novel multi-robot exploration approach based on particle swarm optimization algorithms. In Proceedings of the 2011 IEEE International Symposium on Safety, Security, and Rescue Robotics (SSRR), Kyoto, Japan, 1–5 November 2011; pp. 327–332.
19. Edlinger, R.; Zauner, M.; Rokitansky, W. RRTLAN-A real-time robot communication protocol stack with multi threading option. In Proceedings of the 2013 IEEE International Symposium on Safety, Security, and Rescue Robotics (SSRR), Linkpoping, Sweden, 21–26 October 2013; pp. 1–5.
20. Wiltsche, C.; Lygeros, J.; Ramponi, F.A. Synthesis of an asynchronous communication protocol for search and rescue robots. In Proceedings of the 2013 European Control Conference (ECC), Zurich, Switzerland, 17–19 July 2013; pp. 1256–1261.
21. Pan, Q.W.; Lowe, D. Search and rescue robot team RF communication via power cable transmission line—A proposal. In Proceedings of the International Symposium on Signals, Systems and Electronics, ISSSE'07, Montreal, QC, Canada, 30 July–2 August 2007; pp. 287–290.
22. Araujo, F.; Santos, J.; Rocha, R.P. Implementation of a routing protocol for Ad Hoc networks in search and rescue robotics. In Proceedings of the 2014 IFIP Wireless Days (WD), Rio de Janeiro, Brazil, 12–14 November 2014; pp. 1–7.
23. Nurellari, E.; McLernon, D.C.; Ghogho, M. Distributed two-step quantized fusion rules via consensus algorithm for distributed detection in wireless sensor networks. *IEEE Trans. Signal Inf. Process. Netw.* **2016**, *2*. [CrossRef]
24. Ghassemian, H. A review of remote sensing image fusion methods. *Inf. Fusion* **2016**, *32*, 75–89. [CrossRef]
25. Misra, S.; Vasilakos, A.V.; Obaidat, M.S.; Krishna, P.V.; Agarwal, H.; Saritha, V. A fault-tolerant routing protocol for dynamic autonomous unmanned vehicular networks. In Proceedings of the 2013 IEEE International Conference on Communications (ICC), Budapest, Hungary, 9–13 June 2013; pp. 3525–3529.
26. Chelbi, S.; Duvallet, C.; Abdouli, M.; Bouaziz, R. Event-driven wireless sensor networks based on consensus. In Proceedings of the 2016 IEEE/ACS 13th International Conference of Computer Systems and Applications (AICCSA), Agadir, Morocco, 29 November–2 December 2016; pp. 1–6.
27. Li, X.; Qiao, D.; Li, Y.; Dai, H. A novel through-wall respiration detection algorithm using uwb radar. In Proceedings of the 2013 35th Annual International Conference of the IEEE Engineering in Medicine and Biology Society (EMBC), Osaka, Japan, 3–7 July 2013; pp. 1013–1016.
28. Li, J.; Liu, L.; Zeng, Z.; Liu, F. Advanced signal processing for vital sign extraction with applications in UWB radar detection of trapped victims in complex environments. *IEEE J. Sel. Top. Appl. Earth Obs. Remote Sens.* **2014**, *7*, 783–791.
29. Friedman, M.; Haddad, Y.; Blekhman, A. ACOUFIND: Acoustic ad-hoc network system for trapped person detection. In Proceedings of the 2015 IEEE International Conference on Microwaves, Communications, Antennas and Electronic Systems (COMCAS), Tel-Aviv, Israel, 2–4 November 2015; pp. 1–4.
30. Couceiro, M.S.; Portugal, D.; Rocha, R.P. A collective robotic architecture in search and rescue scenarios. In Proceedings of the 28th Annual ACM Symposium on Applied Computing, Coimbra, Portugal, 18–22 March 2013; pp. 64–69.
31. Alvissalim, M.S.; Zaman, B.; Hafizh, Z.A.; Ma'sum, M.A.; Jati, G.; Jatmiko, W.; Mursanto, P. Swarm quadrotor robots for telecommunication network coverage area expansion in disaster area. In Proceedings of the 2012 SICE Annual Conference (SICE), Akita, Japan, 20–23 August 2012; pp. 2256–2261.
32. Rocha, R.; Dias, J.; Carvalho, A. Cooperative multi-robot systems: A study of vision-based 3-d mapping using information theory. *Robot. Auton. Syst.* **2005**, *53*, 282–311. [CrossRef]
33. Rohmer, E.; Singh, S.P.; Freese, M. V-REP: A versatile and scalable robot simulation framework. In Proceedings of the 2013 IEEE/RSJ International Conference on Intelligent Robots and Systems (IROS), Tokyo, Japan, 3–8 November 2013; pp. 1321–1326.
34. Messina, E.R.; Jacoff, A.S. Measuring the performance of urban search and rescue robots. In Proceedings of the 2007 IEEE Conference on Technologies for Homeland Security, Woburn, MA, USA, 16–17 May 2007; pp. 28–33.

35. Chiou, A.; Wynn, C. Urban search and rescue robots in test arenas: Scaled modeling of disasters to test intelligent robot prototyping. In Proceedings of the UIC-ATC'09—Symposia and Workshops on Ubiquitous, Autonomic and Trusted Computing, Brisbane, Australia, 7–9 July 2009; pp. 200–205.
36. Saeedi, P.; Sorensen, S.A.; Hailes, S. Performance-aware exploration algorithm for search and rescue robots. In Proceedings of the 2009 IEEE International Workshop on Safety, Security & Rescue Robotics (SSRR), Denver, CO, USA, 3–6 November 2009; pp. 1–6.
37. Seeley, T.D.; Buhrman, S.C. Group decision making in swarms of honey bees. *Behav. Ecol. Sociobiol.* **1999**, *45*, 19–31. [CrossRef]
38. Mesbahi, M.; Egerstedt, M. *Graph Theoretic Methods in Multiagent Networks*; Princeton University Press: Princeton, NJ, USA, 2010.
39. Reynolds, C.W. Flocks, herds and schools: A distributed behavioral model. *ACM SIGGRAPH Comput. Gr.* **1987**, *21*, 25–34. [CrossRef]
40. Bullo, F.; Cortes, J.; Martinez, S. *Distributed Control of Robotic Networks: A Mathematical Approach to Motion Coordination Algorithms*; Princeton University Press: Princeton, NJ, USA, 2009; Volume 27.
41. Calderon, J. Mobile Robotics & Intelligent Systems. Robot Swarm Navigation-Obstacles. 2018. Available online: https://youtu.be/6B5TVmT8knI (accessed 0n 21 March 2019).
42. Calderon, J. Mobile Robotics & Intelligent Systems. Robot Swarm Navigation. 2018. Available online: https://youtu.be/F2tsg9jzIoY (accessed on 21 March 2019).

*applied sciences*

MDPI

*Article*

# Multi-Robot Trajectory Planning and Position/Force Coordination Control in Complex Welding Tasks

**Yahui Gan [1,2,*], Jinjun Duan [1,2], Ming Chen [1,2] and Xianzhong Dai [1,2]**

[1] School of Automation, Southeast University, Nanjing 210096, China; duan_jinjun@yeah.net (J.D.); chen_ming@yeah.net (M.C.); xzdai@seu.edu.cn (X.D.)

[2] Key Lab of Measurement and Control of Complex Systems of Engineering, Ministry of Education, Nanjing 210096, China

* Correspondence: ganyahui@yeah.net; Tel.: +86-158-5051-4395

Received: 12 December 2018; Accepted: 27 February 2019; Published: 5 March 2019

**Abstract:** In this paper, the trajectory planning and position/force coordination control of multi-robot systems during the welding process are discussed. Trajectory planning is the basis of the position/force cooperative control, an object-oriented hierarchical planning control strategy is adopted firstly, which has the ability to solve the problem of complex coordinate transformation, welding process requirement and constraints, etc. Furthermore, a new symmetrical internal and external adaptive variable impedance control is proposed for position/force tracking of multi-robot cooperative manipulators. Based on this control approach, the multi-robot cooperative manipulator is able to track a dynamic desired force and compensate for the unknown trajectory deviations, which result from external disturbances and calibration errors. In the end, the developed control scheme is experimentally tested on a multi-robot setup which is composed of three ESTUN industrial manipulators by welding a pipe-contact-pipe object. The simulations and experimental results are strongly proved that the proposed approach can finish the welding task smoothly and achieve a good position/force tracking performance.

**Keywords:** trajectory planning; position/force cooperative control; hierarchical planning; object-oriented; symmetrical adaptive variable impedance

## 1. Introduction

With the complication and diversification of industrial production tasks, multi-robot cooperative systems have demonstrated stronger operational capabilities, more flexible system structures, and stronger collaboration capabilities. Therefore, multi-robot collaboration has become an important challenge for robot control.

Multi-robot collaboration has been applied in many fields such as handing, assembly, welding, etc. The application of this paper is focused on the welding field. In arc welding, the traditional welding workstation consisting of "welding robot + positioner" is not able to meet the current demand for small-volume, customized, flexible, and automated production. Multi-robot cooperative welding adopts a universal and high-degree of freedom handing robot instead of a low-degree of freedom positioner, which can effectively improve the flexibility of welding tasks and welding automation.

Arc welding is a complex system that contains both pose constraints and wrench constraints, the use of multi-robot systems for arc welding has many advantages, but it also brings more complex control problems. Two of the most critical issues are the trajectory planning in the multi-robot collaboration process and the coordinated control of position/force among multi-robot.

For the trajectory planning of multi-robot systems, the current research issues include motion constraints, control methods for cooperative motion, and implementation of multi-robot cooperative systems.

(1) For the motion constraint problem of multi-robot systems, the main task is to derive the end-effector of robot pose, velocity and acceleration constraints. For example, the idea is first pointed out in [1] that when two or more robots grab a common object, the robot end-effector is subject to kinematic constraints, and gives the speed constraint relationship of the multi-robot end-effector when the multi-robot is holding an object and rotates around its center. The pose, speed and acceleration constraints of the end-effector in the case of two robots grasp the common object, operate a pair of pliers and grab an object with a ball joint are deduced in [2,3]. The concept of absolute motion when the two robots cooperatively clamp the workpiece is proposed in [4]. The multi-robot cooperated trajectory planning approach based on the closed kinematic chain model is proposed in [5]; (2) For the control methods of collaborative movements, a method of establishing a constraint model with differential algebraic equations which using feedback linearization is addressed in [6]. The common control methods of general differential algebraic systems is summarized base on the above idea in [7]; (3) The realization for multi-robot collaboration system, the problem of trajectory planning and programming of general multi-robot cooperative system are analyzed. Such as ABB MultiMove function, KUKA RoboTeam function, Yaskawa independence/collaboration function, FANUC cooperative action function, etc.

Although the above research has given a certain impetus to the kinematics constraints and trajectory planning, the current multi-robot coordination tasks are relatively simple and most of them are focus on the coordination of dual robots. The above trajectory planning method is not scalable and feasible for multi-robot collaboration to accomplish specific tasks in a more complex and unstructured environment.

Multi-robot trajectory planning is the basis for completing the welding tasks, there are not only pose constraints, but also the wrench constraints in the process of welding displacement. Therefore, multi-robot position/force coordination control is another key issue in the cooperative welding process. Research methods for multi-robot position/force coordination control include master/slave control, hybrid motion/force control, synchronization control and impedance control. (1) Master/slave control. The control idea is to define one of the robotic arms as the master arm and the other as the salve arm. The master and the slave arm should've meet the certain constraints. The master arm is controlled by the position control mode, and the slave arm follows the motion trend of the master arm detected by the force/torque sensor mounted at the wrist of the slave arm. The master/slave force control approach for the coordination of two arms which carries a common object cooperatively was proposed in [8,9], and the necessity of force control for cooperative multiple robots is also pointed out. However, the slave arm needs to have a fast following response [10], otherwise it will lead to system instability. Force/torque sensor and position controller are difficult to achieve high-speed response in actual control systems. Therefore, the master-slave control strategy is only suitable for low-speed applications; (2) Hybrid motion/force control. The basic idea is that two arms work equally and coordinated by the centralized control. The position control is used in the free space, and the force control is used in the constrained space, such as [11–14]. A difficulty in implementing the hybrid control law in rigid environment is knowing the form of the constraints active at any time. And it also needs to sacrifice some performance by choosing low feedback gains, which makes the motion controller "soft" and the force controller more tolerant of force error; (3) Synchronization control. The basic idea is to track the desired trajectory which is generated by the desired force based on the dynamic model of the manipulators. The control problem is formulated in terms of suitably defined errors accounting for the motion synchronization between the manipulators involved in the cooperative task. The concept of motion synchronization was used in [15,16]. An adaptive control strategy was adopted to track the desired trajectory, ensuring the synchronization position errors converged to zero. In addition, intelligent control strategies were also used in the coordination control of nonlinear cooperative manipulator systems [17–19]. However, the synchronization control is based on the dynamic model. Although the synchronization error at the end effector of the manipulator was considered, due to the difficulties in dynamic modeling, over-complexity control model, strong coupling and nonlinearity, it has not been applied in most actual control systems; (4) Impedance control. The basic idea is to achieve the adjustment of the position based on the force error. Impedance control is a stable and effective method widely used in many fields including coordination. The coordination strategy of the object based on impedance control was studied in [20–24]. The force acting on the object was decomposed into the external force that contributes to the object's motion and the internal force by the end-effector of both arms. Following the guidelines in above references, the external impedance and the internal impedance were combined in a unique control framework [25]. Compared with the first three control strategies, impedance control overcomes the shortcomings of the above control methods, and can effectively control the internal and external force.

In contrast, impedance control has been widely used in multi-robot position/force coordination control. However, a closed-chain system is often subject to external dynamic disturbances and calibration errors in the actual industrial system. These factors can cause time-varying trajectory deviations at the end-effector, and time-varying trajectory deviations can cause unknowns and dynamically changing external forces and internal forces between multi-robot and the operated object. Most of the current impedance control methods that used in the above studies are constant, and the presented control strategies are not feasible for the unknown and dynamically changing trajectory errors. To the best of our knowledge, no research has been reported to solve the trajectory deviation problem during the multi-robot coordination with actual industrial robotic systems.

This paper uses multi-robot systems to complete the pipe-contact-pipe welding task as the research object. A study on the trajectory planning and position/force coordination control in the welding process is conducted.

The remaining of this paper is organized as follows. The system overview of multi-robot cooperative welding system is introduced in Section 2. The object-oriented Multi-robot trajectory planning based on "hierarchical scheme" is proposed in Section 3. The Coordination Strategy based on Symmetrical Adaptive Variable Impedance Control is given in Section 4. A series of experiments are carried out in Section 5, followed by conclusions in Section 6.

## 2. Problem Description

### 2.1. System Description

A multi-robot cooperative welding system refers to the collaboration of many sets of industrial robots on a welding task. In general, three or more robots are necessary. The cooperative system includes at least one welding robot, two or more robots are required which are used to splice the different pieces together. To discuss the key issue in welding process, the multi-robot systems cooperate welding a pipe-connect-pipe are given as an example in this paper. The typical multi-robot cooperation welding system schematic diagram is shown in Figure 1.

Figure 1 shows that the multi-robot welding system that includes three industrial robots, one welding robot, and two transfer robots. When the controller receives the specified task, two transfer robots grasp different welding workpieces, respectively in the initial welding position of relocating and splicing, at the same time, the welding robot begins to welding. After the two transfer robots cooperate with the workpiece, they will change the pose in the whole welding process, and then complete welding with the welding robot.

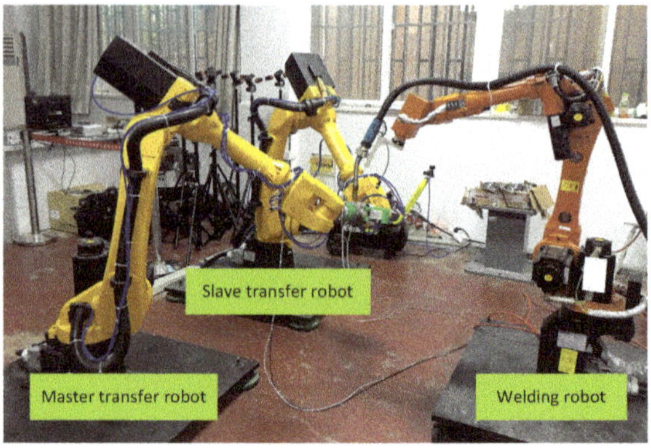


**Figure 1.** The diagram of multi-robot cooperation welding system.

### 2.2. Problem Description

The multi-robot cooperative welding system is shown in Figure 1. To solve the constraint pose and wrench at the same time in welding process, it is difficulty to plan the cooperative motion trajectory and achieve the desired position/force cooperative effect during the transportation.

In order to satisfy the welding process requirements in the continuous welding process, the transfer robots must continuously change the posture of the clamping workpiece to ensure that the welding spot is always in the preset welding position. The welding robot must constantly adjust the position of the welding torch to ensure the welding torch meets the welding requirements. It is required that both the transfer robots and welding robot should satisfy a certain position and posture constraints during the whole welding process.

With the process of welding, the workpiece held by transfer robot is gradually welded, then the transfer robots and the workpiece will be formed as a closed-chain system. In the actual control system, the external disturbance and calibration errors are often existing. These disturbances and calibration errors will cause the time-varying trajectory deviations in the robot end-effector during cooperation, and the dynamic time-varying trajectory deviations can cause a huge internal forces between the robot and the workpiece. Improper control may lead a damage to the workpiece or robot.

From the above analysis, the two key issues in the control of multi-robot cooperative welding system are:

- How to achieve the coordinated motion of multi-robot, which is the problem of planing the multi-robot trajectory;
- How to effectively control the internal force between the robot and the workpiece, which is the problem of position/force cooperation.

## 3. Task-Oriented Multi-Robot Trajectory Planning

### 3.1. Problem Formulation

The schematic diagram of the multi-robot cooperating to complete the welding task is shown in Figure 2.

Where *robot_1* and *robot_2* are used for transfer robots, *robot_3* is used for a welding robot. The transfer robots grab the workpiece $R_a$ and $R_b$, respectively. The welding robot is equipped with a welding torch at the end-effector, and the workpiece $R_a$ and $R_b$ are saddle-shaped after splicing.

The relevant coordinate system shown in Figure 2 is defined as follows. $\boldsymbol{T}_w$ represents the world coordinated system, $\boldsymbol{T}_o$ represents the reference coordinate system of the object. $\boldsymbol{T}_{bi}$ denotes the base coordinate of the $i$-th robot, $\boldsymbol{T}_i$ denotes the coordinate system of the $i$-th robot end-effector. $\boldsymbol{T}_{weld}$ is the weld coordinate system of the workpiece, and $\boldsymbol{T}_{object}$ is the coordinate system of the workpiece.

To facilitate the coordinate conversion, it is usually assumed that the world coordinate system coincides with the base coordinate system of *robot_1*.

The requirements for the multi-robot cooperation to complete the welding task of the tube include:

- The welding torch should meet the requirements of the ship-type welding posture;
- The configuration of each robot is reachable, and the configuration has no singularity in the welding process;
- The motion planning results of each robot are in a common collaborative space;
- There is no collision between the robot itself and another robots in the welding process.

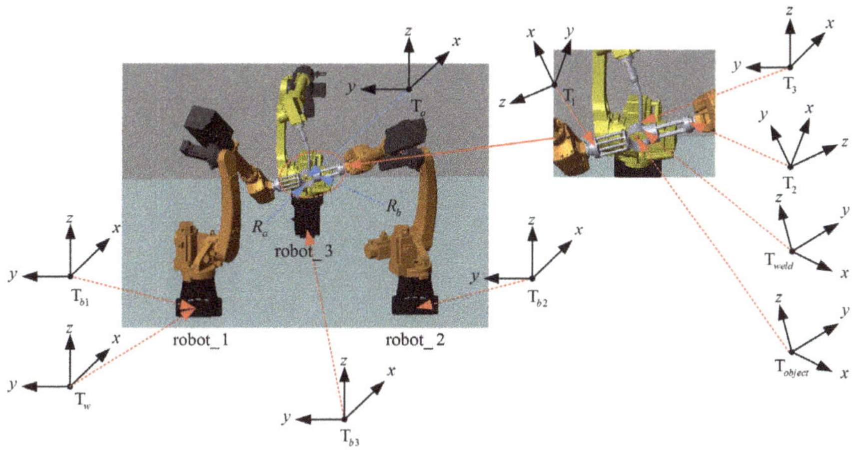


**Figure 2.** The system coordinates of multi-robot pipe-connect-pipe welding task.

### 3.2. Trajectory Planning Strategy

As shown in Figure 2, the coordinate systems involved in the multi-robot cooperative welding process are numerous, and the transformation between coordinate systems is complicated. At the same time, the welding requirements as described above must also be met in order to finish the welding task smoothly. We can conduct that the initial welding position affects whether the entire welding task can proceed smoothly or not. Based on the above constraints and requirements, a multi-robot trajectory planning based on "hierarchical scheme", which considered the optimal initial welding position, is proposed in this paper.

The basic idea of the above scheme is to first determine the optimal initial welding position, that is to determine the position of the reference coordinate system of the object in the world coordinate system. Then according to the welding task, the trajectory of the object in its reference coordinate system is planned. Finally, the robot end-effector trajectory is planned through the constraint relationship between the robot and the object.

The following steps are used to obtain the trajectory of multi-robot.

The first step is to determine the layout of the optimal welding position in the initial state, it is same to determine the position $\boldsymbol{p}$ $(x,y,z)$ which $\boldsymbol{T_o}$ is related to the world coordinate $\boldsymbol{T_w}$.

In combination with the requirements of the multi-robot cooperative welding process, the following indicators are considered to affect the layout selection of initial welding position.

1. The dual-arm task-based directional manipulability measure (DATBDMM) is mainly aimed at the transfer robots.
2. Flexible measure (FM) is mainly for the welding robot.
3. Global joint exercise (GJE) is used for the transfer robots and the welding robot.

According to the above performance indicators, the mathematical model for establishing the optimal initial welding position layout is shown in Equation (1).

$$\begin{cases} \text{DATBDMM}_{cv} = \max \left( \boldsymbol{p}^{\text{T}} \boldsymbol{J}_{vc} \boldsymbol{p} \right)^{-1} \\ \text{FM} = \max \dfrac{\sum\limits_{m-1}^{a} \sum\limits_{n-1}^{b} D_P(\alpha, \beta)}{\sum\limits_{m-1}^{a} \sum\limits_{n-1}^{b} 1} \\ \text{GJE} = \min \sum\limits_{j=1}^{3} \sum\limits_{i=1}^{n} {}^{j}w_i \left( {}^{j}\theta_i(k) - {}^{j}\theta_i(k-1) \right)^2 \end{cases}, \tag{1}$$

where $\boldsymbol{p} \in \boldsymbol{R}^{3\times1}$ represents the velocity unit vector at the point of mass center of the workpiece. $\boldsymbol{J}_{vc} = \left( \boldsymbol{J}_{1cv} \boldsymbol{J}_{1cv}{}^{\text{T}} \right)^{-1} + \left( \boldsymbol{J}_{2cv} \boldsymbol{J}_{2cv}{}^{\text{T}} \right)^{-1}$, $\boldsymbol{J}_{icv}$ represents the speed Jacobian matrix of the robot at the center of operated object. $\alpha, \beta$ indicate the rotation angle of welding torch around the $y$-axis and $z$-axis, respectively. After rotation, it can meet the welding requirements (shipping welding requirements). $D_P(\alpha, \beta)$ denotes a pose reachable function. If there is an inverse solution, it is denoted as $D_P(\alpha, \beta) = 1$. Otherwise, it is denoted as $D_P(\alpha, \beta) = 0$. ${}^{j}w_i$ represents the influence factor of each joint. ${}^{j}\theta_i(k) - {}^{j}\theta_i(k-1)$ indicates the amount of joint change at a certain moment.

The optimal initial welding position $\boldsymbol{p}$ $(x,y,z)$ is determined according to the above performance index, and the mathematical model of the optimal initial welding position for multi-robot cooperative welding is shown in Equation (2), then the optimal solution can be solved.

$$\max f(x, y, z) = \frac{k_{dm} \cdot \text{DATBDMM}_{cv} + k_{fm} \cdot \text{FM}}{k_{gje} \cdot \text{GJE}}, \tag{2}$$

where $f$ represents a multi-objective optimal function, $k_{dm}$,$k_{fm}$,$k_{gje}$ represent the influence factors corresponding to the degree of dual-directional operation of the task, the flexibility, and the amount of joint change, respectively. According to the importance of each parameter and our preliminary experience, the value of $k_{dm}$ is set to 0.5, $k_{fm}$ is set to 0.3 and $k_{gje}$ is set to 0.2.

In the second step, the trajectory of the operated object in the reference coordinate system is planned according to the welding task. The workpiece coordinate system which formed by the pipe connection is shown in Figure 3, the weld formed by the pipe is a saddle-shaped curve.

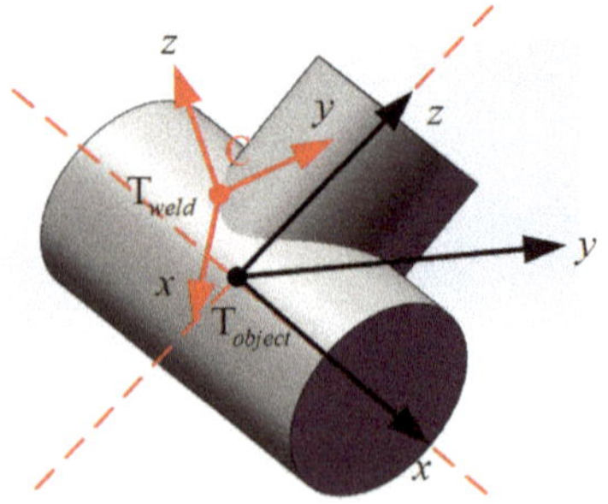


**Figure 3.** The schematic diagram of the object coordinate system and welding coordinate system.

The principle of establishing the reference coordinate system of the operated object is, the intersection point of the pipe center-line is set as the origin of the coordinate system, the $z$-axis is along the center-line of the upper tube and the $x$-axis is along the center-line of the lower tube. In Figure 3, C is a point on the saddle-shaped curve, and the coordinate in the reference coordinate system of the operated object is $C(x, y, z)$. It can use the parametric equation to represent the equation of the saddle-shaped curve in the reference coordinate system as shown in Equation (3).

$$\begin{cases} x = r \cdot \cos\theta \\ y = r \cdot \sin\theta \\ z = \sqrt{R^2 - r^2 \cdot \sin^2\theta} \end{cases}, \tag{3}$$

where $r$ is the radius of the upper tube $R_b$, $R$ is the radius of the lower tube $R_a$, and $\theta$ is the rotation angle parameter.

The solder point $C$ can be adjusted to the ship-type welding posture by two-step, the weld coordinate system rotates $\alpha$ around the $z_{object}$, and rotates $\beta$ around the $x_{object}$. Where $\beta \in (0, 2\pi)$, $\alpha$ is the angle between $z_{weld}$ and the reference coordinate system $z_o$ of the operated object. The formula is shown in Equation (4).

$$\alpha = \cos^{-1}\frac{\sqrt{R^2 - r^2 \cdot \sin^2\theta}}{R \cdot \sqrt{2 + 2 \cdot \frac{r}{R} \cdot \sin^2\theta}}. \tag{4}$$

Based on the above analysis, the trajectory of the operated object in its reference coordinate system can be obtained as shown in Equation (5).

$$\begin{cases} x = 0 \\ y = 0 \\ z = 0 \\ R = f(2\pi \cdot i \cdot t/T) \\ P = 0 \\ Y = 2\pi \cdot i \cdot t/T \end{cases}, \tag{5}$$

where $i$ denotes the $i$-th communication cycle, $t$ denotes the total time which is required for the welding task, and $T$ denotes the interpolation period. $f(\cdot)$ is a variable attitude function, it can be expressed as Equation (6).

$$f(\theta) = \cos^{-1}\frac{\sqrt{R^2 - r^2\sin^2\theta}}{R\sqrt{2 + 2\frac{r}{R}\sin^2\theta}}. \tag{6}$$

In the third step, the trajectory of each robot is planned according to the constraint relationship between the robot's end-effector and the operated object. The movement of the operated object $m$ (assumed coincident with the $\boldsymbol{T}_{object}$) in $\boldsymbol{T}_o$ can be described as $\boldsymbol{T}_m^o(t)$. $\boldsymbol{T}_m^o(t)$ is obtained according to the welding task (Equation (5)), and it can be expressed as Equation (7).

$$\boldsymbol{T}_m^o(t) = \begin{bmatrix} \boldsymbol{R}_m^o(t) & \boldsymbol{p}_m^o(t) \\ 0 & 1 \end{bmatrix}, \tag{7}$$

where $\boldsymbol{R}_m^o(t)$ represents the $(3 \times 3)$ rotation matrix of the centroid with respect to the reference coordinate system of the operated object, and $\boldsymbol{p}_m^o(t)$ represents the $(3 \times 1)$ position matrix of the centroid with respect to the reference coordinate system of the operated object.

According to the closed-chain constraint formed by the transfer robot and the operated object, the pose constraint relationship of the following Equation (8) can be obtained.

$$\begin{cases} \boldsymbol{T}_{bi}^w \cdot \boldsymbol{T}_i^{bi}(q_i) \cdot \boldsymbol{T}_m^i = \boldsymbol{T}_m^w \\ \boldsymbol{T}_o^w \cdot \boldsymbol{T}_m^o(t) = \boldsymbol{T}_m^w \end{cases}, \tag{8}$$

where $T_{bi}^{w}$ is homogeneous transform representing the robot base frame $T_{bi}$ with respect to the world frame $T_w$. $T_i^{bi}(q_i)$ is homogeneous transform representing the end-effector frame of robot $T_i$ with respect to its base frame $T_{bi}$. $T_m^i$ is homogeneous transform representing the mass frame of object $T_m$ with respect to the end-effector frame of robot $T_i$. $T_m^w$ is homogeneous transform representing the mass frame of object $T_m$ with respect to the world frame $T_w$. $T_o^w$ is homogeneous transform representing the object frame $T_o$ with respect to the world frame $T_w$.

From Equation (8), the kinematics of the $i$-th manipulator can be obtained as

$$\begin{aligned} T_i^{bi}(q_i) &= (T_{bi}^{w})^{-1} \cdot T_o^w \cdot T_m^o(t) \cdot (T_m^i)^{-1} \\ &= T_w^{bi} \cdot T_o^w \cdot T_m^o(t) \cdot T_i^m, \end{aligned} \tag{9}$$

where $T_w^{bi}$, $T_o^w$ and $T_i^m$ are constant matrix.

In the entire cooperative welding task, the transfer robots coordinate the workpiece to meet the requirements of the ship-type welding. The welding robot doesn't need to adjust the posture during the whole welding process, only the position of the welding torch in the operated object coordinate system needs to adjust. The position transformation matrix $p_3^o$ of the tip of the welding torch which relative to the reference coordinate system of the operated object can be expressed as Equation (10).

$$p_3^o = \begin{pmatrix} r \cdot \cos\alpha \cdot \cos\beta \cdot \cos\theta - r \cdot \cos\alpha \cdot \cos\beta \cdot \sin\theta + \sin\alpha \cdot \sqrt{R^2 - r^2 \cdot \sin^2\theta} \\ r \cdot \sin\beta \cdot \cos\theta + r \cdot \cos\alpha \cdot \cos\theta \\ -r \cdot \sin\alpha \cdot \cos\beta \cdot \cos\theta + r \cdot \sin\alpha \cdot \sin\beta \cdot \sin\theta + \cos\alpha \cdot \sqrt{R^2 - r^2 \cdot \sin^2\theta} \end{pmatrix}. \tag{10}$$

According to the conversion relationship between the welding robot and the operated object, and the coordinate transform between the welding robot and the world coordinate system, Equation (11) can be obtained.

$$\begin{cases} T_{b3}^{w} \cdot T_3^{b3}(q_i) = T_3^w \\ T_o^w \cdot T_3^o(t) = T_3^w \end{cases} \tag{11}$$

According to the Equation (11), the motion of the welding robot relative to its own coordinate system can be obtained as in Equation (12).

$$T_3^{b3}(q_i) = T_w^{b3} \cdot T_o^w \cdot T_3^o(t). \tag{12}$$

According to Equations (9) and (11), the trajectory of the transfer robot and the welding robot in their own coordinate system can be obtained. Based on the solution formula of inverse kinematics, the trajectory of each robot's joint angles can be obtained.

## 4. Symmetrical Adaptive Variable Impedance Control for Multi-Robot Coordination

Multi-robot trajectory planning is the foundation for the cooperative welding process, because there is no generalized wrench constraint between the welding robot and the transfer robot, so the multi-robot position/force coordination control problems can be converted to the coordination between the transfer robots.

### *4.1. Problem Formulation*

When the transfer robots are operating with a common object, the system becomes a closed-chain system. The closed-chain system which is constitute by the robots and the operated object is shown in Figure 4.

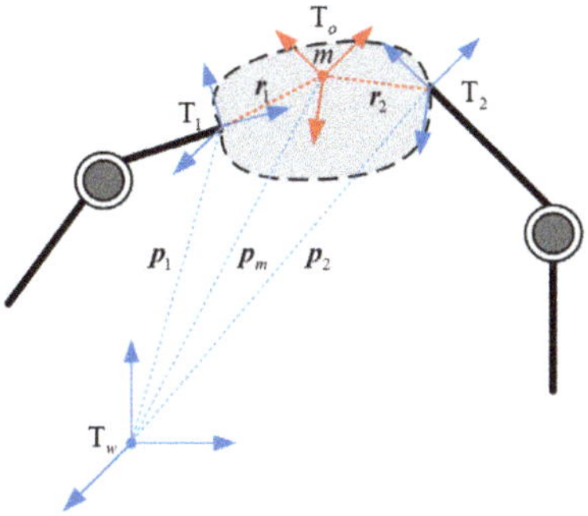


**Figure 4.** The diagram of the closed-chain system.

In Figure 4, $p_i(i=1,2), p_m$ denote the position vector of the robot end-effector with respect to the world coordinate system and the position vector of the operated object centroid with respect to the world coordinate system, respectively. $r_i(i=1,2)$ represents the position vector of the center of mass which is relative to the coordinate system of the robot. The movement of the operated object can be described by the center of mass $m$. And the Newton–Euler equation is shown as Equation (13).

$$\begin{cases} \boldsymbol{f}_m = \boldsymbol{M}\ddot{\boldsymbol{c}} - \boldsymbol{M}\boldsymbol{g} \\ \boldsymbol{n}_m = \boldsymbol{I}\dot{\boldsymbol{\omega}} + \boldsymbol{\omega} \times \boldsymbol{I}\boldsymbol{\omega} \end{cases}, \tag{13}$$

where $\boldsymbol{M}$ and $\boldsymbol{I}$ are the mass and inertia matrices of the object, $\boldsymbol{c}$ and $\boldsymbol{\omega}$ are the position and the angular velocity vector of the object, $\boldsymbol{g}$ is the acceleration of gravity, respectively.

If the movement of the operated object is known (as shown in Equation (5)). According to Equation (13), the wrench which the transfer robots need to exert on the center of operated object can be obtained. The wrench diagram of the force exerting on the operated object is shown in Figure 5.

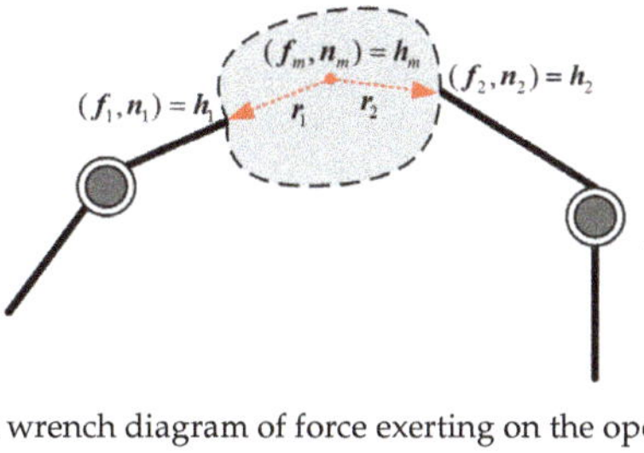


**Figure 5.** The wrench diagram of force exerting on the operated object.

The force formula of the transfer robots exerting on the operated object is shown in Equation (14).

$$\begin{cases} \boldsymbol{f}_m = \sum_{i=1}^{2} \boldsymbol{f}_i \\ \boldsymbol{n}_m = \sum_{i=1}^{2} \boldsymbol{n}_i + \sum_{i=1}^{2} \boldsymbol{r}_i \times \boldsymbol{f}_i \end{cases} \tag{14}$$

where $\boldsymbol{f}_i$ and $\boldsymbol{n}_i$ represent the force and moment of the $i$-th robot end-effector exerting on the operated object, respectively.

According to the concept of "virtual chain", Equation (13) can be expressed in the form of Equation (15).

$$h_m = Wh \tag{15}$$

where $h_m = \begin{bmatrix} f_m \\ n_m \end{bmatrix}$, $h = [{f_1}^T, {n_1}^T, {f_2}^T, {n_2}^T]^T$, $W = \begin{bmatrix} I_3 & 0_3 & I_3 & 0_3 \\ S(r_1) & I_3 & S(r_2) & I_3 \end{bmatrix}$, $S(r_i) = \begin{bmatrix} 0 & -r_{iz} & r_{iy} \\ r_{iz} & 0 & -r_{ix} \\ -r_{iy} & r_{ix} & 0 \end{bmatrix}$, $r_i = [r_{ix} r_{iy} r_{iz}]^T$.

In Equation (15), $W$ denotes the grip matrix and $h$ denotes the wrench matrix which is exerting on the contact point of the operated object. If the wrench is known, the wrench which is exerting on the centroid point of the operated object can be obtained according to the Equation (15). But the actual situation is the inverse problem of Equation (15), it can be attributed to the load distribution problem.

From Equation (15), we can see that once the wrench $h_m$ at the center of the operated object is known, the wrench $h$ which the transfer robots need to exert on the contact point can be obtained by solving the pseudo-inverse matrix. Since $W$ is a row full rank matrix, theoretically there are infinitely many solutions for $h$. According to the conclusion in [20], the general form of the following equation is obtained.

$$h = W^{\dagger} h_m + (I - W^{\dagger} W)\varepsilon, \tag{16}$$

where $W^{\dagger} = AW^T(WAW^T)^{-1}$, $A$ is a positive definite matrix and $\varepsilon$ is an arbitrary vector.

According to the conclusion in [25], Equation (16) can be converted to the form of Equation (17).

$$h = W^{\dagger} h_m + V h_i, \tag{17}$$

where $h_i$ indicates the internal force at the center of mass of the operated object, which can be set according to the actual requirements. According to [26,27], $W^{\dagger}$ and $V$ can be selected as shown in Equations (18) and (19), respectively.

$$W^{\dagger} = \frac{1}{2} \begin{bmatrix} I & 0 \\ -S(r_1) & I \\ I & 0 \\ S(r_2) & I \end{bmatrix}, \tag{18}$$

$$V = \begin{bmatrix} I & 0 \\ -S(r_1) & I \\ -I & 0 \\ S(r_2) & -I \end{bmatrix}. \tag{19}$$

In general, the external forces $h_m$ and the internal forces $h_i$ are given quantity. According to Equation (17), the wrench which needs to exert on the single robot can be obtained.

Although the resultant force of the transfer robots need to exert on the contact point of the object can be operated by the load distribution according to Equation (17). However, when there is the trajectory deviation existing which caused by external distribution or calibration error. The trajectory deviation will affect the motion of the operated object, but also affect the internal force between the transfer robots. If only the resultant force is simply tracked, not only the target trajectory can't be tracked, but also failing to track the desired internal force.

## 4.2. Symmetrical Adaptive Variable Impedance Control for Coordination Control

A symmetrical adaptive variable impedance position/force coordination strategy is proposed to solve the influence of the unknown trajectory deviation which is caused by the external disturbance forces. The basic idea is to first decompose the resultant forces into the internal and external force which are exerting on the contact points of the operated object. And ideally, the desired internal and external force can be obtained. Consider the disturbance force which is caused by the trajectory deviation is dynamically changing, so the adaptive variable impedance is proposed to track the desired position and force.

The first step is to decompose the internal and external force. The desired resultant force of the transfer robots need to exert on the contact point of the operated object is given as shown Equation (17). In order to achieve tracking the desired external force and internal force of the operated object by the transfer robots, Equation (17) can be expressed as a form of force exerting on the contact point of the operated object with transfer robots, it is shown as Equation (20).

$$\boldsymbol{h} = \boldsymbol{h}_E + \boldsymbol{h}_I. \tag{20}$$

where $\boldsymbol{h}_E$ and $\boldsymbol{h}_I$ represent the external force and internal force exerting on the contact point of the operated object with the transfer robots, respectively. They are meet the following equation.

$$\begin{cases} \boldsymbol{h}_E = \boldsymbol{W}^{+}\boldsymbol{h}_m \\ \boldsymbol{h}_I = \boldsymbol{V}\boldsymbol{h}_i \end{cases}. \tag{21}$$

According to the Equation (21), the proposed internal and external force can be known.

In actual control, the wrench of the transfer robots exerting on the contact point of the operated object can be detected by a six-dimensional force/torque sensor which is installed at the end-effector of the robot, and the wrench can be denoted by $\boldsymbol{h}_r$. It is further possible to decompose $\boldsymbol{h}_r$ into the external forces and internal forces as shown in Equation (22).

$$\begin{cases} \boldsymbol{h}_{Er} = \boldsymbol{W}^{+}\boldsymbol{W}\boldsymbol{h}_r \\ \boldsymbol{h}_{Ir} = (\boldsymbol{I} - \boldsymbol{W}^{+}\boldsymbol{W})\boldsymbol{h}_r \end{cases}, \tag{22}$$

where $\boldsymbol{h}_{Er}$ and $\boldsymbol{h}_{Ir}$ represent the actual values of the external force and the internal force which are resolved based on the measured values by the six-dimensional force/torque sensor.

In the second step, position/force coordination control is based on symmetrical adaptive variable impedance. The schematic diagram of multi-robot position/force coordination control based on a symmetrical adaptive variable impedance is shown in Figure 6.

From Figure 6, a symmetric internal and external impedance coordination strategy is used for the transfer robots, and the position control strategy is used for the welding robot to just follow the object's motion. For the transfer robots, the inner impedance controller is composed of transfer robots and the operated object, and the outer impedance controller is composed of the operated object and the environment (external disturbance or external force caused by trajectory deviations). In the actual control, a symmetrical coordination method is adopted to convert the internal and the external force which is exerted on the contact point of the operated object at the end-effector of the single robot. The purpose of the inner impedance controller is to track the desired internal force and modify the movement trajectory of the end of the transfer robot based on the force deviation. The purpose of the outer impedance controller is to track the desired external force and modify the movement trajectory of the operated object according to the force deviation. Then the transfer robots and welding robot update the respective movement trajectories according to the corrected trajectory of the operated object.

Considering that the desired internal forces, external forces, and the uncertain wrench which is caused by external disturbances are time-varying functions, an adaptive variable impedance control strategy is adopted. Therefore, the system is divided into an adaptive variable impedance inner loop controller and an adaptive variable impedance outer loop controller. The diagram of symmetrical internal and external adaptive impedance control is shown in Figure 7.

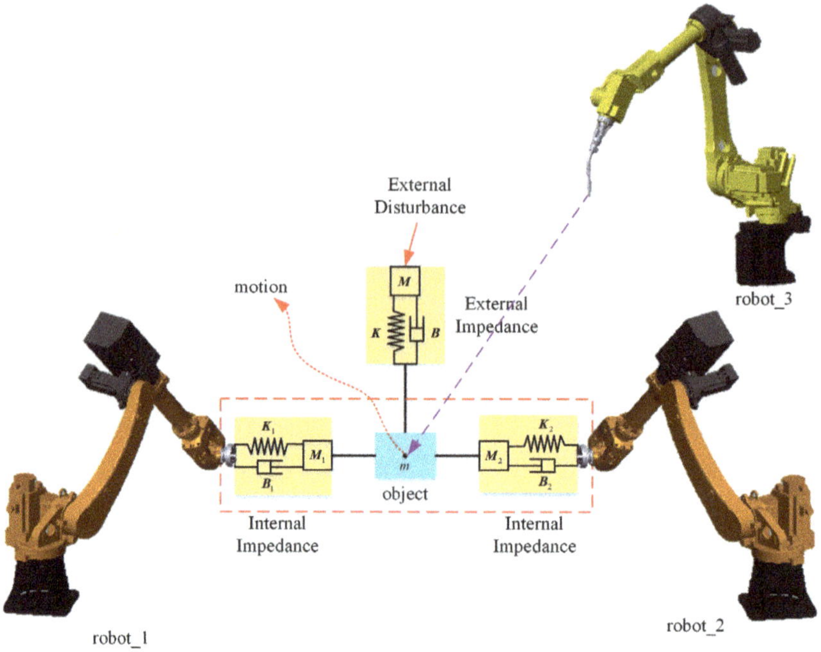


**Figure 6.** The schematic diagram of multi-robot position/force coordination control.

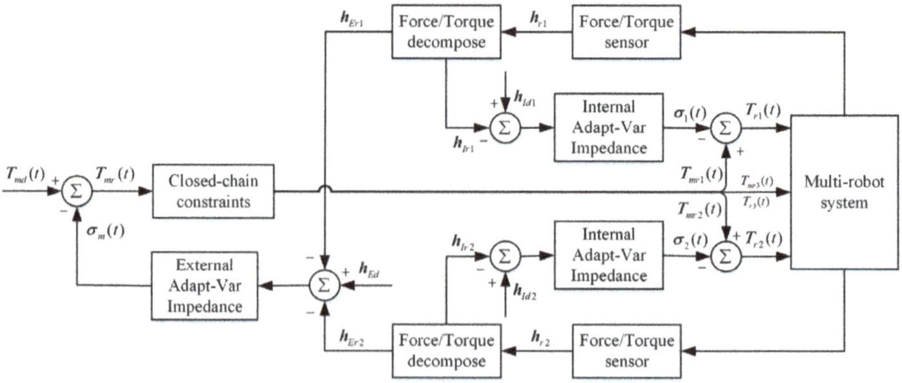


**Figure 7.** The diagram of symmetrical internal and external adaptive impedance control.

As shown in Figure 7, the desired movement trajectory of the operated object which inputs by the system is $\boldsymbol{T}_{md}(t)$. In the ideal conditions, it exists $\boldsymbol{T}_{md}(t) = \boldsymbol{T}_{mr}(t)$. According to the closed-chain constraint conditions, the trajectories ($\boldsymbol{T}_{mr1}(t)$, $\boldsymbol{T}_{mr2}(t)$, $\boldsymbol{T}_{mr3}(t)$) of multi-robot can be obtained, where $\boldsymbol{T}_{mr1}(t) = \boldsymbol{T}_{r1}(t)$, $\boldsymbol{T}_{mr2}(t) = \boldsymbol{T}_{r2}(t)$, $\boldsymbol{T}_{mr3}(t) = \boldsymbol{T}_{r3}(t)$. When uncertainty forces caused by the trajectory deviations exist, they can be obtained by the two six-dimensional force/torque sensor which are denoted by $\boldsymbol{h}_{r1}$ and $\boldsymbol{h}_{r2}$. According to Equation (22), the measurement of the resultant forces can be decomposed into external forces($\boldsymbol{h}_{Er1}$, $\boldsymbol{h}_{Er2}$) and internal forces($\boldsymbol{h}_{Ir1}$, $\boldsymbol{h}_{Ir2}$), respectively.

When there is a force deviation between the measured external force and the desired external force $\boldsymbol{h}_{Ed}$, the system will obtains an error trajectory $\boldsymbol{\sigma}_m(t)$ through the adaptive variable impedance external loop controller, then the corrected trajectory $\boldsymbol{T}_{mr}(t)$ of the operated object is obtained, and the trajectory of each robot is further corrected. When there is a force deviation between the measured internal force and the desired internal force ($\boldsymbol{h}_{Id1}$ and $\boldsymbol{h}_{Id2}$), the system obtains error trajectories ($\boldsymbol{\sigma}_1(t)$ and $\boldsymbol{\sigma}_2(t)$) through the adaptive variable impedance inner loop controller, and then transfer robots's correction trajectories ($\boldsymbol{T}_{r1}(t)$ and $\boldsymbol{T}_{r1}(t)$) can be obtained.

The adaptive variable external impedance law is proposed as Equation (23), as shown in Figure 7.

$$\begin{cases} \boldsymbol{M}[\ddot{\boldsymbol{T}}_{mr}(t) - \ddot{\boldsymbol{T}}_{md}(t)] + [\boldsymbol{B} + \Delta\boldsymbol{B}(t)][\dot{\boldsymbol{T}}_{mr}(t) - \dot{\boldsymbol{T}}_{md}(t)] = \boldsymbol{h}_{Er1}(t) + \boldsymbol{h}_{Er2}(t) - \boldsymbol{h}_{Ed}(t) \\ \Delta\boldsymbol{B}(t) = \dfrac{\boldsymbol{B}}{\dot{\boldsymbol{T}}_{mr}(t) - \dot{\boldsymbol{T}}_{md}(t)}\Phi(t) \\ \Phi(t) = \Phi(t-\lambda) + \sigma\dfrac{\boldsymbol{h}_{Ed}(t-\lambda) - \boldsymbol{h}_{Er1}(t-\lambda) - \boldsymbol{h}_{Er2}(t-\lambda)}{\boldsymbol{B}} \end{cases} \tag{23}$$

$$\begin{cases} \boldsymbol{M}_I[\ddot{\boldsymbol{T}}_{ri}(t) - \ddot{\boldsymbol{T}}_{mri}(t)] + [\boldsymbol{B}_I + \Delta\boldsymbol{B}_I(t)][\dot{\boldsymbol{T}}_{ri}(t) - \dot{\boldsymbol{T}}_{mri}(t)] = \boldsymbol{h}_{Iri}(t) - \boldsymbol{h}_{Idi}(t) \\ \Delta\boldsymbol{B}_I(t) = \dfrac{\boldsymbol{B}_I}{\dot{\boldsymbol{T}}_{ri}(t) - \dot{\boldsymbol{T}}_{mri}(t)}\Phi(t) \\ \Phi(t) = \Phi(t-\lambda) + \sigma\dfrac{\boldsymbol{h}_{Idi}(t) - \boldsymbol{h}_{Iri}(t)}{\boldsymbol{B}_I} \end{cases} \tag{24}$$

where $\boldsymbol{M}$ is the desired inertia matrix, $\boldsymbol{B}$ is the damping matrix, $\lambda$ is the sampling period of the controller and $\sigma$ is the update rate.

The adaptive variable internal impedance law showns in Equation (7) is expressed as Equation (24).

Where $\boldsymbol{M}_I$ denotes the desired inertia matrix of the internal wrench, $\boldsymbol{B}_I$ denotes the damping matrix of the internal wrench, and $i$ denotes $i$-$th$ robot.

In our previous work [28], the adaptive variable impedance control has been proven stable and convergent and been used to track the dynamic force with unknown trajectory deviations.

## 5. Simulations and Experiments

### 5.1. Simulation

This section mainly verifies the feasibility of position/force coordination for multi-robot based on the symmetrical adaptive variable impedance control. Matlab SimMechanics was used to simulate the multi-robot cooperative welding. In order to verify the effectiveness of the algorithm, the simulation is close to the actual physical experiment. During the experiment, it is assumed that there is a certain expected pressure and external disturbances, the purpose is to test the tracking effect of external and internal forces.

The object is composed into two rigid pipes, $R_a$ and $R_b$, respectively. The radius $r_a = 0.051$ m, $r_b = 0.0445$ m, the length $l_a = 0.12$ m, $l_b = 0.08$ m, respectively. The thickness of the pipe is $h = 0.003$ m, and the density is 1000 kg/m$^3$. The center of the 1st robot's end-effector and the center of pipe $R_a$ coincide the center of the 2nd robot's end-effector and the centerline of pipe $R_b$ is coincide.

The optimal initial welding position [0.5 m −0.75 m 0.404 m] can be obtained by solving Equation (2) through genetic algorithm, the optimization process of genetic algorithm is shown in Figure 8. According to the requirements of the welding task, the variable pose trajectory of the operated object is shown as Equations (5) and (6). Furthermore, the motion of each robot with respect to its respective base coordinate system can be solved according to Equations (9) and (12). Assuming that the desired internal wrench is [0 N 0 N 15 N 0 N·m 0 N·m 0 N·m], the desired internal force is among $z$-axis. The external disturbance is operated among $x$ and $y$ axis are shown in Figure 9.

The schematic diagram of the system simulation process for multi-robot completing pipe welding is shown in Figure 10.

The desired motion trajectory of the operated object and the actual motion trajectory after the external disturbance in $x$-axis and $y$-axis are shown in Figure 11, and the internal force tracking effect of the transfer robots is shown in Figure 12.

Combining Figures 11 and 12, it shows that after the external disturbance exerting on the $x$ and $y$ axis, the closed-chain system is flexible. From Figure 12, it shows that during the coordinated welding process, the tracking of the desired internal force can be achieved, and the internal force is maintained within the allowable tolerance range throughout the entire process.

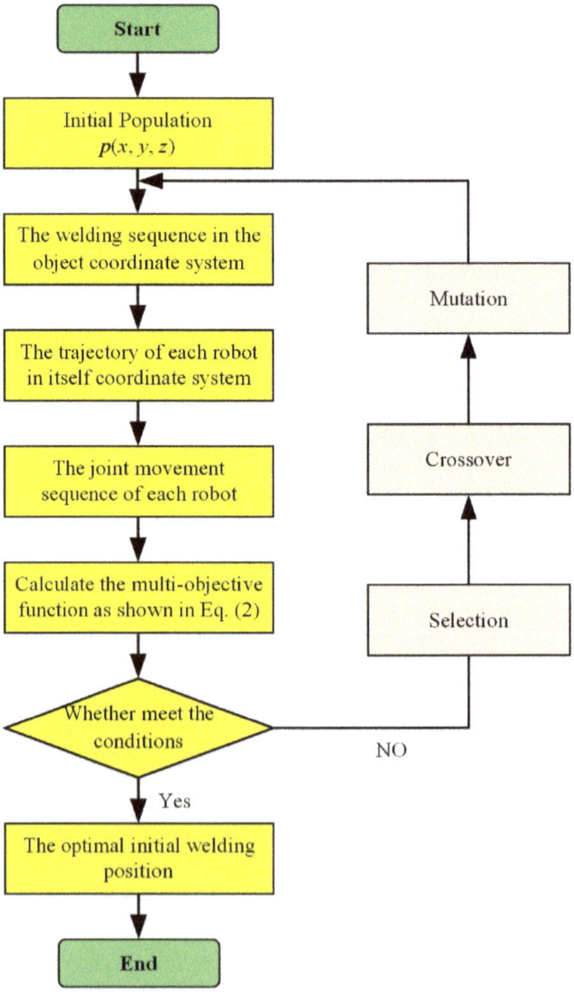


**Figure 8.** The optimization process of genetic algorithm.

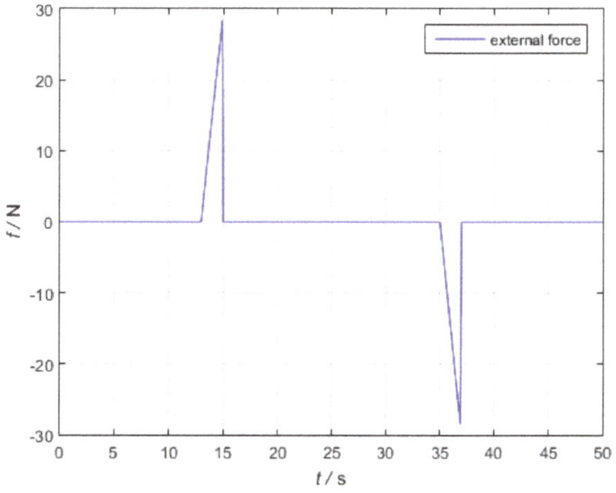


**Figure 9.** The external disturbance exerting on the operated object.

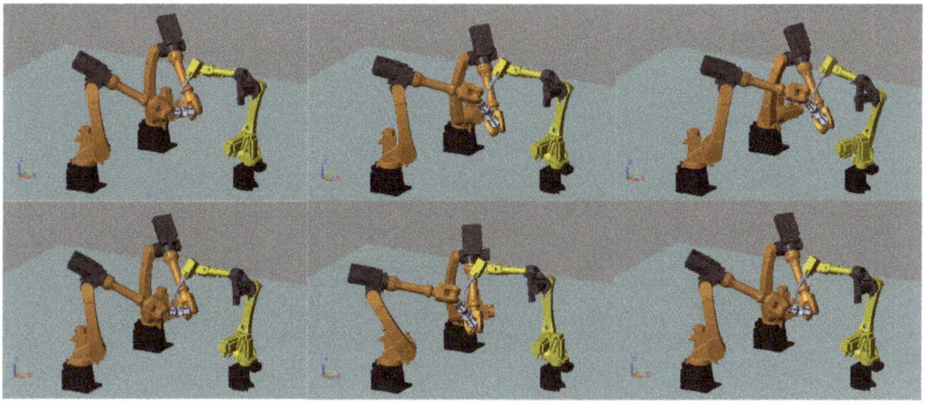

**Figure 10.** The schematic diagram of the system simulation process for multi-robot systems.

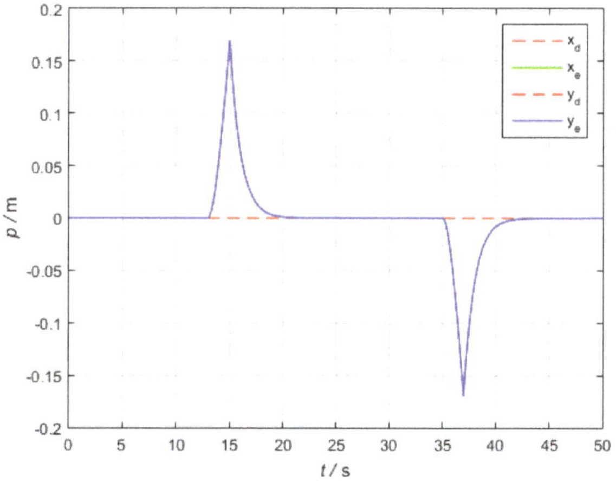


**Figure 11.** The simulation result of desired trajectory and actual trajectory.

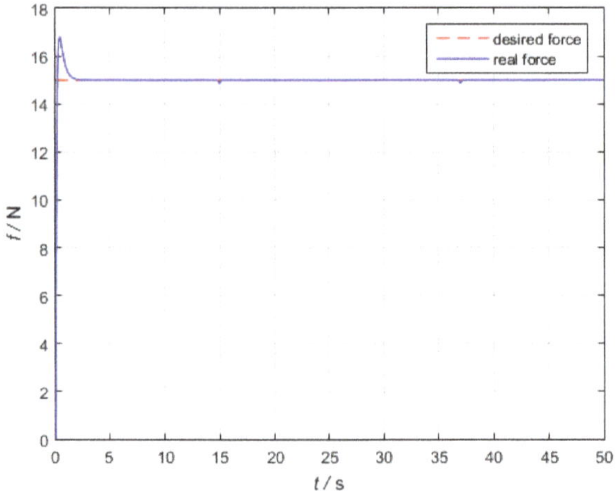


**Figure 12.** The simulation result of the internal force tracking effect.

## 5.2. Experimental Studies

To demonstrate the performance of the proposed algorithm, experiments were conducted using the test-bed as shown in Figure 13. And the logical scheme of the control software is shown in Figure 14.

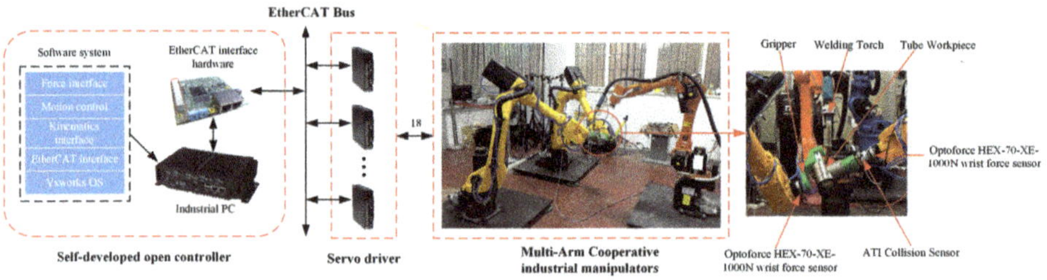


**Figure 13.** Hardware architecture of the test-bed.

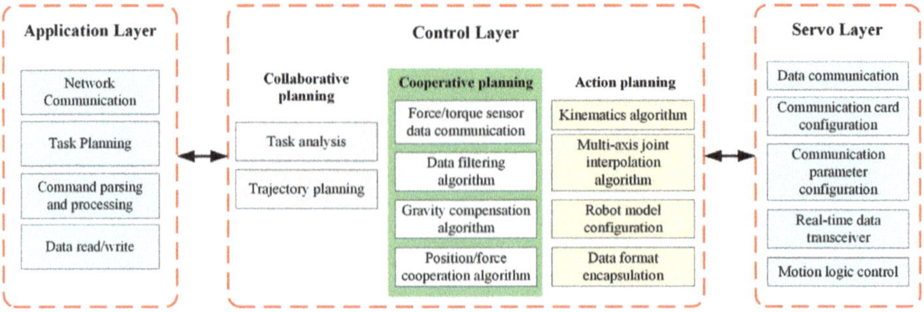


**Figure 14.** The logical scheme of the control software.

The test platform consisted of aself-developed open controller, servo drivers, two ESTUN ER16 industrial manipulators, one ESTUN ER4 industrial manipulators, two force/torque sensors, a collision sensor, and two pairs of gripper. The self-developed open controller used an industrial PC with a configuration of Intel Celeron @1.2 GHz, 512 MB of RAM, VxWorks RTOS. The servo drivers used ESTUN ProNet series, the bandwidth of the servo driver is from 125 Hz to 1000 Hz. Hischer CIFX communication card is used to EtherCAT communication between industrial PC and servo drivers. The force/torque sensors use Optoforce HEX-70-XE-1000N, the collision sensor uses ATI SR-61. The self-developed open controller was used for task coordination, position/force control, motion planning, forward/inverse kinematics, 18-axis cycle synchronization interpolation and human-machine interface. The force/torque sensor was mounted at the wrist of each manipulator. An ATI collision sensor was installed between the end-effector and the gripper of the 1-*st* robot was added for the protection of the whole system. The force/torque sensor provided the UDP protocol with the fastest frequency of 1 kHz, so the force sensor and the controller communicate through UDP. Consider the controller computing power, the bandwidth of the servo drivers and communication frequency of the force sensor at the same time, the communication cycles of the controller and the servo drivers, the controller and the force sensor are both set to 5 ms. Two force sensors were initialized by gravity compensation.

The physics experiment was consistent with the simulation. The transfer robots separately grasp a part of the workpiece to be welded, which was spliced at the initial welding point. The welding robot started arcing at the initial welding point. Then the transfer robots coordinated the workpiece to be displace, and the welding robot completed the welding of the weld seam. The key frames of the whole welding task is shown in Figure 15.

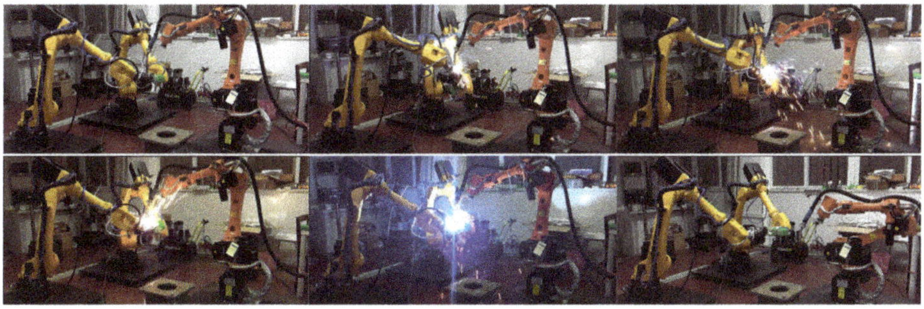

**Figure 15.** The key frames of the whole welding task.

During the welding process, due to the presence of unknown factors such as mechanical calibration error, base coordinate calibration error, and external disturbance, the transfer robots produced an uncertain wrench to the welding workpiece. The internal wrench exerting on the end-effector without force control is shown in Figure 16.

From the above results, we can see that the internal wrench without the force control is large. In order to control the internal forces in a proper range, a symmetrical adaptive variable impedance control is proposed in this paper. To certify the performance of the proposed algorithm, the traditional constant impedance control and the proposed algorithm are compared as shown in Figure 17.

Figure 17 shows that the force control effect has been significantly improved. The desired trajectory of transfer robots' end-effector and the center of workpiece are shown in Figure 18.

From Figure 17, we can conclude that the variable adaptive impedance control can achieve a better effect than the traditional constant impedance control. By using the proposed algorithm, the trajectory deviations of the transfer robots are shown in Figure 19, the trajectory deviations of the welding robot is shown in Figure 20.

Through Figures 17–20, it shows that the internal force of the transfer robots exerting on the workpiece to be welded is within the controllable range and does not cause damage to the welding system. The welding result has been shown in Figure 21. It indicates that a uniform smooth weld seam without cracks has been got by our method. And the welding quality is much better compared with novice welders.

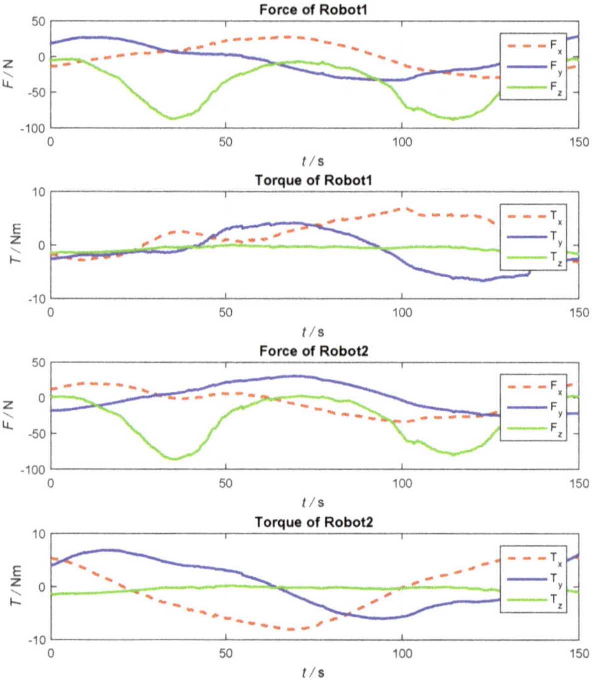


**Figure 16.** The internal wrench exerting on the end-effector without force control.

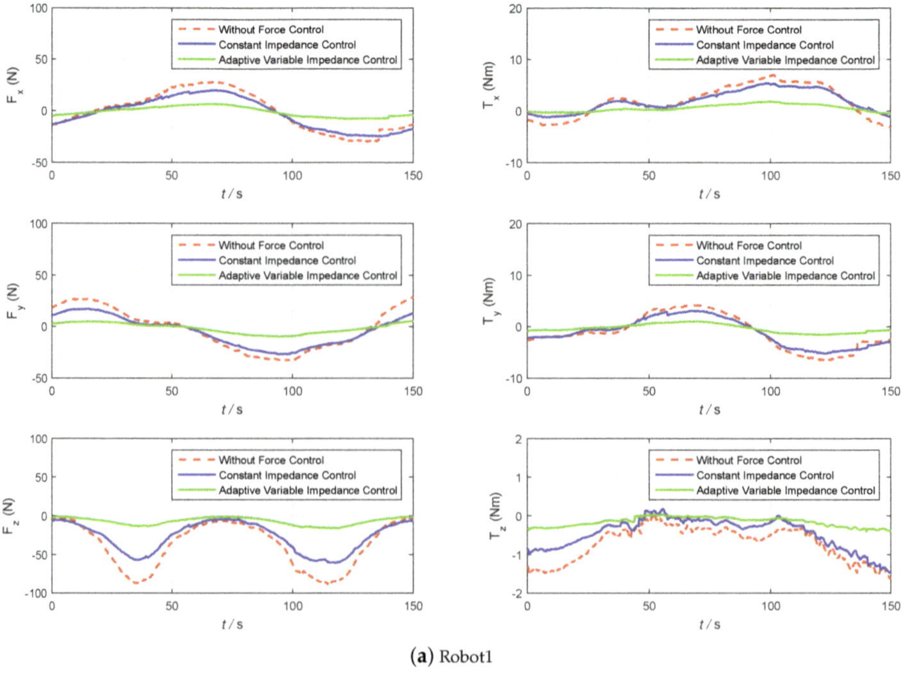


(**a**) Robot1

**Figure 17.** *Cont.*

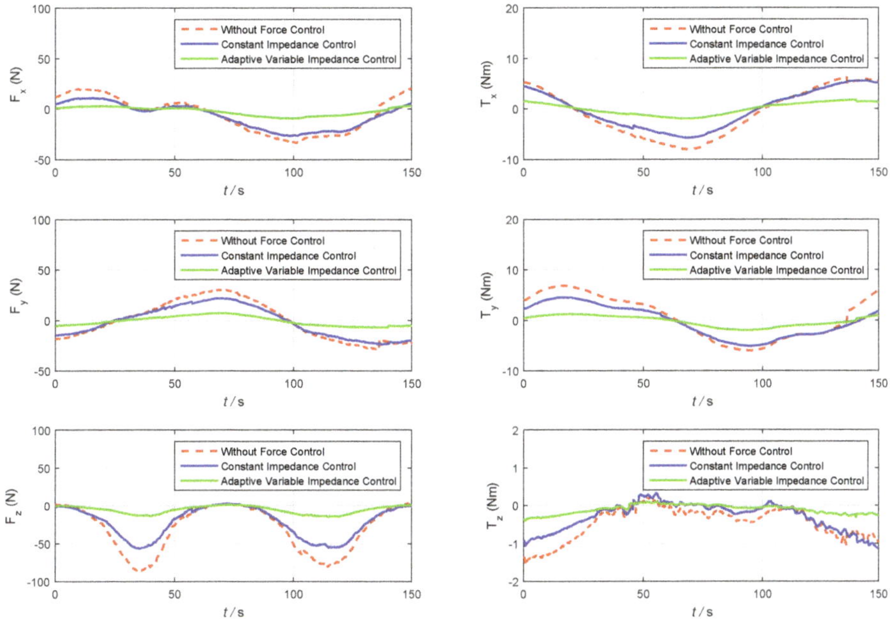


**(b)** Robot2

**Figure 17.** Comparison of two algorithm results.

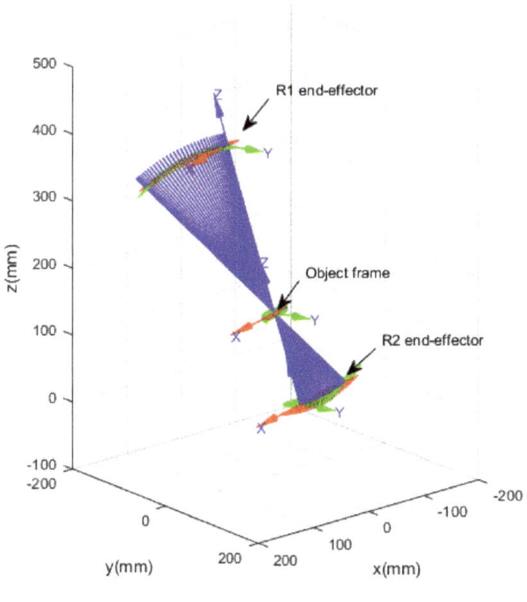


**Figure 18.** The desired trajectory of transfer robots' end-effector and the center of workpiece.

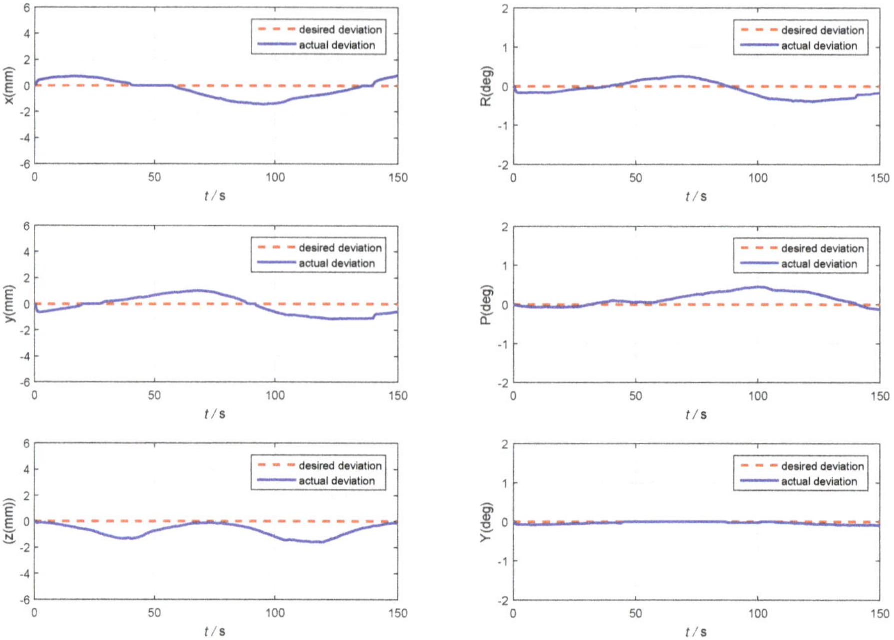


**(a)** The trajectory deviations of the Rotot1.

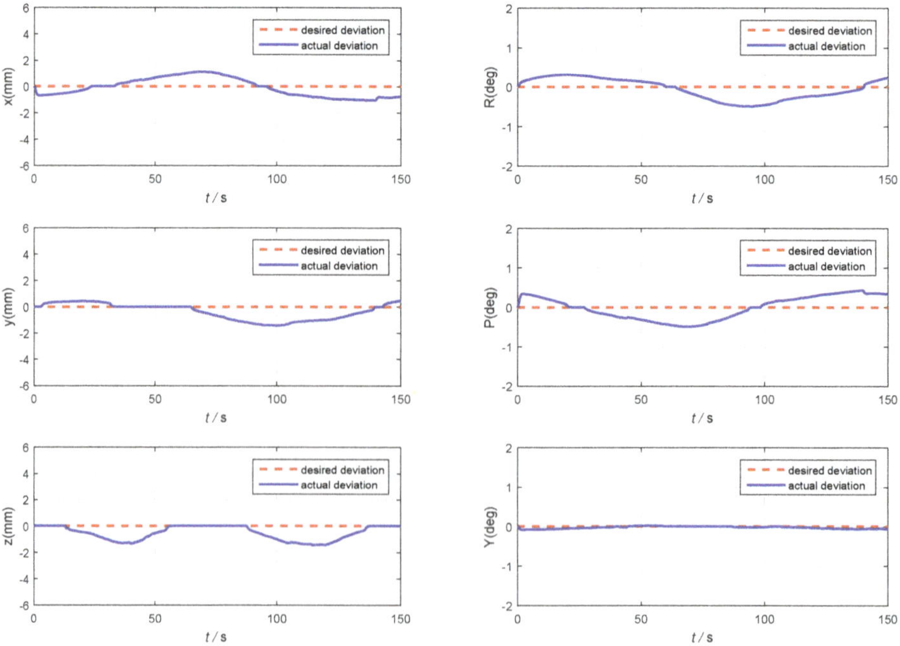


**(b)** The trajectory deviations of the Rotot2.

**Figure 19.** The trajectory deviations of the transfer robots.

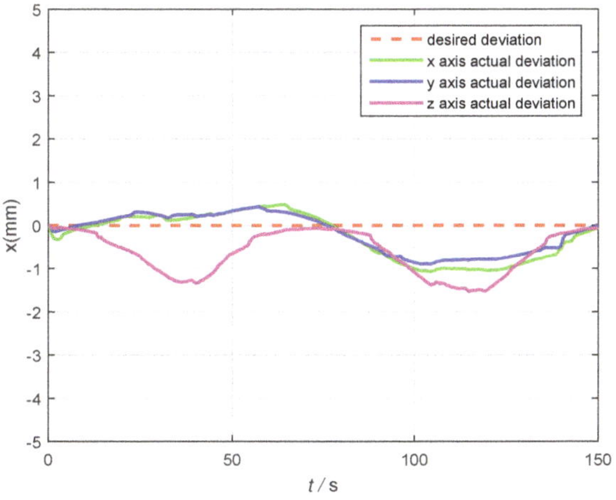


**Figure 20.** The trajectory deviations of the welding robot.

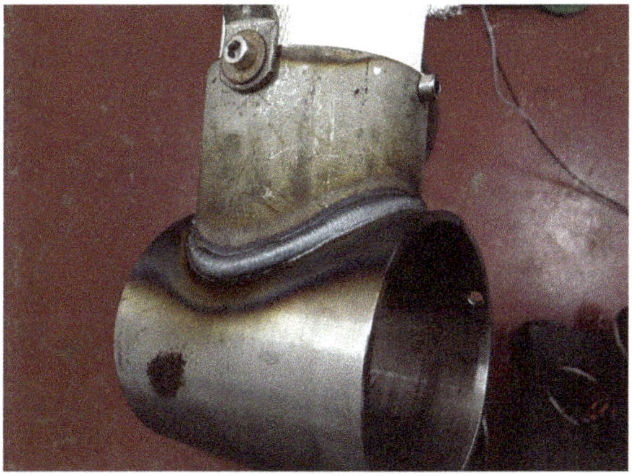

**Figure 21.** Pipe-connect-pipe arc welding result.

In the simulations and experiments, the selection method of the design parameters of the control system ($\boldsymbol{M}$, $\boldsymbol{B}$, $\lambda$ and $\sigma$) can be referenced in [28]. The following parameters can obtain the good performance in the above simulations and experiments, the inertia coefficient $\boldsymbol{M} = diag\{1,1,1,0,0,0\}$, $\boldsymbol{M}_I = diag\{1,1,1,0,0,0\}$, the initial damping coefficient $\boldsymbol{B} = diag\{65,65,65,0,0,0\}$, $\boldsymbol{B}_I = diag\{90,90,90,0,0,0\}$, $\lambda = 0.005s$ and $\sigma = 0.01$.

## 6. Conclusions

In the actual multi-robot cooperative welding system, the trajectory planning and position/force coordination control are the two most critical issues. The trajectory planning is the basis of position/force coordination control. The biggest difference between multi-robot cooperative welding and other types of multi-robot cooperative tasks is that they have more constraints and the coordinate system is more cumbersome and complicated. During the welding of the workpiece, an unknown changing wrench can be generated between the transfer robots and the workpiece due to the external disturbance and calibration errors.

In the face of problems such as complex changes of coordinate system and welding constraints in the process of multi-robot cooperative welding, a planning strategy for "hierarchical planning" is proposed in this paper. Firstly, the optimal initial welding position is determined according to the optimization index, and then the trajectory of the operated object in its coordinate system is planned. Finally, the trajectory of each robot relative to its base coordinate system is obtained.

In actual control systems, due to the calibration errors and external disturbances, the trajectory of robot end-effector is often deviated. In response to this problem, a symmetric adaptive internal and external variable impedance control strategy is used to track the internal and external forces exerting on the operated object. Through simulation and physical tests, it is concluded that the proposed control strategy is applicable to multi-robot cooperative welding systems and can effectively solve the two problems in the process of cooperative welding.

**Author Contributions:** Methodology, Y.G., J.D.; validation and data curation, M.C., J.D.; writing—original draft preparation, Y.G., J.D.; writing—review and editing, Y.G.; visualization, M.C.; supervision, Y.G.; project administration, X.D.; funding acquisition, Y.G., X.D.

**Funding:** This research was funded by National Natural Science Foundation of China under grant number 61873308, 61503076, 61175113 and Natural Science Foundation of Jiangsu Province under grant number BK20150624.

**Conflicts of Interest:** The authors declare no conflict of interest.

## References

1. Mason, M.T. Compliance and Force Control for Computer Controlled Manipulators. *IEEE Trans. Syst. Man Cybern.* **1981**, *11*, 418–432. [CrossRef]
2. Zheng, Y.F.; Luh, J.Y.S. Control of two coordinated robots in motion. In Proceedings of the 1985 24th IEEE Conference on Decision and Control, Fort Lauderdale, FL, USA, 11–13 December 1985; pp. 1761–1766.
3. Luh, J.Y.S.; Zheng, Y.F. Constrained Relations between Two Coordinated Industrial Robots for Motion Control. *Int. J. Robot. Res.* **1987**, *6*, 60–70. [CrossRef]
4. Nagai, K.; Iwasa, S.; Watanabe, K.; Hanafusa, H. Cooperative control of dual-arm robots for reasonable motion distribution. In Proceedings of the 1995 IEEE/RSJ International Conference on Intelligent Robots and Systems. Human Robot Interaction and Cooperative Robots, Pittsburgh, PA, USA, 5–9 August 1995; p. 54.
5. Zhou, B.; Xu, L.; Meng, Z.; Dai, X. Kinematic cooperated welding trajectory planning for master-slave multi-robot systems. In Proceedings of the 2016 35th Control Conference, Chengdu, China, 27–29 July 2016; pp. 6369–6374.
6. Mcclamroch, N.H.; Wang, D. Feedback stabilization and tracking of constrained robots. *IEEE Trans. Autom. Control* **1988**, *33*, 419–426. [CrossRef]
7. Krishnan, H.; Mcclamroch, N.H. *Tracking in Nonlinear Differential-Algebraic Control Systems with Applications to Constrained Robot Systems*; Pergamon Press, Inc.: Oxford, UK, 1994.
8. Nakano, E. Cooperational Control of the Anthropomorphous Manipulator "MELARM". In Proceedings of the 4th International Symposium on Industrial Robots, Tokyo, Japan, 19–21 November 1974; pp. 251–260.
9. Barbieri, L.; Bruno, F.; Gallo, A.; Muzzupappa, M.; Russo, M.L. Design, prototyping and testing of a modular small-sized underwater robotic arm controlled through a Master-Slave approach. *Ocean Eng.* **2018**, *158*, 253–262. [CrossRef]
10. Scaradozzi, D.; Sorbi, L.; Zingaretti, S.; Biagiola, M.; Omerdic, E. Development and integration of a novel IP66 Force Feedback Joystick for offshore operations. In Proceedings of the 22nd Mediterranean Conference on Control and Automation, Palermo, Italy, 16–19 June 2014; pp. 664–669.
11. Hayati, S. Hybrid position/Force control of multi-arm cooperating robots. In Proceedings of the 1986 IEEE International Conference on Robotics and Automation, San Francisco, CA, USA, 7–10 April 1986; pp. 82–89.
12. Uchiyama, M.; Iwasawa, N.; Hakomori, K. Hybrid position/Force control for coordination of a two-arm robo. In Proceedings of the 1987 IEEE International Conference on Robotics and Automation, Raleigh, NC, USA, 31 March–3 April 1987; pp. 1242–1247.
13. Uchiyama, M.; Dauchez, P. A symmetric hybrid position/force control scheme for the coordination of two robots. In Proceedings of the 1988 IEEE International Conference on Robotics and Automation, Philadelphia, PA, USA, 24–29 April 1988; Volume 1, pp. 350–356.
14. Masaru, U.; Pierre, D. Symmetric kinematic formulation and non-master/slave coordinated control of two-arm robots. *Adv. Robot.* **1992**, *7*, 361–383.

15. Sun, D.; Mills, J.K. Adaptive synchronized control for coordination of multirobot assembly tasks. *IEEE Trans. Robot. Autom.* **2002**, *18*, 498–510.
16. Rodriguez-Angeles, A.; Nijmeijer, H. Mutual synchronization of robots via estimated state feedback: A cooperative approach. *IEEE Trans. Control Syst. Technol.* **2004**, *12*, 542–554. [CrossRef]
17. Lian, K.Y.; Chiu, C.S.; Liu, P. Semi-decentralized adaptive fuzzy control for cooperative multirobot systems with H-inf motion/internal force tracking performance. *IEEE Trans. Syst. Man Cybern. Part B Cybern.* **2002**, *32*, 269–280. [CrossRef] [PubMed]
18. Gueaieb, W.; Karray, F.; Al-Sharhan, S. A robust adaptive fuzzy position/force control scheme for cooperative manipulators. *IEEE Trans. Control Syst. Technol.* **2003**, *11*, 516–528. [CrossRef]
19. Gueaieb, W.; Karray, F.; Al-Sharhan, S. A Robust Hybrid Intelligent Position/Force Control Scheme for Cooperative Manipulators. *IEEE/ASME Trans. Mechatron.* **2007**, *12*, 109–125. [CrossRef]
20. Walker, I.D.; Freeman, R.A.; Marcus, S.I. Analysis of Motion and Internal Loading of Objects Grasped by Multiple Cooperating Manipulators. *Int. J. Robot. Res.* **1991**, *10*, 396–409. [CrossRef]
21. Bonitz, R.G.; Hsia, T.C. Force decomposition in cooperating manipulators using the theory of metric spaces and generalized inverses. In Proceedings of the 1994 IEEE International Conference on Robotics and Automation, San Diego, CA, USA, 8–13 May 1994; Volume 2, pp. 1521–1527.
22. Leidner, D.; Dietrich, A.; Schmidt, F.; Borst, C.; Albu-Schäffer, A. Object-centered hybrid reasoning for whole-body mobile manipulation. In Proceedings of the IEEE International Conference on Robotics and Automation, Hong Kong, China, 31 May–7 June 2014; pp. 1828–1835.
23. Dietrich, A.; Wimbock, T.; Albu-Schaffer, A. Dynamic whole-body mobile manipulation with a torque controlled humanoid robot via impedance control laws. In Proceedings of the IEEE/RSJ International Conference on Intelligent Robots and Systems, San Francisco, CA, USA, 25–30 September 2011; pp. 3199–3206.
24. Ott, C.; Hirzinger, G. Comparison of object-level grasp controllers for dynamic dexterous manipulation. *Int. J. Robot. Res.* **2012**, *31*, 3–23.
25. Caccavale, F.; Chiacchio, P.; Marino, A.; Villani, L. Six-DOF Impedance Control of Dual-Arm Cooperative Manipulators. *IEEE/ASME Trans. Mechatron.* **2008**, *13*, 576–586. [CrossRef]
26. Chiacchio, P.; Chiaverini, S.; Sciavicco, L.; Siciliano, B. Global task space manipulability ellipsoids for multiple-arm systems. *IEEE Trans. Robot. Autom.* **1991**, *7*, 678–685. [CrossRef]
27. Erhart, S.; Hirche, S. Internal Force Analysis and Load Distribution for Cooperative Multi-Robot Manipulation. *IEEE Trans. Robot.* **2017**, *31*, 1238–1243. [CrossRef]
28. Duan, J.; Gan, Y.; Chen, M.; Dai, X. Adaptive variable impedance control for dynamic contact force tracking in uncertain environment. *Robot. Auton. Syst.* **2018**, *102*, 54–65. [CrossRef]

applied sciences

MDPI

*Article*

# Pick and Place Operations in Logistics Using a Mobile Manipulator Controlled with Deep Reinforcement Learning

**Ander Iriondo [1,*], Elena Lazkano [2], Loreto Susperregi [1], Julen Urain [1], Ane Fernandez [1] and Jorge Molina [1]**

1 Department of Autonomous and Intelligent Systems, Fundación Tekniker, Iñaki Goenaga, 5-20600 Eibar, Spain; loreto.susperregi@tekniker.es (L.S.); julen.urain@tekniker.es (J.U.); ane.fernandez@tekniker.es (A.F.); jorge.molina@tekniker.es (J.M.)

2 Faculty of Computer Science, P° Manuel Lardizabal, 1-20018 Donostia-San Sebastián, Spain; e.lazkano@ehu.es

* Correspondence: ander.iriondo@tekniker.es

Received: 30 December 2018; Accepted: 17 January 2019; Published: 21 January 2019

**Abstract:** Programming robots to perform complex tasks is a very expensive job. Traditional path planning and control are able to generate point to point collision free trajectories, but when the tasks to be performed are complex, traditional planning and control become complex tasks. This study focused on robotic operations in logistics, specifically, on picking objects in unstructured areas using a mobile manipulator configuration. The mobile manipulator has to be able to place its base in a correct place so the arm is able to plan a trajectory up to an object in a table. A deep reinforcement learning (DRL) approach was selected to solve this type of complex control tasks. Using the arm planner's feedback, a controller for the robot base is learned, which guides the platform to such a place where the arm is able to plan a trajectory up to the object. In addition the performance of two DRL algorithms ((Deep Deterministic Policy Gradient (DDPG)) and (Proximal Policy Optimisation (PPO)) is compared within the context of a concrete robotic task.

**Keywords:** deep reinforcement learning; mobile manipulation; robot learning

## 1. Introduction

Logistics applications demand the development of flexible, safe and dependable robotic solutions for part-handling including efficient pick-and-place solutions.

Pick and place are basic operations in most robotic applications, whether in industrial setups (e.g., machine tending, assembling or bin picking) or in service robotics domains (e.g., agriculture or home). In some structured scenarios, picking and placing is a mature process. However, that is not the case when it comes to manipulating parts with high variability or in less structured environments. In this case, picking systems only exist at laboratory level, and have not reached the market due to factors such as lack of efficiency, robustness and flexibility of currently available manipulation and perception technologies. In fact, the manipulation of goods is still a potential bottleneck to achieve efficiency in the industry and the logistic market.

At the same time, the market demands more flexible systems that allow for a reduction of costs in the supply chain, increasing the competitiveness for manufacturers and bringing a cost reduction for consumers. The introduction of robotic solutions for picking in unstructured environments requires the development of flexible robotic configurations, robust environment perception, methods for trajectory planning, flexible grasping strategies and human–robot collaboration.

This study focused specifically on the development of adaptive trajectory planning strategies for pick and place operations using mobile manipulators. A mobile manipulator is a mobile platform carrying one or more manipulators. The typical configuration of mobile manipulators in industry is an anthropomorphic manipulator mounted on a mobile platform complemented with sensors (laser, vision, and ultrasonic) and tools to perform operations. The mobility of the base extends the work-space of the manipulator, which increases the operational capacity.

Currently, mobile manipulators implementation in industry is limited. One of the main challenges is to establish coordinated movements between the base and the arm in a unstructured environment depending on the application.

Although navigation and manipulation are fields where much work has been done, mobile manipulation is a less known area because it suffers from the difficulties and uncertainties of both previously mentioned fields. We propose to solve the path planning problem using a reinforcement learning (RL) strategy. The aim is to avoid explicit programming of hard-to-engineer behaviours, or at least to reduce it, taking into account the difficulty of foreseeing situations in unstructured environments. RL-based algorithms have been successfully applied to robotic applications [1] and enable learning complex behaviours only driven by a reward signal.

Specifically, this study focused on learning to navigate to such a place that the mobile manipulator's arm is able to pick an object from a table. Due to the limited scope of the arm, not all positions near the table are feasible to pick the object and to calculate those positions analytically is not trivial. The goal of our work was to evaluate the performance of deep reinforcement learning (DRL) [2] to acquire complex control policies such as mobile manipulator positioning by learning only through the interaction with the environment. Specifically, we compared the performance of two model-free DRL algorithms, namely Deep Deterministic Policy Gradient (DDPG) and Proximal Policy Optimisation (PPO) algorithms, in two simulation tests.

The simulated robot we used is the mobile manipulator miiwa, depicted in Figure 1, which has been used in industrial mobile manipulation tasks (e.g., [3]).

**Figure 1.** Kuka miiwa.

## 2. Literature Review

There are different approaches to establishing coordinated movements between the base and the arm depending on the application. Nassal et al. [4] specified three types of cooperation, which differ in the degree of cooperation, the associated complexity and the potential for manipulation capabilities: (1) Loose cooperation is when the mobile base and the manipulator are considered two separate systems, and the base serves as transport. There are several very well-known methods for navigation and trajectory planning [5]. (2) Full cooperation is when the two systems are seen as one (with nine degrees of freedom), and both the base and the arm move simultaneously to position the tool center. There are various approaches to solve the path planning problem [6,7], however, in these cases, the computational cost is very high. (3) Transparent cooperation is the combination of the previous two: the manipulator control compensates for the base and the base moves accordingly to maximise a cost function related to the positioning of the manipulator to perform the task. In [4], an approach to this type of control is proposed.

In general, current robotic solutions follow the loose cooperation approach. The coordination of the two subsystems for mobile manipulation depends on the task that needs to be solved. Berntorp et al. [8] addressed the issue of pick and place where the robotic system must take a can of a known position and place it in another combining the movement of the arm and the base. In [9], the problem of opening doors is considered where the movement of the arm and base are coupled by sensing the forces generated between the arm and the door, and coordinating the forward movement of the base.

One of the principal goals of artificial intelligence (AI) is to enable agents to learn from their experience to become fully autonomous. This experience is obtained by interacting with the environment. Thus, the agent should continue improving its behaviour through trial and error until it behaves optimally. Deep learning enables reinforcement learning to face decision-making problems that were previously infeasible. The union of both of them, namely DRL, makes it possible to learn complex policies guided by a reward signal and has been applied to learn multiple complex decision making tasks in a wide variety of areas. For instance, in [10], the authors learn a chess player agent by self playing that is able to defeat a world champion. In [11], DRL is used to learn an intelligent dialogue generator agent. It also has been successfully applied to computer vision, for example to train a 2D object tracker agent that outperforms state-of-the-art real-time trackers [12].

The field of DRL applied to robotics is becoming more and more popular, and the number of published papers related to that topic is increasing quickly. Applications extend from manipulation [13–16] to autonomous navigation [17,18] and locomotion [19,20].

DRL has been successfully applied in some robotic path planning and control problems. For example, Gu et al. [16] used DRL to learn a low-level controller that is able to control a robotic arm to perform door opening. On the one hand, 25 variables are used to represent the environment's state. More specifically, those variables correspond to the joint angles and their time derivatives, the goal position and the door angle together with the handle angle. On the other hand, the action is represented by the torque for each joint. Robot navigation is another area where DRL also has been applied successfully; for example, in [18], DRL is applied to learn a controller to perform mapless navigation with a mobile robot. In this work, a low dimensionality state representation is also used, specifically 10-dimensional laser range findings, the previous action and the relative target position. The action representation includes linear and angular velocities of the differential robot.

Alternatively, several methods use 2D or 3D images instead of low-dimensional state representations. In [15], for example, 2D images are used with the goal of learning grasping actions. The authors tried to map images directly with low-level control actions and, in addition, the developed system is able to generalise and pick objects that the model has not been trained with. Moreover, other applications (e.g., [21]) use 3D depth images as state representation, but instead of learning directly from pixels, they encode the images to a lower dimensional space before training the model. This encoding enables to accelerate the training process.

Although there are multiple works about learning to control either robotic arms and mobile bases, to the best of our knowledge, references about DRL applied to mobile manipulation are few. The goal of this paper is to show that it is possible to use the arm path planner's feedback to learn a low-level controller for the base. The learned controller is able to guide the robot to a feasible pose in the environment to do the picking trial. To do that, a low-dimensional state representation is used.

In [22], the authors proposed to learn, through interaction with the environment, which place is the best to locate the base of a mobile manipulator, in such a way that the arm is able to pick an object from a table. In that way, the robot takes into account its physical conditions and also the environment's conditions to learn which is the optimal place to perform the grasping action. In this case, they acquire experience off-line, and, after applying some classifiers, such as support vector machines, they are able to learn a model that maps the state of the robot with some feasible places in the environment to do the picking trial. Then, the trained model is used on-line.

In comparison with the previous approach, the goal of the RL-based algorithms is to learn on-line, while the agent interacts with the environment, improving its policy until it reaches to the optimal behaviour. In addition, the goal of our work is to learn a controller, which gives a low-level control signal in each state to drive the mobile robot, while, in [22], instead of learning a controller to drive the robot up to the optimal pose, they tried to find directly which is the optimal pose.

## 3. Methodological Approach

According to Sutton and Barto [23], a reinforcement learning solution to a control problem is defined as a finite horizon Markov Decision Process (MDP). At each discrete time-step $t$, the agent observes the current state of the environment $s_t \in S$, takes an action $a_t \in A(s_t)$, receives a reward $r{:}S \times A \rightarrow \mathbb{R}$ and observes the new state of the environment $s_{t+1}$. At each episode of $T$ time-steps, the environment and the agent are reset to their initial poses. The goal of the agent is to find a policy, deterministic $\pi_\theta(s)$ or stochastic $\pi_\theta(a|s)$, parameterised by $\theta$ under which the expected reward is maximised.

Traditional reinforcement learning algorithms are able to deal with discrete state and action spaces but, in robotic tasks, where both state and action spaces are continuous, discretising those spaces does not work well. Nevertheless, most state-of-the-art algorithms use deep neural networks to map observations with low-level control actions to be able to deal with continuous spaces. In our approach, a DRL algorithm is used to learn where to place the mobile manipulator to make a correct picking trial through the interaction with the environment. Those DRL algorithms need a huge amount of experience to be able to learn complex robotic behaviours and, thus, it is infeasible to train them acquiring experience in real world. In addition, the actions taken by the robot in the initial learning iterations are nearly random and both the robot and the environment might end up damaged as a result. Therefore, DRL algorithms are usually trained in simulation. Learning in simulation enables a faster experience acquisition and avoids material costs. Our study used a simulation-based approach, and we based it on open source robotic tools such as Gazebo simulator, Robot Operating System (ROS) middleware [24], OpenAI Baselines DRL library [25] and Gym/Gym-gazebo toolboxes [26,27].

DRL algorithms follow value-based, policy search or actor–critic architectures [28]. Value based-algorithms estimate the expected reward of an state or state–action pair and are able to deal with discrete action spaces, typically using greedy policies. To be able to deal with continuous action spaces, policy search methods use a parameterised policy and do not need a value function. Usually, those methods have the difficulty of not easily being able to evaluate the policy and they have high variance. Actor–critic architecture is the most used one in the state-of-the-art algorithms and combines the benefits of the two previous algorithm types. On the one hand, actor–critic-based methods use a parameterised policy (actor), and, on the other hand, use a value or action–value function that evaluates the quality of the policy (critic).

The proposed approach follows the actor–critic architecture. On the one hand, a parameterised policy $\pi_\theta$ is used to be able to deal with both continuous state and action spaces in stochastic environments, encoded by the parameter vector $\theta$. On the other hand, a parameterised value function is used to estimate the expected reward at each state or state–action pair, where $w$ is the parameter vector that encodes the critic. The state value function $V_w(s)$ estimates the expected average reward of all actions in the state $s$. The action–value function $Q_w(s,a)$, instead, estimates the expected reward of executing action $a$ in state $s$. Then, the critic's information is used to update both actor and critic. In DRL algorithms, both actor and critic are parameterised by deep neural networks, and the goal is to optimise those parameters to get the agent's optimal behaviour.

In addition, DRL algorithms are divided into two groups, on-policy and off-policy, depending on how they are able to acquire experience. On-policy algorithms expect that the experience used to optimise their behaviour policy is generated by the same policy. Off-policy methods, instead, can use experience generated by another policy to optimise its behaviour policy. Those methods are said to be able to better explore the environment than on-policy methods because they use a more exploratory policy to get experience. In this study, we compared an on-policy algorithm and an off-policy algorithm, to see which type of methods adjusts better to our mobile manipulation behaviour learning. Specifically, PPO and DDPG were used, being those on-policy and off-policy, respectively. The first one learns stochastic policies, which maps states with actions represented by Gaussian probability distributions. DDPG, instead, is able to deterministically map states with actions. Both algorithms follow actor–critic architecture.

### 3.1. *Algorithms*

In this section, we describe the theoretical basis of PPO and DDPG.

#### 3.1.1. PPO

PPO [29] follows the actor–critic architecture and it is based on the trust-region policy optimisation (TRPO) [30] algorithm. This algorithm aims to learn a stochastic policy $\pi_\theta(a_t|s_t)$ that maps states with Gaussian distributions over actions. In addition, the critic is a value function $V_w(s_t)$ that outputs the mean expected reward in state $s_t$. This algorithm has the benefits of TRPO and in general of trust region based methods but it is much simpler to implement it. The intuition behind trust-region based algorithms is that, at each parameter update of the policy, the output distribution cannot diverge too much from the original distribution.

To update the actor's parameters, a clipped surrogate objective is used. Although another loss function is also proposed, using a Kullback–Leibler (KL) divergence [31] penalty on the loss function instead of the clipped surrogate objective, the experimental results obtained are not as good as with the clipped one. Let $r_t(\theta)$ denote the probability ratio defined in Equation (1), so that $r_t(\theta_{old}) = 1$.

$$r_t(\theta) = \frac{\pi_\theta(a_t|s_t)}{\pi_{\theta_{old}}(a_t|s_t)} \tag{1}$$

$\theta_{old}$ is the actor's parameter vector before the update. The objective of TRPO is to maximise the objective function $L(\theta)$ defined in Equation (2). Here, $\mathbb{E}[...]$ indicates the average over a finite batch of samples. The usage of the advantage $\hat{A}_t$ in policy gradient algorithms was popularised by Schulman et al. [32] and indicates how good the performed action is with respect to the average actions performed in each state. To compute the advantages, the algorithm executes a trajectory of $T$ actions and computes them as defined in Equation (3). Here, $t$ denotes the time index $[0, T]$ in the trajectory of length $T$ and $\gamma$ is the discount factor.

$$L(\theta) = \mathbb{E}\left[\frac{\pi_\theta(a_t|s_t)}{\pi_{\theta_{old}}(a_t|s_t)}\hat{A}_t\right] \tag{2}$$

$$\hat{A}_t = -V_w(s_t) + r_t + \gamma r_{t+1} + ... + \gamma^{T-t+1} r_{T-1} + \gamma^{T-t} V_w(s_T) \tag{3}$$

At each policy update, if the advantage has a positive value, the policy gradient is pushed in that direction because it means that the action performed is better than the average. Otherwise, the gradient is pushed in the opposite direction.

Without any constraint, the maximisation of the loss function $L(\theta)$ would lead to big changes in the policy at each training step. PPO modifies the objective function so that penalises big changes in the policy that move $r_t(\theta)$ away from 1. Maintaining $r_t(\theta)$ near to 1 ensures that, at each policy update, the new distribution does not diverge to much from the old one. The objective function is defined in Equation (4).

$$L^{CLIP}(\theta) = \mathbb{E}[min(r_t(\theta)\hat{A}_t, clip(r_t(\theta), 1-\epsilon, 1+\epsilon))\hat{A}_t] \tag{4}$$

Here, $\epsilon$ is a hyper-parameter that changes the clip range.

To update the value function $V_w(s)$ (the critic), the squared-error loss function is used (Equation (5)) between the current state value and a target value. The target value is defined in Equation (6).

$$J(w) = (V_w(s_t) - V_t^{targ})^2 \tag{5}$$

$$V_t^{targ} = \hat{A}_t + V_w(s_t) \tag{6}$$

The PPO algorithm is detailed in Algorithm 1. Although this algorithm is designed to be able to have multiple parallel actors getting experience, only one actor is being used.

**Algorithm 1** Proximal Policy Optimisation (PPO).

**for** $e \in episodes$ **do**
    **for** $a \in actors$ **do**
        Run policy $\pi_{\theta_{old}}$ in environment for T time-steps
        Compute advantage estimates $\hat{A}_1 ... \hat{A}_T$
    **end for**
    Optimise actor's loss function $L^{CLIP}$ with regard to $\theta$, with $K$ epochs and minibatch size N $\leq T \cdot actors$
    $\theta_{old} \leftarrow \theta$
**end for**

#### 3.1.2. DDPG

DDPG [33] combines elements of value function based and policy gradient based algorithms, following the actor–critic architecture. This algorithm aims to learn a deterministic policy $\pi_\theta(s) = a$ and it is derived from the deterministic policy gradient theorem [34].

Following the actor–critic architecture, DDPG uses an action–value function $Q_w(s, a)$ as critic to guide the learning process of the policy and it is based on the deep Q-network (DQN) [35]. Prior to DQN, it was believed that learning value functions with large and nonlinear function approximators was difficult. DQN is able to learn robust value functions due to two innovations: First, the network is trained off-policy getting experience samples from a replay buffer to eliminate temporal correlations. In addition, target networks are used, which are updated more slowly, and this gives consistent targets during temporal difference learning.

The critic is updated according to the gradient of the objective defined in Equation (7).

$$L(w) = \mathbb{E}[(Q_w(s_t, a_t) - y_t)^2] \tag{7}$$

where

$$y_t = r(s_t, a_t) + \gamma Q_w(s_{t+1}, a_{t+1})|_{a_{t+1} = \pi_\theta(s_{t+1})} \tag{8}$$

The actor is updated following the deterministic policy gradient theorem, defined in Equation (9). The intuition is to update the policy in the direction that improves $Q_w(s,a)$ most .

$$\nabla J(\theta) = \mathbb{E}[\nabla_\theta \pi_\theta(s_t) \nabla_a Q_w(s_t, a_t)|_{a=\pi_\theta(s_t)}] \tag{9}$$

As mentioned before, the target value defined in Equation (8) is calculated using the target networks $\pi'_{\theta'}$ and $Q'_{w'}$, which are updated more slowly and this gives more consistent targets when learning the action–value function.

As DDPG is an off-policy algorithm, the policy used to get experience is different from the behaviour policy. Despite the behaviour policy being deterministic, typically, a stochastic policy is used to get experience, being able to better explore the environment. This exploratory policy is usually achieved adding noise to the behaviour policy. Although there are common noises such as normal noise or Ornstein–Uhlenbeck noise [36], which are added directly to the action generated by the policy, in [37], Plapperta et al. proposed adding noise to the neural network's parameter space to improve the exploration and to reduce the training time. The DDPG algorithm is described in Algorithm 2.

**Algorithm 2** Deep Deterministic Policy Gradient (DDPG).

```
Initialise the actor π_θ(s) and the critic Q_w(s, a) networks.
Initialise the target networks Q' y π' with the weights θ' ← θ, w' ← w
Initialise the replay buffer
for e ∈ episodes do
    Initialise noise generation process
    Get the first observation
    for t ∈ steps do
        Select the action a_t = π_θ(s_t) + N
        Execute the action a_t, get the reward r_t and the observation s_{t+1}
        Store the transition < s_t, a_t, r_t, s_{t+1} > in replay buffer
        Get M experience samples < s_i, a_i, r_i, s_{i+1} > from the replay buffer
        y_i = r_i + γQ'_{w'}(s_{i+1}, π'_{θ'}(s_{i+1}))
        Update the critic minimising the loss :
```

$$L = \frac{1}{M}\sum_i (y_i - Q_w(s_i, a_i))^2$$

```
        Update the policy of the actor:
```

$$\nabla_\theta J \approx \frac{1}{M}\sum_i \nabla_a Q_w(s,a)|_{s=s_i, a=\pi_\theta(s_i)} \nabla_\theta \pi_\theta(s)|_{s=s_i}$$

```
        Update target networks:
```

$$\theta' \leftarrow \tau\theta + (1-\tau)\theta'$$

$$w' \leftarrow \tau w + (1-\tau)w'$$

```
    end for
end for
```

### 3.2. Simulated Layout

To model our world, we used the Gazebo model based simulator. The elements that are placed in this simulated world are the robot miiwa, the table and an object, which is on top of the table, as depicted in Figure 2.

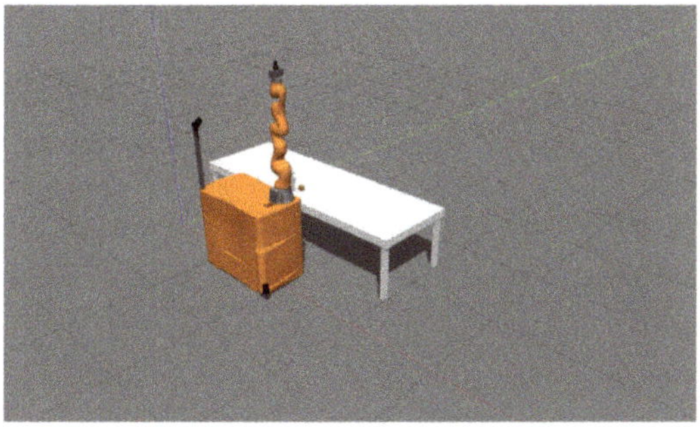

**Figure 2.** Simulated world in Gazebo.

The mobile manipulator miiwa is composed of a 7 DoF arm and a 3 DoF omnidirectional mobile base. To be able to control the mobile base, the *gazebo planar move* plugin was used.

To test if our environment was modelled correctly and to know if the algorithms could learn low-level control tasks in that environment, the learning process was divided into two simulation tests. In both, the algorithm must learn to control the mobile base with velocity commands, such that, at each discrete time-step, the algorithm gets the state of the environment and publishes a velocity command. The objective of those tests was to learn a low-level controller to drive the robot to a place where the arm can plan a trajectory up to the object on the table. To learn those controllers, the feedback of the arm's planner was used. The summary of the tests is listed in Table 1.

**Table 1.** Test setup summary.

| Test | Robot Initial Pose | Box Initial Pose | Objective |
|---|---|---|---|
| Test 1 | Variable | Variable | The arm to be able to plan a trajectory up to the box |
| Test 2 | Variable | Constant | To plan a trajectory with an obstacle in the table |

## 4. Implementation

The application was implemented in a modular way using ROS for several reasons. On the one hand, it enabled us to modularise the application and to parallelise processes. On the other hand, Gazebo is perfectly integrated with ROS and offers facilities to control the simulation using topics, services and actions.

OpenAI Gym is a toolkit to do research on DRL algorithms and to do benchmarks between DRL algorithms. This library includes some simulated environments that are used to test the quality of new algorithms and its use is widespread in the DRL community. Gym offers simple functions to interact with the environments, which are mostly modelled in the Mujoco [38] simulator. Due to the simplicity of the interface that Gym offers, it has become a standard way to interact with environments in DRL algorithms. Gym-gazebo is an extension of Gym that enables the user to create robotic environments in Gazebo and offers the same simple interface to the algorithm to be able to interact with the environment. All the environments we modelled were integrated with Gym-gazebo, which enabled us to straightforwardly use OpenAI Baselines DRL library, which is designed to work with Gym.

Gym-gazebo wraps the used DRL algorithms in ROS nodes and enables interaction with the environment. Nevertheless, another ROS node has been developed that is in charge with controlling all the elements of the simulation and works as bridge between the Gym-gazebo node and the simulator. To be able to control the simulation physic updates and to compute them as fast as possible, we developed a Gazebo plugin, which in turn is a ROS node. Thus, using ROS communication methods, we could control each simulated discrete time-step and we simulated those steps faster than real time.

Specifically, this node takes care of all time-steps being of the same length, executing a fixed number of physic updates at each step.

The *tf broadcaster* node uses the *tf* tool that ROS offers to keep track of all the transformations between frames. We used this node to publish some transformations such as the transformation between the *object* and *world* frames. Consequently, the robot is always aware of where the object is in order to be able to navigate up to it. The implemented architecture is shown in Figure 3.

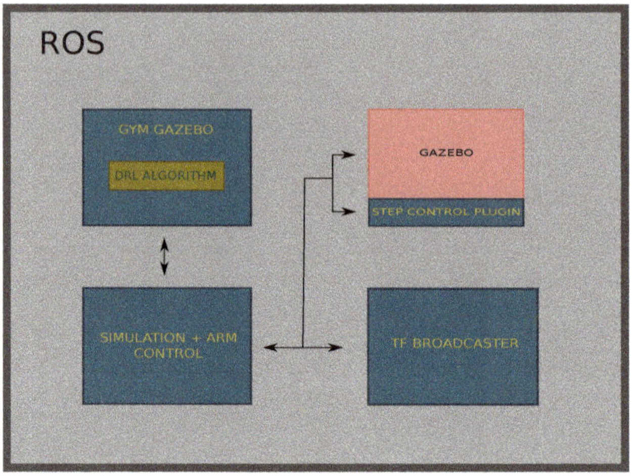


**Figure 3.** Implemented architecture. ROS: Robot Operating System.

### 4.1. Simulation

The learning process was carried out during a fixed number of time-steps, which were divided into episodes of 512 time-steps. Gazebo's *max step size* $T_m$ is an important parameter that indicates the time-step at which Gazebo computes successive states of the model. The default value is 1 ms, but, in this case, $T_m = 2$ ms was used, which gave enough stability and enabled a faster simulation. Thus, each iteration of the physic engine meant 2 ms of simulated time and those iterations could be computed as quickly as possible to be able to accelerate the simulation. Besides, the *real time update rate* $U_r$ parameter indicates how many physic iterations will be tried in one second. The *real time factor* is defined in Equation (10) and indicates how much faster the simulation goes in comparison with the real time. Thus, $rtf = 1$ indicates that the simulation is running on real time.

$$rtf = T_m \cdot U_r \tag{10}$$

In addition, a *frame-skip* $N_{T_m}$ was defined to be able to get a reasonable control rate. In this application, $N_{T_m} = 4$ was used so the discrete time-step size is defined in Equation (11) and doing that we achieve a control rate of 125 Hz.

$$T_s = T_m \cdot N_{T_m} = 8 \text{ ms} \tag{11}$$

Thus, the length of each discrete time-step is 8 ms and those steps are computed faster than real time. Gazebo does not allow the control of the physics engine iterations, so a Gazebo plugin was developed to be able to execute the simulation for some iterations and to be able to compute those iterations as fast as possible. When the plugin was told to run a time-step, it ran $N_{T_m}$ physics iterations and the *real time update rate* was set to 0 to compute those iterations as quickly as possible. Doing this, the *real time factor* increased and enabled us to run the simulation about 5–10 times faster than in real time.

At each step, the learning algorithm generates an action, specifically a velocity command, which is sent to the mobile base. After executing it and at the end of the step, a reward signal is given evaluating the quality of the action performed by the base. The action is the same for all the experiments and is defined in Equation (12).

$$a = [v_x \ \ v_y \ \ \omega_z] \tag{12}$$

To be able to control the mobile base, the *gazebo planar move* plugin was used, which enabled the sending of velocity commands to the omnidirectional base. In addition, it published the odometry information, which was used to know where the robot is respect to a parent coordinate frame. The path planning and control of the arm was made using MoveIt! [39] and enabled us to plan collision free trajectories. Besides that, MoveIt! uses an internal planning scene where objects can be added to be taken into account when the trajectory is planned.

### 4.2. Network Architectures

The library of DRL algorithms used is OpenAI Baselines. This library offers high quality implementations of many state-of-the-art DRL algorithms implemented in Tensorflow [40]. Although the implementation offers complex networks such as convolutional neural networks (CNN) or recurrent neural networks (RNN), we used fully connected multi-layer perceptrons (MLP) to parameterise both policies and value functions.

#### 4.2.1. PPO

The network architecture used in this algorithm is the one proposed by Schulman et al. [29] in the original paper. To parameterise both the policy $\pi_\theta(a_t|s_t)$ and the value function $V_w(s_t)$, a four layered MLP was used. In both actor and critic, the first layer's size depends on the size of the state encoding, which was different in each test. The actor's input layer is followed by two hidden layers of 64 neurons each. The output layer's size depends on the action space's size. In this case, the action was a vector of three elements representing the velocity command, which has to be sent to the mobile base. Specifically, the velocity command is composed of the linear velocities in $x$ and $y$ axes, and the angular velocity in $z$ axis. As mentioned before, this algorithm uses a stochastic policy and, thus, each action is defined by a Gaussian distribution. Each distribution is characterised by a mean and a standard deviation. Hence, the neurons of the last hidden layer are fully connected with the mean of each action. In addition, three variables are used to store and update the standard deviation.

The activation function applied to the output of each neuron of the hidden layers is *Tanh*. Instead, no activation function is applied to the output of the neurons of the last layer.

Regarding the value function $V_w(s_t)$, the first three layers have the same architecture as the actor's first three layers. The output layer, instead, is composed of a single neuron, which is fully connected with each neuron of the last hidden layer and outputs the expected average reward of a state. The activation function used in the hidden layers is *Tanh* as well.

#### 4.2.2. DDPG

The network architecture used is the default implementation that the OpenAI Baselines library offers, which is very similar to the architecture proposed by Lillicrap et al. [33]. As in PPO, to parameterise both the policy $\pi_\theta(s_t)$ and the action–value function $Q(s_t, a_t)$, a four layered MLP was used. This algorithm aims to learn a deterministic policy so each state will be mapped with a concrete action. The actor's input layer also depends on the state encoding and it is followed by two hidden layers, composed of 64 neurons each. Concerning the output layer, it is composed of three neurons, one per action, and each neuron is fully connected with each neuron of the last hidden layer.

With respect to the activation functions, the *Rectified Linear Unit* (ReLU) function is used in each neuron of the hidden layers. In the output layer, instead, *Tanh* is used to bound the actions between $-1$ and 1. In addition, a process called *layer normalisation* [41] is applied to the output of each hidden layer to simplify the learning process and to reduce the training time. DDPG is an off-policy algorithm, hence it uses a more exploratory policy to get experience. As explained in the algorithm's description (Section 3.1.2), there are multiple types of noise but, in this application, the noise is added in the neural network's parameter space.

The action–value function $Q_w(s, a)$ receives as input the state and the action and outputs the expected future reward of executing action $a_t$ in state $s_t$. The input layer depends on the size of the state codification. The input layer is followed by two hidden layers of 64 neurons each and actions are not included until the second hidden layer. The output layer only has one neuron and outputs the $q$-value. The activation function used in the critic network is ReLU for each neuron of the hidden layers. The last layer's neurons do not have any activation function, because the action–value does not have to be bounded.

To minimise the complexity of the critic and to avoid over-fitting, a $L2$ regularisation term is added to the critic's loss function [42]. This penalises the model for having parameters with high values.

### 4.3. Test Setup

Here, we describe the simulation setup in addition to the state codification and the reward function we used in each test. For both PPO and DDPG, the setup and reward functions were the same in all tests.

The robot's work-space is limited and a new episode starts when it goes out of limits. Those limits are defined in Equation (13).

$$x_r^w = [-1.0, 3.0] \quad y_r^w = [-1.5, 0.0] \tag{13}$$

In addition, although the robot was penalised by navigating with high speed, the environment bounds the velocities applied to the mobile base. The maximum linear velocity in $x$ and $y$ axes was 1 m/s and the maximum angular velocity in $z$ axis was 1 rad/s. Thus, if the algorithm predicted velocities higher than those, the limit velocities were applied and the agent was penalised proportionally. Besides, high accelerations were also penalised proportionally, but in this case the environment did not bound them.

The learning process was divided into episodes where the robot had $T$ time-steps to complete the task. The episode length was $T = 512$ time-steps, which is about 4 s, and a discrete time-step $t$ was terminal if:

- The robot collides with the table.
- Robot poses out of limits.
- $t = T$.

The tuning hyper-parameters used in each algorithm are described in Table 2 and were not changed across tests.

**Table 2.** Hyper-parameters.

| **PPO Setup Hyper-Parameters** | |
|---|---|
| Actor/Critic learning rate | $f(step) = step \cdot 3 \times 10^{-4}$ |
| Clip-range | 0.2 |
| Discount factor $\gamma$ | 0.99 |
| Batch size | 512 |
| Mini-batch size | 64 |
| Updates | $\frac{training\ time\_steps}{batch\_size}$ |
| Training epochs per update | 10 |
| **DDPG Setup Hyper-Parameters** | |
| Actor's learning rate | $1 \times 10^{-4}$ |
| Critic's learning rate | $1 \times 10^{-3}$ |
| Batch size | 128 |
| Discount factor $\gamma$ | 0.99 |
| Critic l2 regularisation | $1 \times 10^{-2}$ |
| Running epochs | 500 |
| Cycles per epoch | 10 |
| Rollouts per cycle | 512 |
| Updates | $epochs \cdot cycles$ |
| Training iterations per update | 100 |

#### 4.3.1. Test 1

Here, the objective was to learn a controller that guides the robot to a place near the table so that the arm could plan a trajectory up to the object. The robot and object initial poses were variable, which are defined in Equations (14) and (15). The robot's initial $y_r^w$ coordinate was constant so that the robot always began at the same distance from the table. The state codification is defined in Equation (16) and it is composed of the following 10 variables:

Robot's position in world coordinate system: $x_r^w, x_r^w$
Robot's rotation on z axis in world coordinate system: $yaw_r^w$
Robot's linear velocities on $x$ and $y$ axes: $v_x, v_y$
Robot's angular velocity in z axis: $\omega_z$
Object's position in world coordinate system: $x_{obj}^w, y_{obj}^w$
Distance between the robot and the object: $d(p_r, p_{obj})$
Remaining time steps to end the episode: $t$

$$x_r^w = [0.0, 3.0] \quad y_r^w = -1.5 \quad yaw_r^w = [-\pi, \pi] \tag{14}$$

$$x_{obj}^w = [0.5, 2.5] \quad y_{obj}^w = [0.5, 0.75] \quad z_{obj}^w = 1.2 \tag{15}$$

$$s = [x_r^w \ y_r^w \ yaw_r^w \ v_x \ v_y \ \omega_z \ x_{obj}^w \ y_{obj}^w \ d(p_r, p_{obj}) \ t] \in \mathbb{R}^{10} \tag{16}$$

The linear and the angular velocities included in the state were the velocities sent in the previous time-step to the robot (previous action). The reward function used is defined in Equation (17). A nonlinear function was used to give higher rewards when the robot was close to the object and high velocities and accelerations were penalised to encourage smooth driving. Here, $\Delta v$ means the velocity difference (acceleration) between current and previous action. Instead, $v\_high$ penalised each linear or angular velocity being higher than a threshold. As mentioned before, the maximum linear and angular velocities were 1 m/s and 1 rad/s, respectively. In addition, in the last time-step of the episode, an additional reward was given if the arm could plan a trajectory up to the object. The number of remaining time-steps was included in the state for robot to be aware when this last step was coming. The *collision* variable was equal to 1 when a collision occurred, and 0 otherwise.

$$r(s_t, a_t) = \frac{1}{d(p_r, p_{obj})} \cdot (1 - collision) - 0.5 \cdot \Delta v - 0.5 \cdot v\_high(a_t) + 100 \cdot success \tag{17}$$

$$v\_high(a_t) = \sum_{i=1}^{3} a_{t_i}, \text{ if } a_{t_i} > k \tag{18}$$

4.3.2. Test 2

In this test, as in the previous one, the robot had to navigate to a place near the table such that the arm could plan a trajectory up to the object in the table. In this case, a wall was placed near the object to see if the algorithm could discard poses that were behind the wall. The robot's initial pose was variable (Equation (19)) and the object's initial pose was constant (Equation (20)). As in the first test, the robot's initial $y_r^w$ coordinate was constant so that the robot always began at the same distance from the table. The state codification defined in Equation (21) is composed of the following 8 variables:

Robot's position in object's coordinate system: $x_r^{obj}, x_r^{obj}$
Robot's rotation on $z$ axis in object's coordinate system: $yaw_r^{obj}$
Robot's linear velocities on $x$ and $y$ axes: $v_x, v_y$
Robot's angular velocity in $z$ axis: $\omega_z$
Distance between the robot and object's coordinate system origin: $d(p_r, \mathbb{O})$
Remaining time steps to end the episode: $t$

$$x_r^w = [0.0, 3.0] \quad y_r^w = -1.5 \quad yaw_r^w = [-\pi, \pi] \tag{19}$$

$$x_{obj}^w = 1.7 \quad y_{obj}^w = 0.75 \quad z_{obj}^w = 1.005 \tag{20}$$

$$s = [x_r^{obj} \;\; y_r^{obj} \;\; yaw_r^{obj} \;\; v_x \;\; v_y \;\; \omega_z \;\; d(p_r, \mathbb{O}) \;\; t] \in \mathbb{R}^8 \tag{21}$$

The used reward function is defined in Equation (22). In this case, the distance was computed between the robot's position $p_r$ and object's coordinate system origin $\mathbb{O}$.

$$r(s_t, a_t) = \frac{1}{d(p_r, \mathbb{O})} \cdot (1 - collision) - 0.5 \cdot \Delta v - 0.5 \cdot v\_high(a_t) + 100 \cdot success \tag{22}$$

The environments used in the first and second tests are depicted in Figure 4a,b, respectively.

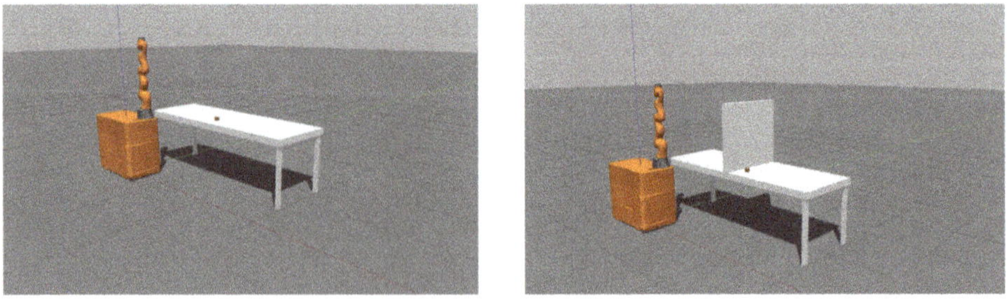

(**a**) Environment 1. (**b**) Environment 2.

**Figure 4.** Simulated environments.

## 5. Results

The learning process was carried out during a fixed number of time-steps and, to be able to evaluate the performance of the algorithms, several evaluation periods of 10 episodes each were made. Specifically, 500 evaluation periods were made uniformly distributed over the learning process and the metrics used to evaluate the performance were the mean accumulated reward and the success rate, which are the most used ones in the DRL community.

### 5.1. Test 1

The goal of this test was to learn a low-level controller that could drive the mobile manipulator's base close to the table, so that the arm could plan a trajectory up to the object. The learning process was carried out during 5M time-steps approximately, and an evaluation period was performed every 20 episodes. The results obtained with both PPO and DDPG algorithms are depicted in Figure 5. Figure 5a shows the mean accumulated rewards obtained in each of the 500 evaluation periods. The obtained success rates are depicted in Figure 5b. Due to the unstable nature of DRL algorithms, the obtained success rates vary considerably. Thus, to better understand the results, the mean value and the maximum/minimum values are shown per 10 test periods.

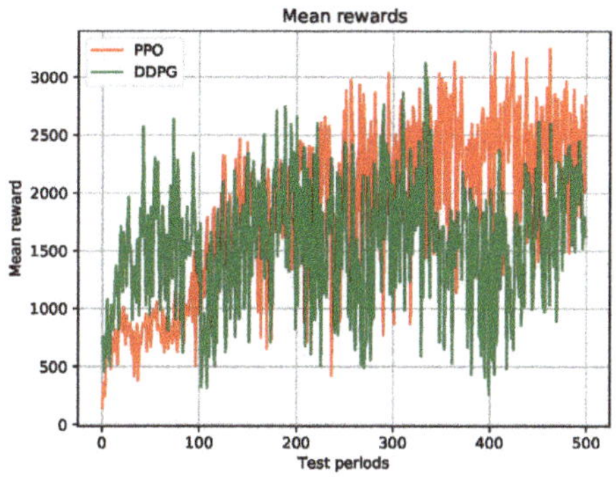


(a) Accumulated mean rewards.

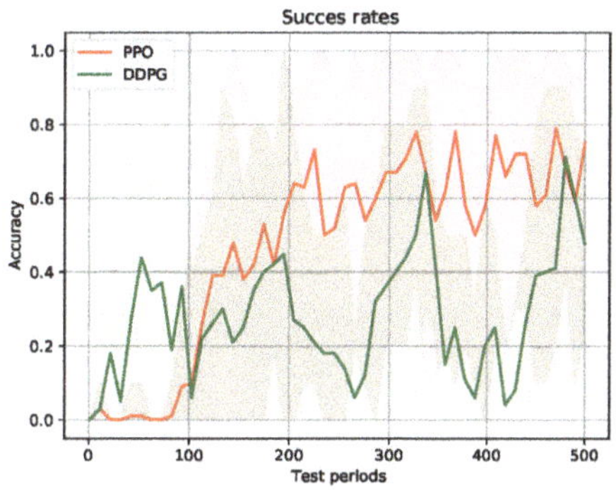


(b) Success rates.

**Figure 5.** Test 1 results.

Concerning the accumulated mean rewards, DDPG converged faster than PPO but, when the learning process moved along, PPO obtained higher rewards. The fact that DDPG obtained higher initial rewards is explained by two main reasons: (1) The policy bounds the actions between $-1$ and 1 so that it is not penalised for high velocities. (2) From the initial learning steps, this algorithm predicts smooth velocity changes across consecutive time-steps and, thus, is not penalised for high accelerations. Instead, PPO does not bound the actions and that is why it is penalised for high velocities and accelerations in the initial learning steps. Nevertheless, this algorithm is able to understand that lower velocities/accelerations are not penalised and, once learned, it obtains higher rewards than DDPG.

Regarding the success rates, it can be seen clearly that DDPG approximated the goal faster than PPO. Although, in some evaluation periods, it succeeded 100% of the time, it presented an unstable behaviour. As explained before, PPO takes more time to approximate the goal but, when it learns to drive smoothly, it shows a more stable behaviour. Besides, it could get 100% success rate considerably more times than DDPG. Off-policy algorithms are said to be able to better explore the environment than on-policy algorithms. In our environment, due to the initial random poses and the limited work-space of the robot, the exploration problem decreased considerably. PPO is an on-policy algorithm and takes more time to explore the environment than DDPG and that is another reason DDPG approaches the goal faster than PPO. Nevertheless, PPO relatively quickly improves enough for its policy to be able to succeed.

DRL has been applied to goal reaching applications either in manipulation or in navigation. Typically, those goals are not surrounded by obstacles and this makes the learning process easier. In this case, the goal (the object) was on top of the table and the robot had to learn to approximate to it without colliding with the obstacle. The reward function defined in this test encouraged the robot approximating to the object as much as possible and the robot had to use the contact information to learn where the table was to not collide with it. Although an additional reward was given when the arm's planning succeeded, it first tried to get as close as possible to the table, sometimes colliding with it and that is one of the reasons that caused the unstable behaviour of both algorithms.

Due to the variable pose of the object, in some episodes, if the object was near to table's edge, it had the possibility to get closer, scoring higher rewards. In addition, the initial poses of both the robot and the object in every test period were totally random and thus the accumulated rewards and/or success rates varied considerably.

### 5.2. Test 2

The goal of this test was to learn a low-level controller that could drive the mobile manipulator's base close enough to the table for the arm to be able to plan a trajectory up to the object. In this case, a wall was placed near the object with the aim of making the decision making problem more difficult. The learning process was carried out during 2.5M time-steps and an evaluation period aws performed every 10 episodes. In Figure 6, the results obtained with both PPO and DDPG algorithms are depicted.

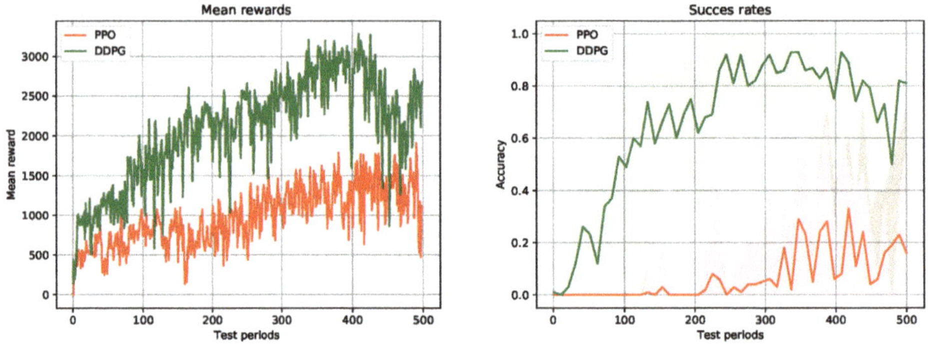


(a) Accumulated mean rewards. (b) Success rates.

**Figure 6.** Test 2 results.

Concerning the accumulated mean rewards (Figure 6a), DDPG's performance was better than PPO's for the overall learning process. Due to the bounded actions and the smooth output of DDPG, it could get higher rewards in the initial learning steps. Thanks to the fact that this algorithm is off-policy, it is able to better explore the environment and, thus, it learns to locate the robot's base much closer to the object than PPO. In addition, the constant pose of the box simplified the exploration problem, since the base always had to navigate to the same area of the environment and, consequently, the mean rewards were not as irregular as in the first test.

The success rates depicted in Figure 6b indicate that DDPG's performance was much better than PPO's, being the former able to score 100% success rate multiple times. DDPG could learn relatively quickly where the grasping zone is, discarding the poses over the wall. After navigating to the grasping zone, the learned policy stoped the robot sending near to zero velocities to the base. In addition, the deterministic behaviour of the algorithm enabled the robot to learn a more stable behaviour near the table, avoiding collisions. Even though PPO could learn to drive the robot near the correct grasping zone, it could not learn to stop it and continued sending velocities high enough to get the robot out of the grasping zone. Besides, due to the stochastic policy, the robot's performance near the table was not as robust as DDPG's, sometimes colliding with it.

## 6. Discussion and Future Work

In this work, we successfully implemented several DRL algorithms for learning picking operations using a mobile manipulator. It is shown that it is possible to use the arm's feedback to learn a low-level controller that drives the base to such a place that the arm is able to plan a trajectory up to the object.

Two state-of-the-art DRL algorithms were applied and compared to learn a mobile manipulation task. Specifically, the arm planner's feedback was used to learn to locate the mobile manipulator's base. To the best of our knowledge, this is the first approach that uses the arm's feedback to acquire a controller for the base.

Although DRL enables the learning of complex policies driven only by a reward signal, the unstable nature of those algorithms makes it difficult to obtain a robust behaviour. In addition, the sensitivity of DRL algorithms to hyper-parameters hinders finding the best parameter combination to get a robust and stable behaviour. Even though an optimisation over those hyper-parameters could be made, the large training times makes this process very expensive and commonly the default values proposed in the literature are used. Concerning the algorithms tested in this work, the results obtained show that the behaviour of the algorithms is dependant on several properties of the environment such as the state/action codification and the reward function definition. In addition, the network architecture used to encode either policies and the value-function has a large effect in the learning process. Even though the same network architecture and hyper-parameters were used in both tests, the results obtained are very different. Therefore, each environment needs the algorithm to be tuned in the best way to solve the problem, which is why the learned policies are not reproducible.

The reward function definition is another key point of the learning process, since it is entirely guided by this signal. A logical approach could be to give a reward to the robot in the last step only if the arm plans a trajectory, but, unfortunately, using only sparse rewards does not work well. In several goal reaching applications, either with arms or mobile bases, a distance dependant reward is proposed. In our application, this encourages the robot to navigate close to the object so the arm can plan a trajectory. After some tests, we saw that a nonlinear distance function accelerates the learning process.

Although DRL algorithms are typically trained in simulation, the experience acquisition is still the bottleneck of DRL based applications applied to robotics. Even though we accelerated the simulation, the entire learning process took several hours. Nevertheless, to be able to transfer the learned policy to a real robot and reduce the reality gap, the robot must be simulated accurately. Thus, a balance between simulation accuracy and training time acceleration should be found.

Moreover, it is complex to tune the algorithms to get a robust performance. We intend to increase the perception capabilities of the robot to be able to navigate in a more secure way and to be aware of the dynamical obstacles placed in the environment, using 2D/3D vision for example. Most of the applications in the literature map observations directly with low-level control actions and this black-box approach is not scalable. To be able to learn multiple behaviours and to combine them, hierarchical DRL proposes to learn a hierarchy of behaviours in different levels. In that vein, our goal is to learn a hierarchy of behaviours and, after training them in simulation, test those behaviours in a real robot.

**Author Contributions:** Conceptualisation, A.I. and L.S.; methodology, A.I., L.S., and E.L.; software, A.I.; formal analysis, A.I., J.U., A.F., and J.M.; data curation, A.I.; writing—original draft preparation, A.I.; writing—review and editing, E.L. and L.S.; and supervision, E.L., L.S.

**Funding:** This Project received funding from the European Union's Horizon 2020 research and Innovation Programme under grant agreement No. 780488.

**Conflicts of Interest:** The authors declare no conflict of interest.

## Abbreviations

The following abbreviations are used in this manuscript:

| | |
|---|---|
| RL | Reinforcement learning |
| DRL | Deep reinforcement learning |
| MDP | Markov Decision Process |
| ROS | Robot Operating System |
| CNN | Convolutional Neural Network |
| RNN | Recurrent Neural Network |
| MLP | Multi Layer Perceptron |
| DDPG | Deep Deterministic Policy Gradient |
| PPO | Proximal Policy Optimisation |

## References


1. Kober, J.; Bagnell, J.A.; Peters, J. Reinforcement learning in robotics: A survey. *Int. J. Robot. Res.* **2013**, *32*, 1238–1274. [CrossRef]
2. Li, Y. Deep reinforcement learning: An overview. *arXiv* **2017**, arXiv:1701.07274.
3. Dömel, A.; Kriegel, S.; Kaßecker, M.; Brucker, M.; Bodenmüller, T.; Suppa, M. Toward fully autonomous mobile manipulation for industrial environments. *Int. J. Adv. Robot. Syst.* **2017**, *14*. [CrossRef]
4. Nassal, U.; Damm, M.; Lüth, T. A mobile platform supporting a manipulator system for an autonomous robot. In Proceedings of the 5th World Conference on Robotics Research, Cambridge, MA, USA, 27–29 Spetember 1994.
5. Siciliano, B.; Khatib, O. *Springer Handbook of Robotics*; Springer: Berlin, Germany, 2016.
6. Padois, V.; Fourquet, J.Y.; Chiron, P. From robotic arms to mobile manipulation: On coordinated motion schemes. In *Intelligent Production Machines and Systems*; Elsevier: Amsterdam, The Netherlands, 2006; pp. 572–577.
7. Tan, J.; Xi, N.; Wang, Y. Integrated task planning and control for mobile manipulators. *Int. J. Robot. Res.* **2003**, *22*, 337–354. [CrossRef]
8. Berntorp, K.; Arzén, K.E.; Robertsson, A. Mobile manipulation with a kinematically redundant manipulator for a pick-and-place scenario. In Proceedings of the 2012 IEEE International Conference on Control Applications (CCA), Dubrovnik, Croatia, 3–5 October 2012; pp. 1596–1602.
9. Meeussen, W.; Wise, M.; Glaser, S.; Chitta, S.; McGann, C.; Mihelich, P.; Marder-Eppstein, E.; Muja, M.; Eruhimov, V.; Foote, T.; et al. Autonomous door opening and plugging in with a personal robot. In Proceedings of the 2010 IEEE International Conference on Robotics and Automation (ICRA), Anchorage, AK, USA, 3–8 May 2010; pp. 729–736.
10. Silver, D.; Hubert, T.; Schrittwieser, J.; Antonoglou, I.; Lai, M.; Guez, A.; Lanctot, M.; Sifre, L.; Kumaran, D.; Graepel, T.; et al. Mastering chess and shogi by self-play with a general reinforcement learning algorithm. *arXiv* **2017**, arXiv:1712.01815.
11. Li, J.; Monroe, W.; Ritter, A.; Galley, M.; Gao, J.; Jurafsky, D. Deep reinforcement learning for dialogue generation. *arXiv* **2016**, arXiv:1606.01541.
12. Yoo, S.; Yun, K.; Choi, J.Y.; Yun, K.; Choi, J. Action-Decision Networks for Visual Tracking with Deep Reinforcement Learning. In Proceedings of the 2017 IEEE Conference on Computer Vision and Pattern Recognition (CVPR), Honolulu, HI, USA, 21–26 July 2017.
13. Levine, S.; Pastor, P.; Krizhevsky, A.; Ibarz, J.; Quillen, D. Learning hand-eye coordination for robotic grasping with deep learning and large-scale data collection. *Int. J. Robot. Res.* **2018**, *37*, 421–436. [CrossRef]
14. Popov, I.; Heess, N.; Lillicrap, T.; Hafner, R.; Barth-Maron, G.; Vecerik, M.; Lampe, T.; Tassa, Y.; Erez, T.; Riedmiller, M. Data-efficient deep reinforcement learning for dexterous manipulation. *arXiv* **2017**, arXiv:1704.03073.
15. Quillen, D.; Jang, E.; Nachum, O.; Finn, C.; Ibarz, J.; Levine, S. Deep Reinforcement Learning for Vision-Based Robotic Grasping: A Simulated Comparative Evaluation of Off-Policy Methods. In Proceedings of the 2018 IEEE International Conference on Robotics and Automation (ICRA), Brisbane, QLD, Australia, 21–25 May 2018.

16. Gu, S.; Holly, E.; Lillicrap, T.; Levine, S. Deep reinforcement learning for robotic manipulation with asynchronous off-policy updates. In Proceedings of the 2017 IEEE International Conference on Robotics and Automation (ICRA), Singapore, 29 May–3 June 2017; pp. 3389–3396, doi:10.1109/ICRA.2017.7989385. [CrossRef]
17. Chen, Y.F.; Everett, M.; Liu, M.; How, J.P. Socially aware motion planning with deep reinforcement learning. In Proceedings of the 2017 IEEE/RSJ International Conference on Intelligent Robots and Systems (IROS), Vancouver, BC, Canada, 24–28 September 2017; pp. 1343–1350.
18. Tai, L.; Paolo, G.; Liu, M. Virtual-to-real deep reinforcement learning: Continuous control of mobile robots for mapless navigation. In Proceedings of the 2017 IEEE/RSJ International Conference on Intelligent Robots and Systems (IROS), Vancouver, BC, Canada, 24–28 September 2017; pp. 31–36.
19. Peng, X.B.; Berseth, G.; Yin, K.; Van De Panne, M. Deeploco: Dynamic locomotion skills using hierarchical deep reinforcement learning. *ACM Trans. Graph. (TOG)* **2017**, *36*, 41. [CrossRef]
20. Heess, N.; Sriram, S.; Lemmon, J.; Merel, J.; Wayne, G.; Tassa, Y.; Erez, T.; Wang, Z.; Eslami, A.; Riedmiller, M.; et al. Emergence of locomotion behaviours in rich environments. *arXiv* **2017**, arXiv:1707.02286.
21. Breyer, M.; Furrer, F.; Novkovic, T.; Siegwart, R.; Nieto, J. Flexible Robotic Grasping with Sim-to-Real Transfer Based Reinforcement Learning. *arXiv* **2018**, arXiv:1803.04996.
22. Stulp, F.; Fedrizzi, A.; Beetz, M.; Autonomous, I.; Group, S. Learning and Performing Place-Based Mobile Manipulation. In Proceedings of the 2009 IEEE 8th International Conference on Development and Learning, Shanghai, China, 5–7 June 2009; pp. 1–7.
23. Sutton, R.S.; Barto, A.G. *Reinforcement Learning: An introduction*; MIT press: Cambridge, UK, 1998; Volume 1.
24. Quigley, M.; Conley, K.; Gerkey, B.; Faust, J.; Foote, T.; Leibs, J.; Wheeler, R.; Ng, A.Y. ROS: An open-source Robot Operating System. In Proceedings of the ICRA Workshop on Open Source Software, Kobe, Japan, May 17 2009; VoLume 3, p. 5.
25. Dhariwal, P.; Hesse, C.; Klimov, O.; Nichol, A.; Plappert, M.; Radford, A.; Schulman, J.; Sidor, S.; Wu, Y. OpenAI Baselines. 2017. Available online: https://github.com/openai/baselines (accessed on 18 January 2019).
26. Brockman, G.; Cheung, V.; Pettersson, L.; Schneider, J.; Schulman, J.; Tang, J.; Openai, W.Z. OpenAI Gym. *arXiv* **2016**, arXiv:1606.01540.
27. Zamora, I.; Gonzalez Lopez, N.; Vilches, V.M.; Hernández Cordero, A.; Robotics, E. Extending the OpenAI Gym for Robotics: A Toolkit for Reinforcement Learning Using ROS and Gazebo. *arXiv* **2017**, arXiv:1608.05742v2.
28. Arulkumaran, K.; Deisenroth, M.P.; Brundage, M.; Bharath, A.A. A brief survey of deep reinforcement learning. *arXiv* **2017**, arXiv:1708.05866.
29. Schulman, J.; Wolski, F.; Dhariwal, P.; Radford, A.; Klimov, O. Proximal policy optimization algorithms. *arXiv* **2017**, arXiv:1707.06347.
30. Schulman, J.; Levine, S.; Moritz, P.; Jordan, M.I.; Abbeel, P. Trust Region Policy Optimization. *arXiv* **2015**, arXiv:1502.05477.
31. Kullback, S.; Leibler, R.A. On information and sufficiency. *Ann. Math. Stat.* **1951**, *22*, 79–86. [CrossRef]
32. Schulman, J.; Moritz, P.; Levine, S.; Jordan, M.; Abbeel, P. High-dimensional continuous control using generalized advantage estimation. *arXiv* **2015**, arXiv:1506.02438.
33. Lillicrap, T.P.; Hunt, J.J.; Pritzel, A.; Heess, N.; Erez, T.; Tassa, Y.; Silver, D.; Wierstra, D. Continuous Control With Deep Reinforcement Learning. *arXiv* **2016**, arXiv:1509.02971v5.
34. Silver, D.; Lever, G.; Heess, N.; Degris, T.; Wierstra, D.; Riedmiller, M. Deterministic Policy Gradient Algorithms. In Proceedings of the 31st International Conference on Machine Learning (ICML 2014), Beijing, China, 21–26 June 2014.
35. Mnih, V.; Kavukcuoglu, K.; Silver, D.; Rusu, A.A.; Veness, J.; Bellemare, M.G.; Graves, A.; Riedmiller, M.; Fidjeland, A.K.; Ostrovski, G.; et al. Human-level control through deep reinforcement learning. *Nature* **2015**, *518*, 529–533. [CrossRef]
36. Uhlenbeck, G.E.; Ornstein, L.S. On the theory of the Brownian motion. *Phys. Rev.* **1930**, *36*, 823. [CrossRef]
37. Plappert, M.; Houthooft, R.; Dhariwal, P.; Sidor, S.; Chen, R.Y.; Chen, X.; Asfour, T.; Abbeel, P.; Openai, M.A. Parameter Space Noise for Exploration. *arXiv* **2018**, arXiv:1706.01905v2.

38. Todorov, E.; Erez, T.; Tassa, Y. Mujoco: A physics engine for model-based control. In Proceedings of the 2012 IEEE/RSJ International Conference on Intelligent Robots and Systems (IROS), Vilamoura, Portugal, 7–12 October 2012; pp. 5026–5033.
39. Chitta, S.; Sucan, I.; Cousins, S. Moveit![ROS topics]. *IEEE Robot. Autom. Mag.* **2012**, *19*, 18–19. [CrossRef]
40. Abadi, M.; Barham, P.; Chen, J.; Chen, Z.; Davis, A.; Dean, J.; Devin, M.; Ghemawat, S.; Irving, G.; Isard, M.; et al. TensorFlow: A System for Large-Scale Machine Learning. In Proceedings of the 12th USENIX Symposium on Operating Systems Design and Implementation (OSDI), Savannah, GA, USA, 2–4 November 2016; Volume 16, pp. 265–283.
41. Ba, J.L.; Kiros, J.R.; Hinton, G.E. Layer Normalization. *arXiv* **2016**, arXiv:1607.06450.
42. Ng, A.Y. Feature selection, L 1 vs. L 2 regularization, and rotational invariance. In Proceedings of the Twenty-First International Conference on Machine Learning, Banff, AB, Canada, 4–8 July 2004; ACM: New York, NY, USA, 2004; p. 78.

*applied sciences*

MDPI

*Article*

# Iterative Learning Method for In-Flight Auto-Tuning of UAV Controllers Based on Basic Sensory Information

**Wojciech Giernacki**

Institute of Control, Robotics and Information Engineering, Electrical Department, Poznan University of Technology, Piotrowo 3a Street, 60-965 Poznan, Poland; wojciech.giernacki@put.poznan.pl; Tel.: +48-61-665-23-77

Received: 31 December 2018; Accepted: 30 January 2019; Published: 14 February 2019

**Abstract:** With an increasing number of multirotor unmanned aerial vehicles (UAVs), solutions supporting the improvement in their precision of operation and safety of autonomous flights are gaining importance. They are particularly crucial in transportation tasks, where control systems are required to provide a stable and controllable flight in various environmental conditions, especially after changing the total mass of the UAV (by adding extra load). In the paper, the problem of using only available basic sensory information for fast, locally best, iterative real-time auto-tuning of parameters of fixed-gain altitude controllers is considered. The machine learning method proposed for this purpose is based on a modified zero-order optimization algorithm (golden-search algorithm) and bootstrapping technique. It has been validated in numerous simulations and real-world experiments in terms of its effectiveness in such aspects as: the impact of environmental disturbances (wind gusts); flight with change in mass; and change of sensory information sources in the auto-tuning procedure. The main advantage of the proposed method is that for the trajectory primitives repeatedly followed by an UAV (for programmed controller gains), the method effectively minimizes the selected performance index (cost function). Such a performance index might, e.g., express indirect requirements about tracking quality and energy expenditure. In the paper, a comprehensive description of the method, as well as a wide discussion of the results obtained from experiments conducted in the AeroLab for a low-cost UAV (Bebop 2), are included. The results have confirmed high efficiency of the method at the expected, low computational complexity.

**Keywords:** UAV; auto-tuning; machine learning; iterative learning; extremum-seeking; altitude controller

## 1. Introduction

### 1.1. Auto-tuning of UAV Controllers—Context and Novelty

Common availability of low-cost, computationally efficient embedded systems and small size sensors directly influence the development of the construction of unmanned aerial vehicles and their applications, the number of which has been increasing in recent years [1–4]. In every UAV prototype, the need to ensure reliability and flight precision, both in manual and autonomous mode, are key aspects and depend directly on the selection of sensors [5], estimation methods [6], and the quality of position and orientation by controllers resulting from the applied control architecture [7,8]. In addition to the advanced control systems that often require precise models of UAV dynamics [9–12], due to their simplicity and versality, fixed-value controllers with a small number of parameters, are commonly and successfully used [13–16]. They determine the safety of operation, maximum flight duration and the UAV's in-flight behavior. That is why it is so important to learn and systematize the mechanisms of optimal self-tuning of their parameters for various environmental disturbances and for a radical change in the dynamics of the UAV itself due to a change in its total mass. Due to the attractive field of applications of such solutions in many areas (transportation and manipulation tasks performed by one or several UAV units [17–19], precision agriculture [20,21], missions requiring the sensory equipment to be re-armed, rescue operations [22], etc.), one is looking for fast solutions with low computational complexity that work in real-time mode.

While the state-of-the-art analysis shows several computationally complex approaches (requiring numerous repetitions and the use of the UAV model) to batch, optimal auto-tuning of controllers (via heuristic bio-inspired [23,24] and deterministic methods [25]), there has been no method reported to optimize the gains of fixed-value UAV controllers so far. No method has also been reported to do the latter in flight, iteratively and exclusively on the basis of available, periodic, basic sensory information (without using the UAV model)—to indirectly increase the flight duration by minimizing the energy expenditure through shaping a smooth flight characteristic. This issue has been selected as the core of the conducted research. The obtained result in the form of effective machine learning method for auto-tuning of gains of UAV controllers is a novelty presented in this article and thoroughly expands the concept of the method presented in [26] (using the weighted sum of the control error and control signal in predefining expectations for time courses and as a measure of tracking quality in the optimization algorithm). In addition, the most important added value also became:

- assessment and systematization (by means of simulation and experimental studies) of the influence of several environmental factors on the process of auto-tuning of UAV controllers during the flight by the proposed extremum-seeking method. The key issue here is the analysis of results in terms of assessing the quality of work of tuned controllers and the work of the optimization mechanism itself in the following test areas: presence of disturbances (wind gusts), UAV mass change, different sensory sources, flight dynamics/optimized performance index,
- outlining the rules for conducting the auto-tuning process of controllers, so that the automatic exploration of the gain space for individual controllers can be as safe as possible (one needs to keep in mind that the proposed method is not based on any stability criterion, which is its main limitation compared with numerous batch solutions based on models).

### 1.2. Motivation

In previous research [26], the author has drawn inspiration from the demanding problems of mobile robotics, which the world research centers have been coping with. Examples of such problems can be found in particular challenges of the Mohamed Bin Zayed International Robotics Challenge (http://www.mbzirc.com) [27], where the common denominator are tasks requiring the use of one or a group of UAV units to conduct autonomous flights with high precision in varied conditions (outdoor and indoor) and varied UAV mass. In preparation for the MBZIRC'2020 edition, it turned out that the only currently available auto-tuning algorithm on commercial auto-pilots (as Pixhawk, Naze32, Open Pilot, CC3D), named `AutoTune`, *"(...) uses changes in flight attitude input by the pilot to learn the key values for roll and pitch tuning. (...) While flying the pilot needs to input as many sharp attitude changes as possible so that the autotune code can learn how the aircraft responds"* [28]. Unfortunately, this solution is problematic due to the tuning safety (especially in prototyping UAV constructions) and control goal set: to provide the most smooth, feasible flight trajectories, which will reduce the control effort to reasonable level, and as a result will be maximally energy efficient. Therefore, in the method considered in this work, a gain tuning of UAV controllers based on dynamic behavior was replaced by more energy-efficient and automatic machine learning technique.

### 1.3. Related Work

Among numerous approaches to machine learning, and apart techniques using neural networks, which require many learning data sets, the mechanisms based on reward and punishment (as in the case of reinforcement learning approaches) are becoming increasingly common. In [29], Rodriguez-Ramos et al. have taught the control system to land autonomously on a moving vehicle, and in [30] Koch et al. trained a flight controller attitude control of a quadrotor through reinforcement learning. Despite the obviously large number of classic approaches to tuning of fixed-value controllers (Panda presents a whole array of such approaches, of which several dozen are practice-oriented [31]), the optimal techniques of iterative learning are invariably gaining on importance [32–34]. Iterative learning techniques have three desirable attributes, namely: automated tuning, low computational complexity (in optimization algorithms, a decision is made only on the basis of current, cyclic information from the selected performance index—cost function), and fast tuning speed [26,35] (in contrast to reinforcement learning approaches, which requires numerous experiments during the learning that makes it unpractical).

While the methods approximating the gradient of the cost function (first- and second-order optimization algorithms) presented in [25,36] can be quite problematic for UAV auto-tuning from noisy measurements (an aspect for careful comparisons in subsequent author's research), the zero-order optimization methods works efficiently because of the speed of calculations. However, it should be remembered and accepted that the obtained solution may be a local (there is no guarantee to obtain global solutions) or a value near it (depending on the declared level of expected accuracy of calculations $\epsilon$).

Among the zero-order optimization methods presented by Chong & Zak in [25], such as Fibonacci-search, golden-search, equal division, and dichotomy algorithms, especially the first two of region elimination methods—developed by Kiefer [37] are effective in optimal control problems [38]. A broad description of the method based on Fibonacci numbers which was used for UAV altitude controller tuning can be found in the mentioned publication [26] of the author—especially mathematical basics and proofs for the region elimination mechanisms. Therefore, for undisturbed presentation of the proposed new method based on the modified golden-search algorithm used in the auto-tuning of the altitude controller during the UAV flight, only necessary mathematical description has been presented in the remaining part of the paper. Instead, the author paid more attention to the application aspects of the method (by placing the necessary pseudocodes) and a wide analysis of the results obtained from the conducted research experiments.

The paper is structured as follows: in Section 2 the UAV description as a control object and measurement system, as well as considered control system, is presented. Therein, the control purpose is highlighted, and the optimization problem is outlined. In the same Section, the proposed auto-tuning method is introduced, and its mathematical basics are explained. Furthermore, the experimental platform is shown. The comprehensive description of simulation and real-world experiments results with discussion are provided in Section 3. Finally, Section 4 presents conclusions and further work plans.

For a better understanding of the presented content, the most important symbols used in the paper are described in Table 1.

**Table 1.** Symbols used in this article.

| Symbol of Variable | Explanation |
|---|---|
| $\alpha, \beta$ | weights (in cost function $J$) |
| $\theta, \phi, \psi$ | *roll*, *pitch*, *yaw* angles |
| $\rho$ | golden-search reduction factor |
| $\epsilon$ | expected accuracy in GLD method |
| $\mathcal{BF}$ | body frame of reference |
| $\mathscr{D}^{(k)}$ | considered range for the optimized parameter at $k$-th iteration |
| $\mathcal{EF}$ | Earth frame of reference |
| $e(t)$ | tracking error (in time domain) |
| $f(\cdot)$ | cost function (in GLD method) |
| $J$ | performance index (cost function in GLD procedure) |

**Table 1.** *Cont.*

| Symbol of Variable | Explanation |
|---|---|
| $k_P, k_I, k_D$ | proportional/integral/derivative gains |
| $N$ | minimal number of iterations required to ensure accuracy $\epsilon$ |
| $N_b$ | number of the predefined bootstrap cycles |
| $N_c$ | number of sampling periods necessary to calculate $J$ at $l$-th iteration of the GLD |
| $N_{max}$ | number of sampling periods related to the length of the tuning procedure |
| $\underline{p}_d$ | vector of desired UAV position |
| $\underline{p}_m$ | vector of measured UAV position |
| $t_a$ | time of gathering information for calculation of $J$ in GLD methods |
| $t_h$ | flight time horizon |
| $T_f$ | time constant of a low-pass filter of transfer function |
| $T_p$ | sampling period for calculation of $J$ |
| $T_s$ | sampling period in low-pass filter |
| $u(t)$ | control signal (in time domain) |
| $x^{(k^-)}$ | lower bound for optimized parameter at $k$-th iteration |
| $x^{(k^+)}$ | upper bound for optimized parameter at $k$-th iteration |
| $\hat{x}$ | candidate point in the optimization procedure |
| $\hat{x}^*$ | iterative estimate of the optimal solution |
| $x_b, y_b, z_b$ | axes of the $\mathcal{BF}$ |
| $x_d, y_d, z_d$ | desired position coordinates |
| $x_e, y_e, z_e$ | axes of the $\mathcal{EF}$ |
| $x_m, y_m, z_m$ | measured position of the UAV |

## 2. Materials and Methods

### 2.1. Multirotor UAV as a Control Object and Its Measurement System

The multirotor UAV can be considered as a multidimensional control plant, being underactuated, strongly non-linear, and highly dynamic with (in general) non-stationary parameters. These features result from its physical structure—especially the use of several propulsion units mounted at the ends of the frame. In addition, measuring, processing, and communication systems are also attached to the middle of this frame—suited for a particular UAV construction. From the perspective of control, the appropriate selection of propulsion units (composed of: brushless direct current motors, electronic speed controllers and propellers) is a key aspect to ensure the expected flight dynamics expressed via thrust ($T$) and torque ($\underline{M}$) generated by the rotational movement of propellers [39]. By changing the rotational speed, it is possible to obtain the expected position and orientation of the UAV in 3D space, i.e., control of its 6 degrees of freedom (DOFs). The obtained control precision also depends on the quality of sensory information to a large extent. Presently, even in the simplest, low-cost UAVs (Figure 1), in order to determine current position and orientation estimates during the flight (e.g., based on more or less advanced modifications of Kalman filters [6]), the sensory data fusion is used (from 3-axes accelerometer, 3-axes gyroscope, 3-axes magnetometer, pressure sensor, optical-flow sensor, GPS, ultrasound sensor, etc.).

In the paper, two sources of measurements are used in the proposed auto-tuning procedure: on-board UAV avionics (for *roll* and *pitch* angles measurements) and external motion capture system (`OptiTrack`) ($X$, $Y$, $Z$ position, and *yaw* angle). In the UAV autonomous control, to ensure unambiguous description of the UAV's position and orientation in 3D space, the North-East-Down (NED) configuration of the reference system is used, since the on-board measurements are expressed in local coordinate system ($\mathcal{BF}$—Body Frame), and the position control, as well as motion capture measurements are defined in the global one ($\mathcal{EF}$—Earth Frame). In the paper of Xia et al. [40], one may find a better known, basic information about the mechanisms of conversions, e.g., how the posture of the multirotor (its rotational and translational motion) can be described by the relative orientation between the $\mathcal{BF}$ and the $\mathcal{EF}$ with the use of the rotation matrix $R \in SO(3)$.

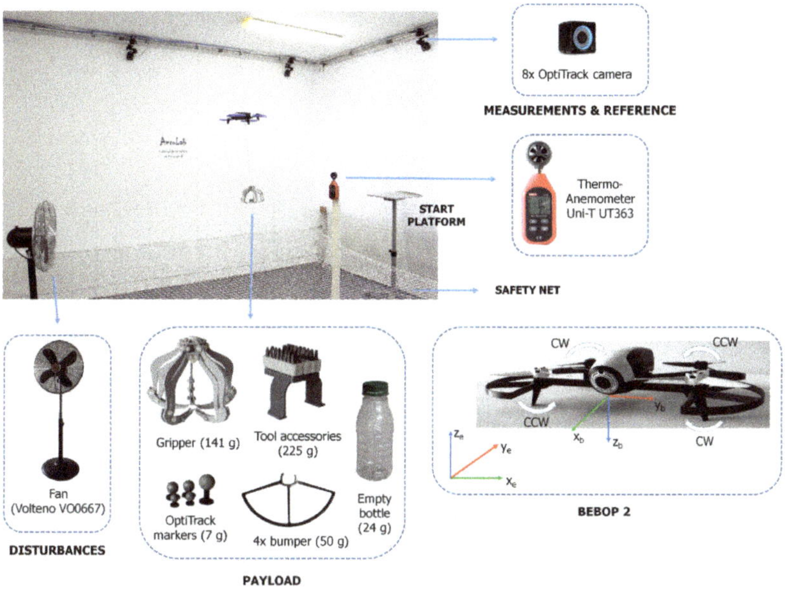


**Figure 1.** The *Bebop 2* quadrotor (and its coordinate system) during one of the initial experiments with the carrying of payload conducted in *AeroLab* of *Poznan University of Technology*.

## 2.2. Considered Control System and Control Purpose (Formulation of Optimization Problem)

The control system of multirotor UAV from Figure 2 considered here is based on cascaded control loops. There is control of angles *roll* ($\theta$) and *pitch* ($\phi$) around the $x_b$ and $y_b$ axes, according to the set (desired) position in the $x_e$ and $y_e$ axes in faster, internal control loops. Their control is performed in slower external loops. The control of $\theta$ and $\phi$ angles occurs indirectly in the realization of autonomous flight trajectory expressed using the vector of desired position trajectory $\underline{p}_d = (x_d, y_d, z_d)^T$ and desired angle of rotation *yaw* ($\psi_d$) around the $z_e$ axis. The purpose of the autonomous control is then to ensure the smallest tracking errors $e(t)$ during the UAV flight, i.e., the difference in the values of the reference signals (desired) and output signals (actual/measured) [41]:

$$\underline{e}_p = \underline{p}_d - \underline{p}_m, \tag{1}$$

$$e_\psi = \psi_d - \psi_m, \tag{2}$$

where the $m$ index refers to the measured values.

Bearing in mind that in UAVs the current tracking error information from (1) and (2) is used as the input of a given fixed-value controllers, in the commonly used proportional-derivative (PD) controller structure or proportional-integral-derivative (PID), it is proposed to use this information (as well as information from the output of a given type of controller with control signal $u(t)$) to formulate a measure of the tracking quality during UAV flight, i.e., the cost function/performance index $J(t)$ (see Figure 3), defined as follows:

$$J(t) = \int_0^{t_a} (\alpha\,|e(t)| + \beta\,|u(t)|)\,dt, \tag{3}$$

where $t_a$ is the time of gathering information (to calculate new controller gains) in the optimization procedure. By introducing the penalty for excessive energy expenditure (expressed in the cost function through actual values of the control signal $u(t)$), it is possible to shape expectations towards transients and the controller's dynamics profile (providing smooth or dynamic flight trajectories). At small values of the $\beta$, the controller works aggressively, using more energy, often at the expense of the appearance of overshoot, which is undesirable in missions and tasks requiring high flight precision.

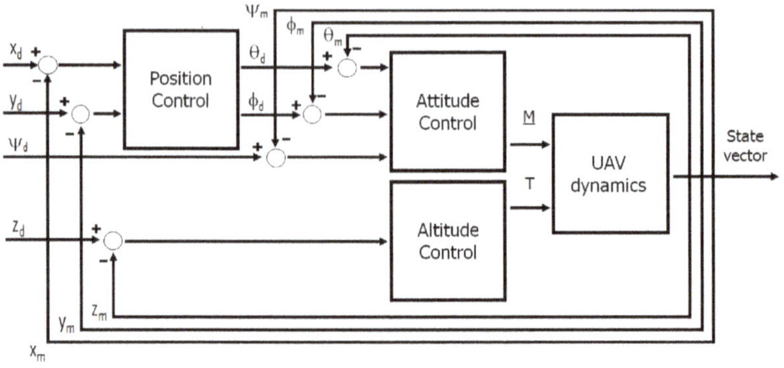


**Figure 2.** Diagram of considered control system.

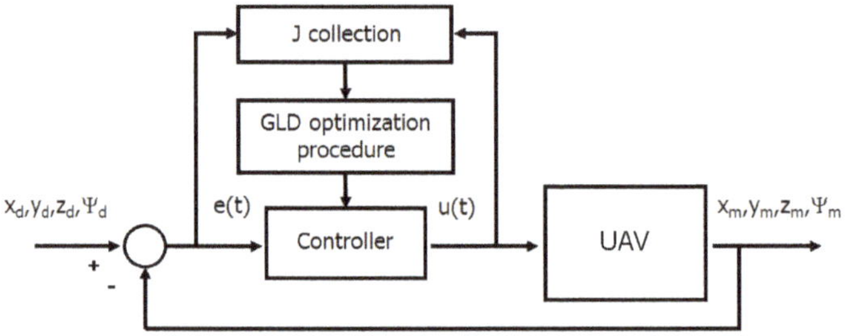


**Figure 3.** General block diagram of the control system with optimization.

Unconstrained control signal $u(t)$ is calculated from the controller's equation, which in the case of PID structure it is given by

$$u(t) = k_P e(t) + k_I \int_0^{t_h} e(t)\,dt + k_D \frac{d}{dt} e(t), \tag{4}$$

where $t_h$ is a flight time horizon, $k_P$ is the proportional gain, $k_I$ represents the integral gain and $k_D$ the derivative gain, respectively. Gains $k_P$ and $k_D$ are expected to be found using the proposed iterative learning method.

**Remark 1.** *In the article, when there is a reference to the PID controller, it should be remembered that only the $k_P$ and $k_D$ gains are tuned automatically, whereas the value $k_I$ (used to eliminate the steady-state error) is selected in a manual manner. The proposed auto-tuning method can be used to optimize the gains of any type of controller with three (or even more) parameters; however, this will result in a longer tuning time. Therefore, from the application point of view, it is better to use the procedure presented further in the article.*

Recalling (4), this work deals with the search for the controller gains $k_P$ and $k_D$, to minimize the cost function (3). That is, the current controller design procedure can be posed as an optimization problem where the solution to the following problem is sought:

$$\begin{array}{ll} \min\limits_{k_1,k_2,\ldots,k_N} & J(t) = \int_0^{t_a} \left(\alpha\,|e(t)| + \beta\,|u(t)|\right) dt, \\ & 0 \le k_1 \le k_1^{max} \\ s.t. & 0 \le k_2 \le k_2^{max} \\ & \ldots \\ & 0 \le k_N \le k_N^{max} \end{array} \tag{5}$$

where $k_1^{max}, k_2^{max}, \ldots, k_N^{max}$ are upper bounds of the predefined ranges of exploration in the optimization procedure of $N$ controller parameters.

**Remark 2.** *In the numerical implementation of optimization problem from (5), to quantify the tracking quality by using the cost function (3), its discrete-time version is used (the integration operation is replaced with the sum of samples). Then the cost function is built from the weighted sum of the absolute values of the tracking error samples and the absolute values of the control signal samples (for a given sampling period $T_p$).*

### 2.3. Procedure for Tuning of Controllers

To increase the safety in the process of tuning UAV controller parameters during the flight, it is proposed to use the procedure from the flowchart (Figure 4), corresponding to the pyramid of subsequent expectations for the work of control system (Figure 5).

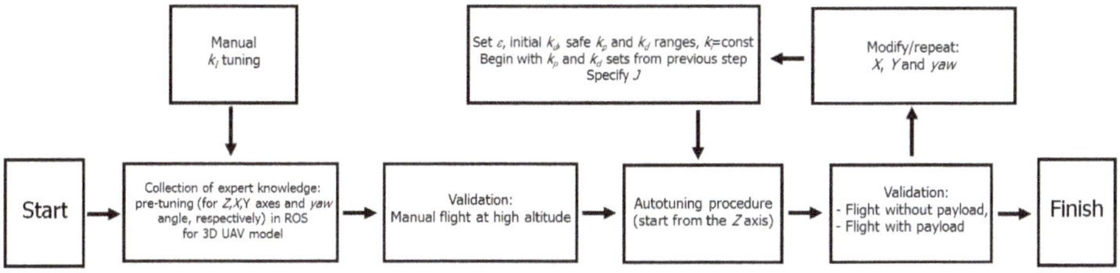


**Figure 4.** Flowchart for the proposed tuning strategy of the UAV controllers.

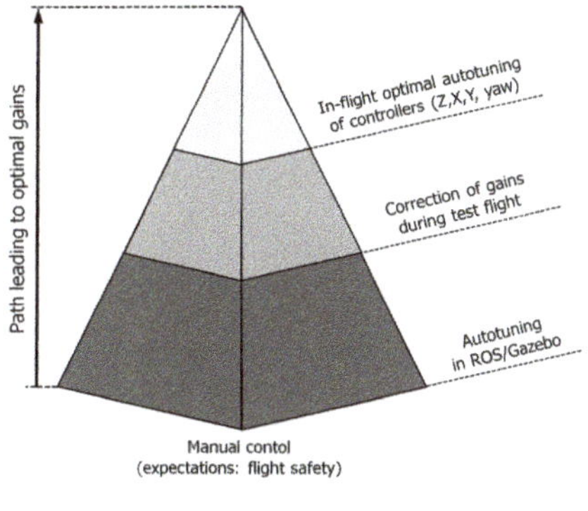


**Figure 5.** Following steps to obtain optimal gains of controllers.

**Remark 3.** *Manual tuning of UAV control system prototype is out of scope of this work (to focus on auto-tuning mechanisms). Some useful information regarding UAV controllers prototyping can be found at well-recognized by the UAV community webpages [42–44].*

### 2.4. Iterative Learning Method for In-Flight Tuning of UAV Controllers—General Idea

Bearing in mind that the search space for the $J_{min}$ for all combinations of controller gains $\underline{k} = (k_1, k_2, \ldots, k_N)^T$ in predefined intervals (ranges) of gains, in the problem outlined in (5) is huge, one needs a fast, effective mechanism for search space exploration. It should be characterized by low computational complexity, and after checking the value of $J(t)$ in a maximum of dozen or several dozen of gain combinations, should be able to provide the value of $J_{min}$ (locally best variant) or a significant improvement compared to the controller's original gains (expressed using e.g., the expected accuracy of the $\epsilon$).

Recalling the publications cited in Section 1.3, iterative learning algorithms are characterized by fast convergence towards the minimum value—especially the region elimination methods (REMs). To be able to use them, one needs to refer to the general idea of iterative learning approaches (proposed by Arimoto et al. in [45]), i.e., minimization of the norm of error (here: cost function) in order to tune particular controller using the periodical repetitiveness of the trials (here: repetitions of the same, predefined trajectory primitives—see Figure 6). Then, to find locally best gains of a particular controller based on a given reference of $x_d$, $y_d$, $z_d$ or $\psi_d$ primitive, and corresponding measured value, the performance index $J(t)$ calculated during the flight with given safe ranges of controller parameters, enables the minimum-seeking procedure to find controller gains with respect to the preferred dynamics, and for a given tolerance of the solution.

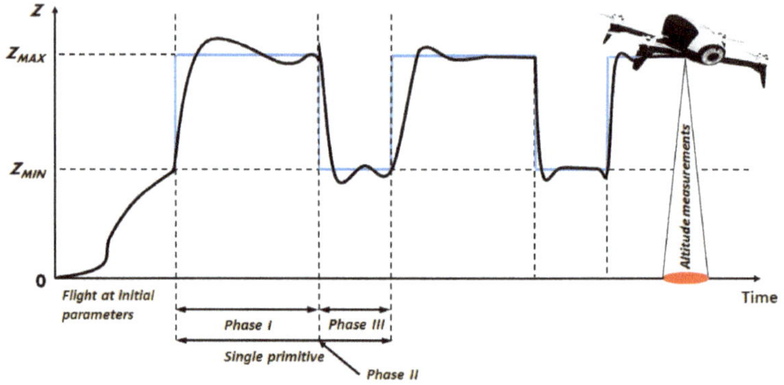


**Figure 6.** Following steps in iterative learning mechanism for altitude controller tuning.

**Remark 4.** *Since the method is based solely on the cyclic collection of measurement data to determine $J(t)$, the need to use the UAV model is reduced. However, its knowledge is advantageous when in the simulation conditions it is possible to roughly estimate/determine the maximum values of the elements of the $\underline{k}$ vector, for which the UAV does not lose its stability.*

For every single primitive being used in the optimization procedure, three phases can be distinguished (see Figure 6):

1. Acquisition of measurement data (current, sampled values: $x_m$, $y_m$, $z_m$ or $\psi_m$) for set controller gains with the assumed $T_p$ and the assumed form of the $J(t)$ function,
2. Determination of new controller gains based on the estimated value of the cost function from the phase no. 1,
3. Adjusting the controller according to iteratively corrected gains and waiting for the time necessary to stop the transient processes caused by it.

Determining the sequence of controller gains is possible by systematically narrowing the search space. For this purpose, the use of region elimination method based on the zero-order deterministic optimization algorithm (GLD), is proposed.

*2.5. Region Elimination Method Based on GLD Algorithm*

Let us consider the problem of iterative searching for a particular controller's gain as a problem of reducing the set range of this gain, in which the criterion of stopping the algorithm is the proximity of following solutions, i.e., the value of the cost function in subsequent solutions (for subsequent controller gains), and the convergence of the algorithm ensures the use of the mechanism based on golden-section search from [25] used in REMs.

Principles and assumptions in the GLD method are similar to those in the modified Fibonacci-search method (FIB) proposed in [26]. The most important are: the unimodality assumption of optimized cost function $f(\cdot)$, lack of knowledge about the global minimum (which gave rise to formulation of stopping criteria in the iterative tuning algorithm, e.g., given tolerance to find the minimizer), successively narrowing the range of values inside which the extremum is known to exist according to the definition (2.1) of the fundamental rule for REMs.

**Definition 1.** *Let us consider an optimization problem of a one-argument unimodal cost function* $f : \mathscr{R} \to \mathscr{R}$ *within in the predefined range* $[x^{(0^-)},x^{(0^+)}]$ *in initial 0th iteration, where* $x^{(0^-)}<x^{(0^+)}$ *of a unimodal function* $f$*. The argument* $x$ *of this function can be interpreted as a gain of controller (here:* $k_P$ *or* $k_D$*), and the value of* $f$ *can be understood as the* $J$ *value (within some horizon) corresponded to it.*

*Now, for a pair of two arguments* $x^{(1^-)}$ *i* $x^{(1^+)}$*, which lie in the range* $[x^{(0^-)},x^{(0^+)}]$*, and which satisfy* $x^{(0^-)}<x^{(1^-)}<x^{(1^+)}<x^{(0^+)}$*, it is true that:*

- *If* $x^{(1^-)} > x^{(1^+)}$*, then the minimum* $\hat{x}^*$ *does not lie in* $(x^{(0^-)}, x^{(1^-)})$*,*
- *If* $x^{(1^-)} < x^{(1^+)}$*, then the minimum* $\hat{x}^*$ *does not lie in* $(x^{(1^+)}, x^{(0^+)})$*,*
- *If* $x^{(1^-)} = x^{(1^+)}$*, then the minimum* $\hat{x}^*$ *does not lie in* $(x^{(0^-)}, x^{(1^-)})$ *and* $(x^{(1^+)}, x^{(0^+)})$*.*

A region elimination fundamental rule is used to find the $\hat{x}^*$ with the minimum value of $f$ within predefined range, based on repeatedly selection of two arguments from the current range according to symmetrically reduction the range of possible arguments:

$$x^{(1^-)} - x^{(0^-)} = x^{(1^+)} - x^{(0^+)} = \rho(x^{(0^+)} - x^{(0^-)}), \tag{6}$$

where $\rho = \frac{3-\sqrt{5}}{2} = 0.381966$ is a golden-search reduction factor.

**Remark 5.** *An advantage of using the golden-search reduction factor (according to Algorithm 1) is the fast exploration of the interval, because following values of x (controller gains) are selected to use one of the values of the cost function calculated in the previous iteration. For this purpose, the interval is divided regarding the golden ratio. As a result, of the use of the golden-search reduction factor for a given interval, two new sub-intervals are obtained. For the new intervals, the ratio of the longer length to the shorter length is equal to the ratio of the length of the divided interval to the length of the longer interval.*

*Due to this mechanism, and by using the golden-search reduction factor, the time of range exploration is shortened (through the reduction the number of points for which* $f$ *function needs to be evaluate) or alternatively the* $f$ *function values can be averaged for the same* $x$ *in following iterations, i.e.,* $x^{((k+1)^+)}$ *and* $x^{(k^-)}$*, which is useful in order to reduce the impact of measurement disturbances during the UAV outdoor flight.*

Based on predefined initial range $x \in \mathscr{D}^{(0)} = \left[x^{(0^-)}, x^{(0^+)}\right]$, the golden-search algorithm can be implemented according to the pseudo-code presented below (Algorithm 1).

**Algorithm 1** Golden-search algorithm.

**Step 1.** Evaluate the minimal number $N$ of iterations required to provide the sufficient (predefined) value of the $\epsilon$:

$$|x^* - \hat{x}^*| \leq \epsilon(x^{(0^+)} - x^{(0^-)}), \tag{7}$$

where $|x^* - \hat{x}^*|$ is the absolute value of the difference between the true (unknown minimum $x^*$) and iterative solution $\hat{x}^*$ (which is assumed to be in the center of $\mathscr{D}^{(N)}$).

**Step 2.** For iteration $k = 1, \ldots, N$,

1) select a pair of intermediate points $\hat{x}^{(k^-)}$ and $\hat{x}^{(k^+)}$ ($\hat{x}^{(k^-)} < \hat{x}^{(k^+)}$, $\left\{\hat{x}^{(k^-)}, \hat{x}^{(k^+)}\right\} \in \mathscr{D}^{(k-1)}$),
2) reduce the range to $\mathscr{D}^{(k)}$ based on REM fundamental rule:
   a) $x^{(k+1)} \in \mathscr{D}^{(k)} = \left[x^{(k-1^-)}, \hat{x}^{(k^+)}\right]$ for $f(\hat{x}^{(k^-)}) < f(\hat{x}^{(k^+)})$,
   b) $x^{(k+1)} \in \mathscr{D}^{(k)} = \left[\hat{x}^{(k^-)}, x^{(k-1^+)}\right]$ for $f(\hat{x}^{(k^-)}) \geq f(\hat{x}^{(k^+)})$,
   c) start next iteration $k := k + 1$.

**Step 3.** Stop the algorithm; put $\hat{x}^* = \frac{1}{2}\left(x^{(N^+)} + x^{(N^-)}\right)$.

For the given value of $\epsilon$, the minimum number $N$ of iteration in the GLD algorithm can be calculated according to:

$$(1 - \rho)^N \leq \epsilon, \tag{8}$$

and for $k = 1, \ldots, N$ one may find the pair of intermediate points using

$$\hat{x}^{(k^-)} = x^{(k-1^-)} + \rho(x^{(k-1^+)} - x^{(k-1^-)}), \tag{9}$$

$$\hat{x}^{(k^+)} = x^{(k-1^-)} + (1 - \rho)(x^{(k-1^+)} - x^{(k-1^-)}). \tag{10}$$

### 2.6. Optimal Gain Tuning of a Two-Parameter Controller Based on Bootstrapping Mechanism

In a two-dimensional space of parameters, the vector of parameters $\underline{x} = \begin{bmatrix} x_1, & x_2 \end{bmatrix}^T$ for the cost function $f(\underline{x})$ (calculated from in-flight measurements) can be interpreted as controller gains (here: $k_P$ and $k_D$). For fast exploration of this space and to give a global character the GLD extremum-seeking procedure, Algorithm 2 is proposed. It is based on the bootstrapping mechanism (see Table 2), for the predefined bootstrap cycles $N_b$. In considered two-parameter controller tuning, in every single bootstrap, two launch of GLD algorithm (for each of controller gains) are executed to obtain expected value of the $\epsilon$. Firstly, the gain no.1 is tuned (while the gain no. 2 is fixed), and then, the gain no. 2 (for fixed value of the no. 1).

**Algorithm 2** Two-parameter controller tuning.

**Step 0.** Put the bootstrap cycles counter to $l = 0$; for initial $\mathscr{D}_i^{(l)}$ $(i = 1, 2)$ define $\epsilon$, $N_b$, and initial value of the second parameter $x_2^{(l)}$ (take $\hat{x}_2^{(l)^*} = x_2^{(l)}$), set $l := l + 1$.

**Step 1.** Find the optimal $\hat{x}_1^{(l)^*}$ using the GLD algorithm, with the second parameter fixed at $\hat{x}_2^{(l-1)^*}$.

**Step 2.** Calculate the optimal $\hat{x}_2^{(l)^*}$ analogously to the method from the Step 1, keeping the first parameter fixed at $\hat{x}_1^{(l)^*}$.

**Step 3.** If $l < N_b$, increase the bootstrap cycles counter $l := l + 1$, and proceed to Step 1, otherwise stop the algorithm—the optimal solution $\underline{\hat{x}}^* = \left[\hat{x}_1^{(l)^*}, \hat{x}_2^{(l)^*}\right]^T$ has been obtained after $N_b$ bootstrap cycles, as desired.

**Table 2.** Steps in the bootstrapping mechanism.

| Bootstrap No. | Gain No.1 | Gain No.2 |
|---|---|---|
| 1 | Tuning (according to the GLD REM) | Kept constant |
| 1 | Kept constant | Tuning |
| 2 | Tuning | Kept constant |
| 2 | Kept constant | Tuning |
| ... | ... | ... |
| $N_b$ | Tuning | Kept constant |
| $N_b$ | Kept constant | Tuning |

To ensure high effectiveness of the proposed method of auto-tuning, one should remember about several important aspects (in configuration and implementation):

- The proposed method requires predefining the initial, admissible ranges for $\underline{x}$, i.e., $\mathscr{D}_i^{(0)} = \left[x_i^{(0^-)},\ x_i^{(0^+)}\right]$ for $i = 1, 2$. It is a crucial choice from the perspective of ensuring the safety of autonomous flight. If there is a such a possibility, it is strongly recommended to use the expert knowledge about the controller gains (from initial flights on the base of analysis of a rise time and the maximum overshoot, prototyping in virtual environment, default settings of on-board controller, detailed analysis of the UAV feedback control system, etc.),
- For the expected tolerance $\epsilon$, the number $N$ is calculated. $2N$ calculations of $f$ are needed in the tuning of a pair of controller parameters of a single bootstrap,
- The algorithm's execution time depends on: $N_b$, $N$, and the time of a single reference primitive, which must be correlated with the expected UAV dynamics and its natural inertia,
- Recalling the most important principles of the zero-optimization method from [26], one needs to have in mind that the proposed method *"(...) is iterative-based and collects information about the performance index (on incremental cost function value) at sampling time instants, equally spaced every $T_p$ seconds"* during the tuning experiments. Thus, for sampling period $T_p$, a single evaluation of $f$ value according to Step 2 of Algorithm 1 with a change of a single parameter of controller is performed using Procedure 1 (for symbols from the Table 1).
- The performance index is calculated as

$$\Delta J^{(n)} = J^{(n+1)} + \Delta J^{(n)}, \tag{11}$$

where $\Delta J^{(n)}$ can be obtained from the discrete-time version of Equation (3), which for $n$-th sample (tracking error and control signal) at time $t = nT_p$ is given by

$$\Delta J^{(n)} = \alpha\,|e_n| + \beta\,|u_n|\,. \tag{12}$$

**Algorithm 3** Evaluation of performance index (with single change of controller parameter) [26]

Recalling defined $N_c$, $N_{max}$, and $n$ for $f(\cdot)$. Then:

- for $n = 1, \ldots, N_c - 1$ with the controller parameters are updated in the previous iteration, the performance index is evaluated using (11) by adding (12); set $J^{(0)} = 0$;
- for $n = N_c$ a single iteration of GLD algorithm is initialized, cost function is stored, and if possible—reduce the range for controller parameters or perform the bootstrap; it results in a transient behavior of the dynamical signal;
- for $n = N_c + 1, \ldots, N_{max}$ tuning is not performed; the controller parameters have been updated; no performance index is collected; transient behavior should decay.

### 2.7. Signals Acquisition and Their Filtration in the Proposed Method

Bearing in mind that in general to determine the performance index, sensory information is used from sources with different precision of estimation of the position and orientation of an UAV, therefore in the auto-tuning procedure it is proposed to use:

- the signals from the UAV odometry—processed using commonly used Kalman filtration. Thanks to that, it is possible to fuse data from several standard UAV on-board sensors,
- low-pass filtration (presented and tested primarily in [26]), expressed by a transfer function of first-order inertia type
$$G(s) = \frac{k}{1 + T_f s}, \tag{13}$$
where $k$ is its gain, and $T_f$ is a chosen time constant (here: $k = 1$, $T_f = 0.1$ sec.
For the implementation of the GLD method, the discretized, recursive version of the low-pass filter (13) for the chosen sampling period $T_s$, is used:
$$y(n) = a(n-1) + (1-a)\,u(n-1), \tag{14}$$
where
$$a = \exp\left(-T_s/T_f\right), \tag{15}$$
and $y(n)$ and $u(n)$ are filtered and pure errors at sample $n$, respectively.
- (optional) measurement information from an external high-precision measurement system—for example, the motion capture system (for indoor flights) or GNSS (outdoor), treated as the ground truth in estimating the difference to UAV avionics measurements.

### 2.8. Experimental Platform

In the real-world experiments, the low-cost, micro quadrotor Bebop 2 from Parrot company, was used (see Figure 1 and [46]). Since it is equipped in P7 dual-core CPU Cortex 9 processor, 1 GB RAM memory, and 8 GB of flash memory, it is possible to perform on-board state estimation of the UAV using Extended Kalman Filter (EKF) for the data gathered from its on-board sensors listed in Table 3. The Bebop 2 uses the Busybox Linux operating system. Compact sizes of the UAV ($33 \times 38 \times 3.6$ cm with hull) and efficient propulsion units ($4 \times 1280$ KV BLDC Motor, 7500-12000 rpm), in combination with 2700 mAh battery provide maximum flight time up to 25 minutes and maximum load capacity up to 550 g (which gives a maximum takeoff mass equal to 1050 g, since the UAV weighs 500 g).

**Table 3.** General characteristic of Bebop 2 sensors.

| Parameter | Value |
|---|---|
| accelerometer & gyroscope | 3-axes MPU 6050 |
| pressure sensor (barometer) | MS5607 (analyses the flight altitude beyond 4.9 m) |
| ultrasound sensor | analyses the flight altitude up to 8 m |
| magnetometer | 3-axes AKM 8963 |
| geolocalization | Furuno GN-87F GNSS module (GPS+GLONASS+Galileo) |
| Wi-Fi Aerials | 2.4 and 5 GHz dual dipole |
| vertical stabilization camera | photo every 16 ms |
| camera | 14 Mpx 3-axis Full HD 1080p with Sunny 180 fish-eye lens: 1/2.3" |

All experimental studies discussed in the article were carried out in `AeroLab` [47], the research space created at the *Institute of Control, Robotics and Information Engineering of Poznan University of Technology* for testing solutions in the field of UAVs flight autonomy, where ground truth is the OptiTrack motion capture system equipped with 8 Prime 13W cameras (with markers placed on the UAV), and a processing unit (PC) equipped with `Motive`—OptiTrack's unified motion capture software platform. The measurement program (Robot Operating System (ROS) node) is executed with the frequency of 100 Hz, control actions with 30 Hz, whereas the tuning methods with 5 Hz. The system is connected to the ground station (Figure 7) to which information about the current position and orientation of the UAV (from motion capture system and UAV) are transmitted. The ground station is the Lenovo Legion Y520 notebook, equipped with Intel Core i7-7700HQ (2.8 GHz frequency), 32 GB DDR4 RAM memory, SSD hard drive and GeForce GTX 1050 2048 MB under Linux Kinetic 16.04 LTS operating system. Such a powerful computer was proposed for the autonomous control of the Bebop 2 UAV, to conduct all necessary calculations at the ground station, including: path planning, data (measurements) processing, autonomous control, auto-tuning of controllers, safety control, etc.

The ground station was also used for tests of the proposed GLD auto-tuning method in simulation environment. These tests were carried out under the control of the ROS, using the open-source flight simulator `Sphinx` [48] and `bebop_autonomy` library [49] extended by models of cascade control system enabling simulation of autonomous flights in $x_e$ and $y_e$ axes(flight for the given coordinates). In the external position control loops, the PID-type controllers have been used.

During the flights, to ensure the safety, Bebop 2 was equipped with 4 bumpers (12.5 g each, made in 3D printing technology) protecting propellers, and in `AeroLab` an additional horizontal safety net was installed to protect it against hard crashes to the ground level. In addition, for security reasons, the priority over the autonomous flight of the drone was allocated to the operator equipped with SkyController 2, enabling manual flight control. Furthermore, a safety button was introduced to cut off the UAV power supply in a situation of imminent danger. It supported initial experiments, where additional safety rope was used.

In experiments on variable mass flights, the UAV was also equipped with a plastic bottle and a gripper (made in 3D printing technology), or alternatively with tool accessories mounted directly on the Bebop (see Figure 1). Additionally, in studies on the influence of environmental disturbances on the auto-tuning process, the UT363 thermo-anometer from Uni-T company was used to measure the air flow speed generated from the Volteno VO0667 fan.

For the simulation and experimental results presented in the next section of the article, a movie clips (available at the webpage http:www.uav.put.poznan.pl), were prepared.

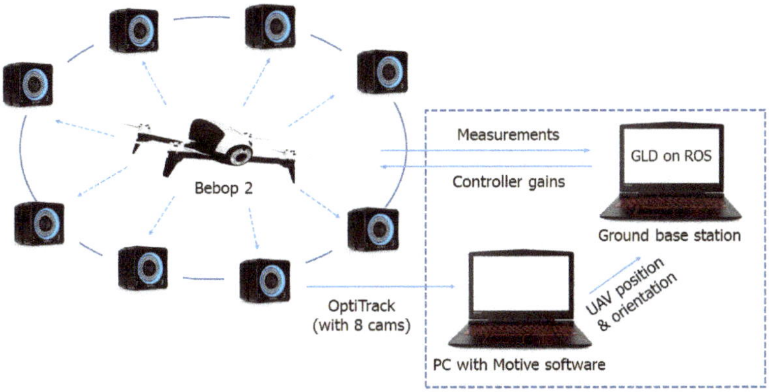


**Figure 7.** Simplified block diagram of measurement and control signal architecture used during the experiments with in-flight tuning of controllers.

## 3. Results and Discussion

### 3.1. Simulation Experiments

Let us consider the problem of searching locally best gains of the altitude PID-type controller of Bebop 2 unmanned aerial vehicle. Default gains are not made available by the Parrot company, hence the problem of finding the best gains (summarized in Table 4) has been treated at the prototyping stage. After development of the 3D model of this UAV (with bumpers) in the Blender software, it was implemented in the ROS/Gazebo environment, giving the physical dimensions, mass and moments of inertia from the real flying robot to its virtual counterpart embedded in the virtual Aerolab scenery. This enabled reliable preliminary experiments to be conducted in the simulator.

The research purposes were set as:

- recognizing the nature of optimized function $J = f(k_P, k_D)$ for its various structures ($\alpha$ = var, $\beta$ = var),
- validation if given gain ranges of $k_P$ and $k_D$ (for a constant, very small value of $k_I = 0.0003$) are safe (i.e., if the closed-loop control system is stable),
- comparative analysis of the effectiveness of GLD and FIB methods.

**Table 4.** Gains of Bebop's controllers used in experiments.

| | *X*-axis | *Y*-axis | *Z*-axis | $\theta$ | $\phi$ | $\psi$ |
|---|---|---|---|---|---|---|
| $k_P$ | 0.69 | 0.69 | 1.32 | default | default | 0.07 |
| $k_I$ | 0.00015 | 0.00015 | 0.0003 | default | default | 0.00001 |
| $k_D$ | 50 | 50 | 10.2 | default | default | 0.9 |

In the first phase of the research, more than 33 hours of simulation tests were conducted. The results are presented in Figure 8. The same dynamics of the desired reference signal was set as in [26] for the FIB method. Every 12 seconds the UAV changed periodically the flight altitude (1.2→1.9→1.2 m). The value of $J$ was being recorded for 10 sec repeatedly. For each combination of $k_P$ and $k_D$ gains, the $J$ value was averaged from 5 trials. The results of 400 combinations of $(k_P, k_D)$ were recorded for three various $J$ functions. In none of the 2000 trials, the UAV model showed dangerous behavior, and as expected: higher values of $k_P$ correspond to a better quality of reference signal tracking (lower values of $J$).

In the second phase of the research, the effectiveness of the GLD and FIB methods was compared for three initial values of the $k_D$ and three $J$ function structures. The very promising results are presented in Figure 8 and Table 5. Both methods effectively explore the gains space ($k_P$, $k_D$) in search for smaller values of $J$, avoiding the local minima (they do not "get stuck" in there)—see Figure 8 (right column). Depending on the set $k_{Dinit}$ gain value, both the methods yield in similar $k_P$ values, but various $k_D$, slowing down the expected tracking dynamics respectively (for larger values of $\beta$). It is particularly noteworthy to compare the signals for subsequent set values of $\beta$ (Figure 9). Bearing in mind the diversity of UAV applications, it is possible to shape the "energy policy", i.e., through an introduction of larger values of $\beta$, one obtain a smooth, slower trajectory of the altitude signal, with a smaller control signal amplitudes (for which the $\beta$ is punishing), which is conducive to extend the flight time.

In relation to the FIB method, an additional time of 96 sec (corresponding to 8 iterations of the auto-tuning algorithm), allows the GLD method in subsequent iterations only to slightly improve the value of the $J$ performance index (respectively by 1.62%, 0.78%, and 2.10%). The introduction of the second bootstrap is justified in the FIB method (improvement by respectively: 11.92%, 4.95%, 1.40%), while in the case of the GLD method, just only one bootstrap provides similar results. The listings from the altitude controller auto-tuning process are available for both methods in the supplementary materials at the AeroLab webpage.

**Table 5.** Results of simulation experiments.

| | FIB | GLD | FIB | GLD | FIB | GLD |
|---|---|---|---|---|---|---|
| $\alpha$ | 1.0 | 1.0 | 0.9 | 0.9 | 0.8 | 0.8 |
| $\beta$ | 0.0 | 0.0 | 0.1 | 0.1 | 0.2 | 0.2 |
| $k_{Dinit}$ | 2.0 | 2.0 | 10.0 | 10.0 | 18.0 | 18.0 |
| $k_P$ range | [0.5,5.0] | [0.5,5.0] | [0.5,5.0] | [0.5,5.0] | [0.5,5.0] | [0.5,5.0] |
| $k_D$ range | [1.0,20.0] | [1.0,20.0] | [1.0,20.0] | [1.0,20.0] | [1.0,20.0] | [1.0,20.0] |
| No. of bootstrap cycles | 2 | 2 | 2 | 2 | 2 | 2 |
| No. of main iterations | 48 | 56 | 48 | 56 | 48 | 56 |
| Tuning time [sec] | 576 | 672 | 576 | 672 | 576 | 672 |
| Low-pass filtration | yes | yes | yes | yes | yes | yes |
| $\epsilon$ | 0.05 | 0.05 | 0.05 | 0.05 | 0.05 | 0.05 |
| Best $k_P$ and $k_D$ values | 3.56/8.70 | 3.70/8.18 | 4.63/5.99 | 4.49/8.58 | 4.23/10.51 | 4.39/15.44 |
| $J_1$ (after the 1st bootstrap) | 6.8235 | 5.4575 | 5.7596 | 5.4865 | 5.9641 | 5.8197 |
| $J_{48}$ (after 48 iter.) | 6.0969 | 5.5024 | 5.4882 | 5.4492 | 5.8815 | 5.9059 |
| $J_{end}$ (after the tuning proc.) | 6.0969 | 5.4149 | 5.4882 | 5.4068 | 5.8815 | 6.0298 |
| $J_{avg}$ (average for tuning proc.) | 6.7264 | 5.4754 | 5.7730 | 5.5643 | 6.0345 | 5.9376 |

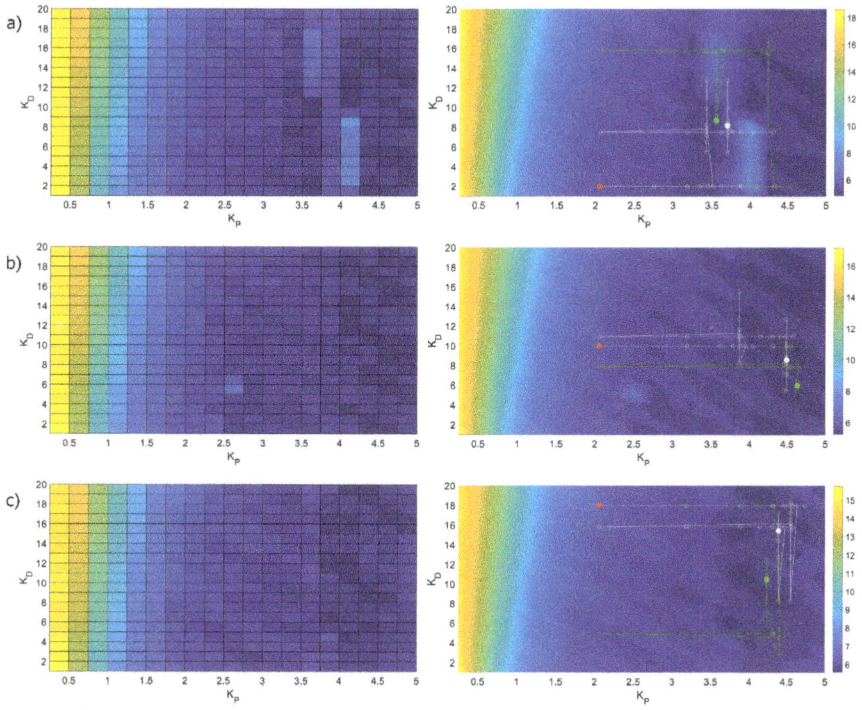


**Figure 8.** Obtained values of the $J$ performance index for $k_P$ and $k_D$ combinations (left column) and $J = f(k_P, k_D)$ approximations (right column) for: (**a**) $\alpha = 1.0$, $\beta = 0.0$, (**b**) $\alpha = 0.9$, $\beta = 0.1$, (**c**) $\alpha = 0.8$, $\beta = 0.2$. FIB (green) and GLD (white) tuning results for: (**a**) $k_{Dinit} = 2$, (**b**) $k_{Dinit} = 10$, (**c**) $k_{Dinit} = 18$ (marked in red).

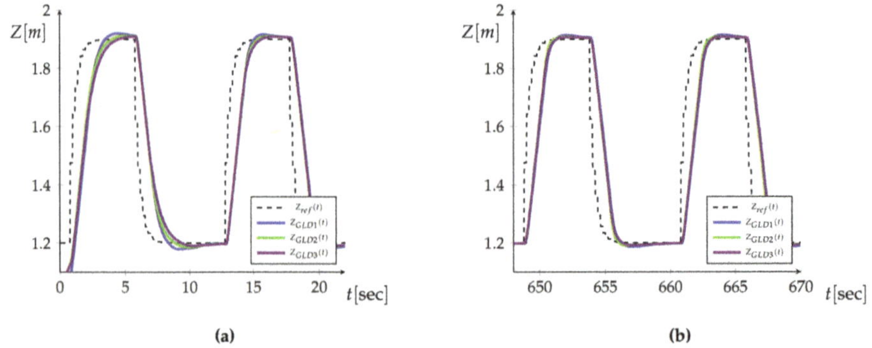


**Figure 9.** Time courses for: (**a**) first two iterations of the GLD method (mistuned gains), (**b**) last two iterations (well-tuned gains) for $Z_{GLD1}$ ($\alpha = 1.0, \beta = 0.0$), $Z_{GLD2}$ ($\alpha = 0.9, \beta = 0.1$), $Z_{GLD3}$ ($\alpha = 0.8, \beta = 0.2$).

## 3.2. Experiments in Flight Conditions

The GLD method was verified in real-world experiments on the same UAV and for the same parameter configuration as in simulation tests. The method was tested with great attention paid to the efficiency of obtaining altitude controller gains and the tracking quality. From variety of conducted experiments, the author decided to present and discussed, a few, which are the most representative. Supplementary materials (video and listings) are available at: http://www.uav.put.poznan.pl.

### 3.2.1. Uncertainty of Altitude Measurements. Change of Sensory Information Sources

The aim of the experiment was to verify how imprecise and non-stationary the altitude measurements of the UAV flight are in the building, based on its basic on-board avionics only. The motion capture system was used as a ground truth. The results are shown in Figure 10. The task for the UAV was to fly to a fixed altitude of 1 m and hover in the air.

As the average error from registered trials is only 0.80%, the range of actual/instantaneous values ranges from 0.85 m to 1.14 m and increases with the passage of time. Such a dispersion of measurements is a problem and major difficulty in the proposed machine learning procedure used in real-world conditions for altitude controller tuning. Therefore, the motion capture system was used for further estimation of the UAV flight altitude. This eliminates the measurement error as a source of additional errors during the $J$ calculation.

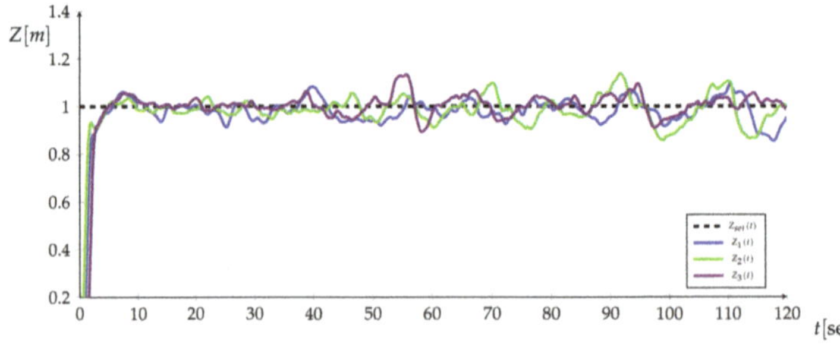


**Figure 10.** Tracking of the reference altitude $Z_{set}$ by the Bebop 2 UAV in three trials ($Z_1$–$Z_3$).

### 3.2.2. Comparison of the Tuning Effectiveness: FIB vs. GLD Method Used in Real-World Conditions

In the Figure 11, the altitude controller gains during auto-tuning procedure using GLD and FIB methods, are presented. Based on simulation results it was decided to terminate both methods after 48 iterations. Final and average values of $J$ (see Figure 12), are lower for the GLD method: 43.97% and

3.39%, for which the tracking quality is better (Figure 13), e.g., lower overshoots were recorded during the tuning time.

Furthermore, it is worth mentioning that both methods here shown convergence in the vicinity of the two local minima of the $J = f(k_P, k_D)$ function, which were estimated based on the preliminary simulation experiments.

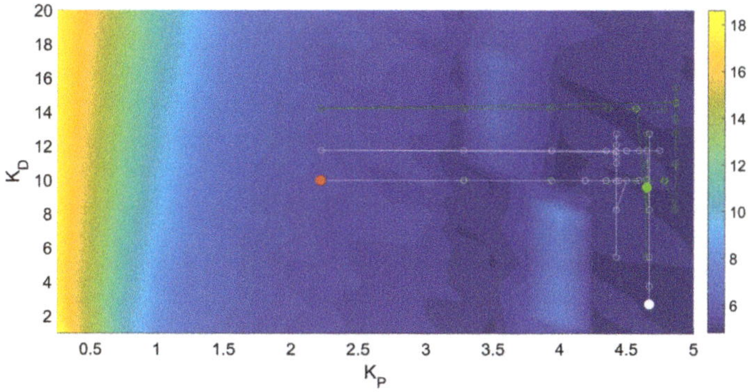


**Figure 11.** The altitude controller gains and $J = f(k_P, k_D)$ values during auto-tuning process using GLD (white color) and FIB (green) methods; $k_{Dinit}$ = 10 (marked in red).

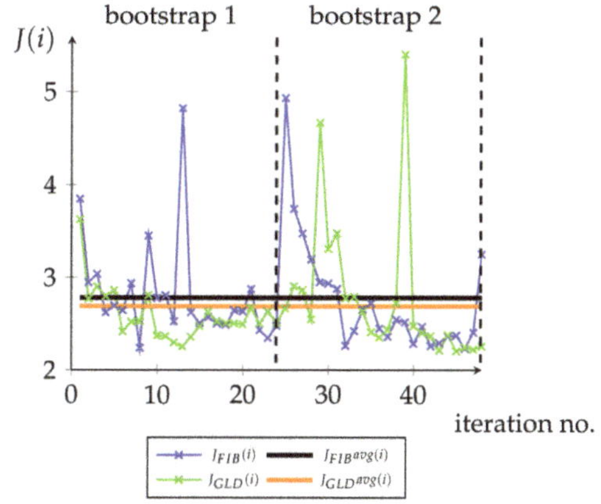


**Figure 12.** Time courses for the GLD and FIB tuning process in real-world conditions.

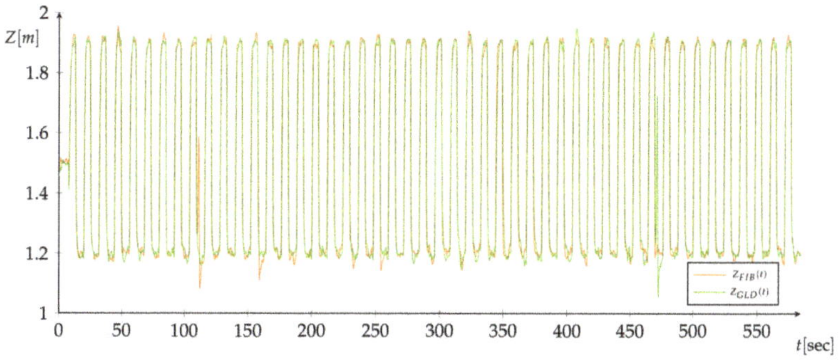


**Figure 13.** Values of $J(i)$ in consecutive steps (i) of GLD and FIB tuning.

3.2.3. Analysis of the Impact of Environmental Disturbances (Wind Gusts) on the Auto-Tuning Procedure

Usually in scientific world literature presented results of research on UAV flights under wind gust conditions concern the case when the air stream is directed towards the UAV frontally. In real flight conditions, this direction is usually random and variable in time. Thus, it was decided to verify the effectiveness of the GLD auto-tuning method with a low-pass filtration, during the UAV flight, in the stream of the air generated from the rotating fan (1.2 m high), at a distance of 1.8 m, behind the UAV on the left, as in Figure 14.

In the auto-tuning procedure, the disturbances were introduced twice (see Figure 15). In the first phase, the maximum air flow speed was 2.7 m/s, in the second—3.7 m/s. It is a severe disturbance referring to the ratio of physical dimensions to the small weight of the UAV. A complete 56-iterative tuning cycle was conducted. The results are summarized in Figure 16 and Table A1 (see Appendix A), and compared with the results of the auto-tuning from the previous Subsection. Very similar, promising final values of the $J$ performance index were obtained—even only slightly smaller for the case of impact of a wind gust during the GLD procedure.

Determinism of the method is illustrated by the results of 10 first iterations in both trials and iterations no. 29-38, where for different values of $J$, the calculated $k_P$ gain values are identical. Similar behavior can be observed in the presence of wind gusts (iterations no. 15–20, and 43–48). In the future research, it is worth considering an approach in which two or several UAV units (agents) could be used to parallel measurements and averaging computations during the auto-tuning procedure, resulting in better tuning precision.

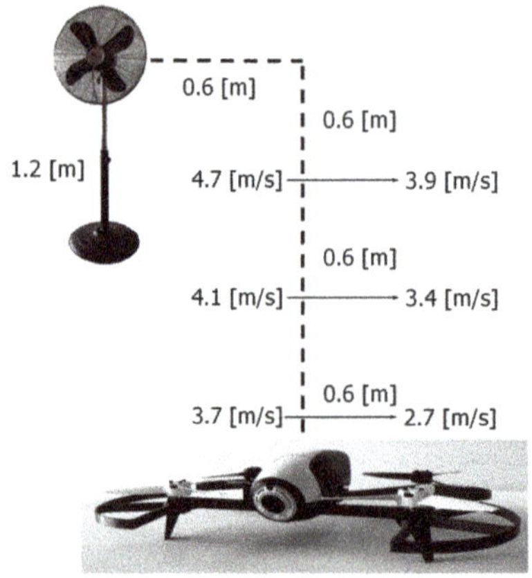


**Figure 14.** Test bed for research on wind gusts impact on the GLD method.

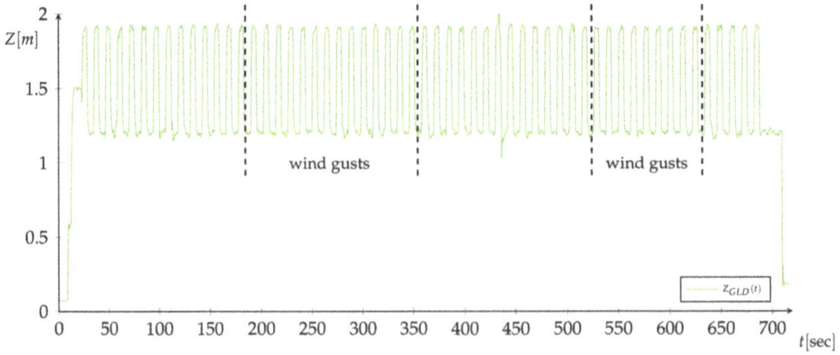


**Figure 15.** Time course for the GLD tuning process in the presence of wind gusts.

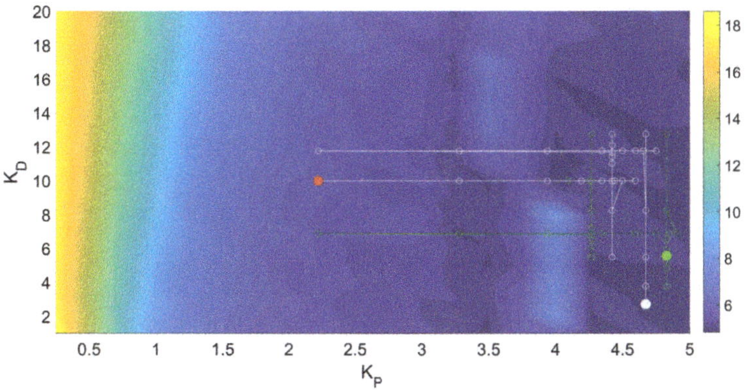


**Figure 16.** The altitude controller gains and $J = f(k_P, k_D)$ values during auto-tuning process using GLD method for nominal case (white color) and at the presence of wind gusts (green); $k_{Dinit}$ = 10 (marked in red).

### 3.2.4. Flights and Auto-Tuning in UAV Mass Change Conditions

The last interesting aspect of the conducted research was to provide knowledge about the quality of the obtained gains in the context of transport tasks and the use of the GLD method to tune the gains of the altitude controller after changing the total takeoff weight of the UAV. A series of experimental studies was conducted for this purpose.

In the simulation experiments, the efficiency of tuning of the UAV altitude controller using the GLD method was verified in conditions of lifting of the additional payload (jar on the gripper and tool accessories attached to the UAV). The gains of the other controllers (for X and Y axes, and for *yaw* angle control), were adopted from Table 4. In subsequent simulations, the values $\alpha$ and $\beta$ of the $J$ function were changed. The results are presented in Table 6, and the search process for the controller's gains is illustrated in the attached video material. Based on the obtained results, it can be noticed that in the case of both payloads tested, the values of $k_P$ were smaller than in the nominal case (flight without payload), and $k_D$ values were larger. Increased starting mass of the UAV forces the use of more thrust to lift the UAV and at the same time—to provide its effective balance, so as not to cause any overshoots (exceeding the given/reference altitude). In the qualitative evaluation of the results of the auto-tuning procedure, the obtained controller using a similar gain value of the proportional part, compensates with a larger gain of $k_D$ the nervous behavior of the UAV (which for particular $J$ function tries to match the dynamics to higher UAV inertia).

**Table 6.** Results of simulation experiments—flying with: gripper & jar (GRIP), and tool accessories (TOOL); for 2 bootstrap cycles, 56 iterations of the GLD method with low-pass filtration and $\epsilon$ = 0.05.

| | GRIP | TOOL | GRIP | TOOL | GRIP | TOOL |
|---|---|---|---|---|---|---|
| $\alpha$ | 1.0 | 1.0 | 0.9 | 0.9 | 0.8 | 0.8 |
| $\beta$ | 0.0 | 0.0 | 0.1 | 0.1 | 0.2 | 0.2 |
| $k_{Dinit}$ | 10.0 | 10.0 | 10.0 | 10.0 | 10.0 | 10.0 |
| $k_P$ range | [0.5,5.0] | [0.5,5.0] | [0.5,5.0] | [0.5,5.0] | [0.5,5.0] | [0.5,5.0] |
| $k_D$ range | [1.0,20.0] | [1.0,20.0] | [1.0,20.0] | [1.0,20.0] | [1.0,20.0] | [1.0,20.0] |
| Best $k_P$ and $k_D$ values | 2.30/15.11 | 2.39/18.94 | 2.20/17.88 | 1.08/18.94 | 0.82/15.77 | 0.67/13.40 |
| $J_1$ (after the 1st bootstrap) | 7.5143 | 7.9741 | 7.6119 | 7.6654 | 7.5773 | 7.5454 |
| $J_{end}$ (after the tuning proc.) | 7.5150 | 7.8334 | 7.5473 | 7.6591 | 7.7385 | 7.4198 |
| $J_{avg}$ (average for tuning proc.) | 7.4306 | 7.9191 | 7.7877 | 7.5485 | 7.5906 | 7.5454 |

In the first real-world experiment (Figure 17), the task of the UAV was to start the autonomous flight from a platform with a plastic bottle attached; then, to fly to the point where the GLD auto-tuning procedure begins; finally, to perform 56 iterations of the algorithm in the presence of wind gusts. The drone, using its on-board avionics (including the optical-flow and ultrasonic sensors) moved vertically after stabilizing the position of the gripper, since it recognized its position as altitude equal to 0, and in effect moved upwards, which created a danger. The same behavior was observed in the second experiment, where the UAV task was to compensate its position in the X, Y, and Z axis (refer to supplementary video material). A decision was made to change the type and manner of payload attachment as shown in Figure 18, which played its role, both in the GLD auto-tuning experiments with additional mass, as well as in transportation tasks at designated nominal gains (see Figure 19). In every conducted trial (Table 7), for subsequent $J$ functions, similar behavior was observed as in the case of simulation tests. For example, let us consider the results obtained for $\alpha = 1.0$ (Figure 20). It can be noticed that the time courses with large overshoots (when the controller forces too hard the UAV, wanting to overcome its increased inertia), result in an increase in the value of $J$ and are effectively rejected in the procedure of seeking the smallest value of this performance index. In addition, by analyzing the subsequent values of this index (Figure 21), it can be seen that the selection of the gain value $k_D$ directly implies the UAV vertical flight dynamics profile. This is particularly seen in the first bootstrap (marked in Figure 20).

Auto-tuning in UAV mass change conditions will be the subject of a separate article, while it is worth stressing that the second problem encountered—mentioned at the beginning of the article—i.e., lack of stability criterion based on which it would be possible to estimate safe gains ranges of $k_P$ and $k_D$ for their exploration in the GLD method. Despite its high efficiency and safe operation in tuning of controllers of UAVs with nominal mass or with low extra mass, in case of large payloads (see Figure 22 for the case of 282 g) one can find examples of unstable flights. Then it is strongly recommended to use preliminary simulation tests based on the model. The introduction of the stability criterion into the proposed GLD method is in the area of further research interest of the author [50].

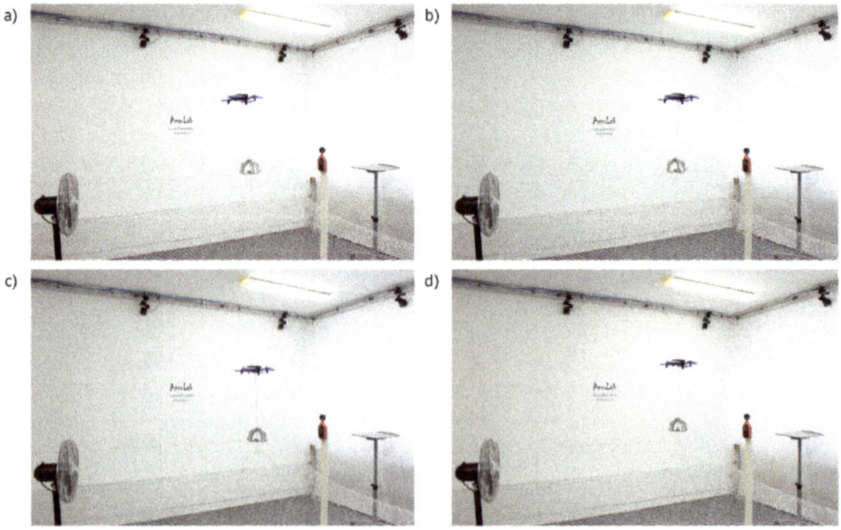


**Figure 17.** Snapshots from one of the initial research experiments with the auto-tuning of the altitude controller during the flight in the presence of wind gusts and with the mass attached to the UAV on a flexible joint.

**Figure 18.** The Bebop 2 with additional payload used for in-flight auto-tuning experiments.

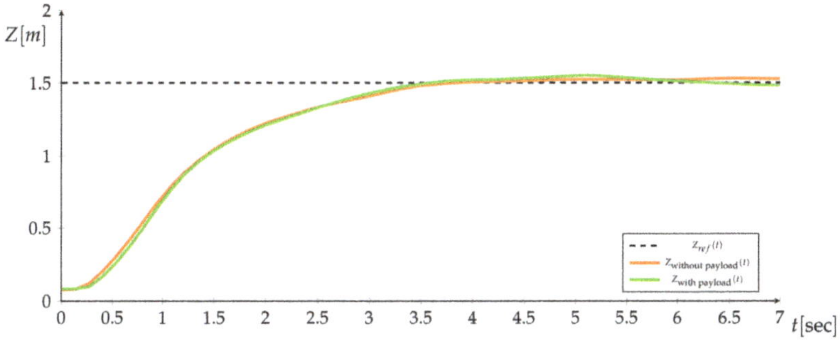


**Figure 19.** Step responses for the Bebop 2 UAV tuned with GLD method—variants: with and without addition mass (225 g tool accessories).

**Table 7.** Results of real-world experiments—flying with the payload (tool accessories); for 2 bootstrap cycles, 56 iterations of the GLD method with low-pass filtration and $\epsilon = 0.05$.

| | Exp.1 | Exp.2 | Exp.3 |
|---|---|---|---|
| $\alpha$ | 1.0 | 0.9 | 0.8 |
| $\beta$ | 0.0 | 0.1 | 0.2 |
| $k_{Dinit}$ | 10.0 | 10.0 | 10.0 |
| $k_P$ range | [0.5,5.0] | [0.5,5.0] | [0.5,5.0] |
| $k_D$ range | [1.0,20.0] | [1.0,20.0] | [1.0,20.0] |
| Best $k_P$ and $k_D$ values | 3.92/7.85 | 4.02/9.57 | 3.20/11.68 |
| $J_1$ (after the 1st bootstrap) | 2.5070 | 2.4992 | 2.6096 |
| $J_{end}$ (after the tuning proc.) | 1.7583 | 2.5779 | 2.5255 |
| $J_{avg}$ (average for tuning proc.) | 2.8091 | 2.4566 | 2.6553 |

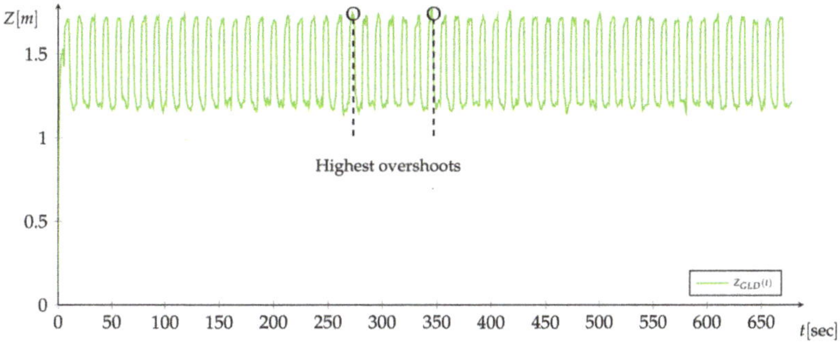


**Figure 20.** Time course from the tuning experiment via GLD method—variant: tuning of the altitude controller during the UAV flight with an additional (heavy) mass (225 g); the experiment interrupted due to the loss of stability.

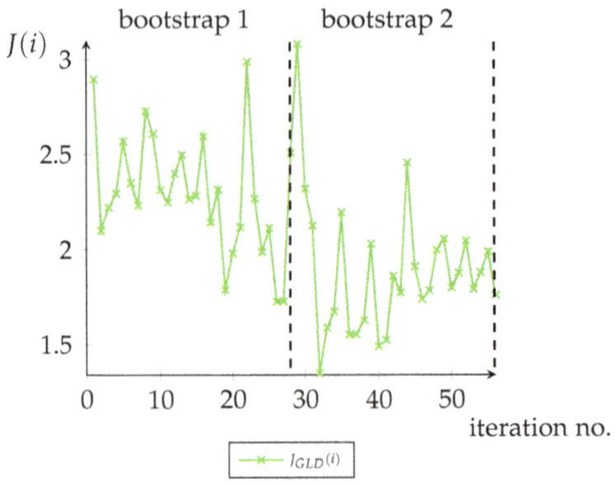


**Figure 21.** Values of $J(i)$ in consecutive steps (i) of the GLD method (Exp. no. 1)—flying with the payload.

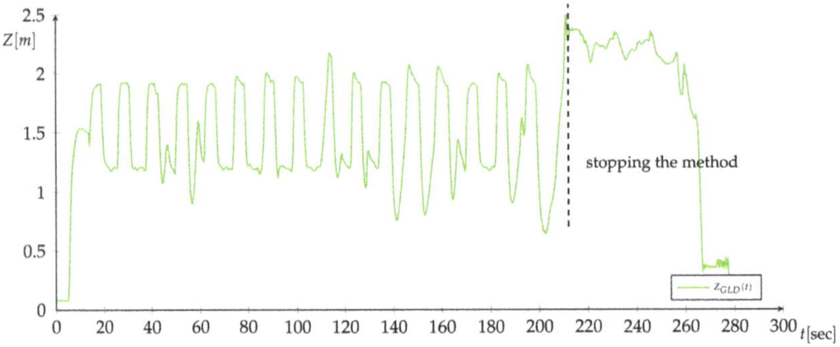


**Figure 22.** Time course from the real-world experiment (Exp. no. 1): tuning of the altitude controller during the UAV flight with an additional (heavy) mass (225 g).

## 4. Conclusions and Further Work

In the paper, a new and efficient real-time auto-tuning method for fixed-parameters controllers based on the modified golden-search (zero-order) optimization algorithm and bootstrapping technique, has been presented. The method ensures fast, iterative behavior, and as a result—returns in the worst case the locally best gains of controller, in the best case—globally optimal. The GLD method is fully automated, and uses a low-pass filtration while working in a stochastic environment. It is a model-free approach, but as it has been articulated in the paper, it is good to combine its advantages with initial model-based prototyping, since the method does not use any stability criterion. The author is interested and looking for the mathematical solutions, i.e., in the area of stochastic analysis and probability, which can be easily adapted into the proposed GLD procedure—without increase of its computational complexity. It will be useful in a context of solving mentioned transportation tasks and problems (especially when flying near to the lifting capacity of the UAV).

**Supplementary Materials:** Recorded videos and data from ROS bags are available online at http://uav.put.poznan.pl.

**Funding:** This research was financially supported as a statutory work of Poznan University of Technology (04/45/DSPB/0196).

**Acknowledgments:** The author would like to thank Bartłomiej Kulecki for his help with software configuration.

**Conflicts of Interest:** The author declares no conflicts of interest.

## Abbreviations

The following abbreviations are used in this manuscript:

| | |
|---|---|
| BF | Body Frame |
| CCW | Counter-Clockwise |
| CW | Clockwise |
| DOF | Degrees of Freedom |
| EF | Earth Frame |
| FIB | Fibonacci-search Method |
| GLD | Golden-search Method |
| GNSS | Global navigation satellite system |
| GPS | Global Positioning System |
| MBZIRC | Mohamed Bin Zayed International Robotics Challenge |
| NED | North-East-Down |
| PD | Proportional-Derivative Controller |
| PID | Proportional-Integral-Derivative Controller |
| REM | Region Elimination Method |
| ROS | Robot Operating System |
| UAV | Unmanned Aerial Vehicle |

## Appendix A

**Table A1.** Comparison of the results of auto-tuning of the UAV's altitude controller using the GLD method—variants: nominal and at the presence of wind gusts.

| | Nominal | Disturbed | Nominal | Disturbed | Nominal | Disturbed |
|---|---|---|---|---|---|---|
| No. of Iter. | $k_P$ | $k_P$ | $k_D$ | $k_D$ | $J$ | $J$ |
| 1 | 2.2190 | 2.2190 | 10.0000 | 10.0000 | 3.6232 | 3.9781 |
| 2 | 3.2810 | 3.2810 | 10.0000 | 10.0000 | 2.7718 | 2.9733 |
| 3 | 3.2813 | 3.2813 | 10.0000 | 10.0000 | 2.9025 | 3.0650 |
| 4 | 3.9377 | 3.9377 | 10.0000 | 10.0000 | 2.7974 | 2.9231 |
| 5 | 3.9379 | 3.9379 | 10.0000 | 10.0000 | 2.8608 | 2.9099 |
| 6 | 4.3435 | 4.3435 | 10.0000 | 10.0000 | 2.4185 | 2.6717 |
| 7 | 4.3436 | 4.3436 | 10.0000 | 10.0000 | 2.5228 | 2.7708 |
| 8 | 4.5943 | 4.5943 | 10.0000 | 10.0000 | 2.5281 | 3.2745 |
| 9 | 4.1886 | 4.1886 | 10.0000 | 10.0000 | 2.8080 | 2.6899 |
| 10 | 4.3435 | 4.3435 | 10.0000 | 10.0000 | 2.3742 | 2.8161 |
| 11 | 4.3436 | 4.0928 | 10.0000 | 10.0000 | 2.3683 | 3.0077 |
| 12 | 4.4393 | 4.1886 | 10.0000 | 10.0000 | 2.3004 | 2.6560 |
| 13 | 4.4393 | 4.1886 | 10.0000 | 10.0000 | 2.2584 | 2.7008 |
| 14 | 4.4985 | 4.2478 | 10.0000 | 10.0000 | 2.3600 | 2.4583 |
| 15 | 4.4210 | 4.2661 | 8.2580 | 8.2580 | 2.4742 | 2.4305 |
| 16 | 4.4210 | 4.2661 | 12.7420 | 12.7420 | 2.6279 | 2.5428 |
| 17 | 4.4210 | 4.2661 | 5.4854 | 5.4854 | 2.5291 | 2.8694 |
| 18 | 4.4210 | 4.2661 | 8.2566 | 8.2566 | 2.5097 | 2.3611 |
| 19 | 4.4210 | 4.2661 | 8.2574 | 8.2574 | 2.5057 | 2.3465 |
| 20 | 4.4210 | 4.2661 | 9.9700 | 9.9700 | 2.4944 | 2.7500 |
| 21 | 4.4210 | 4.2661 | 9.9705 | 7.1985 | 2.6712 | 2.4564 |
| 22 | 4.4210 | 4.2661 | 11.0289 | 8.2569 | 2.5066 | 2.5012 |
| 23 | 4.4210 | 4.2661 | 11.0292 | 6.5441 | 2.6319 | 2.5666 |
| 24 | 4.4210 | 4.2661 | 11.6833 | 7.1982 | 2.5153 | 2.7464 |
| 25 | 4.4210 | 4.2661 | 11.6835 | 6.1397 | 2.6646 | 2.6194 |
| 26 | 4.4210 | 4.2661 | 12.0877 | 6.5439 | 2.9049 | 2.4165 |
| 27 | 4.4210 | 4.2661 | 11.4336 | 6.5441 | 2.8642 | 2.4360 |
| 28 | 4.4210 | 4.2661 | 11.6834 | 6.7939 | 2.5492 | 2.3116 |
| 29 | 2.2190 | 2.2190 | 11.7607 | 6.8711 | 4.6672 | 4.4838 |
| 30 | 3.2810 | 3.2810 | 11.7607 | 6.8711 | 3.3072 | 2.7008 |

**Table A1.** *Cont.*

| | Nominal | Disturbed | Nominal | Disturbed | Nominal | Disturbed |
|---|---|---|---|---|---|---|
| No. of Iter. | $k_P$ | $k_P$ | $k_D$ | $k_D$ | $J$ | $J$ |
| 31 | 3.2813 | 3.2813 | 11.7607 | 6.8711 | 3.4729 | 2.9624 |
| 32 | 3.9377 | 3.9377 | 11.7607 | 6.8711 | 2.7758 | 2.4287 |
| 33 | 3.9379 | 3.9379 | 11.7607 | 6.8711 | 2.7904 | 2.5296 |
| 34 | 4.3435 | 4.3435 | 11.7607 | 6.8711 | 2.6357 | 2.3066 |
| 35 | 4.3436 | 4.3436 | 11.7607 | 6.8711 | 2.4105 | 4.4358 |
| 36 | 4.5943 | 4.5943 | 11.7607 | 6.8711 | 2.3535 | 2.6047 |
| 37 | 4.5943 | 4.5943 | 11.7607 | 6.8711 | 2.4362 | 2.5906 |
| 38 | 4.7493 | 4.7493 | 11.7607 | 6.8711 | 2.7252 | 2.4460 |
| 39 | 4.4986 | 4.7493 | 11.7607 | 6.8711 | 5.4026 | 2.5969 |
| 40 | 4.5943 | 4.8450 | 11.7607 | 6.8711 | 2.4769 | 2.3667 |
| 41 | 4.5943 | 4.8451 | 11.7607 | 6.8711 | 2.4032 | 2.3662 |
| 42 | 4.6535 | 4.9042 | 11.7607 | 6.8711 | 2.3701 | 2.9292 |
| 43 | 4.6718 | 4.8268 | 8.2580 | 8.2580 | 2.2139 | 2.3028 |
| 44 | 4.6718 | 4.8268 | 12.7420 | 12.7420 | 2.3797 | 2.5986 |
| 45 | 4.6718 | 4.8268 | 5.4854 | 5.4854 | 2.2050 | 2.1790 |
| 46 | 4.6718 | 4.8268 | 8.2566 | 8.2566 | 2.2468 | 2.3559 |
| 47 | 4.6718 | 4.8268 | 3.7720 | 3.7720 | 2.2244 | 2.4715 |
| 48 | 4.6718 | 4.8268 | 5.4846 | 5.4846 | 2.2563 | 2.1976 |
| 49 | ... | 4.8268 | ... | 5.4851 | ... | 2.1532 |
| 50 | ... | 4.8268 | ... | 6.5435 | ... | 2.5722 |
| 51 | ... | 4.8268 | ... | 4.8307 | ... | 2.3696 |
| 52 | ... | 4.8268 | ... | 5.4848 | ... | 2.3328 |
| 53 | ... | 4.8268 | ... | 5.4850 | ... | 2.2445 |
| 54 | ... | 4.8268 | ... | 5.8892 | ... | 2.7051 |
| 55 | ... | 4.8268 | ... | 5.2350 | ... | 2.2805 |
| 56 | ... | 4.8268 | ... | 5.4848 | ... | 2.2393 |

## References


1. Valavanis, K.; Vachtsevanos, G.J. (Eds.) *Handbook of Unmanned Aerial Vehicles*; Springer: Dordrecht, The Netherlands, 2015.
2. Jordan, S.; Moore, J.; Hovet, S.; Box, J.; Perry, J.; Kirsche, K.; Lewis, D.; Tsz Ho Tse, Z. State-of-the-art technologies for UAV inspections. *IET Radar Sonar Navig.* **2018**, *12*, 151–164. [CrossRef]
3. Hinas, A.; Roberts, J.M.; Gonzalez, F. Vision-Based Target Finding and Inspection of a Ground Target Using a Multirotor UAV System. *Sensors* **2017**, *17*, 2929. [CrossRef] [PubMed]
4. Sandino, J.; Gonzalez, F.; Mengersen, K.; Gaston, K.J. UAVs and Machine Learning Revolutionising Invasive Grass and Vegetation Surveys in Remote Arid Lands. *Sensors* **2018**, *18*, 605. [CrossRef] [PubMed]
5. Dziuban, P.J.; Wojnar, A.; Zolich, A.; Cisek, K.; Szumiński, W. Solid State Sensors—Practical Implementation in Unmanned Aerial Vehicles (UAVs). *Procedia Eng.* **2012**, *47*, 1386–1389. [CrossRef]
6. Gośliński, J.; Giernacki, W.; Królikowski, A. A nonlinear Filter for Efficient Attitude Estimation of Unmanned Aerial Vehicle (UAV). *J. Intell. Robot. Syst.* **2018**. [CrossRef]
7. Urbański, K. Control of the Quadcopter Position Using Visual Feedback. In Proceedings of the 18th International Conference on Mechatronics (Mechatronika), Brno, Czech Republic, 5–7 December 2018; pp. 1–5.
8. Ebeid, E.; Skriver, M.; Terkildsen, K.H.; Jensen, K.; Schultz, U.P. A survey of Open-Source UAV flight controllers and flight simulators. *Microprocess. Microsyst.* **2018**, *61*, 11–20. [CrossRef]
9. Lozano, R. (Ed.) *Unmanned Aerial Vehicles: Embedded Control*; John Wiley & Sons: New York, NY, USA, 2010.
10. Santoso, F.; Garratt, M.A.; Anavatti, S.G. State-of-the-Art Intelligent Flight Control Systems in Unmanned Aerial Vehicles. *IEEE Trans. Autom. Sci. Eng.* **2018**, *15*, 613–627. [CrossRef]
11. Mahony, R.; Kumar, V.; Corke, P. Multirotor aerial vehicles: Modeling, estimation, and control of quadrotor. *IEEE Robot. Autom. Mag.* **2012**, *19*, 20–32. [CrossRef]

12. Ren, B.; Ge, S.; Chen, C.; Fua, C.; Lee, T. *Modeling, Control and Coordination of Helicopter Systems*; Springer: New York, NY, USA, 2012. [CrossRef]
13. Pounds, P.; Bersak, D.R.; Dollar, A.M. Stability of small-scale UAV helicopters and quadrotors with added payload mass under PID control. *Auton. Robots* **2012**, *33*, 129–142. [CrossRef]
14. Li, J.; Li, Y. Dynamic Analysis and PID Control for a Quadrotor. In Proceedings of the 2011 IEEE International Conference on Mechatronics and Automation (ICMA), Beijing, China, 7–10 August 2011; pp. 573–578. [CrossRef]
15. Espinoza, T.; Dzul, A.; Llama, M. Linear and nonlinear controllers applied to fixed-wing UAV. *Int. J. Adv. Robot. Syst.* **2013**, *10*, 1–10. [CrossRef]
16. Lee, K.U.; Kim, H.S.; Park, J.-B.; Choi, Y.-H. Hovering Control of a Quadrotor. In Proceedings of the 2012 12th International Conference on Control, Automation and Systems (ICCAS), JeJu Island, South Korea, 17–21 October 2012; pp. 162–167.
17. Pounds, P.E.; Dollar, A.M. Aerial Grasping from a Helicopter UAV Platform, Experimental Robotics. *Springer Tracts Adv. Robot.* **2014**, *79*, 269–283. [CrossRef]
18. Kohout, P. A System for Autonomous Grasping and Carrying of Objects by a Pair of Helicopters, Master's Thesis, Czech Technical University in Prague, Prague, Czech Republic, 2017.
19. Spica, R.; Franchi, A.; Oriolo, G.; Bülthoff, H.H.; Giordano, P.R. Aerial grasping of a moving target with a quadrotor UAV. In the Proceedings of the 2012 IEEE/RSJ International Conference on Intelligent Robots and Systems, Vilamoura, Portugal, 7–12 October 2012; pp. 4985–4992. [CrossRef]
20. Yang, F.; Xue, X.; Cai, C.; Sun, Z.; Zhou, Q. Numerical Simulation and Analysis on Spray Drift Movement of Multirotor Plant Protection Unmanned Aerial Vehicle. *Energies* **2018**, *11*, 2399. [CrossRef]
21. Rao Mogili, U.M.; Deepak, B.B.V.L. Review on Application of Drone Systems in Precision Agriculture. *Procedia Comput. Sci.* **2018**, *133*, 502–509. [CrossRef]
22. Imdoukh, A.; Shaker, A.; Al-Toukhy, A.; Kablaoui, D., El-Abd, M. Semi-autonomous indoor firefighting UAV. In Proceedings of the 2017 18th International Conference on Advanced Robotics (ICAR), Hong Kong, China, 10–12 July 2017; pp. 310-315. [CrossRef]
23. Duan, H.; Li, P. *Bio-inspired Computation in Unmanned Aerial Vehicles*; Springer: Berlin, Germany, 2014. [CrossRef]
24. Giernacki, W.; Espinoza Fraire, T.; Kozierski, P. Cuttlesh Optimization Algorithm in Autotuning of Altitude Controller of Unmanned Aerial Vehicle (UAV). In Proceedings of the Third Iberian Robotics Conference (ROBOT 2017), Seville, Spain, 22–24 November 2017; pp. 841–852. [CrossRef]
25. Chong, E.K.P.; Zak, S.H. *An Introduction to Optimization*, 2nd ed.; John Wiley & Sons: Hoboken, NJ, USA, 2001.
26. Giernacki, W.; Horla, D.; Báča, T.; Saska, M. Real-time model-free optimal autotuning method for unmanned aerial vehicle controllers based on Fibonacci-search algorithm. *Sensors* **2018**, *19*, 312. [CrossRef]
27. Spurný, V.; Báča, T.; Saska, M.; Pěnička, R.; Krajník, T.; Loianno, G.; Thomas, J.; Thakur, D.; Kumar, V. Cooperative Autonomous Search, Grasping and Delivering in Treasure Hunt Scenario by a Team of UAVs. *J. Field Robot.* **2018**, 1–24. [CrossRef]
28. Automatic Tuning with AUTOTUNE. Ardupilot.org. Available online: http://ardupilot.org/plane/docs/automatic-tuning-with-autotune.html (accessed on 12 November 2018).
29. Rodriguez-Ramos, A.; Sampedro, C.; Bavle, H.; de la Puente, P.; Campoy, P. A Deep Reinforcement Learning Strategy for UAV Autonomous Landing on a Moving Platform. *J. Intell. Robot. Syst.* **2018**, 1–16. [CrossRef]
30. Koch, W.; Mancuso, R.; West, R.; Bestavros, A. Reinforcement Learning for UAV Attitude Control. Available online: https://arxiv.org/abs/1804.04154 (accessed on 12 November 2018).
31. Panda, R.C. *Introduction to PID Controllers—Theory, Tuning and Application to Frontier Areas*; In-Tech: Rijeka, Croatia, 2012. [CrossRef]
32. Rios, L.; Sahinidis, N. Derivative-free optimization: A review of algorithms and comparison of software implementations. *J. Glob. Optim.* **2013**, *56*, 1247–1293. [CrossRef]
33. Spall, J.C. *Introduction to Stochastic Search and Optimization: Estimation, Simulation, and Control*; Wiley: New York, NY, USA, 2003. [CrossRef]
34. Hjalmarsson, H.; Gevers, M.; Gunnarsson, S.; Lequin, O. Iterative feedback tuning: Theory and applications. *IEEE Control Syst. Mag.* **1998**, *18*, 26–41. [CrossRef]

35. Reza-Alikhani, H. PID type iterative learning control with optimal variable coefficients. In the Proceedings of the 2010 5th IEEE International Conference Intelligent Systems, London, UK, 7–9 July 2010; pp. 1–6. [CrossRef]
36. Ghadimi, S.; Lan, G. Stochastic first-and zeroth-order methods for nonconvex stochastic programming. *SIAM J. Optim.* **2013**, *23*, 2341–2368. [CrossRef]
37. Kiefer, J. Sequential minimax search for a maximum. *Proc. Am. Math. Soc.* **1953**, *4*, 502–506. [CrossRef]
38. Brasch, T.; Byström, J.; Lystad, L.P. *Optimal Control and the Fibonacci Sequence*; Statistics Norway, Research Department: Oslo, Norway, 2012; pp. 1–33. Available online: https://www.ssb.no/a/publikasjoner/pdf/DP/dp674.pdf (accessed on 28 December 2018).
39. Theys, B.; Dimitriadis, G.; Hendrick, P.; De Schutter, J. Influence of propeller configuration on propulsion system efficiency of multi-rotor Unmanned Aerial Vehicles. In Proceedings of the 2016 International Conference on Unmanned Aircraft Systems (ICUAS), Arlington, TX, USA, 7–10 June 2016; pp. 195–201. [CrossRef]
40. Xia, D.; Cheng, L.; Yao, Y. A Robust Inner and Outer Loop Control Method for Trajectory Tracking of a Quadrotor. *Sensors* **2017**, *17*, 2147. [CrossRef] [PubMed]
41. Wang, Y.; Gao, F.; Doyle, F. Survey on iterative learning control, repetitive control, and run-to-run control. *J. Process Control* **2009**, *10*, 1589–1600. [CrossRef]
42. Multicopter PID Tuning Guide. Available online: https://docs.px4.io/en/config_mc/pid_tuning_guide_multicopter.html (accessed on 19 November 2018).
43. How to Tune PID I-Term on a Quadcopter. Available online: https://quadmeup.com/how-to-tune-pid-i-term-on-a-quadcopter/ (accessed on 18 November 2018).
44. Quadcopter PID Explained. Available online: https://oscarliang.com/quadcopter-pid-explained-tuning/ (accessed on 18 November 2018).
45. Arimoto, S.; Kawamura, S.; Miyazaki, F. Bettering operation of robots by learning. *J. Robot. Syst.* **1984**, *1*, 123–140. [CrossRef]
46. Parrot BEBOP 2. The Lightweight, Compact HD Video Drone. Available online: https://www.parrot.com/us/drones/parrot-bebop-2 (accessed on 26 November 2018).
47. AeroLab Poznan University of Technology Drone Laboratory Webpage. Available online: http://uav.put.poznan.pl/AeroLab (accessed on 2 December 2018).
48. What is Sphinx. Available online: https://developer.parrot.com/docs/sphinx/whatissphinx.html (accessed on 18 November 2018).
49. bebop_autonomy—ROS Driver for Parrot Bebop Drone (quadrocopter) 1.0 & 2.0. Available online: https://bebop-autonomy.readthedocs.io/en/latest/ (accessed on 18 November 2018).
50. Giernacki, W.; Horla, D.; Sadalla, T.; Espinoza Fraire, T. Optimal Tuning of Non-integer Order Controllers for Rotational Speed Control of UAV's Propulsion Unit Based on an Iterative Batch Method. *J. Control Eng. Appl. Inform.* **2018**, *24*, 22–31.

*applied sciences*

MDPI

*Article*

# A Trajectory Planning Method for Polishing Optical Elements Based on a Non-Uniform Rational B-Spline Curve

**Dong Zhao and Hao Guo ***

Key Laboratory of Mechanism Theory and Equipment Design of The State Ministry of Education, Tianjin University, Tianjin 300072, China; dongzhao@tju.edu.cn

* Correspondence: guohaohalo@163.com; Tel.: +86-156-2090-9866

Received: 10 July 2018; Accepted: 10 August 2018; Published: 12 August 2018

**Abstract:** Optical polishing can accurately correct the surface error through controlling the dwell time of the polishing tool on the element surface. Thus, the precision of the trajectory and the dwell time (the runtime of the trajectory) are important factors affecting the polishing quality. This study introduces a systematic interpolation method for optical polishing using a non-uniform rational B-spline (NURBS). A numerical method for solving all the control points of NURBS was proposed with the help of a successive over relaxation (SOR) iterative theory, to overcome the problem of large computation. Then, an optimisation algorithm was applied to smooth the NURBS by taking the shear jerk as the evaluation index. Finally, a trajectory interpolation scheme was investigated for guaranteeing the precision of the trajectory runtime. The experiments on a prototype showed that, compared to the linear interpolation method, there was an order of magnitude improvement in interpolation, and runtime, errors. Correspondingly, the convergence rate of the surface error of elements improved from 37.59% to 44.44%.

**Keywords:** hybrid robot; curve fitting; fair optimisation; trajectory interpolation

---

## 1. Introduction

With the rapid development of astronomy, space exploration, and advanced optical instruments, optical elements are being increasingly widely used. The application demands for high-quality and high-efficiency elements present distinct higher requirements for the process technology used in such elements [1]. Computer-controlled optical surfacing (CCOS) has been successfully applied in industrial production, as it can precisely correct the surface error by converting the dwell time of the polishing tool into the feed-rate along the polishing trajectory. Therefore, a trajectory planning method is the key factor affecting high-quality, high-efficiency polishing.

Despite the extent of research on path planning in some fields [2–4], investigation of optical polishing has been rather limited. Although the problem of discontinuous surfaces between the adjacent mm-sized short line segments has attracted wide concern when using parametric curve theory [5–7], frequent acceleration and deceleration result in poor realisation precision of dwell time, i.e., the runtime of trajectory (hereafter referred to as runtime), which further influences the polishing quality of elements and the convergence rate of surfaces [8]. The main focus of this paper is to investigate an interpolation scheme to overcome the above problem, which has two main progressive aspects: fitting and interpolation of parametric curves.

The fitting of parametric curves is a process of converting discrete short line segments into parametric curves. At present, many scholars have carried out research based on the dominant points using a non-uniform rational B-spline (NURBS). Park [9,10] proposed a method for determining the dominant points according to the discrete curvature. Zhou [11] and Xu [12] improved this method by taking concave-convex turning points and extreme points on the curvature curve as dominant points. To improve the fitting precision, Zhao [13] proposed curve fitting taking squared distance minimisation (SDM) as the evaluation index. Although the dominant points based method is easy with regard to calculation and interpolation, it may result in the loss of runtime at non-dominant points. Thus, only the global fitting method is suitable to the optical polishing. To do so, Yang [14] proposed an optimisation algorithm by establishing the evaluation function for deviation of the fitted distance. Li [15] and Lin [16] classified the trajectory into the different forms of NURBS and then employed a piecewise fitting method for real-time implementation. Based on Gaussian elimination and the continuous short block (CSB) look-ahead algorithm, Tsai [17] and Wang [18] realised the on-line transformation from short line segments to NURBS: however, for the global fitting method, the simplified strategy that setting all the weight factors as 1 eliminates the regulating effect on the fairness of curves and easily causes curvature saltation on the trajectory. Therefore, generating a fair trajectory based on NURBS is the first problem facing the polishing trajectory planning technique.

The interpolation of parametric curves is a process that discretises the NURBS into numerical control (NC) commands. Speed planning, as the critical step in the interpolation process, has been an area of research for numerous scholars: this can be classified into the time-optimal approach and the non-time-optimal approach. The time-optimal approach deals with the planning problem by taking the minimisation of the motion time as the objective to promote manufacturing efficiency [19,20]. For example, Timar [21] proposed a speed planning scheme for NC interpolation by the use of the optimal control theory. On this basis, Sencer [22] and Lu [23] considered the driving capacity constraint and trajectory precision constraint in the interpolation, respectively. The non-time-optimal approach usually deals with the planning problem by taking the minimisation of speed fluctuations as the objective [24]. Various methods, such as the feed-rate evolutionary algorithm [25], the equidistance quaternion method [26], and the improved Adams-Malton algorithm [27] are employed to decrease speed fluctuations and guarantee steady, continuous-trajectory operation. Although the effectiveness of the method has been validated experimentally, is cannot be applied directly to optical polishing because the effect of acceleration and deceleration on the realisation precision of runtime is not considered.

Driven by the practical needs to improve the quality of optical polishing, this paper presents a systematic trajectory planning method that particularly enhances the realisation precision of runtime. Following this introduction, Section 2 calculates all the control points of NURBS through numerical solution to overcome the problem of calculation efficiency. In Section 3, an optimisation algorithm of fairing NURBS is established by taking the shear jerk of trajectory as an evaluation index. Section 4 then proposes an interpolation scheme with which to minimise the realisation error of runtime by planning the feed-rate of trajectory according to the given runtime between adjacent NC codes. Section 5 reports experiments on a prototype machine which shows that the proposed trajectory planning method is more accurate than the linear interpolation method. Conclusions are drawn in Section 6.

## 2. Trajectory Fitting Based on the NURBS Curve

In this section, according to the given NC codes, all the control points of NURBS are solved, which provides the necessary mathematical model for the interpolation of parametric curves. The basic settings are as follows:

(1) Considering the stability, ease of use and calculation efficiency, cubic NURBS is employed as the fitting tool. (2) The uniform parametric method is used for trajectory fitting, because the polishing elements have a large radius of curvature and the chord lengths between NC codes are distributed uniformly. (3) All weight factors are set to 1. In this case, the NURBS can be treated as a cubic B-spline to simplify the calculation process.

According to the basic theory of the NURBS, a segment of NURBS $\boldsymbol{C}(u)(0 \leq u \leq 1)$ can be determined based on the four adjacent control points $\boldsymbol{d}_k(k = 0, 1, 2, 3)$. As shown in Figure 1, $\boldsymbol{C}(0)$

and $C(1)$ separately refer to the start point and the end point of the NURBS segment. Based on the aforementioned setting, the NURBS segment can be directly written as:

$$C(u) = \frac{1}{6}\begin{bmatrix} u^3 & u^2 & u & 1 \end{bmatrix}\begin{bmatrix} -1 & 3 & -3 & 1 \\ 3 & -6 & 3 & 0 \\ -3 & 0 & 3 & 0 \\ 1 & 4 & 1 & 0 \end{bmatrix}\begin{bmatrix} d_0 \\ d_1 \\ d_2 \\ d_3 \end{bmatrix} \tag{1}$$

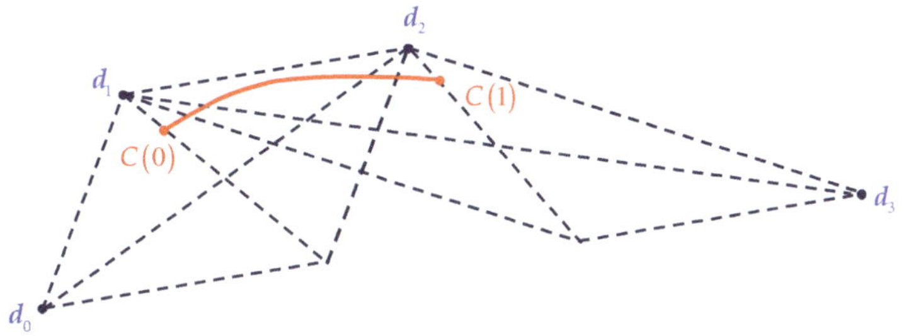


**Figure 1.** Schematic diagram of the parametric curve.

Thus, point $C_i$ can be expressed as:

$$C_i = \frac{1}{6}(d_{i-1} + 4d_i + d_{i+1}) \tag{2}$$

where $C_i$ denotes the $i$th NC code in the $n+1$ lines of NC codes and also is taken as the start point of the $i$th NURBS segment. $d_{i-1}, d_i, d_{i+1}$ respectively denote the three control points corresponding to $C_i$. To calculate $C_1$ and $C_{n+1}$, it is defined that on the condition that $i < 1$, then $d_i = d_1$ and on the condition that $i > n+1$, then $d_i = d_{n+1}$. Equations (3) and (4) can thus be obtained:

$$C_1 = \frac{1}{6}(d_1 + 4d_1 + d_2) = \frac{5}{6}d_1 + \frac{1}{6}d_2 \tag{3}$$

$$C_{n+1} = \frac{1}{6}(d_n + 4d_{n+1} + d_{n+1}) = \frac{1}{6}d_n + \frac{5}{6}d_{n+1} \tag{4}$$

Rewriting Equations (2)–(4) in matrix notation yields:

$$Ad = C$$
$$A = \begin{bmatrix} 5 & 1 & 0 & 0 & 0 & \dots & 0 \\ 1 & 4 & 1 & 0 & 0 & \dots & 0 \\ 0 & 1 & 4 & 1 & 0 & \dots & 0 \\ \vdots & \vdots & \vdots & \vdots & \vdots & \dots & \vdots \\ 0 & 0 & 0 & 0 & 1 & 4 & 1 \\ 0 & 0 & 0 & 0 & 0 & 1 & 5 \end{bmatrix}, d = \begin{bmatrix} d_1 \\ d_2 \\ d_3 \\ \vdots \\ d_n \\ d_{n+1} \end{bmatrix}, C = \begin{bmatrix} 6C_1 \\ 6C_2 \\ 6C_3 \\ \vdots \\ 6C_n \\ 6C_{n+1} \end{bmatrix} \tag{5}$$

If the traditional solution methods such as Gauss elimination method and LU decomposition are directly applied to the Equation (5), it may lead to some undesirable phenomena (such as excessive calculation time) due to the large quantity of NC codes for polishing. Hence, the SOR iterative algorithm [28] is used to find stable numerical solutions of Equation (5) as described below.

The coefficient matrix $A$ can be divided into:

$$A = D - L - U \tag{6}$$

where $\boldsymbol{A} = \text{diag}(a_{11}, a_{22}, \ldots, a_{n+1n+1})$, $-\boldsymbol{L}$ denotes a strictly lower triangular matrix, whose elements below the principal diagonal are corresponding elements of $\boldsymbol{A}$. $-\boldsymbol{U}$ denotes a strictly upper triangular matrix, whose elements above the principal diagonal are corresponding elements of $\boldsymbol{A}$. Owing to the matrix $\boldsymbol{D}$ being invertible, Equation (5) can be modified to:

$$\boldsymbol{d} = \boldsymbol{D}^{-1}(\boldsymbol{L}+\boldsymbol{U})\boldsymbol{d} + \boldsymbol{D}^{-1}\boldsymbol{C} \tag{7}$$

In this case, corner marks $(k)$ and $(k+1)$ are added in Equation (7) to identify the number of iterations. Then, the elementary iterative scheme of $\boldsymbol{d}$ can be written as:

$$\boldsymbol{d}^{(k+1)} = \boldsymbol{M}_1\boldsymbol{d}^{(k)} + \boldsymbol{D}^{-1}\boldsymbol{C} \tag{8}$$

where $\boldsymbol{M}_1 = \boldsymbol{D}^{-1}(\boldsymbol{L}+\boldsymbol{U})$. The component form of Equation (8) can be expressed as:

$$\boldsymbol{d}_i^{(k+1)} = \frac{1}{a_{ii}}\left( - \sum_{\substack{j=1 \\ j\neq i}}^{n+1} a_{ij}\boldsymbol{d}_j^{(k)} + C_i \right), \; i = 1,2,\ldots,n+1 \tag{9}$$

which is also known as the Jacobi iteration scheme, noting that, before calculating $\boldsymbol{d}_i^{(k+1)}$, the iterative values of the first $i-1$ components in $\boldsymbol{d}^{(k+1)}$ have been generated, which are more approximate to the true value than the results obtained by the previous iteration. Therefore, $\boldsymbol{d}_1^{(k)}, \boldsymbol{d}_2^{(k)}, \ldots, \boldsymbol{d}_{i-1}^{(k)}$ can be replaced by $\boldsymbol{d}_1^{(k+1)}, \boldsymbol{d}_2^{(k+1)}, \ldots, \boldsymbol{d}_{i-1}^{(k+1)}$ to make $\boldsymbol{d}_i^{(k+1)}$ closer to the true value. To improve the convergence rate of the iteration further, a proper parameter $\mu$ is selected to conduct the weighted averaging on the aforementioned iterative scheme:

$$\boldsymbol{d}_i^{(k+1)} = \mu\tilde{\boldsymbol{d}}_i^{(k+1)} + (1-\mu)\boldsymbol{d}_i^{(k)}, i = 1,2,\ldots n+1 \tag{10}$$

with:

$$\tilde{\boldsymbol{d}}_i^{(k+1)} = \frac{1}{a_{ii}}\left( -\sum_{j=1}^{i-1} a_{ij}\boldsymbol{d}_j^{(k+1)} - \sum_{j=i+1}^{n+1} a_{ij}\boldsymbol{d}_j^{(k)} + C_i \right) \tag{11}$$

Substituting Equation (11) into Equation (10) leads to the SOR iteration scheme:

$$a_{ii}\boldsymbol{d}_i^{(k+1)} + \mu\sum_{j=1}^{i-1} a_{ij}\boldsymbol{d}_j^{(k+1)} = a_{ii}\boldsymbol{d}_i^{(k)} - \mu\sum_{j=i}^{n+1} a_{ij}\boldsymbol{d}_j^{(k)} + \mu C_i, \; i = 1,2,\ldots,n+1 \tag{12}$$

Note that $\boldsymbol{A}$ is a tridiagonal positive definite matrix, so the optimal value $\mu_{opt}$ of the relaxation factor can be expressed as [29]:

$$\mu_{opt} = \frac{2}{1+\sqrt{1-[\rho(\boldsymbol{M}_1)]^2}} \tag{13}$$

where $\rho(\boldsymbol{M}_1)$ denotes the spectral radius of $\boldsymbol{M}_1$.

It is worth noting that, for a flat element, the aforementioned method can be directly applied to calculate the NURBS trajectory, but for a curved element, the polishing shaft should always lie along the normal direction of the element. In this case, it is necessary to solve two trajectories of end-point and reference-point on the polishing shaft to determine the polishing attitude (for more details, please see [21]).

## 3. Fairing of the NURBS

Although the constructed cubic NURBS by the above method satisfies the G2 continuity characteristics, i.e., the second-order derivative functions of the trajectory is continuous, the motion stability of the trajectory is still influenced by curvature saltation. In this section, a fairing optimisation method is proposed by adjusting the weight factors of NURBS.

The fairing optimisation can be classified into two categories, i.e., global and local fairing according to the number of the adjusted point on NURBS. Considering the significant computational burden of global fairing caused by the large quantity of NC codes for polishing, it is more reasonable to modify the outlier points with curvature saltation by the use of local fairing, which are selected from all NC codes [30]. A filtering process is needed to eliminate the influence of curvature fluctuations caused by discrete calculation.

In the fairing optimisation, the shear jerk of the outlier point is taken as the evaluation index:

$$D_j = \frac{\kappa_{next1} - \kappa_j}{\|C_{next1} - C_j\|} - \frac{\kappa_j - \kappa_{last1}}{\|C_j - C_{last1}\|} \tag{14}$$

where $\kappa_j$ denotes the curvature at the $j$th outlier point $C_j$, $\kappa_{next1}$ and $\kappa_{last1}$ denote the curvature of the $C_{next1}$ and $C_{last1}$, which are on two adjacent sides of the outlier point. The index indicates the curvature changes of the adjacent outlier points.

To generate the weight factors of the various outlier points, the objective function is defined as:

$$L = \sum_{j=1}^{k} D_j^2 + (C_j - C_{j,0})^2 \tag{15}$$

where $C_{j,0}$ denotes the $j$th outlier point before optimisation. It can be seen from Equation (15) that the objective function is composed of two parts: part one is the curvature changes of the outlier point after optimisation, and part two is the adjustment amplitude of outlier points before, and after, optimisation. Thus, the objective function means that fairing optimisation is performed on the premise of modifying the NURBS as little as possible.

According to affine invariant principle of NURBS [31], the four-dimensional (4D) space constructed by the control points and weight factors can be expressed as:

$$C^{\omega}(u) = \sum_{i=1}^{n+1} N_{i,3}(u) \begin{bmatrix} \omega_i \boldsymbol{d}_i \\ \omega_i \end{bmatrix} = \sum_{i=1}^{n+1} N_{i,3}(u) \boldsymbol{d}_i^{\omega} \tag{16}$$

Then, the NURBS defined by Equation (1) can be regarded as the projection of the curve $C^{\omega}(u)$ in 4D space on the centre of the hyperplane $\omega = 1$. Based on Equation (2), it can be seen that:

$$C_j^{\omega} = \frac{1}{6}\boldsymbol{d}_{j,last1}^{\omega} + \frac{2}{3}\boldsymbol{d}_j^{\omega} + \frac{1}{6}\boldsymbol{d}_{j,next1}^{\omega} \tag{17}$$

$$C_{j,last1}^{\omega} = \frac{1}{6}\boldsymbol{d}_{j,last2}^{\omega} + \frac{2}{3}\boldsymbol{d}_{j,last1}^{\omega} + \frac{1}{6}\boldsymbol{d}_j^{\omega} \tag{18}$$

$$C_{j,next1}^{\omega} = \frac{1}{6}\boldsymbol{d}_j^{\omega} + \frac{2}{3}\boldsymbol{d}_{next1}^{\omega} + \frac{1}{6}\boldsymbol{d}_{next2}^{\omega} \tag{19}$$

where $\boldsymbol{d}_j^{\omega}$ denotes the control point corresponding to the $j$th outlier point and $\boldsymbol{d}_{next1}^{\omega}, \boldsymbol{d}_{next2}^{\omega}$ and $\boldsymbol{d}_{last1}^{\omega}, \boldsymbol{d}_{last2}^{\omega}$ denote the control points on the two adjacent sides of the control point $\boldsymbol{d}_j^{\omega}$, respectively.

Furthermore, Equation (15) can be written in 4D space as:

$$L^{\omega} = \sum_{j=1}^{k} \left(D_j^{\omega}\right)^2 + \left(C_j^{\omega} - C_{j,0}^{\omega}\right)^2 \tag{20}$$

where $C_j^\omega - C_{j,0}^\omega$ can be equivalently simplified as $d_j^\omega - d_{j,0}^\omega$ and $D_j^\omega$ can be approximately expressed as [32]:

$$D_j^\omega \approx \frac{\left(C_{j,next1}^\omega\right)'' - \left(C_j^\omega\right)''}{l_{j,next1}^\omega} - \frac{\left(C_j^\omega\right)'' - \left(C_{j,last1}^\omega\right)''}{l_{j,last1}^\omega} \tag{21}$$

where $\left(C_j^\omega\right)''$ denotes the second derivative of the NURBS at $C_j^\omega$, $l_{j,next1}^\omega = \|C_{j,next1,0}^\omega - C_{j,0}^\omega\|$ and $l_{j,last1}^\omega = \|C_{j,0}^\omega - C_{j,last1,0}^\omega\|$. Substituting Equation (21) into Equation (20) yields:

$$L^\omega = \sum_{j=1}^{k}\left(\frac{\left(C_{j,next1}^\omega\right)'' - \left(C_j^\omega\right)''}{l_{j,next1}^\omega} - \frac{\left(C_j^\omega\right)'' - \left(C_{j,last1}^\omega\right)''}{l_{j,last1}^\omega}\right) + \left(d_j^\omega - d_{j,0}^\omega\right)^2 \tag{22}$$

To calculate the minimum value of $L^\omega$, the partial derivative of Equation (22) about $d_j^\omega$ is set to 0:

$$\begin{aligned}\frac{\partial L^\omega}{\partial d_j^\omega} = \sum_{j=1}^{k} 2\left(\frac{d_{j,next2}^\omega - 3d_{j,next1}^\omega + 3d_j^\omega - d_{j,last1}^\omega}{l_{j,next1}^\omega} - \frac{d_{j,next1}^\omega - 3d_j^\omega + 3d_{j,last1}^\omega - d_{j,last2}^\omega}{l_{j,last1}^\omega}\right) \\ \cdot\left(\frac{3}{l_{j,next1}^\omega} + \frac{3}{l_{j,last1}^\omega}\right) + 2\left(d_j^\omega - d_{j,0}^\omega\right) = 0\end{aligned} \tag{23}$$

Then, the equations for $d_1^\omega, d_2^\omega, \ldots, d_k^\omega$ can be expressed as:

$$\begin{aligned} A^\omega d^\omega &= C^\omega \\ A^\omega &= diag\left(\left(\frac{3}{l_{j,next1}^\omega} + \frac{3}{l_{j,last1}^\omega}\right)^2 + 1\right) \\ d^\omega &= (d_1^\omega, d_2^\omega, \ldots, d_k^\omega)^T \\ C^\omega &= (C_1^\omega, C_2^\omega, \ldots, C_k^\omega)^T \\ C_j^\omega &= d_{j,0}^\omega - \left(\frac{d_{j,next2}^\omega - 3d_{j,next1}^\omega - d_{j,last1}^\omega}{l_{j,next1}^\omega} - \frac{d_{j,next1}^\omega + 3d_{j,last1}^\omega - d_{j,last2}^\omega}{l_{j,last1}^\omega}\right)\left(\frac{3}{l_{j,next1}^\omega} + \frac{3}{l_{j,last1}^\omega}\right)\end{aligned} \tag{24}$$

According to Equation (24), the optimised control points and weight factors corresponding to the outlier points can then be generated.

## 4. NURBS Interpolation

The NURBS interpolation is used to discretise the parametric curve to the NC commands based on the planned feed-rate. In this section, an interpolation method for optical polishing is proposed aiming to minimise the realisation error of the trajectory runtime.

### 4.1. Feed-Rate Planning

Feed-rate planning is the main influencing factor in interpolating the NC commands along the trajectory. For the optical polishing, to guarantee the desired runtime of trajectory, the specific method is displayed as follows:

**Step 1:** considering that now there is no analytical solution to calculate the length of NURBS, the Simpson formula is used to obtain the estimation of the length through numerical iteration:

$$p = \frac{up - low}{6}(f(low) + 4f(mid) + f(up)) \tag{25}$$

with:

$$f(u) = \frac{ds}{du} = \sqrt{x'(u) + y'(u) + z'(u)} \tag{26}$$

where $x'(u), y'(u), z'(u)$ respectively denote the first-order derivatives of $x(u), y(u), z(u)$, which are the one-dimensional curves of $C(u)$ along $x, y, z$ axes. $up$ and $low$ denote the upper and lower boundaries of $u$, $mid = (up + low)/2$.

**Step 2**: the length between the two points corresponding to the parameters $up$ and $low$ is calculated:

$$l = \|C(up) - C(low)\| \tag{27}$$

**Step 3:** the error between the aforementioned two lengths is calculated:

$$e = |p - l| \tag{28}$$

The convergence threshold $[e]$ is given. If $e > [e]$, let $up = mid$ and repeat Steps 1 and 2. If $e < [e]$, turn to Step 4.

**Step 4**: the parameter interval $(0, 1)$ was sectioned by the equivalent distance $up - low$ to generate the knot vector $(u_0, u_1, \ldots, u_n)$. Thus, the length of the NURBS is:

$$s \approx \sum_{i=1}^{n} \|C(u_i) - C(u_{i-1})\| \tag{29}$$

Equipped with the length at hand, the S-curve motion law is invoked to guarantee that the feed-rates of the adjacent NURBS segments are changed smoothly. Moreover, the initial sections of the trajectory segments are defined as the feed-rate transition zone. Then, the feed-rates remain constant until the end of the trajectory segments. In this case [33]:

$$s = 2v_{low}t_a + Jt_a^3 + v_{up}t_v \tag{30}$$

$$v_{up} = v_{low} + Jt_a^2 \tag{31}$$

where $v_{low}$ and $v_{up}$ denote the initial and final feed-rates of the trajectory segment. $J$, $2t_a$ and $t_v$ denote the jerk, acceleration (deceleration) time and the uniform motion time, respectively. Thus, the runtime of the trajectory segment is $t_d = 2t_a + t_v$.

It is worth noting that previous studies have shown that the feed-rates, when limited by the runtime of the trajectory segments, are much lower than the maximum value which is constrained by the chord error and the driving capacity. Therefore, there is no need to check the feed-rate again.

### 4.2. NURBS Interpolation

The essence of interpolation is to generate the NC command along the trajectory according to the period $t_s$. As each interpolation point of the NURBS corresponds to one curve parameter, only the curve parameters need to be solved.

Taking $u$ as the function of $t$, the second-order Taylor expansion can be expressed as:

$$u_{i+1} = u_i + \left.\frac{du}{dt}\right|_{t=t_i} t_s + \frac{1}{2}\left.\frac{d^2u}{dt^2}\right|_{t=t_i} t_s^2 \tag{32}$$

The feed-rate $v(u)$ can be written as:

$$v(u) = \|\frac{dC(u)}{du}\| = \|\frac{dC(u)}{du}\frac{du}{dt}\| \tag{33}$$

Furthermore:

$$\frac{du}{dt} = \frac{v(u)}{\|\mathbf{C}'(u)\|} \tag{34}$$

Calculating the derivative of Equation (34):

$$\frac{d^2u}{dt^2} = \frac{\frac{dv(u)}{du}\frac{du}{dt}}{\|\mathbf{C}'(u)\|} - v(u)\frac{\frac{d(\|\mathbf{C}'(u)\|)}{du}\frac{du}{dt}}{\|\mathbf{C}'(u)\|^2} \tag{35}$$

where:

$$\frac{d(\|\mathbf{C}'(u)\|)}{du} = \frac{\mathbf{C}''(u)\cdot\mathbf{C}'(u)}{\|\mathbf{C}'(u)\|}$$

Substituting Equations (33)–(35) into Equation (32) gives:

$$u_{i+1} = u_i + \frac{v_C(u_i)}{\|\mathbf{C}'(u_i)\|}t_s + \frac{1}{2}\left[\frac{v'_C(u_i)}{\|\mathbf{C}'(u_i)\|^2}v_C(u_i) - \frac{\mathbf{C}''(u)\cdot\mathbf{C}'(u)}{\|\mathbf{C}'(u)\|^4}v_C^2(u_i)\right]t_s^2 \tag{36}$$

Substituting $u_{i+1}$ into the curve equation obtained through the fairing optimisation described in Section 3, the next interpolation point can be acquired. Repeating this process until:

$$t_i = \mathrm{ceil}\left(\frac{t_d}{t_s}\right) * t_s \tag{37}$$

where ceil(·) denotes an integer that is rounded up. In this way, interpolation of all trajectory segment can be completed.

## 5. Experiments

Both simulation and experiments were carried out to validate the effectiveness of the presented method on the prototype of the hybrid polishing robot. As shown in Figure 2, it is mainly composed of a 6-DOF (degrees of freedom) hybrid robot, a polishing effector, a magnetic worktable, a column, and a CNC system. The hybrid robot is composed of a 3-DOF (3UPS and UP) parallel mechanism and a 3-DOF wrist. The UP limb and the wrist form a UPS or UPRRR limb. Here, R, U, S, and P represent, respectively, revolute, universal, spherical, and prismatic joints, and the underlined P, S, and R denote the actuated prismatic, spherical, and revolute joints, respectively. The CNC system is built upon an IPC+PMAC open architecture, consisting of a host control computer responsible for reconstruction of the parametric curves, trajectory interpolation, and NC command generation, and a PMAC motion controller for servo-control of the actuated joints.

1. CNC system 2. Column 3. Hybrid robot 4. Polishing effector 5. Magnetic worktable

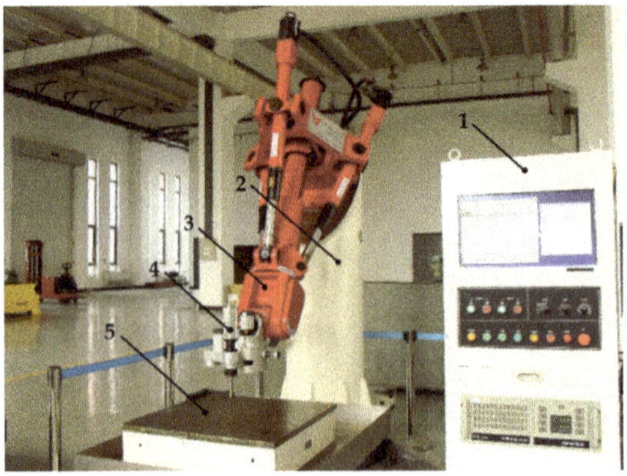


**Figure 2.** The prototype of the polishing robot.

Without loss of generality, a segment of NC codes was taken from the polishing trajectory to validate the effectiveness of the proposed interpolation method. According to the SOR iterative scheme mentioned in Section 2, Figure 3 shows the two fitted trajectories. Then, the optimisation algorithm described in Section 3 is used to smooth the curvature of the trajectory. The change threshold of curvature for judging outlier points is given as 0.01. As shown in Table 1, the fairness of the two trajectories is both significantly improved. The maximum curvatures of the two trajectories are reduced by as much as 79.7% and 63.3% and the maximum absolute values of shear jerks are decreased by 91.2% and 90.2%, correspondingly. Considering the optimisation and interpolation methods for the two trajectories are same, experimental results of the end-point trajectory are just shown in the following discussion for the sake of simplicity.

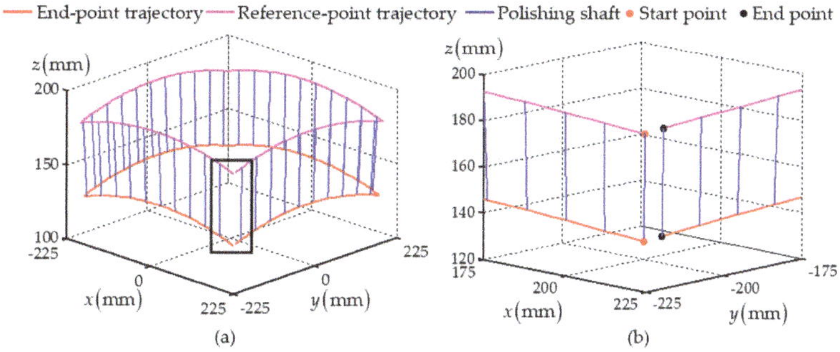


**Figure 3.** The initial fitted polishing trajectory for fairing optimisation and interpolation (**a**) the fitted polishing trajectory (**b**) the partial enlarged view.

Based on the optimisation trajectory shown in Figure 4, the discrete NC command sequences were generated and sent to the PMAC motion controller, which were mapped into the servo-command of actuated joints through an inverse kinematic model. The PMAC motion controller synchronously gathered the positions and velocities fed back from the servo-motors. The interpolation and sampling periods are 10 and 20 ms, respectively. To validate the effectiveness of the method, a comparison experiment was carried out utilizing the linear interpolation method. Figure 5 shows the experimental result, which is computed by the feedback positions of all the actuated joints. It can be seen from the partial enlarged view that the NC commands generated by the proposed method are closer to the NC codes than those found using linear interpolation, which means that the removal position of polishing process can be reached more accurately. Figures 6 and 7 show the runtime between the adjacent NC nodes. Compared with the desired value, the proposed method is able to realise the runtime more precisely, which indicates that the removal quantity during the polishing process can be more precisely controlled, correspondingly. To evaluate the effect of the interpolation method, the indices are defined as the interpolation error ($e_i$) and the runtime error of the trajectory ($e_t$):

$$e_i = \|\mathbf{C}_{i0} - \mathbf{C}_i\|_2,\ i = 1, 2, \ldots, n \tag{38}$$

$$e_{ti} = |t_{i0} - t_i|,\ i = 1, 2, \ldots, n \tag{39}$$

where $\mathbf{C}_{i0}$ and $\mathbf{C}_i$ denote the $i$th desired NC code and actual NC code after interpolation, $t_{i0}$ and $t_i$ denote the desired runtime and the actual runtime between the $i-1$th and $i$th NC codes. It can be seen from Table 2 that, compared to the linear interpolation method, the interpolation error and the runtime error generated through the proposed method are an order of magnitude smaller. These data further imply that the proposed interpolation method has more precision than the linear interpolation method in the polishing process.

**Table 1.** Changes in the two trajectories before, and after, fairing optimisation.

| Trajectory | Before | | After | |
|---|---|---|---|---|
| | Maximum Curvature | Maximum Shear Jerk | Maximum Curvature | Maximum Shear Jerk |
| End-point | 1.249 | 0.285 | 0.253 | 0.025 |
| Reference-point | 1.380 | 0.325 | 0.506 | 0.032 |

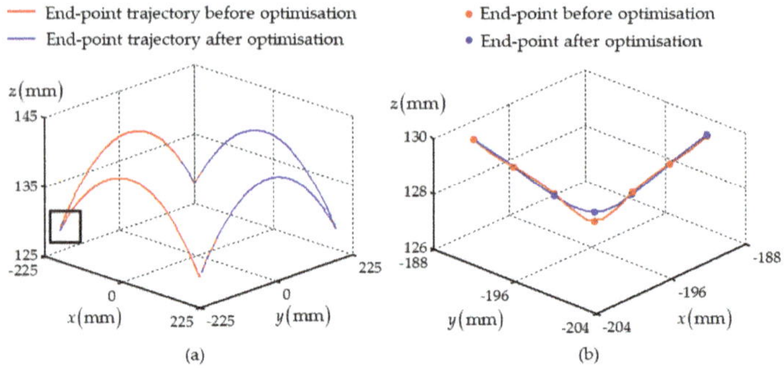


**Figure 4.** The end-point trajectory before, and after, fairing optimization: (**a**) the end-point trajectory; and (**b**) the partial enlarged view

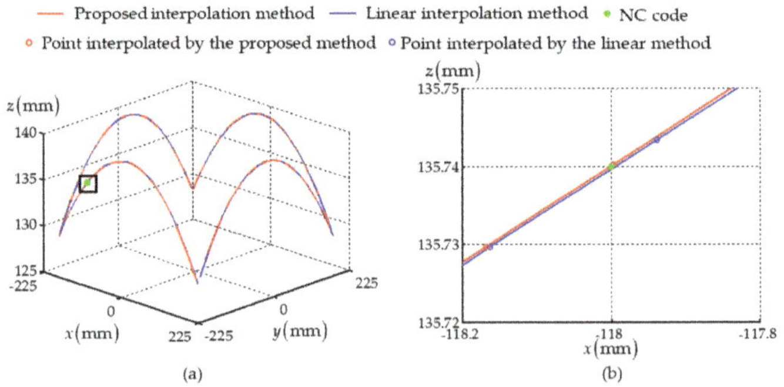


**Figure 5.** The end-point trajectory interpolated by different methods: (**a**) the end-point trajectory; and (**b**) the partial enlarged view.

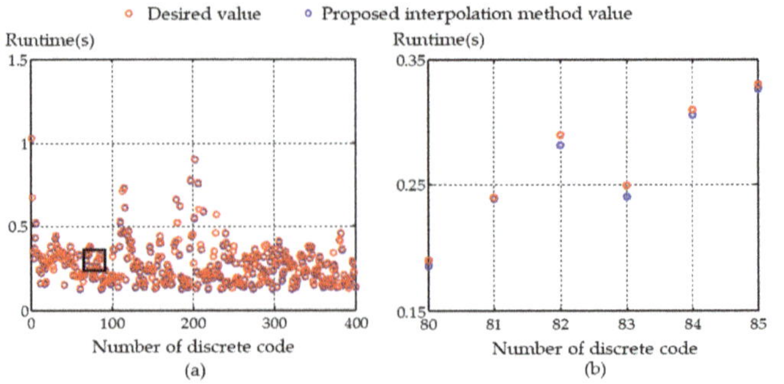


**Figure 6.** Runtime of the trajectory interpolated by the proposed method: (**a**) runtime of the trajectory; and (**b**) the partial enlarged view.

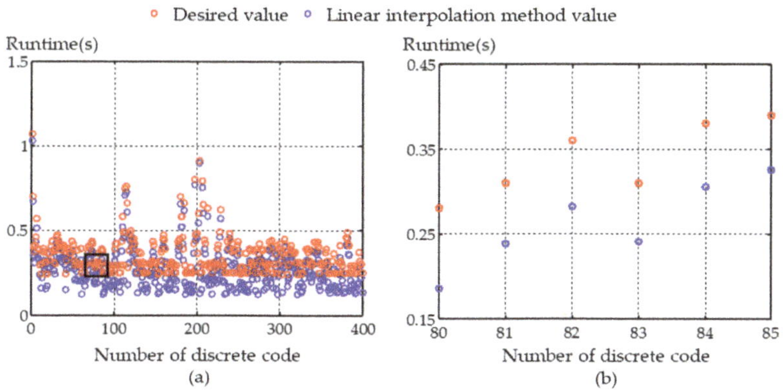


**Figure 7.** Runtime of the trajectory interpolated by the linear method: (**a**) runtime of the trajectory; and (**b**) the partial enlarged view.

**Table 2.** Comparison of the effect of the two interpolation methods.

| Method | Interpolation Error (mm) | | Runtime Error (s) | |
|---|---|---|---|---|
| | Maximum | Mean | Maximum | Mean |
| Proposed | 0.005 | 0.002 | 0.010 | 0.005 |
| Linear | 0.091 | 0.076 | 0.130 | 0.076 |

To validate the modification effect of the interpolation method on the surface error of optical elements, polishing experiments were carried out on fused silica elements by, respectively, using the linear interpolation method and the proposed interpolation method, as shown in Figure 8. Using the Nanovea contour graph, the polished zone (70 mm × 70 mm) of the elements was detected. The experimental results obtained using the two interpolation methods are displayed in Figures 9 and 10, respectively. It can be seen, from the figures, that, after conducting linear interpolation-based polishing, the surface error (PV) decreases from $11.894\lambda(\lambda = 633\text{nm})$ to $7.422\lambda$. In contrast, using the proposed interpolation method, the surface error (PV) is reduced from $11.282\lambda$ to $6.267\lambda$. The convergence rate of the surface error is increased from 37.59% to 44.44%, which further verifies the effectiveness of the proposed method.

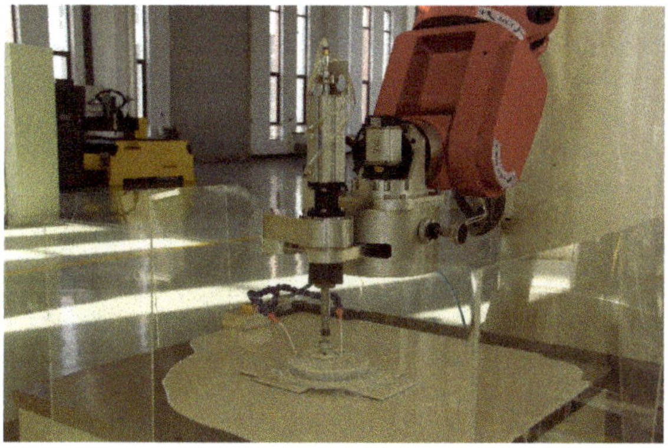

**Figure 8.** Polishing experiment on fused silica elements.

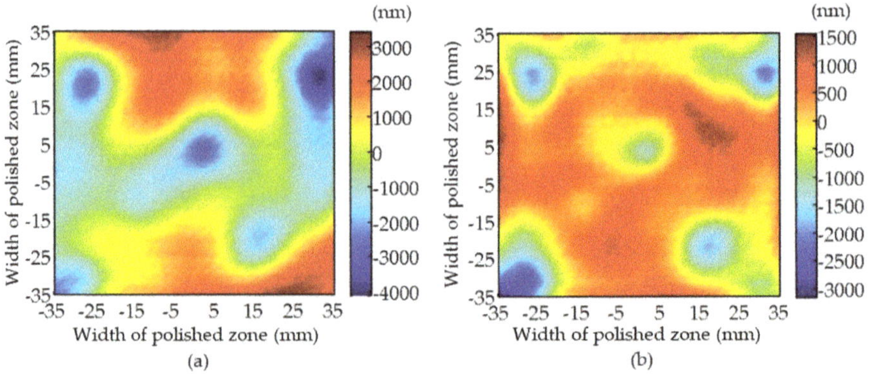


**Figure 9.** Changes in surface errors (**a**) before and (**b**) after conducting the linear interpolation-based polishing.

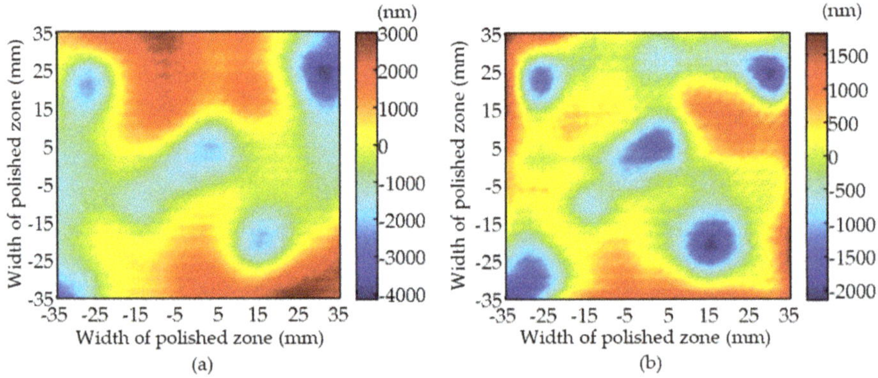


**Figure 10.** Changes in surface errors (**a**) before and (**b**) after conducting the proposed interpolation-based polishing.

## 6. Conclusions

A new trajectory planning method for optical polishing is proposed in this paper. It is developed to generate NC commands that can decrease the interpolation error and the runtime error. First, to obtain the NURBS trajectory without loss of the information about the runtime, a global fitting method is presented based on SOR iteration theory to deal with the problem of the computational burden arising from the large quantity of NC codes for polishing. Then, taking the shear jerk of the NURBS as the evaluation index, a fairing optimisation method is carried out to smooth the curvature saltation of the trajectory. Finally, the feed-rate planning method and the path interpolation scheme are proposed to reduce the realisation error of the trajectory runtime.

Simulation results verify that the fairing optimisation proposed in this research can modify the curvature saltation at the expense of trajectory accuracy compared to the given NC codes; however, the curvature saltation only occurs at the turning point of the trajectory and the trajectory error magnitude generated by the optimisation algorithm is consistent with the linear interpolation. Therefore, the trajectory accuracy is not treated as the index with which to evaluate the smoothing effect in Section 5. The effect of the optimisation algorithm on the optical polishing will be investigated in future work.

It can be seen from Figures 5–7, and Table 2, that the runtime error of the proposed interpolation method arises because the trajectory runtime cannot be divided exactly by the interpolation period and is rounded up to an integer. In contrast, the runtime error of the linear interpolation method arises as a result of the acceleration and deceleration that occurs frequently between adjacent short line segments. Similar to the runtime error, the interpolation error of the proposed interpolation method is a result of the numerical calculation error of the trajectory length and that of the linear interpolation method is due to the transition for the discontinuity of the adjacent short line segments. Then the convergence rate of the surface error is increased with the help of the improvement in the interpolation, and runtime, errors, validating the effectiveness of the proposed interpolation process. For the aforementioned error, their individual effects on the optical polishing need to be further indicated in future work.

**Author Contributions:** D.Z. wrote the manuscript and performed the experiments, H.G. analysed the data and revised the manuscript. All authors discussed the results and commented on the manuscript.

**Funding:** This research was partially funded by the National Science and Technology Major Project of China (grant no. 2017ZX04022001-206), the National Natural Science Foundation of China (grant no. 51420105007), and EU H2020-MSCA-RISE 2016 (grant no. 734272).

**Conflicts of Interest:** The authors declare no conflict of interest.

## References

1. Rupp, W. Conventional optical polishing techniques. *Int. J. Opt.* **1971**, *18*, 1–16. [CrossRef]
2. Kaltsoukalas, K.; Makris, S.; Chryssolouris, G. On generating the motion of industrial robot manipulators. *Rob. Comput. Integr. Manuf.* **2015**, *32*, 65–71. [CrossRef]
3. Makris, S.; Tsarouchi, P.; Matthaiakis, A. Dual arm robot in cooperation with humans for flexible assembly. *CIRP Ann.* **2017**, *66*, 13–16. [CrossRef]
4. Lei, W.; Wang, S. Robust real-time NURBS path interpolators. *Int. J. Mach. Tools Manuf.* **2009**, *49*, 625–633. [CrossRef]
5. Jahanpour, J.; Alizadeh, M. A novel acc-jerk-limited NURBS interpolation enhanced with an optimized S-shaped quintic feedrate scheduling scheme. *Int. J. Adv. Manuf. Technol.* **2015**, *77*, 1889–1905. [CrossRef]
6. Liu, X.; Peng, J.; Si, L. A novel approach for NURBS interpolation through the integration of acc-jerk-continuous-based control method and look-ahead algorithm. *Int. J. Adv. Manuf. Technol.* **2017**, *88*, 961–969.
7. Zhao, J.; Li, L.; Wang, G. Research of NURBS Curve Real-time Interpolation Feed Speed Planning. *Tool Eng.* **2015**, *49*, 7–11.
8. Lin, M.; Tsai, M.; Yau, H. Development of a dynamics-based NURBS interpolator with real-time look-ahead algorithm. *Int. J. Mach. Tools Manuf.* **2007**, *47*, 2246–2262. [CrossRef]
9. Park, H.; Lee, J. B-spline curve fitting based on adaptive curve refinement using dominant points. *Comput.-Aided Des.* **2007**, *39*, 439–451. [CrossRef]
10. Park, H. B-spline surface fitting based on adaptive knot placement using dominant columns. *Comput.-Aided Des.* **2011**, *43*, 258–264. [CrossRef]
11. Zhou, H.; Wang, Y.; Liu, Z. On non-uniform rational B-splines curve fitting based on the least control points. *J. Xian Jiaotong Univ.* **2008**, *42*, 73–77.
12. Xu, J. B-spline Curve Approximation Based on Feature Points Automatic Recognition. *J. Mech. Eng.* **2009**, *45*, 212–217. [CrossRef]
13. Zhao, H.; Lu, Y.; Zhu, L. Look-ahead interpolation of short line segments using B-spline curve fitting of dominant points. *P. I. Mech. Eng. B-J. Eng.* **2014**, *229*, 1–13. [CrossRef]
14. Yang, X.; Hu, Z.; Zhong, Z. Research on the NURBS Curve Fitting for Tool Path Generation. *China Mech. Eng.* **2009**, *20*, 983–984.
15. Li, W.; Liu, Y.; Yamazaki, K. The design of a NURBS pre-interpolator for five-axis machining. *Int. J. Adv. Manuf. Technol.* **2008**, *36*, 927–935. [CrossRef]
16. Lin, K.; Ueng, W.; Lai, J. CNC codes conversion from linear and circular paths to NURBS curves. *Int. J. Adv. Manuf. Technol.* **2008**, *39*, 760–773. [CrossRef]
17. Tsai, M.; Nien, H. Development of a real-time look-ahead interpolation methodology with spline-fitting technique for high-speed machining. *Int. J. Adv. Manuf. Technol.* **2010**, *47*, 621–638. [CrossRef]

18. Wang, J.; Yau, H. Real-time NURBS interpolator: application to short linear segments. *Int. J. Adv. Manuf. Technol.* **2009**, *41*, 1169–1185. [CrossRef]
19. Timar, S.; Farouki, R. Time-optimal traversal of curved paths by Cartesian CNC machines under both constant and speed-dependent axis acceleration bounds. *Rob. Comput. Integr. Manuf.* **2007**, *23*, 563–579. [CrossRef]
20. Sencer, B. Smooth Trajectory Generation and Precision Control of 5-Axis CNC Machine Tools. Ph.D. Thesis, The University of British Columbia, Vancouver, BC, Canada, October 2009.
21. Boyadjieff, C.; Farouki, R.; Timar, S. *Smoothing of Time-Optimal Feed rates for Cartesian CNC Machines*; Springer: Berlin, Germany, 2005; pp. 84–101.
22. Sencer, B.; Altintas, Y.; Croft, E. Feed optimization for five-axis CNC machine tools with drive constraints. *Int. J. Mach. Tools Manuf.* **2008**, *48*, 733–745. [CrossRef]
23. Lu, L.; Zhang, L.; Ji, S. An offline predictive feedrate scheduling method for parametric interpolation considering the constraints in trajectory and drive systems. *Int. J. Adv. Manuf. Technol.* **2016**, *83*, 2143–2157. [CrossRef]
24. Zhao, H.; Zhu, L.; Ding, H. A parametric interpolator with minimal feed fluctuation for CNC machine tools using arc-length compensation and feedback correction. *Int. J. Mach. Tools Manuf.* **2013**, *75*, 1–8. [CrossRef]
25. Sun, Y.; Zhao, Y.; Bao, Y. A novel adaptive-feedrate interpolation method for NURBS tool path with drive constraints. *Int. J. Mach. Tools Manuf.* **2013**, *77*, 74–81. [CrossRef]
26. Zhang, J.; Zhang, L.; Zhang, K. Double NURBS trajectory generation and synchronous interpolation for five-axis machining based on dual quaternion algorithm. *Int. J. Adv. Manuf. Technol.* **2016**, *83*, 2015–2025. [CrossRef]
27. Peng, J.; Liu, X.; Si, L. A Novel Approach for NURBS Interpolation with Minimal Feed Rate Fluctuation Based on Improved Adams-Moulton Method. *Math. Probl. Eng.* **2017**, *2017*, 1–10. [CrossRef]
28. Bai, Z.; Parlett, B.; Wang, Z. On generalized successive over relaxation methods for augmented linear systems. *Numer. Math.* **2005**, *102*, 1–38. [CrossRef]
29. Huang, J. Another Version of SOR Iteration and Its Generalization. Master's Thesis, Dalian University of Technology, Dalian, China, June 2013.
30. Li, Y. Study and Realization on Construction, Fairing and Smooth Joining Between Adjacent of the NURBS Surface. Master's Thesis, Xi'an University of Technology, Xi'an, China, March 2010.
31. Surhone, L.; Timpledon, M.; Marseken, S. *Non-Uniform Rational B-Spline*; Betascript Publishing: Montana, MT, USA, 2010.
32. Zhang, H.; Jiang, D.; Ding, Y. A weight-based optimal faring algorithm for planar cubic NURBS curves. *J. Southwest Minzu Univ.* **2005**, *31*, 351–355.
33. Li, X.; Wu, Y.; Leng, H. Research on a New S-curve Acceleration and Deceleration Control Method. *Modul. Mach. Tool. Autom. Manuf. Technol.* **2007**, 50–53.

applied sciences

MDPI

Article

# Loop Closure Detection Based on Multi-Scale Deep Feature Fusion

**Baifan Chen, Dian Yuan *, Chunfa Liu and Qian Wu**

School of Automation, Central South University, Changsha 410083, China; chenbaifan@csu.edu.cn (B.C.); 15084843844@163.com (C.L.); wuqian945@csu.edu.cn (Q.W.)

* Correspondence: dianyuan@csu.edu.cn; Tel.: +86-185-7310-0034

Received: 25 December 2018; Accepted: 14 March 2019; Published: 17 March 2019

**Abstract:** Loop closure detection plays a very important role in the mobile robot navigation field. It is useful in achieving accurate navigation in complex environments and reducing the cumulative error of the robot's pose estimation. The current mainstream methods are based on the visual bag of word model, but traditional image features are sensitive to illumination changes. This paper proposes a loop closure detection algorithm based on multi-scale deep feature fusion, which uses a Convolutional Neural Network (CNN) to extract more advanced and more abstract features. In order to deal with the different sizes of input images and enrich receptive fields of the feature extractor, this paper uses the spatial pyramid pooling (SPP) of multi-scale to fuse the features. In addition, considering the different contributions of each feature to loop closure detection, the paper defines the distinguishability weight of features and uses it in similarity measurement. It reduces the probability of false positives in loop closure detection. The experimental results show that the loop closure detection algorithm based on multi-scale deep feature fusion has higher precision and recall rates and is more robust to illumination changes than the mainstream methods.

**Keywords:** loop closure detection; convolutional neural network; spatial pyramid pooling

## 1. Introduction

Loop closure detection has become a key problem and research hotspot in the field of mobile robot navigation, particularly in simultaneous localization and mapping (SLAM), because it can reduce the cumulative error of robot pose estimation and achieve accurate navigation in large-scale complex environments. Vision-based loop closure detection, also called visual place recognition, is when the robot identifies the places that have been visited before with images provided by the vision sensor during the navigation. For example, assume there are two images captured at the current time and at an earlier time, the problem of loop closure detection is to judge whether the places at the two moments are the same according to the similarity of these two images. Correct loop closure detection can add an edge constraint in the pose map to help optimize robot motion estimation further and build a consistent map. Wrong loop closure detection will lead to the failure of map building. Therefore, a good loop closure detection algorithm is crucial for consistent mapping and even for the entire SLAM system.

At present, the mainstream methods of visual loop closure detection are based on the Bag of Words (BoW), which cluster the visual features into some "words" and then describe an image in the form of a "words" vector. Thus, the visual loop closure detection problem is transformed into a similarity measure problem with the word vectors of the two images. However, the visual features in the BoW are all artificially designed by researchers in the field of computer vision, and they all belong to the low-level features and are sensitive to illumination changes. With the advent of various visual sensors, different visual features are designed based on the different characteristics of the sensors. However, the design of a new visual feature is often very difficult. In recent years, deep learning methods have developed rapidly. They start from the raw data of the sensor and automatically extract the abstract information of the data through a multi-layer neural network. Compared with traditional image processing, deep learning networks use multiple convolutional layers to extract features and use pooling layers to select features. The extracted image features are more advanced and abstract than traditional artificial visual features. Convolutional Neural Network (CNN) has been widely applied in image retrieval and image classification.

Considering the similarity between visual loop closure detection and image classification (they both need to extract the features of the image, and then complete the related tasks based on the extracted features), this paper applies CNN to loop closure detection and proposes a loop closure detection algorithm based on multi-scale deep feature fusion. The algorithm includes three modules: feature extraction layer, feature fusion layer and decision layer (as shown in Figure 1). We selected the first five convolutional layers of the pre-trained AlexNet network on the ImageNet dataset as the feature extraction layer, which can extract more advanced and more abstract features. In the feature fusion layer, we designed a multi-scale fusion operator with spatial pyramid pooling (SPP) [1] to fuse the deep features with different receptive fields and create a fixed length representation of an image. Finally, in the decision layer, we developed a similarity measurement method by calculating the distinguishability weight of features, which helps reduce the probability of false positives in loop closure detection. The results show that the loop closure detection algorithm has a high precision and recall rate. They also verify the algorithm's robustness to illumination changes.

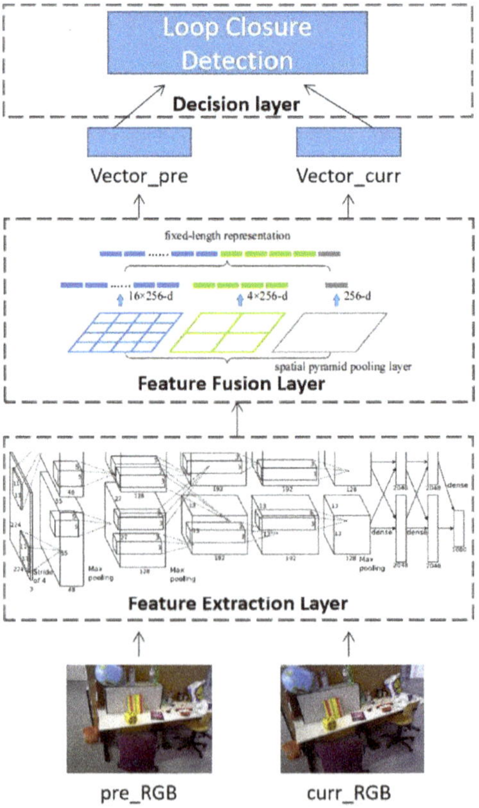


**Figure 1.** The framework of visual loop closure detection algorithm based on multi-scale deep feature fusion.

## 2. Related work

In recent years, scholars have done much research in the direction of loop closure detection algorithms based on vision. The classical algorithms can be roughly divided into two categories: the method based on the BoW (Bag of Word) [2] and the method based on the global descriptor. The first method extracts local features from the scene image and clusters them into multiple "words". Then the whole image is described in the form of vectors based on these "words". Thus, the visual loop closure detection problem is transformed into a similarity measure problem of the description vectors of the two images. BoW is the mainstream method for loop closure detection. A key problem of the BoW method is how to select local features of the image. The common feature points are SIFT [3], SURF [4] and ORB [5]. For example, Mei et al. [6] used the FAST [7] operator to extract the key points and then used the SIFT [3] as feature descriptor. Newman et al. [8] extracted FAST [7] key points and then calculated the descriptors using BRIEFF [9]. For the general case, the images described by the BoW can be compared one-to-one by the histogram or Hamming distance, and the closed loop is detected when the distance is less than a certain threshold. However, in a large-scale scene, search speed is very important, and some researchers have begun to apply the word tree to do efficient loop closure detection. Cummins et al. [10,11] applied Chow-Liu tree approximation to describe the correlation between words and words, and then proposed the classic FAB-MAP method. Glover et al. [12] made public the FAB-MAP development kit based on the work of Cummins et al., which provided convenience for researchers. Maddern et al. [13] proposed the CAT-SLAM method based on FAB-MAP, which combines loop closure detection with a local metric pose filter. Compared with FAB-MAP, the loop closure detection of CAT-SLAM is better. For the second method, the main idea is to describe the entire image with a global descriptor. Ulrich et al. [14] proposed that color histograms provide a compact representation of an image, which results in a system that requires little memory and performs in real-time. But it is very sensitive to changes in illumination. Dalai et al. [15] used histograms of oriented gradients (HOG) as the feature descriptor of the image, which gave very good results for person detection in cluttered backgrounds. GIST [16] had been demonstrated to be a very effective conventional image descriptor, capturing the basic structure of different types of scenes in a very compact way. Based on this, Murillo et al. [17] utilized global gist descriptor computed for portions of panoramic images and a simple similarity measure between two panoramas, which is robust to changes in vehicle orientation, while traversing the same areas in different directions.

Both of the two methods have their own advantages and disadvantages. Furgale et al. [18] proved that the global descriptor method is more sensitive to the camera pose than the BOW method. Milfold [19] and Naseer [20] proposed that the global descriptor method is more robust in the case of illumination changes. Therefore, some researchers have considered combining the two methods and proposed a method of using scene signatures. For example, McManus et al. [21] presented an unsupervised system that produces broad-region detectors for distinctive visual elements, which improved the accuracy of detection. However, the features used in these methods are low-level features and designed artificially in the field of computer vision. They are sensitive to the influence of light, weather and other factors, so these algorithms lack the necessary robustness.

With the disclosure of large-scale datasets (such as ImageNet [22]) and the upgrading of various hardware (such as GPU), deep learning has developed rapidly in recent years. Deep learning can extract abstract and high-level features of the input image through multi-layer neural networks, which is more robust to changes in environmental factors [23,24]. Therefore, it has been widely used in image classification [25] and image retrieval [26]. Considering that visual loop closure detection is similar to image classification and image retrieval, researchers have tried to apply deep learning to loop closure detection. Gao et al. [27,28] took advantage of Autoencoder to extract image features and used the similarity measurement matrix to detect closed loops, which got high accuracy on public datasets. He et al. [29] applied FLCNN (fast and lightweight convolutional neural networks) to extract image features and calculate the similarity matrix, which further improved the real-time and accuracy of loop closure detection. Xia et al. [30] extracted image features by PCANet, and proved that these features are superior to traditional manual design features. Hou et al. [31] used PlaceCNN to extract image features for loop closure detection, which got high accuracy even when the light changed. However, these methods relied on local deep features and ignored the scale problem.

## 3. Loop Closure Detection Algorithm Based on Multi-Scale Deep Feature Fusion

Different from traditional image classification, visual loop closure detection needs to determine whether the two moments are at the same location according to the similarity between the picture collected at the current time and the one taken earlier. Therefore, the algorithm in this paper has paired input, which corresponds to two branches in the algorithm as shown in Figure 1. The algorithm is divided into three layers: feature extraction, feature fusion and decision. The feature extraction layer extracts the deep feature of the input images. The feature fusion layer does multi-scale fusion and normalization of extracted features. The decision layer uses the fusion feature to detect the loop closure. The two branches of the algorithm are identical in the feature extraction layer and the feature fusion layer structure.

### *3.1. Feature Extraction Layer*

The feature extraction layer is composed of two identical CNNs to extract features for two inputs separately. Compared to the early CNN, AlexNet [27] uses a much deeper network model to acquire features. Additionally, it adds modules such as the ReLU activation function, local response normalization (LRN), Dropout, etc., which can reduce the risk of over fitting. Moreover, it takes advantage of multi-GPU to improve the training speed of the network model. In view of these advantages, this paper refers to AlexNet.

The network model has a total of eight layers, consisting of five convolution layers and three fully connected layers. Only the first five layers of AlexNet are needed, as shown in Figure 2. Here, we set the input of the feature extraction layer to be an RGB image of 227 × 227 × 3, and obtained 256 feature maps with a size of 6 × 6 through five convolution layers. (The output data size of each layer is indicated in Figure 2). The convolution layer contains the ReLU activation function and LRN processing, as well as max-pooling. Among them, LRN, which draws on the idea of "lateral inhibition" in neurobiology, is used to locally suppress neurons. When the activation function is ReLU, this "lateral inhibition" is very useful. It can prevent the model from over-fitting prematurely and speed up the training of the model. Its calculation formula is

$$b^i_{x,y} = a^i_{x,y} / \left(k + \alpha \sum\nolimits_{j=\max(0,i-n/2)}^{\min(N-1,i+n/2)} \left(a^i_{x,y}\right)^2\right)^\beta, \tag{1}$$

where $a^i_{x,y}$ denotes the value of the $i$-th convolution kernel after applying the ReLU activation function at position (x, y); n is the number of convolution kernels adjacent in the same position, and N represents the total number of convolution kernels. $k, n, \alpha$ and $\beta$ are tunable parameters; we set $k = 2, n = 5, \alpha = 0.0001, \beta = 0.75$ according to the empirical value.

### *3.2. Feature Fusion Layer*

An RGB image can obtain several feature maps describing different deep features from features extraction layer. For general image classification tasks, a fully connected layer is usually added to weight and sum all the feature maps to obtain the classification result. However, each feature map only corresponds to a small area in the original input image, and its receptive field is small. The deep feature is a local feature. For the loop closure detection problem, we prefer to have different scales of features because this helps to accurately determine whether two images belong to the same scene. In addition, AlexNet [27] requires the fixed-size of the input image to be 227 × 227 × 3. When the size of the input does not meet the requirements, cutting or compressing must happen, which causes the loss or distortion of some image information.

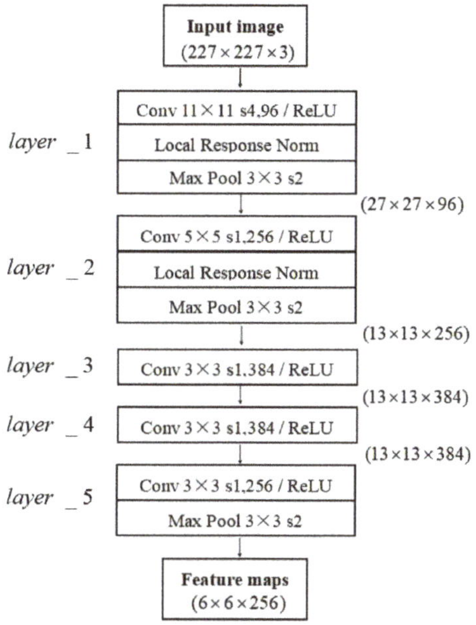


**Figure 2.** Structure of feature extraction layer.

To overcome this problem, we use spatial pyramid pooling (SPP) [1] to fuse multi-scale deep features. SPP divides the feature map into small patches of different sizes by different scales, and then gets each patch's feature. The receptive fields to the input images of these small patches are different, which means the patches correspond to different areas of the input image. Finally, the features extracted from the small patch of different sizes are combined to achieve the fusion of multi-scale features. In addition, the SPP layer can get a fixed-size output and eliminate the restriction that the input image size must be fixed.

Taking a single branch as an example, the feature fusion layer is shown in Figure 3. Here, the input of feature fusion layer is the output from feature extraction layer. After the SPP layer, we can obtain the fixed-length representation of the features, which is finally sent to softmax.

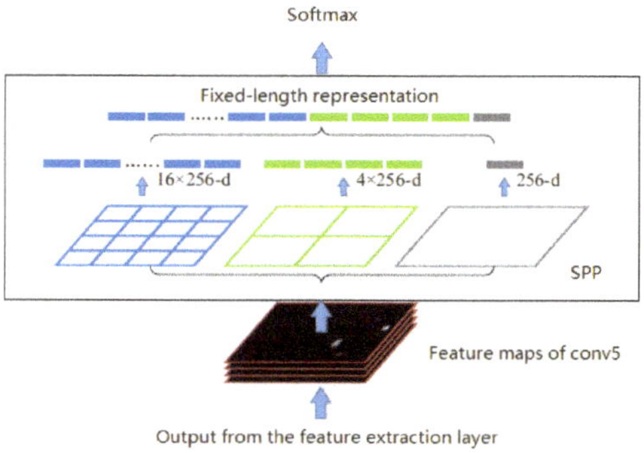


**Figure 3.** Feature fusion layer.

As seen in Figure 3, SPP divides each feature map by using three different scales 4 × 4, 2 × 2, 1 × 1. Each scale corresponds to one layer in the SPP; three scales indicate that the SPP has three layers colored as blue, green and gray. For example, the 4×4 scale divides a feature map into 4 × 4 small patches, and then extracts a feature from each small patch, so each feature map can extract 16 features. There are 256 feature maps as output from the feature extraction layer, and altogether there are 16 × 256 features (blue parts in Figure 3). The features extracted by the different scales of the SPP are concatenated together to obtain a fixed-length feature vector. In order to normalize the results of the SPP, we add a SoftMax layer. Assuming that the feature vector output by the SPP layer is $Z \in R^{n\times 1}$, the output of the SoftMax layer is:

$$Y = \left[\frac{e^{Z_1}}{\sum_{j=1}^{n} e^{Z_j}}, \ldots, \frac{e^{Z_n}}{\sum_{j=1}^{n} e^{Z_j}}\right]^T, Y \in R^{n\times 1}, \quad (2)$$

The number of SPP layers affects the output size of the SoftMax layer. If the number of layers is different, the performance of the corresponding algorithm will be different. In the experiment part, the effects of SPP in different layers on the performance of the algorithm are discussed.

### 3.3. Decision Layer

The decision layer is in charge of loop closure detection. Suppose the inputs of the two branches are Image_1 and Image_2, respectively, and their feature vectors of the SoftMax layer are $f_1$ and $f_2$. The vector dimension is determined by the number of SPP layers and the scale of each layer. Assuming that the dimensions of $f_1$ and $f_2$ are $N$, the similarity of two inputs can be calculated by Equation (3).

$$S_1(f_1, f_2) = 1 - \sum_{i=1}^{N} (f_{1i} - f_{2i}), \quad (3)$$

where $f_1$ and $f_2$ represent the i-th dimension of the feature vector. Setting a threshold T, when $S_1(f_1, f_2) > T$, it means that Image_1 and Image_2 correspond to the same place and a closed loop is detected. Equation (3) treats each feature node fairly, but the distinguishability of each feature node is different. Features such as walls and ground are more common in scenes and their distinguishability is relatively small, while the traffic signs are more distinguishable. If the feature nodes with different distinguishability are treated fairly, more false positives (different places in similar scenes) will occur when detecting the loop closure. Considering this character, we add a weight to each feature node and modify Equation (3) as Equation (4). The weight's value indicates the distinguishability of the feature to the scene.

$$S_1(f_1, f_2) = 1 - \sum_{i=1}^{N} \delta_i (f_{1i} - f_{2i}), \quad (4)$$

where $\delta_i$ represents the weight of the $i$-th feature node, and the larger the value of $\delta_i$, the greater the distinguishability of its corresponding feature in the scene. The weights can be learned by training and accord with Gaussian distribution. The value of $\delta_i$ is calculated as follows:

$$\delta_i = \exp\left(-\frac{(\overline{h}_i - u)^2}{2\delta^2}\right), \quad (5)$$

where $\overline{h}_i$ represents the average response of the $i$-th feature node. If the average response of a feature node is larger, such features are more common (such as ground and sky), and the $\delta_i$ is smaller. If the average response of a feature node is smaller and lower than the mean value of $u$, such features are not common (such as noise), and $\delta_i$ is also smaller. When $\overline{h}_i$ is near the mean value of $u$, the distinguishability is relatively large, and the corresponding weight value is also relatively large. After the network model is trained, the value of $\overline{h}_i$ can be calculated and retained by the test. $u$ and $\delta$ are tunable parameters and set according to experience.

## 4. Parameter Training

### 4.1. Training Method

Since SPP in the feature fusion layer is a special maxing pooling layer, it has no parameters to be trained. Here we only need to train the parameters of the convolution network in the feature extraction layer. The algorithm in this paper consists of two branches with the same structure in the feature extraction layer and the feature fusion layer. For the loop closure detection problem, its positive sample indicates the same location, and the negative sample indicates non-loop closure. Loop closure always happens at different locations and with very few times, which means a large number of classes and small labeled samples. Taking these into consideration, this paper uses the Siamese [32] model to train.

The Siamese network is mainly used in the field of face recognition can solve the classification problem with small sample data well. Figure 4 shows the training model.

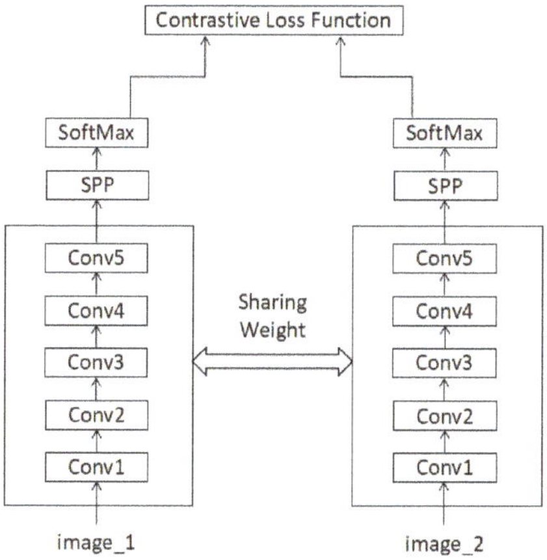


**Figure 4.** Parameter training model for feature extraction layer.

Contrastive loss function is used here. For the $i$-th pair of samples, we assume that the feature vectors of the SoftMax layer are $f_{i1}, f_{i2}$ respectively, and the contrastive loss function is:

$$L = \frac{1}{2M}\sum_{i=1}^{M}[y_i d_i^2 + (1 - y_i)\max(m\arg in - d_i, 0)^2], \quad (6)$$

where $M$ is the number of sample pairs; $d_i = \| f_{i1} - f_{i2} \|_2$ is the Euclidean distance of the two samples in the feature space, and $y_i$ is the label of the $i$-th pair of samples. $y = 1$ is the positive sample, which means that the two images are similar and form a loop closure; $y = 0$ represents a negative sample, which means that the similarity of the two pictures is small and does not form a loop closure. The threshold $margin$ is set to 1 in the experiment. When $y = 1$ and the loss function is $L = \frac{1}{2M}\sum_{i=1}^{M} y_i d_i^2$, if their Euclidean distance d in the feature space is large, the current training model is not good and its loss value increases. When $y = 0$ and the loss function is $L = \frac{1}{2M}\sum_{i=1}^{M}[\max(m\arg in - d_i, 0)^2]$, if their Euclidean distance in the feature space is small, the loss value of the model will become large. The contrastive loss function can express the matching degree of paired samples, and it can be used for the model training.

### 4.2. Model Training

We use the AlexNet model pre-trained by the ImageNet dataset on Caffe [33], and set the parameters of the first five convolutional layers as the initial values of the parameters in the feature extraction layer. Meanwhile, we use the Matterport3D dataset to fine tune the model. The Matterport3D dataset is the world's largest public 3D dataset from 3D scanning solution provider Matterport. It is a large-scale RGB-D dataset containing 10,800 panoramic views from 194,400 RGB-D images of 90 building-scale scenes. The dataset was acquired by a Pro 3D camera. The camera rotates around the center of gravity at each sampling point, and samples the images at six rotation positions, each of which corresponds to 18 sets of pictures. This special way of data acquisition makes the camera cover a wider range of angles, and it provides many loop closure data. Figure 5a–f are some examples of RGB images in the MatterPort3D dataset.

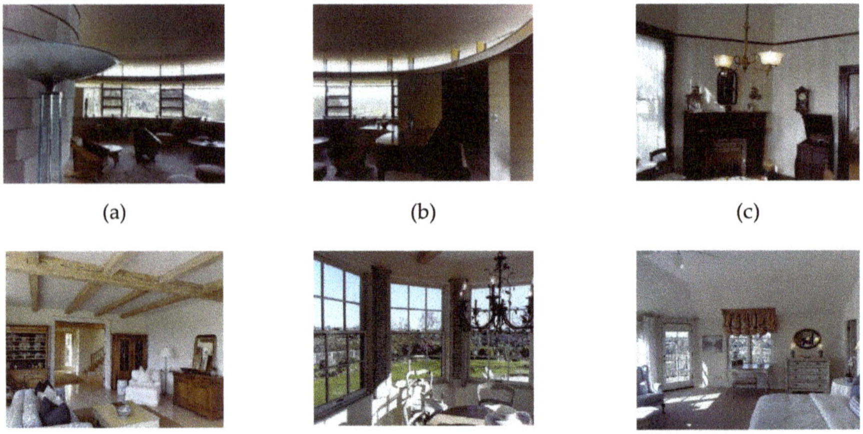

(a) (b) (c)

(d) (e) (f)

**Figure 5.** Examples of an RGB graph in the Matterport3D dataset.

We selected 30,000 pairs of positive samples (loop closure) and 30,000 pairs of negative samples (non-loop closure) in the Matterport3D data set. Then by some data enhancement means such as rotating and translating, the positive and negative samples were each doubled to 60,000. Of these, 10,000 pairs of positive samples and 10,000 pairs of negative samples were selected as test sets, and the other 50,000 pairs of positive samples and 50,000 pairs of negative samples were used for training. Some researchers have found that multi-scale training can improve the accuracy of the network containing SPP layers in the tasks of image classification and object detection. In view of this, this paper uses three different scales of data to train the model: $1280 \times 1024$, $227 \times 227$ and $180 \times 180$. The latter two data are cropped from the original image. In this way, the number of our training set is 300,000, including 150,000 pairs of positive samples and 150,000 pairs of negative samples.

The deep learning framework Caffe requires a fixed size of input images; therefore, we used a combination of single-scale and multi-scale to train the model. In one epoch of model training, single-scale images are input, while in each different epoch of model training, the input batches are of different scales.

In addition, in order to analyze the influence of different SPP layers in the feature fusion layer on the performance of the algorithm, three different SPP are used: a one-layer SPP with a scale of $1 \times 1$, a two-layer SPP with a scale of $1 \times 1, 2 \times 2$, and a three-layer SPP with a scale of $1 \times 1, 2 \times 2, 4 \times 4$. For these three cases, we build three models and train them separately.

## 5. Experiment

In order to verify the performance and effectiveness of the algorithm, we tested the effects of different layers of SPP, the influence of the similarity measurement method in visual loop closure detection and the robustness to illumination changes. An Intel Core i7 processor of 2.8 GHz frequency and an NVIDIA GeForce GTX 1060 with MAX-Q Graphics card were used.

### 5.1. Dataset and Labeling

The dataset was provided by the computer vision group of the Technical University of Munich (TUM) [34]. We used the dataset's RGB-D image sequence, which includes Fr2/rpy, Fr2/large_with_loop, Fr2/pioneer_slam, Fr2/pioneer_slam2 and Fr3/long_office_household. However, the loop closure is not labeled in the TUM dataset and must be manually marked. Because of the 30 Hz/s sample frequency, there are too many images in the same place. Therefore, we selected the key frames with the ground truth of trajectory.

Keyframe selection includes the following steps.

For each image sequence, a key frame list $F = \{\}$ is set, and the first frame is added to $F$.

Sequentially compare the image $f_j$ in the image sequence with the last frame

$$f_i$$

in the key frame list. Assuming that their corresponding poses are $T_j, T_i$ which can be found in the ground truth, the relative translation of the two frames is $M_{j,i} = trans(T_j^{-1} T_i)$, where $trans(\bullet)$ represents the 2 norm of the translational part of the transformation matrix. If $0.1 < M_{j,i} < 1$, the frame $f_j$ is added to the end of the key frame list.

Finally, the loop closure is manually labeled with the selected key frame sequence.

Figure 6a shows the trajectory ground truth of the Fr3/long_office_household sequence, and in Figure 6b red dots are the selected key frames and the black line segments are the loop closure. Figure 6c,d are one pair of loop closure images.

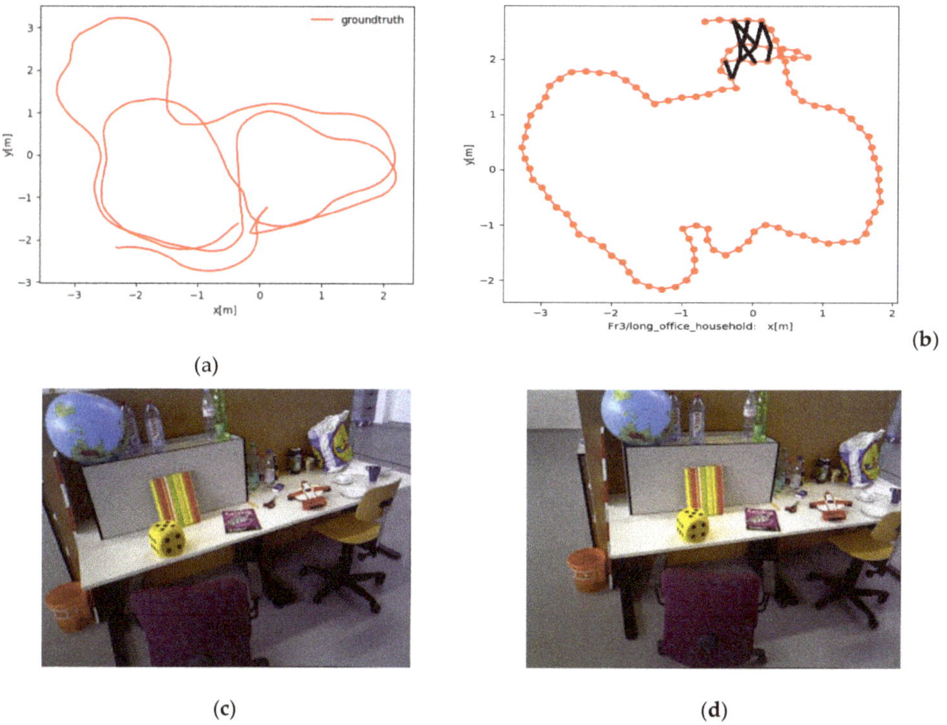


**Figure 6.** Keyframe selection and loop closure example in FR3/long_office_household.

*5.2. Different layers of SPP*

SPP is used in the feature fusion layer to fuse the extracted depth features of different scales. In order to verify how different layers of SPP influence the algorithm results, an SPP with 1 × 1, a two-layer SPP with 1 × 1, 2 × 2, and a three-layer SPP with 1 × 1, 2 × 2, 4 × 4 were used for experiments. The loop closure detection algorithm with these three different SPP layers, noted as spp1, spp12 and spp124, respectively, was compared with FabMap [10].

We selected 689 key frames from the Fr2/rpy, Fr2/large_with_loop and Fr3/long_office_household and labeled 45 loop closure places, which were noted as data_1. The experimental results of the data_1 are shown in Figure 7. It can be seen from Figure 7 that the P-R curves of spp1, spp12 and spp124 are basically on the upper right of the coordinate system, which means they have higher precision and recall rates. Our method can reach 100% precision at 50% recall. In addition, spp124 is better than spp12 and spp1, and spp12 is better than spp1. When the recall rate is 100%, spp124 reaches as high as 78% precision; spp12 is about 62% precision, and spp1 is less than 50% precision. It shows that the greater the layers of SPP, the higher the precision and recall rate of the algorithm.

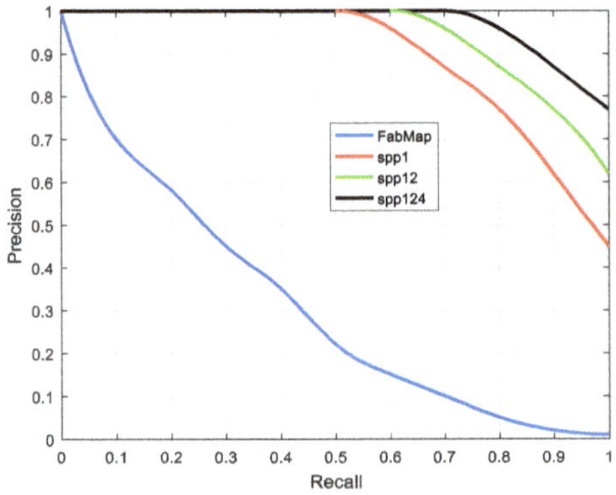


**Figure 7.** P-R curves on data_1.

The above algorithm was also tested in the Fr2/pioneer_slam and Fr2/pioneer_slam2 image sequences which are sampled in a very empty indoor environment and have many similar scenes, such as walls, boards and ground. We selected 435 key frames from the two sequences and labelled 20 loop closure places, which were noted as data_2. The test results on data_2 are shown in Figure 8.

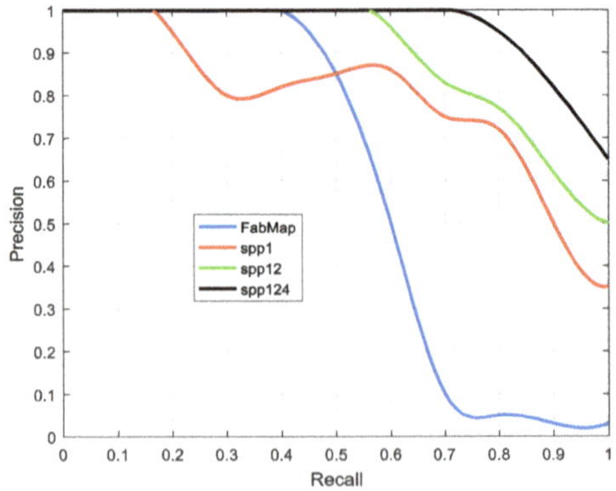


**Figure 8.** P-R curves on data_2.

As can be seen from Figure 8, spp124 has the highest precision-recall rate, followed by spp12. At the 50% recall rate, both spp124 and spp12 can reach 100% precision. For spp1, when the recall rate is less than 50%, the accuracy is higher than the accuracy of FabMap, but when the recall rate is greater than 50%, the accuracy of spp1 is lower than that of FabMap. It is worth mentioning that the performance of all algorithms on data_2 is worse than on data_1. Because there are many similar scenarios in data_2, the algorithms got some false positive results. Overall, the depth features extracted by CNN are more suitable for loop closure detection compared with the traditional artificial design visual features, and increasing the number of SPP layers in the feature fusion layer can improve the accuracy and recall rate.

### *5.3. Similarity Measurement*

In order to verify the effect of weight adjustment in the similarity measurement, we compared the spp1, spp12, and spp124 methods with these added weight adjustments which were denoted as spp1+, spp12+, spp124+. The former directly uses Equation (3) to calculate similarity, and the latter uses Equation (4). In order to calculate $\delta_i$ from Equation (5), we selected 100,000 RGB images from the Matterport3D data set and calculated the average response of each node of the SoftMax layer as $h_i$. Since two branches are identical and share their parameters, only one branch needed to be calculated. The tunable parameter was set as $u = 0.5, \delta = 0.1$. The results of data_1 and data_2 are shown in Figures 9 and 10, respectively.

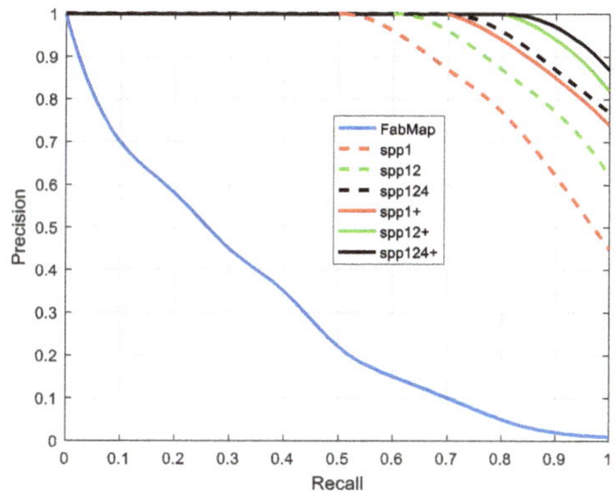


**Figure 9.** P-R curves of seven algorithms on data_1.

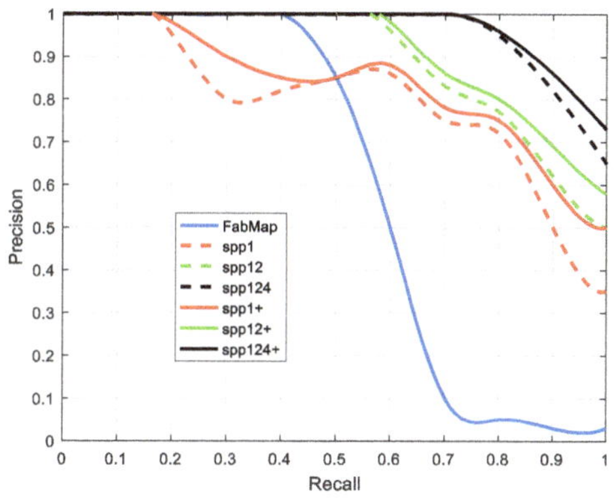


**Figure 10.** P-R curves of seven algorithms on data_2.

At a high recall rate, spp1+, spp12+ and spp124+ have higher precision than spp1, spp12 and spp124. When considering feature distinguishability, the visual loop closure detection algorithm reduced the probability of false positives and improved precision.

### 5.4. Illumination Changes

Illumination changes are the most critical factors affecting the visual loop closure detection. In order to test the robustness of the algorithm to illumination changes, we collected image sequences with different illuminations. We fixed a trajectory and then sampled image sequences at 12:30, 15:00, 17:30 and 19:00. Finally, four image sequences were obtained and named as VS1230, VS1500, VS1730 and VS1900 respectively, as shown in Figure 11. Each image sequence contained 200 frames and 20 closed loops.

(**a**) 12:30

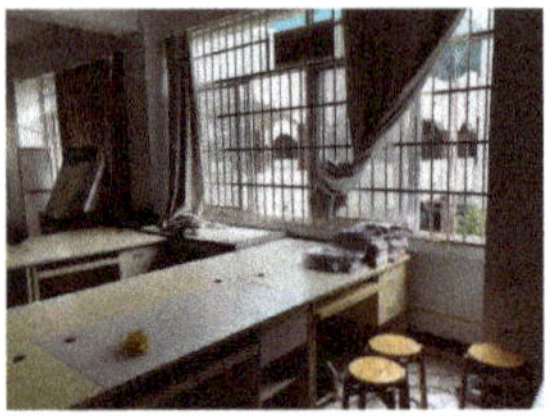

(**b**) 15:00

(**c**) 17:30

(**d**) 19:00

**Figure 11.** Sample images taken at four different times.

The tests were performed on these four image sequences, and the results are shown in Figures 12–15. Since the algorithms of spp1+, spp12+, spp124+ have been proven to be outstanding in 5.3, here we only compared them with FabMap.

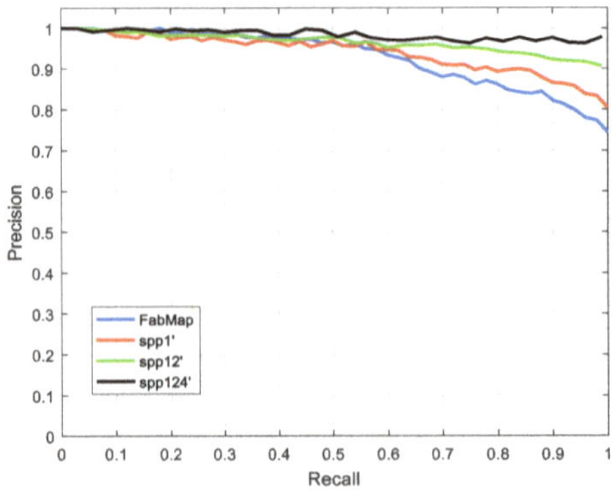


**Figure 12.** P-R curves of four algorithms on VS1230.

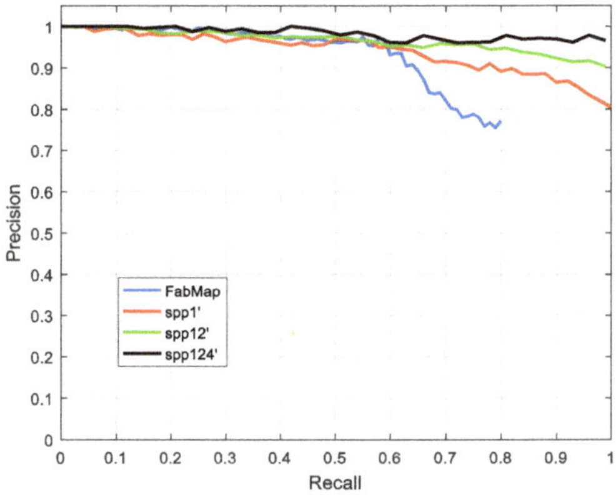


**Figure 13.** P-R curves of four algorithms on VS1500.

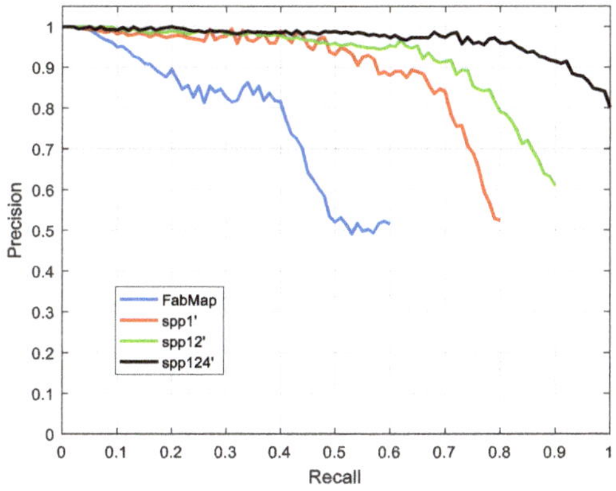


**Figure 14.** P-R curves of four algorithms on VS1730.

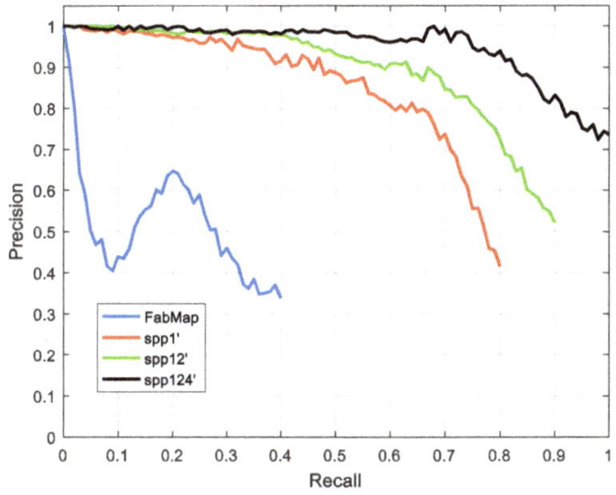


**Figure 15.** P-R curves of four algorithms on VS1900.

At 12:30, the P-R curves of the four algorithms are on the upper right of the coordinate system. However, as the time went by, all P-R values of the algorithms decreased, and the FabMap decreased more obviously because the illumination turned to dim. Especially in the VS1730 and VS1900 image sequences, the accuracy of FabMap was drastically lower at high recall rates. In the VS1900 image sequences, at 100% recall rate, the precision of spp124+ was still as high as 70% or more. This shows that the proposed algorithm is more robust to illumination changes than the traditional BoW method.

In general, for RGB-D SLAM, researchers prefer higher precision because the wrong closed loop leads to a completely wrong pose graph. Therefore, we also did statistical analysis about the average precision of spp1+, spp12+, spp124+, and FabMap on the four image sequences VS1230, VS1500, VS1730, and VS1900.

As can be seen from Figure 16, the average precision of the four algorithms was above 90% at 12:30, and at 17:30, only spp124+ and spp12+ were still above 90% precision. At 17:30, the average precision of FabMap is only 70%. At 19:00, the differences are more obvious. At this time, the average precision of spp124+ was still as high as 90%, and the other three were below 90%, especially FabMap which had an average accuracy of only 42%. It can be concluded that the algorithm of this paper is more robust than the traditional BoW method when the illumination changes.

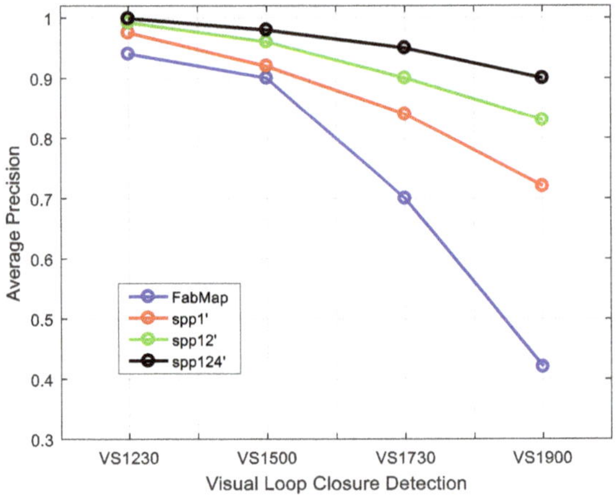


**Figure 16.** Average precision of spp1+, spp12+, spp124+, and FabMap on four image sequences.

## 6. Conclusions

This paper has proposed a loop closure detection algorithm based on multi-scale deep feature fusion. Our proposed algorithm has a higher accuracy and recall rate than the traditional BoW method, and it has better robustness to illumination changes. The key ideas that allowed us to achieve this efficiency are as follows. First, we used CNN to extract features. The features extracted in this way are more advanced and abstract, and have better robustness to illumination changes. Second, we used SPP to fuse the extracted features. By setting the multi-layer SPP with different receptive field sizes to fuse the different scale features in the image, it is more conducive to detecting loop closure. Last but not least, in the similarity calculation process of the decision layer, we added a weight to each feature node according to the distinguishability of the feature nodes to the scene. We have verified the benefits of the above points with experiments. In fact, the neural network used only plays the role of extracting more abstract features. Later, we will consider introducing the difference between the two image features in the middle layer of the deep network, and using the difference feature to detect loop closure.

**Author Contributions:** Conceptualization, B.C. and C.L.; Data curation, D.Y.; Formal analysis, B.C. and D.Y.; Investigation, Q.W.; Methodology, B.C. and C.L.; Project administration, B.C.; Resources, Q.W.; Validation, D.Y.; Writing—original draft, D.Y.; Writing—review & editing, B.C. and D.Y.

**Funding:** This research was supported by the National Natural Science Foundation of China under Grant No. 61403423, the Natural Science Foundation of Hunan Province of China under Grant No. 2018JJ3689, National

Key Research and Development Plan (2018YFB1201602) and Science and Technology Major Project of Hunan Province of China under Grant No. 2017GK1010.

**Conflicts of Interest:** The authors declare no conflict of interest.

## References


1. He, K.; Zhang, X.; Ren, S.; Sun, J. Spatial pyramid pooling in deep convolutional networks for visual recognition. In *European Conference on Computer Vision*; Springer: Cham, Switzerland, 2014; pp. 346–361.
2. Sivic, J. Video Google: A text retrieval approach to object matching in videos. In Proceedings of the Ninth IEEE International Conference on Computer Vision, Nice, France, 13–16 October 2003.
3. Lowe, D.G. Distinctive image features from scale-invariant keypoints. *Int. J. Comput. Vis.* **2004**, *60*, 91–110. [CrossRef]
4. Bay, H. SURF: Speeded up robust features. In *European Conference on Computer Vision*; Springer: Berlin/Heidelberg, Germany, 2006.
5. Mur-Artal, R.; Tardos, J.D. ORB-SLAM2: An Open-Source SLAM System for Monocular, Stereo, and RGB-D Cameras. *IEEE Trans. Robot.* **2017**, *33*, 1255–1262. [CrossRef]
6. Mei, C.; Sibley, G.; Newman, P.; Reid, I. A Constant-Time Efficient Stereo SLAM System. In Proceedings of the 20th British Machine Vision Conference (BMVC), London, UK, 7–10 September 2009; pp. 1–11.
7. Rosten, E.; Drummond, T. Machine learning for high-speed corner detection. In *European Conference on Computer Vision*; Springer: Berlin/Heidelberg, Germany, 2006; pp. 430–443.
8. Churchill, W.; Newman, P. Experience-based navigation for long-term localisation. *Int. J. Robot. Res.* **2013**, *32*, 1645–1661. [CrossRef]
9. Calonder, M.; Lepetit, V.; Ozuysal, M.; Strecha, C.; Fua, P. BRIEF: Computing a local binary descriptor very fast. *IEEE Trans. Pattern Anal. Mach. Intell.* **2012**, *34*, 1281–1298. [CrossRef] [PubMed]
10. Schindler, G.; Brown, M.; Szeliski, R. City-Scale Location Recognition. In Proceedings of the IEEE Conference on Computer Vision and Pattern Recognition (CVPR), Minneapolis, MN, USA, 17–22 June 2007; pp. 1–7.
11. Cummins, M.; Newman, P. Probabilistic appearance-based navigation and loop closing. In Proceedings of the IEEE International Conference on Robotics and Automation, Roma, Italy, 10–14 April 2007; pp. 2042–2048.
12. Glover, A.; Maddern, W.; Warren, M.; Stephanie, R.; Milford, M.; Wyeth, G. Openfabmap: An open source toolbox for appearance-based loop closure detection. In Proceedings of the IEEE International Conference on Robotics and Automation (ICRA), Saint Paul, MN, USA, 14–18 May 2007; pp. 4730–4735.
13. Maddern, W.; Milford, M.; Wyeth, G. CAT-SLAM: Probabilistic localisation and mapping using a continuous appearance-based trajectory. *Int. J. Robot. Res.* **2012**, *31*, 429–451. [CrossRef]
14. Dalai, N.; Triggs, B. Histograms of oriented gradients for human detection. In Proceedings of the IEEE Conference Computer Vision and Pattern Recognition, San Diego, CA, USA, 20–25 June 2005; pp. 886–893.
15. Ulrich, I.; Nourbakhsh, I. Appearance-based place recognition for topological localization. In Proceedings of the IEEE International Conference on Robotics and Automation (ICRA), San Francisco, CA, USA, 24–28 April 2000; Volume 2, pp. 1023–1029.
16. Oliva and, A. Torralba. Building the gist of a scene: The role of global image features in recognition. *Vis. Percept. Prog. Brain Res.* **2006**, *155*, 23–36.
17. Murillo, A.C.; Kosecka, J. Experiments in place recognition using gist panoramas. In Proceedings of the IEEE International Conference on Computer Vision Workshops (ICCV), Kyoto, Japan, 27 September–4 October 2009; pp. 2196–2203.
18. Furgale, P.; Barfoot, T.D. Visual teach and repeat for long-range rover autonomy. *J. Field Robot.* **2010**, *27*, 534–560. [CrossRef]
19. Milford, M.J.; Wyeth, G.F. SeqSLAM: Visual route-based navigation for sunny summer days and stormy winter nights. In Proceedings of the IEEE International Conference on Robotics and Automation (ICRA), Saint Paul, MN, USA, 14–18 May 2012; pp. 1643–1649.
20. Naseer, T.; Spinello, L.; Burgard, W.; Stachniss, C. Robust visual robot localization across seasons using network flows. In Proceedings of the Twenty-Eighth AAAI Conference on Artificial Intelligence, Québec City, QC, Canada, 27–31 July 2014; pp. 2564–2570.
21. Mcmanus, C.; Upcroft, B.; Newmann, P. Scene Signatures: Localised and Point-less Features for Localisation. *Image Proc.* **2014**. [CrossRef]

22. Deng, J.; Dong, W.; Socher, R.; Li, L.J.; Li, K.; Fei-Fei, L. ImageNet: A large-scale hierarchical image database. In Proceedings of the IEEE Conference on Computer Vision and Pattern Recognition (CVPR), Miami, FL, USA, 20–25 June 2009; pp. 248–255.
23. Chatfield, K.; Simonyan, K.; Vedaldi, A.; Zisserman, A. Return of the devil in the details: Delving deep into convolutional nets. *Comput. Sci.* **2014**, 1–11.
24. Wan, J.; Wang, D.Y. Deep learning for content-based image retrieval: A comprehensive study. In Proceedings of the 22nd ACM International Conference on Multimedia, Orlando, FL, USA, 3–7 November 2014; pp. 157–166.
25. Krizhevsky, A.; Sutskever, I.; Hinton, G.E. ImageNet classification with deep convolutional neural networks. In Proceedings of the 25th International Conference on Neural Information Processing Systems, Lake Tahoe, NV, USA, 3–6 December 2012; pp. 1097–1105.
26. Babenko, A.; Slesarev, A.; Chigorin, A.; Lempitsky, V. Neural codes for image retrieval. In *European Conference on Computer Vision*; Springer: Cham, Switzerland, 2014; pp. 584–599.
27. Gao, X.; Zhang, T. Unsupervised learning to detect loops using deep neural networks for visual SLAM system. *Auton. Robot.* **2017**, *41*, 1–18. [CrossRef]
28. Gao, X.; Zhang, T. Loop closure detection for visual slam systems using deep neural networks. In Proceedings of the Chinese Control Conference, Hangzhou: Chinese Association of Automation, Hangzhou, China, 28–30 July 2015; pp. 5851–5856.
29. He, Y.L.; Chen, J.T.; Zeng, B. A Fast loop closure detection method based on lightweight convolutional neural network. *Comput. Eng.* **2018**, *44*, 182–187.
30. Xia, Y.; Li, J.; Qi, L.; Fan, H. Loop closure detection for visual SLAM using PCANet features. In Proceedings of the IEEE International Joint Conference, Vancouver, BC, Canada, 24–29 July 2016; pp. 2274–2281.
31. Hou, Y.; Zhang, H.; Zhou, S. Convolutional neural network-based image representation for visual loop closure detection. In Proceedings of the IEEE International Conference, Lijiang, China, 20–25 April 2015; pp. 2238–2245.
32. Chopra, S.; Hadsell, R.; Lecun, Y. Learning a similarity metric discriminatively, with application to face verification. In Proceedings of the IEEE Computer Society Conference on Computer Vision and Pattern Recognition, San Diego, CA, USA, 20–25 June 2005; pp. 539–546.
33. Jia, Y.; Shelhamer, E.; Donahue, J.; Donahue, J.; Karayev, S.; Long, J.; Girshick, R. Caffe: Convolutional Architecture for Fast Feature Embedding. In Proceedings of the 22nd ACM International Conference on Multimedia, Orlando, FL, USA, 3–7 November 2014.
34. Sturm, J.; Engelhard, N.; Endres, F.; Burgard, W.; Cremers, D. A benchmark for the evaluation of RGB-D SLAM systems. In Proceedings of the IEEE/RSJ International Conference on Intelligent Robots and Systems, Vilamoura, Portugal, 7–12 October 2012; pp. 573–580.

*applied sciences*

MDPI

*Article*

# IMU-Assisted 2D SLAM Method for Low-Texture and Dynamic Environments

**Zhongli Wang** [1,2,*], **Yan Chen** [1], **Yue Mei** [1], **Kuo Yang** [1] **and Baigen Cai** [1,2]

1 School of Electronic Information and Engineering, Beijing Jiaotong University, Beijing 100044, China; 17120213@bjtu.edu.cn (Y.C.); 16120249@bjtu.edu.cn (Y.M.); 15120302@bjtu.edu.cn (K.Y.); bgcai@bjtu.edu.cn (B.C.)

2 Beijing Engineering Research Center of EMC and GNSS Technology for Rail Transportation, Beijing 100044, China

* Correspondence: zlwang@bjtu.edu.cn; Tel.: +86-10-5168-4361

Received: 14 October 2018; Accepted: 4 December 2018; Published: 7 December 2018

**Featured Application: The work in this paper is about the core techniques of mobile robots, especially for the localization and navigation in indoor or structured environments, such as the home service robot, AGV in a factory, et al.**

**Abstract:** Generally, the key issues of 2D LiDAR-based simultaneous localization and mapping (SLAM) for indoor application include data association (DA) and closed-loop detection. Particularly, a low-texture environment, which refers to no obvious changes between two consecutive scanning outputs, with moving objects existing in the environment will bring great challenges on DA and the closed-loop detection, and the accuracy and consistency of SLAM may be badly affected. There is not much literature that addresses this issue. In this paper, a mapping strategy is firstly exploited to improve the performance of the 2D SLAM in dynamic environments. Secondly, a fusion method which combines the IMU sensor with a 2D LiDAR, based on framework of extended Kalman Filter (EKF), is proposed to enhance the performance under low-texture environments. In the front-end of the proposed SLAM method, initial motion estimation is obtained from the output of EKF, and it can be taken as the initial pose for the scan matching problem. Then the scan matching problem can be optimized by the Levenberg–Marquardt (LM) algorithm. For the back-end optimization, a sparse pose adjustment (SPA) method is employed. To improve the accuracy, the grid map is updated with the bicubic interpolation method for derivative computing. With the improvements both in the DA process and the back-end optimization stage, the accuracy and consistency of SLAM results in low-texture environments is enhanced. Qualitative and quantitative experiments with open-loop and closed-loop cases have been conducted and the results are analyzed, confirming that the proposed method is effective in low-texture and dynamic indoor environments.

**Keywords:** dynamic environment; closed-loop detection; sparse pose adjustment (SPA); inertial measurement unit (IMU); simultaneous localization and mapping (SLAM)

---

## 1. Introduction

Simultaneous localization and mapping (SLAM) provides the mobile robot the ability to set up a model of the working space and to localize itself, and it is the most important ability for a truly autonomous robot able to operate within real-world environments. There are several main kinds of sensors widely used for SLAM research or applications, such as cameras [1], RGBD sensors [2], LiDAR [3], and even the fusion of different kinds of sensors [4]. Each kind of sensor has its own advantages and limitations. The camera-based methods are easy to be affected by illumination or seasons changing. Comparatively, LiDAR has the advantages of high precision, good real-time performance, and strong anti-interference ability, so the LiDAR-based SLAM has been widely used in many practical applications, such as autonomous vehicles, home service robots, and automatic guided vehicles in civilian areas.

Generally, LiDAR-based SLAM can be separated into two stages according to the information processing: the front-end stage and the back-end stage.

In the front-end stage, currently observed scanning data is matched with the previously scanned one, which is the very fundamental step in SLAM—data association (DA). There are usually two strategies for data association: scan-to-scan matching-based [5,6] and scan-to-map matching-based [7]. Scan-to-scan matching-based DA is commonly used to compute the relative motion between two consecutive scanning results. The matching process is to adjust the pose of the current scan to make sure an overlap exists as much as possible between current scan and reference scan. For example, the HG-SLAM method in reference [8] adopts a scan-to-scan matching strategy and Iterative Closest Point (ICP) [9] is used to compute the rigid transformation between adjacent scanning frames. The advantage of scan-to-scan matching strategy is that it has a lower computational cost because only two frames are considered. However, due to the noise data and many other factors, the result is prone to quickly accumulating errors. On the other hand, scan-to-map matching-based DA is to align the current scan with the existing map. For example, Hector SLAM [7] takes a scan-to-map matching strategy to solve the data association problem, where the current frame is aligned with the entire map to achieve the rigid transformation from the current scanning set to the built map, which uses the Gauss–Newton method to solve the nonlinear optimization problems. Usually, to further improve the stability, the features extracted from scanned points can be used for matching [6]. The cumulative error is limited in this case and a higher precision mapping result can be achieved for small-scale environments. Because the current scan will be matched with the whole or part of the existing map, it has a higher computational cost compared with scan-to-scan matching.

Based on the initial pose estimation results obtained in the front-end stage, some optimization methods are applied to improve the accuracy and robustness of SLAM results in the back-end stage. Particle filter-based [10] or graph-based optimization are two popular methods used for nonlinear optimization in SLAM. The graph-based method [11,12] uses a collection of nodes to represent the poses and features; the edges in the graph are the results generated by DA from observations, also regarded as the constraints and different optimization methods can be applied to minimize the error expressed by the constraints. Loop closure detection (LCD) when a robot revisits a place plays a very important role in SLAM [13,14]. It will be used as a constraint for global optimization and can reduce the accumulated error. LCD is heavily dependent on the DA results. In HG-SLAM [8], a hierarchical loop closure method based on a local map [15–17] is proposed. The optimization in the back-end stage can usually enhance the results, but it needs a good initial value to start over, or it will fall into the local minimum or even be unable to converge.

Though the initial estimated pose can be optimized in the second stage (back-end stage), the accumulated errors of DA may give rise to big problems for the results of SLAM, and the accuracy and the stability will be badly affected, sometimes even failing to obtain the results. This is often the case in low-texture environments for a 2D scanner, for example, the mobile robot moves along the long corridor in an indoor environment, and due to the measuring limitation of the scanner, there may be no obvious changes between two consecutive scanning outputs, making the DA process difficult. Additionally, when there are moving objects in a dynamic environment, the scanned points located on the dynamic target are taken as noise data and badly impact the DA process. Both cases bring great challenges for the DA task. If a large error is generated in the early stage of the DA process, the following steps, including the LCD and the back-end optimization, will be affected. The closed-loop detection is actually a scan-to-map matching process, and can help to limit the error accumulation generated in the DA process. However, it needs a good initial pose estimate. Unfortunately, a good initial pose is hard to obtain in a low-texture or dynamic environment. From this point of view, the DA and LCD are still an open challenging problem.

To solve the aforementioned problems, some approaches based on sensor fusion are put forward for data association. With the particle filter (PF) framework [18,19], the GMapping approach [10] introduces the adaptive resampling technique, calculating the particle distribution not only relying on the current observation of the LiDAR, but also the odometry information; the uncertainty of the robot pose is reduced and the particle dissipation problem is minimized. But no LCD is included in their method. The IMU sensor in the Hector SLAM method [7] can provide an initial pose estimation to solve the least squares problems. By making use of the extended Kalman filter (EKF) framework, Hector SLAM uses the rigid transformation from the current point cloud to the existing map as input to update the pose. Marco Baglietto [20] integrated a 6DOF IMU with a 2D LiDAR to explore the rescue scene and to find a safe route. Jian [21] uses IMU data to project the single-line LiDAR data to the horizontal plane to achieve the purpose of drawing the tree distribution map in the forest. However, in this method, IMU is only used to assist in mapping, and is not involved in positioning, so is not suitable for low-texture environments. Hesch [22] and others connect a single-line LiDAR with an inertial measurement unit. The system can provide real-time, complex 3D indoor environment tracking. However, the implementation of the method is based on the assumption that all vertical walls are orthogonal to each other.

All the above methods are based on the assumption of a static environment, however, this comes into conflict with the real applications because there are always other moving objects around, such as people moving around, or even other moving robots. Dynamic environments will inevitably affect the reliability of data association, which results in inaccurate maps. Currently, there are two ways to deal with the dynamic environment. One way is integrating the moving object detecting and tracking with the traditional SLAM method, finally building a dynamic map and static map to provide comprehensive information of the whole environment. Another way is to detect and eliminate the observed information caused by dynamic objects. Holz [23] decomposes the task into the SLAM problem in the static environment and the navigation problem of the dynamic environment in the known map. Li [24] proposes a mapping strategy for occupancy grid map in the dynamic environment. Montemerlo [25] and others identify the moving human body in the process of positioning, thus improving the robustness of pose estimation. Hahnel [26] uses a probabilistic model to describe the movement of human bodies, and the observation data caused by the human body are ignored during the map building. Avots [27] and others estimate the state of the door in the environment through a particle filter. Duckett [28] uses a special map representation to merge changes in the map. Wang [29] proposes a method to track and predict dynamic obstacles.

Motivated by the existing fusion methods, in this paper, a fusion-based method which combines an IMU sensor with 2D LiDAR for low-texture and dynamic environments is proposed. Compared with existing literature, the main contributions of this work include:

(1) Based on EKF framework, the information from the IMU sensor is integrated with the 2D LiDAR sensor, and an initial motion estimation can be obtained by the fusion, which can be taken as the initial pose for the scan matching problem. This greatly improves the accuracy and stability of the DA results under the low-texture and dynamic environment.
(2) By generating static local maps, a map-updating strategy is exploited to improve the accuracy of DA and closed-loop detection in the dynamic environment.
(3) With scan-to-map matching methods and periodic back-end optimization with the sparse pose adjustment (SPA) method, the accuracy and stability of the SLAM are improved obviously for the low-texture environment. Furthermore, quantitative experiments are conducted to evaluate the proposed method.

## 2. The Proposed Method

The data flow diagram of the proposed method is illustrated in Figure 1; the rectangle box in the graph represents the data, and the oval box represents the operation. The data flow is as follows:

(1) Based on previously estimated pose at time T-1, which is obtained by the fusion of 6DOF IMU data and LiDAR, and the current output of IMU sensor, the initial pose of LiDAR at time T is estimated based on EKF estimation. The output of EKF is then forwarded to participate in scan matching. The results of the scan matching can be involved in the EKF prediction at the next time. Here, a scan-to-submap strategy is employed to greatly reduce the time consumption.
(2) After a scan matching, the system will carry out a closed-loop detection. If a loop closure is found by a suitable matching, the result will be added as a constraint to the back-end optimizer. The back-end optimizer will run one time every 5 s, and output the LiDAR pose at all moments. We can get all optimized static maps by making use of all point cloud data.
(3) The scan-to-submap matching strategy is used to solve the data association problem. A local map (submap) is composed by a number of consecutive LiDAR data frames. When a frame is inserted into the corresponding local map, we will estimate the best LiDAR pose with the existing frames in the local map. The estimation is actually to align the current point cloud with the local map to find the optimal matching, which is a nonlinear least squares problem. To solve the problem, the occupancy grid map is continuous with the bicubic interpolation method.
(4) To further improve the map accuracy, a sparse pose adjustment (SPA) algorithm [30] is periodically activated for the back-end optimization. Owing to the merits of the SPA algorithm, for example, it is robust and tolerant to initialization value, with very low failure rates (getting stuck in local minima) for both incremental and batch processing, and the convergent rate is very fast as it requires only a few iterations of the LM method (this is one of the key factors in our application). With the periodic optimization process, the accumulated error can be limited. It can improve the success rate of closed-loop detection in low-texture environments.

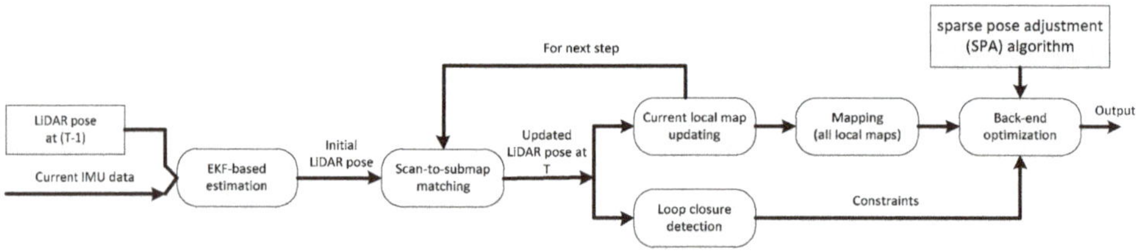


**Figure 1.** The data flow diagram of the proposed method.

### 2.1. Mapping

We use the grid map to represent the environment, which has been first proposed by Elfes and Moravec [31–33] in the 1980s, and further extended by Elfes [34,35] later. Their fundamental works established the theoretical framework of the grid map. As shown in Figure 2, it is a 2D occupancy grid map, and the value of the cell in the map represents the possibility of being occupied with the objects/obstacles and ranges from $-1$ to 1. The value $-1$ means the cell is empty or has available space, and is indicated as white cells in the grid map. The value 1 means the cell is an obstacle or has no available space, and is indicated as black. The value 0 usually represents an unknown area and is indicated as the gray in the map. The map is intuitive, with easy maintenance and an easy introduction to navigation algorithm. How to choose a suitable grid size is empirically determined, which should balance the map accuracy and computational cost. According to the environment scale of the application, the size of the grid is set to 5 cm in this paper.

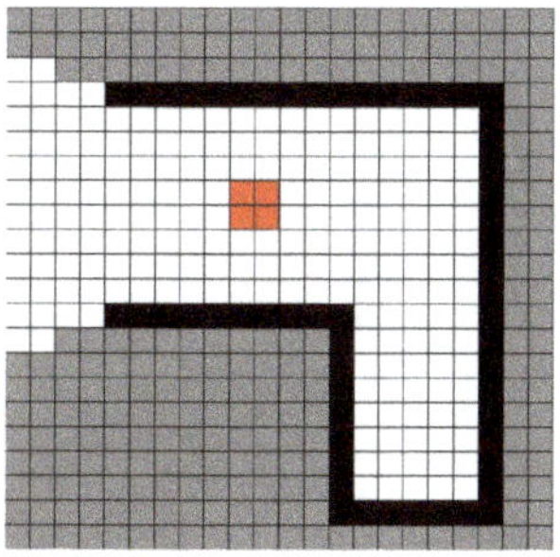

**Figure 2.** 2D grid-based map of occupancy.

The updating model of the grid map can be expressed as follows:

$$M_{new}(\mathrm{p_m}) = M_{old}(\mathrm{p_m}) + lomeas \quad (1)$$

In Equation (1), $M_{new}(\mathrm{p_m})\in[-1,1]$ represents the grid value that has been updated by the point cloud, $M_{old}(\mathrm{p_m})$ represents the previous value that has not been updated, *lomeas* represents the measured value. There are two cases when a grid map is updated by a point cloud. The first is the laser point falls on the grid, and the grid can be regarded as occupied with *lomeas* = 0.3. The second is the laser beam can pass through the grid, and the grid can be regarded as free with *lomeas* = −0.3. Based on Equation (1), a modified updating strategy is used to improve the accuracy of scan matching and closed-loop detection for dynamic environments.

For the first case, the updating strategy can be expressed in Equation (2), it can restrict the updating of the free grid to eliminate the impact of the arrival of the dynamic target.

$$\begin{cases} M_{new}(\mathrm{p_m}) = M_{old}(\mathrm{p_m}) + 0.3 & -0.9 \le M_{old}(\mathrm{p_m}) \le 1 \\ M_{new}(\mathrm{p_m}) = M_{old}(\mathrm{p_m}) & -1 \le M_{old}(\mathrm{p_m}) < -0.9 \end{cases} \quad (2)$$

For the second case, the updating strategy can be expressed in Equation (3). The updating strategy can accelerate the updating process of the occupied grid to eliminate the effect of the departure of the moving target.

$$\begin{cases} M_{new}(\mathrm{p_m}) = M_{old}(\mathrm{p_m}) - 0.3 & -1 \le M_{old}(\mathrm{p_m}) \le 0.9 \\ M_{new}(\mathrm{p_m}) = -0.95 & 0.9 < M_{old}(\mathrm{p_m}) \le 1 \end{cases} \quad (3)$$

Figure 3 is the simulation comparison between the proposed updating strategy and the existing method. In this figure, the upper three from left to right are the grid maps at time T, T+1, and T+2 using the proposed updating strategy. The bottom row is the results from the existing method. The red point in each figure represents the LiDAR, which goes straight along a corridor. Two dynamic targets enter into the LiDAR scanning range from the unknown area. One target is approaching the LiDAR, and the other one crosses the corridor from left to right. We can find that in the map of the existing method, a trajectory of a moving target is left, although this trajectory will slowly become shallower with the continuous updating, but it will affect the accuracy of scan matching, especially in a low-texture environment. On the contrary, in the map generated by the proposed updating strategy, there is no moving target trajectory left, which eliminates the effects of the dynamic target.

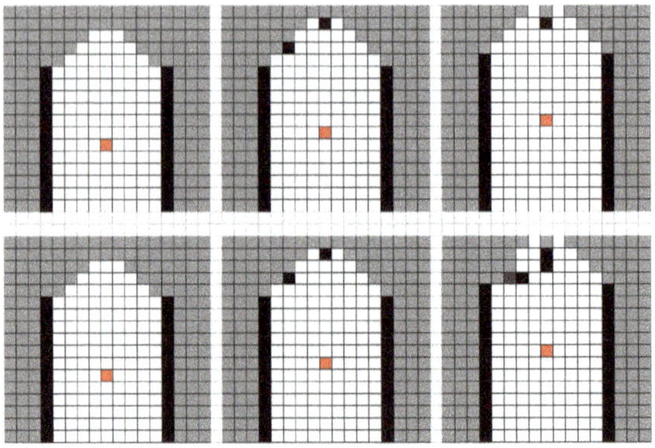

**Figure 3.** Simulation comparisons of map updating between the proposed strategy and the existing one.

The grid map is processed with the bicubic interpolation method for derivative computing. As shown in Figure 4, $P_{\mathrm{m}}$(x,y) is the target point to be estimated, and there are sixteen grid points $P_{ij}(i = 0, 1, \ldots, 3, j = 0, 1, \ldots, 3)$ around $P_m$. The main idea is to calculate the probability of target point $P_{\mathrm{m}}$ through the probability of the 16 grid points with a weighted sum strategy. The weight is expressed with $W(z)$ and defined in Equation (4); it is a bicubic function used here, and $z$ is the distance from the current point to the target $P_m$.

$$W(z) = \begin{cases} (a+2)|z|_3 - (a+3)|z|_2 + 1 & for\ |z| \leq 1 \\ a|z|_3 - 5a|z|_2 + 8a|z| - 4a & for\ 1 < |z| < 2 \\ 0 & otherwise \end{cases} \quad (4)$$

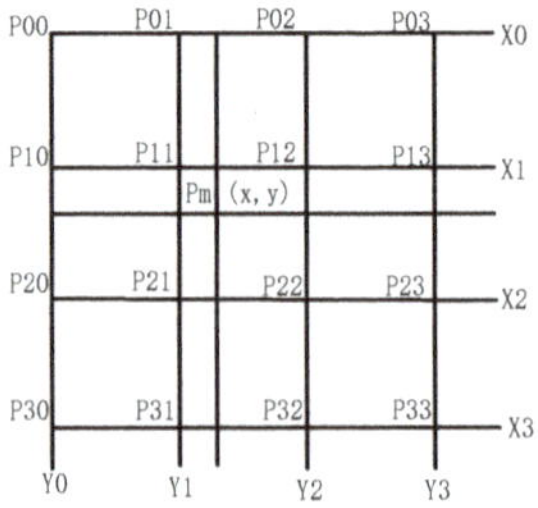


**Figure 4.** The bicubic interpolation of the grid map.

The grid at line $i$ and column $j$ is expressed as:

$$\mathrm{k}_{ij} = W(x - x_i)W(y - y_j) \quad (5)$$

The value of $P_{\mathrm{m}}$ is defined as:

$$M(\mathrm{p_m}) = \sum_{j=0}^{3}\sum_{i=0}^{3} M(\mathrm{p}_{ij}) * \mathrm{k}_{ij} \quad (6)$$

Then we can calculate the derivative of $M(\mathrm{p_m})$ relative to x or y.

### 2.2. Coordinate Transformation

To improve the success rate of closed-loop detection, a local map which is composed of the latest 10 continuous frames is defined. A coordinate $X^R O^R Y^R$ is defined for the local map. The scanned points are represented in LiDAR coordinates and defined as:

$$b^i = (b^i{}_x, b^i{}_y)_{\ i=0,1,\ldots I} \tag{7}$$

The pose of the LiDAR at the time $n$ related to the local map coordinate of $m$ is expressed as:

$$\xi_m{}^n = (\xi^n{}_x, \xi^n{}_y, \xi^n{}_\theta) \tag{8}$$

The pose of the local map $m$ related to the world coordinate system is defined as:

$$\psi^m = (\psi^m{}_x, \psi^m{}_y, \psi^m{}_\theta) \tag{9}$$

Then the transformation from the LiDAR coordinate to the local map coordinate is expressed as:

$$G^i(\xi^n) = \begin{pmatrix} \cos\xi^n{}_\theta & -\sin\xi^n{}_\theta \\ \sin\xi^n{}_\theta & \cos\xi^n{}_\theta \end{pmatrix} \begin{pmatrix} b^i{}_x \\ b^i{}_y \end{pmatrix} + \begin{pmatrix} \xi^n{}_x \\ \xi^n{}_y \end{pmatrix} = \begin{pmatrix} b^i{}_x \\ b^i{}_y \end{pmatrix}_{X^R O^R Y^R} \tag{10}$$

The transformation between the pose of LiDAR from the local map coordinate to the world coordinate system is defined as:

$$L^n(\psi^m) = \begin{pmatrix} \cos\psi^m{}_\theta & -\sin\psi^m{}_\theta & 0 \\ \sin\psi^m{}_\theta & \cos\psi^m{}_\theta & 0 \\ 0 & 0 & 1 \end{pmatrix} \begin{pmatrix} \xi^n{}_x \\ \xi^n{}_y \\ \xi^n{}_\theta \end{pmatrix} + \begin{pmatrix} \psi^m{}_x \\ \psi^m{}_y \\ \psi^m{}_\theta \end{pmatrix} = \begin{pmatrix} \xi^n{}_x \\ \xi^n{}_y \\ \xi^n{}_\theta \end{pmatrix}_{X^W O^W Y^W} = \xi^n{}_{X^W O^W Y^W} \tag{11}$$

### 2.3. EKF-Based Sensor Fusion

The 6DOF pose of the mobile robot can be expressed as $x = [p^T, \Omega^T, v^T]^T$ where $p = (p_x, p_y, p_z)^T$ denotes the position information of the mobile robot, $\Omega = (\phi, \theta, \psi)T$ denotes orientation in roll, pitch, and yaw, $v = (v_x, v_y, v_z)^T$ denotes the velocity. The output information of IMU is expressed as $u = [\omega^T, a^T]^T$, as $\omega = (\omega_x, \omega_y, \omega_z)^T$ is the angular velocity, and $a = (a_x, a_y, a_z)^T$ is the acceleration. Then, the state transition of the 6-DOF motion estimation system can be expressed as:

$$\dot{p} = v \tag{12}$$

$$\dot{\Omega} = E_\Omega \cdot \omega \tag{13}$$

$$\dot{v} = R_\Omega \cdot a + g \tag{14}$$

where $R_\Omega$ denotes cosine matrix vector from mobile robot to global coordinate, $E_\Omega$ transform angle velocity into the derivative of Euler angle derivative, and g is gravity constant. Since a low-cost IMU sensor is used, the effect of the earth rotation is not considered.

The prediction model of EKF is provided by the output of the IMU sensor. In order to alleviate the error caused by the IMU drift, additional preprocessing or correction is necessary. The observation model of the EKF-based system is the pose information of the 3-DOF in the plane provided by the 2D SLAM system. The error of the IMU is corrected using the 3-DOF pose information, and the pose of the 6-DOF is updated. The process of prediction and update of EKF is shown as follows:

$$(P^+)^{-1} = (1-k)\cdot P^{-1} + k\cdot C^T R^{-1} C \tag{15}$$

$$\hat{x}^+ = P^+[(1-k)\cdot P^{-1}\hat{x} + k\cdot C^T R^{-1}\xi^*]^{-1} \tag{16}$$

$$\mathrm{K} = \mathrm{PC}^{\mathrm{T}}\left(\frac{1-k}{k}\cdot\mathrm{R} + \mathrm{C}^{\mathrm{T}}\mathrm{PC}\right)^{-1} \tag{17}$$

$$\mathrm{P}^{+} = \mathrm{P} - (1-k)^{-1}\cdot\mathrm{KCP} \tag{18}$$

$$\hat{\mathrm{x}}^{+} = \hat{\mathrm{x}} + \mathrm{K}(\xi^{*} - \mathrm{C}\hat{\mathrm{x}}) \tag{19}$$

where $\xi^{*}$ represents 3-DOF pose information provided by the 2D SLAM system, $k$ represents weight coefficient of updating pose, the 6-DOF pose information generated by EKF is reduced dimension by projection matrix C, and the result is used as the initial pose of the 2D SLAM system.

### 2.4. Data Association

For the low-texture environment, such as in a corridor, it is possible to get several adjacent frames which are almost the same. If there is no assistance from other information, the scan matching may cause a large error. Therefore, the output of EKF is taken to participate in the scan matching. When a LiDAR frame is inserted into the corresponding local map, we will estimate the best LiDAR pose in the local map based on the previous frames in the local map. The method is to find the LiDAR pose to make sure the total probability of the point cloud in the local map is large enough, a nonlinear least square method can be used.

The goal is to solve $\xi^{n*}$ with the cost function as follows:

$$\xi^{n*} = \underset{\xi}{\operatorname{argmin}} \sum_{i=1}^{I} \left[1 - M(\mathrm{G}^{\mathrm{i}}(\xi^{\mathrm{n}}))\right]^{2} \tag{20}$$

The function $M(.)$ defines the probability value in the local submap. The function $G(.)$ is a coordinate transformation. The above optimization function can be rewritten as:

$$\sum_{i=1}^{I} \left[1 - M(\mathrm{G}^{\mathrm{i}}(\xi^{\mathrm{n}} + \Delta\xi^{\mathrm{n}}))\right]^{2} \to 0 \tag{21}$$

With the first-order Taylor expansion of the above equation, we can get:

$$\sum_{i=1}^{I} \left[1 - M(\mathrm{G}^{\mathrm{i}}(\xi^{\mathrm{n}})) - \nabla M(\mathrm{G}^{\mathrm{i}}(\xi^{\mathrm{n}}))\frac{\partial \mathrm{G}^{\mathrm{i}}(\xi^{\mathrm{n}})}{\partial \xi^{\mathrm{n}}}\Delta\xi^{\mathrm{n}})\right)\right]^{2} \to 0 \tag{22}$$

With the initial pose $\xi^{\mathrm{n}} = \mathrm{C}x$, taking derivative of the above equation with respect to $\Delta\xi^{\mathrm{n}}$, then we have

$$2\sum_{i=1}^{I} \left[\nabla M(\mathrm{G}^{\mathrm{i}}(\xi^{\mathrm{n}}))\frac{\partial \mathrm{G}^{\mathrm{i}}(\xi^{\mathrm{n}})}{\partial \xi^{\mathrm{n}}}\right]^{T} \left[1 - M(\mathrm{G}^{\mathrm{i}}(\xi^{\mathrm{n}})) - \nabla M(\mathrm{G}^{\mathrm{i}}(\xi^{\mathrm{n}}))\frac{\partial \mathrm{G}^{\mathrm{i}}(\xi^{\mathrm{n}})}{\partial \xi^{\mathrm{n}}}\Delta\xi^{\mathrm{n}})\right)\right] = 0 \tag{23}$$

Solving for $\Delta\xi^{\mathrm{n}}$ yields the Levenberg–Marquardt [36] equation for the minimization problem:

$$\Delta\xi^{\mathrm{n}} = \mathrm{H}^{-1}\sum_{i=1}^{I} \left[\nabla M(\mathrm{G}^{\mathrm{i}}(\xi^{\mathrm{n}}))\frac{\partial \mathrm{G}^{\mathrm{i}}(\xi^{\mathrm{n}})}{\partial \xi^{\mathrm{n}}}\right]^{T} \left[1 - M(\mathrm{G}^{\mathrm{i}}(\xi^{\mathrm{n}}))\right] = 0 \tag{24}$$

where

$$\mathrm{H} = \left[\nabla M(\mathrm{G}^{\mathrm{i}}(\xi^{\mathrm{n}}))\frac{\partial \mathrm{G}^{\mathrm{i}}(\xi^{\mathrm{n}})}{\partial \xi^{\mathrm{n}}}\right]^{T} \left[\nabla M(\mathrm{G}^{\mathrm{i}}(\xi^{\mathrm{n}}))\frac{\partial \mathrm{G}^{\mathrm{i}}(\xi^{\mathrm{n}})}{\partial \xi^{\mathrm{n}}}\right] + \lambda \tag{25}$$

Because LM optimization introduces a damping coefficient in the Gauss–Newton method here, it is used to control the step to prevent divergence. We can tune the dynamic damping coefficient by

LM algorithm. In each iteration step, the error caused by new change will be monitored. If the new error is smaller than before, $\lambda$ will decrease in the next iteration, otherwise, it will increase.

*2.5. Closed-Loop Detection and Back-End Optimization*

Though a scan-to-map matching strategy is used which can limit the accumulated error, with the continuous movement of the robot, the error will still continue to accumulate. Moreover, in order to reduce the error of closed-loop detection in the low-texture environment, the local map is applied to eliminate the accumulated error. The sparse pose adjustment (SPA) [30] method for back-end optimization is used. When a new frame is added to the local map, a 3D search window along the direction of $x$, $y$, and $\theta$ around the estimated LiDAR pose is defined first. Then, searching the window step by step is carried out in the local map by calculating the total probability of all points in the frame. If the total probability exceeds the threshold, then it is regarded as a successful closed-loop detection. The relative pose between the LiDAR and the corresponding local map is used as a constraint to optimize pose. When a constraint is added, the pose in the trajectory of LiDAR and all local maps are optimized.

The optimization is formulated as a nonlinear least square problem; we can add constraints to it at any moment.

$$\underset{\Psi,\Omega}{\operatorname{argmin}} \frac{1}{2}\sum_{mn}\rho(E^2(\psi^{\mathrm{m}},\xi^{\mathrm{n}}{}_{\mathrm{X^W O^W Y^W}};\Sigma_{mn},\xi_m{}^{\mathrm{n}})) \tag{26}$$

where

$$E^2(\psi^{\mathrm{m}},\xi^{\mathrm{n}}{}_{\mathrm{X^W O^W Y^W}};\Sigma_{mn},\xi_m{}^{\mathrm{n}}) = e(\psi^{\mathrm{m}},\xi^{\mathrm{n}}{}_{\mathrm{X^W O^W Y^W}};\xi_m{}^{\mathrm{n}})^T\Sigma_{mn}^{-1}e(\psi^{\mathrm{m}},\xi^{\mathrm{n}}{}_{\mathrm{X^W O^W Y^W}};\xi_m{}^{\mathrm{n}}) \tag{27}$$

$$\Psi = \left\{\psi^1,\ldots,\psi^{\mathrm{m}},\ldots,\psi^{\mathrm{M}}\right\} \tag{28}$$

$$\Omega = \left\{\xi^1{}_{\mathrm{X^W O^W Y^W}},\ldots,\xi^{\mathrm{n}}{}_{\mathrm{X^W O^W Y^W}},\ldots,\xi^{\mathrm{N}}{}_{\mathrm{X^W O^W Y^W}}\right\} \tag{29}$$

$$e(\psi^{\mathrm{m}},\xi^{\mathrm{n}}{}_{\mathrm{X^W O^W Y^W}};\xi_m{}^{\mathrm{n}}) = \xi_m{}^{\mathrm{n}} - \begin{pmatrix} R_{\psi^{\mathrm{m}}}^{-1}(t_{\psi^{\mathrm{m}}} - t_{\xi^{\mathrm{n}}{}_{\mathrm{X^W O^W Y^W}}}) \\ \psi^{\mathrm{m}}{}_{\theta} - \xi^{\mathrm{n}}{}_{\theta\mathrm{X^W O^W Y^W}} \end{pmatrix} \tag{30}$$

These constraints have the form of relative poses $\xi_{\mathrm{m}}{}^{\mathrm{n}}$ and associated covariance matrices $\Sigma_{\mathrm{mn}}{}^{-1}$ which can be estimated by [37]. The final step is to estimate the local map pose $\Psi$ set and LiDAR pose set $\Omega$. Considering the low-texture environment may give rise to very similar scanning data, which may lead to a mismatch and result in a large error, Huber loss function is introduced to reduce the influence of outliers in the quadratic term of the objective function.

$$\rho_\delta(a) = \begin{cases} \frac{1}{2}a^2 & for\ |a| \leq \delta \\ \delta(\left|a\right| - \frac{1}{2}\delta) & otherwise \end{cases} \tag{31}$$

## 3. Experiments

*3.1. The Platform*

The mobile robot platform used in the experiments is shown in Figure 5. Though the embedded controller can get the IMU and odometry data simultaneously, only IMU data is currently considered in the experiment. A 2D LiDAR sensor UTM-30LX-EW was used [38]. The IMU used in the experiments is an ADIS16365 from Analog Device, which contains a three-axis gyroscope and a three-axis accelerometer. Each sensor has its own dynamic compensation equation, and can provide accurate measurement. The sampling frequency of IMU is 500 HZ. We ran the algorithm under ubuntu14.04 with ROS Indigo [39], with a processor Intel (R) Core (TM) i7-6700HQ @ 2.60 GHZ and 8 GB memory.

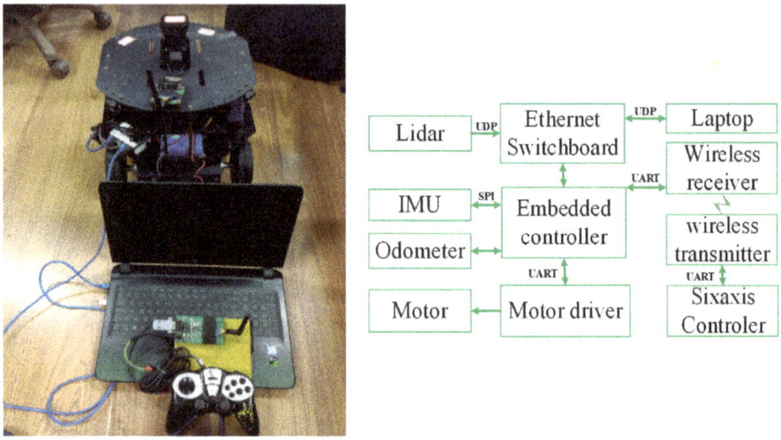


**Figure 5.** The mobile robot used for the experiment and its architecture. (**a**) The mobile robot used for the experiments; (**b**) The architecture of the mobile robot.

To verify the performance of the proposed method, we conducted some quantitative and qualitative experiments. For the quantitative comparison, because of the limitation of ground truth data, we conducted the experiments in the constrained environment, where we could measure the length with an accurate handheld laser range finder. The comparison was made among the Cartographer (LiDAR-only) [40] and Cartographer with IMU fusion-based method, and the proposed fusion-based method, which is discussed in Section 3.2.

There are some state-of-the-art 2D LiDAR-based SLAM methods, such as Hector SLAM [7], GMapping [10], KartoSLAM [41], CoreSLAM [42], LagoSLAM [43], and Cartographer [40], which have been proposed, and some papers on the performance evaluation are published [44–47]. As mentioned in reference [44], among the former five methods (Cartographer was proposed later), KartoSLAM, GMapping, and HectorSLAM show better performance than the other two methods. Both KartoSLAM and LagoSLAM are graph-based optimizations, but the computation load of LagoSLAM is higher than that of KartoSLAM. In particular, KartoSLAM showed the best performance in the real world because the SPA solver is employed and it is a full SLAM approach. In this paper, we focus on the accuracy of the map and real-time application, so the methods including KartoSLAM and Cartographer are used for the evaluation. Since Cartographer is the most state-of-the-art method, we take it as the baseline method in this paper. The experiment results will be introduced in Section 3.3.

The dataset used in Section 3.3 are from reference [48–50]. The performance of the algorithms can be evaluated by the accuracy of the map, which is usually defined by the distance between the obtained map and ground truth map or the accuracy of the robot trajectory, but it is usually difficult to generate the ground truth. Additionally, many open standard datasets were recorded with only one sensor. All these factors make the comparison a little hard. In this paper, we apply the proportion of occupied and free cells to the quantitative comparison, which has been proposed in reference [44] for quality evaluation. As shown in Figure 6, the perceptible feature of the picture of a map is the accuracy of the walls or edges. That is, for two maps which are similar in most of the areas, the more blurs in a map means the lower its quality. It is obviously straightforward and no ground truth is needed.

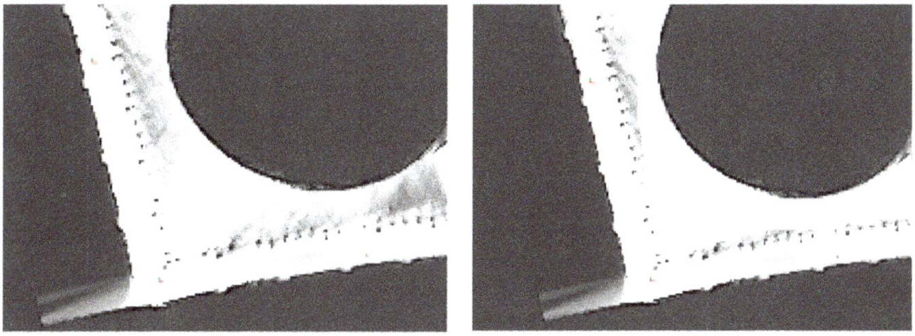

(**a**) (**b**)

**Figure 6.** The edge and wall representation in a grid map (the more blurriness in a map, the lower quality it is). (**a**) A submap with more blurries (lower quality); (**b**) A submap with less blurries (higher quality).

### 3.2. Quantitative Evaluation

In order to verify the performance of the proposed method in the low-texture environment with dynamic target interference, we elaborately selected two low-texture scenes: the first one was an open-loop scene, the other one was a closed-loop scene. The open-loop scene was a classroom corridor in our university. As shown in Figure 7, the corridor was 35.3 m long and 2.8 m wide. The closed-loop experiment was carried out in a looped classroom corridor 65 m long and 3.9 m wide, which is presented in Figure 8.

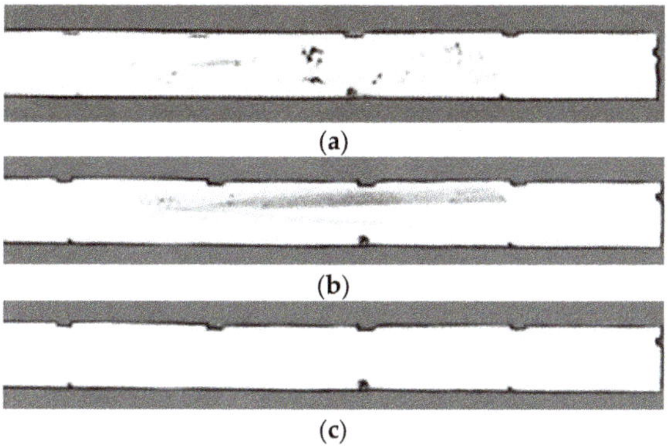

**Figure 7.** Maps built in open-loop scene (inside the corridor area, the darker the area was, the more blurry it was, that is, the lower quality it will be). (**a**) A map built with Cartographer with LiDAR only; (**b**) a map built with Cartographer with LiDAR+IMU fusion; (**c**) a map built with the proposed method.

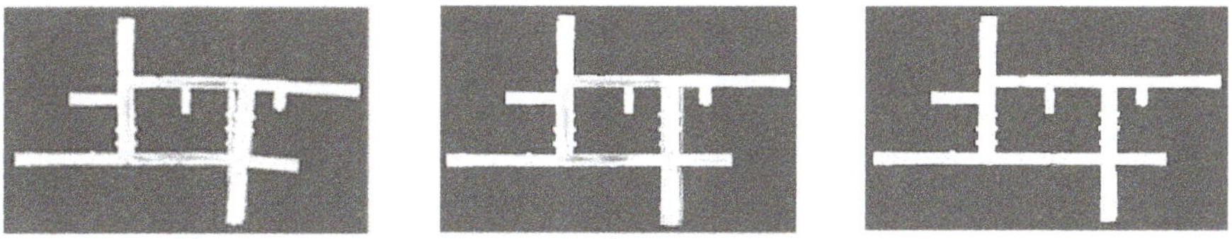

(**a**) (**b**) (**c**)

**Figure 8.** Evaluation experiments in closed-loop scene. (**a**) A map created by Cartographer with LiDAR only; (**b**) a map created by Cartographer with LiDAR+IMU fusion; (**c**) a map created by the proposed method.

We operated the mobile robot along a predefined trajectory to build the map. At the same time, some students moved around the robot elaborately to simulate the interference of moving obstacles. We carried out three kinds of experiments in the open-loop and closed-loop environment.

Figure 7 shows the mapping results with the open-loop scene. The first map was built by Cartographer with LiDAR only [40]. The second map was built by Cartographer with LiDAR+IMU fusion. The third map was built by the proposed method. The results of the open-loop scenes are shown in Figure 7. The results for the closed-loop scene are shown in Figure 8.

As shown in Figure 7, the experiments in the open-loop scene show that though the map built with LiDAR and IMU is improved, because of the existence of the moving obstacles, there are still many noise data in the final map. However, with the proposed method, these noise data are removed. In order to compare the accuracy of three maps for the open-loop scene, we measured the length of the same segment in the corridor, and the estimated lengths with three methods are shown in Table 1. We can see from Figure 7 and Table 1 that the accuracy and stability of the map are greatly improved with the proposed method.

The experiments for the closed-loop scene in Figure 8 show the performances of the three methods (Cartographer without IMU, Cartographer with IMU, and the proposed). In Figure 8a, because the low texture of the environment leads to large error drifting in the front-end stage, after the optimization in the back-end stage, the result map still has inconsistencies. There are also many blurs inside the corridor caused by the moving persons during the experiments. Because the LiDAR-only method cannot recognize the moving objects, it takes all the data for scan matching, then the points located on the moving objects are noise data, which lead to the DA error and map updating error too. Comparatively, as shown in Figure 8b, by fusing with IMU, the fusion-based Cartographer method improves the DA error greatly, but it still cannot remove the moving object effectively, which leads to the blurs inside the corridor of the map. For the proposed method, benefiting from the prediction of IMU and SPA algorithm in the front-end stage, the closed-loop optimization, and large error removing in the back-end stage, the proposed method outperformed the others, as illustrated in Figure 8c. In order to make a quantitative comparison in this case, we selected six segments with different lengths in the closed-loop scene, which are marked as segments A to F, as shown in Figure 9, and the estimated results with three methods are shown in Table 2.

**Table 1.** Error Comparison in Open Loop.

| Real Length of the Corridor (m) | Length Obtained by LiDAR-Only Method | Relative Error | Length Obtained by LiDAR & IMU Fusion Method | Relative Error | Length Obtained by the Proposed Method | Relative Error |
|---|---|---|---|---|---|---|
| 28.9 | 26.5 | −8.3% | 27.1 | −6.2% | 27.8 | −3.8% |

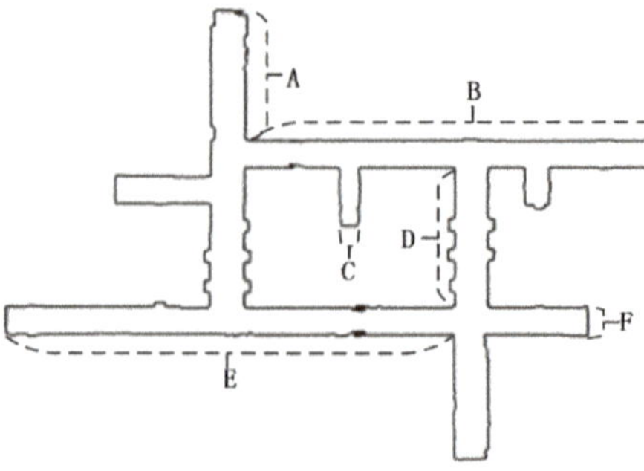


**Figure 9.** The selected segments in closed-loop scene.

We can see the accuracy is improved by the proposed method in Table 2. This trend will be more obvious with longer segments, as for the segments B and E. We also found that the lengths of the segments estimated by Cartographer with the LiDAR-only method are often shorter than the actual lengths in both open-loop and closed-loop scenes, which is caused by the similarity of two adjacent frames in the low-texture environment. That means that because there are not many differences between two frames in a low-texture environment, for the optimization process, the LiDAR-only method does the scan matching around the previous pose, and it tends to converge earlier to the error tolerance. However, the fusion-based method does the scan matching with the initial pose estimated by the fusion of IMU and the previous pose, and the result is closer to the actual value. The gap between the two kinds of methods (with IMU or not) will be greater if a low-frequency output of LiDAR is used.

**Table 2.** Error Comparison in Closed Loop.

| Segment | Actual Value | LiDAR | Relative Error (LiDAR) | LiDAR & IMU | Relative Error (LiDAR & IMU) | LiDAR & IMU & Static Map/ Proposed Method | Relative Error (LiDAR & IMU & Static Map) |
|---|---|---|---|---|---|---|---|
| A | 11.0 | 10.4 | −5.5% | 10.8 | −1.8% | 11.1 | +0.9% |
| B | 42.8 | 38.3 | −10.5% | 42.9 | +0.2% | 42.9 | +0.2% |
| C | 1.8 | 1.6 | −11.1% | 1.7 | −5.6% | 1.7 | −5.6% |
| D | 12.4 | 12 | −3.2% | 11.5 | −7.3% | 12.1 | −2.4% |
| E | 42.8 | 38.3 | −10.5% | 40 | −6.5% | 43.1 | +0.7% |
| F | 2.4 | 2.6 | +8.3% | 2.5 | +4.2% | 2.6 | +8.3% |

### 3.3. *Qualitive Evaluation*

In this section, some state-of-the-art 2D LiDAR-based SLAM, such as HectorSLAM and Cartographer, are evaluated qualitatively with the proposed method.

Figure 10 shows the maps of Cartographer (LiDAR-only) and the proposed fusion methods in the outdoor environment, which were conducted at the outdoor square road behind the Siyuan building in the university. The left is the map with LiDAR-only used, the right is obtained with the fusion-based method. Figure 11 shows the maps with Cartographer; the experiments were conducted on the first floor of teaching buildings No. 9 and No. 1. There were some persons passing by during the experiment process, which led to the blurs in the result map. As aforementioned in Figure 6, many blurs remain in the map with LiDAR-only; comparatively, the map generated by the proposed method is sharper and clearer.

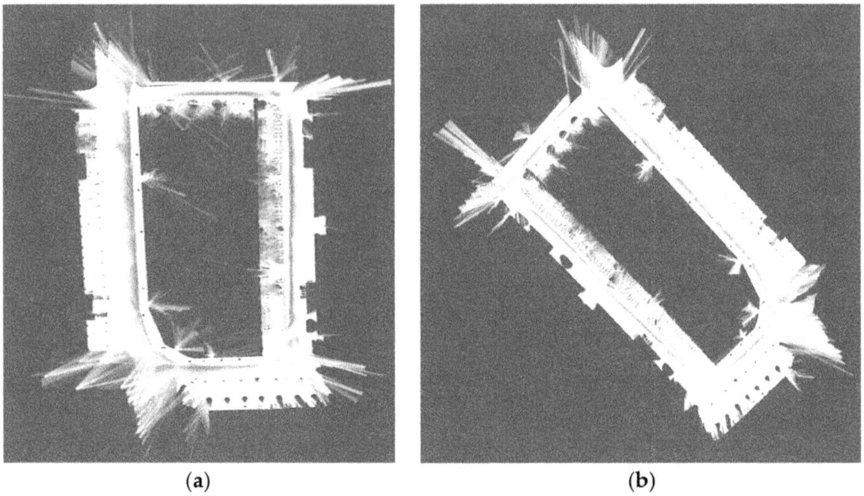

Figure 10. Closed-loop experiment in outdoor environment. (a) Cartographer (LiDAR only); (b) the proposed method.

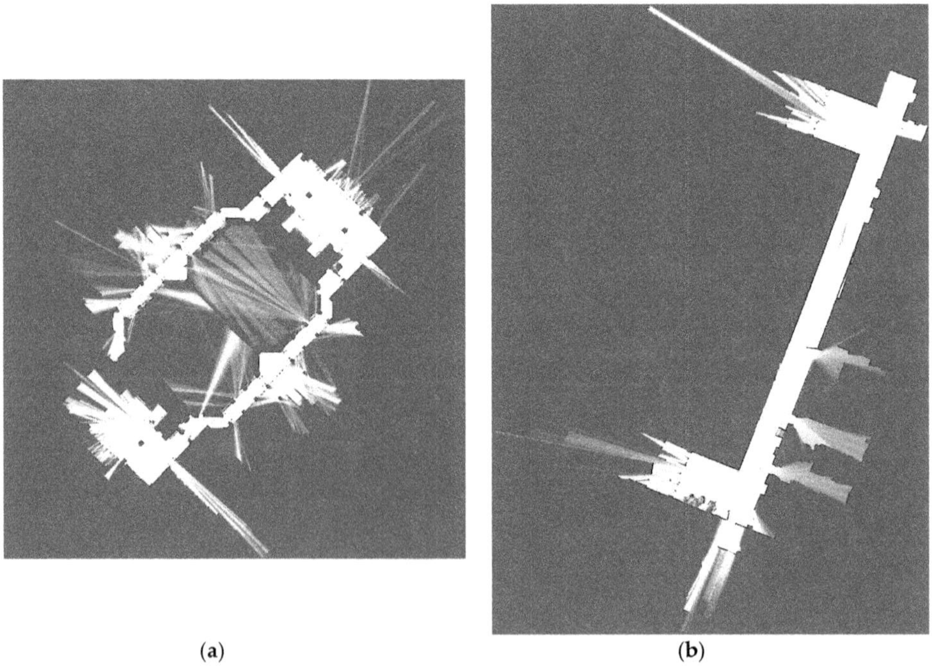

Figure 11. The map generated with the proposed method in indoor environment. (a) mapping result of first floor of teaching building No. 9; it is a closed-loop classroom corridor, but we deliberately arranged to not use the last section data, and the mapping result is acceptable. (b) mapping results of the first floor of teaching building No.1.

We used some open datasets [48] to evaluate the proposed fusion-based method, and the results are presented in Figures 12–14. In all these three figures, the left one shows the results with Cartographer of LiDAR only, and the right one shows the results with the proposed method. In Figure 12 the left map shows larger accumulated error indicated by large ellipse. In Figure 12, there is more blurriness in the map generated by Cartographer of LiDAR only. In Figure 13, the map generated by the proposed method is sharp with less blurriness.

Figure 15 shows the experiment results obtained by HectorSLAM, Cartographer (LiDAR-only), and the proposed method with "Team_Hector_MappingBox_Dagstuhl_Neubau.bag" (first raw in the figure) and "Team_Hector_MappingBox_L101_Building.bag" (second raw in the figure) respectively; other datasets obtained the same results. In these experiments, the map with HectorSLAM seems not convergent and leads to inconsistencies. On the other hand, the maps built by Cartographer (LiDAR only) (Figure 15e) and the proposed method (Figure 15f) show not much difference in this case. This is because the environment is not a low-texture one.

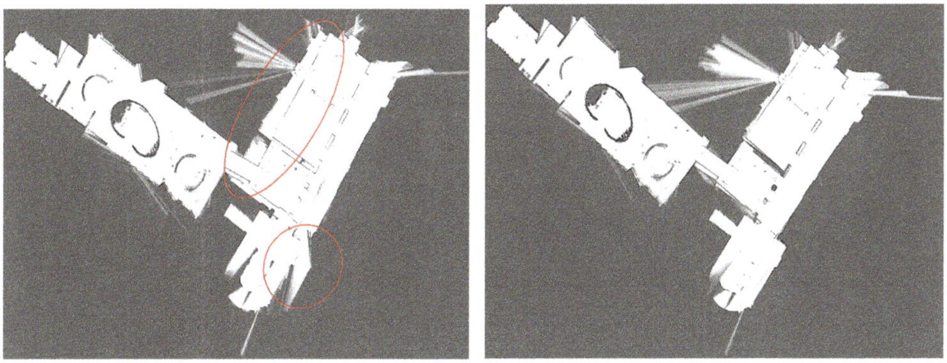

(a) (b)

**Figure 12.** Map built by Cartographer (LiDAR only) (**a**) and the proposed method (**b**) with file b0-2014-07-21-12-42-53.bag [48].

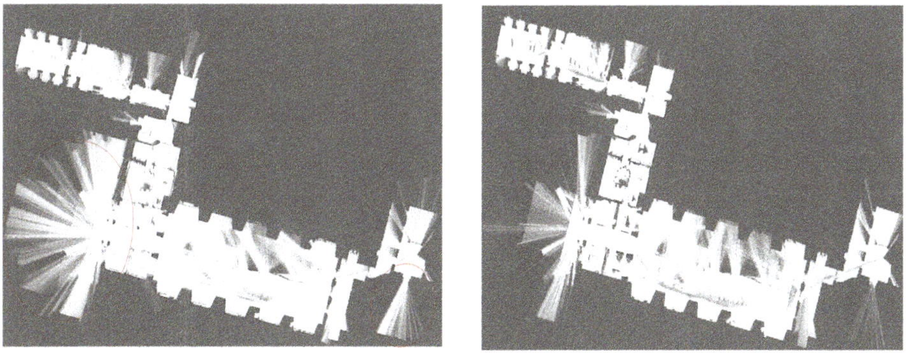

(a) (b)

**Figure 13.** Map built by Cartographer (LiDAR only) (**a**) and the proposed method (**b**) with file b0-2014-10-07-12-43-25.bag [48].

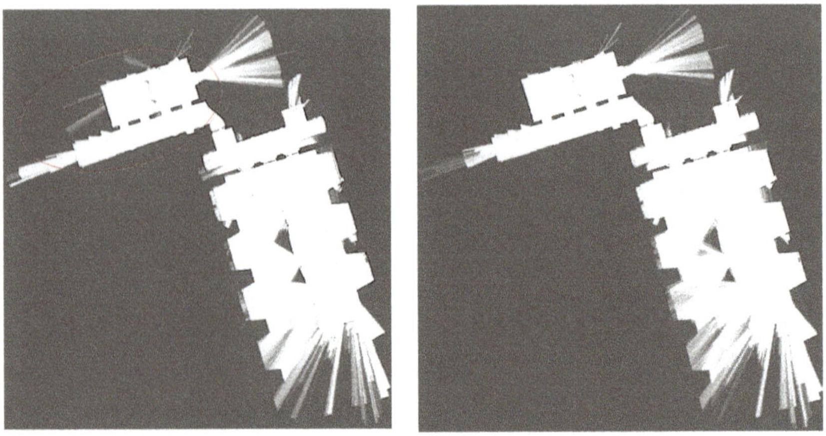

(a) (b)

**Figure 14.** Map built by Cartographer (LiDAR only) (**a**) and the proposed method (**b**) with file b0-2014-08-14-13-36-48.bag [48].

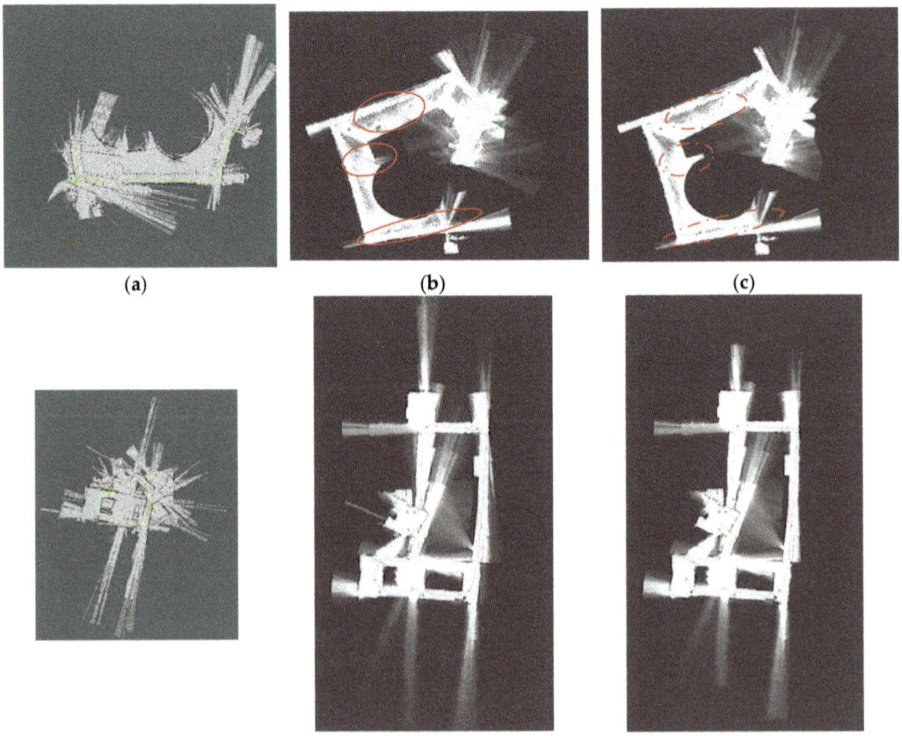

(a) (b) (c)

(d) (e) (f)

**Figure 15.** Evaluation results with HectorSLAM, Cartographer (LiDAR only) and the proposed method. The first raw is the generated maps with "Team_Hector_MappingBox_Dagstuhl_Neubau.bag" file. They are (**a**) HectorSLAM; (**b**) Cartographer (LiDAR only); (**c**) the proposed method respectively. The second raw is the generated maps with "Team_Hector_MappingBox_L101_Building.bag" file. They are (**d**) HectorSLAM; (**e**) Cartographer (LiDAR only); (**f**) the proposed method respectively.

## 4. Conclusions

In this paper, a mapping strategy is firstly exploited to improve the performance of the 2D SLAM in dynamic and low-texture environments, and a fusion method which combines the IMU sensor with a 2D LiDAR is used. The initial motion estimation obtained from the output of EKF is used as the initial pose for the scan matching. Some measurements are taken to remove the noise or large error data during the matching process. The optimization with sparse pose adjustment (SPA) method is adopted at the back-end stage. With the cooperation of both the front-end and back-end, the accuracy and stability of SLAM in low-texture and dynamic environments are greatly improved. The experiments with both open-loop and closed-loop scenes have been conducted, which demonstrated the performance of the proposed method. The evaluation experiments with some state-of-the-art methods show the improvement of the proposed method.

Though there are many excellent works on evaluation of the 2D LiDAR SLAM, such as the references [44–47], to name a few, and some open datasets for the evaluation are presented, we found that, because the ground truth is limited, and many datasets recorded only one sensor, some differences of the input data format make the dataset not able to run on all state-of-the-art methods. All these factors make the evaluation still a challenging work. The next works include the evaluation framework for the state-of-the-art methods and the dataset with ground truth, and making the proposed method run on an embedded hardware platform and applying it to real applications.

**Author Contributions:** The work presented here was carried out in collaboration with all authors. Z.W. made the research theme, designed the methods, and wrote the paper. Y.C., Y.M. and K.Y. carried out the experiments, analyzed the data, interpreted the results, and prepared the draft. B.C. worked on reviewing the manuscript.

**Funding:** This work was supported in part by the Natural Science Foundation of China under Grant No.61573057, partly by the Fundamental Research Funds of BJTU (2017JBZ002), and partly by National Natural Science Foundation of China under Grant 61702032.

**Conflicts of Interest:** The authors declare no conflict of interest.

## References

1. Valiente, D.; Payá, L.; Jiménez, L.M.; Sebastián, J.M.; Reinoso, Ó. Visual information fusion through Bayesian inference for adaptive probability-oriented feature matching. *Sensors* **2018**, *18*, 2041. [CrossRef] [PubMed]
2. Meng, X.; Gao, W.; Hu, Z. Dense RGB-D SLAM with multiple cameras. *Sensors* **2018**, *18*, 2118. [CrossRef] [PubMed]
3. Christopher, G.; Matthew, D.; Christopher, R.H.; Daniel, W.C. Enabling off-road autonomous navigation-simulation of LIDAR in dense vegetation. *Electronics* **2018**, *7*, 154.
4. Qian, C.; Liu, H.; Tang, J.; Chen, Y.; Kaartinen, H.; Kukko, A.; Zhu, L.; Liang, X.; Chen, L.; Hyyppä, J. An integrated GNSS/INS/LiDAR-SLAM positioning method for highly accurate forest stem mapping. *Remote Sens.* **2016**, *9*, 3. [CrossRef]
5. Olson, E. M3RSM: Many-to-many multi-resolution scan matching. In Proceedings of the IEEE International Conference on Robotics and Automation (ICRA), Seattle, WA, USA, 26–30 May 2015.
6. Martin, F.; Triebel, R.; Moreno, L.; Siegwart, R. Two different tools for three-dimensional mapping: DE-based scan matching and feature-based loop detection. *Robotica* **2014**, *32*, 19–41. [CrossRef]
7. Kohlbrecher, S.; Meyer, J.; von Stryk, O.; Klingauf, U. A flexible and scalable SLAM system with full 3D motion estimation. In Proceedings of the IEEE International Symposium on Safety, Security and Rescue Robotics (SSRR), Kyoto, Japan, 1–5 November 2011.
8. Ratter, A.; Sammut, C. Local Map Based Graph SLAM with Hierarchical Loop Closure and Optimization. In Proceedings of the Australasian Conference on Robotics and Automation (ACRA), Canberra, Australia, 2–4 December 2015.
9. Zhang, Z. Iterative point matching for registration of free-form curves and surfaces. *Int. J. Comput. Vis.* **1994**, *13*, 119–152. [CrossRef]
10. Grisetti, G.; Stachniss, C.; Burgard, W. Improved techniques for grid mapping with rao-blackwellized particle filters. *IEEE Trans. Robot.* **2007**, *23*, 34–46. [CrossRef]

11. Grisetti, G.; Kummerle, R.; Stachniss, C.; Burgard, W. A Tutorial on Graph-Based SLAM. *Intell. Transp. Syst. Mag. IEEE* **2010**, *2*, 31–43. [CrossRef]
12. Latif, Y.; Cadena, C.; Neira, J. Robust loop closing over time for pose graph SLAM. *Int. J. Robot. Res.* **2013**, *32*, 1611–1626. [CrossRef]
13. Lourakis, M.; Argyros, A. SBA: A software package for generic sparse bundle adjustment. *ACM Trans. Mathe. Softw. (TOMS)* **2009**, *36*, 2. [CrossRef]
14. Kummerle, R.; Grisetti, G.; Strasdat, H.; Konolige, K.; Burgard, W. g2o: A general framework for graph optimization. In Proceedings of the IEEE International Conference on Robotics and Automation, Shanghai, China, 9–13 May 2011; pp. 3607–3613.
15. Bosse, M.; Zlot, R. Map matching and data association for large-scale two dimensional laser scan-based slam. *Int. J. Robot. Res.* **2008**, *27*, 667–691. [CrossRef]
16. Ratter, A.; Sammut, C.; McGill, M. GPU accelerated graph SLAM and occupancy voxel based ICP for encoder-free mobile robots. In Proceedings of the International Conference on Intelligent Robots and Systems, Tokyo, Japan, 3–7 November 2013; pp. 540–547.
17. Olson, E.B. Real-time correlative scan matching. In Proceedings of the IEEE International Conference on Robotics and Automation (ICRA'09), Kobe, Japan, 12–17 May 2009; pp. 4387–4393.
18. Doucet, A.; de Freitas, J.; Murphy, K.; Russel, S. Rao-Blackwellized particle filtering for dynamic Bayesian networks. In Proceedings of the 16th Conference on Uncertainty in Artificial Intelligence, Stanford, CA, USA, 30 June–3 July 2000; pp. 176–183.
19. Murphy, K. Bayesian map learning in dynamic environments. In Proceedings of the 12th International Conference on Neural Information Processing Systems, Denver, CO, USA, 29 November–4 December 1999; pp. 1015–1021.
20. Baglietto, M.; Sgorbissa, A.; Verda, D.; Zaccaria, R. Human navigation and mapping with a 6DOF IMU and a laser scanner. *Robot. Autonom. Syst.* **2011**, *59*, 1060–1069. [CrossRef]
21. Tang, J.; Chen, Y.; Kukko, A.; Kaartinen, H.; Jaakkola, A.; Khoramshahi, E.; Hakala, T.; Hyyppä, J.; Holopainen, M.; Hyyppä, H. SLAM-Aided Stem Mapping for Forest Inventory with Small-Footprint Mobile LiDAR. *Forests* **2015**, *6*, 4588–4606. [CrossRef]
22. Hesch, J.A.; Mirzaei, F.M.; Mariottini, G.L.; Roumeliotis, S.I. A Laser-Aided Inertial Navigation System (L-INS) for Human Localization in Unknown Indoor Environments. In Proceedings of the IEEE International Conference on Robotics and Automation, Anchorage, AK, USA, 3–7 May 2010; pp. 5376–5382.
23. Holz, D.; Lorken, C.; Surmann, H. Continuous 3D sensing for navigation and SLAM in cluttered and dynamic environments. In Proceedings of the International Conference on Information Fusion, Cologne, Germany, 30 June–3 July 2008; pp. 1–7.
24. Li, L.; Yao, J.; Xie, R.; Tu, J.; Feng, C. Laser-Based Slam with Efficient Occupancy Likelihood Map Learning for Dynamic Indoor Scenes. *ISPRS Ann. Photogramm. Remote Sens. Spat. Inf.* **2016**, *III-4*, 119–126. [CrossRef]
25. Montemerlo, M.; Thrun, S.; Whittaker, W. Conditional particle filters for simultaneous mobile robot localization and people-tracking. In Proceedings of the IEEE International Conference on Robotics & Automation, Washington, DC, USA, 11–15 May 2002; Volume 1, pp. 695–701.
26. Hahnel, D.; Schulz, D.; Burgard, W. Map building with mobile robots in populated environments. In Proceedings of the IEEE/RSJ International Conference on Intelligent Robots & Systems, Lausanne, Switzerland, 30 September–4 October 2002; Volume 1, pp. 496–501.
27. Avots, D.; Lim, E.; Thibaux, R.; Thrun, S. A probabilistic technique for simultaneous localization and door state estimation with mobile robots in dynamic environments. In Proceedings of the IEEE/RSJ International Conference on Intelligent Robots & Systems, Lausanne, Switzerland, 30 September–4 October 2002; Volume 1, pp. 521–526.
28. Duckett, T. Dynamic maps for long-term operation of mobile service robots. In Proceedings of the Robotics: Science and Systems, Massachusetts Institute of Technology, Cambridge, MA, USA, 8–11 June 2005; pp. 17–24.
29. Wang, C.C.; Thorpe, C.; Thrun, S. Online Simultaneous Localization and Mapping with Detection and Tracking of Moving Objects: Theory and Results from a Ground Vehicle in Crowded Urban Areas. In Proceedings of the IEEE International Conference on Robotics & Automation, Taipei, Taiwan, 14–19 September 2003; Volume 1, pp. 842–849.
30. Konolige, K.; Grisetti, G.; Kummerle, R.; Burgard, W.; Limketkai, B.; Vincent, R. Sparse pose adjustment for 2D mapping. In Proceedings of the IROS, Taipei, Taiwan, 18 October 2010.

31. Moravec, H.P.; Elfes, A. High resolution maps from wide angle sonar. In Proceedings of the 1985 IEEE International Conference on Robotics and Automation, St. Louis, MO, USA, 25–28 March 1985; IEEE Comput. Soc. Press: Silver Spring, MD, USA, 1985; pp. 116–121.
32. Elfes, A. Sonar-based real-world mapping and navigation. *IEEE J. Robot. Autom.* **1987**, *3*, 249–265. [CrossRef]
33. Moravec, H.P. Sensor fusion in certainty grids for mobile robots. *Ai Mag.* **1988**, *9*, 14.
34. Elfes, A. Occupancy Grids: A Stochastic Spatial Representation for Active Robot Perception. In Proceedings of the Sixth Conference Annual Conference on Uncertainty in Artificial Intelligence, Cambridge, MA, USA, 27–29 July 1990; Morgan Kaufmann: San Francisco, CA, USA, 1990; pp. 136–146.
35. Elfes, A. Dynamic control of robot perception using multi-property inference grids. In Proceedings of the 1992 IEEE International Conference on Robotics and Automation, Nice, France, 12–14 May 1992; IEEE Comput. Soc. Press: Los Alamitos, CA, USA, 1992; pp. 2561–2567.
36. Moré, J.J. The Levenberg-Marquardt algorithm: Implementation and theory. *Numer. Anal.* **1978**, *630*, 105–116.
37. Olson, E.; Leonard, J.; Teller, S. Fast iterative alignment of pose graphs with poor initial estimates. In Proceedings of the IEEE International Conference on Robotics and Automation, Orlando, FL, USA, 15–19 May 2006.
38. HOKUYO. Available online: https://www.hokuyo-aut.jp/search/single.php?serial=170 (accessed on 29 October 2018).
39. Indigo. Available online: http://wiki.ros.org/indigo (accessed on 29 October 2018).
40. Hess, W.; Kohler, D.; Rapp, H.; Andor, D. Real-time loop closure in 2D LIDAR SLAM. In Proceedings of the IEEE International Conference on Robotics & Automation, Stockholm, Sweden, 16–21 May 2016.
41. Vincent, R.; Limketkai, B.; Eriksen, M. Comparison of indoor robot localization techniques in the absence of GPS. In Proceedings of the SPIE: Detection and Sensing of Mines, Explosive Objects, and Obscured Targets XV of Defense, Security, and Sensing Symposium, Orlando, FL, USA, 5–9 April 2010.
42. Steux, B.; El Hamzaoui, O. tinySLAM: A SLAM algorithm in less than 200 lines C-language program. In Proceedings of the International Conference on Control Automation Robotics & Vision (ICARCV), Singapore, 7–10 December 2010.
43. Carlone, L.; Aragues, R.; Castellanos, J.A.; Bona, B. A linear approximation for graph-based simultaneous localization and mapping. In Proceedings of the International Conference Robotics: Science and Systems, Los Angeles, CA, USA, 27–30 June 2011.
44. Filatov, A.; Filatov, A.; Krinkin, K.; Chen, B.; Molodan, D. 2D SLAM Quality Evaluation Methods. Available online: https://arxiv.org/pdf/1708.02354.pdf (accessed on 26 November 2018).
45. Santos, J.M.; Portugal, D.; Rocha, R.P. An evaluation of 2D SLAM techniques available in Robot Operating System. In Proceedings of the 2013 IEEE International Symposium on Safety, Security, and Rescue Robotics (SSRR), Linkoping, Sweden, 21–26 October 2013.
46. Rainer, K.; Steder, B.; Dornhege, C.; Ruhnke, M.; Grisetti, G.; Stachniss, C.; Kleiner, A. On measuring the accuracy of SLAM algorithms. *Auton. Robots* **2009**, *27*, 387–407.
47. Markus, K.; Rainer, K.; Martin, F.; Stefan, M. Benchmarking the Pose Accuracy of Different SLAM Approaches for Rescue Robotics. Available online: https://www.ultrakoch.org/Work/Publications/arc15.pdf (accessed on 28 November 2018).
48. Cartographer ROS. Available online: https://google-cartographer-ros.readthedocs.io/en/latest/data.html# (accessed on 10 November 2018).
49. tu-darmstadt-ros-pkg. Available online: http://code.google.com/p/tu-darmstadt-ros-pkg/downloads/list (accessed on 25 November 2018).
50. slam benchmarking. Available online: http://ais.informatik.uni-freiburg.de/slamevaluation/datasets.php (accessed on 25 November 2018).

applied sciences

MDPI

Article

# MIM_SLAM: A Multi-Level ICP Matching Method for Mobile Robot in Large-Scale and Sparse Scenes

**Jingchuan Wang [1,2,*], Ming Zhao [1,2] and Weidong Chen [1,2]**

1 Department of Automation, Shanghai Jiao Tong University, Shanghai 200240, China; mzhao1993@sjtu.edu.cn (M.Z.); wdchen@sjtu.edu.cn (W.C.)
2 Laboratory of System Control and Information Processing, Ministry of Education of China, Shanghai 200240, China
* Correspondence: jchwang@sjtu.edu.cn; Tel: +86-21-3420-4513

Received: 19 October 2018; Accepted: 27 November 2018; Published: 30 November 2018

**Abstract:** In large-scale and sparse scenes, such as farmland, orchards, mines, and substations, 3D simultaneous localization and mapping are challenging matters that need to address issues such as maintaining reliable data association for scarce environmental information and reducing the computational complexity of global optimization for large-scale scenes. To solve these problems, a real-time incremental simultaneous localization and mapping algorithm called MIM_SLAM is proposed in this paper. This algorithm is applied in mobile robots to build a map on a non-flat road with a 3D LiDAR sensor. MIM_SLAM's main contribution is that multi-level ICP (Iterative Closest Point) matching is used to solve the data association problem, a Fisher information matrix is used to describe the uncertainty of the estimated pose, and these poses are optimized by the incremental optimization method, which can greatly reduce the computational cost. Then, a map with a high consistency will be established. The proposed algorithm has been evaluated in the real indoor and outdoor scenes as well as two substations and benchmarking dataset from KITTI with the characteristics of sparse and large-scale. Results show that the proposed algorithm has a high mapping accuracy and meets the real-time requirements.

**Keywords:** data association; 3D-SLAM; localization; mapping

---

## 1. Introduction

In recent years, the simultaneous localization and mapping (SLAM) based on a 3D LIDAR sensor have become an important topic in robotics due to the rise of autonomous driving technology. Meanwhile, the building of a map has become the basis for an autonomous mobile robot to complete tasks such as inspection and autonomous navigation [1] in some harsh environments.

The research on SLAM can be traced back to Smith et al. [2] of Stanford University in the 1980s. They published a seminal paper on SLAM, and later generations have done considerable work on it. Among them, the algorithms such as Extended Kalman Filters [3], Extended Information Filters [4], and Rao-Blackwellized Particle Filters [5] based on a filtering idea all adopt the Bayesian state estimation theory to estimate the posterior probability of the system's state. However, the essence of this kind of algorithm is only optimizing the local state of the system. As the distance traveled by the robot increases, it is difficult to establish a high consistency map. Another kind of algorithm is the graph optimization that was first proposed by Lu and Milio [6]. The essence of this algorithm is maintaining all the observation and space constraints between observations, then using the maximum likelihood method to estimate the pose of the robot. Better mapping results can be obtained in large-scale scenes using the global optimization method.

Although the SLAM problem has been thoroughly researched, there are still many problems to be solved, including the fact that computational complexity will increase from two-dimensions to three-dimensions, and how to ensure a consistent map with the increase of the size of mapping and reliable data association with less information in sparse environments. Combining these issues, Zlot [7] utilizes a rotating 2D LIDAR sensor to achieve high-precision map construction in a wide mine, but it cannot be applied online. Olufs [8] proposes a data association method in sparse environments, but it is only suitable for solving the localization problem under known maps. Huang [9] proposes a sparse local submap joining filter (SLSJF) for map-building in large-scale environments. However, this depends on environmental features. Moosmann [10] and Nüchter [11] propose a real-time SLAM framework, and the experiments in some scenarios have achieved good results, but the results of the mapping are not tested for large-scale and sparse environments. Zhang [12] estimates the pose of the robot by extracting the feature points in the scene, but it is difficult to establish a closed-loop map as the map size expands. Wang [13] proposes a multi-layer matching SLAM in large-scale and sparse environments, but it only solves the data association in a 2D scene. The algorithm will be ineffective when the robot drives on a non-flat road. Therefore, this paper is dedicated to solving the following two problems:

- In sparse environments, due to the scarce environmental information perceived by a 3D LIDAR sensor and the non-flat road, data association is easy to fall into a local minimum. Liu [14] proposes a LiDAR SLAM method in natural terrains which fuses multiple sensors including two 3D LiDAR. Therefore, the robustness of data associated with a lower accumulated error should be strengthened.
- In large-scale environments, it is difficult to obtain a reasonable pose estimation when the robot returns to a region it has previously explored and then generates an inconsistent map. Liang [15] addresses the laser-based loop closure problem by fusing visual information. Hess [16] can achieve real-time mapping in indoor scenes, and it may fail in the closing loop due to the incremental computation in large-scale outdoor scenes. How to reduce the computational complexity of graph optimization is also a problem that needs to be improved.

To solve these problems, we propose a real-time incremental SLAM algorithm called the MIM_SLAM based on a multi-level ICP matching method. This paper is organized as follows. We continue in the next section with an overview of the algorithm. Section 3 details the process of the MIM_SLAM algorithm. Then, we follow up with the experimental results and analysis in Section 4 and finally provide a summary in Section 5.

## 2. Algorithm Overview

Figure 1 shows the overall MIM_SLAM algorithm framework. The MIM_SLAM algorithm simplifies the SLAM problem into the data association and the incremental pose graph optimization. Firstly, the data association problem is solved by the multi-level ICP matching method which includes matching the time-neighbor frame, matching the current frame with the map, and matching the current frame with an area-neighbor keyframe. Then the uncertainty of the estimated pose is described via the Fisher information matrix. After the above steps, this algorithm can obtain the transformation matrix $T_{ij}$ and covariance matrix $\Sigma_{ij}$ between pose $x_i$ and pose $x_j$, which can be saved in the pose graph. Finally, the incremental QR decomposition method [14] is used to optimize the pose graph. In the next section, the MIM_SLAM algorithm will be described in detail.

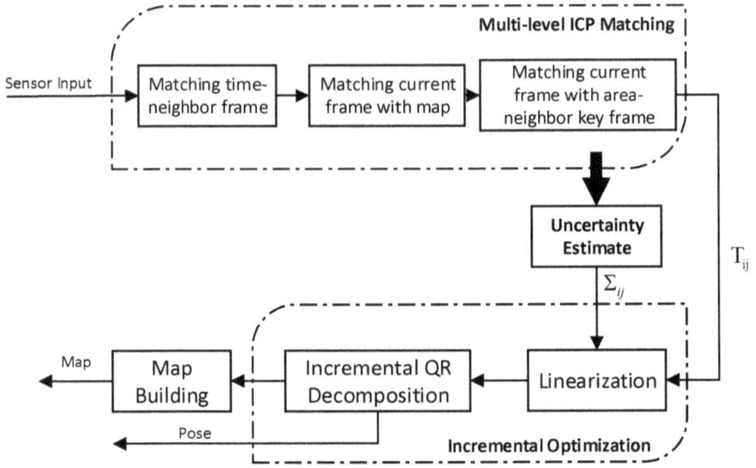


**Figure 1.** The proposed MIM_SLAM algorithm framework.

## 3. Algorithm Description

### *3.1. SLAM as an Incremental Optimization Problem*

This part analyzes the SLAM problem from the probabilistic model and transforms the problem into a least squares problem based on the maximum a posteriori estimate. Then, the incremental QR decomposition method [17] is used to solve the least squares problem.

We denote robot pose by $X = \{x_i\}$ with $i \in 0 \ldots M$, the landmark by $L = \{l_j\}$ with $j \in 1 \ldots N$, control input given through wheeled odometry and IMU by $U = \{u_i\}$ with $i \in 1 \ldots M$, landmark measurement by $Z = \{z_i\}$ with $i = 1 \ldots M$. The SLAM problem is equivalent to estimating the posterior probability distribution:

$$\begin{aligned} &P(X, L|Z, U) \\ &= P(L|Z, X, U)P(X|Z, U) \\ &= P(L|Z, X)P(X|Z, U) \end{aligned} \quad (1)$$

From Equation (1), it can be seen that the landmark $L$ depends on the observations and robot pose sequences, so the SLAM problem is further simplified to the pose estimation problem utilizing Baye's rule $P(X|Z, U) = P(Z|X, U)P(X|U)/P(Z|U)$. In this paper, we just consider the online SLAM problem [4], which only involves estimating the posterior over the momentary pose, this is $P(x_M|z_{1:M}, u_{1:M})$.

$$\begin{aligned} &P(x_M|z_{1:M}, u_{1:M}) \\ &= \frac{P(z_{1:M}|x_M, u_{1:M})P(x_M|u_{1:M})}{P(z_{1:M}|u_{1:M})} \\ &= \frac{P(z_M|x_M, z_{1:M-1}, u_{1:M})P(x_M|z_{1:M-1}, u_{1:M})}{P(z_M|z_{1:M-1}, u_{1:M})} \\ &= \eta P(z_M|x_M)P(x_M|z_{1:M-1}, u_{1:M}) \\ &= \eta P(z_M|x_M)\int P(x_M|x_{M-1}, z_{1:M-1}, u_{1:M})P(x_{M-1}|z_{1:M-1}, u_{1:M})dx_{M-1} \\ &= \eta P(z_M|x_M)\int P(x_M|x_{M-1}, u_M)P(x_{M-1}|z_{1:M-1}, u_{1:M-1})dx_{M-1} \\ &= \eta P(x_0)\prod_{i=1}^{M} P(x_i|x_{i-1}, u_i)P(z_i|x_i) \end{aligned} \quad (2)$$

In Equation (2) line 4 to 5, it is derived using the theorem of total probability. $P(x_0)$ is a prior on the initial state, $P(x_i|x_{i-1}, u_i)$ is the motion model, and $P(z_i|x_i)$ is the measurement model.

To estimate the pose of the robot, we transform the SLAM problem into the least-squares problem based on the maximum a posteriori estimate. The maximum a posteriori estimate $X^*$ and $L^*$ for the trajectory and map are obtained by minimizing the negative log of the joint probability from Equation (2):

$$X^* = -\underset{X}{\operatorname{argmin}}\eta \left\{ \sum_{i=1}^{M} \log P(x_i|x_{i-1}, u_i) + \sum_{i=1}^{M} \log P(z_i|x_i) \right\} \tag{3}$$

$$L^* = \{(x_1^*, z_1), \ldots, (x_i^*, z_i)\} \tag{4}$$

Both of the motion model and measurement model are generally assumed such that they meet the Gaussian distribution. Then, both of them can be converted to the following form:

$$x_j \sim N(x_i \oplus T_{ij}, \Sigma_{ij}) \tag{5}$$

where the operator $\oplus$ denotes the coordinate transformation. $T_{ij}$ and $\Sigma_{ij}$ represent respectively the transformation matrix and covariance matrix between pose $x_i$ and pose $x_j$, which will be obtained through the multi-level ICP matching method and uncertainty estimation method in the next two paragraphs. Additionally, it can be written as follows:

$$\begin{cases} x_i = f(x_{i-1}, u_i) + m_i \\ z_i = g(x_i) + n_i \end{cases} \tag{6}$$

where $m_i$ is the motion noise, and $n_i$ is the measurement noise. This leads to the following nonlinear least squares problem:

$$X^* = \underset{X}{\operatorname{argmin}} \left\{ \sum_{i=1}^{M} \|f(x_{i-1}, u_i) - x_i\|_{m_i}^2 + \sum_{i=1}^{M} \|g(x_i) - z_i\|_{n_i}^2 \right\} \tag{7}$$

Since the above model has non-linear functions that are not easy to solve, they must be linearized, as Gauss-Newton [18] and Levenberg-Marquardt [19] has done. It can eventually be transformed into the general least-squares problem as follows, and the specific derivation can refer to the literature [14].

$$\theta^* = \underset{\theta}{\operatorname{argmin}} \|A\theta - b\|^2 \tag{8}$$

where the vector $\theta$ contains all pose variables, and the matrix $A$ is a large and sparse Jacobian matrix.

The problem in Equation (8) is apparently solved by the Cholesky decomposition method, but the essential problem is the number of calculations involved in solving the information matrix $A^T A$. Thus, applying standard QR decomposition to matrix $A$:

$$A = Q \begin{bmatrix} R \\ 0 \end{bmatrix} \tag{9}$$

Apply Equation (9) to the least squares problem in Equation (8). As $Q$ is an orthonormal matrix, $Q^TQ = I$ ($I$ is identity matrix). Thus, an additional term $Q^T$ is added to the second line in (10), which doesn't change the length of the vector $A\theta - b$.

$$\begin{aligned} \|A\theta - b\|^2 &= \|Q\begin{bmatrix} R \\ 0 \end{bmatrix}\theta - b\|^2 \\ &= \|Q^TQ\begin{bmatrix} R \\ 0 \end{bmatrix}\theta - Q^Tb\|^2 \\ &= \|\begin{bmatrix} R \\ 0 \end{bmatrix}\theta - \begin{bmatrix} d \\ e \end{bmatrix}\|^2 \\ &= \|R\theta - d\|^2 + \|e\|^2 \end{aligned} \tag{10}$$

where $[d,e]^T = Q^Tb$ is defined, and (10) obtain the minimum $\|e\|^2$ if and only if $R\theta = d$. Now, the SLAM problem has been solved by the above analysis. However, as a new observation arrives, the Jacobian matrix $A$ does not significantly change. Thus the results of the QR decomposition at previous times can be used as the iterative initial value and the process is incremental, which can significantly reduce computational complexity.

### *3.2. The Multi-Level ICP Matching Method*

Before detailing this method, it is useful to understand the ICP algorithm. The standard ICP [20] (Iterative Closest Point) algorithm firstly establishes the point-to-point correspondence between the two point clouds by the nearest neighbor principle and then establishes the matching error function.

Finally, it computes a transformation matrix so that the error function is minimized. In the past 30 years, many ICP variant algorithms [21] have been proposed. In this paper, an ICP variant algorithm that calls point-to-plane ICP [22] is used. Compared with the standard ICP, the robustness and accuracy are better. This algorithm is listed as Algorithm 1.

**Algorithm 1:** Point-to-plane ICP.

**Input:** Two point cloud: $A = \{a_i\}$, $B = \{b_i\}$; An initial transformation: $T_0$
**Output:** Transformation $T_{AB}$ which aligns $A$ and $B$; Fitness Score: $Score_{fitness}$

1: $T \leftarrow T_0$
2: **while** not converged **do**
3:     **for** $i \leftarrow 1$ *to* $N$ **do**
4:         $m_i \leftarrow$ FindClosestPointInA$(T \cdot b_i)$
5:         **if** $\|m_i - T \cdot b_i\| \leq d_{\max}$ **then** $w_i \leftarrow 1$
6:         **else** $w_i \leftarrow 0$
7:         **end**
8:     **end**
9:     $T \leftarrow \text{argmin}\left\{\sum_i w_i \|\eta_i \cdot (T \cdot b_i - m_i)\|^2\right\}$
10: **end**
11: $T_{AB} \leftarrow T$
12: $Score_{fitness} = \sum_i w_i \|\eta_i \cdot (T \cdot b_i - m_i)\|^2$

The observation of the 3D LiDAR sensor is denoted as $z_k$ at time $k$. There are data associations between all previous observations $z_{1:k-1}$ and the current observation included in $z_k$. A traditional scan matching method associates $z_k$ with $z_{k-1}$ at last time, or associates $z_k$ with all previous observations $z_{1:k-1}$. These two methods either have a substantial accumulated error, or have large amounts of calculations which cannot meet online applications. In this paper, the data association is divided into two categories: continuous in time and continuous in area.

For observations that are continuous in the time, the method firstly utilizes the ICP algorithm to associate $z_k$ and $z_{k-1}$, and a rough estimation of the current robot's pose can be obtained. The initial value of ICP is obtained by using a Kalman filter to merge the wheeled odometry with IMU. Because the inter-frame matching cannot obtain the accurate pose estimation, the next step will match the current frame with the map to further eliminate the accumulated error.

As shown in Figure 2, $M_i$ denotes the global map at time $i$ and $T_i^W$ denotes the pose of the robot in the world coordinate system at time $i$. $T_{i+1}^L$ denotes the transformation matrix from time $i+1$ to time $i$ (that is, the output of inter-frame matching). $z_{i+1}$ denotes the observation at time $i+1$. The process for matching the current frame with the map is as follows:

- Finding the nearest neighbor in $M_i$ for each point in $z_{i+1}$, and saving it as $m_{i+1}$. We use the combination of octree and approximate nearest neighbor algorithm in PCL [23] for speeding up.
- Taking $z_{i+1}$ and $m_{i+1}$ as the input of Algorithm 1, the transformation matrix $T_{opt}$ can be obtained after registering the two point clouds.
- The pose of the robot at time $k+1$ is $T_{i+1}^W = T_i^W \cdot T_{i+1}^L \cdot T_{opt}$.

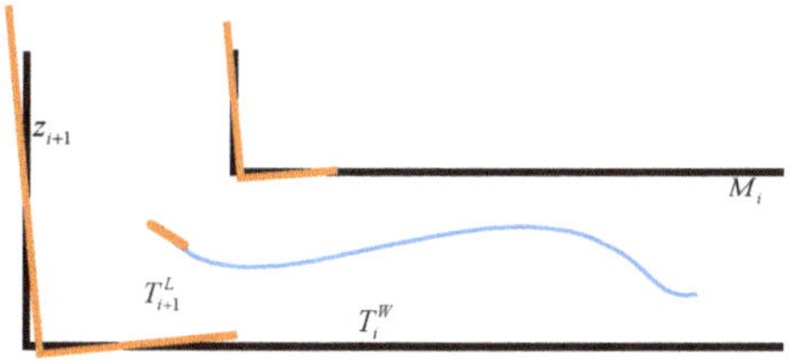


**Figure 2.** The matching process of the current frame and map.

The effectiveness of the above scan matching method is verified as shown in Figure 3. The white point clouds denote the map that SLAM has established. The blue point clouds denote the point clouds after inter-frame matching. The redpoint clouds denote the point clouds after matching the current frame with the map. It can be seen that the accumulated error in the process of inter-frame matching is obviously eliminated.

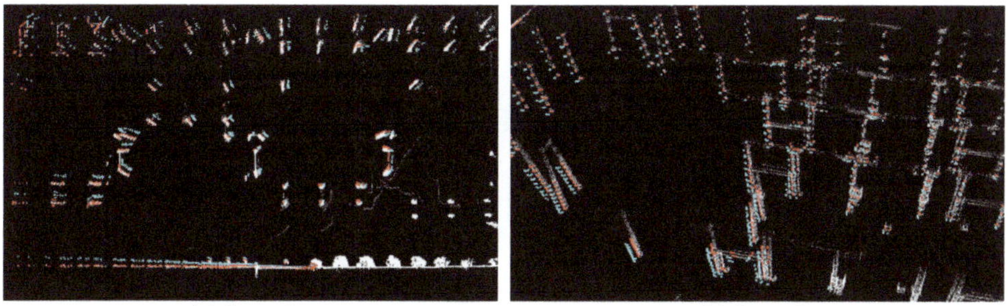

(**a**) Scene 1 (**b**) Scene 2

**Figure 3.** An example of the matching process.

For observations that are continuous in the area, the robot returns to a region that it has previously explored. In particular, in large-scale and sparse environments, the accumulative error cannot be completely eliminated when the robot travels a long distance if only the above two matching processes are considered. So the following method matches the current frame $z_i$ with the area-neighbor keyframe

$z_\tau$ based on a priori of pose estimation. If the matching score is less than a certain threshold, the pose constraint between $x_i$ and $x_\tau$ will be established. The process is listed as Algorithm 2. The initial parameter is: $L = \{(0, x_0, z_0)\}, d = 0$.

**Algorithm 2:** Matching the current frame with an area-neighbor keyframe.

**Input:** Current pose and observation: $x_i, z_i$
**Output:** Transformation $T_{i\tau}$ which aligns $z_i$ and $z_\tau$

```
calculate d = d + ||x_i − x_{i−1}||
if d < d_thre then
        return null
end
foreach (k_τ, x_τ, z_τ) in L do
    if k − k_τ > k_thre then
        if ||x_t − x_τ|| < x_thre then
            put z_t and z_τ into Algorithm 1, get T and Score_fitness
            if Score_fitness < S_thre then
                T_iτ ← T
                return T_iτ
            end
        end
    end
end
L = L ∪ (k_i, x_i, z_i), d = 0, k = k + 1
return null
```

### 3.3. Uncertainty Estimation

After multi-level ICP matching, the transformation matrix $T_{ij}$ between pose $x_i$ and pose $x_j$ can be obtained, but we still need to know the exact uncertainty estimation, namely the covariance matrix $\Sigma_{ij}$. However, an incorrect covariance matrix may damage the established map.

In this paper, the inversion of the Fisher information matrix is used as the covariance. The Fisher information matrix is defined as the function of the expected measurement and the surface slope scanned by the laser sensor [24]. Liu [24] and Wang [25] give the derivation of the Fisher information matrix based on a two-dimensional probability grid. It is now extended to three-dimensions. The Fisher information matrix is discretized:

$$\hat{L}(p) = \sum_{i=1}^{N} \frac{1}{\sigma_i^2} \left(\frac{\Delta r_{iE}}{\Delta p}\right)^T \left(\frac{\Delta r_{iE}}{\Delta p}\right) \tag{11}$$

$$\frac{\Delta r_{iE}}{\Delta p} = \left[\frac{\Delta r_{iE}}{\Delta x}, \frac{\Delta r_{iE}}{\Delta y}, \frac{\Delta r_{iE}}{\Delta z}, \frac{\Delta r_{iE}}{\Delta \varphi}, \frac{\Delta r_{iE}}{\Delta \psi}, \frac{\Delta r_{iE}}{\Delta \theta}\right] \tag{12}$$

where $p = [x, y, z, \varphi, \psi, \theta]$ is the pose of the robot, and $\sigma_i^2$ is the noise variance of the $i$-th 3D laser range finder scan ray. $N$ is the total number of scan rays. $r_{iE}$ is the expected distance from the robot to the nearest obstacle along the $i$-th scan ray. To calculate Equation (12) efficiently, the point cloud map is converted into a three-dimensional grid map. Thus $r_{iE}$ can be computed as follows:

$$r_{iE} = \frac{\sum_{j=1}^{s} r_{ij}\mu_{ij}}{\sum_{j=1}^{s} \mu_{ij}} \tag{13}$$

where $r_{ij}$ is the distance between the robot and the $j$-th voxel along the direction of an $i$-th scan ray. $\mu_{ij}$ is the occupancy value of the corresponding voxel. $s$ is the sequence number of the ending voxel. By combining Equation (11) and Equation (12), we obtain:

$$\hat{L}(p) = \sum_{i=1}^{N} \frac{1}{\sigma_i^2} \begin{bmatrix} \frac{\Delta r_{iE}^2}{\Delta x^2} & \frac{\Delta r_{iE}^2}{\Delta x \Delta y} & \frac{\Delta r_{iE}^2}{\Delta x \Delta z} & \frac{\Delta r_{iE}^2}{\Delta x \Delta \varphi} & \frac{\Delta r_{iE}^2}{\Delta x \Delta \psi} & \frac{\Delta r_{iE}^2}{\Delta x \Delta \theta} \\ \frac{\Delta r_{iE}^2}{\Delta x \Delta y} & \frac{\Delta r_{iE}^2}{\Delta y^2} & \frac{\Delta r_{iE}^2}{\Delta y \Delta z} & \frac{\Delta r_{iE}^2}{\Delta y \Delta \varphi} & \frac{\Delta r_{iE}^2}{\Delta y \Delta \psi} & \frac{\Delta r_{iE}^2}{\Delta y \Delta \theta} \\ \frac{\Delta r_{iE}^2}{\Delta x \Delta z} & \frac{\Delta r_{iE}^2}{\Delta y \Delta z} & \frac{\Delta r_{iE}^2}{\Delta z^2} & \frac{\Delta r_{iE}^2}{\Delta z \Delta \varphi} & \frac{\Delta r_{iE}^2}{\Delta z \Delta \psi} & \frac{\Delta r_{iE}^2}{\Delta z \Delta \theta} \\ \frac{\Delta r_{iE}^2}{\Delta x \Delta \varphi} & \frac{\Delta r_{iE}^2}{\Delta y \Delta \varphi} & \frac{\Delta r_{iE}^2}{\Delta z \Delta \varphi} & \frac{\Delta r_{iE}^2}{\Delta \varphi^2} & \frac{\Delta r_{iE}^2}{\Delta \varphi \Delta \psi} & \frac{\Delta r_{iE}^2}{\Delta \varphi \Delta \theta} \\ \frac{\Delta r_{iE}^2}{\Delta x \Delta \psi} & \frac{\Delta r_{iE}^2}{\Delta y \Delta \psi} & \frac{\Delta r_{iE}^2}{\Delta z \Delta \psi} & \frac{\Delta r_{iE}^2}{\Delta \varphi \Delta \psi} & \frac{\Delta r_{iE}^2}{\Delta \psi^2} & \frac{\Delta r_{iE}^2}{\Delta \psi \Delta \theta} \\ \frac{\Delta r_{iE}^2}{\Delta x \Delta \theta} & \frac{\Delta r_{iE}^2}{\Delta y \Delta \theta} & \frac{\Delta r_{iE}^2}{\Delta z \Delta \theta} & \frac{\Delta r_{iE}^2}{\Delta \varphi \Delta \theta} & \frac{\Delta r_{iE}^2}{\Delta \psi \Delta \theta} & \frac{\Delta r_{iE}^2}{\Delta \theta^2} \end{bmatrix} \quad (14)$$

According to the Cramér–Rao Bound theory [26], the lower bound of covariance can be determined by the inversion of the Fisher information matrix. Finally, the uncertainty of the pose estimation can be obtained:

$$\mathrm{cov}(p) = \hat{L}^{-1}(p) \quad (15)$$

## 4. Experimental Results and Analysis

As shown in Figure 4, the mobile robot called SmartGuard [27] is used to verify the proposed algorithm. It is a completely autonomous robotic system that can inspect substation equipment. SmartGuard is equipped with a wheeled odometry, an IMU, and a 3D LIDAR (RS-Lidar-16). The RS-Lidar-16 can measure out to 150 m with a high precision, ±2 cm, and it has a +15 to −15-degree vertical field of view. It continuously scans the 360-degree surrounding environment at a 10 Hz frame rate and at 300,000 points/sec. The capability of the computer that runs the SLAM algorithm is as follows: Intel Core i5-6300HQ CPU 2.3 Hz and 8 G DDR3 RAM.

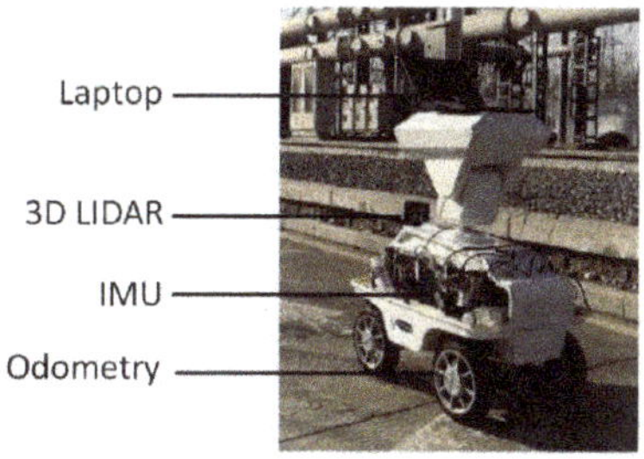


**Figure 4.** The mobile robot (SmartGuard) for experiments.

### *4.1. Indoor and Outdoor Mapping Test*

To verify the mapping performance, the proposed algorithms have been tested in several different indoor and outdoor environments (Figure 5). Figure 5a shows that the proposed MIM_SLAM algorithm can build a consistency map with lower accumulated errors after a 180 m long loop. Figure 5b shows that the consistency map can also be generated from a narrow and long corridor. As shown in Figure 5c,d, in the wide range of the parking lot, the robot traveled approximately 1600 m at a speed of 0.6 m/s; the trajectory is shown in Figure 5c. The map shown in Figure 5c is still accurate and clear.

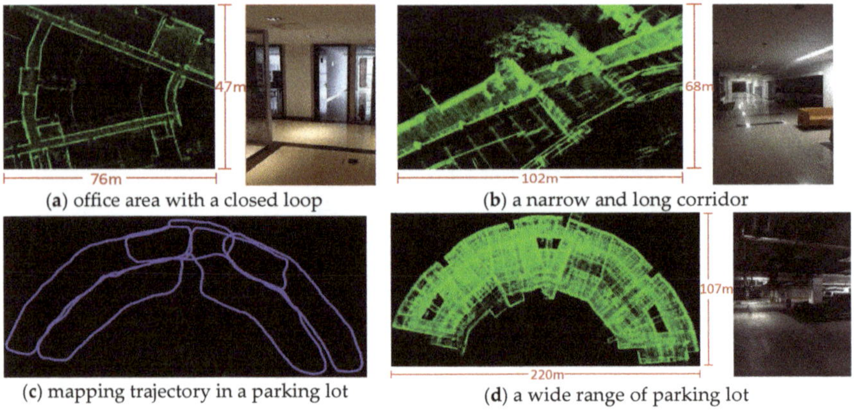


(**a**) office area with a closed loop (**b**) a narrow and long corridor

(**c**) mapping trajectory in a parking lot (**d**) a wide range of parking lot

**Figure 5.** The maps generated in the indoor and outdoor environments.

### 4.2. Large-Scale and Sparse Scenes Test

Two substations are selected to verify the validity of the MIM_SLAM algorithm in the larger-scale and sparse environments. A general substation comprises thousands of square meters, and its electrical equipment needs to include a sufficient distance for electrical safety. We first tested in substation A shown in Figure 6. During the experiment, the robot traveled approximately 600 m at a speed of 0.6 m/s and the 3D LIDAR sensor collected a total of 10,080 frames. The plan and the environment of substation A are shown in Figure 6. The red line denotes the robot's trajectory ABCDA. The point cloud maps established by the MIM_SALM and the LOAM [12] which has been considered state-of-the-art in LiDAR SLAM are shown in Figure 7.

Intuitively, the MIM_SLAM algorithm has a high consistency map compared with the LOAM. Due to the accumulated error, the estimated pose has a great deviation when the robot travels from area A along the red track shown in Figure 6, and then returns back to area A. The LOAM cannot effectively handle the data association at this moment, which leads to the wrong association and damage to the established map. In this paper, the multi-level ICP matching method takes into account this situation and generates a reliable constraint between the two poses. The incremental optimization method is used to optimize the global pose. Thus the final map has a high consistency.

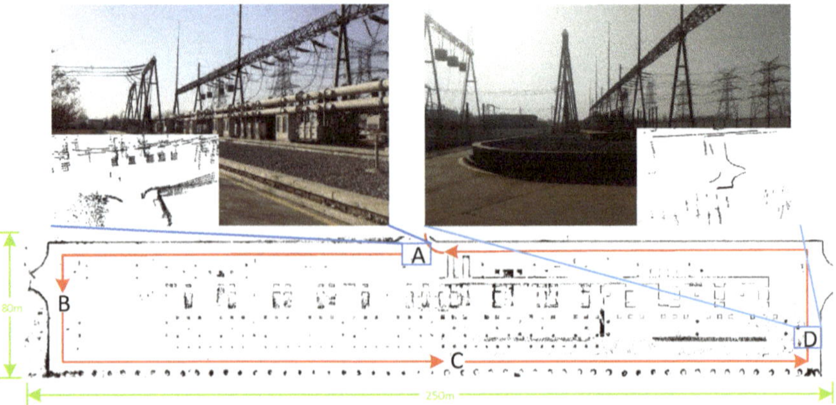


**Figure 6.** The plan and the environment of substation A.

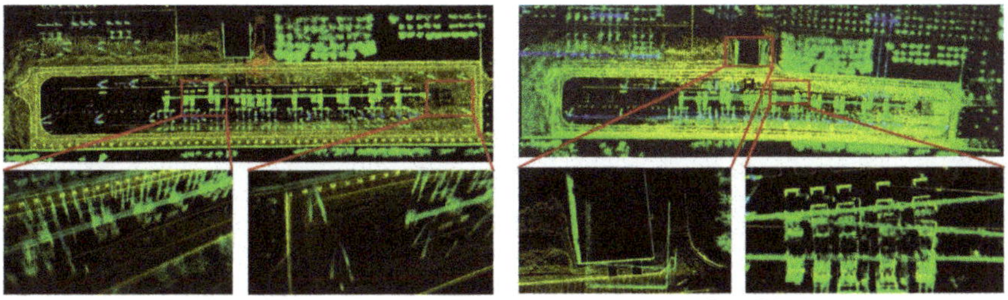

(**a**) MIM_SLAM (**b**) LOAM

**Figure 7.** The point cloud maps established by the MIM_SLAM and the LOAM [12] algorithms.

Next, we tested it in substation B. Compared with substation A, this was larger. Meanwhile, the electrical equipment is placed more sparsely, and less information is gathered by the 3D LIDAR sensor. During the experiments, the robot traveled approximately 1450 m at a speed of 0.6 m/s, and the LIDAR sensor collected a total of 24,220 frames. The plan and the environment of substation B are shown in Figure 8. The red line denotes the robot's trajectory ABCDEABFGDC. The point cloud maps established by the MIM_SLAM algorithm and the LOAM are shown in Figure 9.

It can be seen from the observation of the two scenes given in Figure 8 that the electrical equipment is distributed more sparsely, and more than 70% of the laser rays reach the maximum measurement range. As shown in Figure 9a, both contour lines of the map established by the MIM_SLAM algorithm are at right angles at areas F and G. Additionally, when the robot returned back to the area C, the details of the point cloud map at area C showed that the telegraph poles and the road edge are clearly visible, and there is no incorrect point cloud accumulation. Because the LOAM cannot effectively process the data association of the closed-loop area, the performance is very poor in this situation, which has more closed-loop areas in large-scale environments, as shown in Figure 9b.

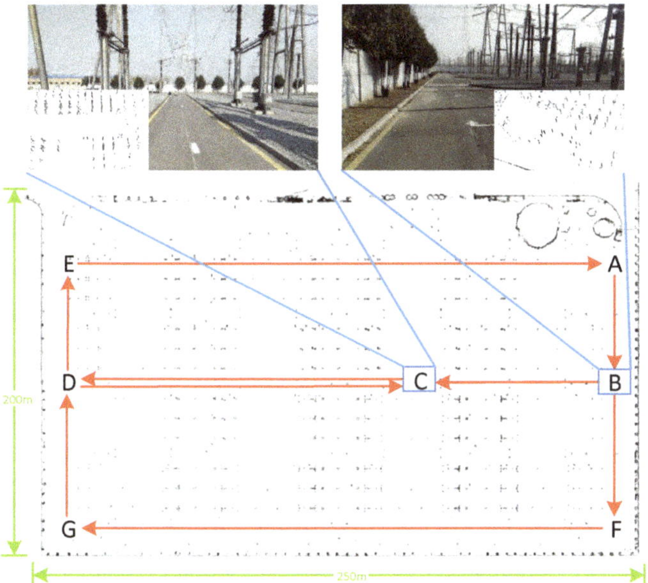


**Figure 8.** The plan and the environment of substation B.

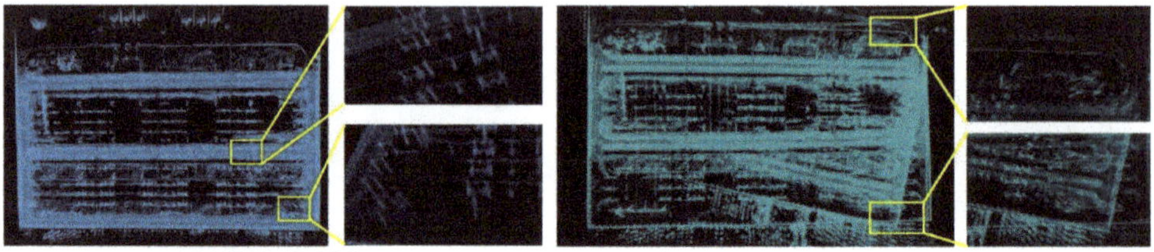

(**a**) MIM_SLAM (**b**) LOAM

**Figure 9.** The point cloud maps established by the MIM_SLAM and the LOAM algorithms.

### *4.3. Mapping Accuracy Test*

In order to analyze the mapping accuracy quantitatively, we selected two substations with large-scale and sparse characteristics which are shown in Figures 6 and 8, also the standard ICP [20] and the LOAM [12] algorithm was selected to compare with the MIM_SLAM algorithm. In the experiments, the robot is controlled running at a speed of 0.6 m/s on a flat road along a straight line with a length of 10 m. The accuracy of mapping is defined as the robot's localization error $\varepsilon$ [28], expressed as follows:

$$\begin{aligned} \varepsilon &= \tfrac{1}{N}\sum_{i=0}^{N-1} \|e_i\| \\ e_i &= (x_0 \oplus T_i) - (x_0^* \oplus T_i^*) \end{aligned} \tag{16}$$

where $x_0$ and $x_0^*$ denote the initial pose. $\oplus$ is the standard motion composition operator. $T_i$ denotes the transformation matrix of the estimated pose $x_i$ relative to the initial pose $x_0$. $T_i^*$ denotes the transformation matrix of the true pose $x_i^*$ relative to the initial pose $x_0^*$. $\|\cdot\|$ is the 2-Norm used in this paper. For the robot to navigate in three-dimensional environments, $e_i$ is represented as $(x_e, y_e, z_e, \varphi_e, \psi_e, \theta_e)$. Then $\varepsilon$ can be divided into two parts: translation error and rotation error.

$$\varepsilon = \frac{1}{N}\sum_{i=0}^{N-1} \|trans(e_i)\| + \frac{1}{N}\sum_{i=0}^{N-1} \|rot(e_i)\| \tag{17}$$

Since it is difficult to obtain the true pose of the robot, 20 points are selected from the robot's trajectory. The first point is taken as the initial pose. For the next 19 points, the difference value of the localization results relative to the initial pose is calculated. The difference value of the true pose can be measured by the tape. In Table 1, the localization error is tested with the standard ICP, the LOAM and the MIM_SLAM algorithms under substation A and B. It can be seen that the MIM_SLAM algorithm is more precise. If there is a closed-loop area on the robot's path, the MIM_SLAM algorithm will be better than the LOAM. The processing time per frame using different approaches is calculated and shown in Table 2. MIM_SLAM has a similar time consuming compared to the LOAM and meets the real-time requirements. In Section 4.4, this will be further analyzed in the benchmark dataset.

**Table 1.** The localization error results of different approaches/scenes.

| | **Standard ICP** | **LOAM** | **MIM_SALM** |
|---|---|---|---|
| | | Trans. Error (unit: m) | |
| A | $0.213 \pm 0.148$ | $0.064 \pm 0.057$ | $0.052 \pm 0.043$ |
| B | $0.282 \pm 0.177$ | $0.075 \pm 0.066$ | $0.066 \pm 0.062$ |
| | | Rot. Error (unit: deg) | |
| A | $2.5 \pm 1.9$ | $1.8 \pm 1.2$ | $1.6 \pm 1.4$ |
| B | $3.1 \pm 2.2$ | $2.5 \pm 2.1$ | $2.2 \pm 1.8$ |

**Table 2.** The average time consumption per frame for different approaches/scenes.

| | Standard ICP | LOAM | MIM_SALM |
|---|---|---|---|
| | | Processing Time (unit: s) | |
| A | 0.0405 | 0.1036 | 0.0886 |
| B | 0.0416 | 0.1067 | 0.0932 |

### *4.4. Benchmarking Datasets Test*

Then, datasets from the KITTI odometry benchmark [29] are used to evaluated MIM_SLAM. The datasets take advantage of the autonomous driving platform Annieway to develop novel challenging real-world computer vision benchmarks. The autonomous driving platform is equipped with a 360° Velodyne HDL-64E laser scanner and two high-resolution color and grayscale video cameras. Accurate ground truth is provided by a GPS localization system. We selected three typical types of environments: "urban" with building around (sequence 07), "country" on small roads with vegetation in this scene (sequence 03), and "highway" where roads are wide and lower dynamic (sequence 06).

(1) Sequence 03: This dataset is designed to verify that MIM_SLAM can achieve a low drift pose estimation in sparse vegetation environment. The mapping result is shown in Figure 10a and the trajectory and the ground truth are shown in Figure 11a. Due to the scarce stable features in this scene, there is a bit of position deviation after 420 m of traveling compared with the ground truth in Figure 11a. To evaluate the pose estimation accuracy, we use the evaluation method in the KITTI odometry benchmark which calculated the translational and rotational errors for all possible subsequences of length (100, 200, … , 800) meters. As is shown in Figure 12a, both LOAM and MIM_SLAM achieve lower drift values compared with the standard ICP method, and MIM_SLAM is slightly worse than LOAM.

(2) Sequence 06: We use this dataset to verify that MIM_SLAM can effectively process the data association of the closed-loop area shown in Figures 10 and 11. From Figure 12b, it can be seen that the translational and rotational errors are lower compared to the standard ICP, and slightly better than LOAM.

(3) Sequence 07: This urban road scene is highly dynamic and large-scale. The vehicle travels approximately 660 m at a speed of 6.2 m/s. The mapping result is shown in Figure 10c and the trajectory and the ground truth are shown in Figure 11c. Intuitively, the details of the point cloud map are clear. The estimated pose and the ground truth are almost overlapping. The pose estimation error is shown in Figure 12c quantitatively.

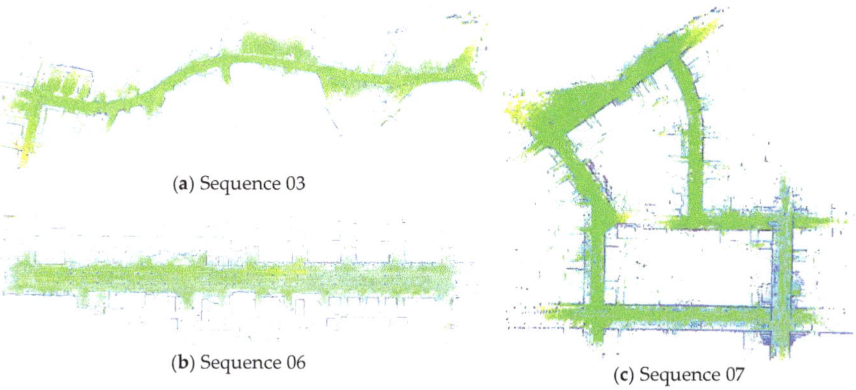

(**a**) Sequence 03

(**b**) Sequence 06

(**c**) Sequence 07

**Figure 10.** The mapping results in the three sequences.

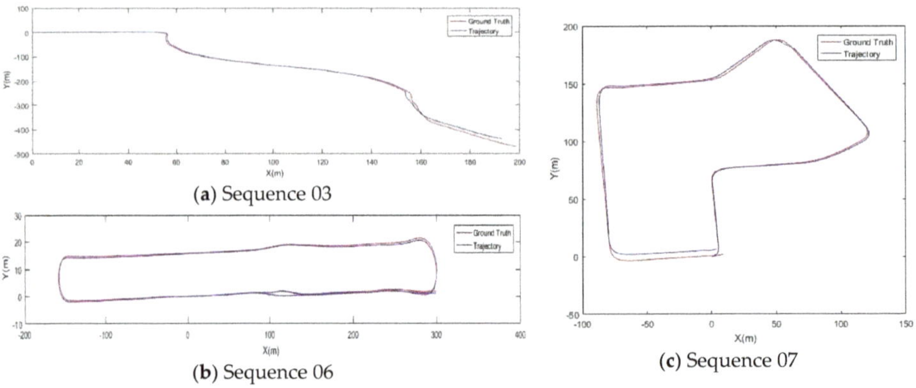


(**a**) Sequence 03

(**b**) Sequence 06

(**c**) Sequence 07

**Figure 11.** The trajectory produced by MIM_SLAM in the three sequences.

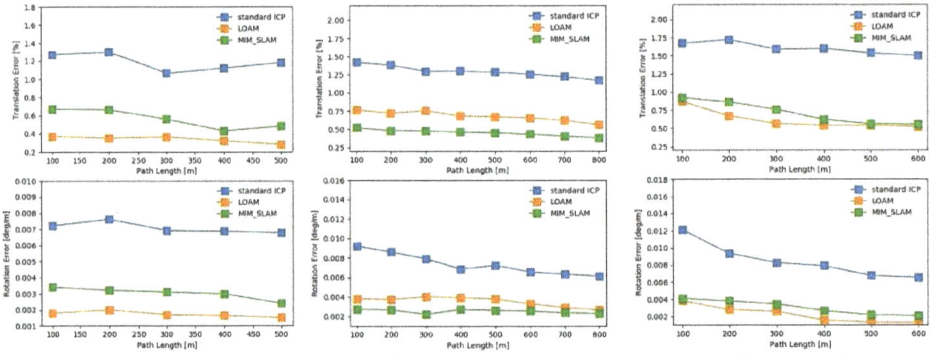


(**a**) Sequence 03 (**b**) Sequence 06 (**c**) Sequence 07

**Figure 12.** The pose estimation error in the three sequences.

## 5. Conclusions

For large-scale and sparse environments, we propose a novel MIM_SLAM algorithm in this paper, which simplifies the SLAM problem to the least-squares optimization problem. The contribution of this paper is that the multi-level ICP matching method is proposed to solve the data association problem and the uncertainty estimation is handled by the Fisher information matrix. It considers the accumulated errors in two aspects: matching between time-neighbor frames (building a low-drift LiDAR odometry) and matching between area-neighbor frames (dealing with the data association at a revisited area). Moreover, multi-level matching and incremental optimization reduce the computational complexity while ensuring mapping accuracy, we can achieve accurate mapping and real-time applications. Additionally, its application is not only limited to mobile robots, and it can potentially be extended to other vehicles, e.g., UAVs. Experimental results show that it can effectively build a high consistency map with a smaller amount of environmental information and the large-scale scenes, and it also achieves similar or better accuracy compared with the standard ICP and the state-of-the-art LOAM algorithm in the KITTI dataset. Additionally, there are some limitations in high dynamic scenes. Our future work will focus on the construction of a more robust data association method, we will integrate the output of our multi-level ICP matching with an IMU in a Kalman filter to further reduce the accumulated error.

**Author Contributions:** This study was completed by the co-authors. J.W. conceived and led the research. The major experiments and analyses were undertaken by J.W. and M.Z. W.C. supervised and guided this study. J.W., M.Z. and W.C. wrote the paper. All the authors have read and approved the final manuscript.

**Funding:** This research was partly funded by the National Natural Science Foundation of China (Grant No. 61773261) and the National Key R&D Program of China (Grant No. 2017YFC0806501).

**Conflicts of Interest:** The authors declare no conflicts of interest.

## References

1. Martínez, J.L.; Morán, M.; Morales, J.; Reina, A.J.; Zafra, M. Field Navigation Using Fuzzy Elevation Maps Built with Local 3D Laser Scans. *Appl. Sci.* **2018**, *8*, 397. [CrossRef]
2. Smith, R.; Cheeseman, P. On the Estimation and Representations of Spatial Uncertainty. *Int. J. Robot. Res.* **1987**, *5*, 56–68. [CrossRef]
3. Shinsuke, M.M.D.; Noboru, I.M.D.; Yoshito, I.M.D. Estimating Uncertain Spatial Relationships in Robotics. *Mach. Intell. Pattern Recognit.* **2013**, *5*, 435–461.
4. Thrun, S.; Liu, Y.; Koller, D.; Ng, A.Y.; Ghahramani, Z.; Durrant-Whyte, H. Simultaneous localization and mapping with sparse extended information filters. *Int. J. Robot. Res.* **2004**, *23*, 693–716. [CrossRef]
5. Montemerlo, M.S. Fastslam: A factored solution to the simultaneous localization and mapping problem with unknown data association. *Arch. Environ. Contam. Toxicol.* **2015**, *50*, 240–248.
6. Gutmann, J.S.; Konolige, K. Incremental mapping of large cyclic environments. In Proceedings of the 1999 IEEE International Symposium on Computational Intelligence in Robotics and Automation (CIRA '99), Monterey, CA, USA, 8–9 November 2002; pp. 318–325.
7. Zlot, R.; Bosse, M. Efficient large-scale 3D mobile mapping and surface reconstruction of an underground mine. In *Field and Service Robotics*; Springer: Berlin/Heidelberg, Germany, 2014; pp. 479–493.
8. Olufs, S.; Vincze, M. An efficient area-based observation model for monte-carlo robot localization. In Proceedings of the 2009 IEEE/RSJ International Conference on Intelligent Robots and Systems (IROS 2009), St. Louis, MO, USA, 10–15 October 2009; pp. 13–20.
9. Huang, S.; Wang, Z.; Dissanayake, G. Sparse local submap joining filter for building large-scale maps. *IEEE Trans. Robot.* **2008**, *24*, 1121–1130. [CrossRef]
10. Moosmann, F.; Stiller, C. Velodyne slam. In Proceedings of the 2011 IEEE Intelligent Vehicles Symposium (IV), Baden, Germany, 5–9 June 2011; pp. 393–398.
11. Nüchter, A.; Lingemann, K.; Hertzberg, J.; Surmann, H. 6D SLAM-3D mapping outdoor environments. *J. Field Robot.* **2007**, *24*, 699–722. [CrossRef]
12. Zhang, J.; Singh, S. LOAM: Lidar Odometry and Mapping in Real-time. In Proceedings of the 2014 Robotics: Science and Systems, Berkeley, CA, USA, 12–16 July2014; Volume 2.
13. Wang, J.; Li, L.; Zhe, L.; Weidong, C. Multilayer matching SLAM for large-scale and spacious environments. *Int. J. Adv. Robot. Syst.* **2015**, *12*, 124. [CrossRef]
14. Liu, Z.; Chen, H.; Di, H.; Tao, Y.; Gong, J.; Xiong, G.; Qi, J. Real-Time 6D Lidar SLAM in Large Scale Natural Terrains for UGV. In Proceedings of the 2018 IEEE Intelligent Vehicles Symposium (IV), Changshu, China, 26–30 June 2018; pp. 662–667.
15. Liang, X.; Chen, H.; Li, Y.; Liu, Y. Visual laser-SLAM in large-scale indoor environments. In Proceedings of the IEEE International Conference on Robotics and Biomimetics, Qingdao, China, 3–7 December 2017; pp. 19–24.
16. Hess, W.; Kohler, D.; Rapp, H.; Andor, D. Real-time loop closure in 2D LIDAR SLAM. In Proceedings of the IEEE International Conference on Robotics and Automation, Stockholm, Sweden, 16–21 May 2016; pp. 1271–1278.
17. Kaess, M.; Ranganathan, A.; Dellaert, F. iSAM: Incremental smoothing and mapping. *IEEE Trans. Robot.* **2008**, *24*, 1365–1378. [CrossRef]
18. Hartley, H.O. The modified Gauss-Newton method for the fitting of non-linear regression functions by least squares. *Technometrics* **1961**, *3*, 269–280. [CrossRef]
19. Moré, J.J. The Levenberg-Marquardt algorithm: Implementation and theory. In *Numerical Analysis*; Springer: Berlin/Heidelberg, Germany, 1978; pp. 105–116.
20. Besl, P.J.; Mckay, N.D. Method for registration of 3-D shapes. *IEEE Trans. Pattern Anal. Mach. Intell.* **2002**, *14*, 239–256. [CrossRef]

21. Li, L.; Liu, J.; Zuo, X.; Zhu, H. An Improved MbICP Algorithm for Mobile Robot Pose Estimation. *Appl. Sci.* **2018**, *8*, 272. [CrossRef]
22. Chen, Y.; Medioni, G. Object modelling by registration of multiple range images. *Image Vis. Comput.* **1992**, *10*, 145–155. [CrossRef]
23. Rusu, R.B.; Cousins, S. 3D is here: Point Cloud Library (PCL). In Proceedings of the 2011 IEEE International Conference on Robotics and Automation, Shanghai, China, 9–13 May 2011; pp. 1–4.
24. Liu, Z.; Chen, W.; Wang, Y.; Wang, J. Localizability estimation for mobile robots based on probabilistic grid map and its applications to localization. In Proceedings of the 2012 IEEE Conference on Multisensor Fusion and Integration for Intelligent Systems (MFI), Hamburg, Germany, 13–15 September 2015; pp. 46–51.
25. Wang, Y.; Chen, W.; Wang, J.; Wang, H. Active global localization based on localizability for mobile robots. *Robotica* **2015**, *33*, 1609–1627. [CrossRef]
26. Bobrovsky, B.; Zakai, M. A lower bound on the estimation error for certain diffusion processes. *IEEE Trans. Inf. Theory* **1976**, *22*, 45–52. [CrossRef]
27. Wang, B.; Guo, R.; Li, B.; Han, L.; Sun, Y.; Wang, M. SmartGuard: An autonomous robotic system for inspecting substation equipment. *J. Field Robot.* **2012**, *29*, 123–137. [CrossRef]
28. Burgard, W.; Stachniss, C.; Grisetti, G.; Steder, B.; Kümmerle, R.; Dornhege, C.; Ruhnke, M.; Kleiner, A.; Tardös, J.D. A comparison of SLAM algorithms based on a graph of relations. In Proceedings of the 2009 IEEE/RSJ International Conference on Intelligent Robots and Systems (IROS 2009), St. Louis, MO, USA, 10–15 October 2009; pp. 2089–2095.
29. Geiger, A. Are we ready for autonomous driving? The KITTI vision benchmark suite. In Proceedings of the 2012 Computer Vision and Pattern Recognition, Providence, RI, USA, 16–21 June 2012; pp. 3354–3361.

applied sciences

MDPI

Article

# A Graph Representation Composed of Geometrical Components for Household Furniture Detection by Autonomous Mobile Robots

**Oscar Alonso-Ramirez [1,*], Antonio Marin-Hernandez [1,*], Homero V. Rios-Figueroa [1], Michel Devy [2], Saul E. Pomares-Hernandez [3] and Ericka J. Rechy-Ramirez [1]**

1 Artificial Intelligence Research Center, Universidad Veracruzana, Sebastian Camacho No. 5, Xalapa 91000, Mexico; hrios@uv.mx (H.V.R.-F.); erechy@uv.mx (E.J.R.-R.)
2 CNRS-LAAS, Université Toulouse, 7 avenue du Colonel Roche, F-31077 Toulouse CEDEX, France; devy@laas.fr
3 Department of Electronics, National Institute of Astrophysics, Optics and Electronics, Luis Enrique Erro No. 1, Puebla 72840, Mexico; spomares@inaoep.mx
* Correspondence: oscalra_820@hotmail.com (O.A.-R.); anmarin@uv.mx (A.M.-H.); Tel.: +52-228-817-2957 (A.M.-H.)

Received: 30 September 2018; Accepted: 8 November 2018; Published: 13 November 2018

**Abstract:** This study proposes a framework to detect and recognize household furniture using autonomous mobile robots. The proposed methodology is based on the analysis and integration of geometric features extracted over 3D point clouds. A relational graph is constructed using those features to model and recognize each piece of furniture. A set of sub-graphs corresponding to different partial views allows matching the robot's perception with partial furniture models. A reduced set of geometric features is employed: horizontal and vertical planes and the legs of the furniture. These features are characterized through their properties, such as: height, planarity and area. A fast and linear method for the detection of some geometric features is proposed, which is based on histograms of 3D points acquired from an RGB-D camera onboard the robot. Similarity measures for geometric features and graphs are proposed, as well. Our proposal has been validated in home-like environments with two different mobile robotic platforms; and partially on some 3D samples of a database.

**Keywords:** service robot; graph representation; similarity measure

## 1. Introduction

Nowadays, the use of service robots is more frequent in different environments for performing tasks such as: vacuuming floors, cleaning pools or mowing the lawn. In order to provide more complex and useful services, robots need to identify different objects in the environment; but also, they must understand the uses, relationships and characteristics of objects in the environment.

The extraction of an object's characteristics and its spatial relationships can help a robot to understand what makes an object useful. For example, the largest surfaces of a table and a bed differ in planarity and height; then, modeling object's characteristics can help the robot to identify their differences. A robot with a better understanding of the world is a more efficient service robot.

A robot can reconstruct the environment geometry through the extraction of geometrical structures on indoor scenes. Wall, floor or ceiling extractions are already widely used for environment characterization; however, there are few studies on extracting the geometric characteristics of furniture. Generally, the extraction is performed on very large scenes, composed of multiple scans and points of view, while we extract the geometric characteristics of furniture from only a single point of view (Figure 1).

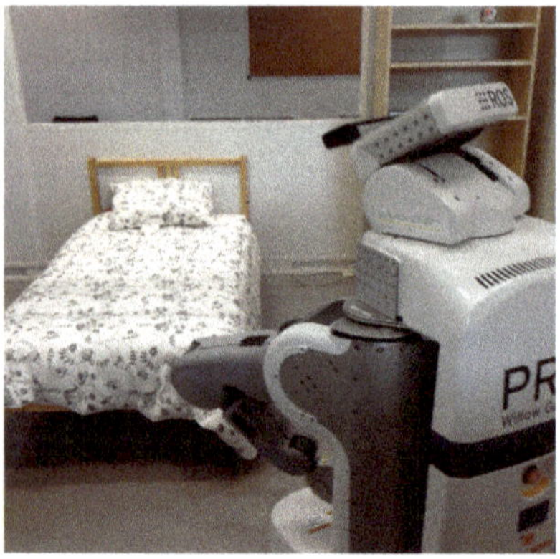

**Figure 1.** A service robot in a home-like environment.

Human environments are composed of many type of objects. Our study focuses on household furniture that can be moved by a typical human and that is designed to be moved during normal usage.

Our study omits kitchen furniture, fixed bookcases, closets or other static or fixed objects; because they can be categorized in the map as fixed components, therefore, the robot will always know their position. The frequency each piece of furniture is repositioned will vary widely: a chair is likely to move more often than a bed. Furthermore, the magnitude of repositioning a piece of furniture will vary widely: the intentional repositioning of a chair will likely move farther than the incidental repositioning of a table.

When an object is repositioned, if the robot does not extract the new position of the object and update its knowledge of the environment, then the robot's localization will be less accurate.

The main idea is to model these pieces of furniture (offline) in order to detect them on execution time and add them to the map simultaneously, not as obstacles, but as objects with semantic information that the robot could use later.

In this work, the object's horizontal or quasi-horizontal planes are key features (Figure 2). These planes are common in home environments: for sitting at a table, for lounging, or to support another object. The main difference between various planes is whether the horizontal plane is a regular or irregular flat surface. For example, the horizontal plane of a dining table or the horizontal plane of the top of a chest of drawers is different from the horizontal plane of a couch or the horizontal plane of a bed.

Modeling these planes can help a robot to detect those objects and to understand the world. It can help a robot, for example, to know where it can put a glass of water.

This work proposes that each piece of furniture is modeled with graphs. The nodes represent geometrical components, and the arcs represent the relationships between the nodes. Each graph of a piece of furniture has a main node representing a horizontal plane, generally the horizontal plane most commonly used by a human. Furthermore, this principle vertex is normally the horizontal plane a typical human can view from a regular perspective (Figure 2). In our framework, we take advantage of the fact that the horizontal plane most easily viewed by a typical human is usually the horizontal plane most used by a human. The robot has cameras positioned to provide the robot with a point of view similar to the point of view of a typical human.

**Figure 2.** Common scene of a home environment. Each object's horizontal plane is indicated with green.

To recognize the furniture, the robot uses an RGB-D camera to acquire three-dimensional data from the environment. After acquiring 3D data, the robot extracts the geometrical components and creates a graph for each object in the scene. Because the robot maintains the relationships between all coordinates of its parts, the robot can transform the point cloud to a particular reference frame, which simplifies the extraction of geometric components; specifically, to the robot's footprint reference frame.

From transformed point clouds of a given point of view, the robot extracts graphs corresponding to partial views for each piece of furniture in the scene. The graphs generated are later compared with the graphs of the object's models contained in a database.

The main contributions of this study are:

- a graph representation adapted to detect pieces of furniture using an autonomous mobile robot;
- a representation of partial views of furniture models given by sub-graphs;
- a fast and linear method for geometric feature extraction (planes and poles);
- metrics to compare the partial views and characteristics of geometric components; and
- a process to update the environment map when furniture is repositioned.

## 2. Related Work

RGB-D sensors have been widely used in robots; therefore, they are excellent for extracting information about diverse tasks in diverse environments. These sensors have been employed to solve many tasks; e.g., to construct 3D environments, for object detection and recognition and in human-robot interaction. For many tasks, but particularly when mobile robots must detect objects or humans, the task must be solved in real time or near real time; therefore, the rate at which the robot processes information is a key factor. Hence, an efficient 3D representation of objects that can quickly and accurately detect them is important.

Depending on the context or the environment, there are different techniques to detect and represent 3D objects. The most common techniques are based on point features.

The extraction of some 3D features is already available in libraries like Point Cloud Library (PCL) [1], including: spin images or fast point feature histograms. Those characteristics provide good results as the quantity of points increases. An increase in points, however, increases the computational time. A large object, such as furniture, requires computational times that are problematic for real-time tasks. A comparative evaluation of PCL 3D features on point clouds was given in [2].

The use of RGB-D cameras for detecting common objects (e.g., hats, cups, cans, etc.) has been accomplished by many research teams around the world. For example, a study [3] presented an approach based on depth kernel features that capture characteristics such as size, shape and edges. Another study [4] detected objects by combining sliding window detectors and 3D shape.

Others works ([5,6]) have followed a similar approach using features to detect other types of free-form objects. For instance, in [5], 3D-models were created and objects detected simultaneously by using a local surface feature. Additionally, local features in a multidimensional histogram have been combined to classify objects in range images [6]. These studies used specific features extracted from the objects and then compared the extracted features with a previously-created database, containing the models of the objects.

Furthermore, other studies have performed 3D object-detection based on pairs of points from oriented surfaces. For example, Wahl et al. [7] proposed a four-dimensional feature invariant to translation and rotation, which captures the intrinsic geometrical relationships; whereas Drost et al. [8] have proposed a global-model description based on oriented pairs of points. These models are independent from local surface-information, which improves search speed. Both methods are used to recognize 3D free-form objects in CAD models.

In [9], the method presented in [8] was applied to detect furniture for an "object-oriented" SLAM technique. By detecting multiple repetitive pieces of furniture, the classic SLAM technique was extended. However, they used a limited range of types of furniture, and a poor detection of furniture was reported when the furniture was distant or partially occluded.

On the other hand, Wu et al. [10] proposed a different object representation to recognize and reconstruct CAD models from pieces of furniture. Specifically, they proposed to represent the 3D shape of objects as a probability distribution of binary variables on a 3D voxel grid using a convolutional deep belief network.

As stated in [11], it is reasonable to represent an indoor environment as a collection of planes because a typical indoor environment is mostly planar surfaces. In [12], using a 3D point cloud and 2D laser scans, planar surfaces were segmented, but those planes are used only as landmarks for map creation. In [13], geometrical structures are used to describe the environment. In this work, using rectangular planes and boxes, a kitchen environment is reconstructed in order to provide to the robot a map with more information about, i.e., how to use or open a particular piece of furniture.

A set of planar structures to represent pieces of furniture was presented in [14], which stated that their planar representations "have a certain size, orientation, height above ground and spatial relation to each other". This method was used in [15] to create semantic maps of furniture. This method is similar to our framework; however, their method used a set of rules, while our method uses a probabilistic framework. Our method is more flexible, more able to deal with uncertainty, more able to process partial information and can be easily incorporated into many SLAM methods.

In relation to plane extraction, a faster alternative than the common methods for plane segmentation was presented in [16]. They used integral images, taking advantage of the structured point cloud from RGB-D cameras.

Another option is the use of semantic information from the environment to improve the furniture detection. For example in [17], geometrical properties of the 3D world and the contextual relations between the objects were used to detect objects and understand the environment. By using a Conditional Random Field (CRF) model, they integrated object appearance, geometry and relationships with the environment. This tackled some of the problems with feature-based approaches, including pose variation, object occlusion or illumination changes.

In [18], the use of the visual appearance, shape features and contextual relations such as object co-occurrence was proposed to semantically label a full 3D point cloud scene. To use this information, they proposed a graphical model isomorphic to a Markov random field.

The main idea of our approach is to improve the understanding of the environment by identifying the pieces of furniture, which is still a very challenging task, as stated in [19]. These pieces of furniture will be represented by a graph structure as a combination of geometrical entities.

Using graphs to represent environment relations was done in [20,21]. Particularly, in [20], a semantic model of the scene based on objects was created; where each node in the graph represented an object and the edges represented their relationship, which were also used to improve the object's detection. The work in [21] used graphs to describe the configuration of basic shapes for the detection of features over a large point cloud. In this case, the nodes represent geometric primitives, such as planes, cylinders, spheres, etc.

Our approach uses a representation similar to [21]. Each object is decomposed into geometric primitives and represented by a graph. However, our approach differs because it processes only one point cloud at a time, and it is a probabilistic framework.

## 3. Furniture Model Representation and Similarity Measurements

Our proposal uses graphs to represent furniture models. Specifically, our approach focuses on geometrical components and relationships in the graph instead of a complete representation of the shape of the furniture.

Each graph contains the furniture's geometrical components as nodes or a vertex. The edges or arcs represent the adjacency of the geometrical components. These geometrical components are described using a set of features that characterize them.

### 3.1. Furniture Graph Representation

A graph is defined as an ordered pair $G = (V, E)$ containing a set $V$ of vertices or nodes and a set $E$ of edges or arcs. A piece of furniture $F^i$ is represented by a graph as follows:

$$F^i = (V^i, E^i), \qquad \text{with } i = [1, ..., N_f] \tag{1}$$

where $F^i$ is an element from the set of furniture models $\mathcal{F}$; the sets $V^i$ and $E^i$ contain the vertices and edges associated with the $i$th class; and $N_f$ is the number of models in the set $\mathcal{F}$, i.e., $|\mathcal{F}| = N_f$.

The set of vertices $V^i$ and the set of edges $E^i$ are described using lists as follows:

$$\begin{aligned} V^i &= \{v_1^i, v_2^i, ..., v_{n_v^i}^i\} \\ E^i &= \{e_1^i, e_2^i, ..., e_{n_e^i}^i\} \end{aligned} \tag{2}$$

where $n_v^i$ and $n_e^i$ are the number of vertices and edges, respectively, for the $i$th piece of furniture.

The functions $V(F^i)$ and $E(F^i)$ are used to recover the corresponding lists of vertex and edges of the graph $F^i$.

An edge $e_j^i$ is the $j$th link in the set, joining two nodes for the $i$th piece of furniture. As connections between nodes are a few, we use a simple list to store them. Thus, each edge $e_j^i$ is described as:

$$e_j^i = (a, b) \tag{3}$$

where $a$ and $b$ correspond to the linked vertices $v_a^i, v_b^i \in V^i$, such that $a \neq b$; and since the graph is an undirected graph, $e_j^i = (a, b) = (b, a)$.

### 3.2. Geometric Components

The vertices on a furniture graph model are geometric components, which roughly correspond to the different parts of the furniture. For instance, a chair has six geometric components: one horizontal plane for sitting, one vertical plane for the backrest and four tubes for the legs. Each component has different characteristics to describe it.

Generally, a geometric component $Gc^k$ is a non-homogeneous set of characteristics:

$$Gc^k = \{ft_1^k, ft_2^k, ..., ft_{n_{ft}^k}^k\} \tag{4}$$

where $k$ designates an element from the set $\mathcal{G}c$, which contains $N_k$ different types of geometric components, and $n_{ft}^k$ is the number of characteristics of the $k$th geometric component. Characteristics or features of a geometrical component can be of various types or sources. A horizontal plane, for example, includes characteristics of height, area and relative measures.

Each vertex $v_j^i$ is then a geometric component of type $k$. The function $G_c(v_j^i)$ returns then the type and the set of features of geometric component $k$.

Figure 3 shows an example of a furniture model $F^i$. It is composed of four vertices ($n_v^i = 4$) and three edges ($n_e^i = 3$). There are three different types of vertices because each geometric component ($N_k = 3$) is represented graphically with a different shape of the node.

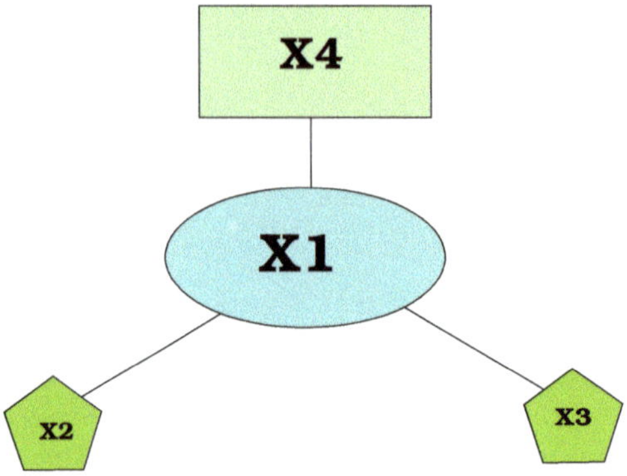


**Figure 3.** Example of a furniture graph with four nodes and three edges. The shape of the node correspond to a type of geometric component.

Despite the simplified representation of the furniture, the robot cannot see all furniture components from a single position.

### 3.3. Partial Views

At any time, the robot's perspective limits the geometric components the robot can observe. For each piece of furniture, the robot has a complete graph including every geometric component of the piece of furniture, then some subgraphs should be generated corresponding to different views that a robot can have. Each sub-graph contains the geometric components the robot would observe from a hypothetical perspective.

Considering the following definition, a graph $H$ is called a subgraph of $G$, such that $V(H) \subseteq V(G)$ and $E(H) \subseteq E(G)$, then a partial view $Fp^i$ for a piece of furniture is a subgraph from $F^i$, which is described as:

$$Fp^i = (\tilde{V}^i, \tilde{E}^i) \tag{5}$$

such that $\tilde{V}^i \subseteq V^i$ and $\tilde{E}^i \subseteq E^i$. The number of potential partial views is equal to the number of possible subsets in the set $F^i$; however, not all partial views are useful. See Section 4.3.

In order to match robot perception with the generated models, similarity measurements for graphs and geometric components should be defined.

### 3.4. Similarity Measurements

#### 3.4.1. Similarity of Geometric Components

Generally, a similarity measure $s_{Gc}$ of two type $k$ geometric components $Gc^k$ and $Gc^{k\prime}$ is defined as:

$$s_{Gc}^k(Gc^k, Gc^{k\prime}) = 1 - \sum_i^{n_{ft}^k} w_{Gi}^k d(ft_i^k, ft_i^{k\prime}) \tag{6}$$

where $k$ represents the type of geometric component, $w_{Gi}$ are weights and $d(ft_i^k, ft_i^{k\prime})$ is a function of the difference of the $i$th feature of the geometric components $Gc^k$ and $Gc^{k\prime}$, defined as follows:

$$\delta\varphi = \frac{|ft_i^k - ft_i^{k\prime}| - \epsilon_{ft_i}}{ft_i^k} \quad (7)$$

then:

$$d(ft_i^k, ft_i^{k\prime}) = \begin{cases} 0, & \delta\varphi < 0 \\ \delta\varphi, & 0 \leq \delta\varphi \leq 1 \\ 1, & \delta\varphi > 1 \end{cases} \quad (8)$$

where $\epsilon_{ft_i}$ is a measure of the uncertainty of the $i$th characteristic. $\epsilon_{ft_i}$ is considered a small value that specifies the tolerance between two characteristics. Equation (7) normalizes the difference; whereas Equation (8) equals zero if the difference of two characteristics is within the acceptable uncertainty $\epsilon_{ft_i}$ and equals one when they are totally different.

### 3.4.2. Similarity of Graphs

Likewise, the similarity $s_F$ of two furniture graphs (or partial graphs) $F^i$ and $F^{i\prime}$ is defined as:

$$s_F(F^i, F^{i\prime}) = \sum_{j}^{n_v^i} w_{Fj} s_{Gc}^k(Gc_j^k, Gc_j^{k\prime}) \quad (9)$$

where $w_{Fj}$ are weights, corresponding to the contribution of the similarity $s_{Gc}$ between the corresponding geometric components $j$ to the graph model $F^i$.

It is important to note that:

$$\sum_{i}^{n_{ft}^k} w_{Gi}^k = 1 \qquad \sum_{j}^{n_v^i} w_{Fj} = 1 \quad (10)$$

In the next section, values for Equations (6) and (9), in a specific context and environment, will be provided.

## 4. Creation of Models and Extraction of Geometrical Components

In order to generate the proposed graphs, 3D models per furniture are required; so that geometrical components can be extracted.

Nowadays, it is possible to find online 3D models for a wide variety of objects and in many diverse formats. However, those 3D models contain many surfaces and components, which are never visible from a human (or a robot) perspective (e.g., the bottom of a chair or table). It could be possible to generate the proposed graph representation from those 3D models; nevertheless, it would be necessary to make some assumptions or eliminate components not commonly visible.

At this stage of our proposal, the particular model of each piece of furniture is necessary is necessary, not a generic model of the type of furniture. Furthermore, it was decided to construct the model for each piece of furniture from real views taken by the robot; consequently, visible geometrical components of the furniture can be extracted, and then, an accurate graph representation can be created.

Furniture models were generated from point clouds obtained with an RGB-D camera mounted on the head of the robot, in order to have a similar perception to a human being. The point clouds were merged together with the help of an Iterative Closest Point (ICP) algorithm. In order to make an accurate registration, the ICP algorithm finds and uses the rigid transformation between two point clouds. Finally, a downsample was performed to get an even distribution of the points on the 3D model. This is achieved by dividing the 3D space into voxels and combining the points that lie within into one output point. This allow reducing the number of points in the point cloud while maintaining the characteristics as a whole. Both the ICP and the downsampling algorithm were used from the PCL library [1] In Figure 4, some examples are shown of the 3D point cloud models.

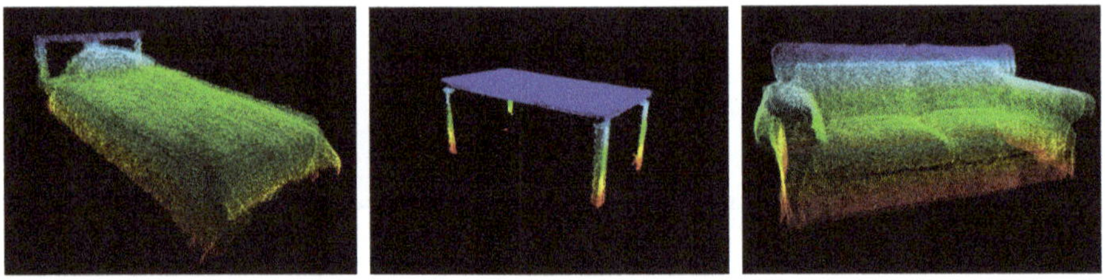

**(a)** **(b)** **(c)**

**Figure 4.** Example of the created models, for (**a**) a bed, for (**b**) a table and for (**c**) a couch.

### *4.1. Extraction of Geometrical Components*

Once 3D models of furniture are available, the type of geometric components should be chosen and then extracted. At this stage of the work, it has been decided to use a reduced set of geometrical components, which is composed of horizontal and vertical planes and legs (poles). As can been seen, most of the furniture is composed of flat surfaces (horizontal or vertical) and legs or poles. Conversely, some surfaces are not strictly flat (e.g., the horizontal surface of a bed or the backrest of a coach); however, many of them can be roughly approximated to a flat surface with some relaxed parameters. For example, the main planes for a bed and a table can be approximated by a plane equation, but with different dispersion; a small value for the table and a bigger value for the bed. Currently, curve vertical surfaces have not been considered in our study. Nevertheless, they can be incorporated later.

#### 4.1.1. Horizontal Planes' Extraction

Horizontal plane detection and extraction is achieved using a method based on histograms. For instance, a table and a bed have differences in their horizontal planes. In order to capture the characteristics of a wide variety of horizontal planes, three specific tasks are performed:

1. Considering that a robot and its sensors are correctly linked (i.e., between all reference frames, fixed or mobile), it is possible to obtain a transformation for a point cloud coming from a sensor in the robot's head into a reference frame to the base or footprint of the robot (see Figure 5). The TF package in ROS (Robotic Operation System) performs this transformation at approximately 100 Hz.
2. Once transformation between corresponding references frames is performed, a histogram of heights of the points in the cloud with reference to the floor is constructed.

   Over this histogram, horizontal planes (generally composed of a wide set of points) generate a peak or impulse. By extracting all the points that lie over those regions (peaks), the horizontal planes can be recovered. The form and characteristics of the peak or impulse in the histogram refer to the characteristics of the plane. For example, the widths of the peaks in Figure 6 are different; these correspond to: (1) a flat and regular surface for a chest of drawers (Figure 6b) and (2) a rough surface of a bed (Figure 6d).

   Nevertheless, there can be several planes merged together in a peak in a scene, i.e., two or more planes with the same height.
3. To separate planes merged together in a peak in a scene, first, all the points of the peak are extracted, and then, they are projected to the floor plane. A clustering algorithm is then performed in order to separate points corresponding to each plane, as shown in Figure 7.

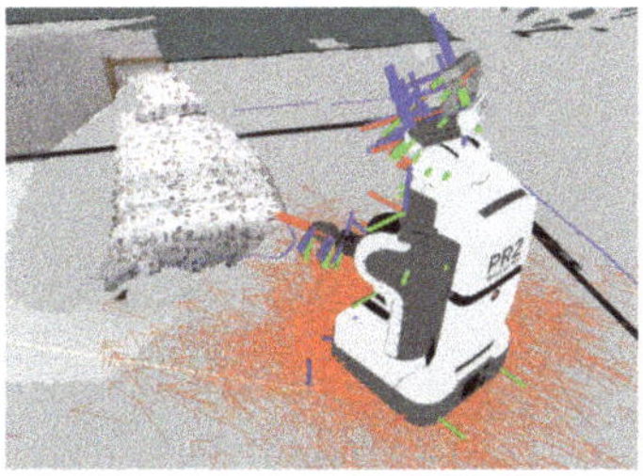

**Figure 5.** Visualization of a PR2 robot with its coordinate frames and the point cloud from the scene transformed to the world reference frame.

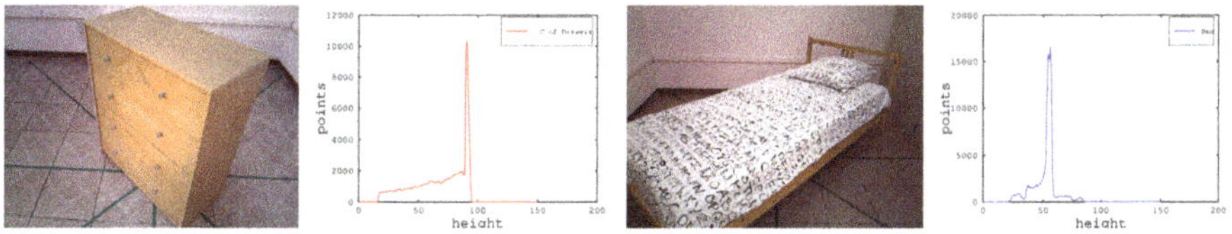


(**a**) (**b**) (**c**) (**d**)

**Figure 6.** Example of a height histogram of two different pieces of furniture. In (**a**), the RGB image of a chest of drawers, and in (**b**), its height histogram. In (**c**,**d**), the RGB image of a bed and its height histogram, respectively.

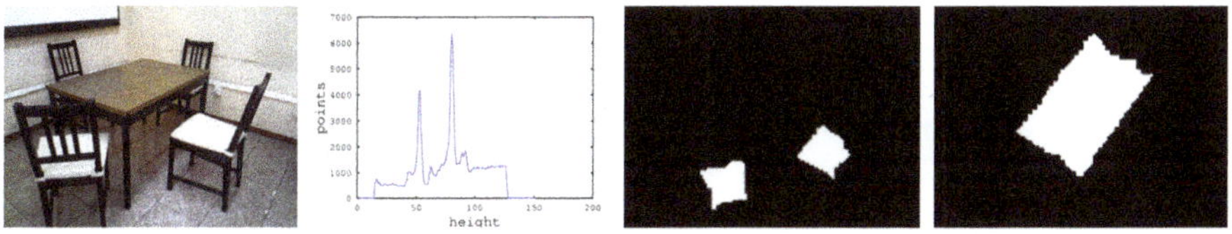


(**a**) (**b**) (**c**) (**d**)

**Figure 7.** Example of a height histogram and the objects segmentation. In (**a**), the RGB from the scene, in (**b**), the height histogram, and (**c**,**d**) show the projections for each peak found on the histogram.

#### 4.1.2. Detection of Vertical Planes and Poles

Vertical planes' and poles' extraction follows a similar approach. Specifically, they are obtained as follows:

1. In these cases, the distribution of the points is analyzed on a 2D histogram generated by projecting all the points into the floor plane.
2. Considering this 2D histogram as a grayscale image, all the points from a vertical plane will form a line on the image, and the poles will form a spot; then extracting the lines and spots, and their corresponding projected points, will provide the vertical planes and poles (Figure 8).

Finally, image processing algorithms can be applied in order to segment those lines on the 2D histogram and then recover points corresponding to those vertical planes.

In the cases where two vertical planes are projected to the same line, they can be separated by a similar approach mentioned in the previous section for segmenting horizontal planes; however in this case, the points are projected to a vertical plane. Then, it is possible to cluster them and perform some calculations on them like the area of the projected surface. For the current state of the work, we are more interested in the form of the furniture rather than its parts, so adjacent planes belonging to the same piece of furniture, i.e., the drawers of a chest of drawers, are not separated. If the separation of these planes were necessary, an approach similar to the one proposed in [22] could be used; where they segmented the drawers from a kitchen.

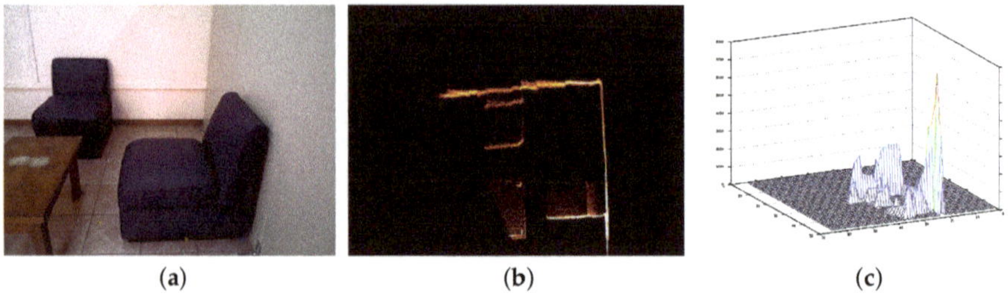


**Figure 8.** Example of a floor projection and its 2D histogram. In (**a**), the RGB image from the scene, in (**b**), the floor projection, and in (**c**), the 2D histogram.

With this approach, it is also possible to detect small regions or dots that correspond to the legs of the chairs or tables. Despite the small sizes of the regions or dots, they are helpful to characterize the furniture.

Our proposal can work with full PCD scenes without any requirement of a downsampling, which must be performed in algorithms like RANSAC in order to maintain a low computational cost. Moreover, histograms are computed linearly.

### 4.2. Characteristics of the Geometrical Components

Every geometrical component must be characterized in order to complete the graph for every piece of furniture. For simplicity, at this point, all the geometrical components have the same characteristics. However, more geometrical components can be added or replaced in the future. The following features or characteristics of the geometrical components are considered:

- Height: the average height of the points belonging to the geometric component.
- Height deviation: standard height deviation of the points in the peak or region.
- Area: area covered by the points.

Figure 9 shows the values of each characteristic for the main horizontal plane of some furniture models. The sizes of the boxes were obtained by a min-max method. More details will be given in Section 5.2. The parallelepiped represents the variation (uncertainty) for each variable.

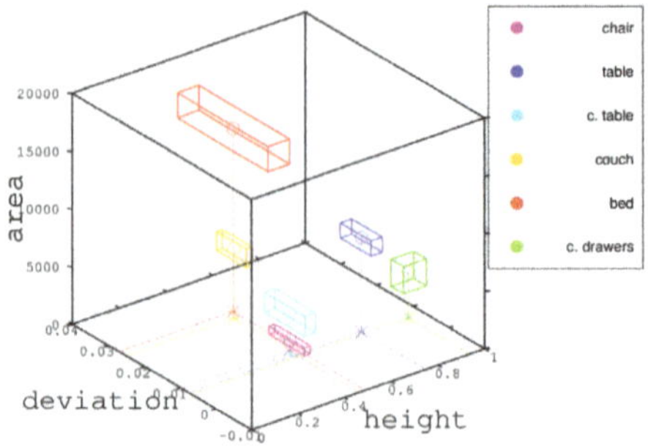


**Figure 9.** Characteristics of the main geometrical component (horizontal plane) for diverse pieces of furniture.

### 4.3. Examples of Graph Representation

Once geometric components have been described, it is possible to construct a graph for each piece of furniture.

Let $F^*$ be a piece of furniture (e.g., a table); therefore, as stated in Equation (1), the graph is described as:

$$F^* = (V^*, E^*)$$

where $V^*$ has five vertices and $E^*$ four edges (i.e., $n_v^* = 5$ and $n_e^* = 4$). The five vertices correspond to: one vertex for the horizontal surface of the table and one for each of the four legs. The main vertex is the horizontal plane, and an edge will be added whenever two geometrical components are adjacent, so in this particular case, the edges correspond to the union between each leg and the horizontal plane.

Figure 10a presents the graph corresponding to a dinning table. Similarly, Figure 10b shows the graph of a chair. It can be seen that the chair's graph has one more vertex than the table's graph. This extra vertex is a vertical component corresponding to the backrest of the chair.

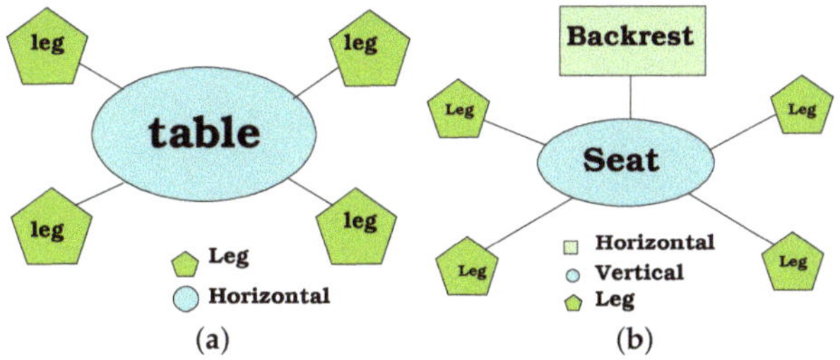


**Figure 10.** Example of complete graph models for: (**a**) a dinning table and (**b**) a chair; different shapes represent different types of geometric components.

As mentioned earlier, a graph serves as a complete model for each piece of furniture because it contains all its geometrical components. However, as described in Section 3.3, the robot cannot view all the components of a given piece of furniture because of the perspective. Therefore, subgraphs are created in order to compare robot perception with models.

To generate a small set of sub-graphs corresponding to partial views, several points of view have been grouped into four quadrants. These are: two subgraphs for the front left and right views and two more for the back view left and right (Figure 11). However, due to symmetry and occlusions, the set of subgraphs can be reduced. For example, in the case of a dining table, there is only one graph without sub-graphs because its four legs can be seen from many viewpoints.

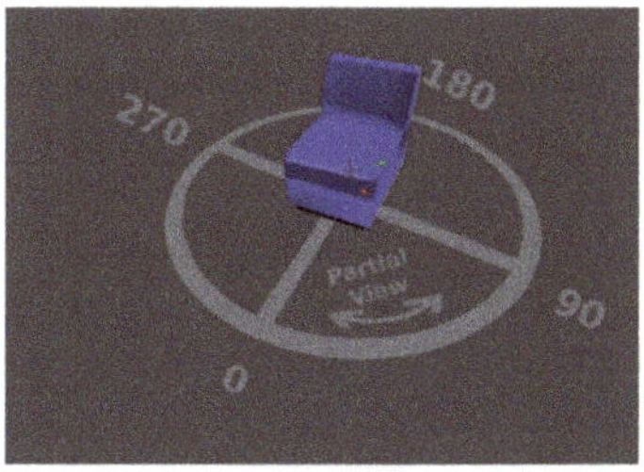


**Figure 11.** Visualizing the different points of view used to generate partial views.

Consequently, graphs require also to specify which planes are on opposite sides (if there are any), because this information is important to specify which components are visible from every view.

The visibility of a given plane is encoded at the vertex. For example, for a chest of drawers, it is not possible to see the front and the back at the same time.

Figure 12 shows an example of a graph model and some subgraphs for a couch graph model. It can be seen from Figure 12a that the small rectangles to the side of the nodes indicate the opposite nodes. The sub-graphs (Figure 12b,c) represent two sub-graphs of the left and right frontal views, respectively. The reduction of the number of vertex of the graph is clear. Specifically, the backrest and the front nodes are shown, whereas the rear node is not presented because it is not visible for the robot from a frontal view of the furniture. Thus, a subgraph avoids comparing components that are not visible from a given viewpoint.

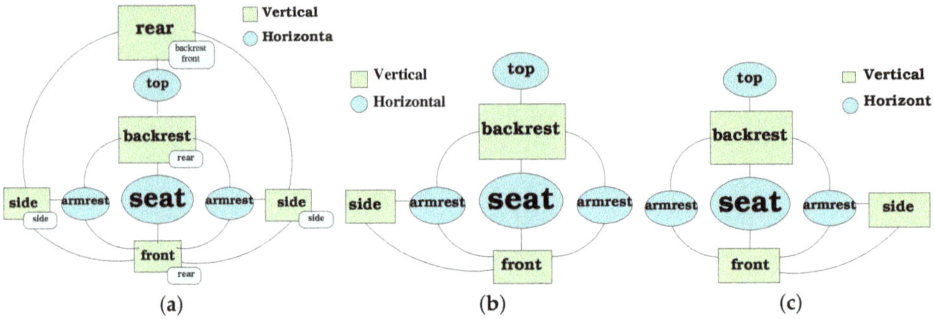


**Figure 12.** In (**a**), the graph corresponding to the complete couch graph, and in (**b**,**c**), two sub-graphs for the front left and right views.

## 5. Determination of Values for Models and Geometric Components

In order to validate our proposal, a home environment has been used. This environment is composed of six different pieces of furniture ($N_f = 6$), which are:

$$\mathcal{F} = \{dinning\ table, chair, couch, center\ table,\\ bed, chest\ of\ drawers\}$$

To represent the components of those pieces of furniture, the following three types of geometrical components have been selected:

$$\mathcal{G}c = \{horizontal\ plane, vertical\ plane, legs\}$$

and the features of the geometrical components are described in Section 4.2.

### 5.1. Weights for the Geometrical Components' Comparison

In order to compute the proposed similarity $s_{Gc}$ between two geometrical components, it is necessary to determine the corresponding weights $w_{Gi}$ in Equation (6). Those values have been determined empirically, as follows:

From a set of scenes taken by the robot, a set of them where each piece of furniture was totally visible. Then, geometrical components were extracted following the methodology proposed in Section 4.1. The weights were selected according to the importance of each feature to a correct classification of the geometrical component.

Table 1 shows the corresponding weights for the three geometrical components.

**Table 1.** Weights for similarity estimation.

| | $w^k_{G1}$ (Height) | $w^k_{G2}$ (h. Deviation) | $w^k_{G3}$ (Area) |
|---|---|---|---|
| $w^H_{Gi}$ (horizontal) | 0.65 | 0.15 | 0.25 |
| $w^V_{Gi}$ (vertical) | 0.5 | 0.2 | 0.3 |
| $w^L_{Gi}$ (legs) | 0.5 | 0.2 | 0.3 |

### 5.2. Uncertainty

Uncertainty values in Equation (7) were estimated using an empirical process. From some views selected for each piece of furniture fully observable, the difference with its correspondent model was calculated; in order to have an estimation of the variation of corresponding values, with the complete geometrical component. Then, the highest difference for each characteristic was selected as the uncertainty.

As can be seen from Figure 9, the use of characteristics (height, height deviation and area) is sufficient to classify the main horizontal planes. Moreover, over this space, characteristics and uncertainty from each horizontal plane make regions fully classifiable.

There are other features of the geometrical components that are useful to define the type of geometrical component or their relations, so they have been added to the vertex structure. These features are:

- Center: the 3D point center of the points that makes the geometrical component.
- PCA eigenvectors and eigenvalues: eigenvector and eigenvalues resulting from a PCA analysis for the region points.

The center is helpful to establish spatial relations, and the PCA values help to discriminate between vertical planes and poles. By finding and applying an orthogonal transformation, PCA converts a set of possible correlated variables to a set of linearly uncorrelated variables called principal components. Since, in PCA, the first principal component has the largest possible variance, then, in the case of poles, the first component should be aligned with the vertical axis, and the variance of other components should be significantly smaller. This not so in the case of planes, where two first components can have similar variance values. Components are obtained by the eigenvector and eigenvalues from PCA.

### 5.3. Weights for the Graphs' Comparison

Conversely, as weights were determined using Equation (7), the weights for the similarity between graphs (Equation (9)) were calculated based on the total area of models for each piece of furniture.

Given the projected area of each geometrical component of the graph model, the total area is calculated. Then, the weights for each vertex (geometrical component) have been defined as the percentage of its area in comparison to the total area. Moreover, when dealing with a subgraph from a model, the total area is determined by the sum of areas from the nodes from that particular view (subgraph). Thus, there is a particular weight vector for each graph and subgraph in the environment.

Table 2 shows the values of computed areas for the chest of drawers corresponding to the graph and subgraph of the model (Figure 13).

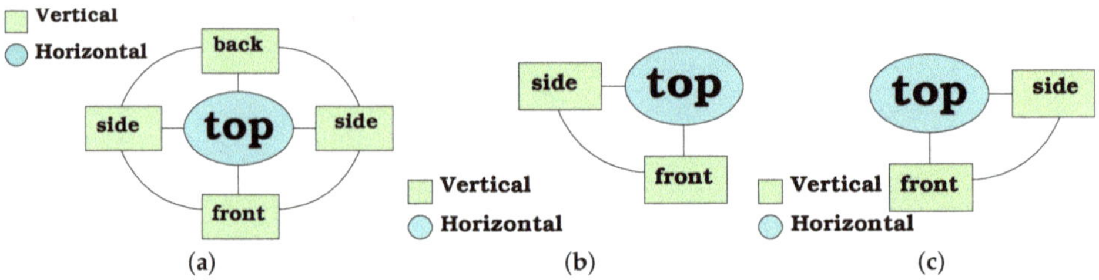


**Figure 13.** Graph models for the chest of drawers: in (**a**), the graph for the full model, and in (**b**,**c**), graphs for partial front views, left and right, respectively.

**Table 2.** Example of the weights for the comparison of the chest of drawers graph, based on the area of each geometrical component.

| | Side | Front | Side | Back | Top |
|---|---|---|---|---|---|
| area | 1575.5 | 9155.5 | 1575.5 | 9155.5 | 1914 |
| Model % | 6.74 | 39.16 | 6.74 | 39.16 | 8.19 |
| Partial View % | 12.45 | 72.41 | | | 15.13 |
| Partial View % | | 72.41 | 12.45 | | 15.13 |

Table 3 shows the weights in the case of the dinning table, which does not have sub-graphs (Figure 10a).

**Table 3.** Example of the weights for the comparison of the dining table graph, based on the area of each geometrical component.

| | Table | Leg | Leg | Leg | Leg |
|---|---|---|---|---|---|
| area | 8159 | 947 | 947 | 947 | 947 |
| Model % | 68.31 | 7.92 | 7.92 | 7.92 | 7.92 |
| Partial View % | 68.31 | 7.92 | 7.92 | 7.92 | 7.92 |

## 6. Evaluations

Consider an observation of a scene, where geometrical components have been extracted, by applying the methods described in Section 4.1.

Let $O$ be the set of all geometrical components observed on a scene, then:

$$O = \{O^1, ..., O^{N_k}\} \tag{11}$$

where $O^k$ is the subset of geometrical components of the type $k$.

In this way, observed horizontal geometrical components found on the scene are in the same subset, lets say $O^*$. Consequently, it is possible to extract each one of them in the subset and then compare them to the main nodes for each furniture graph.

Once the similarity between the horizontal components on the scene and the models has been calculated, all the categories with a similarity higher than a certain threshold are chosen as probable models for each horizontal component. A graph is then constructed for each horizontal component, where adjacent geometrical components are merged with it. Then, this graph is compared with the subgraphs of the probable models previously selected.

The first column in Figure 14 shows some scenes from the environment, where different pieces of furniture are present. Only four images with the six types of furniture are shown. The scene in Figure 14a is composed of a dinning table and a chair. After the geometrical components are extracted (Figure 14b), two horizontal planes corresponding to the dinning table and the chair are selected.

A comparison of those horizontal components to each one of the main nodes of the furniture graphs is performed. This results in two probable models (table and chest of drawers) for the plane labeled as "H0", which actually corresponds to the table, and three probable models (chair, center table and couch) for the plane labeled "H01" (which corresponds to a chair). The similarities computed can be observed as "Node Sim." in Table 4.

Next, graphs are constructed for each horizontal plane (the main node) and adding its adjacent components. In this case, both graphs have only one adjacent node.

Figure 15a shows the generated Graph ("G0") from the scene and the partial-views graphs from the selected probable models. It can be observed that "G0" has an adjacent leg node, so it can only be a sub-graph for the table graph since the chest of drawers graph has only adjacent vertical nodes.

For "G1" (Figure 16a), its adjacent node is a vertical node with a higher height than the main node, so it is matched to the backrest node of the chair and of the couch (Figure 16b,d). Moreover, there is no match with the center table graph (Figure 16c).

The similarity for adjacent nodes is noted in Table 4. Graph similarity is calculated with Equation (9) and shown in the last column of the table. "G0" is selected as a table and "G1" as a chair (Figure 14c).

Figure 14 shows the results of applying the described procedure to different scenes with different types of furniture. The first column (Figure 14a,d,g,j) shows the point clouds from the scenes. The column at the center (Figure 14b,e,h,k) shows the geometrical components found on the corresponding scene. Finally, the last column (Figure 14c,f,i,l) shows the generated graphs classified correctly.

**Table 4.** Example for graph classification.

| Main Node | Main Node Sim. | Adjacent Nodes | Adjacent Node Sim. | Graph Similarity |
|---|---|---|---|---|
| H00 | Table<br>0.8743 | V02 | Table leg<br>0.7015 | 0.0.6670 |
| H00 | Chest of Drawers<br>0.8014 | V02 | No Match | X |
| H01 | Chair<br>0.9891 | V01 | Backrest<br>0.8878 | 0.5740 |
| H01 | Center Table<br>0.8895 | V01 | No match | X |
| H01 | Couch<br>0.7277 | V01 | Backrest<br>0.7150 | 0.3204 |

As our approach is probabilistic, it can deal with the noise from the sensor; as well as partially occluded views, at this time, with occlusions no greater than 50% of the main horizontal plane, as this plane is the key factor in the graph.

While types of furniture can have the same graph structure, values in their components are particular for a given instance. Therefore, it is not possible to recognize with the same graph different instances of a type of furniture. In other words, a particular graph for a bed cannot be used for beds with different sizes; however, the graph structure could be the same.

Additionally, to test the approach on more examples, we have tested on selected images from the dataset SUNRGB-D [23]. Figure 17 shows the results of the geometrical components' extraction and the graph generation for the selected images. The color images corresponding to house environments similar to the test environments scenes are presented at the top of the figure; the corresponding geometrical component extraction is shown at the center of the figure; and the corresponding graphs are presented at the bottom of the figure.

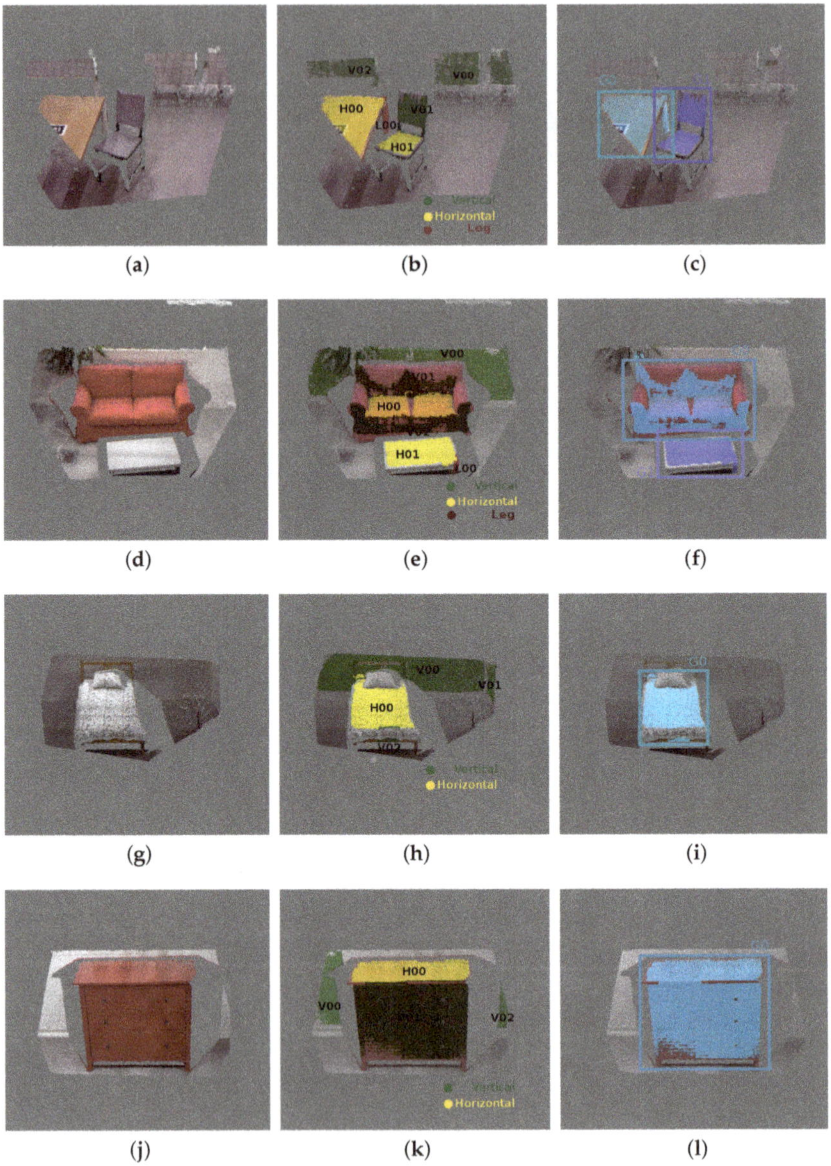


**Figure 14.** Results obtained for the furniture detection; each row shows the results for one scene. In (**a**,**d**,**g**,**j**), the original point clouds from the scenes. In (**b**,**e**,**h**,**k**) are shown the geometrical components found in the scene; the vertical planes are in green color, the horizontals in yellow and the legs in red. The bounding boxes in (**c**,**f**,**i**,**l**) show the graphs generated that were correctly identified as a piece of furniture.

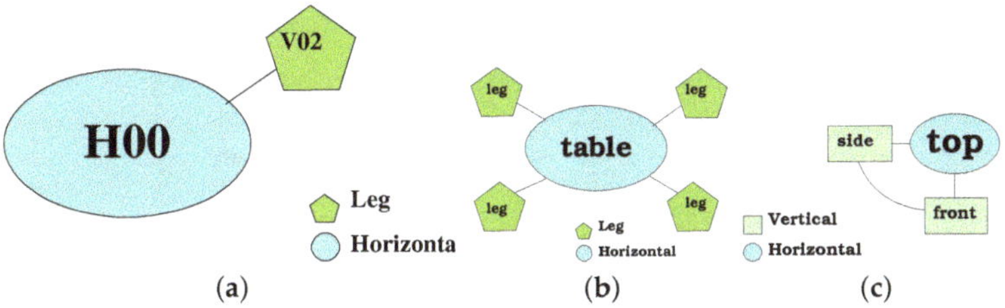


**Figure 15.** Graph comparison: in (**a**), one of the graphs generated from the scene in Figure 14b; in (**b**,**c**), partial graphs selected for matching with the graph in (**a**), corresponding to the table and the chest of drawers.

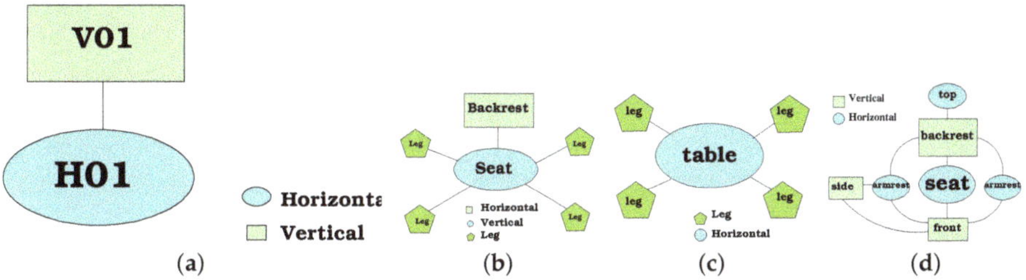


**Figure 16.** Graph comparison: in (**a**), one of the graphs generated from the scene in Figure 14b, and in (**b**–**d**), partial graphs selected for matching corresponding to the chair, the center table and the couch.

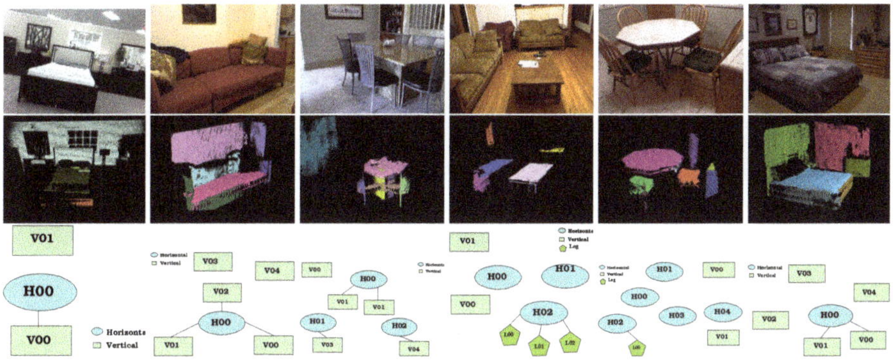


**Figure 17.** Tests over images from the dataset SUNRGB-D [23]: at the top, the RGB images, in the middle, the corresponding extracted geometrical components, and at the bottom, the corresponding graphs.

It is important to note that to detect the geometrical components over these scenes, the point cloud from each selected scene was treated using the intrinsic and extrinsic parameters provided in the dataset. Consequently, a similar footprint reference frame is obtained such as the frame provided by the robots in this work. This manual adjustment was necessary to simulate the translation needed to align the coordinate frame of the image with the environment, as described in Section 4.1.

Since the proposed approach requires a complete model for each piece of furniture, which were not provided by the dataset, it was not possible to match the graphs generated to a given piece of furniture. However, the approach has worked as expected (Figure 17).

## 7. Conclusions and Future Work

This study proposed a graph representation useful for detecting and recognizing furniture using autonomous mobile robots. These graphs are composed of geometrical components, where each geometrical component corresponds roughly to a different part of a given piece of furniture. At this stage of our proposal, only three different types of geometrical components were considered; horizontal planes, vertical planes and the poles. A fast and linear method for extraction of geometrical components has been presented to deal with 3D data from the scenes. Additionally, similarity metrics have been proposed in order to compare the geometrical components and the graphs between the models in the learned set and the scenes. Our graph representation allows performing a directed search when comparing the graphs. To validate our approach, evaluations were performed on house-like environments.

Two different environments were tested with two different robots provided with different brands of RGB-D cameras. The environments presented the same type of furniture, but not completely the same instances; and robots had their RGB-D camera in the head to provide a similar point of view to a human. These preliminary evaluations have proven the efficiency of the presented approach.

As future work, other geometrical components (spheres and cylinders) and other characteristics will be added and assessed to improve our proposal. Furthermore, other types of furniture will be evaluated using our proposal.

It will be desirable to bring the robots to a real house environment to test their performance.

**Author Contributions:** Conceptualization, O.A.-R. and A.M.-H. Investigation, O.A.-R., Supervision, A.M.-H., H.V.R.-F. and M.D. Formal analysis, H.V.-R.F. and M.D. Validation, S.E.P.-H. and E.J.R.-R.

**Funding:** This research received no external funding.

**Acknowledgments:** The lead author thanks CONACYT for the scholarship granted during his Ph.D.

**Conflicts of Interest:** The authors declare no conflict of interest.

## References

1. PCL—Point Cloud Library (PCL). Available online: http://www.pointclouds.org (accessed on 10 October 2017).
2. Alexandre, L.A. 3D descriptors for object and category recognition: A comparative evaluation. In Proceedings of the Workshop on Color-Depth Camera Fusion in Robotics at the IEEE/RSJ International Conference on Intelligent Robots and Systems (IROS), Vilamoura, Portugal, 7–12 October 2012; Volume 1, p. 7.
3. Bo, L.; Ren, X.; Fox, D. Depth kernel descriptors for object recognition. In Proceedings of the IEEE/RSJ International Conference on Intelligent Robots and Systems, San Francisco, CA, USA, 25–30 September 2011; pp. 821–826. [CrossRef]
4. Lai, K.; Bo, L.; Ren, X.; Fox, D. Detection-based object labeling in 3D scenes. In Proceedings of the IEEE International Conference on Robotics and Automation, Saint Paul, MN, USA, 14–18 May 2012; pp. 1330–1337. [CrossRef]
5. Guo, Y.; Bennamoun, M.; Sohel, F.; Lu, M.; Wan, J. An Integrated Framework for 3-D Modeling, Object Detection, and Pose Estimation From Point-Clouds. *IEEE Trans. Instrum. Meas.* **2015**, *64*, 683–693. [CrossRef]
6. Hetzel, G.; Leibe, B.; Levi, P.; Schiele, B. 3D object recognition from range images using local feature histograms. In Proceedings of the IEEE Computer Society Conference on Computer Vision and Pattern Recognition (CVPR 2001), Kauai, HI, USA, 8–14 December 2001; Volume 2, pp. 394–399. [CrossRef]
7. Wahl, E.; Hillenbrand, U.; Hirzinger, G. Surflet-pair-relation histograms: A statistical 3D-shape representation for rapid classification. In Proceedings of the Fourth International Conference on 3-D Digital Imaging and Modeling, Banff, AB, Canada, 6–10 October 2003; pp. 474–481. [CrossRef]
8. Drost, B.; Ulrich, M.; Navab, N.; Ilic, S. Model globally, match locally: Efficient and robust 3D object recognition. In Proceedings of the IEEE Computer Society Conference on Computer Vision and Pattern Recognition (CVPR'10), San Francisco, CA, USA, 13–18 June 2010; pp. 998–1005. [CrossRef]

9. Salas-Moreno, R.F.; Newcombe, R.A.; Strasdat, H.; Kelly, P.H.J.; Davison, A.J. SLAM++: Simultaneous Localisation and Mapping at the Level of Objects. In Proceedings of the IEEE Conference on Computer Vision and Pattern Recognition (CVPR), Long Beach, CA, USA, 15–21 June 2013; pp. 1352–1359. [CrossRef]
10. Wu, Z.; Song, S.; Khosla, A.; Yu, F.; Zhang, L.; Tang, X.; Xiao, J. 3D ShapeNets: A deep representation for volumetric shapes. In Proceedings of the IEEE Conference on Computer Vision and Pattern Recognition (CVPR), Long Beach, CA, USA, 15–21 June 2015; pp. 1912–1920. [CrossRef]
11. Swadzba, A.; Wachsmuth, S. A detailed analysis of a new 3D spatial feature vector for indoor scene classification. *Robot. Auton. Syst.* **2014**, *62*, 646–662. [CrossRef]
12. Trevor, A.J.B.; Rogers, J.G.; Christensen, H.I. Planar surface SLAM with 3D and 2D sensors. In Proceedings of the IEEE International Conference on Robotics and Automation (ICRA'12), Saint Paul, MN, USA, 14–18 May 2012; pp. 3041–3048. [CrossRef]
13. Rusu, R.B.; Marton, Z.C.; Blodow, N.; Dolha, M.; Beetz, M. Towards 3D Point cloud based object maps for household environments. *Robot. Auton. Syst.* **2008**, *56*, 927–941. [CrossRef]
14. Günther, M.; Wiemann, T.; Albrecht, S.; Hertzberg, J. Building semantic object maps from sparse and noisy 3D data. In Proceedings of the IEEE/RSJ International Conference on Intelligent Robots and Systems, Tokyo, Japan, 3–7 November 2013; pp. 2228–2233. [CrossRef]
15. Günther, M.; Wiemann, T.; Albrecht, S.; Hertzberg, J. Model-based furniture recognition for building semantic object maps. *Artif. Intell.* **2017**, *247*, 336–351. [CrossRef]
16. Holz, D.; Holzer, S.; Rusu, R.B.; Behnke, S. Real-Time Plane Segmentation Using RGB-D Cameras. In *RoboCup 2011: Robot Soccer World Cup XV*; Röfer, T., Mayer, N.M., Savage, J., Saranlı, U., Eds.; Springer: Berlin/Heidelberg, Germany, 2012; pp. 306–317.
17. Lin, D.; Fidler, S.; Urtasun, R. Holistic Scene Understanding for 3D Object Detection with RGBD Cameras. In Proceedings of the International Conference on Computer Vision (ICCV), Seoul, Korea, 27 October–3 November 2013; pp. 1417–1424. [CrossRef]
18. Koppula, H.S.; Anand, A.; Joachims, T.; Saxena, A. Semantic Labeling of 3D Point Clouds for Indoor Scenes. In *Advances in Neural Information Processing Systems 24*; Shawe-Taylor, J., Zemel, R.S., Bartlett, P.L., Pereira, F., Weinberger, K.Q., Eds.; Curran Associates, Inc.: Dutchess County, NY, USA, 2011; pp. 244–252.
19. Wittrowski, J.; Ziegler, L.; Swadzba, A. 3D Implicit Shape Models Using Ray Based Hough Voting for Furniture Recognition. In Proceedings of the International Conference on 3D Vision—3DV 2013, Seattle, WA, USA, 29 June–1 July 2013; pp. 366–373. [CrossRef]
20. Chen, K.; Lai, Y.K.; Wu, Y.X.; Martin, R.; Hu, S.M. Automatic Semantic Modeling of Indoor Scenes from Low-quality RGB-D Data Using Contextual Information. *ACM Trans. Graph.* **2014**, *33*, 208:1–208:12. [CrossRef]
21. Schnabel, R.; Wessel, R.; Wahl, R.; Klein, R. Shape recognition in 3d point-clouds. In Proceedings of the 16th International Conference in Central Europe on Computer Graphics, Visualization and Computer Vision in co-operation with EUROGRAPHICS, Pilsen, Czech Republic, 4–7 February 2008; pp. 65–72.
22. Rusu, R.B.; Marton, Z.C.; Blodow, N.; Holzbach, A.; Beetz, M. Model-based and learned semantic object labeling in 3D point cloud maps of kitchen environments. In Proceedings of the IEEE/RSJ International Conference on Intelligent Robots and Systems, St. Louis, MO, USA, 10–15 October 2009; pp. 3601–3608. [CrossRef]
23. Song, S.; Lichtenberg, S.P.; Xiao, J. SUN RGB-D: A RGB-D scene understanding benchmark suite. In Proceedings of the IEEE Conference on Computer Vision and Pattern Recognition (CVPR), Long Beach, CA, USA, 15–21 June 2015; Volume 5, p. 6.

*applied sciences*

MDPI

*Article*

# Multiellipsoidal Mapping Algorithm

**Carlos Villaseñor, Nancy Arana-Daniel *, Alma Y. Alanis, Carlos Lopez-Franco and Javier Gomez-Avila**

Centro Universitario de Ciencias Exactas e Ingenierías, Universidad de Guadalajara, Blvd Marcelino García Barragán 1421, Guadalajara 44430, Mexico; cavp@outlook.com (C.V.); almayalanis@gmail.com (A.Y.A.); clzfranco@gmail.com (C.L.-F.); javier.ega@hotmail.com (J.G.-A.)

* Correspondence: nancyaranad@gmail.com; Tel.: +52-331-547-3877

Received: 16 June 2018; Accepted: 25 July 2018; Published: 27 July 2018

**Abstract:** The robotic mapping problem, which consists in providing a spatial model of the environment to a robot, is a research topic with a wide range of applications. One important challenge of this problem is to obtain a map that is information-rich (i.e., a map that preserves main structures of the environment and object shapes) yet still has a low memory cost. Point clouds offer a highly descriptive and information-rich environmental representation; accordingly, many algorithms have been developed to approximate point clouds and lower the memory cost. In recent years, approaches using basic and "simple" (i.e., using only planes or spheres) geometric entities for approximating point clouds have been shown to provide accurate representations at low memory cost. However, a better approximation can be implemented if more complex geometric entities are used. In the present paper, a new object-mapping algorithm is introduced for approximating point clouds with multiple ellipsoids and other quadratic surfaces. We show that this algorithm creates maps that are rich in information yet low in memory cost and have features suitable for other robotics problems such as navigation and pose estimation.

**Keywords:** object mapping; Geometric Algebra; Differential Evolution

## 1. Introduction

Many algorithms have been developed for the robotic mapping problem [1], which arises when a robot is provided with a spatial model of the environment. Unlike many 3D reconstruction algorithms that use interpolation to obtain the best object rendering, the principal aim of robotic mapping is to create maps that are rich in information and low in memory cost. The environment and objects could be acquired through 3D sensor like Laser Rangefinder or multiples camera view with algorithms like Structure from Motion [2]. With these techniques we obtain dense point clouds.

On the one hand, there exist 3D mapping algorithms, e.g., Simultaneous Localization and Mapping (SLAM), that model key point maps. The principal features of such algorithms are representations with low memory cost and ease of use. However, the maps are not information-rich; consequently, such algorithms cannot be used in many applications, such as object manipulation and aerial navigation. On the other hand, many algorithms create extensive object representations as 3D reconstructions based on the Radial Basis Function (RBF) [3] or other functions [4], resulting in richer information maps, albeit at a high memory cost; additionally, such algorithms are difficult to use. Finding a good balance between representation and memory cost is an important current research topic.

In the last decade, algorithms based on object mapping with geometric entities have become very popular for many applications [1], such as urban and office environment mapping. Object mapping refers to algorithms that approximate the volume shape of a point cloud (it is to say approximate a point cloud) with multiply geometric entities, this is analogous to use curve fitting in a regression problem, but we use geometric entities to describe and preserve the volume shapes instead of functions because there is not dependent variables. The quality of the approximation could be measured using the mean distance of the points to the geometric entity.

Geometric entities have been used before to approximate point clouds, for instance, we observe the use of multiplanar representations [5] and spherical and linear representations [6–9]. Each of these approaches fits the parameters of their geometric entities to obtain the best approximation of sensor data. However, if more complex entities are used, we can obtain approximations of point clouds with better accuracy, and using fewer entities than those obtained by using mulitiplanar, spherical or linear representations.

In this paper, a new object-mapping algorithm is presented. This algorithm is capable of approximating the point cloud with multiple ellipsoids and other quadratic surfaces such as spheres, pairs of planes, and pseudocylinders, which allows us to reduce the number of geometric entities used to represent the objects and complete scenes of the environment, this feature is shown in a controlled experiment, but to automate this process a hierarchical clustering must be implemented, as discussed in the future work section.

This algorithm can also be related to Hyperellipsoidal Neurons (HNs) [10] based on Geometric Algebra (GA) [11], where every neuron is trained with k-means++ [12] and Differential Evolution (DE) [13] for adapting the point cloud implicit surface.

In the following sections, we show that the multiellipsoidal mapping algorithm is capable of creating information-rich maps with low memory cost and that the entities obtained with the algorithm are even capable of deforming into other quadratic surfaces due to their representation in GA as multivectors. The entities are described in GA; then, the map is suitable for developing new algorithms that work in this algebra, such as path planning [14], pose estimation [15,16], and other tasks [17].

The paper is organized as follows. In Section 2, we introduce several mathematical and algorithmic tools used in this paper. Section 3 presents the adaptation strategy of the ellipsoidal surfaces to a point cloud. In Section 4, we explore various experiments to show the performance of the algorithm. In Section 5, the differences from object-mapping algorithms based on function approximations are discussed. Finally, conclusions are presented in Section 6, and in Section 7 we discuss the future work.

## 2. Mathematical Background

In this section, we introduce the mathematical framework of GA to represent the geometric entities. We also introduce the basic notions of k-means++ and DE algorithms to optimize the representation. The notation shown here will be used in the development of the proposed algorithm.

### 2.1. Geometric Algebra

GAs are Clifford Algebras [11,18] constructed over a bilinear form of the vector space $\mathbb{R}^{p,q,r}$, where $(p,q,r)$ is the algebraic signature. This GA is denoted by $\mathbb{G}_{p,q,r}$ and it is an equivalent notation for $C\ell_{p,q,r}(\mathbb{R})$. The elements that belong to this algebra are called multivectors. Let us denote by "$\circ$" the Clifford product, by "$\bullet$" the inner product and by "$\wedge$" the outer product. Then, for two basis vectors, (1) states the Clifford product behavior, where $e_i \wedge e_j$ is a bivector or a 2-vector element; all vectors in $\mathbb{R}^{p,q,r}$ are included in $\mathbb{G}_{p,q,r}$ as 1-vectors (a $k$-vector describe a multivector span by $k$ basis vectors). In addition, the properties of the bilinear form shown in (2) are present, where "$\cdot$" is the common dot product used in linear algebra, thus $\mathbb{R}^{p,q,r} \subset \mathbb{G}_{p,q,r}$.

$$e_i \circ e_j = \begin{cases} e_i \wedge e_j & \text{if } i \neq j \\ 1 & \text{if } 1 \leq i = j \leq p \\ -1 & \text{if } p < i = j \leq q \\ 0 & \text{if } q < i = j \leq r \end{cases} \tag{1}$$

$$a \circ a = a \bullet a = a \cdot a, \quad a \in \mathbb{R}^{p,q,r} \tag{2}$$

GAs are associative and anticommutative algebras; additionally, due to the generality of the Clifford product, many properties of other mathematical frameworks are present [11] such as complex numbers, quaternions, and other Cayley-Dickson algebras, in addition to Pauli matrices and spinor algebras.

Nevertheless, GAs are not only useful for integrating multiple mathematical frameworks but also present attractive features for geometric entity representation. In a GA, multivectors by themselves could represent geometric entities in their inner or outer product representation instead of the traditional geometric locus. The importance of this feature is that the geometric entities, as well as their operators, are represented as elements of the same algebra (both are represented as multivectors).

## 2.2. *Hyperconformal Geometric Algebra*

The most used GA is perhaps the Conformal Geometric Algebra (CGA) for the 3D case $\mathbb{G}_{4,1}$, where the algebraic signature is $(4, 1, 0)$, and its extension to n-dimensions $\mathbb{G}_{n+1,1}$ to represent the $\mathbb{R}^n$ vector space. In this algebra, it is possible to represent points, pairs of points, lines, planes, circles and spheres. Many algorithms for robotics and machine vision [19], e.g., finding path planning algorithms [14], pose estimation [15], structure extraction from motion [16], geometric entity detection [6,7], and robotic mapping algorithms [8,9], have been developed with this algebra.

However, to use more complex geometric entities as algebra elements, other algebras must be used. The GA $\mathbb{G}_{6,3}$ [20] is a generalization of $\mathbb{G}_{4,1}$, in which such geometric entities like ellipsoids, planes, pair of planes, pseudocylinders, spheres and others deformed quadratic surfaces can be represented as multivectors. In [10], $\mathbb{G}_{6,3}$ was extended to any dimension in the so-called Hyperconformal GA (HGA) with a notation $\mathbb{G}_{2n,n}$ for representing the vector space $\mathbb{R}^n$. This algebra is constructed by using a homogeneous stereographic projection in (3) over every coordinate. Figure 1 describes a graphical representation of the projection.

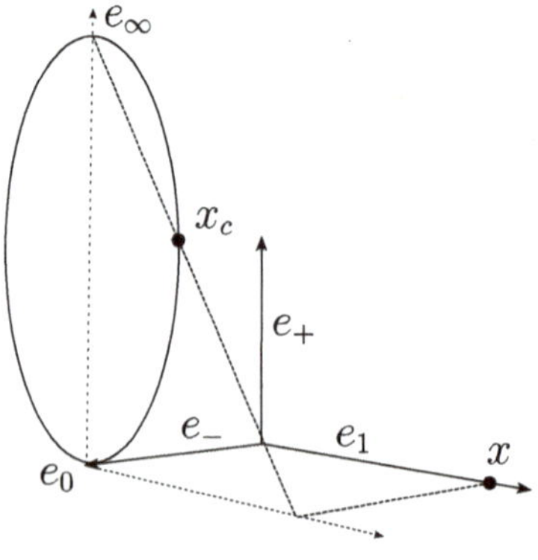


**Figure 1.** Homogeneous stereographic projection.

The null basis are related to (3) by $e_0 = e_+ + e_-$ and $e_\infty = e_+ - e_-$ and they have the property $e_0^2 = e_\infty^2 = 0$. Given a three-dimensional space with basis $(e_x, e_y, e_z)$, the previous notation must be extended. Consider that the stereographic projection is applied in every axis, e.g., the null basis of the coordinate $e_x$ are denoted with $e_{0x}$ and $e_{\infty x}$.

$$x_c = 2\frac{x}{x^2+1}e_1 + \frac{x^2-1}{x^2+1}e_+ + e_- = xe_1 + \frac{1}{2}x^2 e_\infty + e_0 \tag{3}$$

Hence, the entities in $\mathbb{G}_{6,3}$ can be calculated using the null basis. Let $x = p_x e_x + p_y e_y + p_z e_z$ be a point in $\mathbb{R}^3$, the its representation in $\mathbb{G}_{6,3}$ is denoted by $X$ as is shown in (4), where $e_{i\infty}$ is the point at infinity in the $i$ homogeneous stereographic projection and $e_0 = 1/3(e_{0x} + e_{0y} + e_{0z})$ is the point at zero.

$$X = p_x e_x + p_y e_y + p_z e_z + \frac{1}{2}(p_x^2 e_{\infty x} + p_y^2 e_{\infty y} + p_z^2 e_{\infty z}) + e_0 \tag{4}$$

In $\mathbb{G}_{6,3}$, $H$ denotes a 1-vector that represents an fixed-axes ellipsoid with center $(c_x, c_y, c_z)$ and semiaxis $(r_x, r_y, r_z)$, as can be seen in (5), where $e_\infty = e_{\infty z} + e_{\infty y} + e_{\infty z}$.

$$E = \frac{c_x}{r_x^2}e_x + \frac{c_y}{r_y^2}e_y + \frac{c_z}{r_z^2}e_z + \frac{1}{2}\left(\frac{c_x^2}{r_x^2} + \frac{c_y^2}{r_y^2} + \frac{c_z^2}{r_z^2} - 1\right)e_\infty + \frac{1}{r_x^2}e_{0x} + \frac{1}{r_y^2}e_{0y} + \frac{1}{r_z^2}e_{0z} \tag{5}$$

These notations are different from the ones shown in [20], but already presented in [10]. Other geometric entities are derived from the ellipsoid, as shown in Table 1. In Section 3, this theory is used to develop the optimization of the ellipsoidal surfaces.

**Table 1.** Geometric entities that can be represented by a deformed ellipsoid.

| Entity | Representation |
|---|---|
| Sphere | $S = E$ if $r_x = r_y = r_z$ |
| Pseudocylinder | $C = \lim_{r_i \to \infty} E$, $i \in \{x, y, z\}$ |
| Pair of planes | $P_p = \lim_{r_i, r_j \to \infty} E$, $i, j \in \{x, y, z\}$ and $i \neq j$ |

## 2.3. k-Means Algorithm

The k-means algorithm is a popular algorithm for clustering and is frequently used in unsupervised techniques. For a set of points $P = \{p_1, p_2, \cdots, p_n\}$, the k-means algorithm aims to find the partition $S = \{S_1, S_2, \cdots, S_k\}$ with $k \leq n$, as shown in (6), where $c_i$ is the centroid of the cluster $S_i$.

$$S = \underset{S}{\operatorname{argmin}} \sum_{i=1}^{k} \sum_{p \in S_i} p - c_i^2 \tag{6}$$

For this paper, this optimization problem is solved using Lloyd's algorithm with a variant of initialization known as k-means++ [12].

## 2.4. Differential Evolution

DE is an evolutionary algorithm for multivariate functions optimization with many highly successful applications [13]. DE proposes a set of Candidate Solutions (CS) that compete for the best performance in the objective function $f(\cdot)$. In Figure 2, the basic DE scheme is described.

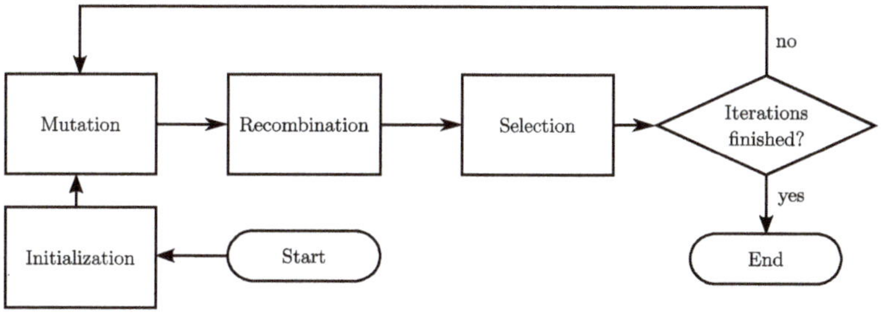


**Figure 2.** Differential Evolution scheme.

A CS is a vector whose elements match the variables in $f(\cdot)$. The population is randomly initialized inside predefined bounds that limit the search space. Afterwards in the mutation process, a donor vector $v_i$ is calculated for every CS $x_i$, by randomly selecting three different CSs, $\{r_1, r_2, r_3\}$ and applying (7), where $F$ is the mutation factor $F \in [0, 2]$.

$$v_i = r_1 + F(r_2 - r_3) \tag{7}$$

In the recombination process, the CS $x_i$ is recombined with the donor vector $v$ in every dimension $j$ to obtain a new CS $u_i$ as is shown in (8), where $rand() \sim U[0, 1]$, and $CR$ is the crossover rate.

$$u_i(j) = \begin{cases} v_i(j) & \text{if } rand() \leq CR \\ x_i(j) & \text{if } rand() > CR \end{cases} \tag{8}$$

Finally, in the selection process the new CS $u_i$ is compared to the actual CS $x_i$, and use (9) (for a minimization problem) to choose whether to keep the actual solution or replace it with the new one.

$$x_i = \begin{cases} u_i & \text{if } f(u_i) \leq f(x_i) \\ x_i & \text{otherwise} \end{cases} \tag{9}$$

The mutation, recombination, and selection function for a certain number of iterations is continuously performed, and ultimately, the CS with the best performance in $f(\cdot)$ is returned.

## 3. Ellipsoidal Surfaces Optimization

In 2017, we presented the Hyperellipsoidal Neuron (HN) [10], where the neuron represents an hyperellipsoidal decision surface. The HN is capable of deforming the decision surface into geometric entities, such as a pair of planes, and spheres and pseudocylinders. In this paper, we use the same propagation of the HN for representing the ellipsoidal surfaces.

We use the parametrization functions $\psi_1(\cdot)$ and $\psi_2(\cdot)$ defined in (10) and (11) respectively.

$$\psi_1(x) = \left[x_1, \cdots, x_n, 1, -\frac{1}{2}x_1^2, \cdots, -\frac{1}{2}x_n^2\right]^T \tag{10}$$

$$\psi_2(E) = \left[\frac{c_1}{r_1^2}, \cdots, \frac{c_n}{r_n^2}, -\frac{1}{2}\left(\frac{c_1^2}{r_1^2} + \cdots + \frac{c_n^2}{r_n^2} - 1\right), \frac{1}{r_1^2}, \cdots, \frac{1}{r_n^2}\right]^T \tag{11}$$

Note that the product $\psi_1(X)^T \psi_2(E)$ is a parameterization in $\mathbb{R}^{2n+1}$ of the the inner product in the algebra $\mathbb{G}_{2n,n}$, and by the definition of an ellipsoid in the inner product null space, it is possible to ensure that a point lies on the ellipsoid surface if $\psi_1(X)^T \psi_2(E) = 0$.

In [10] was presented a training algorithm of HNs for classification. However, the aim of this paper is not to classify points but to approximate surfaces of cloud points with ellipsoids. Hence, in what follows, the development of a new training algorithm to solve the problem of obtaining object maps of environments is presented.

*Training Algorithm for 3D Mapping*

Two sets of parameters, the center $\{c_x, c_y, c_z\}$ and the semi-axes $\{r_x, r_y, r_z\}$, are adapted to represent an ellipsoid. Similar to the case of RBF networks, the ellipsoid center is trained with k-means++, taking the centroid of the cluster as the center of the ellipsoid. Then, for a point cloud with $n$ points $\{X_1, X_2, \cdots, X_n\}$, $k$ clusters $\{S_1, S_2, \cdots, S_k\}$ and $k$ centroids $\{c_1, c_2, \cdots, c_k\}$ for the ellipsoids are found.

For training the semiaxes, the inner product of each point and ellipsoid $\psi_1(X)^T \psi_2(E)$ is minimized, consequently the distance between the points and the ellipsoid surface is minimized. Additionally, the volume of the ellipsoid, defined as $V = \frac{4}{3}\pi r_x r_y r_z$, must be penalized to avoid trivial solutions, e.g., an ellipsoid contained all of the point cloud or a ellipsoid approximating just one point.

Finally, the fitness function for the DE algorithm in (12) is designed, where every cluster $S_i$, calculated by k-means++, is used for adapting the semiaxes $\{r_x, r_y, r_z\}$ of an ellipsoid with center $c_i$. The first term penalizes the distance from every point in the cluster to the ellipsoid surface (outside or inside) by applying the root mean square error (RMSE), and the second term penalizes the density of each cluster $S_i$ by computing the ratio of the volume of the entity to the number of points contained in $S_i$, i.e., $S_i$.

$$(r_x, r_y, r_z) = \underset{(r_x, r_y, r_z)}{\arg\min} \quad \alpha \sqrt{\frac{1}{S_i} \sum_{X \in S_i} [\psi_1(X)^T \psi_2(E)]^2} + \frac{4(1-\alpha)}{3S_i} \pi r_x r_y r_z \tag{12}$$

The free parameter $\alpha$ controls how much the ellipsoid can grow; $\alpha$ and the parameter $k$ control the granularity of the ellipsoidal map. The granularity refers to the number and size of the ellipsoids that represent an object.

## 4. Experiments

To show the capabilities of the proposed algorithm, in this section, the results of experiments are presented. For all of the following experiments, DE is used with parameter settings: $F = 1.2$ and $CR = 0.7$ and a fixed population of 10 particles and 50 iterations. The objective function has the parameter $\alpha = 0.8$, that was chosen heuristically for a good granularity balance. The point clouds are provided by an SRI-500 Laser Rangefinder from Acuity Technologies capable of scanning 800,000 points per second at distances of up to 500 feet (150 meters approximately).

### 4.1. Object Mapping

In experiment 1, as shown in Figure 3, the point cloud (left) is composed of 10,916 three-dimensional points and the multiellipsoidal map (right) contains 150 ellipsoids.

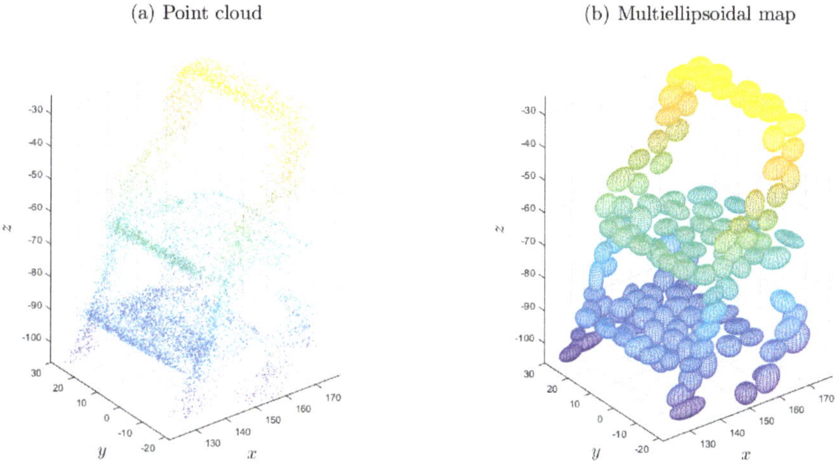


**Figure 3.** Experiment 1.

Similarly, in experiment 2, shown in Figure 4, a map of a tree of 31,049 points is adapted with 400 ellipsoids. Notably, the ellipsoids that are on the floor of the approximation are projected onto the 2D plane defined by the x and y axes of the figure.

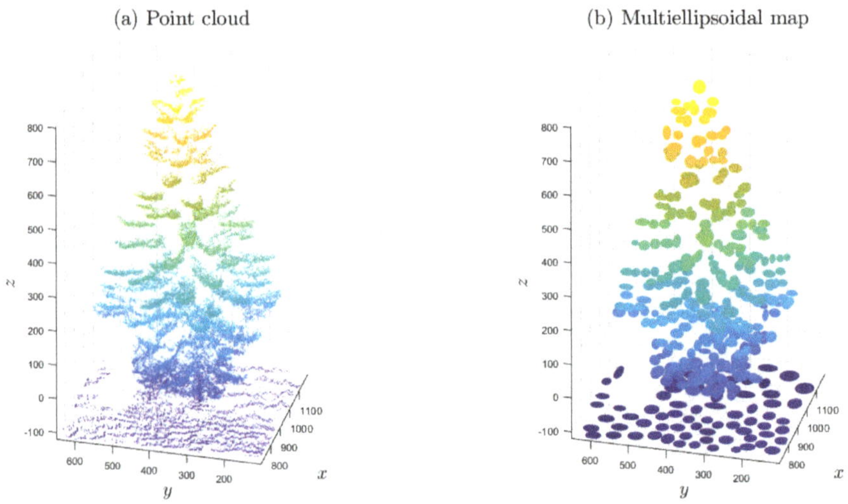


**Figure 4.** Experiment 2.

In Figure 5, the results of experiment 3 are shown. This is an example of a person with opened arms. As can be seen, concave areas are represented accurately by several ellipsoids. The point cloud has 40,883 points, and the multiellipsoidal map contains only 350 ellipsoids.

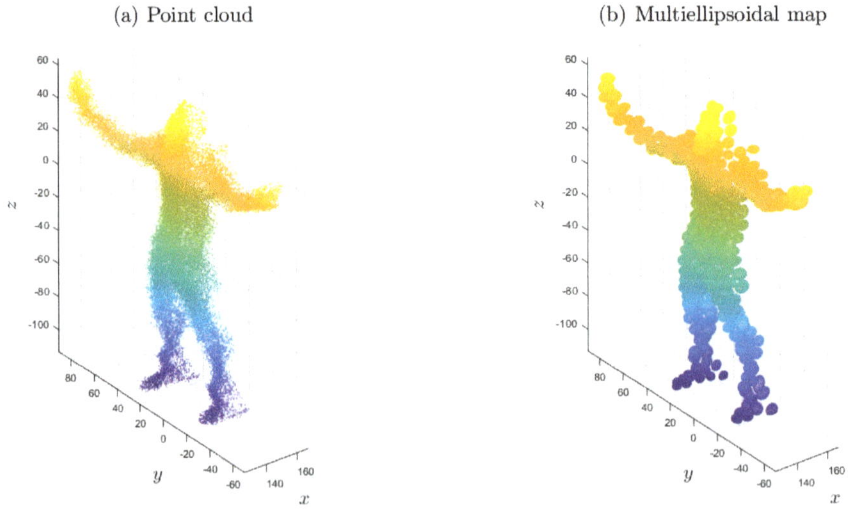


**Figure 5.** Experiment 3.

In experiment 4 in Figure 6, a man is standing with his back toward the observer. The point cloud is formed by 37,142 points, and the multiellipsoidal map contains 350 ellipsoids.

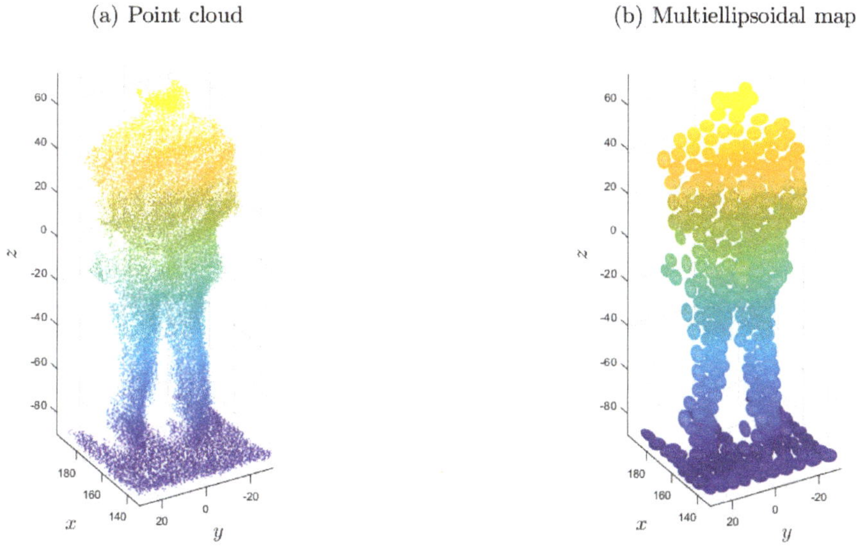


**Figure 6.** Experiment 4.

In Figure 7, another human figure is considered, in this case, a man with open arms sitting in a chair. The point cloud is formed by 42,272 points, and the multiellipsoidal map has 350 ellipsoids.

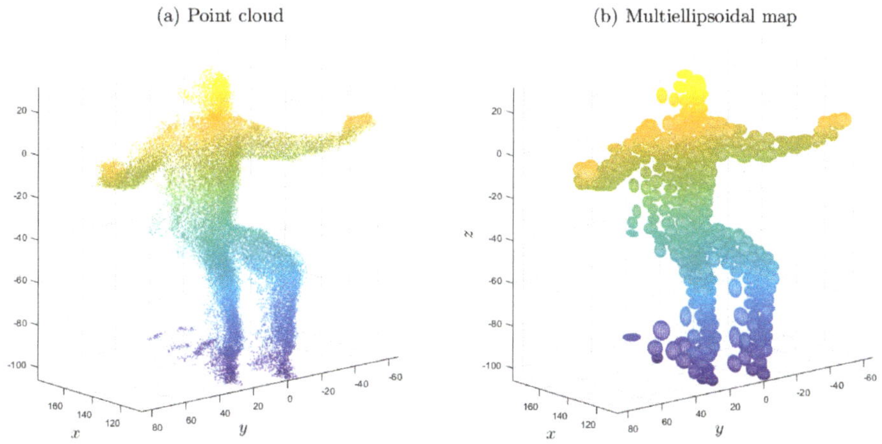


**Figure 7.** Experiment 5.

Summarizing these five experiments, it is observed that a good approximation and representation of the objects are obtained even when a small number of ellipsoids is used. Table 2 shows the memory cost. Consider a four-byte floating-point number; then, the memory cost of the cloud point (three floating-point numbers for each point) and of the multiellipsoidal map (six floating-point numbers for each ellipsoid) is calculated. Finally, a percentage cost of each map is shown, where it is easy to observe that an information-rich map with low memory cost has been obtained.

**Table 2.** Memory cost of the representation—part 1.

| Experiment | Point Cloud | | Multiellipsoidal Map | | Percentage Cost |
|---|---|---|---|---|---|
| | Points | Bytes | Ellipsoids | Bytes | |
| Experiment 1 | 10,916 | 130,992 | 150 | 3600 | 2.7482% |
| Experiment 2 | 31,049 | 372,588 | 400 | 9600 | 2.5765% |
| Experiment 3 | 40,883 | 490,596 | 350 | 8400 | 1.7122% |
| Experiment 4 | 37,142 | 445,704 | 350 | 8400 | 1.8846% |
| Experiment 5 | 42,272 | 507,264 | 350 | 8400 | 1.6559% |

### 4.2. Varying the Number of Ellipsoids

The percentage cost shown in Table 2 depends on the chosen number of ellipsoids ($k$). To show how the representation changes with $k$, consider experiment 6 in Figure 8, where a point cloud of an office chair is approximated with various numbers of ellipsoids. The number of ellipsoids that are shown depends on the application.

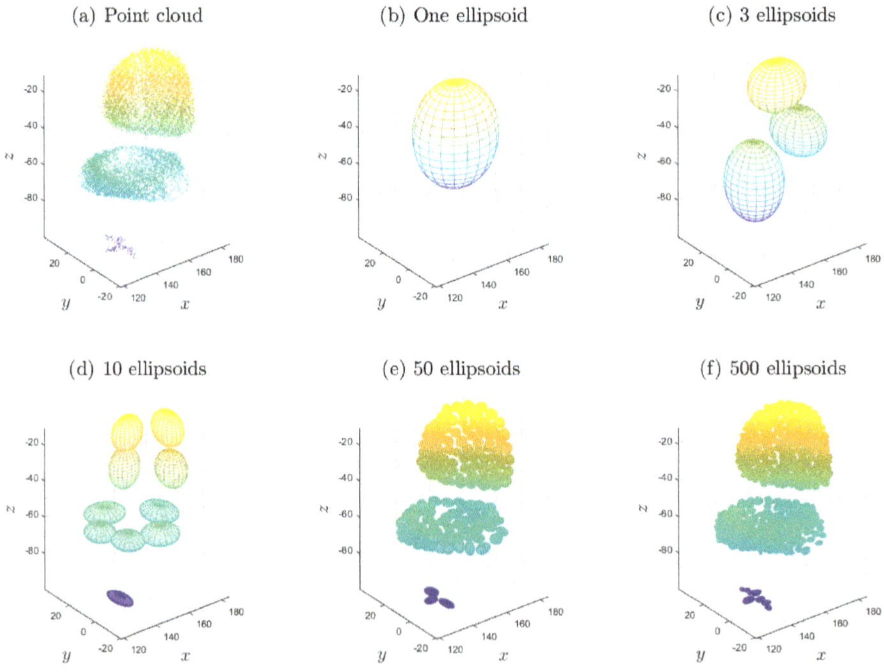


**Figure 8.** Experiment 6.

### 4.3. Ellipsoid Deformation

Table 1 presents the geometric entities that can be obtained when an ellipsoid is deformed using the GA framework. Figure 9 shows the results of experiment 7, where a trash can is approximated with two ellipsoids. We use a threshold of four meters for deforming the ellipsoids. It can be observed that an ellipsoid is deformed into a pseudocylinder for adapting all of the points of the sides of the trash can; additionally, to represent the floor, another ellipsoid is deformed into a pair of planes. This representation is very useful for robotic navigation; however, because a partition clustering technique is used, the ellipsoid size is also controlled by the number of ellipsoids. For a more general way of segmenting, it is always possible to choose to change to a hierarchical clustering algorithm which would allow us to obtain the mentioned deformations of the ellipsoids into pseudo cylinders or pair of planes.

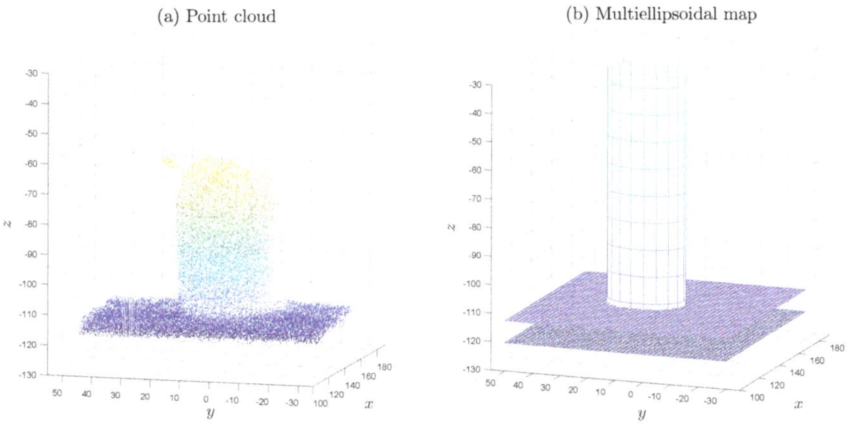


**Figure 9.** Experiment 7.

## *4.4. Environment Mapping*

The multiellipsoidal mapping algorithm presented is useful for mapping not only for objects but also environments. Figures 10 and 11 presents two experiments that show the proposed algorithm's capabilities for mapping environments. In Table 3, the memory cost of representing these environments is shown.

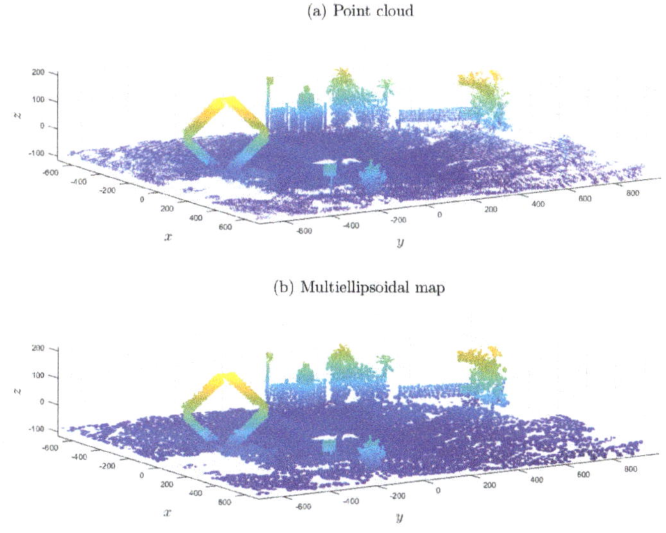


**Figure 10.** Experiment 8.

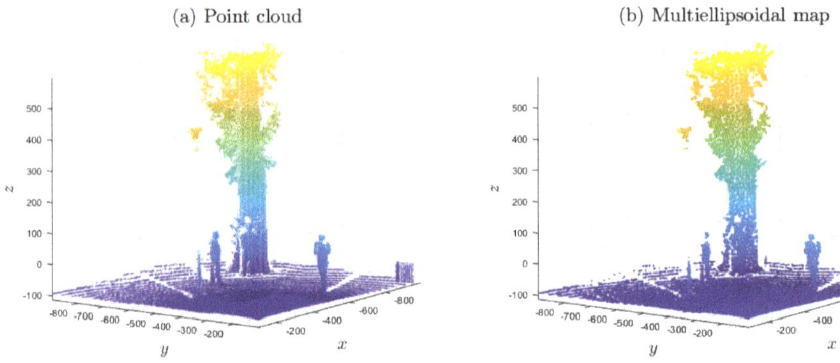


**Figure 11.** Experiment 9.

**Table 3.** Memory cost of the representation—part 2.

| Experiment | Point Cloud | | Multiellipsoidal Map | | Percentage Cost |
|---|---|---|---|---|---|
| | Points | Bytes | Ellipsoids | Bytes | |
| Experiment 8 | 714,446 | 8,573,352 | 8880 | 213,120 | 2.4858% |
| Experiment 9 | 70,447 | 845,364 | 5920 | 142,080 | 16.8069% |

## 4.5. Comparison to Spherical Mapping

Multiplanar mapping algorithms have been used for many applications such as urban mapping and office-like environments representation; however, these algorithms are not suitable to mapping free form objects. To solve this problem, dense multiplanar representations have been developed [5]; however, another problem then arises because, in such cases, the planes boundaries must be defined.

Spherical mapping [8,9] solves both problems by approximating the cloud point with spheres. Because the proposed approach is an extension of the spherical mapping, both approaches are empirically compared to show that ellipsoids can more accurately represent a point cloud.

Let us consider the model error for the point cloud of experiment 1. Let $x \in S_i$ represent the points in the cluster $S_i$ and $d$ be the closest to $x$ three-dimensional points on the spherical or ellipsoidal surface. Then, the RMSE defined by (13) is shown, where $i$ is the number of the cluster. Figure 12 shows the ellipsoidal and spherical approximations with $k = 100$. The spherical mapping is created in the exactly same way as ellipsoidal, except for adapting only one radius.

$$e_i = \sqrt{\frac{\sum_{x \in S_i} (d(x) - x)^2}{S_i}} \tag{13}$$

In Figure 13, we present the histograms of error of the ellipsoidal and the spherical approaches. In the left, the error $e_i$ calculated using Equation (13) of each ellipsoid is shown. In the same way, in the right we have the calculated error for each sphere. In Table 4, their statistical measures are provided.

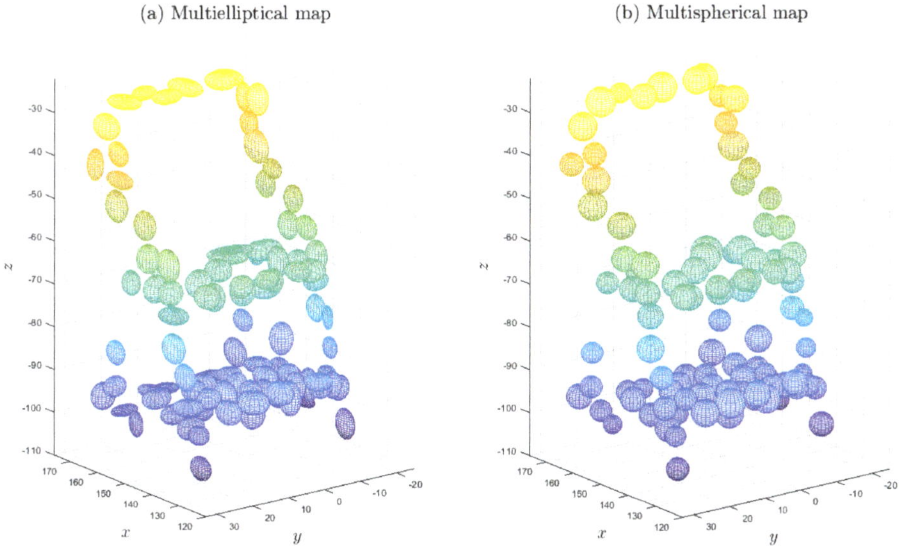


**Figure 12.** Maps of experiment 10.

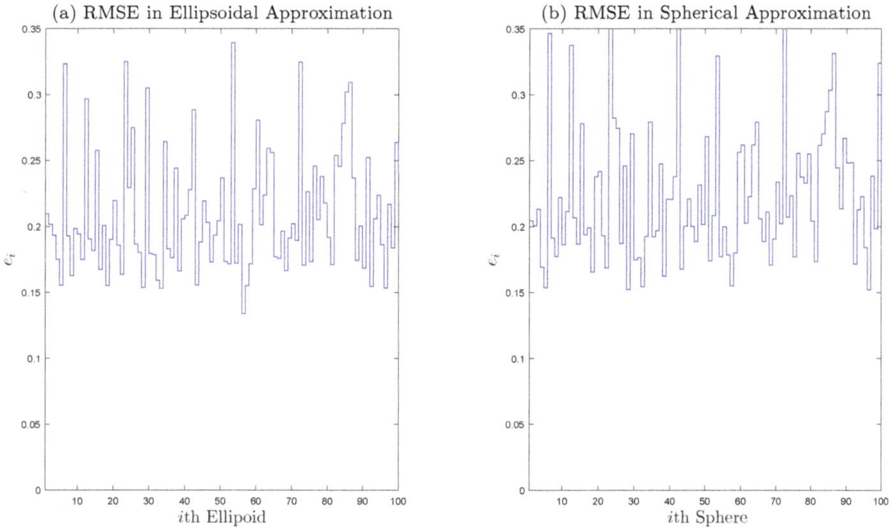


**Figure 13.** Errors of experiment 10.

**Table 4.** Errors of ellipsoidal and spherical approximations.

| Approximation | Mean | STD | Maximum | Minimum | Total |
|---|---|---|---|---|---|
| Ellipsoidal | 0.2079 | 0.0451 | 0.3604 | 0.1351 | 20.7936 |
| Spherical | 0.2258 | 0.0540 | 0.4647 | 0.1475 | 22.5769 |

Note that the multiellipsoidal mapping has a smaller error than the spherical representation due to the degrees of freedom of the ellipsoid. Hence, in this experiment, it has been empirically shown that the ellipsoidal mapping represents an improvement in the spherical mapping algorithm.

## 5. Discussion of Other Mapping Algorithms

In this section, the differences between mapping algorithms based on the approximation of geometric entities and those based on function approximation are discussed; in particular, the difference between the approximation of geometric entities represented in GA as multivectors [6–9] and approximations of functions such as the Gaussian approximation with RBF networks [3] or generalized distance functions [4].

### 5.1. The Best Approximation

Regarding the best approximation, the RBF is a successful technique because, in similarity with the Multi-layer Perceptron, the universal approximation theorem [21] shows that RBF can be used for approximating any free form object. However, there is no theorem that shows that planes, spheres, and ellipsoids are in fact universal approximators. Consequently, it is possible to claim that the approximation of functions will result in the best fit of the point cloud. This is a good property if the application is point cloud interpolation.

### 5.2. Compactness and Heuristics

As can be seen, function approximation results in the best approximation; however, because functions with the domain $(-\infty, \infty)$ are considered, such functions have to be bounded, and a heuristic is needed to determine the plane position (and the points transformation) where the 3D function is defined. This problem implies that more parameters for the approximation must be used. In contrast, using geometric entities of GA, self-bounded entities such as spheres and ellipsoids and infinite entities such as pseudocylinders, planes, and pairs of planes are obtained. Additionally, because all the entities belong to the same algebra and the geometric locus is defined by their null space, heuristics to fix the representation are not required. Hence, we conclude that geometric entities are useful for easy implementation and compactness. We can also argue that the best information compression can be obtained as shown in Tables 2–4.

### 5.3. The Best Mathematical Framework

For most robotic applications, obtaining the map is merely an initial step. Hence, it is important that the map be suitable for subsequent tasks. In the case of functions, it is computationally expensive to apply a rigid transformation as is needed in many applications. In contrast, geometric entities represented by multivectors are suitable for such transformations, because the same operators can be used in every entity. These features make GA suitable to pose estimation [15], movement estimation [16], navigation over rough terrain [14], computation of inverse kinematics [22], object manipulation [23], 3D reconstruction of buildings [6], and other applications in computer graphics [24]. In conclusion, the mapping algorithms based on GA offer a suitable framework for algorithms development.

## 6. Conclusions

In this paper, a new mapping algorithm based in $\mathbb{G}_{6,3}$ GA has been presented, capable of adapting multiples ellipsoids to obtain an object representation that is more compact than the point cloud. This multiellipsoidal mapping offers information-rich maps suitable for representation and approximation at low memory cost. Furthermore, the use of the GA framework allows us to work with an algebraic representation of ellipsoids that is capable of deforming the ellipsoids by themselves into other quadratic surfaces, such as spheres, pair of planes and pseudocylinders; this feature is valuable for robotic mapping.

As our results show, compared to other object-mapping algorithms like multiplanar mapping [5] and spherical volume registration [8,9], ellipsoids could adapt better free form objects. In the Section 5, we discussed that in contrast to mapping algorithms based in RBF [3] or other functions [4], the ellipsoid is an element of GA $\mathbb{G}_{6,3}$, and is hence easier to manipulate than functions. This property will provide a better framework for other robotic tasks such as path planning and navigation.

## 7. Future Work

The proposed algorithm is strongly related to the clustering algorithms that segment the cloud point to get other quadratic surfaces; a hierarchical clustering methodology is necessary. This improvement will reduce the number of geometric entities. Consequently, no $k$ parameter should be found.

The presented algorithm can also be abstracted as a one-layer neural network consisting in HNs. The extension of this neural network could be trained for on-line capabilities (e.g., trained with Extended Kalman Filter), this could lead to a dynamic multielliptical mapping algorithm. Furthermore, the parameters $k$ and $\alpha$ were selected heuristically; for a better understanding of how these parameters affect the granularity of the mapping algorithm, a new study has to be done.

**Author Contributions:** Conceptualization, C.V. and N.A.-D.; Formal analysis, C.V. and J.G.-A.; Funding acquisition, A.Y.A. and C.L.-F.; Investigation, C.V., N.A.-D.; Methodology, C.V.; Project administration, N.A.-D.; Software, C.V. and J.G.-A.; Supervision, N.A.-D. and C.L.-F.; Validation, J.G.-A.; Writing—original draft, C.V.; Writing—review and editing, N.A.-D. and A.Y.A.

**Funding:** This work has been supported by CONACYT Mexico, through Projects CB256769 and CB258068 ("Project supported by Fondo Sectorial de Investigación para la Educación").

**Conflicts of Interest:** The authors declare no conflict of interest.

## References

1. Thrun, S. Robotic mapping: A survey. *Explor. Artif. Intell. New Millenn.* **2002**, *1*, 1–35.
2. Szeliski, R. *Computer Vision: Algorithms and Applications*; Springer Science & Business Media: Berlin, Germany, 2010.
3. Carr, J.C.; Beatson, R.K.; Cherrie, J.B.; Mitchell, T.J.; Fright, W.R.; McCallum, B.C.; Evans, T.R. Reconstruction and representation of 3D objects with radial basis functions. In Proceedings of the 28th Annual Conference on Computer Graphics and Interactive Techniques, Los Angeles, CA, USA, 12–17 August 2001; ACM: New York, NY, USA, 2001; pp. 67–76.
4. Poranne, R.; Gotsman, C.; Keren, D. *3D Surface Reconstruction Using a Generalized Distance Function. Computer Graphics Forum*; Wiley Online Library: Hoboken, NJ, USA, 2010; Volume 29, pp. 2479–2491.
5. Argiles, A.; Civera, J.; Montesano, L. Dense Multi-Planar Scene Estimation From a Sparse Set of Images. In Proceedings of the 2011 IEEE/RSJ International Conference on Intelligent Robots and Systems, San Francisco, CA, USA, 25–30 September 2011; pp. 4448–4454.
6. Bayro-Corrochano, E.; Bernal-Marin, M. Generalized Hough transform and conformal geometric algebra to detect lines and planes for building 3D maps and robot navigation. In Proceedings of the 2010 IEEE/RSJ International Conference on Intelligent Robots and Systems, Taipei, Taiwan, 18–22 October 2010; pp. 810–815.
7. López-González, G.; Arana-Daniel, N.; Bayro-Corrochano, E. *Conformal Hough Transform for 2D and 3D Cloud Points. Iberoamerican Congress on Pattern Recognition*; Springer: Berlin, Germany, 2013; pp. 73–83.
8. Rivera-Rovelo, J.; Bayro-Corrochano, E. Segmentation and volume representation based on spheres for non-rigid registration. In *International Workshop on Computer Vision for Biomedical Image Applications*; Springer: Berlin, Germany, 2005; pp. 449–458.
9. Rivera-Rovelo, J.; Bayro-Corrochano, E.; Dillmann, R. Geometric neural computing for 2d contour and 3d surface reconstruction. In *Geometric Algebra Computing*; Springer: Berlin, Germany, 2010; pp. 191–209.
10. Villaseñor, C.; Arana-Daniel, N.; Alanís, A.Y.; López-Franco, C. Hyperellipsoidal Neuron. In Proceedings of the International Joint Conference on Neural Networks (IJCNN), Anchorage, Alaska, 14–19 May 2017; pp. 788–794.

11. Hestenes, D.; Sobczyk, G. *Clifford Algebra to Geometric Calculus: A Unified Language for Mathematics and Physics*; Springer Science & Business Media: Berlin, Germany, 2012; Volume 50.
12. Arthur, D.; Vassilvitskii, S. k-means++: The advantages of careful seeding. In Proceedings of the Eighteenth Annual ACM-SIAM Symposium on Discrete Algorithms, New Orleans, LA, USA, 7–9 January 2007; Society for Industrial and Applied Mathematics: Philadelphia, PA, USA, 2007; pp. 1027–1035.
13. Das, S.; Suganthan, P.N. Differential evolution: A survey of the state-of-the-art. *IEEE Trans. Evolut. Comput.* **2011**, *15*, 4–31. [CrossRef]
14. Valencia-Murillo, R.; Arana-Daniel, N.; López-Franco, C.; Alanís, A.Y. Rough Terrain Perception Through Geometric Entities for Robot Navigation. In Proceedings of the 2nd International Conference on Advances in Computer Science and Engineering, Los Angeles, CA, USA, 1–2 July 2013.
15. Rosenhahn, B.; Sommer, G. Pose estimation in conformal geometric algebra part i: The stratification of mathematical spaces. *J. Math. Imaging Vis.* **2005**, *22*, 27–48. [CrossRef]
16. Arana-Daniel, N.; Villaseñor, C.; López-Franco, C.; Alanís, A.Y. Bio-inspired aging model-particle swarm optimization and geometric algebra for structure from motion. In *Iberoamerican Congress on Pattern Recognition*; Springer: Berlin, Germany, 2014; pp. 762–769.
17. Bayro-Corrochano, E. *Geometric Computing: For Wavelet Transforms, Robot Vision, Learning, Control and Action*; Springer Publishing Company: Berlin, Germany, 2010.
18. Dorst, L.; Fontijne, D.; Mann, S. *Geometric Algebra for Computer Science: An Object-Oriented Approach to Geometry*; Morgan Kaufmann Publishers Inc.: Burlington, MA, USA, 2009.
19. Bayro-Corrochano, E.; Reyes-Lozano, L.; Zamora-Esquivel, J. Conformal geometric algebra for robotic vision. *J. Math. Imaging Vis.* **2006**, *24*, 55–81. [CrossRef]
20. Zamora-Esquivel, J. G6,3 geometric algebra; description and implementation. *Adv. Appl. Clifford Algebras* **2014**, *24*, 493–514. [CrossRef]
21. Hornik, K. Approximation capabilities of multilayer feedforward networks. *Neural Netw.* **1991**, *4*, 251–257. [CrossRef]
22. Hildenbrand, D.; Zamora, J.; Bayro-Corrochano, E. Inverse kinematics computation in computer graphics and robotics using conformal geometric algebra. *Adv. Appl. Clifford Algebras* **2008**, *18*, 699–713. [CrossRef]
23. Hildenbrand, D.; Bayro-Corrochano, E.; Zamora, J. Advanced geometric approach for graphics and visual guided robot object manipulation. In Proceedings of the 2005 IEEE International Conference on Robotics and Automation, Barcelona, Spain, 18–22 April 2005; pp. 4727–4732.
24. Vince, J. *Geometric Algebra for Computer Graphics*; Springer Science & Business Media: Berlin, Germany, 2008.

*applied sciences*

MDPI

*Article*

# Topological Map Construction Based on Region Dynamic Growing and Map Representation Method

**Fei Wang *, Yuqiang Liu, Ling Xiao, Chengdong Wu and Hao Chu**

Faculty of Robot Science and Engineering, Northeastern University, Shenyang 110819, China; yuqiang0616@163.com (Y.L.); 1870719@stu.neu.edu.cn (L.X.); wuchengdong@mail.neu.edu.cn (C.W.); chuhao@mail.neu.edu.cn (H.C.)

* Correspondence: wangfei@mail.neu.edu.cn; Tel.: +86-139-400-58702

Received: 25 December 2018; Accepted: 3 February 2019; Published: 26 February 2019

**Featured Application: It is suitable for robots in the human-machine collaboration category, especially service robots.**

**Abstract:** In the human–machine interactive scene of the service robot, obstacle information and destination information are both required, and both kinds of information need to be saved and used at the same time. In order to solve this problem, this paper proposes a topological map construction pipeline based on regional dynamic growth and a map representation method based on the conical space model. Based on the metric map, the construction pipeline can initialize the region growth point on the trajectory of the mobile robot. Next, the topological region is divided by the region dynamic growth algorithm, the map structure is simplified by the minimum spanning tree, and the similar region is merged by the region merging algorithm. After that, the parameter TM (topological information in the map) and the parameter OM (occupied information in the map) are used to represent the topological information and the occupied information. Finally, a topological map represented by the colored picture is saved by converting to color information. It is highlighted that the topological map construction pipeline is not limited by the structure of the environment, and can be automatically adjusted according to the actual environment structure. What's more, the topological map representation method can save two kinds of map information at the same time, which simplifies the map representation structure. The experimental results show that the map construction method is flexible, and that resources such as calculation and storage are less consumed. The map representation method is convenient to use and improves the efficiency of the map in preservation.

**Keywords:** human–machine interactive navigation; mobile robot; topological map; regional growth

## 1. Introduction

Human–machine interactive navigation refers to the process through which machines and operator cooperate with each other to control the movement of devices and realize interactive navigation [1]. This type of navigation has been widely used in indoor service robots [2,3], self-driving cars [4,5], automatic guided vehicles (AGV) [6,7], and so on [8,9]. Among them, the most important part between the operator and machines is the map. The map needs to combine continuous spatial environment information with human abstract intentions, and discover the relationships between intentions and real spatial areas; finally, these connections are abstracted into a sequence of events (that is, a topological map) [10].

Many researchers began to study topological maps very early. Nodes are jointly represented by the sector feature of a laser and a proportional invariant feature of vision, which do not depend on any artificial landmark, and the global location of robots in the process of map creation [11]. However, the laser sector feature is required in the intersection area among different channels, and the visual feature is also required to be proportional invariant. Therefore, the application effect will be affected in such large-scale and featureless scenarios as a living room. The final topological map is sparse, and cannot achieve accurate navigation obstacle avoidance [12]. A self-organizing method of hierarchical clustering (Map-TreeMaps) is proposed in [13]. Each unit of the map represents the structured data of the tree, while the treemap method provides a global view of the local hierarchy. This method enhances the generality of the construction of topological maps and solves the problem that some environmental geometric structures are constrained. Similar to the Chow–Liu tree model in [14], as long as the number of searching layers is big enough, various spatial structures can be detected and clustered separately. In addition, the segmentation results can be easily optimized by thresholding the weights of local subgraphs. Both methods abandon the details of the local subgraph, which makes it difficult to achieve accurate navigation obstacle avoidance. An auxiliary graph is used to solve the problem of local subgraph association in [15], which improves the efficiency of segmentation and doesn't solve the problem of subgraph details.

Moreover, the above clustering segmentation cannot deal with large-scale and open space, which is not human-friendly for human–machine interactive navigation. For example, in the application of service robots, large living rooms and long corridors cannot be regarded as a region, but should be divided into several regions according to the actual situation [16]. A novel and efficient method for updating Voronoi diagrams was proposed in [17], which only updates those units that are actually affected by the environment, and finally lower the number of visits and computing time. In addition, a skeleton-based Voronoi diagram method is also proposed, which is particularly effective for noise removal. A simultaneous location and mapping algorithm (VorSLAM) based on Voronoi map representation is proposed in [18]. One of the basic features of this algorithm is that the features correspond to the local map one by one, and each feature is associated with a local map defined on the feature. This not only retains the details of the local map, it also alleviates the problem of large scene segmentation by Voronoi partitioning. However, the Voronoi graph is based on the principle of distance or special structure to divide the space, so that the area obtained is "basically the same size", and the special structure also limits the applicability of the algorithm.

From the research history of topological map construction, a generalized Voronoi map [19] and spectral clustering [20] are the two main methods for topological map construction at present [21]. For example, a lightweight method is proposed to create maps by combining metric maps and topological information in [22]. By combining the information of two maps, the robot can realize autonomous navigation and obstacle avoidance in a large area. In paper [23], spectral clustering and an extended Voronoi graph are used to construct a topological graph from a metric graph. The specific idea is to use spectral clustering to segment the metric graph and get the center of the cluster. After determining the first vertex, other vertices are established by an extended Voronoi graph, and vertices are divided into connection points and tail nodes. Although the combination of two maps achieves navigation avoidance, the difficulty of map preservation is increased; the combination of two segmentation methods improves the efficiency and scope of application of topological segmentation, and the redundancy and uncontrollability of segmentation results is coming, which easily leads to the accumulation of topological vertices in some areas.

To solve the above problems, this paper proposes a novel pipeline of constructing a topological map and an efficient way of expressing a topological map. The main contributions of this paper are as follows:

- A complete system of building, saving, and loading topographic maps is proposed, which can make the topographic maps readily applied.
- A topological segmentation method based on region dynamic growth is proposed, which makes the region segmentation no longer limited by the geometrical structure of the environment, and also more in line with the actual needs of human–machine interactive navigation.

- A representation method of a topological map based on the conical space model is proposed, which makes the map retain not only the information of the topological relationship, but also the information regarding the obstacle occupied.

Finally, several comparative experiments are carried out in the Gazebo simulator provided by ROS (Robot Operating System, ROS) and the author's lab. The experimental results verify the effectiveness of the proposed system and method. The overall block diagram of the topological map building system is shown in Figure 1.

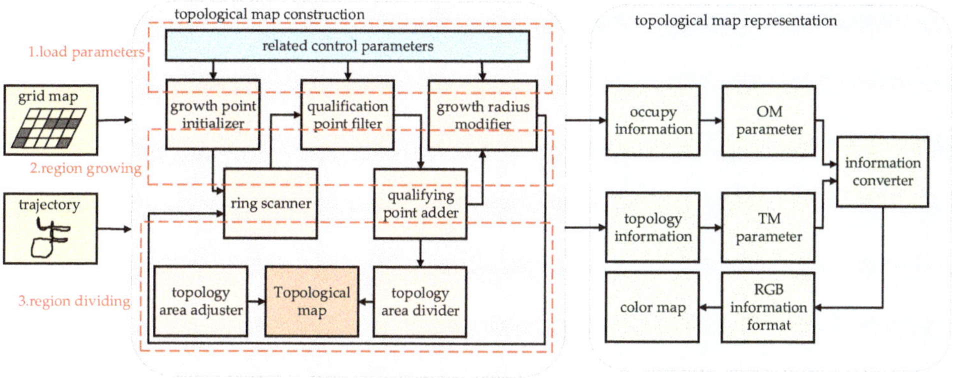


**Figure 1.** The architecture of the topological map construction system.

## 2. Topological Map Construction Based on Regional Dynamic Growth

The process of building a topological map based on region dynamic growth is shown in the left part of Figure 1. The system first randomly selects the initial growth point on the trajectory of the robot, and then grows dynamically according to the control parameters. When the regional characteristics meet the requirements, the growth will stop. This method is inspired by the spherical expansion of voxel clusters in paper [24]. Then, after each region has grown, it will be segmented from the meter map, and the color identification of the topological region will be established. The center of gravity of the topological region will be used as the center of the region, and the neighborhood of the region will be scanned. Finally, after all the topological areas have been established, the region adjustment part will delete, merge, and grow the unqualified areas, and finally complete the construction of the whole environment's topological map.

### 2.1. Regional Growth Process

Based on the metric map and the trajectory of the robot, the map building system will randomly initialize the region growing point on the trajectory (which is in the same coordinate system as the metric map). It is worth mentioning that this point is not the final center of the region (also called the topological vertex). Then, the circular growth of the region is achieved by using a circular scanner with the same degree of growth in all directions on the plane, rather than depending on the geometric structure or special characteristics of the environment.

There are several important parameters in the regional dynamic growth algorithm, which are described below:

- $r_a$: The points' addition ratio in the circular scan, which refers to the ratio of qualified points in each circular scan. It can prevent malformation, making the growth area approximately circular or elliptical rather than elongated.
- $r_o$: The obstacle ratio in the circular scan, which refers to the ratio of unqualified points to the total qualified points in the convex hull region composed of qualified points in the added points

and areas in each annular scanning. It can prevent the region from growing into a concave region, and ensure the convexity of the topological region.

- $r_p$: The points pass ratio of the topological region, which refers to the ratio of all the points in the topological region to all the points in the circular region (ideal region). It can reflect the contrast between the growth area and the ideal area in the current state. It is used to modify the growth radius in real time, so its role is to prevent growth deficiency.
- $R_w$: The control weight of dynamic modification of the growth radius, which refers to the extent of radius modification in each region.

The concept of the qualified point is mentioned above. If one of them is not satisfied, it is called an unqualified point. The qualified point needs to satisfy all the following conditions:

1. The metric map area that corresponded to the point must be in a free space, and cannot be an occupied space or an unknown space.
2. The topological map area corresponding to the point must be a non-topological identifier area, which can only be a free space or a local area identifier.
3. Conditions 1 and 2 must be satisfied for all points through which the ray emitted from the region vertex passes.

As the location of the robot's trajectory must be free and can basically traverse the whole environment, it is better to sample the growing points from the robot's trajectory. Then, the main steps of the region dynamic growth algorithm are as follows (Figure 2, the symbols in the Figure 2d will be described later):

(1) Determine the growth point and scan all the adjacent points around the current region vertex in a circular way.
(2) Rays from the growth point to each point were calculated using the Bresenham algorithm [25], and all of the qualified points were screened.
(3) Calculate the convex hull after each qualified point is added using the Graham scanning algorithm [26], and retain the point where the obstacle ratio satisfies the requirements.
(4) Add qualifying points to the topological area. At the same time, the points pass ratio and the points add ratio in the current state are calculated. If the points' addition ratio is less than the threshold, the growth of the current region is exited, indicating that the region growth is completed ahead of time. If the addition rate is normal, the points pass ratio will be used to modify the maximum growth radius. The modified formula is shown in Equation (1):

$$R_{\max}^{t+1} = R_{\max}^{t} + R_w \cdot r_p \cdot R_{\max}^{t} \tag{1}$$

where $R_{\max}^{t}$ is the maximum growth radius at the current time $t$, and $R_{\max}^{t+1}$ is the maximum growth radius at the time $t + 1$.

After that, the regional dynamic growth algorithm will grow in other areas in the same way until the topological area covers most of the environment.

The center of each region will be updated after the end of each region growth. This paper considers that every topological region is an irregular convex polygon composed of convex hull points in the region, and the polygon is approximately circular or elliptic, so the center of gravity of the polygon can be used as a new center.

As shown in (d) of Figure 2, the polygon is considered to be composed of multiple triangles of area $S(S_0, S_1, S_2, \ldots)$, where $p_c$ represents a vertex and $p_0p_1, p_1p_2, p_2p_3, \ldots$ represent the other two points in triangles. Then, if the coordinates of three vertices of the triangle $\Delta p_1p_2p_3$ are known: $p_1(x_1, y_1)$,$p_2(x_2, y_2)$,$p_3(x_3, y_3)$, the common center of gravity coordinates can be obtained, as shown in Equation (2).

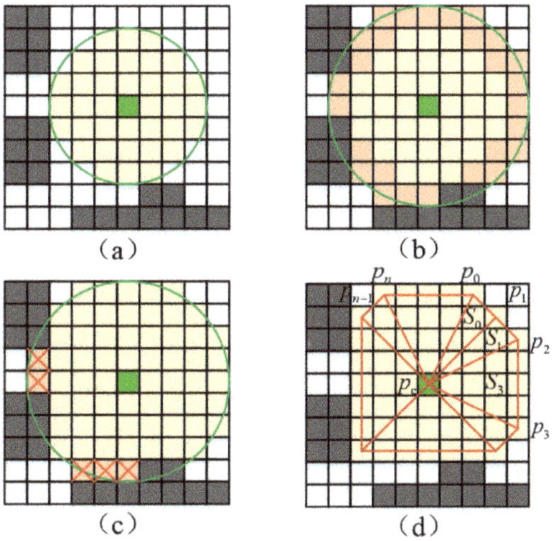


**Figure 2.** Main steps of the regional dynamic growth algorithm. (**a**) Growth state at a certain moment. (**b**) The next moment is grown in a circular expansion manner to obtain adjacent candidate points. (**c**) Remove the unqualified points in the adjacent candidate points. (**d**) Add qualified candidate points to the topological region.

However, Equation (2) is not suitable for multi-triangle calculation, so the triangle area calculation method is used in this paper. Firstly, the area of the triangle is calculated by vector cross-multiplication, as shown in Equation (3). In addition, it is not necessary to consider whether the traversal order of the three points is clockwise or counter-clockwise in the calculation process, because it can be cancelled in the subsequent calculations.

$$\begin{cases} x_g = \frac{x_1+x_2+x_3}{3} \\ y_g = \frac{y_1+y_2+y_3}{3} \end{cases} \tag{2}$$

$$S = \frac{(x_2 - x_1) * (y_3 - y_1) - (x_3 - x_1) * (y_2 - y_1)}{2} \tag{3}$$

Assuming that an irregular convex polygon is composed of $n$ triangles, in which each triangle has an area of $S_i$ and a center of gravity of $G_i(x_i, y_i)$, the integral can be transformed into an accumulation sum by using the formula for calculating the center of gravity of a planar thin plate, as shown in Equation (4):

$$\begin{cases} x_g = \frac{\iint\limits_D x\mathrm{d}S}{S} = \frac{\sum\limits_{i=1}^{n} x_i S_i}{\sum\limits_{i=1}^{n} S_i} \\ y_g = \frac{\iint\limits_D y\mathrm{d}S}{S} = \frac{\sum\limits_{i=1}^{n} y_i S_i}{\sum\limits_{i=1}^{n} S_i} \end{cases} \tag{4}$$

Then, in any irregular convex $n$ polygon $(p_0, p_1, p_2, \ldots, p_n)$ with $p_i(x_i, y_i)(i = 0, 1, 2, 3, \ldots, n)$ as a vertex, it is divided into $n$ triangles composed of a center point $p_c(x_c, y_c)$ as a vertex and any two points $p_i$. Then, the coordinates of the center of gravity of the polygon are calculated as shown in the following Equation (5):

$$\begin{cases} x' = \dfrac{\sum\limits_{i=1}^{n}(x_c+x_i+x_{i-1})S_i}{3\sum\limits_{i=1}^{n}S_i} \\ y' = \dfrac{\sum\limits_{i=1}^{n}(y_c+y_i+y_{i-1})S_i}{3\sum\limits_{i=1}^{n}S_i} \end{cases} \quad (5)$$

Finally, the principal semi-axial length and the short semi-axis length of the region are calculated by principal component analysis. The pseudo code of the regional dynamic growth algorithm is shown in Algorithm 1.

**Algorithm 1:** Regional dynamic growth

**Objectives:** The growing point is regarded as a temporary topological vertex, and then all the neighboring points are scanned annularly to select eligible points and add them to the current topological region.
**Input:** Growth point $p_c(x, y)$, maximum growth radius $R_{max}$.
**Output:** Central point $p'_c$ of the topological region, radius $R'_{new}$ of the topological region.

initial growth radius $r$
initialization of eligible point set $P_p = \{\}$
initialization of unqualified point set $P_f = \{\}$
**for** $(r = 1; r < R_{max}; r++)$
    $P_t = \{\}$
    $P_t \leftarrow$ ring scanner $(p_c, r)$
    **for each** $p_i$ **in** $P_t$
        **if** ($p_i$ is failure)
            $P_f \leftarrow$ fail point filter $(p_i)$
        **continue**
        $P_r \leftarrow \{\}$
        $P_r \leftarrow$ calculate the point at which ray $(p_c \rightarrow p_i)$ passes
        **if** ($P_r$ is failure)
            $P_f \leftarrow$ fail point filter$(p_i)$
            **continue**
        $P_h \leftarrow$ calculate the current convex hull $(P_p, p_i)$
        calculate the number of unqualified points containing $P_f$ in convex hull $P_h$
        **if** (obstacle ratio $r_a$ doesn't satisfies the requirements)
            $P_f \leftarrow$ fail point filter $(p_i)$
    $P_p \leftarrow P_t$, and add a topological area identifier
    **if** (addition rate $r_a$ satisfies the requirements)
        $R'_{max} \leftarrow$ modify the growth radius $(R_{max}, r_p, R_w)$
        $R_{max} \leftarrow R'_{max}$
    **else break**
$P'_c \leftarrow$ recalculate the regional center $(P_c, P_p)$
$R'_{new} \leftarrow$ recalculate the radius information of the area $(P_p)$
**return** $R'_{max}, P'$

### 2.2. Regional Adjustment Process

After the growth of the previous section, the whole environment will be divided into regions of different sizes because of the dynamic growth. The small areas not only waste the vertex resources of the topological map, but also cannot be used for the interactive navigation of the actual scene. Therefore, it is necessary to merge small areas and optimize the topological map.

In order to improve the coverage of topographic maps, the region can be regenerated. During the secondary growth process, if the radius of the current region is less than the threshold value and no new points are added, or the maximum radius of secondary growth is meeted, the growth process will quit. The method is similar to regional growth in Section 2.1, which is not discussed here again.

In addition, the Kruskal algorithm is used to adjust the topological map before region merging, and the minimum spanning tree form of the topological map is obtained, which further simplifies the map structure.The next step is to merge some areas. The region merging algorithm first counts the topological vertices that meet the merge requirements. It needs to satisfy the following two conditions:

1. The radius of the main vertex area is less than the threshold.
2. The total area of the merged area is less than the threshold.
3. The obstacle ratio of the merged area is less than the threshold.

---

**Algorithm 2:** Region Merging

**Objectives:** To merge small areas, delete the original topological vertices and generate new topological vertices, and change the color identification of the topological area.
**Input:** Topological region vertex sequence set $P_m$ to be merged.
**Output:** The status of this subarea merge: true means the merge was successful; false means no merge occurred.

initialize the set of topological vertex ordinals $V_{com} = \{\}$ that need to be merged
initialize the point set $P_{com}= \{\}$ of the points contained in the merged region
initialize the point set $P_f= \{\}$ of the disqualified points around the merged area
initialize the first vertex information ( $p_{first} \leftarrow P_m^0$ ) of the merge process
calculate the weighted center coordinate $p_c \leftarrow P_m$
calculate the average scan radius $r_c \leftarrow P_m$
**for** $r$ **in** $[0, r_c]$
    $P \leftarrow$loop traversal of all the points contained within the scan radius.
    **for** $P_i$ **in** $P$
        **if** ($P_i$ belongs to the topological area)
            $P_{com} \leftarrow P_i$
        **else** $P_f \leftarrow P_i$
**if** ($P_{com}$ is empty)
    **return false**
$P_h \leftarrow$calculate the current convex hull ($P_{com}$)
calculate the number of unqualified points containing $P_f$ in convex hull $P_h$
**if** (obstacle ratio $r_a$ doesn't satisfies the requirements)
    **return false**
delete all of the vertex information in $P_m$
Merge the vertex regions in $P_m$ and add the region identifier at the same time
$P_c' \leftarrow$recalculate the center of the area
$r_n \leftarrow$recalculate the radius of the merged area
Add a new topological vertex ($r_n$, $P_c'$, $P_{com}$)
**return true**

---

After the vertex numbers that need to be merged are obtained, the region merge can be performed. The main steps are as follows:

(1) The points in the corresponding topological region of all the vertices are extracted, and the qualified points (identical with the vertex color marker) and the failure points (different from the vertex color) are counted.
(2) The convex hulls of qualified points are calculated, the unqualified points of convex hulls are counted from the failure points, and the obstacle ratio is calculated. If the obstacle ratio is greater than the threshold, then exit.
(3) If the obstacle ratio is less than the threshold, the region merging is carried out, the original vertex is deleted, and a new vertex is established.

The pseudo code of the algorithm is shown in Algorithm 2.

## 3. Topological Map Representation

The process of map preservation is shown in the block diagram on the right side of Figure 1, from which it can be seen that the representation method of the topological map is a process of combining the metric map with the topological map. The method is inspired by the ROS package [27], which expands from a gray value to an RGB color, so that the occupied information and the topological information is saved efficiently while saving and reusing the topological map. In this paper, the parameter TM and the parameter OM are used to represent the topological information and the occupied information, respectively; then, the information converter is used to convert the parameters into the RGB value of the color picture, so that the obtained RGB value can not only be used to display and identify the topology area in real time, but can also contribute to save the map information as a color image.

### 3.1. Online Representation of the Map

The metric map divides the space into a finite number of grid cells $M = \{m_i | i = 0, 1, 2, \ldots n\}$, each $m_i$ corresponding to an occupied variable. If the grid cell is completely occupied by "1" and is not occupied as "0", then $p(m_i = 1)$ or $p(m_i)$ indicates the possibility that the grid cell is occupied [28]. Therefore, the smallest unit of the metric is the grid cell, and a grid cell is represented by only one variable.

This only represents the information-occupied obstacles; although it can achieve accurate navigation and obstacle avoidance, it cannot achieve a higher level of control. For example, the user tells the robot "I want to go to the kitchen!", and the robot does not know how to perform unless the robot knows the specific location of the kitchen [29]. In order to solve this problem, a scheme combining the metric map and the topological map has been proposed.

A topological map is an environment representation method based on an adjacent graph, which can be represented by an adjacency list structure. Similar to the concept of graphs, topological maps have two basic elements: nodes (which are called vertices in this paper) and edges. Nodes represent different locations that can be distinguished in the environment or various states of distinguishable robots; edges represent relationships between nodes, such as distance or motion control commands. Such representations have low storage requirements, and also support efficient path planning, especially for large-scale unknown environments or outdoor environments.

In this paper, the vertices mainly include the position of the vertices, the vertex number, the adjacent vertices, and so on. The edges mainly include starting and ending vertices, weights, etc. In addition, metric maps and topological maps exist in the form of a "hierarchical map" in the framework proposed in this paper. The two maps interact with each other, learn from each other, and complement each other, providing different map services for the system.

### 3.2. Topological Map Preservation

The preservation of topological map is the process of integrating topological information and occupied information to form RGB image information. In order to describe the two kinds of map information conveniently, a conical space model is proposed, which is controlled by the parameter TM, the parameter OM, and integer one. The height of the cone is determined by the integer one, and the height is always one. The radius of the bottom surface of the cone is determined by the parameter OM, and the value of the polar coordinates of the bottom surface is determined by the parameter TM, which is inspired by the HSV color space theory [30]. An example process is shown in Figure 3 below.

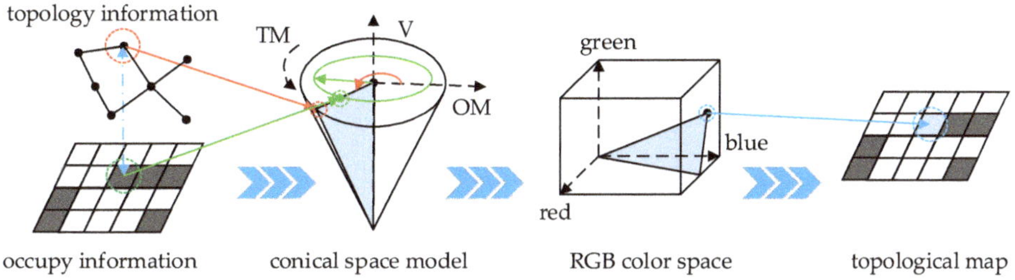


**Figure 3.** Topological map preservation process.

In the figure, the red line represents the conversion process of the topological information, the green line represents the conversion process of the occupied information, and the blue line represents the synthesis process between the two pieces of information. It should be noted that the relationship between the grid cell and topological vertex is "many-to-one", which means that a topological vertex corresponds to multiple grid elements. In other words, multiple grid cells form a region, and a region is represented by a topological vertex.

In the process of preservation, the information of the topological vertex is extracted from the topological map firstly; then, the occupied information of the grid cell corresponding to the topological vertex is found in the metric map, and the two pieces of information are converted into the parameters TM and OM. In the conical space model, a cross-section can be obtained by these two parameters. The cross-section represents the result of the synthesis of two kinds of information. However, such information is still unusable, and needs to be converted again. The blue triangle cross-section is shown in Figure 3. Therefore, the information converter is proposed to convert the cross-section information into RGB information, which is often used in the field of vision. The RGB value is used to represent a grid cell, and then each grid cell is operated in the same way. Finally, the color map containing the topological information and occupancy information is saved. In the following, the TM and OM parameters are explained in detail:

TM is measured by the angle, and the range of values is [0, 360]. It is calculated counterclockwise from red. This value is used to represent the topological neighborhood, which can represent at most 360 topological neighborhoods. In order to show a higher degree of discrimination, values can be taken at intervals, such as 20 color values for the identification of the topological region, and 18 kinds of topological region identifiers can be used.

OM represents the degree of proximity to the spectral color, and the range of values is [0, 1]. This value is used to represent whether the obstacle is occupied at a point in the map. The larger the value, the greater the likelihood that there is an obstacle at that point; the smaller the value, the greater the likelihood that the point may be free.

Then, in the beginning of topological map preservation, the occupied information and topological information need to be converted. The conversion formulas are as shown in Equation (6):

$$\begin{cases} TM = a + b \times P_i \\ OM = k \times p(m_i) \end{cases} \tag{6}$$

where $a$ is the start value, $b$ is the interval value, $k$ is the certain coefficient, and $a = 0, b = 20, k = 1$ is taken in the paper.

After that, some intermediate variables should be computed from these two parameters, and the formulas are shown in Equation (7), where $V = 1$ is taken in the paper. Then, with the help of intermediate variables, the RGB information can be obtained by Equation (8):

$$\begin{cases} h = \left\lfloor \frac{TM}{60} \right\rfloor \pmod{6} \\ f = \frac{TM}{60} - h \\ p = V \times (1 - OM) \\ q = V \times (1 - f \times OM) \\ g = V \times (1 - (1 - f) \times OM) \end{cases} \quad (7)$$

In addition, the yaml file and the picture in pgm format are used to save the metric map traditionally, where the picture in pgm picture is a grayscale picture, and only the gray level data can be saved. Therefore, the picture in ppm picture is used to save the topological map, because the picture in ppm format can save RGB color data. Moreover, the ppm image format is divided into the ASCII encoding format (file descriptor is P3) and the binary-encoding format (file descriptor is P6), because the binary encoding format consumes less memory than the ASCII encoding format, so the binary encoding format is used.

$$(R, G, B) = \begin{cases} (255, g, p) & \text{if h} = 0 \\ (q, 255, p) & \text{if h} = 1 \\ (p, 255, g) & \text{if h} = 2 \\ (p, q, 255) & \text{if h} = 3 \\ (g, p, 255) & \text{if h} = 4 \\ (255, p, q) & \text{if h} = 5 \end{cases} \quad (8)$$

Except for the ppm image, a yaml file that contains the pixel coordinates of topological vertices, adjacent vertices, and other information is also saved, and the details saved in this yaml file will be explained in the next section. In order to simplify the calculation, in the following experiments, the area indicated by red is occupied, the area indicated by green is free and the area indicated by blue is unknown.

### 3.3. Topological Map Reading

The reading of the topological map means that the map file is parsed first, and then the obstacle occupied information and topological information are restored. The reading process takes two steps: reading the yaml file and the ppm image.

Here, to explain the contents of the yaml file, the file contains the following main parts:

- Image file path: Refers to the saved path of the ppm image file.
- Resolution: Refers to the resolution of the map, and is used to represent the scale of a pixel in the real world, with a unit (meters/pixel).
- Origin: Refers to the two-dimensional (2D) pose of the lower left pixel in the map, as (x, y, yaw), with yaw as the counter-clockwise rotation (yaw = 0 means no rotation). The yaw is ignored in this paper.
- Free thresh: Pixels with an occupancy probability less than this threshold are considered completely free.
- Occupied thresh: Pixels with an occupancy probability greater than this threshold are considered completely occupied.
- Mode: The way the file is saved, which can be one of three values: trinary, scale, or raw. Trinary is the default.
- Number of topological vertices.

◆ Topological information: A series of lists that contain the pixel coordinates of topological vertices, adjacent vertices, and other information (such as semantic information; this is empty in the paper).

In the map representation method proposed in this paper, the metric map is a two-dimensional map, so the map coordinates, pixel coordinates, initial points, and other relationships can be shown in Figure 4 below. In the image, different color regions represent different topological regions, and dark color regions are more likely to be occupied.

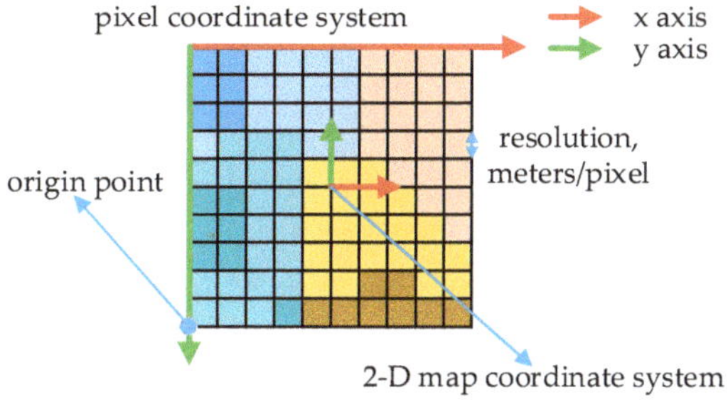


**Figure 4.** Relations between coordinate systems in the picture in ppm format.

Traditionally, the origin of the pixel coordinate system is in the upper left corner of the picture, while the origin of the map coordinate system is in the center of the picture, and the pixel coordinates of the center can be calculated by the origin point. Then, with the help of the resolution of the map, all the pixels can be restored to grid cells.

In the first place, the pixel coordinates of the pixel points in the picture need to be converted into map coordinates by the map coordinates of the lower left pixel, and the specific conversion formula is as shown in Equation (9):

$$\begin{cases} x_{\text{map}} = x_{\text{orign}} + x_{\text{pixel}} * res \\ y_{\text{map}} = \text{y}_{\text{orign}} + \text{h}_{\text{map}} * res - \text{y}_{\text{pixel}} * res \end{cases} \tag{9}$$

where $(x_{\text{pixel}}, \text{y}_{\text{pixel}})$ is the pixel coordinate, $(x_{\text{map}}, y_{\text{map}})$ is the map coordinate, $res$ is the map resolution, and $\text{h}_{\text{map}}$ is the height of the map.

At the same time, the RGB image information in the ppm image file needs to be converted into two parameters in the conical space model, and the conversion formula is shown in Equation (10).

$$\begin{aligned} & V \leftarrow \max(R,G,B) \\ & OM \leftarrow \begin{cases} \frac{V-\min(R,G,B)}{V} & \text{if } V \neq 0 \\ 0 & \text{otherwise} \end{cases} \\ & TM \leftarrow \begin{cases} \frac{60(G-B)}{V-\min(R,G,B)} & \text{if V} = \text{R} \\ \frac{120+60(B-R)}{V-\min(R,G,B)} & \text{if V} = \text{G} \\ \frac{240+60(R-G)}{V-\min(R,G,B)} & \text{if V} = \text{B} \end{cases} \end{aligned} \tag{10}$$

where if $TM < 0$, then $TM = TM + 360$. After the above calculation, the final range of the three values is: $0 \leq V \leq 1, 0 \leq OM \leq 1, 0 \leq TM \leq 360$.

Afterwards, the topological information can be obtained from the parameter TM, and the topological map is established. In addition, the occupied information of the point is read from the parameter OM, and the metric map is established. In this way, the topological information and occupied information can be completely recovered.

## 4. Experiment and Result

Finally, several experiments will be carried out to illustrate the effectiveness of the proposed algorithm and the rationality of the topological map representation method. Relevant experiments were completed in the simulation environment, and the metric map was obtained by using the gmapping algorithm [31].

In the Gazebo simulation environment, four environments were built representing an indoor home, office, pillar, and open space. The experimental environment is shown in Figure 5. In the experiment, the turtlebot2 robot equipped with a 2D laser sensor was used as an experimental platform. All of the experiments were performed on a notebook with a memory 8G, i7 processor, and a GTX1050 graphics card.

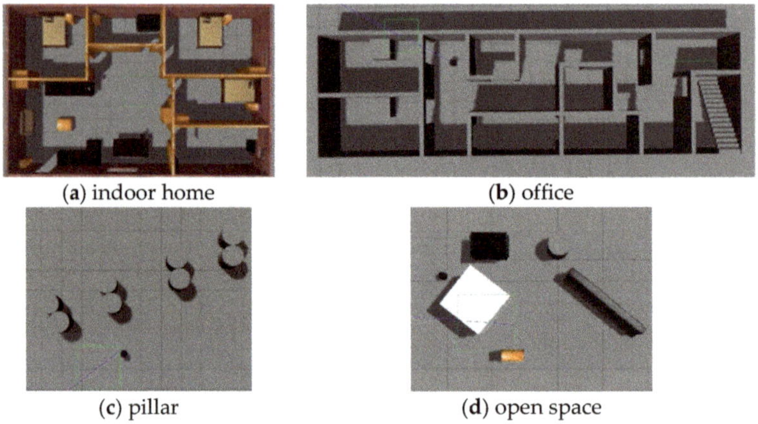

(**a**) indoor home (**b**) office (**c**) pillar (**d**) open space

**Figure 5.** Simulation environment.

The first experiment was carried out in the indoor home environment, and each part of the whole system was tested. The experimental results are shown in Figure 6, which includes the construction of the metric map, the construction of the topological map, the adjustment of the topological map, and the preservation of the topological map. These continuous processes truly implement the process of topological mapping from build to save.

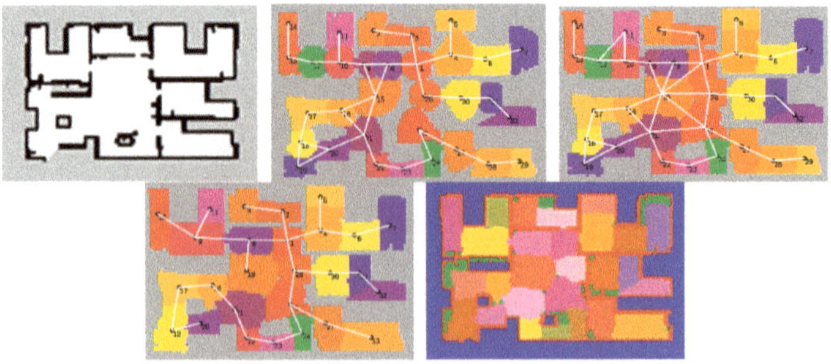

**Figure 6.** Topological map construction process.

In the following, relevant experiments were carried out in the office environment, pillar environment, and open space. The results are shown in Figure 7, which shows the metric map and topological map constructed in different environments. Among them, the growth radius is between 1.2–3 m, the map resolution is 0.20 m, and the control weight is one.

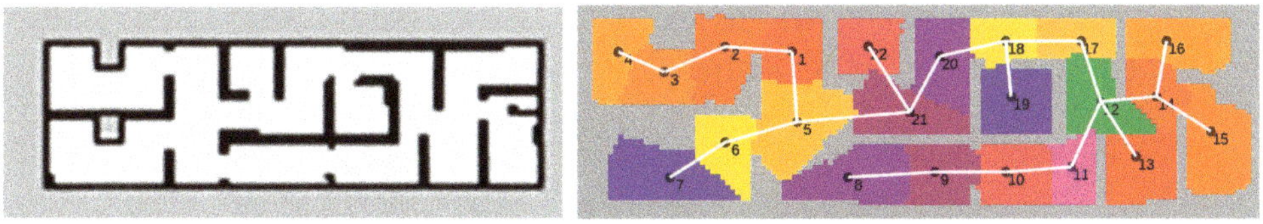

(**a**) office environment, where $r_a$ =0.10, $r_o$ =0.01, $r_p$ =0.80

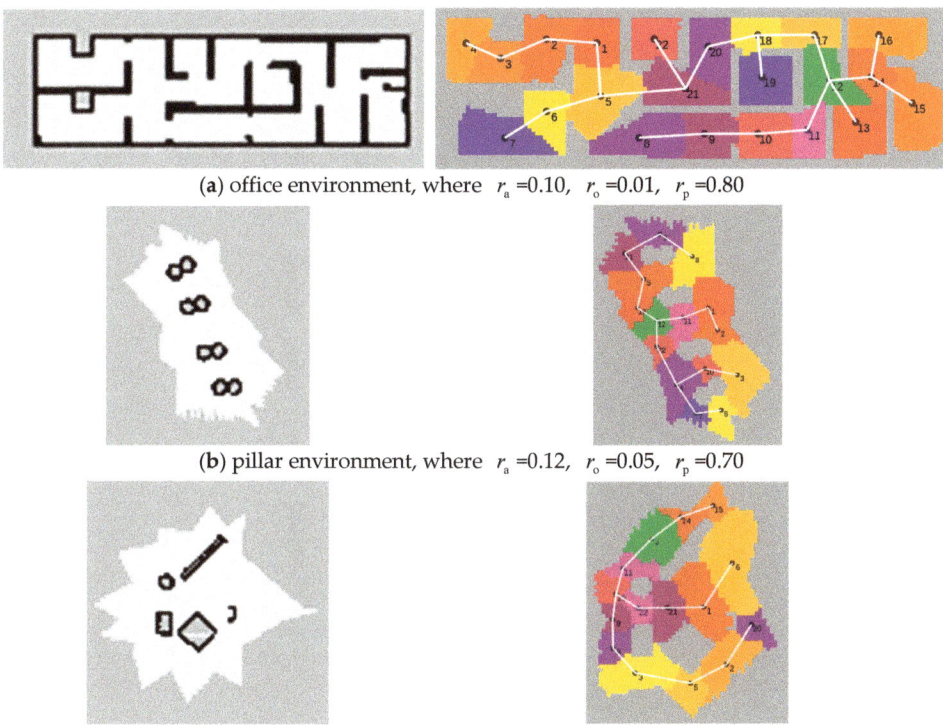

(**b**) pillar environment, where $r_a$ =0.12, $r_o$ =0.05, $r_p$ =0.70

(**c**) open space, where $r_a$ =0.13, $r_o$ =0.06, $r_p$ =0.65

**Figure 7.** Occupied map (**left**) and its corresponding topological map (**right**).

The office environment consists of many small spaces and long corridors. It can be seen that the map construction algorithm performs poorly at the door and the area is somewhat deformed. However, in the latter two examples, the map construction algorithm is more perfect: the region is more similar to the ellipse, and the dynamic growth and merging process makes the small region fully covered.

Next, in the indoor environment, this paper chooses a special location (next to the dining table in the living room) to build the map, and discusses the impact of these three parameters on the map construction under the circumstances of changing the obstacle ratio, the point addition ratio, and the points pass ratio. Here, the growth radius is set to 1.5 m, the map resolution is 0.2 m, and the control value $R_W$ is set to one. The details of regional growth are shown in Figure 8.

The obstacle ratio is to prevent the region from growing into a concave region, and ensure the convexity of the topological region. As can be seen in group (**a**) of Figure 7, with the increase of the obstacle ratio, the area will grow toward a small space (such as a door) next to it, and a few obstacle points will be added continuously.

The points addition ratio prevents malformation, making the growth area approximately circular or elliptical rather than elongated. It can be seen in group (**a**) of Figure 8 that when the point addition ratio is too small or too large, it is easy to cause growth malformation or insufficient growth.

The points pass ratio can reflect the contrast between the growth area and the ideal area in the current state. It is used to modify the growth radius in real time, so its role is to prevent growth deficiency. When the growth area is far from the ideal area, it can make the region grow seriously. This effect is well reflected in the (**c**) group, but it also destroys the convex nature of the area. Therefore, the above three parameters need to be adjusted according to the actual situation, and the topological map will represent the environment better.

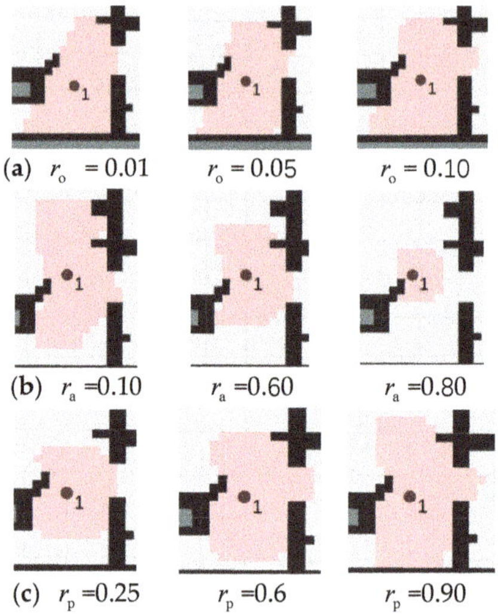


**Figure 8.** The influence of three parameters on the construction of the topological map. (**a**) Impact of changes in the obstacle ratio on the region. (**b**) Impact of changes in the point addition ratio on the region. (**c**) Impact of changes in the points pass ratio on the region.

In addition, map resolution is another important factor affecting the construction of topological maps. In the indoor home environment, a large number of experiments were carried out for different resolutions in order to obtain an optimal topological map. Figure 9 shows the effect of map resolution on topological map connectivity and occupancy.

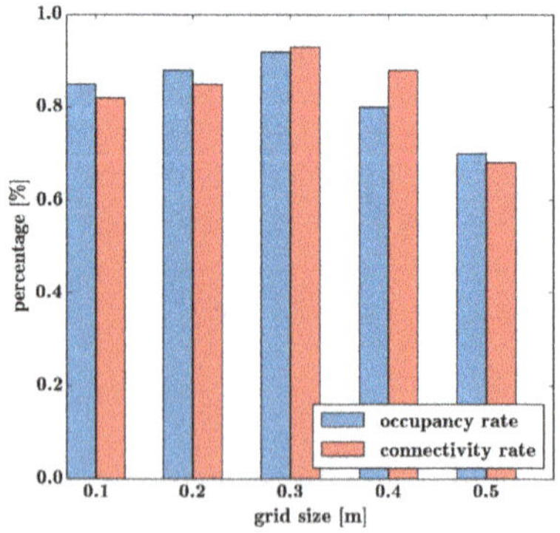


**Figure 9.** The impact of map resolution on topological map construction.

Finally, topological maps built in the office environment, pillar environment, and open space are saved, as shown in Figure 10. Experiments show that the topological map representation proposed in this paper can save the occupied information and topological information well.

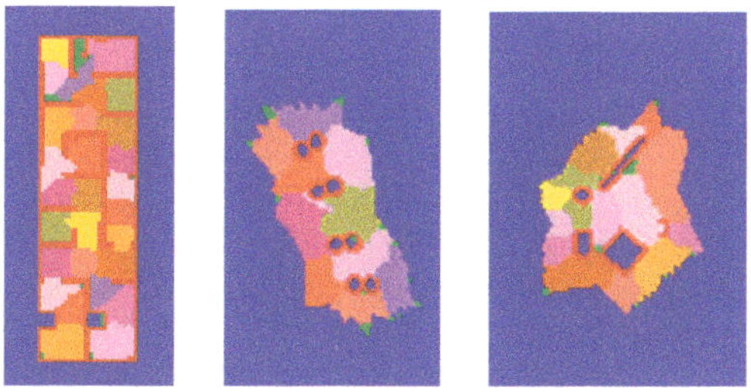

**Figure 10.** Topological maps saved in three environments.

What's more, the system proposed in this paper was integrated into the turtlebot2 equipped with the hokuyo laser sensor and NVidia TK1, and the topological map construction was successfully completed in the author's laboratory, refer to the Supplementary Materials. This experiment proves the practicality of the pipeline proposed in this paper on mobile platforms such as service robots. The configuration and map of the turtlebot2 experimental system is shown in Figure 11.

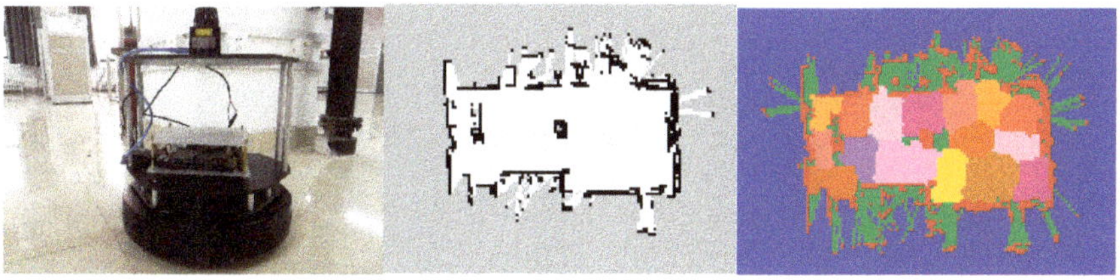

**Figure 11.** Setup of the turtlebot2 experiments and maps.

Not only is the topological map construction method not limited by the environment geometry, but the topological map preservation method can also simultaneously save the occupied information and topological information to achieve the map conditions required for accurate navigation. In order to better illustrate that, Table 1 summarizes some of the information about the topological map creation in various environments. As can be seen from the table, the topological map can cover the entire environment, and the more complex the environment, the more topological vertices the map construction method will use to describe it. Although the topological map storage requirements are larger than the metric maps, more information will be saved to provide the same precision navigation environment.

**Table 1.** Topological map information for each scene.

| | Topological Map Information | | Storage Requirements (kB) | |
|---|---|---|---|---|
| | Number of Vertexes | Occupied Ratio | Metric Map | Topological Map |
| Indoor (171.224 $m^2$) | 31 | 0.922444 | 36.0183 | 109.46 |
| Office (115.389 $m^2$) | 22 | 0.985809 | 30.0172 | 91.27 |
| Pillar (129.821 $m^2$) | 15 | 0.984625 | 20.0172 | 72.061 |
| Open space (171.261 $m^2$) | 15 | 0.936123 | 16.0172 | 48.0853 |
| Laboratory (277.999 $m^2$) | 16 | 0.883049 | 30.0181 | 90.0903 |

## 5. Conclusions

In this paper, a new framework for creating high-precision topological maps and efficient map representations for diverse environments is presented. In various structured environments, the regional dynamic growth algorithm proposed in this paper can divide the free space into multiple convex regions, each forming a topological region to represent the environment. This greatly reduces the use requirements compared to the most advanced methods, and extends the range of use. The topological map representation method proposed in this paper uses the parameter TM and the parameter OM of the conical space model to represent the occupation information and topological information of the map efficiently, which satisfies the conditions of high-precision navigation.

Through a large number of experiments, the topological map construction method proposed in this paper can construct a topological map that conforms to a specific environment without being limited by the environment geometry. In addition, the topological map can be made to better describe the entire environment by modifying the corresponding control parameters. It also verifies that the topological map preservation method proposed in this paper can save the occupied information and topological information at the same time by occupying a small amount of storage space. The universality of the construction method and the efficiency of the preservation method provide conditions for human–machine interactive navigation, so that the topological map can be truly applied in real life.

For the future research, since this paper only studies two-dimensional topological maps, three-dimensional topological maps will be the direction of future work. In addition, the semantic information of the actual scene can be added to the topological node to improve the robot's understanding of the actual environment.

## 6. Patents

A patent named "Human–machine interactive navigation system and method based on brain–computer interface" is pending.

**Supplementary Materials:** A video is available online at https://youtu.be/XUidY4vnslU, Video title: A topology map construction experiment based on dynamic growth algorithm.

**Author Contributions:** This study was completed by the co-authors. F.W., C.W., and H.C. conceived and led the research. The major experiments and analyses were undertaken by Y.L. and L.X.; F.W. supervised and guided this study. Y.L. and L.X. wrote the paper. All the authors have read and approved the final manuscript. Conceptualization, F.W.; Formal analysis, Y.L.; Investigation, Y.L.; Methodology, Y.L.; Project administration, F.W., C.W. and H.C.; Resources, F.W.; Software, Y.L.; Validation, L.X.; Writing—original draft, Y.L.; Writing—review & editing, Y.L. and L.X.

**Funding:** This research was funded by Fundamental Research Funds for the Central Universities of China, grant number N172608005 and Liaoning Provincial Natural Science Foundation of China, grant number 20180520007.

**Conflicts of Interest:** The authors declare no conflict of interest.

## References

1. Mavridis, N. A review of verbal and non-verbal human–robot interactive communication. *Robot. Auton. Syst.* **2015**, *63*, 22–35. [CrossRef]
2. Luo, R.C.; Shih, W. Autonomous Mobile Robot Intrinsic Navigation Based on Visual Topological Map. In Proceedings of the 2018 IEEE International Symposium on Industrial Electronics (ISIE), Cairns, Australia, 13–15 June 2018; pp. 541–546.
3. Chung, M.J.Y.; Pronobis, A.; Cakmak, M.; Fox, D.; Rao, R.P.N. Autonomous question answering with mobile robots in human-populated environments. In Proceedings of the 2016 IEEE/RSJ International Conference on Intelligent Robots and Systems (IROS), Daejeon, Korea, 9–14 October 2016; pp. 823–830.
4. Jo, K.; Kim, C.; Sunwoo, M. Simultaneous localization and map change update for the high definition map-based autonomous driving car. *Sensors* **2018**, *18*, 3145. [CrossRef] [PubMed]
5. Lin, H.Y.; Yao, C.W.; Cheng, K.S. Topological map construction and scene recognition for vehicle localization. *Auton. Robot.* **2018**, *42*, 65–81. [CrossRef]
6. Qing, G.; Zheng, Z.; Yue, X. Path-planning of automated guided vehicle based on improved Dijkstra algorithm. In Proceedings of the 2017 Chinese Control and Decision Conference (CCDC), Chongqing, China, 28–30 May 2017; pp. 7138–7143.
7. Papoutsidakis, M.; Kalovrektis, K.; Drosos, C.; Stamoulis, G. Design of an Autonomous Robotic Vehicle for Area Mapping and Remote Monitoring. *Int. J. Comput. Appl.* **2017**, *167*, 36–41. [CrossRef]

8. Chandrasekaran, B.; Conrad, J.M. Human-robot collaboration: A survey. In Proceedings of the 2015 Southeast Conference, Fort Lauderdale, FL, USA, 9–12 April 2015; pp. 1–8.
9. Alitappeh, R.J.; Pereira, G.A.S.; Araújo, A.R.; Pimenta, L.C.A. Multi-robot deployment using topological maps. *J. Intell. Robot. Syst.* **2017**, *86*, 641–661. [CrossRef]
10. Johnson, C. Topological Mapping and Navigation in Real-World Environments. Master's Thesis, University of Michigan, Ann Arbor, MI, USA, 12 December 2017.
11. Li, X.; Qiu, H. An effective laser-based approach to build topological map of unknown environment. In Proceedings of the 2015 IEEE International Conference on Robotics and Biomimetics (ROBIO), Zhuhai, China, 6–9 December 2015; pp. 200–205.
12. Konolige, K.; Marder-Eppstein, E.; Marthi, B. Navigation in hybrid metric-topological maps. In Proceedings of the 2011 IEEE International Conference on Robotics and Automation (ICRA), Shanghai, China, 9–13 May 2011; pp. 3041–3047.
13. Azzag, H.; Lebbah, M.; Arfaoui, A. Map-TreeMaps: A New Approach for Hierarchical and Topological Clustering. In Proceedings of the 2010 International Conference on Machine Learning and Applications (ICMLA), Washington, DC, USA, 12–14 December 2010; pp. 873–878.
14. Liu, M.; Colas, F.; Siegwart, R. Regional topological segmentation based on mutual information graphs. In Proceedings of the 2011 IEEE International Conference on Robotics and Automation (ICRA), Shanghai, China, 9–13 May 2011; pp. 3269–3274.
15. Bandera, A.; Sandoval, F. Spectral clustering for feature-based metric maps partitioning in a hybrid mapping framework. In Proceedings of the 2009 IEEE International Conference on Robotics and Automation (ICRA), Kobe, Japan, 12–17 May 2009; pp. 1868–1874.
16. Yu, W.; Amigoni, F. Standard for Robot Map Data Representation for Navigation. In Proceedings of the 2014 IEEE/RSJ International Conference on Intelligent Robots and Systems (IROS), Chicago, IL, USA, 14–18 September 2014; pp. 3–4.
17. Lau, B.; Sprunk, C.; Burgard, W. Improved updating of Euclidean distance maps and Voronoi diagrams. In Proceedings of the 2010 IEEE/RSJ International Conference on Intelligent Robots and Systems (IROS), Taipei, Taiwan, 18–22 October 2010; pp. 281–286.
18. Guo, S.; Ma, S.; Li, B.; Wang, M.; Wang, Y. Simultaneous Location and Mapping Through a Voronoi-diagram-based Map Representation. *Acta Autom. Sin.* **2011**, *37*, 1095–1104.
19. Ramaithitima, R.; Whitzer, M.; Bhattacharya, S.; Kumar, V. Automated creation of topological maps in unknown environments using a swarm of resource-constrained robots. *IEEE Robot. Autom. Lett.* **2016**, *1*, 746–753. [CrossRef]
20. Kaleci, B.; Senler, C.M.; Parlaktuna, O.; Gürel, U. Constructing Topological Map from Metric Map Using Spectral Clustering. In Proceedings of the 2016 International Conference on TOOLS with Artificial Intelligence (ICTAI), San Jose, CA, USA, 6–8 November 2016; pp. 139–145.
21. Liu, M.; Colas, F.; Pomerleau, F.; Siegwart, R. A Markov semi-supervised clustering approach and its application in topological map extraction. In Proceedings of the 2012 International Conference on Intelligent Robots and Systems (2012), Vilamoura, Algarve, Portugal, 7–12 October 2012; pp. 4743–4748.
22. Ravankar, A.A.; Ravankar, A.; Emaru, T.; Kobayashi, Y. A hybrid topological mapping and navigation method for large area robot mapping. In Proceedings of the 2017 Society of Instrument and Control Engineers of Japan (SICE), Kanazawa, Japan, 19–22 September 2017; pp. 1104–1107.
23. Kaleci, B.; Parlaktuna, O.; Gurel, U. A comparative study for topological map construction methods from metric map. In Proceedings of the 2018 Signal Processing and Communications Applications Conference (SIU), Izmir, Turkey, 2–5 May 2018; pp. 1–4.
24. Blochliger, F.; Fehr, M.; Dymczyk, M.; Schneider, T.; Siegwart, R. Topomap: Topological mapping and navigation based on visual slam maps. In Proceedings of the 2018 IEEE International Conference on Robotics and Automation (ICRA), Brisbane, QLD, Australia, 21–25 May 2018; pp. 1–9.
25. The Bresenham Line-Drawing Algorithm. Available online: https://www.cs.helsinki.fi/group/goa/mallinnus/lines/bresenh.html (accessed on 9 December 2018).
26. Graham, R.L. An efficient algorithm for determining the convex hull of a finite planar set. *Inf. Process. Lett.* **1972**, *1*, 132–133. [CrossRef]
27. map_server. Available online: http://wiki.ros.org/map_server (accessed on 9 December 2018).

28. Colleens, T.; Colleens, J.J.; Ryan, D. Occupancy grid mapping: An empirical evaluation. In Proceedings of the 2007 Mediterranean Conference on Control & Automation, Athens, Greece, 27–29 June 2007; pp. 1–6.
29. Saunders, J.; Syrdal, D.S.; Koay, K.L.; Burke, N.; Dautenhahn, K. "Teach Me–Show Me"—End-User Personalization of a Smart Home and Companion Robot. *IEEE Trans. Hum.-Mach. Syst.* **2016**, *46*, 27–40. [CrossRef]
30. Hanbury, A. Circular statistics applied to colour images. In Proceedings of the 2003 Computer Vision Winter Workshop, Valtice, Czech Republic, 3–6 February 2003; pp. 53–71.
31. Grisettiyz, G.; Stachniss, C.; Burgard, W. Improving Grid-based SLAM with Rao-Blackwellized Particle Filters by Adaptive Proposals and Selective Resampling. In Proceedings of the 2005 IEEE International Conference on Robotics and Automation, Barcelona, Spain, 18–22 April 2005; pp. 2432–2437.

*applied sciences*

MDPI

*Article*

# Navigating a Service Robot for Indoor Complex Environments

**Jong-Chih Chien [1], Zih-Yang Dang [2] and Jiann-Der Lee [2,3,4,*]**

[1] Degree Program of Digital Space and Product Design, Kainan University, Taoyuan 33857, Taiwan; jcchien@mail.knu.edu.tw
[2] Department of Electrical Engineering, Chang Gung University, Taoyuan 33302, Taiwan; yang1227@hotmail.com
[3] Department of Neurosurgery, Chang Gung Memorial Hospital at LinKou, Taoyuan 33302, Taiwan
[4] Department of Electrical Engineering, Ming Chi University of Technology, New Taipei City 24301, Taiwan
* Correspondence: jdlee@mail.cgu.edu.tw

Received: 30 December 2018; Accepted: 29 January 2019; Published: 31 January 2019

**Featured Application: Navigation for a service robot with facial and gender recognition capabilities in an indoor environment with static and dynamic obstacles.**

**Abstract:** This paper investigates the use of an autonomous service robot in an indoor complex environment, such as a hospital ward or a retirement home. This type of service robot not only needs to plan and find paths around obstacles, but must also interact with caregivers or patients. This study presents a type of service robot that combines the image from a 3D depth camera with infrared sensors, and the inputs from multiple sonar sensors in an Adaptive Neuro-Fuzzy Inference System (ANFIS)-based approach in path planning. In personal contacts, facial features are used to perform person recognition in order to discriminate between staff, patients, or a stranger. In the case of staff, the service robot can perform a follow-me function if requested. The robot can also use an additional feature which is to classify the person's gender. The purpose of facial and gender recognition includes helping to present choices for suitable destinations to the user. Experiments were done in cramped but open spaces, as well as confined passages scenarios, and in almost all cases, the autonomous robots were able to reach their destinations.

**Keywords:** robot; obstacle avoidance; facial and gender recognition

## 1. Introduction

Today's workers in care-taking facilities have their hands full taking care of patients, thus require help for everyday routines. There are many branches of investigation into using machines to help out with daily routines in the health-care industry, such as gesture recognition [1], or pedestrian movement prediction [2]. We choose to investigation helpful tasks, such as delivering proper medicines to target patients, or helping to carry heavy loads in follow-me mode, performed by an autonomous robot with face and gender recognition abilities. For these purposes, this paper presents a tri-wheeled autonomous service robot equipped with a camera, an RGB-Depth (RGB is acronym for Red, Green, and Blue) sensor, and sonar sensors. It has built-in facial recognition ability, ability to separate staff or patients from visitors, and it can also distinguish the gender of a visitor for record-keeping purposes. By recognizing staff, it can offer the staff functions not available to patients or visitors, such as the follow-me function. By recognizing patients, it can help dispense the proper medication to each individual patient. A possible scenario for this robot could be using it to dispense medicine to selected patients who may be walking around the hallways. In navigation, it can avoid static as well as dynamic obstacles while moving toward its objective, using a dual-level Adaptive Neuro-Fuzzy Inference

System (ANFIS)-based fuzzy controller. It uses a depth-map to detect obstacles ahead, then uses image processing techniques to extract information as input into an ANFIS-based fuzzy system for analysis in order for the service to be able to avoid obstacles. The robot also incorporates sonar information using another ANFIS-based fuzzy system when it determines that depth-map information is insufficient.

In 1998, Yamauchi and Schultz [3] proposed an idea to include extra distance sensors on robots, such as laser distance sensors and ultrasound distance sensors. The distance sensors can aid in detecting surrounding objects or obstacles in real time so that the robot can locate or correct itself using known map data. However, the high costs of the precision sensors make them unsuitable for popular use. Even though ultrasonic sensors are relatively cheap by comparison, their deviation of errors is wider and they have detection blind spots; the detection distance is only proportional to the volume of the sound generator. Prahlad [4] proposed a driver-less car that has face detection and tracking capabilities. Similarly, we added onto our architecture the ability to perform face and gender recognition. Correa [5] proposed the development of a sensing system in an indoor environment, allowing the robot to have autonomous navigation and the ability for identification of its environment. His proposed system consisted of two parts: The first part is a reactive navigation system where the robot uses the RGB-Depth sensor to receive depth information and uses the arrangement of obstacles as the basis to determine the path to avoid the obstacles indoors; the second part uses an artificial neural network to identify different configurations of the environment. Csaba [6] introduced an improved version based on fuzzy rules. His system uses 16 rules, three inputs and one output, one Mamdani-type fuzzy controller, and obtained acceptable results in real-time experiments. In 2016, Algredo-Badillo [7] presented the possibility of a fuzzy control system, with the output from depth sensor as its input, for an autonomous wheelchair as a possible design. In 2018, de Silva [8] and Jin et al. [9] show that fusion of data from multiple sensors can perform better than a single sensor for a driver-less vehicle.

Our study builds an autonomous robot with a camera, a RGB-depth sensor, and sonar sensors. Prior to moving, the robot scans the entire room with a pre-trained pedestrian classifier in order to determine if a targetable person or persons exist that need to be tracked. If the robot locates its target and moves toward the person, the person's face is located and facial and gender recognitions would be performed simultaneously in order to determine if the target person's identity is in the database or if the person is a stranger. While moving, the dual sensor inputs to the fuzzy-based real-time obstacle avoidance system are used for path planning. The target person can then input target location and follows the robot or activates the robot's follow-me function instead. Using the follow-me function, the robot can help carry heavy loads for the user. The flowchart of the system is shown below in Figure 1.

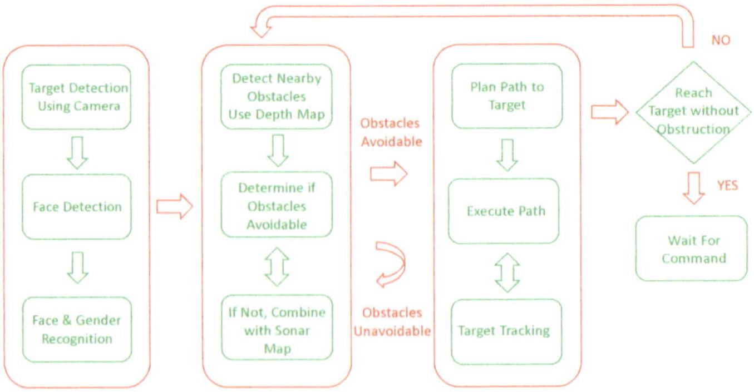


**Figure 1.** The system flowchart.

## 2. Method

### 2.1. Facial Recognition

We used OpenCV's LPBH (Local Binary Patterns Histogram) face recognizer method [10], which uses Local Binary Patterns Histogram as descriptors. It has shown to work better than the EigenFaces [11] or the FisherFaces [12] face recognizer methods under different environments and lighting conditions, and can reach an accuracy rate of more than 94% when 10 or more faces per person are used during its training [13]. The training and testing datasets were taken from Aberdeen [14], GUFD [15], and Utrecht ECVP [16]. Pictures of laboratory personnel were later insert into the training dataset. The purpose of adding face and gender recognition is so that the robot can locate an assigned target or targets, then performs its assigned tasks accordingly. The flowchart for face and gender recognition is shown in Figure 2.

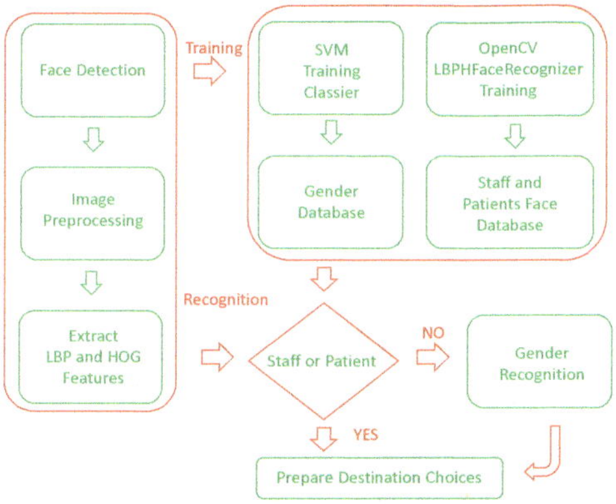


**Figure 2.** Facial and gender recognition flowchart.

### 2.2. Gender Recognition

In the gender identification part, the features with higher classification accuracy are selected from LBP [17] and HOG (Histogram of Oriented Gradients) [18] of different scales, and then screened using *p*-values before being combined. The advantage of this method is that it uses the differences between LBP and HOG methods of calculation to increase the classification accuracy. Simple experiments show that the results of combining both is higher than using either one alone. In addition, using the *p*-value to filter can be used to pick out features that are more prominent, so that the number of classification features required would be greatly reduced, and the time required for training or testing could also be improved.

After *p*-value is used to filter the statistical values of the texture features of HOG and LBP, the more robust features in the training samples are used to train a SVM (Support Vector Machine) [19] model, which is used as the basis for the classification of the test samples. The training and testing data were taken from the same online databases for facial recognition. We compared the accuracy rates of LBP, HOG, and LPB + HOG + *p*-value filtering. The results are shown below in Table 1. The rate of success of HOG + LBP + *p*-value filtering has reached 92.6%.

**Table 1.** Comparison of Gender Classification Accuracy Using Different Features.

| Feature (s) | Classification Accuracy | Feature Points |
|---|---|---|
| LBP | 91.8862% | 2891 |
| HOG | 90.0464% | 1548 |
| LBP + HOG + *p*-value | 92.6829% | 1365 |

## 2.3. Object Segmentation

Object segmentation is performed using map from the depth sensor. An example of depth map vs. standard RGB camera is shown below in Figure 3.

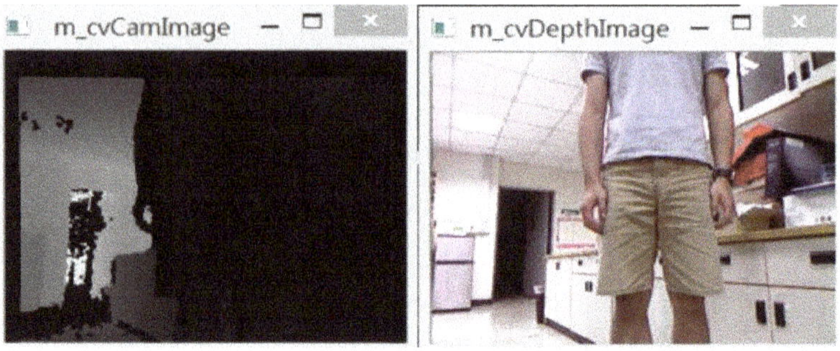


**Figure 3.** Example of depth map vs. standard camera image.

A depth map includes all objects nears and far, and contains too much information for accurate processing, so we decided that the threshold of the depth map should first be limited by a depth value less or equal to 1.5 m. An example result is shown below in Figure 4:

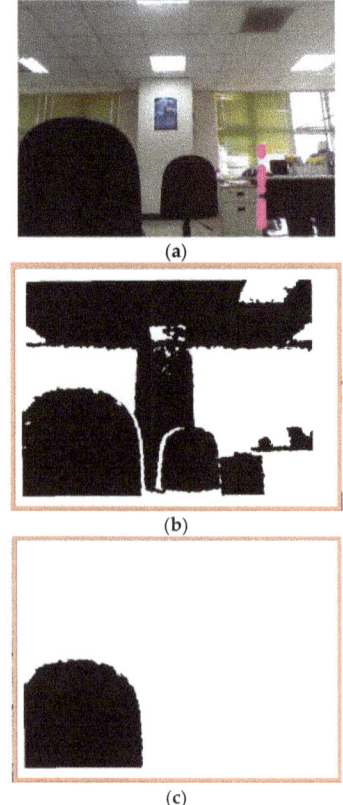

(a)

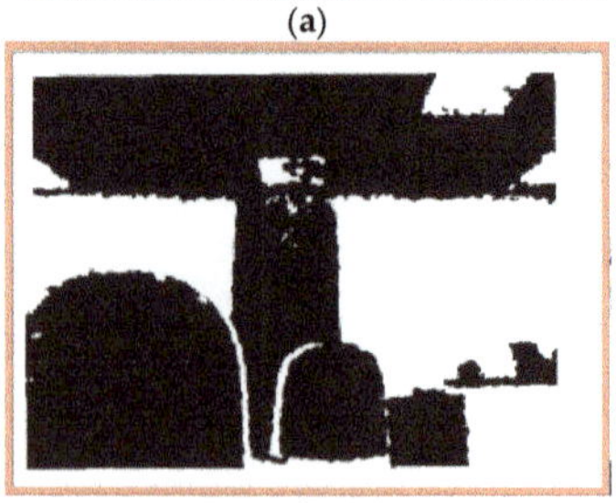

(b)

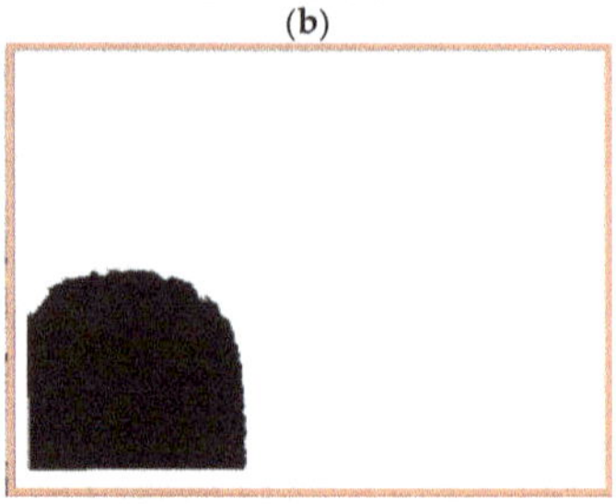

(c)

**Figure 4.** (a) Camera image, (b) depth map, (c) depth map threshold of 1.5 m.

At this range, generally the obstacles tend to be at the bottom of the depth image. In order to speed up the process, we extract the regions of interest (ROI) from the entire image by disregarding the depth information from above the middle of the height of the image. This is also to avoid the effects of lighting from above. In Correa's paper [5], he listed eight possible scenarios for obstacle arrangement; where the scene is divided into five parts, and if any of the space is occupied by an obstacle then it is marked, as shown in Figure 5. For example, the obstacle arrangement in Figure 4c is the scenario of Figure 5d.

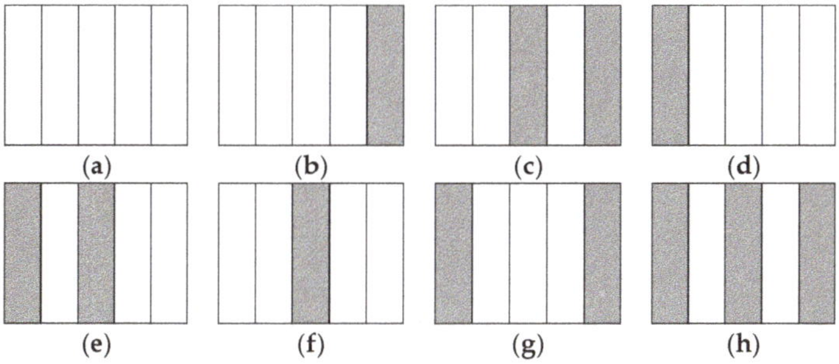


**Figure 5.** Correa's eight obstacles scenarios. (**a**) No obstacle; (**b**) One obstacle at the far right; (**c**) Two obstacles at the right, separated by a gap; (**d**) One obstacle at the far left; (**e**) Two obstacles at the left, separated by a gap; (**f**) An obstacle at the middle; (**g**) Two obstacles at the far right and far left; (**h**) Three obstacles separated by two gaps.

In the most likely cases are cases (f–h), scenarios where the robot uses Correa's method randomly move left or right, and can easily make the wrong decision. So instead of detecting obstacle arrangement first before moving, we decided that the robot should actively seek gaps between obstacles while moving in order to find the largest gap and judge if there is a chance to pass through. Figure 6 shows the various gaps detected using this method. The proposed method would be more proactive than that proposed by Correa.

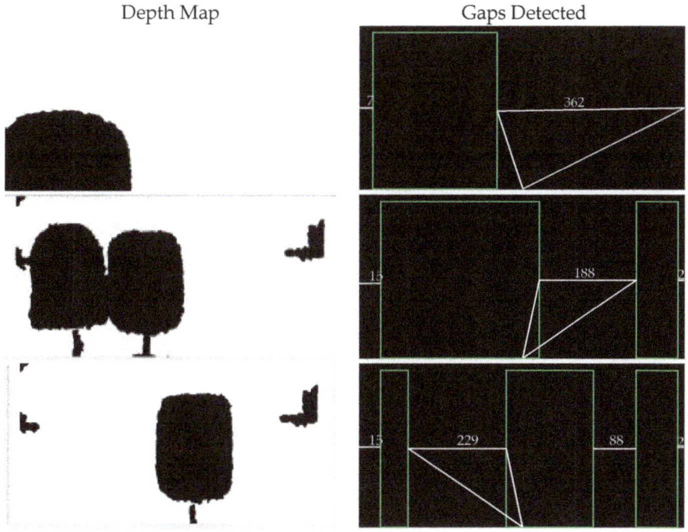


**Figure 6.** Threshold depth maps and detected gaps.

## *2.4. Robot Movements in the Presence of Obstacles*

There are five possible commands for the robots to move: forward, toward left, toward right, pause, and turn around. Forward command is issued when fuzzy analysis of the obstacles shows that there is a gap in the middle that is passable. There is a necessary initial condition for the robot to move forward: that within its depth sensor's field-of-view, at least one passable gap exists. If this initial condition does not exist, then the robot can turn around, move forward a little, then turn around again in order to increase its field-of-view. The turn towards left command is issued if the fuzzy analysis of the obstacles determines that there is a gap wide enough to pass and the center of the gap is toward the left of the depth map, but it is not a hard right, rather at an angle that causes the robot to move toward the center of the gap; similarly for the turn towards right command. The turnaround command is issued when fuzzy analysis determines that there is no gap wide enough to pass through. The pause command is issued if additional information from the sonar is required. The flowchart for obstacle avoidance is shown below in Figure 7.

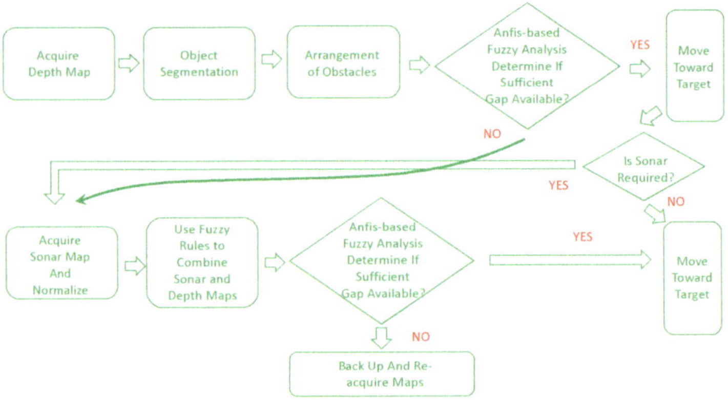


**Figure 7.** The flowchart for robot's obstacle avoidance.

## *2.5. ANFIS-Based Fuzzy System*

ANFIS is the acronym for Adaptive Neuro-Fuzzy Inference System. An ANFIS [20] system structure for two variables and two rules is shown in Figure 8, where

Rule 1: If input x is $A_1$ and input y is $B_1$,
Then $f_1 = p_1x + q_1y + r_1$.
Rule 2: If input x is $A_2$ and input Y is $B_2$,
Then $f_2 = p_2x + q_2y + r_2$.

Then the output f is a linear combine of weighted $f_1$ and $f_2$. In Figure 8, the first layer is input layer, which contains the membership functions of variables. The second layer is the rule layer, which gets fuzzy rules from the combinations of the membership function of each variable. The third layer is the normalization layer, which normalizes the results from the previous layer. The fourth layer is the inference layer. The fifth layer is output layer, which calculates the sum from previous layer's output values. De-fuzzification is then performed on the output.

In our system, there are two fuzzy systems, as shown in Figure 7, where the first system uses depth map alone in determining whether a crossable gap exists between obstacles. In the second system, where depth map alone is determined to be insufficient for accurate judgement of the closeness of obstacles, the input from sonars are then used in addition to the depth map as inputs. In the following subsections, we will discuss these two fuzzy systems.

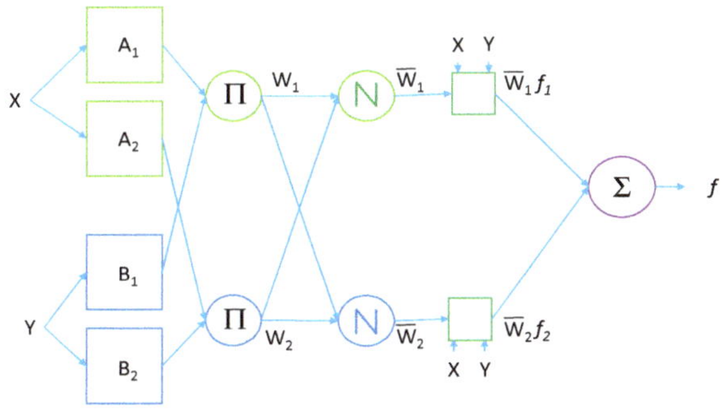


**Figure 8.** The ANFIS-based fuzzy system for two rules.

### 2.5.1. Fuzzy System Using Depth Map Alone

Because our RGB-Depth sensor is only accurate for obstacles up to 1.5 m away, there is a possibility of mis-judging gap width, and that a crossable gap may be judged as not crossable. Therefore, this is where the first fuzzy system is used to judge whether a gap is crossable. We use as input the value of the depth map, and the absolute difference between the depth values of neighboring obstacles. COG (Center of Gravity) is used as the de-fuzzification method. The 9 rules are shown in Table 2. The threshold values of 400 and 700 were determined experimentally first.

**Table 2.** Fuzzy rules for using depth map alone.

| | |
|---|---|
| 1 | When gap width is less than robot width, and the absolute depth difference between neighbors is less than 400, then output LOW. |
| 2 | When gap value is less than robot width, and the absolute depth difference between neighbors is between 400 and 700, then output LOW. |
| 3 | When gap value is less than robot width, and the absolute depth difference between neighbors is greater than 700, then output LOW. |
| 4 | When gap value is around robot width, and the absolute depth difference between neighbors is less than 400, then output LOW. |
| 5 | When gap value is around robot width, and the absolute depth difference between neighbors is between 400 and 700, then output LOW. |
| 6 | When gap value is around robot width, and the absolute depth difference between neighbors is greater than 700, then output HIGH. |
| 7 | When gap value is greater than robot width, and the absolute depth difference between neighbors is less than 400, then output HIGH. |
| 8 | When gap value is greater than robot width, and the absolute depth difference between neighbors is between 400 and 700, then output HIGH. |
| 9 | When gap value is greater than robot value, and the absolute depth difference between neighbors is greater than 700, then output HIGH. |

An example using these rules is shown in Figure 9, where the gap value is 241, around the robot's width, and the depth difference between neighbors is 663; the 9 rows in Figure 9 represent inputs and outputs for each of the 9 fuzzy rules. The bottom right graph is the visualization for the final output. The red lines for the input columns are visual representation of the input values.

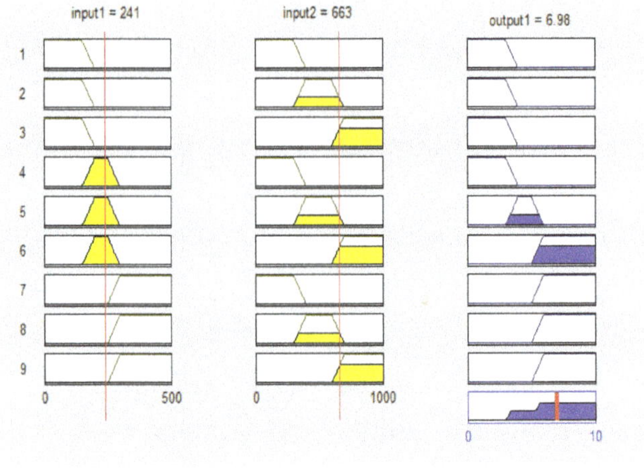


**Figure 9.** An example of fuzzy-rules-based decision.

2.5.2. Combining Sensors

One drawback using the depth map is the problem that if the obstacles are too low and too close to the robot, then they are undetectable by using the depth map alone. In this case, the sonars will be activated as aid in obstacle-avoidance. However, judging how much weight should be given to the depth map or the sonar map in order to yield the optimal map so as to find the largest available gap is a problem. It requires solving the weights in Equation (1) at each instance of decision making:

$$\text{(Result Map)} = W_d * \text{(Depth Map)} + W_s * \text{(Normalized Sonar Map)} \quad (1)$$

The field of research of combining sonar map with other sensors, such as depth map, is still an area that requires exploration [21,22]. The sonar map can only be used for obstacles that are close, so using the sonar map alone may cause the robot to spin continuously in order to acquire more information. Because of this, our research uses the depth map to move the robot until it moves too close to the gap or obstacles and determines that additional information would be required, then it activates the sonar system in order to acquire the sonar map. There are eight sonar sensors installed in a semi-circular fashion, as illustrated in Figure 10.

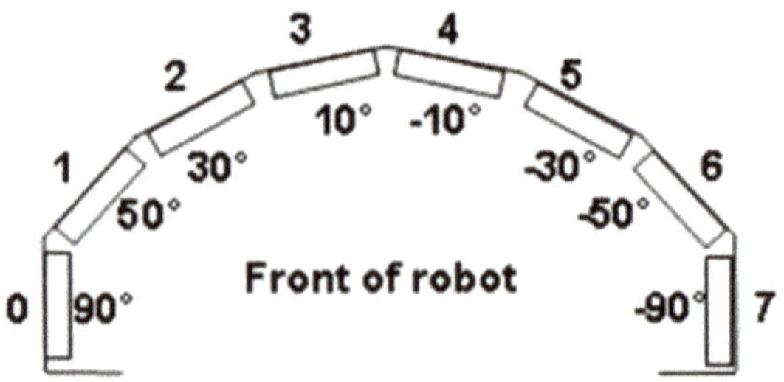


**Figure 10.** Placement of sonar sensors near our robot's base.

However, the leftmost and rightmost sensors are not used because they seem to cause erroneous decisions. The sonar signals are first normalized between 0 and 1.0, and it is determined that signal strength less than 0.4 indicates that obstacles are close, while greater than 0.6 indicates that obstacles are far away. If each of the six sonar signals are used as individual fuzzy inputs, then the system would become overly complicated. Therefore, in order to reduce the number of fuzzy rules, the signals for the left three sensors are averaged into a single input, and the signals for the right three sensors are averaged into another single input. This would effectively reduce the number of fuzzy rules. We designed a total of 8 fuzzy rules for combining the sensors. They are listed in Table 3.

**Table 3.** Fuzzy rules for combining sensors.

| | |
|---|---|
| 1 | If the gap within depth map is greater than robot width, and the left sonar input is high, and the right sonar input is high, then $W_d = 1.0$ $W_s = 0.0$. |
| 2 | If the gap within depth map is greater than robot width, and the left sonar input is high, and the right sonar input is low, then $W_d = W_s = 0.5$. |
| 3 | If the gap within depth map is greater than robot width, and the left sonar input is low, and the right sonar input is high, then $W_d = W_s = 0.5$. |
| 4 | If the gap within depth map is greater than robot width, and the left sonar input is low, and the right sonar input is low, then $W_d = W_s = 0.5$. |
| 5 | If the gap within depth map is less than robot width, and the left sonar input is low, and the right sonar input is low then, $W_d = 0.0$, $W_s = 0.5$. |
| 6 | If the gap within depth map is less than robot width, and the left sonar input is high, and the right sonar input is low then, $W_d = W_s = 0.5$. |
| 7 | If the gap within depth map is less than robot width, and the left sonar input is low, and the right sonar input is low then, $W_d = W_s = 0.5$. |
| 8 | If the gap within depth map is less than robot width, and the left sonar input is high, and the right sonar input is high then, $W_d = W_s = 0.5$. |

An example of using these rules as shown in Figure 11, where gap width is around the robot's width, the normalized and averaged value for left sonar is 0.259, and the normalized and averaged value for right sonar is 0.729. The eight rows in Figure 11 represents the inputs and outputs for the 8 rules. The red lines are symbolic representations of the input values.

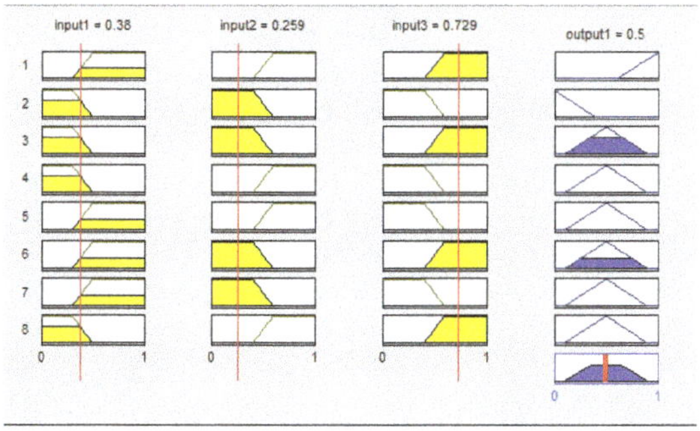


**Figure 11.** An example of using fuzzy rules to combine sensors input.

Figure 12 shows the paths the robot takes when decisions are made based on using fuzzy rules to combine sonar map and depth map versus using only the sonar map alone. In situations like this, we can see the advantage of combining the inputs of sensors.

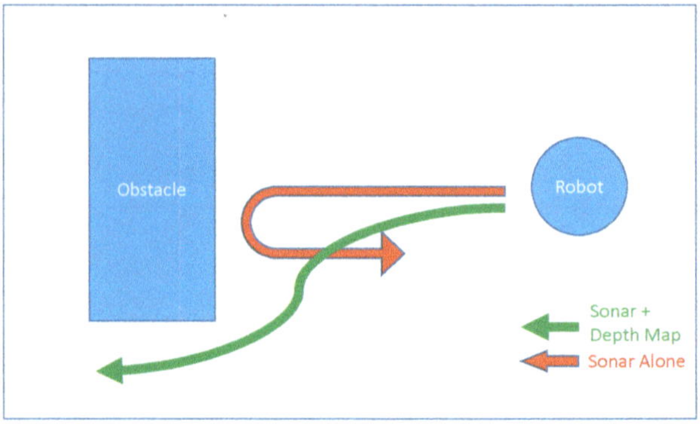


**Figure 12.** The path our robot takes when using sonars alone vs. combining sensors.

## 3. Experiment

### 3.1. Setup

The experimental robot is shown in Figure 13, with camera and RGB-Depth sensor on its top and middle, respectively. The sonar sensors are located below. Its width is 381 mm, and it has a swing radius of 26.7 cm. Its software is running on a laptop located in the middle.

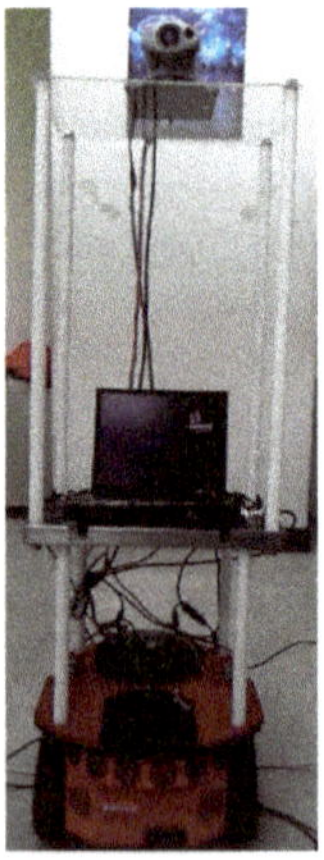

**Figure 13.** The experimental robot.

The experiments are divided into 2 subsections. The first is the face recognition plus gender determination. The second subsection is to test the obstacle avoidance capability of the robot. This particular subsection is further divided into situations when the obstacles are static, and when the obstacles are moving (e.g., pedestrians). These experiments were performed in real-time. These experiments were performed in an area of about 7 m by 3 m, with office chairs as obstacles.

### 3.2. Face and Gender Recognition

In the first experiment, the robot is placed a little distance away from the human person. The robot would adjust its orientation and distance to the human person so as to place the human face, detected using Viola's method [23], square in the middle of its field-of-view, and tracked using the KLT feature tracker [24]. The training sample included 400 male photos and 300 female photos. The test samples included 265 male photos and 227 female photos. The correct gender recognition rate was 92.6829%. Later, 10 photos of each laboratory personnel were added to the training set, and the robot was able to perform on-line face recognition during the test, and recognize the human persons in front as laboratory personnel and correctly classified their gender. After recognition and classification, choices for destinations are then presented to the user via the monitor based on the result, including a Follow-Me choice for lab workers. Examples of classification results are shown in Figure 14.

(**a**) Female Lab Worker (**b**) Male Lab Worker

**Figure 14.** Examples of facial recognition and gender classification.

### 3.3. *Obstacles Avoidance*

#### 3.3.1. Static Obstacle Avoidance

In the static obstacles experiment, the testing site is a hallway. The obstacles are placed along the hallway, as shown in Figure 15.

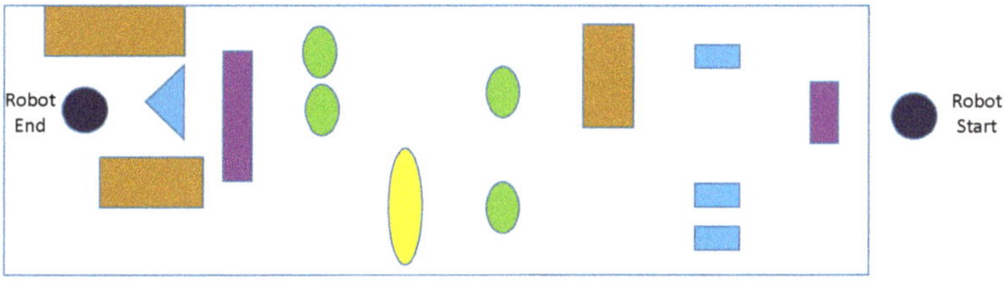


**Figure 15.** Static obstacles for the robot.

Sufficient gaps are left between the obstacles so that the robot should be able to reach its final destination. The following plot, Figure 16, shows the path the robot took. The robot's position was recorded at an interval of every 10 s.

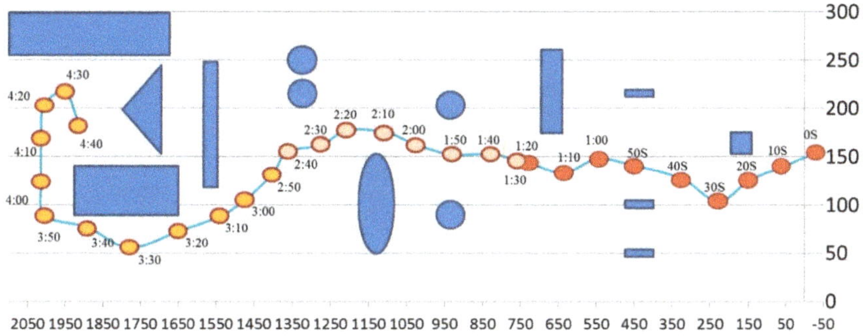


**Figure 16.** The path the robot took with positions taken at 10 s interval.

There are a total of 17 possible scenarios designed along the path that we think that the robot would need to make critical decisions about whether to incorporate sonar sensors in its decisions. The following figure, Figure 17, shows where each of the 17 scenarios, $T_1$–$T_{17}$, took place, and Table 4 shows the input values and the robot's decisions for the 17 scenarios.

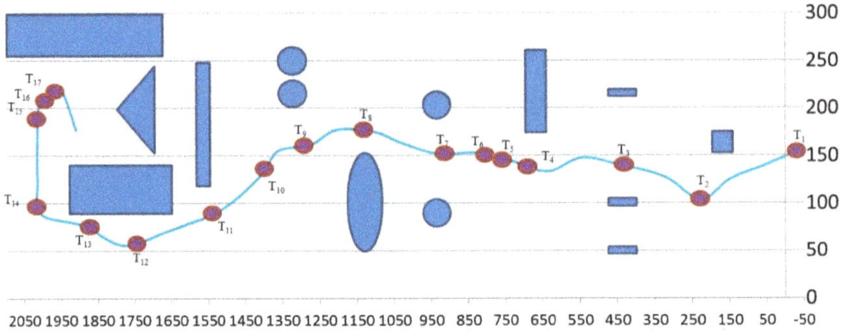


**Figure 17.** The seventeen scenarios ($T_1$–$T_{17}$) where decisions of the robot were recorded.

**Table 4.** Sensors Inputs and Decision for Scenarios $T_1$–$T_{17}$.

| Scenario | Left Obstacle Depth Value | Right Obstacle Depth Value | Gap Normalized to Robot Width | Left Sonar Normalized | Right Sonar Normalized | Final Decision |
|---|---|---|---|---|---|---|
| $T_1$ | N/A | 317 | 1.41 | 1 | 1 | Forward |
| $T_2$ | 97.6 | 10,000 | 1.67 | 0.2 | 1 | Right Turn |
| $T_3$ | N/A | 106.3 | 1.20 | 0.8 | 0.44 | Left Turn |
| $T_4$ | N/A | N/A | 0.0 | 0.22 | 0.22 | Forward |
| $T_5$ | 128.9 | N/A | 0.0 | 0.2 | 0.62 | Right Turn |
| $T_6$ | 122.1 | 125.6 | 1.64 | 0.4 | 0.4 | Forward |
| $T_7$ | 139.9 | 176.3 | 1.56 | 0.58 | 0.66 | Forward |
| $T_8$ | 272.2 | N/A | 2.24 | 0.4 | 0.36 | Forward |
| $T_9$ | N/A | N/A | 0.0 | 0.3 | 0.22 | Forward |
| $T_{10}$ | N/A | 188.8 | 1.67 | 1 | 0.5 | Forward |
| $T_{11}$ | 163.0 | 160.3 | 1.05 | 0.5 | 0.38 | Forward |
| $T_{12}$ | 100.1 | N/A | 1.33 | 0.08 | 0.13 | Right Turn |
| $T_{13}$ | 193.1 | 1000.0 | 0.84 | 0.4 | 1 | Right Turn |
| $T_{14}$ | 163.5 | 196.4 | 0.83 | 0.54 | 0.4 | Forward |
| $T_{15}$ | N/A | N/A | 0.0 | 0.04 | 0.22 | Forward |
| $T_{16}$ | N/A | N/A | 0.0 | 0.22 | 0.22 | Forward |
| $T_{17}$ | 115.9 | N/A | 1.13 | 0.22 | 0.42 | Right Turn |

Figure 18 shows the fuzzy inputs and outputs at $T_1$–$T_{17}$.

Table 5 below illustrates the comparisons between our method, Correa's method [3], and Csaba's method [4], using single-point analysis at each scenario point. Success is defined as being able to avoid collision.

**Table 5.** Comparisons between Correa, Csaba, and our proposed method on $T_1$–$T_{17}$ scenarios.

| Scenario | Correa | Csaba | Our Method |
|---|---|---|---|
| $T_1$ | Fail! | Success | Success |
| $T_2$ | Fail! | Fail! | Success |
| $T_3$ | Fail! | Success | Success |
| $T_4$ | Success | Success | Success |
| $T_5$ | Fail! | Fail! | Success |
| $T_6$ | Fail! | Fail! | Success |
| $T_7$ | Success | Success | Success |
| $T_8$ | Success | Success | Success |
| $T_9$ | Success | Success | Success |
| $T_{10}$ | Fail! | Success | Success |
| $T_{11}$ | Success | Success | Success |
| $T_{12}$ | Fail! | Success | Success |

**Table 5.** *Cont.*

| Scenario | Correa | Csaba | Our Method |
| --- | --- | --- | --- |
| $T_{13}$ | Fail! | Fail! | Success |
| $T_{14}$ | Fail! | Success | Success |
| $T_{15}$ | Fail! | Success | Success |
| $T_{16}$ | Success | Success | Success |
| $T_{17}$ | Fail! | Success | Success |
| $T_{1}$ | Success | Success | Success |

**Figure 18.** Fuzzy inputs and decisions for combining sensor maps at $T_1$–$T_{17}$.

### 3.3.2. Dynamic Obstacle Avoidance

Eleven scenarios were setup using one or two moving obstacles for this experiment. The obstacles are to simulate pedestrians in a service environment, and none was the robot's target. Table 6 illustrates these scenarios.

**Table 6.** Eleven scenarios for dynamic moving obstacles.

| Scenario # | Scenario |
|---|---|
| 1 | Single Obstacle Moving Towards the Robot |
| 2 | Single Obstacle Moving Fast from Right of the Robot |
| 3 | Single Obstacle Moving Slow from Right of the Robot |
| 4 | Single Obstacle Moving Fast from Left of the Robot |
| 5 | Single Obstacle Moving Slow from Left of the Robot |
| 6 | Dual Obstacles Moving Towards the Robot Then Separates |
| 7 | Dual Obstacles Moving From Left and Right of the Robot Then Crisscross |
| 8 | Dual Obstacles Moving From Left of the Robot At The Same Speed |
| 9 | Dual Obstacles Moving From Left of the Robot At Different Speeds |
| 10 | Dual Obstacles Moving From Right of the Robot At The Same Speed |
| 11 | Dual Obstacles Moving From Right of the Robot At Different Speeds |

Figure 19 illustrates how the robot responded at each of these scenarios. The red line represents the path of the robot, the blue line represents the path of the first obstacle, and orange line represents the path of the second obstacle. Each dot represents sampled locations taken between fixed time intervals. In each of these scenarios, except for the third, the robot was able to pass through the obstacles and reach the other side. In the third scenario, the robot was able to avoid collision by turning around.

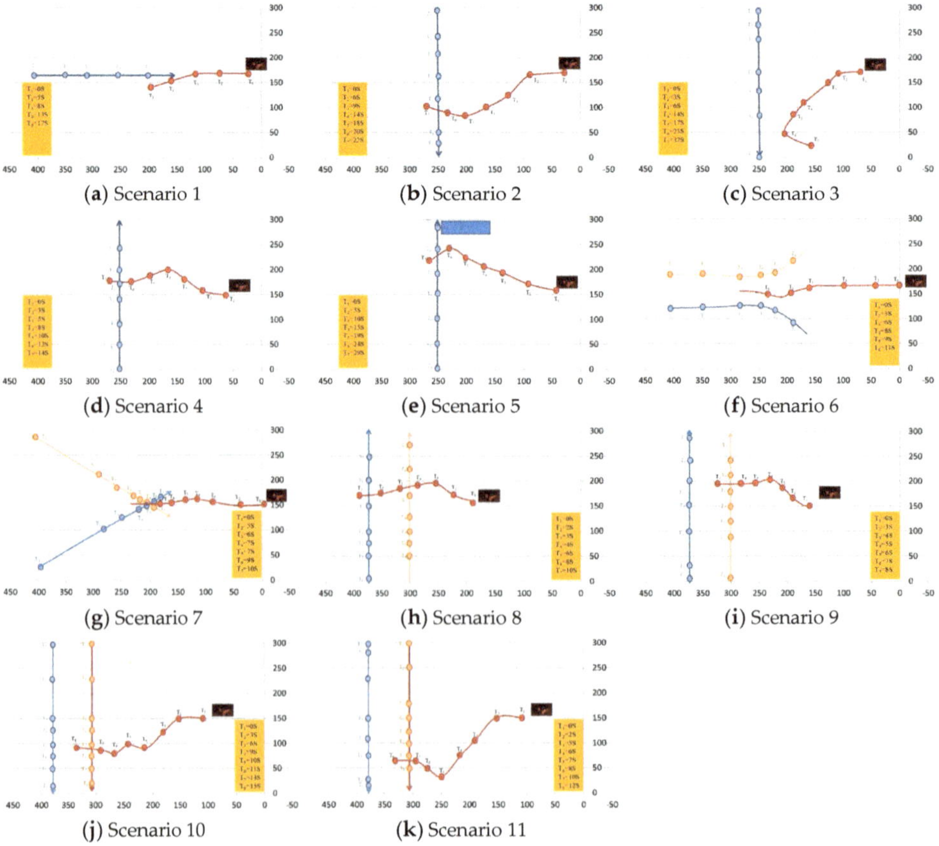

(**a**) Scenario 1 (**b**) Scenario 2 (**c**) Scenario 3
(**d**) Scenario 4 (**e**) Scenario 5 (**f**) Scenario 6
(**g**) Scenario 7 (**h**) Scenario 8 (**i**) Scenario 9
(**j**) Scenario 10 (**k**) Scenario 11

**Figure 19.** The paths of the moving obstacles (blue, orange) and the robot (red) for the 11 scenarios.

## 4. Discussion and Conclusions

In this paper, we presented a service robot with face recognition and gender classification abilities, with accuracy reaching 92.68%. The service robot can perform different tasks based on the classification results, such as activating the Follow-Me function for laboratory staff only. Possible uses of the robots including delivering medicines to target patients, assist visitors to find patients, and other tasks. We also developed a dual-level ANFIS-based fuzzy obstacle-avoidance system based on the inputs from two different types of sensors: RGB-Depth and sonars. It is found that the obstacle-avoidance capability using both types of sensors surpasses the performance of using just a single type sensor. We performed the experiment testing the obstacle avoidance capability of the robot under both static and dynamic environments and found that the robot can successfully maneuver around obstacles in almost all cases. In the future, we hope to improve the robot's navigation abilities by including advanced path planning using indoor maps of the environment, or different hardware configurations for other purposes.

**Author Contributions:** Conceptualization, methodology, analysis, writing, J.-C.C. and J.-D.L.; Experiments, Z.-Y.D.

**Funding:** The work was partly supported by Ministry of Science and Technology (MOST) and Chang Gung Memorial Hospital, Taiwan, Republic of China, under Grant MOST107-2221-E-182-026-MY2 and CMRPD2G0121.

**Conflicts of Interest:** The authors declare no conflict of interest.

## References


1. Yang, G.; Lv, H.; Chen, F.; Pang, Z.; Wang, J.; Yang, H.; Zhang, J. A Novel Gesture Recognition System for Intelligent Interaction with a Nursing-Care Assistant Robot. *Appl. Sci.* **2018**, *8*, 2349. [CrossRef]
2. Chen, Z.; Song, C.; Yang, Y.; Zhao, B.; Hu, Y.; Liu, S.; Zhang, J. Robot Navigation Based on Human Trajectory Prediction and Multiple Travel Modes. *Appl. Sci.* **2018**, *8*, 2205. [CrossRef]
3. Yamauchi, B.; Shultz, A.; Adams, W. Mobile robot exploration and map-building with continuous localization. In Proceedings of the Internet Content Rating Association (ICRA), Leuven, Belgium, 20 May 1998; Volume 4, pp. 3175–3720.
4. Vadakkepat, P.; Lim, P.; de Silva, L.C.; Jing, L.; Ling, L.L. Multimodal Approach to Human-Face Detection and Tracking. *IEEE Trans. Ind. Electron.* **2008**, *55*, 1385–1393. [CrossRef]
5. Correa, D.S.O.; Sciotti, D.F.; Prado, M.G.; Sales, D.O.; Wolf, D.F.; Osorio, F.S. Mobile Robots Navigation in Indoor Environments Using Kinect Sensor. In Proceedings of the 2012 Second Brazilian Conference on Critical Embedded Systems, São Paulo, Brazil, 21–25 May 2012; pp. 36–41.
6. Csaba, G. Fuzzy Based Obstacle Avoidance for Mobil Robots with Kinect Sensor. In Proceedings of the 2012 4th IEEE International Symposium on Logistics and Industrial Informatics (LINDI), Smolenice, Slovakia, 5–7 September 2012; pp. 135–144.
7. Algredo-Badillo, I.; Morales-Raales, L.A.; Hernández-Gracidas, C.A.; Cortés-Pérez, E.; Pimentel, J.J.A. Self-navigating Robot based on Fuzzy Rules Designed for Autonomous Wheelchair Mobility. *Int. J. Comput. Sci. Inf. Secur.* **2016**, *14*, 11–19.
8. de Silva, V.; Roche, J.; Kondoz, A. Fusion of LiDAR and Camera Sensor Data for Environment Sensing in Driverless Vehicles. 2018. Available online: https://dspace.lboro.ac.uk/2134/33170 (accessed on 10 November 2018).
9. Jin, X.-B.; Su, T.-L.; Kong, J.-L.; Bai, Y.-T.; Miao, B.-B.; Dou, C. State-of-the-Art Mobile Intelligence: Enabling Robots to Move Like Humans by Estimating Mobility with Artificial Intelligence. *Appl. Sci.* **2018**, *8*, 379. [CrossRef]
10. OpenCV. Available online: http://www.opencv.org/ (accessed on 2 September 2018).
11. Turk, M.; Pentland, A. Eigenfaces for recognition. *J. Cogn. Neurosci.* **1991**, *3*, 71–86. [CrossRef] [PubMed]
12. Belhumeur, P.N.; Hespanha, J.P.; Kriengman, D.J. Eigenfaces vs. fisherfaces: Recognition using class specific linear projection. *IEEE Trans. Pattern Anal. Mach. Intell.* **1997**, *19*, 711–720. [CrossRef]
13. Chien, L.W.; Ho, Y.F.; Tsai, M.F. Instant Social Networking with Startup Time Minimization Based on Mobile Cloud Computing. *Sustainability* **2018**, *10*, 1195. [CrossRef]

14. Aberdeen Facial Database. Available online: http://pics.psych.stir.ac.uk/zips/Aberdeen.zip (accessed on 10 February 2017).
15. GUFD Facial Database. Available online: http://homepages.abdn.ac.uk/m.burton/pages/gfmt/Glasgow%20Face%20Recognition%20Group.html (accessed on 10 February 2017).
16. Utrecht ECVP Facial Database. Available online: http://pics.psych.stir.ac.uk/zips/utrecht.zip (accessed on 10 February 2017).
17. Ullah, I.; Aboalsamh, H.; Hussain, M.; Muhammad, G.; Mirza, A.; Bebis, G. Gender Recognition from Face Images with Local LBP Descriptor. *Arch. Sci. J.* **2012**, *65*, 353–360.
18. Ren, H.; Li, Z. Gender Recognition Using Complexity-Aware Local Features. In Proceedings of the 2014 22nd International Conference on Pattern Recognition, Stockholm, Sweden, 24–28 August 2014; pp. 2389–2394.
19. Moghaddam, B.; Yang, M.-H. Gender Classification with Support Vector Machines. In Proceedings of the Fourth IEEE International Conference on Automatic Face and Gesture Recognition, Grenoble, France, 28–30 March 2000; p. 306.
20. Jang, J.-S.R. ANFIS: Adaptive-network-based fuzzy inference system. *IEEE Trans. Syst. Man Cybern.* **1993**, *23*, 665–685. [CrossRef]
21. Flynn, A.M. Combining Sonar and Infrared Sensors for Mobile Robot Navigation. *Int. J. Robot. Res.* **1988**, *7*, 5–14. [CrossRef]
22. Elfes, A.; Matthies, L. Sensor integration for robot navigation: Combining sonar and stereo range data in a grid-based representataion. In Proceedings of the 26th IEEE Conference on Decision and Control, Los Angeles, CA, USA, 9–11 December 1987; pp. 1802–1807.
23. Viola, P.; Jones, M.J. Rapid Object Detection using a Boosted Cascade of Simple Features. In Proceedings of the IEEE Computer Society International Conference on Computer Vision and Pattern Recognition, Kauai, HI, USA, 8–14 December 2001; Volume 1, pp. 511–518.
24. Tomasi, C.; Kanade, T. *Detection and Tracking of Point Features*; Technical Report CMU-CS_91-132; Carnegie Mellon University: Pittsburgh, PA, USA, 1991.

*applied sciences*

MDPI

*Article*

# Automated Enemy Avoidance of Unmanned Aerial Vehicles Based on Reinforcement Learning

**Qiao Cheng *, Xiangke Wang *, Jian Yang and Lincheng Shen**

College of Intelligence Science and Technology, National University of Defense Technology, Changsha 410073, China; yj_ntx@163.com (J.Y.); lcshen@nudt.edu.cn (L.S.)

* Correspondence: qiao.cheng@nudt.edu.cn (Q.C.); xkwang@nudt.edu.cn (X.W.)

Received: 23 November 2018; Accepted: 8 February 2019; Published: 15 February 2019

**Abstract:** This paper focuses on one of the collision avoidance scenarios for unmanned aerial vehicles (UAVs), where the UAV needs to avoid collision with the enemy UAV during its flying path to the goal point. Such a type of problem is defined as the enemy avoidance problem in this paper. To deal with this problem, a learning based framework is proposed. Under this framework, the enemy avoidance problem is formulated as a Markov Decision Process (MDP), and the maneuver policies for the UAV are learned based on a temporal-difference reinforcement learning method called Sarsa. To handle the enemy avoidance problem in continuous state space, the Cerebellar Model Arithmetic Computer (CMAC) function approximation technique is embodied in the proposed framework. Furthermore, a hardware-in-the-loop (HITL) simulation environment is established. Simulation results show that the UAV agent can learn a satisfying policy under the proposed framework. Comparing with the random policy and the fixed-rule policy, the learned policy can achieve a far higher possibility in reaching the goal point without colliding with the enemy UAV.

**Keywords:** enemy avoidance; reinforcement learning; decision making; hardware-in-the-loop simulation; unmanned aerial vehicles

---

## 1. Introduction

Unmanned Aerial Vehicles (UAVs) have received considerable attention in many areas [1], such as commercial, search and rescue, military, and so on. In the military area, there are applications such as the surveillance [2,3], target tracking [4,5], target following [6,7], and so on. Among these applications, collision avoidance is one of the most important concerns [8], especially in unsafe environment. In such cases, a UAV should keep safe separation with various kinds of objects, such as static obstacles [9,10], teammates [11], and moving enemies. The strategies toward different approaching objects are different due to specific requirements in dealing with those objects. This paper focuses on avoiding the collision with moving enemies. There are many researches on collision avoidance problems. However, relatively fewer works are on the avoidance of moving enemies, comparing with those on the avoidance of static obstacles and flying teammates. Furthermore, the uncertain motion of enemies and the necessity to attack enemies create more challenges on the avoidance of moving enemies than avoiding other objects. Besides, the mission of the UAV, such as reaching a specific goal destination, should also be considered. Therefore, the avoidance of moving enemies is a challenge problem, and such a problem is defined as the enemy avoidance problem in this paper.

There are many approaches to handle the collision avoidance problem in different stages [12]. Many of those approaches rely on models for the dynamic of the environment and UAVs. However, the accuracy of these models can sometimes greatly affect the performance of those methods. Moreover, building these models is not easy work, and is even impractical. On the other hand, a complex model means heavy computation load when making decisions. Therefore, learning methods are increasingly used in collision avoidance problems, which are based on collected data. However, most of these learning methods are used to predict the effect of the decision, not directly used for the decision making. Different from other learning methods, reinforcement learning is a very popular method for sequential decision making problems [13]. It can learn to make decisions incrementally based on feedback from the environment. Therefore, it can generate a good policy even if the models of the environment are unknown. Since a sequence of appropriate actions are required to avoid the enemy UAV, the enemy avoidance problem can be regarded as a sequential decision making problem. Therefore, this paper proposes a framework which incorporates the reinforcement learning to deal with the enemy avoidance problem.

Many methods for the collision avoidance problem discretize the state space to make decisions [14,15]. However, this paper studies the enemy avoidance problem in continuous state space. Therefore, the function approximation technique, which can handle continuous space, is also embodied in the proposed framework.

As for the UAVs, most researches [16–18] focus on quadrotors rather than fixed-wing UAVs in collision avoidance problems. However, the dynamics of quadrotors and fixed-wing UAVs are different. In addition, the UAV in the enemy avoidance problem has the mission to reach the goal point, and thereby needs to keep away from the enemy UAV and even attack enemy UAVs. In practical application, the fixed-wing UAVs are more suitable for such a problem scenario for their better mission fulfillment properties, such as higher endurance and greater speeds. Therefore, this paper focuses on the enemy avoidance problem of fixed-wing UAVs.

Since learning the policy on the real UAV platform would bring about great consumption, a hardware-in-the-loop (HITL) simulation system is constructed. With hardware-in-the-loop, the simulation system can provide very consistent properties to that of the real environment, which highly respects the kinecmatic and maneuver constraints of the UAVs. Furthermore, it saves the energy to build a model for the related hardware, which is usually very hard to build accurately. Comparing with the real UAV platform, the HITL simulation system can repeat the experiments as many times as needed without worrying about UAV costs.

The contributions of this paper are summarized as follows.

(i) An interesting new problem called the enemy avoidance problem is defined, which can be a good adding up scenario to the collision avoidance problem. The newly defined problem is different from most of the existing collision avoidance problems, for it is to avoid the collision with the enemy UAV rather than static obstacles or moving teammates.

(ii) A novel framework is proposed to learn the policy for the decision making UAV. The proposed framework formulates the enemy avoidance problem as a Markov Decision Process (MDP) problem, and solves the MDP problem by a temporal-difference reinforcement learning method called Sarsa. The Cerebellar Model Arithmetic Computer (CMAC, [19]) technique is also embodied in the proposed framework for the generalization of the continuous state space. With this framework, such a decision making problem is transformed from the usual computational problem to a learning problem. Besides, it can learn the policy with an unknown environment model, and can make decisions based on continuous state space rather than discrete ones like most existing works do.

(iii) A hardware-in-the-loop (HITL) simulation environment for the enemy avoidance problem is constructed, which is used for the policy learning and policy testing experiments. Different from the simulation environment in most of the existing works, this HITL simulation system saves a lot of model designing trouble, and has better consistency to the real environment, such as the environment noise. When comparing with real environment platforms, the HITL simulation system has the advantage of saving experimental cost.

The remainder of this paper is outlined as follows. Section 2 gives some reviews on the related literature. The enemy avoidance problems are presented in Section 3. The proposed framework for the enemy avoidance problem is elaborated in Section 4. Section 5 details the construction of the hardware-in-the-loop simulation environment. Simulation experiments and results are illustrated in Section 6. Finally, Section 7 concludes the whole work and discusses future works.

## 2. Literature Review

There are many researches on collision avoidance problems, and different methods are used to solve different collision avoidance problems. Therefore, this section will give a summary about several widely used methods in collision avoidance problems, as well as a comparison between our work and these existing works.

One of the most widely used methods is to formulate the collision avoidance problem as an optimization problem, while considering all kinds of constrains. Therefore, to avoid collision is to solve the optimization problem with appropriate methods under different constrains. The work in [20] formulates the collision avoidance problem as a convex optimization problem, and seeks for a suitable control constraint set for participating UAV based on reachable sets and tubes for UAVs. This method is limited to linear systems. The collision avoidance in work [21] is formulated as a set of linear quadratic optimization problems, which are solved with an original geometric based formulation. To handle flocking control with obstacle avoidance, work [22] proposes a UAV distributed flocking control algorithm based on the modified multi-objective pigeon-inspired optimization (MPIO), which considers both the hard constraints and the soft ones. Our previous works [23,24] formulate the conflict avoidance problem as a nonlinear optimization problem, and then use different methods to solve such an optimization problem. The work in [23] proposes a two-layered mechanism to guarantee safe separation, which finds the optimal heading change solutions with the vectorized stochastic parallel gradient descent-based method, and finds the optimal speed change solutions with a mixed integer linear programming model. The work in [24] uses the stochastic parallel gradient descent (SPGD) method to find the feasible initial solutions, and then uses the Sequential quadratic programming (SQP) algorithm to compute the local optimal solution. Even for the obstacle avoidance problem in other areas, the optimization methods are also used. For example, two swarm based optimization techniques are used in work [25] to offer obstacle-avoidance path planning for mobility-assisted localization in wireless sensor networks (WSN), which are grey wolf optimizer and whale optimization algorithm. The main difference between the collision avoidance for UAVs and the obstacle-avoidance path planning in WSN lies in the constrains and objective of the optimization model. Usually, solving the optimization problem requires a lot of computation. Therefore, our work does not formulate the enemy avoidance problem as an optimization problem, but formulates it as an MDP problem and solves the MDP incrementally by interaction with the environment.

Another kind of method for solving collision avoidance problems is to predict the potential collision with certain techniques. The work in [26] proposes an approach based on radio signal strengths (RSS) measurements to obtain position estimation of the UAV, and to detect the potential collisions based on the position estimations, and then to distribute the UAVs at different altitudes to avoid collision. The work in [27] proposes a model-based learning algorithm that enables the agent to learn an uncertainty-aware collision prediction model through deep neural networks, so as to avoid the collision with unknown static obstacles. The work in [28] proposes a data-driven end-to-end motion planing approach which helps the robot navigate to a desired target point while avoiding collisions with static obstacles without the need of a global map. This approach is based on convolutional neural networks (CNNs), and the robot is provided with expert demonstrations about navigation in a given virtual training environment. One of the problems for such kind of methods is that high capacity learning algorithms like deep learning tend to overfit when little training data is available. Therefore, the work in [29] collects a lot of crash samples to build a dataset by crashing their drone 11,500 times. The used data driven approach demonstrates such negative data is also crucial for learning how to navigate without collision. However, to collect both positive data and negative data for prediction is very costly. Different from these works, our work aims to obtain a policy with the proposed framework.

The policy is a mapping from the state directly to the action, therefore, no prediction of the collision is needed.

As reinforcement learning gains its popularity in decision making problems, there are works that use different reinforcement learning to solve the collision avoidance problem. The work in [30] proposes to combine Model Predictive Control (MPC) with reinforcement learning to learn obstacle avoidance policies for the UAV in a simulation environment. In this method, the MPC is used to generate data at the training time, and the deep neural network policies are trained with an off-policy reinforcement learning method called guided policy search based on the generated data. The work in [31] proposes a geometric reinforcement learning algorithm for UAV path planning, which constructs a specific reward matrix to include the geometric distance and risk information. This algorithm considers the obstacles as risk and builds a risk model for the obstacles, which is used in constructing the reward matrix. The work in [32] models the UAV collision avoidance problem as a Partially Observable Markov Decision Process (POMDP) and uses Monte Carlo Value Iteration (MCVI) to solve the POMDPs, which can cope with high-dimensional continuous-state space in a collision avoidance problem. The work in [33] formulates the problem of collision avoidance as an MDP and a POMDP, and uses generic MDP/POMDP solvers to generate avoidance strategies. Though the framework proposed in our work is also based on reinforcement learning, many details are different from the these works. For example, this paper uses neither special data generating process, nor complex reward function designing. Besides, the environment transition model is unknown in this paper, and a different reinforcement learning method is adopted.

Another big difference between the existing works and our work is that the collision avoidance problem in this paper is not exactly the same as those in previous works. First, the UAV in this paper needs to avoid collision with a moving enemy UAV, not static obstacles [25] or teammates [11]. Besides, the actions the UAV uses to avoid collision include both heading angle change and velocity change. The work in [34] investigated strategies for multiple UAVs to avoid collision with moving obstacles, which is a little similar with collision with enemy UAV. However, their work assumes all UAVs and all obstacles have constant ground speeds, and the direction of the velocity vector of an obstacle is constant. Therefore, they only consider change in direction of the UAV for collision avoidance, and do not consider change in velocity of the UAV. Furthermore, this paper does not attempt to build an environment model, but approximates it by continuous interaction with the environment. Similarlly, work [35] also approximates the unmodeled dynamics of the environment, but it uses back propagation neural networks and proposes a tree search algorithm to find the near optimal conflict avoidance solutions. In addition, this paper considers continuous state space in the decision making process, not the discrete one like many other related works do.

## 3. Problem Definition

We call the problem posed in this paper the enemy avoidance problem, which is different from the usual collision avoidance problems or path planning problems. Before proposing methods to solve this problem, we first give a detailed description for the enemy avoidance problem, as well as the related assumptions and definitions.

### *3.1. Problem Description*

For the convenience of the research, we define the enemy avoidance problem in a fixed region. There are two UAVs flying toward each other in the region, namely the decision making UAV $f$ and the enemy UAV $e$. The fixed region is where the two UAVs may collide with each other. The decision making UAV $f$ enters the region from the left side, while the enemy UAV $e$ enters the region from the right side. Both UAVs are flying toward their own goal points. Let $G_f$ and $G_e$ denote the goal point of the decision making UAV $f$ and that of the enemy UAV $e$, respectively. The goal point $G_f$ for the decision making UAV $f$ is located near the right edge of the region, while the goal point $G_e$ for the enemy UAV $e$ is located near the left edge of the region. Both goal points are on the middle line of the region, as shown in Figure 1.

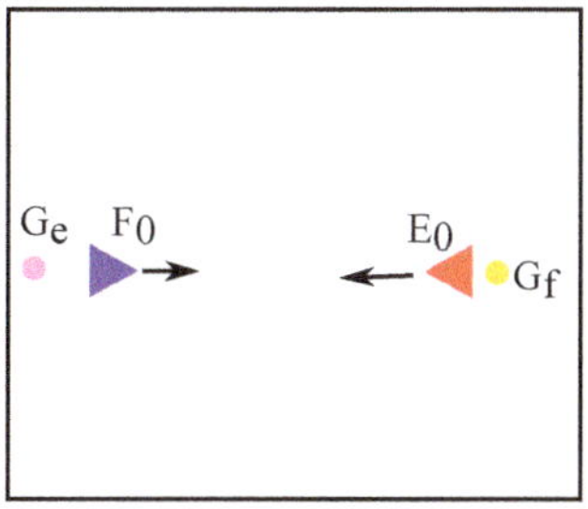


**Figure 1.** The enemy avoidance scenarios.

In such a region, there is no hovering requirement or taking off and landing requirements, therefore the fixed-wing UAV can be easily applied in such problem settings.

The mission of the decision making UAV $f$ is to reach the goal point $G_f$ safely and with as little cost as possible. However, the decision making UAV $f$ and the enemy UAV $e$ are flying in opposite directions along the middle line of the region toward their own goal points, which poses the decision making UAV $f$ the danger of collision with the enemy UAV $e$. Therefore, the decision making UAV $f$ needs to avoid the enemy UAV $e$ during its flight towards the goal point $G_f$. Ways to avoid collision with the enemy UAV include changing the heading angle or the velocity, and attacking the enemy UAV. Though changing the heading angle can let the decision making UAV avoid flying directly into the enemy UAV if the the enemy UAV happens to be in the heading direction of the decision making UAV, inappropriate heading angle change may make the decision making UAV fly too far away from the goal point $G_f$. Similarly, changing the velocity at an appropriate time can also avoid collision with the enemy UAV, such as accelerating to pass the enemy UAV before the collision or decelerating to wait for the enemy UAV to pass. Successfully attacking the enemy UAV can also provide good insurance for the decision making UAV to fulfill its mission. However, the attacking action may fail in destroying the enemy UAV, and the decision making UAV suffers certain losses when using attacking action. Therefore, the decision making UAV should not use the attacking action too often. To achieve the mission requirements, the decision making UAV cannot use just one single avoiding action, but should arrange all the actions in an appropriate sequence.

On the other hand, the enemy UAV simply flies toward the goal point $G_e$ with constant velocity and heading angle if the decision making UAV does not collide with it or attack it successfully. The scenario is supposed to end as soon as the decision making UAV has collided with or successfully attacked the enemy UAV, or the decision making UAV has reached the goal point $G_f$ successfully or has been out of the region.

To arrange an appropriate action sequence is the process of decision making, which is also the main focus of this work. In decision making, such an action sequence is called the policy. Based on the action chosen by the on-board agent at each decision making step, the decision making UAV can adjust its flying attitude or attack the enemy UAV. The agent makes decisions based on both its own information from its on-board sensor system and the enemy UAV's information from the ground station. The action executed by the decision making UAV makes the UAV changes its state in the environment. On the other hand, the enemy UAV also updates its states in the environment and senses its own states from the environment with its on-board sensor system. The ground station captures all UAVs' information, and then transmits the information to the decision making UAV. Figure 2 presents the overall decision making process of the enemy avoidance problem. Therefore, how the decision making agent uses the gathered information to make decisions for avoiding collision with the enemy UAV is what this paper is going to solve. Furthermore, the time and the location at which the enemy

UAV enters the region are not fixed each time. Therefore, the decision making ability of the decision making UAV should be able to generalize to different enemy avoidance cases.

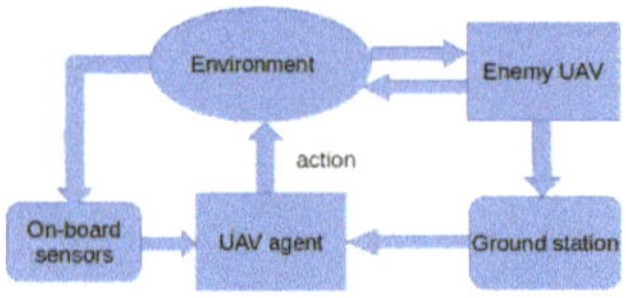


**Figure 2.** The decision making framework.

### 3.2. Assumptions and Definitions

At first, we need to make some assumptions about the posed enemy avoidance problem.

**Assumption 1.** *The enemy UAV has constant desired velocity and desired heading angle, while the actual velocity and heading angle of the enemy UAV oscillate a little around the desired ones.*

**Assumption 2.** *The decision making UAV can change its attitude and attack the enemy UAV during its flight according to the action decided by the UAV agent.*

**Assumption 3.** *The heights of the UAVs are not considered. Therefore, all the distances in the problem are simply computed by two dimensions.*

**Assumption 4.** *Each UAV obtains its own position and attitude (velocity and heading angle) with its on-board sensor system. The decision making UAV can obtain information about all the UAVs through the ground station.*

**Assumption 5.** *There are no other UAVs in the region except the decision making UAV and the enemy UAV, as well as no obstacles in the region.*

In these assumptions, Assumptions 1 and 3 are used to simplify the enemy avoidance problem, so that we can focus more on other more important factors in the enemy avoidance problem. The researched results then can be used as the basis of more practical problems. With Assumption 5, this paper can focus on the collision avoidance of the enemy UAV, and does not need to consider collision with other UAVs and obstacles.

Furthermore, we give some definitions that will be used in solving the enemy avoidance problem.

**Definition 1.** *(Distance). The distance between two points $a = (x_a, y_a)$ and $b = (x_b, y_b)$ is calculated with the following equation:*

$$d_{ab} = \sqrt{(x_a - x_b)^2 + (y_a - y_b)^2} \tag{1}$$

**Definition 2.** *(Reaching Goal). A UAV $f$ is regarded as having reached a goal G when the following condition is satisfied:*

$$d_{fg} < r_g \tag{2}$$

*where $d_{fg}$ denotes the distance between the UAV $f$ and the goal point G, and $r_g$ is the specified goal radius.*

**Definition 3.** *(Collision). A UAV $f$ is regarded as having collided with the enemy UAV $e$ when the following condition is satisfied:*

$$d_{fe} < r_c \tag{3}$$

*where $d_{fe}$ denotes the distance between the UAV $f$ and the enemy UAV $e$, and $r_c$ is the specified collision radius.*

**Definition 4.** *(Attacking Probability). The success of an attacking action is defined by the attacking probability P, which is specified by the following equation:*

$$P = e^{1-\frac{d_{fe}}{30}} \tag{4}$$

## 4. Problem Sovling

This paper proposes a new framework to solve the enemy avoidance problem, which formulates the enemy avoidance problem as the Markov Decision Process (MDP) and learns the decision making policy for the enemy avoidance problem based on reinforcement learning. Firstly, the detail of formulating the enemy avoidance problem as the Markov Decision Process (MDP) is presented, which is the basis of the reinforcement learning. Secondly, how to learn the policy based on a temporal-difference reinforcement learning method called Sarsa is elaborated, as well as the embodied function approximator called CMAC for the generalization of the continous state.

### *4.1. Formulate the Problem as the MDP*

Reinforcement learning has been widely used in sequential decision making problems which are formulated as the Markov Decision Process (MDP). Typically, an MDP comprises of four elements: the state set $\mathcal{S}$, the action set $\mathcal{A}$, the transition function $\mathcal{T}$, and the reward function $\mathcal{R}$. When an agent is in a state $s \in \mathcal{S}$, it can choose an action $a \in \mathcal{A}$ according to a policy $\pi$. After executing the action $a$, the agent will enter into the next state $s' \in \mathcal{S}$ according to the transition function $\mathcal{T}$, and will receive an immediate reward $r$ according to the reward function $\mathcal{R}$. In this enemy avoidance problem, the environment transition function $\mathcal{T}$ is unknown, and will be learned by interaction with the environment. To formulate the enemy avoidance problem as the MDP, the state space $\mathcal{S}$, the action space $\mathcal{A}$, and the reward function $\mathcal{R}$ are defined as follows.

#### 4.1.1. State Space

In this enemy avoidance problem, the design of the state space mainly considers the positions and attitudes of the UAVs, as well as the position of the goal point $G_f$. However, these raw data are not used directly as the state variables. Instead, higher-level variables based on these data are defined. To be specific, the state space contains three sets of variables:

(i) Variables about the status of the decision making UAV $f$:
- $v_f$: The velocity of the decision making UAV $f$.
- $\psi_f$: The heading angle of the decision making UAV $f$.

(ii) Variables about the goal point $G_f$:
- $d_{fg}$: The distance from the decision making UAV $f$ to the goal point $G_f$.
- $\omega_g$: The angle between the north direction and the line from the decision making UAV $f$ to the goal point $G_f$.

(iii) Variables about the status of the enemy UAV $e$:
- $v_e$: The velocity of the enemy UAV $e$.
- $\psi_e$: The heading angle of the enemy UAV $e$.
- $d_{fe}$: The distance from the decision making UAV $f$ to the enemy UAV $e$.
- $\omega_e$: The angle between the north direction and the line from the decision making UAV $f$ to the enemy UAV $e$.

As we can see, there are 8 state variables in total. Figure 3 illustrates these state variables. The value ranges for all the velocity variables are $[10, 20]$, while the value ranges for all the angle variables are $[0, 360)$. Suppose the length and width of the region are $l$ and $w$, respectively. Thus, the value ranges of all the distance variables are $(0, d)$, where $d = \sqrt{l^2 + w^2}$. The reference frame is set in this way: the $X$ axis points to the North, and the $Y$ axis points to the East. Besides, the system neglects the rotation and acceleration of the earth, and the earth is assumed to be flat.

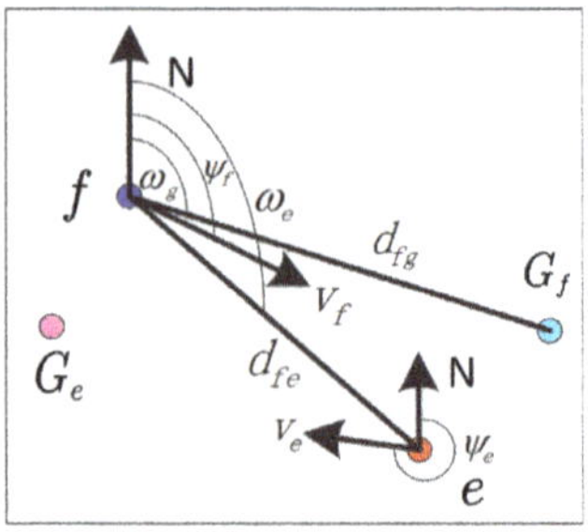


**Figure 3.** The denotations for the state variables.

### 4.1.2. Action Space

At each decision making step, the agent needs to decide an action for execution, so as to change the altitude of the decision making UAV. Six actions are defined in this paper: flying toward the goal, accelerating, decelerating, increasing the heading angle, decreasing the heading angle, and attacking the enemy. Denote these actions with $A = \{a_0, a_1, ..., a_5\}$. Each action corresponds to a way to change the desired attitude. The details are listed as follows.

- $a_0$: Fly toward the goal straightly, while keeping the desired velocity as the same in the previous step.
- $a_1$: Increase the desired velocity with $\Delta V$ and fly with the same desired heading angle as the previous step.
- $a_2$: Decrease the desired velocity with $\Delta V$ and fly with the same desired heading angle as the previous step.
- $a_3$: Increase the desired heading angle with $\Delta\phi$ and fly with the same desired velocity as the previous step.
- $a_4$: Decrease the desired heading angle with $\Delta\phi$ and fly with the same desired velocity as the previous step.
- $a_5$: Attack the enemy UAV, while the desired attitude changes as that of $a_0$.

The desired height is set with a fixed value $h$ in all cases, for the flying height is not considered in this problem.

### 4.1.3. Reward Function

There are two types of rewards in this enemy avoidance problem, denoted by $r_1$ and $r_2$, respectively. The first type of reward $r_1$ is the usual reward given at each decision making step, which aims to encourage the agent to reach the goal point in as few steps as possible. The other reward $r_2$ is the reward given at the end of the episode for different ending reasons. There are four situations that will end an episode. Denote these four situations by $S_T = \{s_{T1}, s_{T2}, s_{T3}, s_{T4}\}$, which are defined as follows.

- $s_{T1}$: The decision making UAV has collided with the enemy UAV.
- $s_{T2}$: The decision making UAV has reached the goal point.
- $s_{T3}$: The decision making UAV has attacked the enemy UAV successfully.
- $s_{T4}$: The decision making UAV has been out of the region.

Since the mission of the decision making UAV is to reach the goal point successfully, the agent will be rewarded with a very big value when the mission is fulfilled. Cases that the decision making

UAV needs to avoid, such as colliding with the enemy UAV or out of the region, should be punished. Successfully attacking the enemy UAV can guarantee the fulfillment of the mission for the decision making UAV, therefore it should be rewarded. However, since the attacking action is costly and may fail, it will be better to limit the use of the attacking action. Considering this, a small punishment is given when attacking action is used. Therefore, we define the reward $r = r_1 + r_2$, where $r_1 = -1$, and $r_2$ is defined as below:

$$r_2 = \begin{cases} -20 & a = a_5; \\ -100 & s = s_{T1}; \\ -100 & s = s_{T4}; \\ 2 & s = s_{T3}; \\ 500 & s = s_{T2}. \end{cases} \quad (5)$$

### 4.2. Learning Policy with the Sarsa Method

In the first part of the new framework, the enemy avoidance problem has been formulated as an MDP. For the second part of the new framework, the agent of the decision making UAV is allowed to learn the policy based on the Sarsa reinforcement learning method.

There are mainly three classes of methods for solving the reinforcement learning problem [13]: dynamic programming, Monte Carlo methods, and temporal-difference (TD) learning. Dynamic programming methods require a complete and accurate model of the environment (the transition function $\mathcal{T}$), while the Monte Carlo methods are not suitable for step-by-step incremental computation. Only the temporal-difference methods require no model and are fully incremental. In the enemy avoidance problem, the environment transition function $\mathcal{T}$ is unknown, and the policies need to be learned by continuous interaction with the environment. Therefore, the temporal-difference methods are more suitable for the enemy avoidance problem. As one of the temporal-difference reinforcement learning methods, the Sarsa method is chosen to be used in the proposed framework to learn the policy for the enemy avoidance problem.

Since the state space in the enemy avoidance problem is continuous, it is impractical to visit each state with each action infinitely often. Therefore, certain function approximations are needed to generalize the state space from relatively sparse interaction samples and with fewer variables than there are states. In this paper, the CMAC (cerebellar model arithmetic computer, [19]) technique is also embodied in the proposed framework to approximate the $Q$-value function when learning with the Sarsa method.

The details of how the proposed framework embodies the Sarsa method and the CMAC function approximator to learn the policy for the enemy avoidance problem are elaborated as follows.

#### 4.2.1. Sarsa Method

The Sarsa method is an on-policy temporal-difference learning method, where the agent attempts to update the policy that is used to make decisions for the decision making UAV $f$ at the same time. Different from most reinforcement learning methods where the main goal is to estimate the optimal value function, the Sarsa agent learns an action-value function $Q(s,a)$ rather than a state-value function $V(s)$. For the enemy avoidance problem, the Sarsa agent updates its action-value function $Q(s,a)$ after every transition from a state $s \in S$, where $s$ is not an episode ending situation. That is to say, $s \notin S_T$. If the state $s' \in S_T$, then $Q(s',a') = 0$.

The updating rule is given in Equation (6), where $\gamma$ is the discount rate, and $\alpha$ is a step-size parameter. Every element of the quintuple of the enemy avoidance events, $(s,a,r,s',a')$, are used in this updating rule. Such a quintuple makes up a transition from one state-action pair of the enemy avoidance problem to the next, and therefore gives rise to the name Sarsa for the algorithm.

$$Q(s,a) \leftarrow Q(s,a) + \alpha(r + \gamma Q(s',a' - Q(s,a)) \quad (6)$$

In the proposed framework, the Sarsa algorithm [13] is adapted into the enemy avoidance problem for the agent to learn the policy, which is presented in Algorithm 1.

**Algorithm 1** Sarsa Algorithm for the Enemy Avoidance Problem

```
Initialize Q(s, a) arbitrarily
for each episode do
    Initialize s as all the UAVs having entered the problem region
    Choose a ∈ A for s using policy derived from Q with ε−greedy
    for each step of episode do
        Take action a, observe r, s'
        Choose a' ∈ A for s' using policy derived from Q with ε−greedy
        Q(s, a) ← Q(s, a) + α(r + γQ(s', a') − Q(s, a))
        s ← s'; a ← a'
    end for
    until s ∈ S_T
end for
```

4.2.2. CMAC Function Approximation

In the proposed framework for the enemy avoidance problem, the state space is continuous. Therefore, the learning agent for the decision making UAV needs to use function approximation to generalize from limited experienced states. With function approximation, the action-value function $Q(s, a)$ of the enemy avoidance problem is maintained in a parameterized functional form with parameter vector $\overrightarrow{\theta}$. In this framework, the linear function form is used, as presented by Equation (7), where $\overrightarrow{\phi}_{(s,a)}$ is the feature vector of the function approximation.

$$Q(s, a) = \overrightarrow{\theta}^T \overrightarrow{\phi}_{(s,a)} \tag{7}$$

The CMAC (cerebellar model arithmetic computer, [19]) is one of those linear function approximators, and thus is used to construct the feature vector $\overrightarrow{\phi}_{(s,a)}$ in the proposed framework. To update the parameter vector $\overrightarrow{\theta}$, the gradient-descent method is adopted in the proposed framework as well.

The CMAC discretizes the continuous state space of the enemy avoidance problem by laying infinite axis-parallel tilings over all the eight state variables and then generalizes them via multiple overlapping tilings with some offset. Each element of a tiling is called a tile, which is a binary feature, as shown by Equation (8).

$$\phi_{(s,a)}(i) = \begin{cases} 1 & \text{tile } i \text{ is activated.} \\ 0 & \text{otherwise.} \end{cases} \tag{8}$$

Therefore, the CMAC maintains $Q(s, a)$ of the enemy avoidance problem in the following form:

$$Q(s, a) = \sum_{i=1}^{n} \theta(i)\phi_{(s,a)}(i) = \sum_{i \in I(\overrightarrow{\phi}_{(s,a)})} \theta(i) \tag{9}$$

where $I(\overrightarrow{\phi}_{(s,a)})$ is the set of tiles that are activated by the pair $(s, a)$ in the enemy avoidance problem, whose tile values are 1.

The parameter vector $\overrightarrow{\theta}$ is adjusted by the gradient-descent method, whose updating rule is as follows:

$$\overrightarrow{\theta}_{t+1} = \overrightarrow{\theta}_t + \alpha\delta_t \tag{10}$$

where $\delta_t$ is the usual TD error,

$$\delta_t = r_{t+1} + \gamma Q_t(s_{t+1}, a_{t+1}) - Q_t(s_t, a_t) \tag{11}$$

## 5. Simulation Environment

Since it is hard to build an accurate environment model for the enemy avoidance problem, and the RL agent needs to approximate the environment model through continuous interaction with the environment, this paper builds a hardware-in-the-loop (HITL) simulation system for the enemy avoidance problem. In this HITL simulation system, the kinecmatic and dynamic of UAVs are modeled by the X-plane simulators, while the maneuver and control properties of the system are confined by the hardware controller called Pixhawk. With such a HITL simulation system, the simulation can be more consistent to real flying, and can reduce the energy of building a complex environment model and save the cost of a real flying test.

In this section, how the HITL simulation system is constructed will be detailed. After this, the simulation process for an episode will be given.

### 5.1. System Construction

The X-plane flight simulator is used in this paper to simulate the flying dynamics of both the decision making UAV and the enemy UAV, and each X-plane is controled by a Pixhawk (PX4) flight controller. The Pixhawk is a hardware which is also used in controlling the real UAVs. In the simulation system, there is a ground station which can broadcast the information of all the UAVs to every UAV. The PX4 can control the flying of the UAVs according to the desired attitude. The desired attitude is composed of the desired velocity, the desired heading angle, and the desired height. In each UAV, there is an on-board sensor system which is used to sense the position and the attitude information of the UAV. Besides, each UAV has an agent for the communication and the decision control. To be specific, the agent for the decision making UAV has three modules: communication module, decision making module, and translator module. The on-board sensor system sends the sensed information to the agent through the communication module, while the communication module also sends the received information to the ground station and receives the enemy UAV's information from the ground station. Based on all the information received by the communication module, the decision making module then decides which action the decision making UAV should take, while the translator module interprets the action into desired attitude and sends it to the PX4 for the UAV flying control. Since the enemy UAV in this paper is assumed to fly towards the goal point $G_e$ directly all along the process, there is no decision making module in the enemy agent. Therefore, the enemy agent is composed of two modules: communication and control module. The communication module of the enemy agent has the same function as that of the agent for the decision making UAV. In the enemy agent, the control module sends the desired attitude to the PX4, where the desired velocity and the desired height are fixed at the initialization of the simulation, and the desired heading angle is calculated based on the relationship between the goal point and the position of the enemy UAV.

The structure of the simulation system is illustrated by Figure 4.

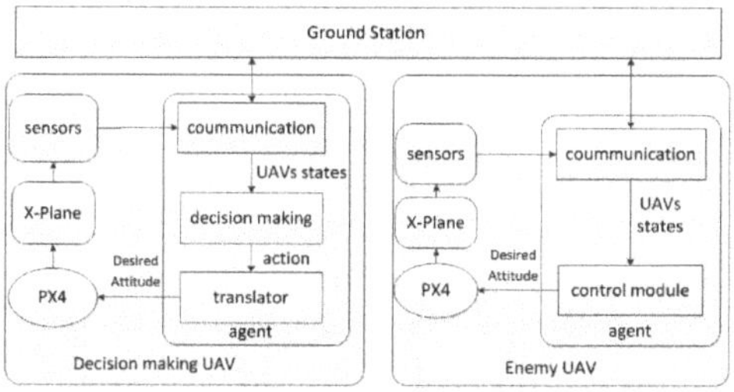


**Figure 4.** The structure of the simulation system.

Since another purpose of this paper is to explore the collision avoidance solution for the fixed-wing UAV, the X-plane simulator is set to use the fixed-wing simulation model. The simulation environment is shown as in Figure 5.

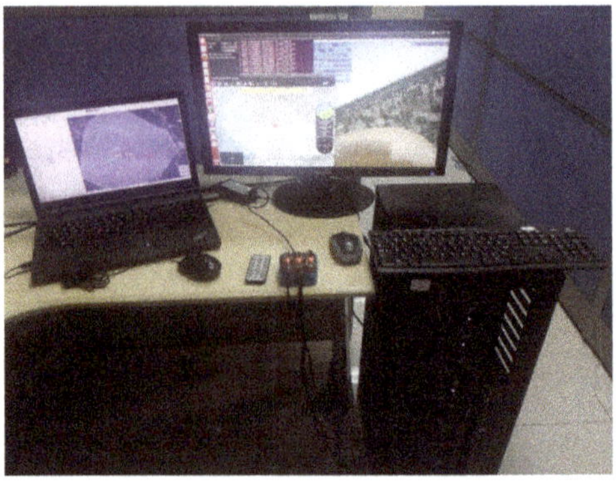

**Figure 5.** The simulation environment.

Beside receiving and sending the flying information of the UAVs, the ground control station also needs to send commands to all the UAVs. With these commands, the simulation process can be well controlled by the ground station.

### *5.2. Simulation Process*

In order to collect as many samples as possible for the policy learning, the simulation should be carried for many episodes. Each episode is run with the same process, as shown by Figure 6. There are five steps for an episode, listed as follows.

(i) Both PX4s are set on the mission mode, so as to control the two UAVs to loiter around their own loiter points outside of the region. The loiter points are send to the PX4 from the ground station before the simulation begins.
(ii) The ground station sends a command asking all the UAVs to fly to their goal points. Both PX4s are set to the offboard mode, so that the UAV agents can guide their own UAVs to fly towards their goal points. In this stage, the agents use the simple-fly pattern to guide the flying of the UAVs. In the simple-fly pattern, the desired heading angle of the UAV is the angle toward the goal point directly, while the desired velocity and the desired height stay unchanged.
(iii) When both UAVs have entered the region for the enemy avoidance problem, the ground station sends a command to inform both UAV agents that the new episode begins. Upon receiving this command, the decision making agent changes into its decision making pattern from the simple-fly pattern, during which the agent can either learn the policy with reinforcement learning or make decisions with certain policy. On the other hand, the enemy agent still guide the enemy UAV fly straightly toward its goal point $G_e$ with simple-fly pattern during this stage.
(iv) When the simulation meets one of those ending conditions, the ground station sends a command to both UAVs to inform the ending of the episode. Both PX4s are set back to the mission mode after receiving this command from the ground station.
(v) Under the mission mode, both PX4s control their own UAVs fly back to their loiter points again, since the loiter points remain as their next point in the mission mode. When both UAVs have returned to their loiter points, the ground station sends the command to make both UAVs fly into the region again for the start of the next episode.

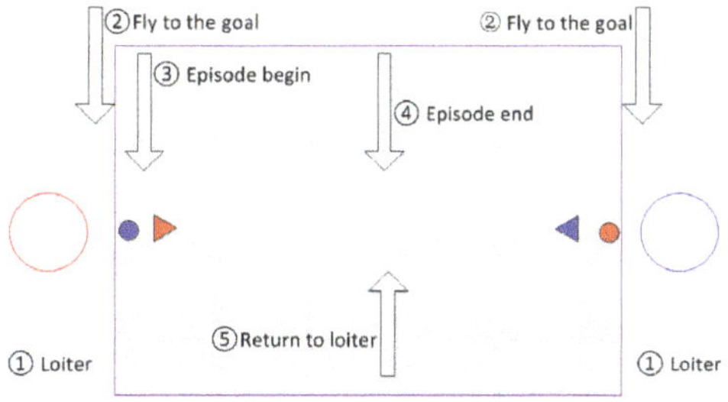


**Figure 6.** The simulation process.

## 6. Implementation and Results

Based on the HITL simulation environment built in Section 5, the decision policy for the enemy avoidance problem will be learned with the new framework constructed in Section 4. Furthermore, a fixed-rule policy and a random policy are designed to compare with the policy learned by the Sarsa based framework.

The experiment settings are as follows. The region for the enemy avoidance problem has a length of $l = 600$ m and a width of $w = 450$ m. The outside loiter center of each UAV is 100 meters away from the nearest region edge, and the loiter radius is 100 m. The goal point $G_f$ is inside the region, which is 150 m away from the right edge of the region. The decision making UAV loiters around the goal point $G_f$ with a loiter radius of 50 m when it has arrived the goal point $G_f$, and the goal radius is $r_g = 100$ m. The collision radius is $r_c = 40$ m. Set $\Delta V = 1.0$ m/s and $\Delta\phi = 5.0$ degree.

### 6.1. Learning with the Sarsa Based Framework

First, we use the newly constructed framework based on the Sarsa method and the CMAC function approximator to learn the policy for the enemy avoidance problem in the HITL simulation environment.

As in reinforcement learning, the goal is to maximize the expected accumulate rewards, thus the action-value function $Q$ is an effective metric to measure the performance of the policy learning.

Considering the mission of the decision making UAV, it has to reach the goal point $G_f$ successfully as soon as possible. On the other hand, the number of steps that an episode lasts can partially indicates how long it takes the decision making UAV to reach the goal point $G_f$. Therefore, the number of steps for an episode can be used as a metric to measure the performance to some extend.

However, there are several ending reasons. It can take very few steps to end the episode if the decision making UAV collides with or attacks the enemy UAV at a very early stage of an episode. Therefore, the number of steps alone is not enough to measure the performance of the learning in the enemy avoidance problem. To this end, the numbers of episodes that ending for different reasons are calculated, so as to measure more accurately how the learning changes the episode ending reasons.

The experiment for learning the policy with the proposed framework has been run for 14,362 episodes. The related parameter settings are: $\epsilon = 0.01$, $\alpha = 0.1$, $\gamma = 0$, and the number of tiles laid for each variable is $n = 32$. The results of the above three metrics are presented as follows.

Figure 7 illustrates how the $Q$ -value changes with the increase of the episodes during the policy learning process. The figure is drawn with data filtered by a window of 1000 episodes. From Figure 7, it can be seen that the $Q$-value increases as the number of episodes increases, and becomes stable after around 10,000 episodes.

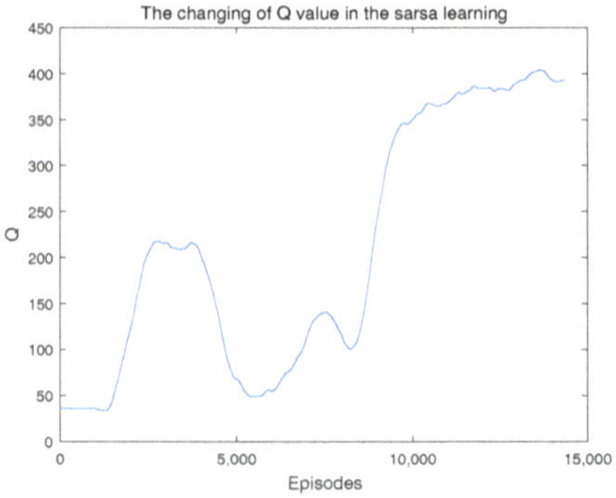


**Figure 7.** The *Q*-value for each episode in the policy learning.

Figure 8 shows how the number of steps for each episode changes as the number of policy learning episodes increases. The figure is also drawn with data filtered by a window of 1000 episodes. In Figure 8, the number of steps fluctuates very much at the beginning, but converges to a relatively small number as the number of episodes increases, and it comes to a plateau after around 10,000 episodes.

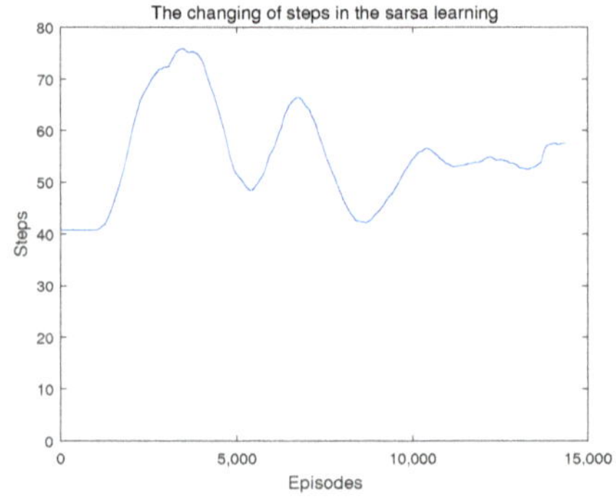


**Figure 8.** The number of steps for each episode in the policy learning.

Figure 9 presents how the accumulated numbers of different ending reasons change during the policy learning process. From Figure 9, it can be see that the episodes are mostly ended for collision with the enemy UAV at the initial period of the learning process. After about 1500 episodes, the number of episodes ending for reaching the goal begins to increase rapidly, which surpasses those of other ending reasons very soon. The number of episodes ending for successfully attacking the enemy UAV has a rise from about the 4000th episode to about the 8000th episode, and surpasses the number of episodes ending for collision with the enemy UAV during this period. Except for ending for reaching the goal point, the numbers of other ending reasons all stop to increase after about 9000 episodes. These results mean that the learning agent gradually learns to avoid the enemy UAV and attack the enemy UAV, and then it learns to limit the usage of the attacking action, and finally it has learned a very good policy to reach the goal point.

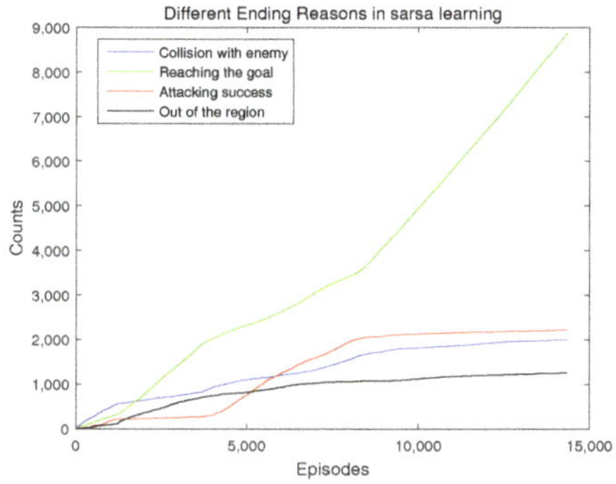


**Figure 9.** The numbers of different ending reasons in the policy learning.

From the above results for three metrics, it can be seen that the policy learning process under the proposed framework converges after about 10,000 episodes. Such episode number for converging is a reflection of the sample complexity of this learning based method. The proposed framework is to reduce the computation cost on the price of the sample complexity. However, such policy learning can be done before the real-time decision making, while decision making depending on some computation methods need to bear those computation costs at each real-time decision making step. Therefore, the proposed framework still has its advantages on these points.

### 6.2. Designing Comparison Policies

In order to show the effectiveness of the policy learned by the Sarsa based framework, we design a set of rules to guide the flying of the decision making UAV based on human experience, which is called the fixed-rule policy. As a more baseline comparison, a random policy is also used for the comparison. The detail of these two comparison policies are introduced in the following.

#### 6.2.1. Fixed-Rule Policy

The fixed-rule policy is designed with human experience. It composes of several if-then rules. At each decision making step, the agent examines the current state and decides which rule can be used. The detail of the designed rules are listed as in Algorithm 2.

The designing of these rules aims at guiding the UAV flying toward the goal as straightforward as possible and trying to avoid the enemy UAV at the same time. However, these if-then rules are very simple ones due to the limitation of the human knowledge. Therefore, the hypothesis here is that the Sarsa based framework can help the agent discover more latent rules to guide the flying of the decision making UAV.

**Algorithm 2** Fixed-rule Policy

```
if the distance to the nearest enemy<100 m then
    if decision steps from the last attacking action>10 decision steps then
        execute the attacking action
    else
        fly to the goal
    end if
else
    if the enemy UAV in the flying direction (within 2.5 degree) then
        if the enemy is on the right side then
            increase the flying angle
        else
            decrease the flying angle
        end if
    else
        fly to the goal
    end if
end if
```

#### 6.2.2. Random Policy

The purpose of designing this random policy is to set a fundamental baseline policy for the effectiveness comparison of all other policies. The hypothesis is that those effective policies should all have better performance than this random policy.

In this random policy, the agent chooses action at each decision making step randomly. The probabilities of choosing each action are equal, so that no special favor is given to any action.

### *6.3. Policy Comparison Results*

To test the effectiveness of the policy learned with the Sarsa based framework, another set of experiments are carried on the same HITL simulation platform. For comparison, the fixed-rule policy and the random policy are also tested with the same experimental settings. Each policy is run for 1200 episodes, and the results are summarized as follows.

Firstly, the numbers of steps taken in each episode for using different policies are compared in Figure 10. The figure is drawn with data filtered with a window of 100 episodes. It can be seen that the number of steps are stable in the testing process, no matter which policy is used. However, it takes different numbers of steps to end an episode when different policies are used. Overall, the random policy takes the least number of steps, while the policy learned with the Sarsa based framework takes the most. Though more number of steps means more cost, it is also a reflection of higher possibility to reach the goal point and better policy to avoid the enemy UAV. Such understanding is from the following two aspects. First, the episode ending for reaching the goal point takes more steps than the episode ending before reaching the goal point for other reasons, because of the longer distance. On the other hand, it needs more steps to fly away to avoid the collision with the enemy UAV than to fly straightly towards the goal. To more accurately show the performance of different policies, more results are presented below.

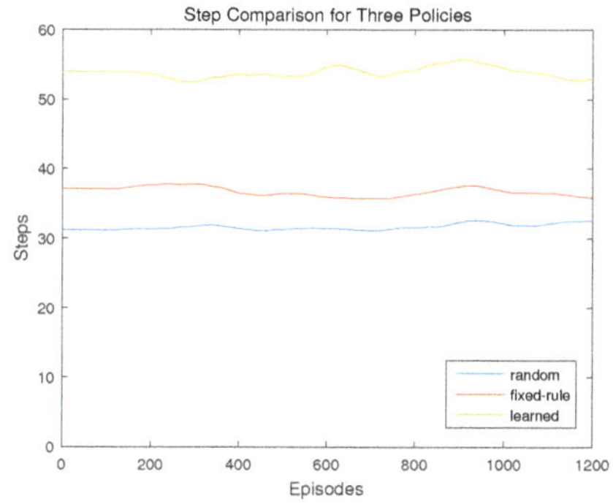


**Figure 10.** The steps of three policies.

### 6.3.1. Results of the Random Policy

Figure 11 illustrates how the numbers of episodes ending for different reasons change as the number of the total episode increases under the random policy. From Figure 11, it can be seen that there are mainly three ending reasons for these episodes, which are collision with the enemy UAV, reaching the goal point successfully, and attacking the enemy UAV successfully. Besides, there is nearly no case for flying out of the region. The number of the episodes ending for reaching the goal point increases slowest among the three main ending reasons, while the episode number for successfully attacking of the enemy UAV increases the quickest. This indicates that it is easier to end the episode by attacking the enemy UAV. The reason for this should be that ending the episode by successfully attacking takes only an attacking action as long as the attacking action is executed at a relatively near distance from the enemy UAV. On the contrast, to reach the goal point successfully is much harder, for it needs to avoid the moving enemy UAV by taking many decision steps in an appropriate sequence.

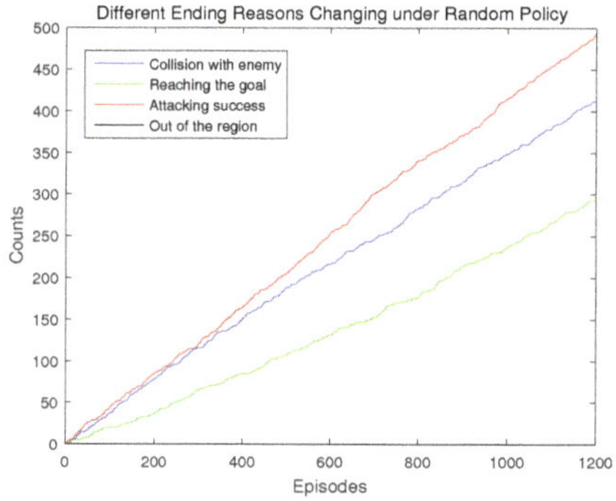


**Figure 11.** The numbers of different ending reasons when using the random policy.

Figure 12 presents the total episode numbers of different ending reasons when the random policy is used. It can be seen more clearly that the number of episodes ending for reaching the goal point is smaller than those for the other two ending reasons (collision with the enemy UAV and attacking the enemy UAV successfully). Therefore, the random policy is unable to guide the decision making UAV to fulfill its mission very successful, and the results of this policy only reflect different challenges for achieving different ending reasons.

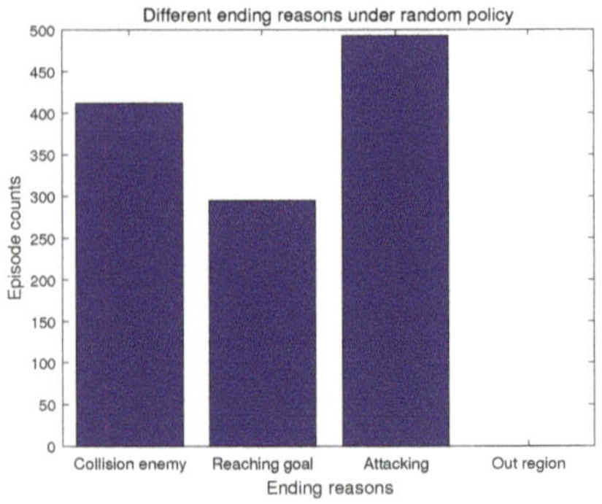


**Figure 12.** The total numbers of different ending reasons when using the random policy.

6.3.2. Results of the Fixed-Rule Policy

When the fixed-rule policy is used, the numbers of episodes ending with different reasons increase during the testing process, as shown in Figure 13. Still, there are three main ending reasons, which are collision with the enemy UAV, reaching the goal point successfully, and attacking the enemy UAV successfully. The number of episodes ending for attacking the enemy UAV is less than that of collision with the enemy UAV, while the number of the episodes ending for reaching the goal point successfully is nearly equal to that of collision with the enemy UAV. Compared with the random policy, these results show that the fixed-rule policy can reduce the unnecessary use of the attacking action, and can increase the use of some effective enemy avoidance actions so as to increase the probability of reaching the goal point successfully.

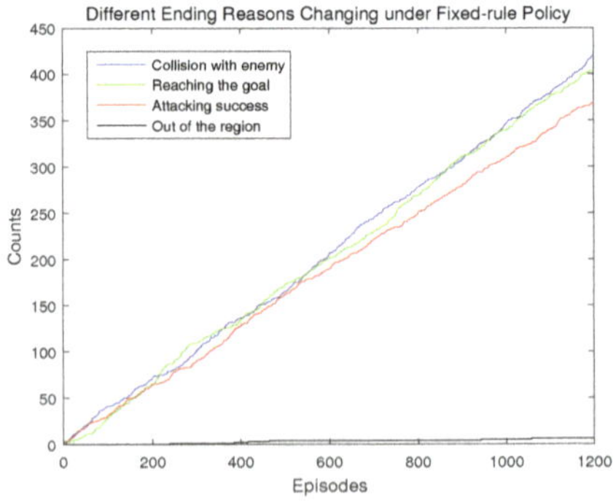


**Figure 13.** The numbers of different ending reasons when using the fixed-rule policy.

Figure 14 gives the total number of episodes for each ending reason in the fixed-rule policy experiment. Compared with that of the random policy, the fixed-rule policy can obviously reduce the number of episodes ending for attacking the enemy UAV, but is not effective in reducing the number of episodes for colliding with the enemy UAV.

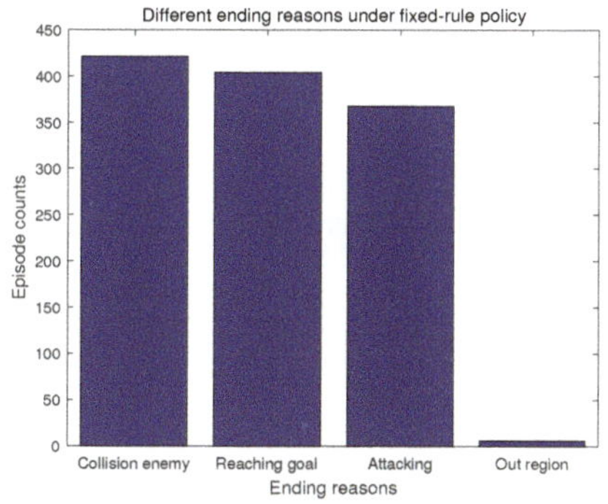


**Figure 14.** The total numbers of different ending reasons when using the fixed-rule policy.

### 6.3.3. Results of the Sarsa Learned Policy

Finally, the policy learned with the Sarsa based framework is also tested in the platform. In this experiment, the agent only takes action at each decision making step by following the learned policy, without any further learning.

Figure 15 presents how the numbers of episodes ending for different reasons change when the policy learned with the Sarsa based framework is adopted. Different from the other two policies, there is only one main ending reason: reaching the goal point. The numbers of episodes ending for both collision with the enemy UAV and attacking the enemy UAV are sharply reduced in the learned policy, and can be neglected when compared with that of the reaching goal case. Obviously, the policy learned with the Sarsa based framework can successfully guide the decision making UAV reaching the goal point.

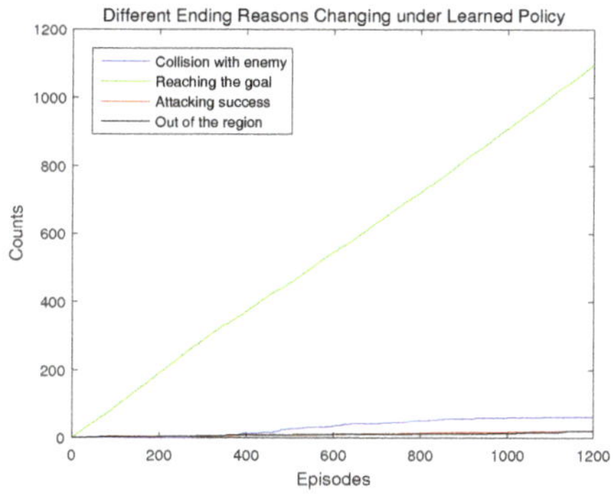


**Figure 15.** The numbers of different ending reasons when using the policy learned with the Sarsa based framework.

Figure 16 gives the total numbers of episodes ending for different reasons when the policy learned with the Sarsa based framework is used. It can be seen that the number of episodes for collision with the enemy UAV and the number of episodes for attacking the enemy UAV are both reduced to quite small numbers, which indicates the effectiveness of the policy learned with the Sarsa based framework.

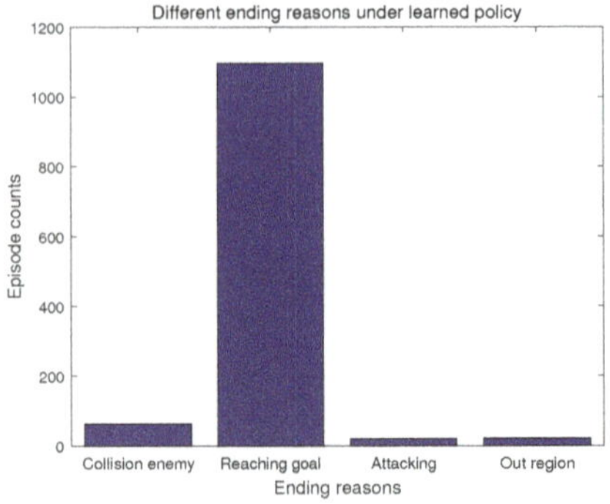


**Figure 16.** The total numbers of different ending reasons when using the learned policy.

## 7. Conclusions and Future Work

In this paper, we have defined the enemy avoidance problem, which is quite different from most collision avoidance problems. To solve the enemy avoidance problem, we have proposed a new framework, which formulates the enemy avoidance problem as the Markov Decision Process, and learns policy for the decision making UAV based on the Sarsa reinforcement learning method, and generalizes the continuous state space with the CMAC function approximation technique. Furthermore, we have constructed a HITL simulation system to learn the policy with the proposed framework and to verify the effectiveness of different policies for the enemy avoidance problem. The carried experiments show that the HITL simulation system is valid in presenting the enemy avoidance problem, and also suitable for the RL based framework. The Sarsa based framework can learn a satisfying policy for the enemy avoidance problem. Further experiments also show that the policy learned with the Sarsa based framework outperforms both the random policy and the fixed-rule policy in solving the enemy avoidance problem.

There are several advantages in our proposed method. Firstly, it does not require a mathematically built environment model to be computed when making decisions. Secondly, it learns the policy off-line, and the learned policy is a simple mapping from the state to the action. Therefore, little computation is needed for the on-line decision making. Thirdly, the policy is learned in the continuous space, and has good generalization ability. However, there are still some weaknesses in our proposed method. Firstly, samples are needed for the learning, which could also be costly. Secondly, there are a lot of simplifications in the researched problem, so inconsistency may occur if the method is applied directly into some more practical real problems.

Therefore, some future works can be explored. Firstly, a more practical problem setting could be studied. For example, equip the enemy UAV with more practical properties like changeable desired attitude and the attacking ability, or extend the two dimensional problem into three dimensions, or add noise like wind disturbance to the environment. Secondly, more efficient reinforcement learning methods and even some other techniques could be applied, so as to shorten the learning process. Thirdly, ways to combine the learning method with human experience could be explored, so as to make use of human experience to facilitate the learning and to discover latent knowledge by agent learning. Fourthly, methods which consider the collision with both obstacles and other UAVs would be more useful in improving the autonomy of the UAVs, and therefore could be studied in the future. Lastly, extending the problem to scenarios with more UAVs can be a very worth research area.

**Author Contributions:** Conceptualization, Q.C., X.W. and J.Y.; Funding acquisition, X.W. and L.S.; Investigation, Q.C. and J.Y.; Methodology, Q.C., X.W. and J.Y.; Project administration, L.S.; Supervision, X.W. and L.S.; Validation, Q.C.; Visualization, Q.C. and X.W.; Writing—original draft, Q.C.; Writing—review & editing, Q.C., X.W., J.Y. and L.S.

**Funding:** This work is supported by the National Natural Science Foundation of China under Grants 61403406.

**Conflicts of Interest:** The authors declare no conflict of interest.

## References

1. Yu, X.; Zhang, Y. Sense and avoid technologies with applications to unmanned aircraft systems: Review and prospects. *Prog. Aerosp. Sci.* **2015**, *74*, 152–166. [CrossRef]
2. Motlagh, N.H.; Bagaa, M.; Taleb, T. UAV-based IoT platform: A crowd surveillance use case. *IEEE Commun. Mag.* **2017**, *55*, 128–134. [CrossRef]
3. Gu, J.; Su, T.; Wang, Q.; Du, X.; Guizani, M. Multiple moving targets surveillance based on a cooperative network for multi-UAV. *IEEE Commun. Mag.* **2018**, *56*, 82–89. [CrossRef]
4. Liu, Y.; Wang, Q.; Hu, H.; He, Y. A Novel Real-Time Moving Target Tracking and Path Planning System for a Quadrotor UAV in Unknown Unstructured Outdoor Scenes. *IEEE Trans. Syst. Man Cybern. Syst.* **2018**. [CrossRef]
5. Yadav, I.; Eckenhoff, K.; Huang, G.; Tanner, H.G. Visual-Inertial Target Tracking and Motion Planning for UAV-based Radiation Detection. *arXiv* **2018**, arXiv:1805.09061.
6. Vanegas, F.; Campbell, D.; Roy, N.; Gaston, K.J.; Gonzalez, F. UAV tracking and following a ground target under motion and localisation uncertainty. In Proceedings of the IEEE Aerospace Conference, Big Sky, MT, USA, 4–11 March 2017; pp. 1–10.
7. Li, S.; Liu, T.; Zhang, C.; Yeung, D.Y.; Shen, S. Learning Unmanned Aerial Vehicle Control for Autonomous Target Following. *arXiv* **2017**, arXiv:1709.08233.
8. Mahjri, I.; Dhraief, A.; Belghith, A. A review on collision avoidance systems for unmanned aerial vehicles. In *International Workshop on Communication Technologies for Vehicles*; Springer: Berlin, Germany, 2015; pp. 203–214.
9. Gottlieb, Y.; Shima, T. UAVs task and motion planning in the presence of obstacles and prioritized targets. *Sensors* **2015**, *15*, 29734–29764. [CrossRef]
10. Ramasamy, S.; Sabatini, R.; Gardi, A.; Liu, J. LIDAR obstacle warning and avoidance system for unmanned aerial vehicle sense-and-avoid. *Aerosp. Sci. Technol.* **2016**, *55*, 344–358. [CrossRef]
11. Liu, Z.; Yu, X.; Yuan, C.; Zhang, Y. Leader-follower formation control of unmanned aerial vehicles with fault tolerant and collision avoidance capabilities. In Proceedings of the International Conference on Unmanned Aircraft Systems (ICUAS), Denver, CO, USA, 9–12 June 2015; pp. 1025–1030.
12. Jenie, Y.I.; van Kampen, E.J.; Ellerbroek, J.; Hoekstra, J.M. Taxonomy of conflict detection and resolution approaches for unmanned aerial vehicle in an integrated airspace. *IEEE Trans. Intell. Transp. Syst.* **2017**, *18*, 558–567. [CrossRef]
13. Sutton, R.S.; Barto, A.G. *Reinforcement Learning: An Introduction*; MIT Press: Cambridge, MA, UK, 1998.
14. Radmanesh, M.; Kumar, M.; Nemati, A.; Sarim, M. Dynamic optimal UAV trajectory planning in the national airspace system via mixed integer linear programming. *Proc. Inst. Mech. Eng. G J. Aerosp. Eng.* **2016**, *230*, 1668–1682. [CrossRef]
15. Ong, H.Y.; Kochenderfer, M.J. Short-term conflict resolution for unmanned aircraft traffic management. In Proceedings of the IEEE/AIAA 34th Digital Avionics Systems Conference (DASC), Prague, Czech Republic, 13–17 September 2015.
16. Alonso-Mora, J.; Naegeli, T.; Siegwart, R.; Beardsley, P. Collision avoidance for aerial vehicles in multi-agent scenarios. *Auton. Robot.* **2015**, *39*, 101–121. [CrossRef]
17. Dentler, J.; Kannan, S.; Mendez, M.A.O.; Voos, H. A real-time model predictive position control with collision avoidance for commercial low-cost quadrotors. In Proceedings of the IEEE Conference on Control Applications (CCA), Buenos Aires, Argentina, 19–22 September 2016; pp. 519–525.
18. Alvarez, H.; Paz, L.M.; Sturm, J.; Cremers, D. Collision avoidance for quadrotors with a monocular camera. In *Experimental Robotics*; Springer: Cham, Switzerland, 2016; pp. 195–209.
19. Albus, J.S. *Brains, Behaviour, and Robotics*; Byte Books: Perterborough, NH, USA, 1981.
20. Zhou, Y.; Baras, J.S. Reachable set approach to collision avoidance for UAVs. In Proceedings of the IEEE 54th Annual Conference on Decision and Control (CDC), Osaka, Japan, 15–18 December 2015; pp. 5947–5952.
21. D'Amato, E.; Mattei, M.; Notaro, I. Bi-level Flight Path Planning of UAV Formations with Collision Avoidance. *J. Intell. Robot. Syst.* **2018**, *93*, 193–211. [CrossRef]

22. Qiu, H.; Duan, H. A multi-objective pigeon-inspired optimization approach to UAV distributed flocking among obstacles. *Inf. Sci.* **2018**, in press. [CrossRef]
23. Yang, J.; Yin, D.; Cheng, Q.; Shen, L. Two-Layered Mechanism of Online Unmanned Aerial Vehicles Conflict Detection and Resolution. *IEEE Trans. Intell. Transp. Syst.* **2017**. [CrossRef]
24. Yang, J.; Yin, D.; Shen, L.; Cheng, Q.; Xie, X. Cooperative Deconflicting Heading Maneuvers Applied to Unmanned Aerial Vehicles in Non-Segregated Airspace. *J. Intell. Robot. Syst.* **2018**, *92*, 187–201. [CrossRef]
25. Alomari, A.; Phillips, W.; Aslam, N.; Comeau, F. Swarm Intelligence Optimization Techniques for Obstacle-Avoidance Mobility-Assisted Localization in Wireless Sensor Networks. *IEEE Access* **2017**, *6*, 22368–22385. [CrossRef]
26. Masiero, A.; Fissore, F.; Guarnieri, A.; Pirotti, F.; Vettore, A. UAV positioning and collision avoidance based on RSS measurements. *Int. Arch. Photogramm. Remote Sens. Spat. Inf. Sci.* **2015**, *40*, 219. [CrossRef]
27. Kahn, G.; Villaflor, A.; Pong, V.; Abbeel, P.; Levine, S. Uncertainty-aware reinforcement learning for collision avoidance. *arXiv* **2017**, arXiv:1702.01182.
28. Pfeiffer, M.; Schaeuble, M.; Nieto, J.; Siegwart, R.; Cadena, C. From perception to decision: A data-driven approach to end-to-end motion planning for autonomous ground robots. In Proceedings of the IEEE International Conference on Robotics and Automation (ICRA), Singapore, 29 May–3 June 2017; pp. 1527–1533.
29. Gandhi, D.; Pinto, L.; Gupta, A. Learning to fly by crashing. In Proceedings of the IEEE/RSJ International Conference on Intelligent Robots and Systems (IROS), Vancouver, BC, Canada, 24–28 September 2017; pp. 3948–3955.
30. Zhang, T.; Kahn, G.; Levine, S.; Abbeel, P. Learning deep control policies for autonomous aerial vehicles with mpc-guided policy search. In Proceedings of the IEEE International Conference on Robotics and Automation (ICRA), Stockholm, Sweden, 16–21 May 2016; pp. 528–535.
31. Zhang, B.; Mao, Z.; Liu, W.; Liu, J. Geometric reinforcement learning for path planning of UAVs. *J. Intell. Robot. Syst.* **2015**, *77*, 391–409. [CrossRef]
32. Bai, H.; Hsu, D.; Kochenderfer, M.J.; Lee, W.S. Unmanned aircraft collision avoidance using continuous-state POMDPs. *Robot. Sci. Syst. VII* **2012**, *1*, 1–8.
33. Temizer, S.; Kochenderfer, M.; Kaelbling, L.; Lozano-Pérez, T.; Kuchar, J. Collision avoidance for unmanned aircraft using Markov decision processes. In Proceedings of the AIAA Guidance, Navigation, and Control Conference, Toronto, ON, Canada, 2–5 August 2010; p. 8040.
34. Seo, J.; Kim, Y.; Kim, S.; Tsourdos, A. Collision avoidance strategies for unmanned aerial vehicles in formation flight. *IEEE Trans. Aerosp. Electron. Syst.* **2017**, *53*, 2718–2734. [CrossRef]
35. Yang, J.; Yin, D.; Cheng, Q.; Shen, L.; Tan, Z. Decentralized cooperative unmanned aerial vehicles conflict resolution by neural network-based tree search method. *Int. J. Adv. Robot. Syst.* **2016**, *13*. [CrossRef]

*applied sciences*

MDPI

*Article*

# Predictable Trajectory Planning of Industrial Robots with Constraints

**Youdong Chen * and Ling Li**

School of Mechanical Engineering and Automation, Beihang University, Beijing 100191, China; liuyunmuzi@126.com

* Correspondence: chenyd@buaa.edu.cn; Tel.: +86-1369-312-7687

Received: 6 November 2018; Accepted: 7 December 2018; Published: 17 December 2018

**Abstract:** Trajectory prediction is currently attracting considerable attention. This paper proposes geodesic trajectory planning with end-effector and joint constraints to predict the trajectory properties of the end-effector, such as velocities, accelerations, and smoothness. The prediction of the trajectory properties is independent of the joint trajectories. The prediction makes it possible to adjust the trajectory properties in line with a light computational burden. To demonstrate the effectiveness of the proposed method, experiments were conducted using the Efort robot. The experiments show that the proposed method can predict the properties of the trajectory and modify the trajectory to meet the constraints.

**Keywords:** predictable trajectory planning; geodesic; constrained motion

## 1. Introduction

There are growing interests in trajectory prediction. Trajectory prediction can be classified into two kinds: trajectory prediction for future time and trajectory prediction for future tasks. Future actions have been predicted for obstacle avoiding and trajectory re-planning [1,2]. Appropriate trajectories for future tasks have been predicted by using machine learning [3,4]. Fast motion planning has been obtained from experience. The transfer of previously optimized trajectories to a new situation cannot be made in the joint space. Little attention has been paid to Cartesian trajectory prediction that predicts the motion performed by the robot end-effector under joint velocity/acceleration constraints.

Many practical tasks impose constraints on robot motions, such as machining, welding, and cutting and so on. RRT (rapidly-exploring random tree)/PRM (probabilistic roadmap)-based approaches were used to plan the path with end-effector constraints [5–9]. The Cartesian positions of the end-effector at the via-points were achieved, but the motions of the end-effector between via points are not predictable, in view of the nonlinear effects introduced by the direct kinematics.

There are limits of joint positions, velocities, and accelerations. To obtain smooth motions, the joint constraints should be bounded. Joint constraints are often met by optimizing appropriate objective functions, such as keeping the joints close to their range centers [10], using the joint ranges in a weighted pseudo-inversion [11], or defining an infinity norm to be minimized at the velocity levels [12]. However, this method did not track the assigned paths. Inverse kinematics was used to enable joint velocity and acceleration not exceeding their limits [13], which was achieved by scaling the task time. However, these methods cannot meet joint constraints.

Generally, there are three questions. (1) The motions of the end-effector between via points are not predictable in trajectory planning with end-effector constraints. (2) There is no guarantee that the end-effector will track the assigned paths in trajectory planning with joint constraints. (3) End-effector constraints and joint constraints have seldom been considered simultaneously in the previous studies. In this paper, we attempt to solve these problems.

Geodesic is an important concept in Riemannian geometry and can be used in robot trajectory planning [14,15] and machine tool interpolation [16]. By selecting suitable Riemannian metrics and local coordinates, the direct output of the proposed method is joint trajectories.

In this paper, we present a predictable trajectory planning with constraints using geodesics. The end-effector position, velocities, and accelerations can be predicted without forward kinematics. The prediction of end-effector motion properties can be obtained before the calculation of the geodesic method. While tracking the constrained end-effector paths, the joint velocity/acceleration constraints can be met by scaling the motion time. The trajectory based on this method is predictable and has a light computational burden. Furthermore, the end-effector motions are of high precision.

This paper is organized as follows. The predictable trajectory planning with constraints is presented in Section 2. Using null constraints and a virtual end-effector constraint, predictable linear and circular trajectory planning are described in Section 3. Experiments using the Efort robot are illustrated in Section 4. Finally, the conclusion is given.

## 2. Predictable Trajectory Planning with Constraints

### *2.1. Predictable Trajectory Planning with End-Effector Constraints*

The robot end-effector moves under some constraints to accomplish some tasks in practical applications. The end-effector position constraint $H$ and orientation constraint $F$ can be described by

$$\begin{cases} H(p) = 0 \\ F(n, o, a) = 0 \end{cases} \tag{1}$$

where $p$ and $\begin{Bmatrix} n & o & a \end{Bmatrix}$ are the end-effector position and orientation respectively.

The end-effector trajectories in Cartesian space can be predicted without forward kinematics. The prediction is independent of the joint trajectories and can be made before the calculation of the proposed geodesic method. The prediction can adjust the end-effector motion properties and has light computational burdens.

First of all, the corresponding Cartesian paths meet the end-effector constraints defined in Equation (1). In Cartesian space, the end-effector moves the shortest paths that are usually not linear paths. Because the linear paths may not satisfy the end-effector constraints the shapes of the Cartesian paths are determined by the constraints.

Secondly, the Cartesian velocity is constant. Because the velocity of a geodesic is constant, the velocity always equal to the initial velocity. The velocity of a geodesic trajectory $v$ is

$$v = \frac{L_d}{t_w} = v_0, \tag{2}$$

where $L_d$, $t_w$ and $v_0$ are the path length, the motion time, and the initial velocity, respectively.

Because the velocity of a geodesic is constant, the end-effector performs a uniform curved motion. The tangential acceleration is zero. The centripetal acceleration $a$ is

$$a = \frac{v^2}{r}, \tag{3}$$

where $r$ is the radius of the curvature of a geodesic path.

Finally, the joint and Cartesian trajectories are both smooth. Because the solutions of the geodesic equations are $C^3$ continuous [10], the joint trajectories are smooth as are the Cartesian trajectories.

The joint trajectories are calculated by the geodesic method. The position and orientation are the functions of the joint variables,

$$\begin{array}{l} p = p(q_1 \cdots q_k) \\ n = n(q_{k+1} \cdots q_n) \\ o = o(q_{k+1} \cdots q_n) \\ a = a(q_{k+1} \cdots q_n) \end{array} \tag{4}$$

where $q_i$ are joint variables. By the implicit function theorem [17], there are $q_{i_0}$ and $q_{j_0}$ which are derived from Equation (1)

$$\begin{array}{l} q_{i_0} = q_{i_0}(q_1, \cdots q_{i_0-1}, q_{i_0+1}, \cdots q_{k_0}) \\ q_{j_0} = q_{j_0}(q_{k_0+1}, \cdots q_{j_0-1}, q_{j_0+1}, \cdots q_n) \end{array} \tag{5}$$

The position and orientation vectors can be described as,

$$\begin{array}{l} p = p(q_1, \cdots q_{i_0-1}, q_{i_0+1}, \cdots q_{k_0}) \\ \mathrm{O} \cong n + o + a = \mathrm{O}(q_{k_0+1}, \cdots q_{j_0-1}, q_{j_0+1}, \cdots q_n) \end{array} \tag{6}$$

In this way, the end-effector constraints are implicit in the position and orientation vector. For simplicity, they can be denoted as,

$$\begin{array}{l} u_i = \begin{cases} q_i, i \leq i_0 - 1 \\ q_{i+1}, k_0 \geq i \geq i_0 + 1 \end{cases} \\ v_i = \begin{cases} q_{k+i}, i \leq j_0 - k_0 - 1 \\ q_{k+i+1}, n \geq i \geq j_0 + 1 \end{cases} \\ p = (p_1, p_2, p_3) \\ \mathrm{O} = (\mathrm{O}_1, \mathrm{O}_2, \mathrm{O}_3) \end{array} \tag{7}$$

Define Riemannian metrics as

$$\begin{array}{l} {}^d g = \sum\limits_{\alpha,i,j} \frac{\partial p_\alpha}{\partial u_i} \frac{\partial p_\alpha}{\partial u_j} du_i du_j = \sum\limits_{i,j} {}^d g_{ij} du_i du_j \\ {}^o g = \sum\limits_{\alpha,i,j} \frac{\partial \mathrm{O}_\alpha}{\partial v_i} \frac{\partial \mathrm{O}_\alpha}{\partial v_j} dv_i dv_j = \sum\limits_{i,j} {}^o g_{ij} dv_i dv_j \end{array} \tag{8}$$

where $du_i$ is the differential of $u_i$, and $\Sigma_{\alpha,i,j}$ is the summation notation with the index $\alpha, i, j$. The quantities relevant position and orientation with superscript are $d$ and $o$, respectively. The geodesic equations are

$$\begin{cases} \frac{d^2 u_i}{dt^2} + \sum\limits_{k,j} {}^d\Gamma^i_{kj} \frac{du_k}{dt} \frac{du_j}{dt} = 0 \\ \frac{d^2 v_l}{dt^2} + \sum\limits_{m,b} {}^o\Gamma^l_{bm} \frac{dv_m}{dt} \frac{dv_b}{dt} = 0 \end{cases} \tag{9}$$

The Christoffel symbols ${}^d\Gamma^i_{kj}$ and ${}^o\Gamma^l_{mb}$ are given as

$$\begin{cases} {}^d\Gamma^i_{kj} = \frac{1}{2} \sum\limits_{m} {}^d g^{ei} \left( \frac{\partial^d g_{ke}}{\partial u_j} + \frac{\partial^d g_{je}}{\partial u_k} - \frac{\partial^d g_{kj}}{\partial u_e} \right) \\ {}^o\Gamma^l_{bm} = \frac{1}{2} \sum\limits_{f} {}^o g^{fl} \left( \frac{\partial^o g_{nf}}{\partial v_m} + \frac{\partial^o g_{mf}}{\partial v_b} - \frac{\partial^o g_{bm}}{\partial v_f} \right) \end{cases} \tag{10}$$

where ${}^d g^{ei}$ and ${}^o g^{fl}$ are the elements of inverse Riemannian metric coefficient matrices ${}^d G^{-1} = \left({}^d g_{ij}\right)^{-1}$ and ${}^o G^{-1} = ({}^o g_{mn})^{-1}$ respectively [16]. $\partial$ is the notation of partial differential. The Christoffel symbols are derived from the position and orientation vectors with end-effector constraints. The joint trajectories are obtained by Equation (9) together with Equations (5) and (7) with boundary conditions

$$\begin{cases} u_i(t_0) = {}^0u_i \\ v_l(t_0) = {}^0v_l \\ u_i(t_0 + t_w) = {}^fu_i \\ v_l(t_0 + t_w) = {}^fv_l \end{cases} \tag{11}$$

where $t_0$, $\{{}^0u_i, {}^0v_l\}$ and $\left\{{}^fu_i, {}^fv_l\right\}$ are the start time, the start and end joint positions, respectively. In other words, the direct output of the geodesic method is joint trajectories.

As we can see above, trajectory planning with the end-effector constraints is given. The prediction of the end-effector motion properties can be made before the calculation of the geodesic method. The prediction is demonstrated in Section 3.1 in a simple way. The reason why the trajectories can be predicted is that they are geodesic trajectories. Geodesic has many merits, such as the shortest path, constant velocity.

### 2.2. *Joint Velocity/Acceleration Limits Avoidance*

The joint velocity and acceleration limits must be taken into account. Assume that the trajectory planned by the proposed method is,

$$\left\{t_w, q_i, \dot{q}_i, \ddot{q}_i (i = 1, \cdots, n)\right\} \tag{12}$$

Introducing a set of joint velocity/acceleration limits as follows,

$$\begin{cases} \left|\dot{q}_i\right| \le \dot{q}_{i,\lim} \\ \left|\ddot{q}_i\right| \le \ddot{q}_{i,\lim} \end{cases} \tag{13}$$

where $\dot{q}_{i,\lim}$ and $\ddot{q}_{i,\lim}$ are the velocity and acceleration limits of the $i$th joint. Define a scaling factor

$$\alpha = \max\left\{\frac{\left|\dot{q}_i\right|}{\dot{q}_{i,\lim}}, \frac{\left|\ddot{q}_i\right|}{\ddot{q}_{i,\lim}}, i = 1, \cdots, n\right\} \tag{14}$$

If $\alpha \ge 1$, there is at least one of the joint velocities or accelerations out of its limit. Update the motion time by the scaling factor,

$$t_{new} = \alpha t_w \tag{15}$$

The end-effector motions are predictable. The end-effector moves along the desired paths. The end-effector velocity is renewed as,

$$v_{new} = \frac{1}{\alpha} v \tag{16}$$

Along the geodesic path, the tangential acceleration is still zero. The centripetal acceleration is

$$a_{new} = \frac{{v_{new}}^2}{r} = \frac{1}{\alpha^2} a \tag{17}$$

The new joint trajectories are obtained by solving the geodesic equations, Equation (9) under the updated motion time with the boundary conditions,

$$\begin{cases} u_i(t_0) = {}^0u_i \\ v_l(t_0) = {}^0v_l \\ u_i(t_0 + t_{new}) = {}^fu_i \\ v_l(t_0 + t_{new}) = {}^fv_l \end{cases} \tag{18}$$

Actually, the joint velocity and acceleration limits avoidance is achieved by scaling the motion time. Our method can track the assigned paths, and the velocities while the accelerations can be predicted in advance.

### 2.3. Acceleration/Deceleration Profile

Because the geodesic is constant-speed, the planning geodesic trajectory is constant speed. The acceleration/deceleration profile should be considered at the endpoints in the practical applications.

Generally, the acceleration/deceleration profile is given as

$$\frac{d^2q_i}{dt^2} + \sum_{k,j} \Gamma^i_{kj} \frac{dq_k}{dt}\frac{dq_j}{dt} = \frac{acc}{\sqrt{\sum_j (\frac{dq_j}{dt})^2}} \frac{dq_i}{dt} \tag{19}$$

where $\Gamma^i_{kj}$ and $acc$ are the Christoffel symbols and the acceleration respectively [10]. In fact, $acc$ is the tangential acceleration. For linear motions, the centripetal acceleration equals zero. The tangential acceleration is the acceleration $acc$. The acceleration of curved motions $acc$ is actually the tangential acceleration.

## 3. Predictable Linear and Circular Trajectory Planning

### 3.1. Predictable Linear Trajectory Planning

If the constraint defined in Equation (1) is zero, the method illustrated in Section 2 degenerates into linear trajectory planning. Because the shortest path in the Cartesian space is a linear path, the Riemannian metrics ${}^dg$ and ${}^og$ become [9],

$$\begin{cases} {}^dg = (dp)^2 \\ {}^og = (dn)^2 + (do)^2 + (da)^2 \end{cases} \tag{20}$$

The geodesic equations are

$$\begin{cases} \frac{d^2q_i}{dt^2} + \sum_{k,j} {}^d\Gamma^i_{kj} \frac{dq_k}{dt}\frac{dq_j}{dt} = 0 \\ \frac{d^2q_l}{dt^2} + \sum_{m,b} {}^o\Gamma^l_{bm} \frac{dq_b}{dt}\frac{dq_n}{dt} = 0 \end{cases} \tag{21}$$

We predict that the end-effector performs linear paths with the joint trajectories derived by Equation (19). The velocity and acceleration are constant and zero respectively. The Cartesian trajectories are smooth. These predictions are validated as follows.

The joint variables $\theta_1, \cdots, \theta_{k_0}$ are regarded as local coordinates of the position space. The components of the position vector $p_x$, $p_y$ and $p_z$ can also be chosen as coordinates. The Riemannian metric defined in Equation (19) can be rewritten as

$${}^dg = (dp)^2 = \begin{pmatrix} dp_x & dp_y & dp_z \end{pmatrix} I \begin{pmatrix} dp_x & dp_y & dp_z \end{pmatrix}^T, \tag{22}$$

where $I$ is a $3 \times 3$ identity matrix. The geodesic equations become

$$\begin{cases} \frac{d^2 p_x}{dt^2} = 0 \\ \frac{d^2 p_y}{dt^2} = 0 \\ \frac{d^2 p_z}{dt^2} = 0 \end{cases} \tag{23}$$

which are equations of the linear path. Since geodesic has no relation to the coordinates, Equations (19) and (21) represent the same geodesic. The robot end-effector will move linearly under the joint trajectories determined by Equation (19). The velocity and acceleration are constant and zero respectively from Equation (21). The end-effector motion properties can be made before the calculation of the geodesic method. The Cartesian trajectories are smooth by the theory of the ordinary differential equation.

### 3.2. Predictable Circular Trajectory Planning

A geodesic of a spherical surface is a big circle. To obtain circular geodesic trajectory planning, we can impose a virtual end-effector constraint as

$$(p - p_0)^2 = R^2, (R > 0) \tag{24}$$

where $p_0$ and $R$ are the center and the radius of a spherical surface respectively. The corresponding Riemannian metrics and geodesic equations can be obtained as in Section 2.

It is predicted that the end-effector performs circular paths with joint trajectories derived by the geodesic equations. The velocity, tangential acceleration, and centripetal acceleration are constant $v$, zero, and $\frac{v^2}{R}$ respectively. The Cartesian trajectories are smooth. Similar validation can be obtained.

## 4. Experiment

The proposed predictable trajectory planning method and joint velocity/acceleration limits avoidance schemes were validated in the following experiment. The experiment used a 6-DOF industrial robot Efort robot (ER3A-3C (HD)) produced by Anhui Efort intelligent equipment Co., Ltd. The link parameters of the Efort robot are shown in Table 1. The joint velocity limits given by the manufacture are presented in Table 2. The joint acceleration limit is $2 \times 10^4$ (rad/s$^2$).

**Table 1.** Link parameters of the Efort robot.

| $i$ | $\alpha_{i-1}$ | $a_{i-1}$ (mm) | $d_i$(mm) | $\theta_i$ |
|---|---|---|---|---|
| 1 | 0 | 0 | 0 | $\theta_1$ |
| 2 | $-\pi/2$ | 50 | 0 | $\theta_2$ |
| 3 | 0 | 270 | 0 | $\theta_3$ |
| 4 | $-\pi/2$ | 70 | 299 | $\theta_4$ |
| 5 | $\pi/2$ | 0 | 0 | $\theta_5$ |
| 6 | $-\pi/2$ | 0 | 0 | $\theta_6$ |

**Table 2.** The joint velocity limits of the Efort robot.

| Joint | 1 | 2 | 3 | 4 | 5 | 6 |
|---|---|---|---|---|---|---|
| limits(rad/s) | 4.01 | 4.01 | 4.36 | 5.59 | 5.59 | 7.33 |

An actual machining process is performed on the Efort robot. The coordinates and time stamps of the via points on the machined path are listed in Table 3. The paths are shown in Figure 1. The desired Cartesian velocity is 800 mm/s. To reduce the vibration, the Cartesian velocities at the corner points are set at 200 mm/s.

Before the experiment, we can predict the trajectories. The robot will move along the path shown in Figure 1 with the end-effector constraints. There will be an acceleration zone, a constant-velocity zone and a deceleration zone in a normal block. In a short block, there will be only an acceleration/deceleration zone or a constant-velocity zone. After avoiding the joint limits, the end-effector will move the desired path. The motion time and velocities will be scaled.

In the constant-velocity zones, the trajectories are generated by Equation (19). The trajectories in the acceleration and deceleration zones are generated by Equation (17). The corresponding results are shown in Figure 2. In the first acceleration zone, the velocity creeps up from 0 to 200 mm/s in 0.2 s. The velocity runs to 800 mm/s within 0.04 s. The total motion time grows from 6.18 s to 6.49 s with the joint avoidance scheme. Corresponding velocities are also changed.

The joint trajectories are shown in Figure 3. As shown in Table 4, the maximum velocity and the maximum acceleration of joint 3 are 5.06 rad/s and 2.32 × 104 rad/s2, respectively. Both of the values exceed their limits. The method of joint velocity/acceleration limits avoidance introduced in Section 2.2 is employed to adjust the joint velocity and acceleration trajectories. Actually, there are three areas where the joint velocities exceed their limits. These trajectories can be modified to comply with the velocity and acceleration constraints by scaling the motion time. Scaling the neighboring zones using the scaling factor defined in Equation (12), the motion time is updated by the scaling factor as in Equation (13). The maximum velocity and the maximum acceleration of joint 3 reduced to 4.36 rad/s and 1.85 × 104 rad/s2. The maximum joint velocities and accelerations are shown in Table 4. Table 4 shows that most of the values are reduced and all values are in the allowable range. The modified joint trajectories are shown in Figure 3. To see the details clearly, the first modified zone is taken as an example, as shown in Figure 4. In this zone, the motion time (0.48–0.76) (s) is scaled to (0.48–0.85) (s). The velocity of the constant-velocity zone is reduced from 800 mm/s to 592.57 mm/s, as shown in Figure 2. The Cartesian paths marked red in Figure 1 are the joint modified zones. The Cartesian path marked green in Figure 1 is above the plane. As is shown, the predicted Cartesian path coincides with the modified test one. The Cartesian velocity shown in Figure 2 is consistent with the predictions that are acceleration, constant velocity, and deceleration zones. It can be observed that the proposed method allows us to keep the joint velocities and accelerations below their limits while precisely tracking the desired paths. The machined workpiece is shown in Figure 5. The computational time of each interpolation point is about 1ms, which meets the real time demands. The Cartesian trajectories move as is predicted.

**Table 3.** The coordinates and time stamps of the via points.

| | (px, py, pz) (mm) | Time (s) | | (px, py, pz) (mm) | Time (s) |
|---|---|---|---|---|---|
| 1 | (500, −260, 100) | 0 | 14 | (310, −70, 80) | 2.89 |
| 2 | (500, −260, 80) | 0.2 | 15 | (310, −70, 100) | 2.99 |
| 3 | (500, −60, 80) | 0.48 | 16 | (355, −115, 100) | 3.31 |
| 4 | (300, −60, 80) | 0.76 | 17 | (355, −115, 80) | 3.41 |
| 5 | (300, −260, 80) | 1.04 | 18 | (355, −205, 80) | 3.86 |
| 6 | (500, −260, 80) | 1.32 | 19 | (310, −160, 80) | 4.18 |
| 7 | (500, −260, 100) | 1.42 | 20 | (310, −160, 100) | 4.28 |
| 8 | (490, −250, 80) | 1.49 | 21 | (445, −205, 100) | 4.99 |
| 9 | (490, −250, 100) | 1.59 | 22 | (445, −205, 80) | 5.09 |
| 10 | (490, −70, 80) | 1.83 | 23 | (400, −250, 80) | 5.41 |
| 11 | (310, −70, 80) | 2.07 | 24 | (355, −205, 80) | 5.72 |
| 12 | (490, −250, 80) | 2.41 | 25 | (490, −70, 80) | 5.98 |
| 13 | (310, −250, 80) | 2.65 | 26 | (490, −70, 100) | 6.18 |

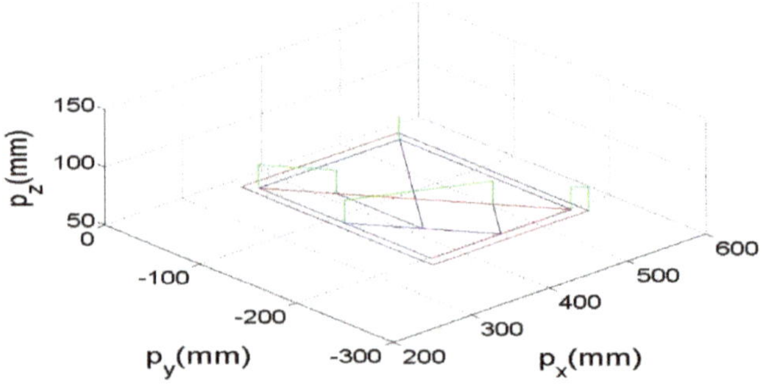


**Figure 1.** The Cartesian path of the Efort robot.

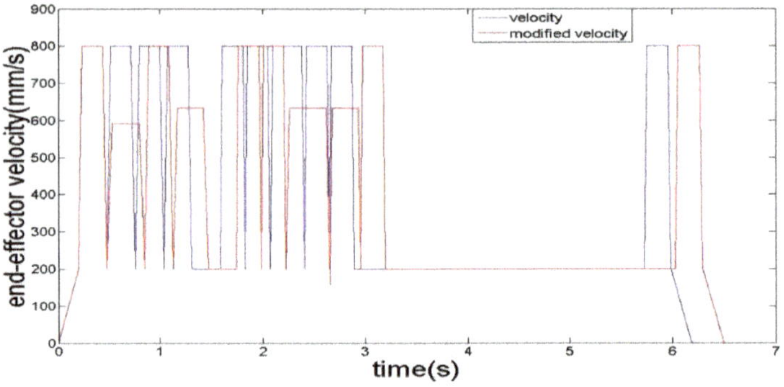


**Figure 2.** The Cartesian velocity of the Efort robot.

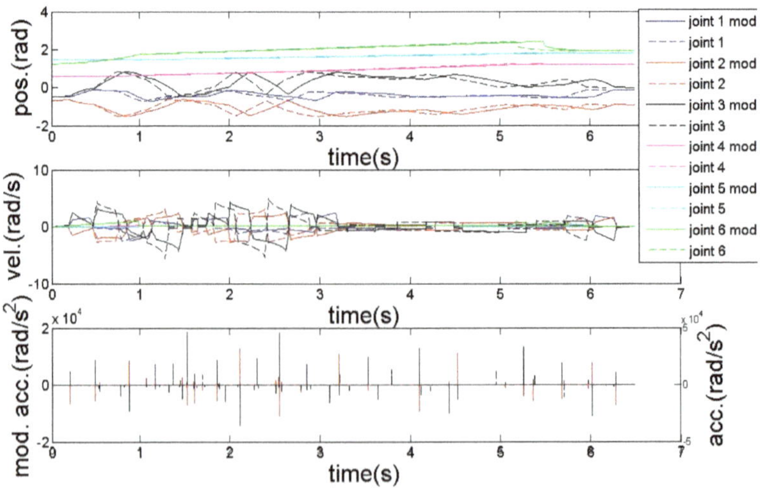


**Figure 3.** The joint trajectories (Here, pos., vel., acc., mod. are the abbreviation of position, velocity, acceleration and modified, respectively).

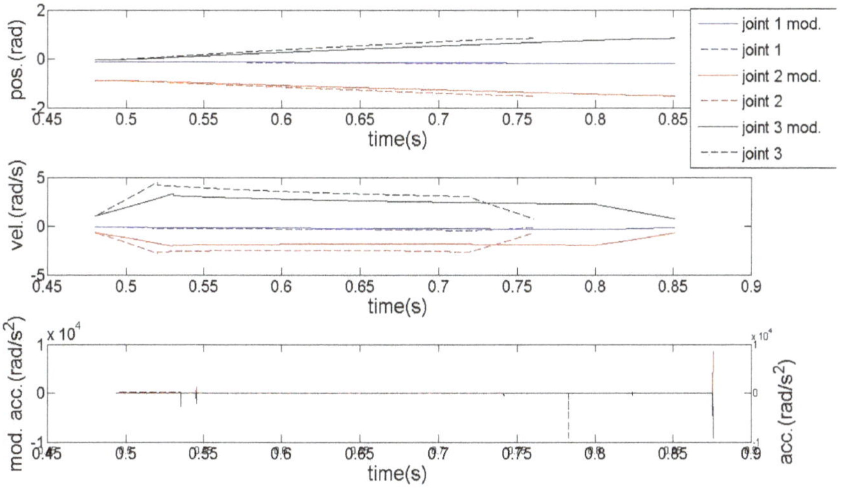


**Figure 4.** The local joint trajectories of first three joints (Here, pos., vel., acc., mod. are the abbreviation of position, velocity, acceleration and modified, respectively.).

**Table 4.** Comparisons of joint velocities and accelerations.

| | Max. Vel. (rad/s) | Max. Modified Vel. (rad/s) | Max. Acc. (rad/s$^2$) | Max. Modified Acc. (rad/s$^2$) |
|---|---|---|---|---|
| joint 1 | 2.49 | 2.49 | 7.12 × 103 | 7.10 × 103 |
| joint 2 | 3.21 | 2.73 | 1.37 × 104 | 1.28 × 104 |
| joint 3 | 5.51 | 4.36 | 2.32 × 104 | 1.85 × 104 |
| joint 4 | 0.22 | 0.22 | 450.47 | 214.25 |
| joint 5 | 0.1 | 0.1 | 3.25 × 103 | 3.16 × 103 |
| joint 6 | 0.1 | 0.1 | 1.67 × 104 | 1.61 × 104 |

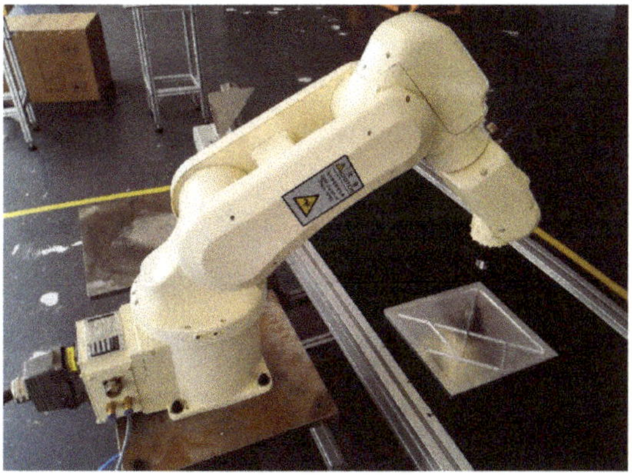

**Figure 5.** A machined workpiece.

## 5. Conclusions

In this paper, a predictable trajectory planning of industrial robots with constraints is presented. The end-effector and joint velocity/acceleration constraints are all considered. The end-effector motions, velocities, and accelerations can be predicted. The prediction is embedded in two aspects. (1) The direct output of the method is the joint trajectories while the Cartesian trajectories even between via points can be predicted. (2) The end-effector paths can be predicted with velocity and acceleration constraints. Experiments with this method using the Efort robot are given. The results show that the trajectory can be predicted and modified on-line to meet the robot constraint.

There are also some limitations. In practice, the end-effector constraints may not be described easily. In the paper, we give two specific end-effector constraints—linear and circular end-effector motions. There are two ways to deal with complicated end-effector constraints. One is to explore the equations of end-effector constraints as in Equation (1). The other is to fit complex end-effector motions by linear and circular motions.

There are also questions to be studied, such as the accuracy of the prediction. Related problems include the size, influence factor, and improvements of the accuracy of the prediction, and so on.

**Author Contributions:** Y.C. and L.L. conceived and designed the experiments, L.L. performed the experiments, Y.C. and L.L. analyzed the data, L.L. contributed materials/analysis tools, L.L. wrote the original draft preparation, Y.C.; wrote the review and editing, Y.C. acquired the funding.

**Funding:** This research was funded by the Beijing Science and Technology Plan (D161100003116002).

**Conflicts of Interest:** The authors declare no conflict of interest.

## References

1. Vendrell, E.; Mellado, M.; Crespo, A. Robot planning and re-planning using decomposition, abstraction, deduction, and prediction. *Eng. Appl. Artif. Intell.* **2001**, *14*, 505–518. [CrossRef]
2. Mora, M.C.; Tornero, J. Predictive and multirate densor-based planning under uncertainty. *IEEE Trans. Intell. Transp. Syst.* **2015**, *16*, 1493–1504. [CrossRef]
3. Jetchev, N.; Toussaint, M. Fast motion planning from experience: Trajectory prediction for speeding up movement generation. *Auton. Robots* **2013**, *34*, 111–127. [CrossRef]
4. Jetchev, N.; Toussaint, M. Trajectory prediction in cluttered voxel environments. In Proceedings of the 2010 IEEE International Conference on Robotics and Automation (ICRA), Anchorage, AK, USA, 3–7 May 2010; pp. 2523–2528.
5. Yao, Z.; Gupta, K. Path planning with general end-effector constraints. *Robot. Auton. Syst.* **2007**, *55*, 316–327. [CrossRef]
6. Jaillet, L.; Porta, J.M. Path planning under kinematic constraints by rapidly exploring manifolds. *IEEE Trans. Robot.* **2013**, *29*, 105–117. [CrossRef]
7. Berenson, D.; Srinivasa, S.S.; Ferguson, D.; Kuffner, J.J. Manipulation planning on constraint manifolds. In Proceedings of the 2010 IEEE International Conference on Robotics and Automation (ICRA), Kobe, Japan, 12–17 May 2009; pp. 625–632.
8. Suh, C.; Taewoong, T.; Kim, B.; Noh, H.; Kim, M.; Park, F.C. Tangent space RRT: A randomized planning algorithm on constraint manifolds. In Proceedings of the 2010 IEEE International Conference on Robotics and Automation (ICRA), Shanghai, China, 9–13 May 2011; pp. 4968–4973.
9. Berenson, D.; Srinivasa, S.S. Probabilistically complete planning with end-effector pose constraints. In Proceedings of the 2010 IEEE International Conference on Robotics and Automation (ICRA), Anchorage, AK, USA, 3–7 May 2010; pp. 2724–2730.
10. Samson, C.; Borgne, M.L.; Espiau, B. *Robot Control: The Task Function Approach*; Clarendon: Oxford, UK, 1991.
11. Chanand, T.; Dubey, R. A weighted least-norm solution based scheme for avoiding joint limits for redundant joint manipulators. *IEEE Trans. Robot. Autom.* **1995**, *11*, 286–292.
12. Deo, A.S.; Walker, I.D. Minimum effort inverse kinematics for redundant manipulators. *IEEE Trans. Robot. Autom.* **1997**, *13*, 767–775. [CrossRef]
13. Antonelli, G.; Chiaverini, S.; Fusco, G. A new on-line algorithm for inverse kinematics of robot manipulators ensuring path tracking capability under joint limits. *IEEE Trans. Robot. Autom.* **2003**, *19*, 162–167. [CrossRef]
14. Chen, Y.; Li, L.; Ji, X. Smooth and Accurate Trajectory Planning for Industrial Robots. *Adv. Mech. Eng.* **2014**, *6*, 342137. [CrossRef]
15. Zefran, M.; Kumar, V.; Croke, C.B. On the generation of smooth three-dimensional rigid body motions. *IEEE Trans. Robot. Autom.* **1998**, *14*, 576–589. [CrossRef]

16. Chen, Y.; Li, L. Smooth geodesic interpolation for five-axis machine tools. *IEEE/ASME Trans. Mechatron.* **2016**, *21*, 1592–1603. [CrossRef]
17. Rudin, W. *Principles of Mathematical Analysis*, 3rd ed.; McGraw-Hill Companies, Inc.: New York, NY, USA, 1976.

*applied sciences*

MDPI

*Article*

# Comparison of Spray Deposition, Control Efficacy on Wheat Aphids and Working Efficiency in the Wheat Field of the Unmanned Aerial Vehicle with Boom Sprayer and Two Conventional Knapsack Sprayers

**Guobin Wang [1], Yubin Lan [1,*], Huizhu Yuan [2], Haixia Qi [1], Pengchao Chen [1], Fan Ouyang [1] and Yuxing Han [1,*]**

1 National Center for International Collaboration Research on Precision Agricultural Aviation Pesticides Spraying Technology (NPAAC), South China Agricultural University, Guangzhou 510642, China; guobinwang@stu.scau.edu.cn (G.W.); qihaixia@scau.edu.cn (H.Q.); pengchao@stu.scau.edu.cn (P.C.); ouyangfan@scau.edu.cn (F.O.)

2 State Key Laboratory for Biology of Plant Disease and Insect Pests, Institute of Plant Protection, Chinese Academy of Agricultural Sciences, Beijing 100193, China; hzhyuan@ippcaas.cn

* Correspondence: ylan@scau.edu.cn (Y.L.); yuxinghan@scau.edu.cn (Y.H.); Tel.: +86-020-8528-1421 (Y.L.); +86-020-8528-8202 (Y.H.)

Received: 4 December 2018; Accepted: 24 December 2018; Published: 9 January 2019

**Abstract:** As a new low volume application technology, unmanned aerial vehicle (UAV) application is developing quickly in China. The aim of this study was to compare the droplet deposition, control efficacy and working efficiency of a six-rotor UAV with a self-propelled boom sprayer and two conventional knapsack sprayers on the wheat crop. The total deposition of UAV and other sprayers were not statistically significant, but significantly lower for run-off. The deposition uniformity and droplets penetrability of the UAV were poor. The deposition variation coefficient of the UAV was 87.2%, which was higher than the boom sprayer of 31.2%. The deposition on the third top leaf was only 50.0% compared to the boom sprayer. The area of coverage of the UAV was 2.2% under the spray volume of 10 L/ha. The control efficacy on wheat aphids of UAV was 70.9%, which was comparable to other sprayers. The working efficiency of UAV was 4.11 ha/h, which was roughly 1.7–20.0 times higher than the three other sprayers. Comparable control efficacy results suggest that UAV application could be a viable strategy to control pests with higher efficiency. Further improvement on deposition uniformity and penetrability are needed.

**Keywords:** unmanned aerial vehicle; pesticide application; deposition uniformity; droplets penetrability; control efficacy; working efficiency

## 1. Introduction

Wheat is one of the major food crops in China; the planting area and yield account for 21.4% and 20.9%, respectively, in 2017 (Data from National Bureau of Statistics of China). However, there are many kinds of pests and diseases which harm the production of wheat. These pests and diseases are controlled basically by the application of chemical products.

The selection of the equipment to be used is a critical factor for chemical pest control. In China, more than 88% of sprayers are manually operated [1], which include electric or manual air-pressure knapsack sprayer and knapsack mist-blower sprayer. The quality of the application depends mainly on the skill of the operators. These types of equipment are of low cost, easily maintained and adequate to control periodic and localized problems. However, applications with knapsack sprayer generally lead to high chemical exposure of the operators [2,3] and postural discomfort [4]. The operational farm size is increasing with the growth of agricultural co-operatives, land leasing and contract farming, while the labor force is declining by urbanization and rural–urban migration [1], leading to these low-efficiency and labor-intensive equipment types no longer being suitable for crop protection. As one of the alternative equipment types, self-propelled boom sprayer appeared on the market, equipped with horizontal spray boom. These machines have relatively higher working efficiency, lower chemical exposure and higher deposition [5]. However, the complicated terrain and small farm size with separated plots limit the use of boom sprayer in China. In recent decades, to adapt to this unique operating environment and meet the shortage supply of the crop protection equipment, unmanned aerial vehicles (UAV) for pesticide application have been developed quickly in China. Comparing with the manned agricultural aircraft, UAVs do not require navigation station or airport, and the edge of field can be its landing site [6]. The low rate of no-load flight and less flight crew reduce the expenditure of operations and administration [6]. Meanwhile, UAVs have short turning radius due to hover and turn around flexibly in the air, which are suitable for working in rough terrain and small plots with high efficiency [1,7,8]. Comparing with the conventional ground crop protection machinery, UAVs operate with lower labor intensity, operator exposure and have a higher working efficiency, especially in rough terrain and small plots [1,6,9]. According to the statistics data by the Chinese Ministry of Agriculture, nearly 14,000 crop protection UAVs are used in the country. The spraying area approached 5.5 million hectares in 2017.

Because of the broad prospect of application, UAVs have attracted plenty of scholar's attention. In the aspect of optimizing operational altitudes and speeds, Qin et al. [7] optimized the flying parameters for preventing plant hoppers, showing that a flight height of 1.5 m and a flying velocity of 5 m/s achieved the maximum lower layer deposition and the most uniform distribution for HyB-15L UAV sprayer (Gao Ke Xin Nong Co. Ltd., Shenzhen, Guangdong, China). The optimal parameters change with the type of UAV and the crop. In a spraying test of different shape (open center shape and round head shape) of circus trees, the 3W-LWS-Q60S UAV (Zhuhai Crop Guardian Aerial Plant Protection Co., Zhuhai, Guangdong, China) performs better when the working height is 1.0 m compared with 0.5 and 2 m [10]. In the aspect of deposition uniformity, a multi-spraying swath test is conducted with different UAV sprayers [9]. There is an obviously inconsistent amount of deposition in the longitudinal and lateral direction and this phenomenon has been reported in many studies [7,10,11]. The control efficacy on pests and diseases is one of the most important evaluation indices of chemical application. Quite different from the conventional large volume application, the UAV sprayer belongs to low volume (LV, 4.7–46.7 L/ha) or ultra-low volume (ULV, 0–4.7 L/ha) [12] spraying equipment with the spray volume in the range of 1–40 L/ha [6,11]. Meanwhile, with the same active ingredient applied per acre, the chemical concentration of UAV is particularly high. Qin et al. [7] studied the control efficacy of HyB-15L UAV with spraying Chlorpyrifos·Regent EC against plant hoppers and found that the insecticidal efficacy is 92% and 74% at 3 and 10 days after application, respectively. On the premise of guaranteed control efficacy, the working efficiency is another important evaluation index of application. Currently, UAVs mostly rely on semi-autonomous control with the flight altitude belonging to autonomous control. The control range is 200–300 m in visible distance with manual control [6]. The payload capacity of the aerial application UAVs is generally 5–25 kg [6,13,14]. Considering the limited payload and the flight range, the effective spray work rates of 2–5 ha/h can be achieved in a vineyard with a gasoline-powered helicopter (RMAX, Yamaha motor Co., Cypress, CA, USA) [11]. The working efficiency of different UAVs in the grain-filling stage of wheat was studied by Wang et al. [9], with the daily working area ranging from 13.4 to 18.0 ha in 8 h. Pesticide spray drift is an important environmental problem for aerial application. Compared with the manned agricultural aircraft [6,15,16], the droplet drift of UAV is effectively reduced with the lower flight height [6,7] and the downwash wind [17,18]. According to Xue et al. [19], under the wind speed of 3 m/s, 90% of drift droplets of Z-3 UAV are located within a range of 8 m of the target area. Similar to Xue et al., Wang et al. [17] measured that 90% of drift droplets of a fuel powered single-rotor UAV are within 9.3–14.5 m under the wind speed of 0.76–5.5 m/s. Compared with the conventional boom sprayer [20,21], the droplets drift distance of UAV sprayer would be further.

Despite these preceding studies, research is focused mainly on parameter optimization, droplet deposition and biological efficacy of one equipment. Few studies compare different kinds of crop protection equipment, especially including UAV sprayer. Under the same working condition, the comparison of different crop protection equipment on deposition, control efficacy and working efficiency is very important for equipment selection and application quality analysis. The main objective of this research was to compare the application quality of a battery motive 3WTXC8-5 six-rotor UAV with a 3WX-280H self-propelled boom sprayer and two conventional knapsack sprayers (3WBS-16A2 electric air-pressure knapsack sprayer and WFB-18 knapsack mist-blower sprayer). The comparison items included the spray deposition on the plants and run-off, uniformity and penetrability of the deposition, deposition characterization (including droplet size, number of spray deposits and the area of coverage), pesticide efficacy on wheat aphid and working efficiency. The experimental results show that the control efficacy of the UAV on wheat aphid was comparable to other spraying equipment with the working efficiency significantly higher than others. Unfortunately, UAV still have many problems on deposition uniformity and droplets penetrability.

## 2. Materials and Methods

To compare the advantage and shortcoming of the UAV with other spraying equipment, we selected three typical types of crop protection equipment including a self-propelled boom sprayer and two conventional knapsack sprayers for field spray deposition, control efficacy on wheat aphid and working efficiency tests. The spray deposition was compared from four aspects: the total amount of deposition and losses to the ground, deposition uniformity, droplets penetration in the canopy and characterization of the deposition (including droplet size, number of spray deposits and the area of coverage).

### 2.1. Spray Equipment

A battery motive 3WTXC8-5 six-rotor UAV (Henan Tianxiucai Aviation protection machinery Co., Ltd., Kaifeng, China) (Figure 1A) was used in this study. The UAV was powered by two 12,000 mAh Li-Po batteries (Shenzhen Grepow battery Co., Ltd., Shenzhen, China). The flying time was 15–20 min with full tank. The flight speed was 12.6–14.4 km/h with two rotary cup atomizer arranged on both sides. The interval of nozzles was 0.85 m and the installation angle was vertically downward. The rotate speed of the disk was 10,000 rotations per minute. The chemicals were transferred from the tank to the nozzles by a HXB600 micro liquid pump (Shanghai Hallya Electric Co., Ltd., Shanghai, China) and the flow rate was 1.24 L/min. The accuracy of the flight height and flight velocity were controlled by the well-trained operator. The flight height was 1.0 m and the effective spraying width was 4.0 m. The spray volume of one sortie was close to 10 L/ha, which was equal to two times the tank capacity.

One of the comparison systems was the 3WX-280H self-propelled boom (SPB) sprayer produced by Sino-Agri Fengmao Plant Protection Machinery Company (Beijing, China) with the tank capacity of 280 L (Figure 1B). There are 12 ISO 04 nozzles (Spraying system Co.) installed vertically on the spray boom with the same interval of 0.5 m on the 6-m boom. The spraying height of the boom was 0.5 m from the top of the canopy. The spray pressure was 4 bar and the flow rate was 18.2 L/min. The operation speed was 6.0–6.5 km/h. Under these application conditions, the spray volume was close to 300 L/ha.

**Figure 1.** Four test sprayers: (**A**) 3WTXC8-5 six-rotor unmanned aerial vehicle (UAV) sprayer; (**B**) 3WX-280H self-propelled boom (SPB) sprayer; (**C**) WFB-18 knapsack mist-blower (KMB) sprayer; and (**D**) 3WBS-16A2 electric air-pressure knapsack (EAP) sprayer.

The two other conventional sprayers were WFB-18 knapsack mist-blower (KMB) sprayer (Sino-agri Fengmao, China) (Figure 1C) and 3WBS-16A2 electric air-pressure knapsack (EAP) sprayer (Chuangxing sprayer factory, Xinxiang, China) (Figure 1D). EAP sprayer was equipped with twin hollow cone nozzles and a pressure pump provided a maximum pressure of 4 bar and a flow rate of 1.6 L/min. The tank capacity was 16 L and the length of the lance of the EAP sprayer was 81 cm. The spray swath width was close to 2.5 m. The traveling speed in the test was approximately 1.1–1.3 km/h and the spray volume under these application conditions was close to 300 L/ha. The KMB sprayer was developed to improve the spraying efficiency of air-pressure knapsack sprayer, which was equipped with a tank, a spray pipe, a nozzle and a gasoline engine. The droplets were sprayed from the nozzle, which were further atomized by the high speed air flow. The high speed air flow was produced by high speed whirling impeller driven by the gasoline engine. The flow rate in this test was 2.0 L/min with the tank capacity of 18 L. The spraying swath width was close to 6.0 m. The traveling speed was approximately 2.7–3.0 km/h and the spray volume was close to 75 L/ha. The working height of the nozzle of two conventional sprayers was 0.5 m from the top of the canopy. The spraying patterns of two knapsack were both swinging spraying.

All working parameters and spray volumes for each sprayer were established taking into account local farmers practices. Before tests, all spray equipment performed a preliminary test to calibrate the equipment to ascertain the flow rate of the nozzles. After the flow rate was ascertained, the traveling speed was also calculated to obtain the stated application rate. To achieve the velocity, the operator needed to repeat several times before undertaking each trial until the desired traveling speed was reached. Each spraying treatment was done by a well-trained applicator.

## 2.2. Experiment Design

### 2.2.1. Field Plots

The tests were conducted at the agricultural experiment station of the Chinese Academy of Agricultural Science located at Xinxiang, Henan Province, China (latitude 35°8′8″, longitude 113°46′58″) (Figure 2). The experimental farm is a trapezoidal field nearly 50 ha in area and it consists of many square fields approximately 200 m on a side (Figure 2). The tested material was "Bainong AK58" wheat in the booting-filling stage on 27 April, which was sown on 10 October, 2014. The plant spacing, plant height, leaf area of the flag leaf, and planting density were 12 cm, 84.8 ± 4.1 cm, 1927.1 $cm^2$ and $4.1 \times 10^6$ plant/ha, respectively. The wheat in the whole test area grew well and consistently.

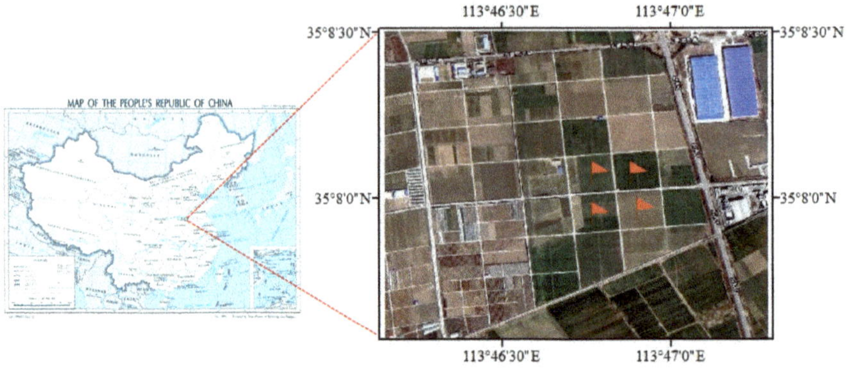


**Figure 2.** The test location and the brief overview of the studied wheat field. The experimental fields are marked with red flags.

### 2.2.2. Spray-Deposition Measurements

The experiment consists of five treatments: four kinds of spray equipment treatment and a blank control. The spray deposition, the control efficacy on wheat aphids and working efficiency were tested. The spray deposition and the control efficacy were tested in a 170 m × 190 m area (Figure 3A). In the test field, treatments were arranged within the location as a randomized complete block design with three replications, resulting in three blocks each with five plots corresponding to different treatments. Each plot was a 30 m × 50 m area. Ten-meter buffer zones [17,19] between plots were set to avoid the drift pollution (Figure 3A).

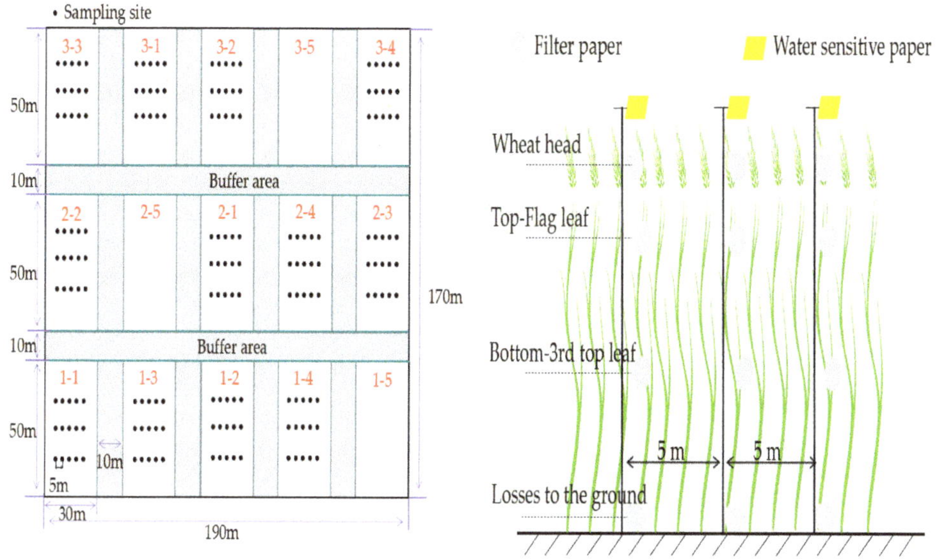


(A) (B)

**Figure 3.** (**A**) Experimental layout of each treatment; and (**B**) placement of filter papers and water-sensitive paper at each sampling position within the wheat canopy.

Prior to the application in each treatment, sample collectors were placed at each plot in five equally-spaced sample sites. Each sample site was 5 m apart and spanned a total of 20 m. The sample sites were repeated three times with an interval of 15 m between each repetition. To avoid cross contamination between plots, sampling was arranged at the center of the plots, as shown in Figure 3A.

Sample collectors at each sample site consisted of one water-sensitive paper (WSP) (25 mm × 75 mm) and four filter papers (90 mm in diameter) (Figure 3B). The WSP was used to evaluate the characteristic of deposition such as an area of coverage, number of spray deposits and droplet size. The filter papers were used to measure the deposition distribution in the canopy. The WSPs were fixed horizontally on plastic rods through double-headed clamps. The heights of these clamps were adjusted to assist in positioning at a height equivalent to the head of the wheat canopy. The filter papers were used to simulate leaves to collect foliar deposition at different heights and losses to the ground. On each sampling site, three filter papers were attached on the wheat head, flag leaf and third top leaf of one wheat plant. To evaluate the losses to ground, a filter paper was placed on the ground.

In all tests, 70% Imidacloprid·Regent Water dispersible granule (WDG) at 85.7 g a.i./ha (active ingredient) (Zhejiang Sega Science and Technology Co., Ltd.) and allure red (80% purity, purchased from Beijing Oriental Care Trading Ltd.) were added into the tank. The pesticides were prepared according to the recommended dose. Allure red was used as the tracer according to the spraying area at 450 g/ha. Allure red, a water-soluble colorant, is frequently used in these types of studies [3,9,15]. It presents high recovery rate, high photostability and low acute toxicity in different species of animals according to the Joint FAO/WHO Expert Committee on Food Additives as well as the European Union's Scientific Committee for Food.

Nearly 30 s after spraying, five randomly selected wheat plants of over ground part, WSPs and filter papers of different canopy positions at each sampling site were collected and placed in labeled plastic zip-lock bags. Five random wheat plants were combined and bagged as one sample. The sample of wheat plants was used to measure the total deposition per plant. For each plot, there were 15 wheat plant samples, 60 filter papers (4 from each sampling site) and 15 WSPs. All samples were placed in zip-lock bags along with a label describing the treatment, replication, and location information. Samples were placed into light-proof seal box immediately after collection and transported to the laboratory for analysis.

Each filter paper and the wheat sample were washed in 0.02 L and 0.2 L of distilled water in the collection bags, respectively. Samples were agitated and vibrated for 10 min to allow the dye to dissolve into the water solution. Previous tests have shown that this methodology results in near total recovery of dye deposited on samples [22]. After vibration and elution, a sample portion of the wash effluent was filtered through a 0.22 μm membrane and the filtered wash effluent was poured into a cuvette to measure the absorbance value by a UV2100 ultraviolet and visible spectrophotometer (LabTech, Co. Ltd., Beijing, China) at an absorption wavelength of 514 nm. Spray deposits were quantified by comparison with similarly determined dye concentrations from spray tank samples and the area of the respective samples. The data were expressed as a quantity of dye (μg) deposited per unit area of the sample ($cm^2$) or quantity of dye (μg) deposited per plant. Note that throughout the text whenever the term "deposition" is used, it refers to the mass of dye (which would correspond to amount of active ingredient), not the mass of total spray mix, deposited on specified sampling surface. The distribution uniformity of the spray deposition in the canopy was analyzed using the value corresponding to the coefficient of variation (CV), calculated as the quotient between the standard deviation and the average of the spray deposits on the crop. In addition, for each application, the recovery rate was calculated using Equation (1):

$$R = (D \times P/A) \times 10^6 \quad (1)$$

where R is the recovery rate (%), D is the average deposited tracer solution per plant (μg/plant), P is the plant density ($4.1 \times 10^6$ plant/ha), A is the additive amount of allure red (450 g/ha), and $10^6$ is the unit conversion factor.

In the laboratory, the WSPs were scanned at a resolution of 600 dpi with a scanner. After that, imagery software DepositScan [23] (USDA, USA) was utilized to extract droplet deposits in the digital image and analyzed the droplet size, number of spray deposits and the area of coverage. The Volume Median Diameter (VMD) is a key index for reflecting the droplet size, which was used in this study.

The climatic conditions were recorded using a Kestrel 5500 digital meteorograph (Loftopia, LLC, USA), which recorded temperatures of 18.3–22.4 °C, a relative humidity of 46.3–53.7% and wind velocities of 3.6–10.8 km/h.

#### 2.2.3. Investigation of Control Efficacy

To analyze the field efficacy of different spraying equipment with different spray deposition characteristics, we selected wheat aphid as the field test target according to the occurrence of diseases and pests. The wheat aphid was investigated and recorded four times according to pesticide field efficacy test criteria. Before the pesticide application on 27 April, the base number of the wheat aphids per hundred plants was more than 500, which meet the control criteria. The assessment was made by sampling five locations per plot on wheat aphid. The aphid number of 10 strains of wheat per location was investigated before spraying and the wheats were marked with a red string. After application on Days 1, 3, and 7, the number of aphid in the same location and plant was investigated again. The overall control effect against wheat aphid was calculated without regard to the types or instars of the wheat aphid. The mortality and control effects were obtained based on the population numbers of live insects in each zone before and after spraying. The control effect was calculated according to Equations (2) and (3):

$$\text{Mortality (\%)} = (\text{The number of pests before application} - \text{The number of pests after application})/\text{The number of pests before application} \times 100 \quad (2)$$

$$\text{Control effect (\%)} = [\text{Observed mortality (\%)} - \text{Control mortality (\%)}]/[100 - \text{Control mortality (\%)}] \times 100 \quad (3)$$

#### 2.2.4. Working Efficiency Test

To better reflect the work efficiency (ha/h) of different equipment, the spraying area and the total spraying time of five filled tanks of chemical for each sprayer were recorded. The spray equipment was operated by experienced operators. The preparation time of mixing chemicals and other preparations, such as replacing the batteries or adding chemicals, did not count in the spraying time because they are greatly influenced by the cooperation and organization of the large-scale application. The spraying parameters are described in Section 2.2.1.

#### 2.2.5. Statistical Analysis

Before the significant difference analysis, the percentage of the area of coverage on WSPs and the mortality of wheat aphids were transformed using $y = \arcsin\sqrt{X/100}$. The total deposition and losses to the ground, deposition on different canopies, number of spray deposits and droplet size were $\log(x + 1)$ transformed to stabilize wide variances and meet normality assumptions. After transformation, the data were analyzed for normality using the Kolmogorov–Smirnov test and for equal variances across the treatments and repeats using Levene's test ($p < 0.05$). The significant difference for the transformed data was conducted using analysis of variance (ANOVA) by Duncan's test at a significance level of 95% with SPSS v22.0 (SPSS Inc, an IBM Company, Chicago, IL, USA).

## 3. Results and Discussion

### 3.1. Spray Deposition

#### 3.1.1. Total Deposition on the Crops and Losses to the Ground

The samples of the wheat plant were used to measure the total spray deposition per plant. The filter papers were used to measure the spray deposition in different wheat canopy and losses to the ground, and the recovery rate was calculated from the amount of the total deposition and additive. The total deposition of the UAV was not significantly different from the other sprayers (Table 1). Among the four sprayers, the EAP sprayer achieved the highest total deposition and the KMB sprayer achieved the lowest (Table 1). Compared with the SPB and KMB sprayer, the UAV and EAP sprayer had significantly higher recovery rates. The dose transfer process studied showed the transfer of pesticide from the spray tank to the target organism and in this process, losses of drift and run-off had great influence on deposition [24]. In this study, the SPB and KMB sprayer had relatively lower depositions and, correspondingly, those two sprayers had the higher run-offs, which were 0.39 and 0.50 μg/cm$^2$ (Table 1). Similar to other studies [25,26], it could be that high-volume spraying easily leads to run-off. In Table 1, Columns 4 and 5, compared with other sprayers, UAV had the lowest losses to the ground.

**Table 1.** The average (Avg.) and coefficient of variation (CV) of the total deposition on the plants, losses to the ground and the recovery rate of four sprayers.

| Spray Equipment | Total Deposition | | Losses to the Ground | | Recovery Rate |
|---|---|---|---|---|---|
| | Avg. (μg/Plant) | CV (%) | Avg. (μg/cm$^2$) | CV (%) | Avg. (%) |
| UAV sprayer | 76.8 a | 87.2 | 0.13 b | 78.2 | 70.0 a |
| SPB sprayer | 68.7 a | 32.1 | 0.39 a | 39.0 | 62.7 a |
| EAP Sprayer | 84.8 a | 84.4 | 0.14 b | 118.2 | 77.3 a |
| KMB Sprayer | 61.9 a | 81.2 | 0.50 a | 97.1 | 56.5 a |

Note: Values followed by the same letter in the column do not differ statistically ($p < 0.05$; Duncan's Test).

#### 3.1.2. The Uniformity of the Deposition

In addition to the total deposition, the uniformity of the deposition is also very important for controlling pests and diseases. The uniformity of the SPB sprayer was better than the others. The CV of the deposition was only 32.1%, which was significantly lower than the others (Table 1). This means it had a better deposition uniformity. However, the CVs of the deposition in this study was much greater than the value of Chinese National Standard requirement (10%) [27]. This may be due to the different measuring methods. In the study by Yang et al. [28], the CV of the deposition distribution ranged from 5% to 10% with different spray pressures and nozzle heights, which are far lower than the result of our study. This is because the Teejet pattern check was used by Yang et al., which is not influenced by the environment and the crop canopy. By comparison, the results in our study were a reflection of the actual uniformity of the deposition distribution on the crops.

Compared with the SPB sprayer, the UAV sprayer had a significant lower uniformity of the deposition distribution with the CV of 87.2% (Table 1). This result was larger than the Civil Aviation of China General Aviation Operation Quality and Technology Standard for ultra-low volume spraying, which is 60% [8]. The uniformity of the deposition distribution of the UAV was influenced by many factors, such as the types [9], the flight accuracy, the flight parameters [7], the spraying system, the biased downwash wind [18] and the meteorological condition. Precise flight route control and auto navigation are essential for chemicals application with improving the deposition distribution uniformity [14]. Xue et al. [8] developed an automatic navigation unmanned spraying system for the N-3 unmanned helicopter, which can significantly improve the deposition uniformity. The flight parameters also have a great influence on the deposition distribution uniformity. Various UAV sprayers under different flight parameters have been tested to find optimal spraying heights and speeds [7,10,18]. Qin et al. [7] found that the flight parameters affect not only the distribution uniformity on rice canopy but also the control efficacy on wheat hoppers. Bae and Koo [29] pointed out that most agricultural helicopters exhibit biased downwash, resulting in an uneven spray pattern. To address this problem, a roll-balanced agricultural helicopters with an elevated-pylon tail rotor system was developed. In their study, the uniformity of the spray patterns and area of coverage was improved.

The CVs of depositions of the EAP and KMB sprayer were 84.4% and 81.2%, respectively, which indicate a nonuniform deposition (Table 1), mainly due to the manual operation of the sprayer. The deposition uniformity was worse because it depended on the stability of the traveling speed along the spraying route and on the regularity of the arm movement of the operators.

### 3.1.3. Droplets Penetrability

In the test, filter papers were used to measure the tracer deposition on the different canopies of wheat. An analysis of the deposition on the wheat heads showed that the deposition of the low volume sprayer of the UAV and KMB sprayer were significantly higher than the SPB and EAP sprayer (Figure 4). The greatest deposition was achieved with the conventional KMB sprayer with 1.46 $\mu g/cm^2$, which was 18.7%, 80.2% and 111.6% more than the UAV, EAP and SPB sprayer, respectively (Figure 4). From the droplet size test results presented in Section 3.1.4, the VMD of droplets of KMB and UAV sprayer were lower than the other two sprayers. The main reason for higher wheat head deposition of UAV and KMB sprayer may be that fine droplets were better retained in the upper canopy [30].

The deposition of four sprayers on the flag leaf (top canopy) ranged from 0.57 to 1.02 $\mu g/cm^2$ (Figure 4). The UAV sprayer had the highest deposition on the flag leaf, which was 41.7–78.9% higher than the other sprayers. However, influenced by the great variability, the depositions on the flag leaf were not significantly different. In the study by Zhu et al. [31], the spray deposits decreased dramatically from the top to the bottom of the canopies and also tended to linearly decrease as the leaf area index increase. Due to the overlap and the block of the blades at the later growth stage of the wheat, the depositions on the lower parts were far less than top parts. The depositions of different sprayers were influenced by many factors, such as spraying pressure, spraying height and spraying pattern. The deposition of the UAV sprayer on the third top leaf (bottom canopy) was only 0.26 $\mu g/cm^2$, which was not significantly different from the other two conventional sprayers. However, it was 50.0% compared to the SPB sprayer and the deposition of SPB sprayer on the bottom was 0.52 $\mu g/cm^2$.

To improve the adhesive rate of solution and control efficacy, it is crucial to enhance the droplets penetrability and obtain a homogeneous deposition distribution [31], especially since many pests and diseases occur on the bottom of plants. Although many studies [18,19] have proven that the downwash airstream generated by the UAV is conducive to the disturbance of leaves and droplets penetration, the results from our study proved that the penetration of the droplets of UAV is still worse than boom sprayer. This result is similar to the results of Wang et al. [9] and Qin et al. [7]. There are many factors accounting for the poor droplet penetrability. One of the most important reasons is that the droplets were sprayed by the rotary cup atomizers, which lack the downward kinetic energy in comparison with hydraulic nozzles. Wang et al. [9] compared four kinds of UAV sprayer and found the flight height and the flight speed had a pronounced impact on droplet penetrability. In their study, the penetrability of the droplets was inversely proportional to the flight speed and the flight height. The flight parameter had an effect on the downwash flow, which further affected the droplets penetrability. Chen et al. [18] testified that the vertically downward wind had a significant effect on the penetrability. To improve the droplet penetrability of the UAV sprayer, the fly height and speed should be lower, thus ensuring operating efficiency, and suitable nozzles should be chosen to improve the droplets penetrability.

The deposition uniformity on different parts of the canopy was consistent with the total deposition. The SPB sprayer had the best deposition uniformity (CVs < 56%) on different canopies and other sprayers had lower uniformity (CVs > 70%).

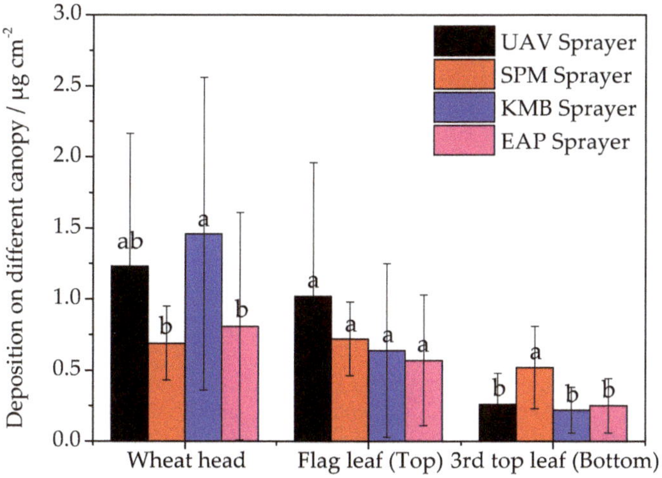


**Figure 4.** Deposition of four sprayers on the different canopies of wheat. Bars with different letters are significant different, Duncan's test, $p < 0.05$.

### 3.1.4. Characterization of Deposition

In the test, the WSPs were used to evaluate the characteristics of the deposition, which included droplet size, number of spray deposits and the area of coverage. Influenced by the spray volume and spraying system, the deposition characteristics of different sprayers were quite different (Figure 5). The spray volume of SPB sprayer approached 300 L/ha, which was identical to EAP sprayer belonging to the large volume application. Furthermore, the nozzles of SPB sprayer are similar to EAP sprayer, which are hydraulic nozzles. From the test results of WSPs, the VMD of the droplets of the SPB and EAP sprayer were 272.3 and 254.1 μm, respectively. They were not significantly different from each other (Figure 5). Although the VMD of droplets and the spray volume of SPB sprayer were similar to EAP sprayer, the area of coverage and the number of spray deposits of SPB sprayer were 75.9% and 73.9% greater than the EAP sprayer with no significant difference (Figure 5). From the CVs results of the total deposition, the greater area of coverage and number of spray deposits of SPB sprayer may be due to the better deposition uniformity. As a conventional knapsack sprayer, the spray volume of KMB sprayer was quite lower than the EAP sprayer of 75 L/ha. With the air assistance, the VMD of droplets of KMB sprayer was significantly lower than the EAP sprayer, which was only 154.7 μm. Results from previous studies had proved that the area of coverage is proportional to the spray volume [32]. The area of coverage of KMB sprayer was lower than the EAP sprayer with no significant difference (Figure 5). However, with finer droplets, the number of spray deposits of the KMB sprayer was quite similar to the EAP sprayer (Figure 5).

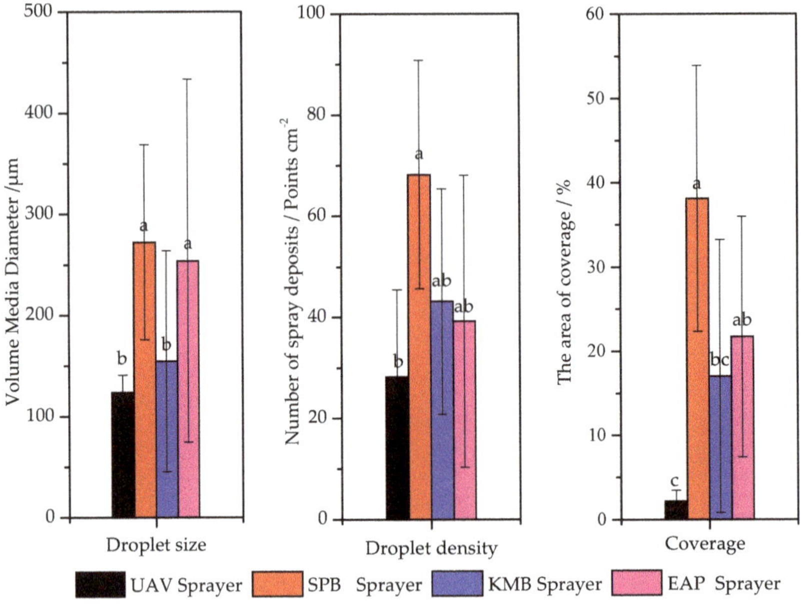


**Figure 5.** Characterization of the deposition including droplet size, number of spray deposits and the area of coverage.

The spray volume of UAV sprayer was 5 L/ha, which was lower than the other three sprayers. With the unique atomization method, the droplets were smashed by the high rotational speed of the rotary cup (10,000 rpm). The VMD of the droplets was 124.0 μm, which was finer than the three other sprayers (Figure 5). Another characteristic of this nozzle, which differed from hydraulic nozzle, is that the width of the droplet spectrum of this nozzle was narrower and the droplet size was more uniform with high rotational speed [33]. The result in this study verified that the CV of the VMD of the droplets was only 17.1%, which was far less than the other sprayers (CV > 35%) (Figure 5). Because of the lower spray volume, the area of coverage of the UAV was 2.2%, which were quite lower than the other sprayers and was only 5.8% to 12.8% of the other three sprayers. The number of spray deposits of the UAV was 28.2 points/cm$^2$, which was also significantly lower than others (Figure 5). This may go against the control of pests or diseases, especially when the contact pesticides were applied. Therefore, the focus of further studies will be improving the spreading coefficient of the droplets and increasing the area of the coverage, such as adding adjuvant in the tank [34,35], or changing into electrostatic-charged sprays [36,37].

Spraying systems had a great effect on the characteristics of deposition, including the area of coverage, number of spray deposits and droplet size. It is difficult to judge the application quality by a single index. Although the lower spray volume of UAV could lead to lower area of coverage and fewer deposits, the dose applied per area was not significant lower than other sprayers, because of the higher concentration of each droplets. Syngenta Crop Protection AG (Basel, Switzerland) recommends that the satisfactory results can be obtained by a number of spray deposits of at least 20–30 points/$cm^2$ for insecticide or pre-emergence herbicide applications, 30–40 points/$cm^2$ for contact post-emergence herbicide applications and 50–70 points/$cm^2$ for fungicide applications [23]. The number of spray deposits only needs to reach a certain threshold to achieve a good control efficacy [38].

### 3.2. Control of Wheat Aphids

The control efficacy of the four sprayers applying 70% Imidacloprid·Regent WDG on wheat aphids are indicated in Figure 6. Although the spray deposition characteristics of the four sprayers were significantly different from each other, the differences of control efficacy were not remarkable. The control efficacies on Day 7 after application were all beyond 70%. From the comparison of four sprayers, it was found that the SPB and KMB sprayers achieved the best control efficacy, while the efficacies were intermediate for EAP and UAV sprayer. Deposit structure plays a major role in toxin efficacy [24,38]. According to the results of the deposition, the SPB sprayer had the best deposition uniformity and penetrability, which benefit the control of wheat aphids, especially since wheat aphids tend to favor the lower portions of plants. The larger area of coverage and a larger number of spray deposits increased the odds of interaction between active ingredients and pests.

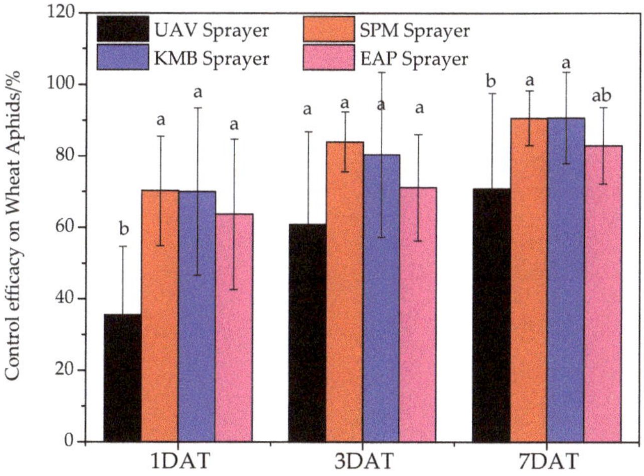


**Figure 6.** Control efficacy (%) of four sprayers against wheat aphids in the field tests at Days 1, 3, 7 after treatment. DAT, days after treatment.

Because UAV sprayers are still in the initial development stage in China, there are still many problems on the spraying system and working parameter. These problems lead to unevenness deposition and poor penetrability. The lower area of coverage and fewer spray deposits with the nonuniform deposition reduced the chance of aphids to be exposed to the insecticides. The control efficacy of UAV sprayer on Day 1 after application was significantly lower than other sprayers, which was only equal to 50.5% of the SPB sprayer. However, in time, the control efficacy increased. The reason could be that, with the movement of the aphids and the systemic action of the active ingredients, the chance of wheat aphids being exposed to the active ingredients increased. On Day 3 after application, the control efficacy increased to 60.9% and, on Day 7 after application, the control efficacy increased to 70.9%. Although it was also significantly lower than the SPB and KMB sprayers, it was still an acceptable result for farmers, especially in complex small plots or rice fields, where the SPB sprayer is hard to work and the KMB and EAP sprayer have a low work efficiency. Although the UAV sprayer, especially for the electrical multi-rotor UAV, used in the field, is an innovation, the low volume application is not a new technology. Many studies had been conducted to evaluate the feasibility of low volume application for pests and diseases control. In the laboratory, Maczuga and Mierzejewski [39] found that several spray deposits of 5–10 points/cm$^2$ on foliage after spraying were effective (90% mortality) against second and third instars. Washington et al. [40] investigated the effect of fungicide spray number of spray deposits (points/cm$^2$), droplet size and proximity of the spray deposits to fungal spores on the banana leaf surface. The test results suggest that the inhibition zones of two contact fungicides on the leaf surface extend beyond the visible edge of the spray droplet deposit and a mean droplet deposit density of 30 points/cm$^2$ can inhibit the germination of the ascospore below 1%. Latheef et al. [36] evaluated the efficacy of aerial electrostatic-charged sprays for season-long control of sweet potato whiteflies, and they concluded that the control efficacy of electrostatic-charged sprays with spray volume of 4.68 L/ha was comparable with those on cotton treated with conventional applications of 46.8 L/ha.

### 3.3. Working Efficiency

Working efficiency is also an important evaluation index for equipment selection. The UAV sprayer used in the test belongs to semi-autonomous control, which is the most common in China. By recording the spraying time of five sorties with the filled tank, the average spraying time of each sortie of the UAV sprayer was calculated as 0.095 h (Table 2). The average spraying area of each sortie was 0.39 ha with the spray volume of 5 L (Table 2). Calculated from the spraying time and area, the work efficiency of the UAV sprayer was 4.11 ha/h (Table 2). The work efficiency test results were consistent with Wang et al. [9], who reported that the working efficiency of UAV was at 13.4–18.0 hectare per 8 h, i.e., 1.68–2.25 ha/h. In his test, the operation items included the time of preparation, route planning, failure maintenance, and ground service, which accounted for 50% of the whole process. With micro-electronic technology development, the multi-sensor data fusion and real-time kinematic positioning technology and product will be applied in agriculture UAV, with which the UAV can achieve fully autonomous flight, significantly improving the working efficiency [41].

**Table 2.** Working efficiency (ha/h) of four sprayers.

| Sprayer | Tank Capacity (L) | Spray Area (Means ± Standard Error, ha) | Spray Time (Means ± Standard Error, h) | Working Efficiency (ha/h) |
|---|---|---|---|---|
| UAV sprayer | 5 | 0.39 ± 0.04 | 0.095 ± 0.01 | 4.11 |
| SPB sprayer | 280 | 0.93 ± 0.06 | 0.39 ± 0.03 | 2.38 |
| KMB Sprayer | 18 | 0.22 ± 0.02 | 0.14 ± 0.01 | 1.57 |
| EAP Sprayer | 16 | 0.039 ± 0.004 | 0.19 ± 0.01 | 0.21 |

Note: Spray area in the table means the area per spray with full tank. Spray time in the table means the time per spray with full tank.

Compared with the UAV sprayer, the SPB sprayer had a relatively lower working efficiency. The average spraying time was 0.39 h and the spraying area was 0.93 ha with the tank capacity of 280 L (Table 2). Although there are many other kinds of boom sprayers with a much longer boom, the complexity of the terrain and the farm size limit the usage of larger boom sprayers in China. The average farm size in China is amongst the smallest in the world of 0.67 ha and more than half (57.5%) of farms are small (<2 ha) [1]. Thus, this kind of small boom sprayer is widespread with the working efficacy of 2.38 ha/h (Table 2).

From the test results, the working efficiency of two conventional knapsack sprayers were 1.57 and 0.21 ha/h, respectively (Table 2). Conventional knapsack sprayers need to be carried on the back, which easily leads to exhausting and lower work efficiency. Especially for the EAP sprayer, the working efficacy was only 5.1% of the UAV sprayer. Although these knapsack sprayers also hold very large market share (88%, from China Agriculture Yearbook Editorial Committee 2016) in China, with the development of technology and the growth of farming size, these lower working efficiency sprayers will be tapered and the studies on high efficiency sprayers will be necessary.

## 4. Conclusions

In this study, four typical sprayers were used for pesticide application in the wheat field. The total deposition on the plants and losses to the ground, the uniformity of the deposition, droplets penetrability, characteristics of the deposition, control efficacy on wheat aphids and working efficiency were compared in this research. The conclusions are shown as follows:

1) The total deposition of the UAV sprayer was not significantly different from the other three sprayers, but the losses to the ground were the lowest.
2) The UAV sprayer had a poor deposition uniformity (CV = 87.2%) and droplets penetrability (0.26 $\mu g/cm^2$ on the third top leaf), which need to further improve in the future. By comparison, the SPB sprayer has the best deposition uniformity (CV = 32.1%) and droplets penetrability (0.52 $\mu g/cm^2$ on the third top leaf).
3) The area of coverage, number of spray deposits and droplet size varied with spray volume and the nozzle type of the different sprayers. Compared with other sprayers, the deposition characterizations of the UAV sprayer were a lower area of coverage (2.2%), a lfewer spray deposits (28.2 points/$cm^2$), finer droplet size (VMD = 124.0 $\mu m$), higher concentration and lower spray volume.
4) Although the spray deposition characterizations of the UAV sprayer were different from the three other sprayers, the control efficacy of applying 70% Imidacloprid·Regent WDG on wheat aphids was comparable with other sprayers. The control efficacy of UAV sprayer on Day 7 after the application was 70.9%.
5) The working efficiency of the UAV sprayer was 4.11 ha/h, which was 1.7, 2.6, and 20.0 times those of SPB, KMB and EAP sprayer, respectively. This is the greatest advantage of the UAV sprayer.

The experiment demonstrated the feasibility and high efficiency of the UAV sprayer. The deposition uniformity and the droplets penetrability of the UAV also need to be improved. Due to the lower coverage and poor deposition uniformity, the effective measures, such as optimizing the spraying system or adding adjuvant in the tank to improve deposition uniformity and penetrability would be needed in the future.

**Author Contributions:** G.W., Y.L., H.Y., H.Q., P.C., F.O. and Y.H. conceived the idea of the experiment. G.W. performed the experiments and analyzed the data. G.W. and Y.H. wrote and revised the paper.

**Funding:** The study was funded by The National Key Research and Development Plan: High Efficient Ground and Aerial Spraying Technology and Intelligent Equipment (2016YFD0200700), Science and Technology Planning Project of Guangdong Province (2017B010117010), The leading talents of Guangdong province program (2016LJ06G689), and Key science and technology plan of Guangdong Province (2017B010116003).

**Conflicts of Interest:** The authors declare no conflict of interest.

## References

1. Yang, S.; Yang, X.; Mo, J. The application of unmanned aircraft systems to plant protection in China. *Precis. Agric.* **2018**, *19*, 278–292. [CrossRef]
2. Zhang, X.; Zhao, W.; Jing, R.; Wheeler, K.; Smith, G.A.; Stallones, L.; Xiang, H. Work-related pesticide poisoning among farmers in two villages of Southern China: A cross-sectional survey. *BMC Public Health* **2011**, *11*, 429. [CrossRef] [PubMed]
3. Cao, L.; Cao, C.; Wang, Y.; Li, X.; Zhou, Z.; Li, F.; Yan, X.; Huang, Q. Visual determination of potential dermal and inhalation exposure using allura red as an environmentally friendly pesticide surrogate. *ACS Sustain. Chem. Eng.* **2017**, *5*, 3882–3889. [CrossRef]
4. Ghugare, B.D.; Adhaoo, S.H.; Gite, L.P.; Pandya, A.C.; Patel, S.L. Ergonomics evaluation of a lever-operated knapsack sprayer. *Appl. Ergon.* **1991**, *22*, 241–250. [CrossRef]
5. Sánchez-Hermosilla, J.; Rincón, V.J.; Páez, F.; Fernández, M. Comparative spray deposits by manually pulled trolley sprayer and a spray gun in greenhouse tomato crops. *Crop Prot.* **2012**, *31*, 119–124. [CrossRef]

6. Xiongkui, H.; Bonds, J.; Herbst, A.; Langenakens, J. Recent development of unmanned aerial vehicle for plant protection in East Asia. *Int. J. Agric. Biol. Eng.* **2017**, *10*, 18–30. [CrossRef]
7. Qin, W.C.; Qiu, B.J.; Xue, X.Y.; Chen, C.; Xu, Z.F.; Zhou, Q.Q. Droplet deposition and control effect of insecticides sprayed with an unmanned aerial vehicle against plant hoppers. *Crop Prot.* **2016**, *85*, 79–88. [CrossRef]
8. Xue, X.; Lan, Y.; Sun, Z.; Chang, C.; Hoffmann, W.C. Develop an unmanned aerial vehicle based automatic aerial spraying system. *Comput. Electron. Agric.* **2016**, *128*, 58–66. [CrossRef]
9. Wang, S.; Song, J.; He, X.; Song, L.; Wang, X.; Wang, C.; Wang, Z.; Ling, Y. Performances evaluation of four typical unmanned aerial vehicles used for pesticide application in China. *Int. J. Agric. Biol. Eng.* **2017**, *10*, 22–31. [CrossRef]
10. Zhang, P.; Deng, L.; Lyu, Q.; He, S.; Yi, S.; Liu, Y.; Yu, Y.; Pan, H. Effects of citrus tree-shape and spraying height of small unmanned aerial vehicle on droplet distribution. *Int. J. Agric. Biol. Eng.* **2016**, *9*, 45–52. [CrossRef]
11. Giles, D.; Billing, R. Deployment and performance of an unmanned aerial vehicle for spraying of specialty crops. In Proceedings of the International Conference of Agricultural Engineering, Zurich, Switzerland, 6–10 July 2014.
12. Law, S.E. Embedded-electrode electrostatic-induction spray-charging nozzle: Theoretical and engineering design. *Trans. ASAE* **1978**, *21*, 1096–1104. [CrossRef]
13. Zhang, C.; Kovacs, J.M. The application of small unmanned aerial systems for precision agriculture: A review. *Precis. Agric.* **2012**, *13*, 693–712. [CrossRef]
14. Huang, Y.; Hoffmann, W.C.; Lan, Y.; Wu, W.; Fritz, B.K. Development of a spray system for an unmanned aerial vehicle platform. *Appl. Eng. Agric.* **2009**, *25*, 803–809. [CrossRef]
15. Fritz, B. Meteorological effects on deposition and drift of aerially applied sprays. *Trans. ASAE* **2006**, *49*, 1295–1301. [CrossRef]
16. Bird, S.L.; Perry, S.G.; Ray, S.L.; Teske, M.E. Evaluation of the AgDISP aerial spray algorithms in the AgDRIFT model. *Environ. Toxicol. Chem.* **2002**, *21*, 672–681. [CrossRef]
17. Wang, X.; He, X.; Wang, C.; Wang, Z.; Li, L.; Wang, S.; Jane, B.; Andreas, H.; Wang, Z. Spray drift characteristics of fuel powered single-rotor UAV for plant protection. *Trans. ASAE* **2017**, *33*, 117–123, (In Chinese with English abstract). [CrossRef]
18. Chen, S.; Lan, Y.; Li, J.; Zhou, Z.; Liu, A.; Mao, Y. Effect of wind field below unmanned helicopter on droplet deposition distribution of aerial spraying. *Int. J. Agric. Biol. Eng.* **2017**, *10*, 67–77. [CrossRef]
19. Xue, X.; Tu, K.; Qin, W.; Lan, Y.; Zhang, H. Drift and deposition of ultra-low altitude and low volume application in paddy field. *Int. J. Agric. Biol. Eng.* **2014**, *7*, 23–28. [CrossRef]
20. Zhao, H.; Xie, C.; Liu, F.; He, X.; Zhang, J.; Song, J. Effects of sprayers and nozzles on spray drift and terminal residues of imidacloprid on wheat. *Crop Prot.* **2014**, *60*, 78–82. [CrossRef]
21. Wolters, A.; Linnemann, V.; Zande, J.C.v.d.; Vereecken, H. Field experiment on spray drift: Deposition and airborne drift during application to a winter wheat crop. *Sci. Total Environ.* **2008**, *405*, 269–277. [CrossRef] [PubMed]
22. Qiu, Z.; Yuan, H.; Lou, S.; Ji, M.; Yu, J.; Song, X. The research of water soluble dyes of Allura Red and Ponceau-G as tracers for determing pesticide spray distribution. *Agrochemicals* **2007**, *46*, 323–325. (In Chinese with English abstract) [CrossRef]
23. Zhu, H.; Salyani, M.; Fox, R.D. A portable scanning system for evaluation of spray deposit distribution. *Comput. Electron. Agric.* **2011**, *76*, 38–43. [CrossRef]
24. Ebert, T.A.; Taylor, R.A.J.; Downer, R.A.; Hall, F.R. Deposit structure and efficacy of pesticide application. 2: Trichoplusia ni control on cabbage with fipronil. *Pestic. Sci.* **1999**, *55*, 793–798. [CrossRef]
25. Rincón, V.J.; Sánchez-Hermosilla, J.; Páez, F.; Pérez-Alonso, J.; Callejón, Á.J. Assessment of the influence of working pressure and application rate on pesticide spray application with a hand-held spray gun on greenhouse pepper crops. *Crop Prot.* **2017**, *96*, 7–13. [CrossRef]
26. Sánchez-Hermosilla, J.; Rincón, V.J.; Páez, F.; Agüera, F.; Carvajal, F. Field evaluation of a self-propelled sprayer and effects of the application rate on spray deposition and losses to the ground in greenhouse tomato crops. *Pest Manag. Sci.* **2011**, *67*, 942–947. [CrossRef] [PubMed]
27. *Agricultural and Forestry Machinery–Inspection of Sprayers in Use–Part 2: Horizontal Boom Sprayers*; ISO/TC 23/SC 6. ISO 16122-2:2015; International Organization for Standardization (ISO): Geneva, Switzerland, 2015.

28. Yang, D.B.; Zhang, L.N.; Yan, X.J.; Wang, Z.Y.; Yuan, H.Z. Effects of droplet distribution on insecticide toxicity to asian corn borers (Ostrinia furnaealis) and spiders (Xysticus ephippiatus). *J. Integr. Agric.* **2014**, *13*, 124–133. [CrossRef]
29. Bae, Y.; Koo, Y.M. Flight attitudes and spray patterns of a roll-balanced agricultural unmanned helicopter. *Appl. Eng. Agric.* **2013**, *29*, 675–682. [CrossRef]
30. Hislop, E.C.; Western, N.M.; Butler, R. Experimental air-assisted spraying of a maturing cereal crop under controlled conditions. *Crop Prot.* **1995**, *14*, 19–26. [CrossRef]
31. Zhu, H.; Dorner, J.W.; Rowland, D.L.; Derksen, R.C.; Ozkan, H.E. Spray penetration into peanut canopies with hydraulic nozzle tips. *Biosyst. Eng.* **2004**, *87*, 275–283. [CrossRef]
32. Ferguson, J.C.; Chechetto, R.G.; Hewitt, A.J.; Chauhan, B.S.; Adkins, S.W.; Kruger, G.R.; O'Donnell, C.C. Assessing the deposition and canopy penetration of nozzles with different spray qualities in an oat (*Avena sativa* L.) canopy. *Crop Prot.* **2016**, *81*, 14–19. [CrossRef]
33. Craig, I.P.; Hewitt, A.; Terry, H. Rotary atomiser design requirements for optimum pesticide application efficiency. *Crop Prot.* **2014**, *66*, 34–39. [CrossRef]
34. Xu, L.; Zhu, H.; Ozkan, H.E.; Bagley, W.E.; Krause, C.R. Droplet evaporation and spread on waxy and hairy leaves associated with type and concentration of adjuvants. *Pest Manag. Sci.* **2011**, *67*, 842–851. [CrossRef] [PubMed]
35. Van Zyl, S.A.; Brink, J.-C.; Calitz, F.J.; Fourie, P.H. Effects of adjuvants on deposition efficiency of fenhexamid sprays applied to Chardonnay grapevine foliage. *Crop Prot.* **2010**, *29*, 843–852. [CrossRef]
36. Latheef, M.A.; Carlton, J.B.; Kirk, I.W.; Hoffmann, W.C. Aerial electrostatic-charged sprays for deposition and efficacy against sweet potato whitefly (*Bemisia tabaci*) on cotton. *Pest Manag. Sci.* **2009**, *65*, 744–752. [CrossRef] [PubMed]
37. Yanliang, Z.; Qi, L.; Wei, Z. Design and test of a six-rotor unmanned aerial vehicle (UAV) electrostatic spraying system for crop protection. *Int. J. Agric. Biol. Eng.* **2017**, *10*, 68–76. [CrossRef]
38. Ebert, T.A.; Taylor, R.A.J.; Downer, R.A.; Hall, F.R. Deposit structure and efficacy of pesticide application. 1: Interactions between deposit size, toxicant concentration and deposit number. *Pestic. Sci.* **1999**, *55*, 783–792. [CrossRef]
39. Maczuga, S.A.; Mierzejewski, K.J. Droplet size and density effects of bacillus thuringiensis kurstaki on gypsy moth (Lepidoptera: Lymantriidae) Larvae. *J. Econ. Entomol.* **1995**, *88*, 1376–1379. [CrossRef]
40. Washington, J.R. Relationship between the spray droplet density of two protectant fungicides and the germination of mycosphaerella fijiensis ascospores on banana leaf surfaces. *Pest Manag. Sci.* **1997**, *50*, 233–239. [CrossRef]
41. Lan, Y.; Chen, S.; Fritz, B.K. Current status and future trends of precision agricultural aviation technologies. *Int. J. Agric. Biol. Eng.* **2017**, *10*, 1–17. [CrossRef]

*applied sciences*

MDPI

*Article*

# Design and Experiment of a Variable Spray System for Unmanned Aerial Vehicles Based on PID and PWM Control

**Sheng Wen [1,2], Quanyong Zhang [2,3], Jizhong Deng [2,3,*], Yubin Lan [2,3], Xuanchun Yin [2,3] and Jian Shan [2,3]**

1 Engineering Foundation Teaching and Training Center, South China Agricultural University, Guangzhou 510642, China; vincen@scau.edu.cn
2 National Center for International Collaboration Research on Precision Agriculture Aviation Pesticides Spraying Technology, Guangzhou 510642, China; qy_zhang@stu.scau.edu.cn (Q.Z.); ylan@scau.edu.cn (Y.L.); xc_yin@scau.edu.cn (X.Y.); shanj@stu.scau.edu.cn (J.S.)
3 Engineering College, South China Agricultural University, Guangzhou 510642, China
* Correspondence: jz-deng@scau.edu.cn

Received: 12 October 2018; Accepted: 30 November 2018; Published: 3 December 2018

**Abstract:** Unmanned aerial vehicle (UAV) variable-rate spraying technology, as the development direction of aviation for plant protection in the future, has been developed rapidly in recent years. In the actual agricultural production, the severity of plant diseases and insect pests varies in different locations. In order to reduce the waste of pesticides, pesticides should be applied according to the severity of pests, insects and weeds. On the basis of explaining the plant diseases and insect pests map in the target area, a pulse width modulation variable spray system is designed. Moreover, the STMicroelectronics-32 (STM32) chip is invoked as the core of the control system. The system combines with sensor technology to get the prescription value through real-time interpretation of prescription diagram in operation. Then, a pulse square wave with variable duty cycles is generated to adjust the flow rate. A closed-loop Proportional-Integral-Derivative (PID) control algorithm is used to shorten the time of system reaching steady state. The results indicate that the deviation between volume and target traffic is stable, which is within 2.16%. When the duty cycle of the square wave is within the range of 40% to 100%, the flow range of the single nozzle varies from 0.16 L/min to 0.54 L/min. Variable spray operation under different spray requirements is achieved. The outdoor tests of variable spray system show that the variable spray system can adjust the flow rapidly according to the prescription value set in the prescription map. The proportion of actual droplet deposition and deposition density in the operation unit is consistent with the prescription value, which proves the effectiveness of the designed variable spray system.

**Keywords:** UAV; variable spray; prescription map translation; PID algorithm

---

## 1. Introduction

Pests and diseases of crops are a major factor affecting the yield and quality of crops, and chemical pesticides are the main means for their prevention and control. The pesticide application method currently used in China is mainly based on uniform spraying. Therefore, the utilization rate of pesticides and herbicides is still low [1]. According to statistics, as of 2017, the pesticide utilization rate in china was only 36.6%, which was lower than the level of 50% in developed countries [2–5]. The extensive use of pesticides directly endangers the ecological environment and human health. Therefore, pesticide reduction and efficiency have been realized worldwide. In the field of plant protection, the variable spray technology can be applied on demand. It has definite prospects and potentialities in improving the utilization rate of pesticides and reducing pesticide residues [6]. Chen et al. [7] developed a field information processing system based on a Beidou positioning embedded vehicle variable sprayer. The system was able to complete the spray operation according to the generated job prescription map. However, the performance of field information processing system was not completely verified by the field testing. Perez-Ruiz et al. [8] used a geospatial prescription map prepared for olive trees, along with Real-Time Kinematic-Global Positioning System (RTK-GPS)-based position information to control the spray rate. This system only implements variable spray based on tree shape, and does not combine the degree of disease and pest with tree shape. Qiu et al. [9] used a variable-pressure spray system that controls the flow rate with an electric control valve. The step responses of the five target flows were measured experimentally. The results showed that the rise time, peak time and overshoot of the nonlinear system have amplitude correlation. However, the way of regulating flow by an electronic control valve is only applicable to ground machinery. Compared with the high-speed flight of plant protection unmanned aerial vehicles (UAV), the response time of the electronic control valve has a greater impact on the system. Gonzalez et al. [10] designed a nonlinear variable spray system based on pressure regulation. By establishing the transfer function of the open-loop system, the nonlinear control of the variable spray system was realized. However, for the low volume spray of plant protection UAV, the droplet size and the effect of spraying operation will be affected by the pressure. Shahemabadi and Moayed [11] proposed an algorithm to improve the Pulse width modulation (PWM) algorithm. By controlling the rising or falling state of the valve opening corresponding to the high and low pulse levels, an adjustment range of the flow rate from 0% to 100% can be realized according to the adjustment precision of 2.5%. The response time by using pulse to adjust valve opening cannot meet the requirements of UAV high-speed flight.

At present, the precision variable spraying technology for ground plant protection machinery in China is developing more rapidly. However, the technology of agricultural aviation plant protection is still in its infancy, mainly because the plant protection UAV is faster than the ground plant protection machinery. UAV has the characteristics of high control precision and fast response speed to the variable spray system. In this study, a plant protection UAV variable spray system is designed based on the interpretation of the work prescription map. The system obtains prescription information for real-time location through graphic interpretation of plant protection UAV, and PWM-Proportional–Integral–Derivative (PID) control is used to adjust the spray volume quickly and accurately. The stability and feasibility of the system are verified by experiments, which provide a theoretical reference for the research of plant protection UAV variable spray system and extend to the field of modern agricultural aviation for plant protection.

The paper is organized as follows: Section 2 introduces the working principle of the plant protection UAV variable spray system based on prescription diagram interpretation. Section 3 introduces the components, design and simulation of PID algorithm in detail. Section 4 presents the results of the experiments and discussion. Finally, Section 5 provides the concluding remarks.

## 2. System Composition and Working Principle

The plant protection drone variable spray system designed by the research team is presented in Figure 1. It comprises a prescription graphic translation subsystem and a variable spraying subsystem. At first, the prescription value of the medication is interpreted by the prescription graphic translation subsystem from the prescription map of the work area. Subsequently, the prescription value is transmitted to the variable spray system. At last, the closed-loop PID control algorithm is utilized to adjust the duty cycle according to the prescription value received. By assigning a value to the single-chip timer counter, a square wave signal with an adjustable duty cycle is generated to adjust the rotation speed of the micro-diaphragm pump. Therefore, the way of variable spraying is realized. In the spray system pipeline, the flow information in the spray pipeline is fed back to the STM32 controller by the Hall flow sensor. Meanwhile, the instantaneous flow rate, real-time position, flight parameters and prescription values are displayed through the liquid crystal display (LCD).

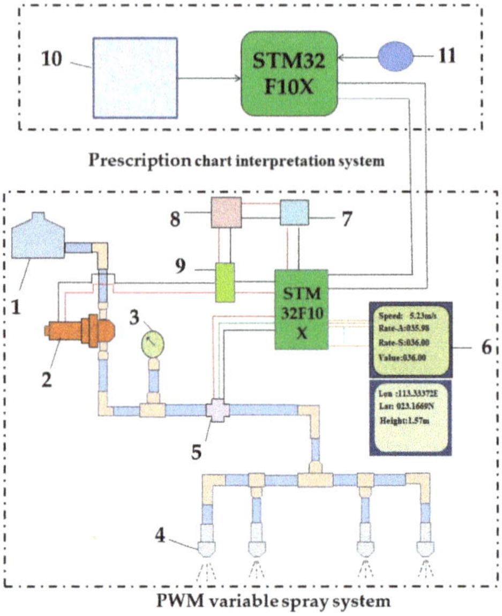


**Figure 1.** Schematic diagram of the structure of a prescription graphical interpretation of a variable spray system. PWM: pulse width modulation. **Note:** 1. Medicine box; 2. micro-diaphragm pump; 3. digital pressure gague; 4. pressure nozzle; 5. hall flow sensor; 6. liquid crystal display (LCD); 7. buck module; 8. 12 V direct-current (DC) power; 9. drive amplification module; 10. prescription figure; 11. GPS.

## 3. System Design

### *3.1. Prescription Map Generation and Interpretation*

In order to get the prescription map, the ArcMap software (Environment System Research Institute, ESRI) was used to generate a prescription map with different prescription values, so that the variable spray system can be guided by the prescription map. At the same time, in order to verify the effectiveness of the system, a 40 m × 60 m field experimental field was selected in the Zengcheng Experimental Teaching Base of South China Agricultural University in Guangzhou, China (113°38′15″ E, 23°14′37″N). The target area was divided into small slices of 10 m × 10 m. The UAV flew over the center of each operation unit. The flight path and unit division are shown in Figure 2.

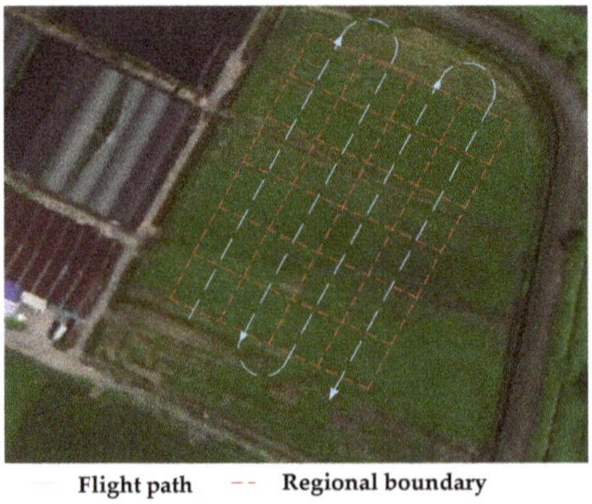


**Figure 2.** Sketch map of target plot.

Each 10 m × 10 m area is an operational unit, and a spraying amount is presented in each operational unit. According to the performance of each component of the variable spray system, five different levels of dosage (7.5 L/hm$^2$, 15 L/hm$^2$, 22.5 L/hm$^2$, 30 L/hm$^2$, and 37.5 L/hm$^2$) were set up. The prescription value of each operation unit was selected randomly from the above 5 gradient prescription values, which are shown in Figure 3.

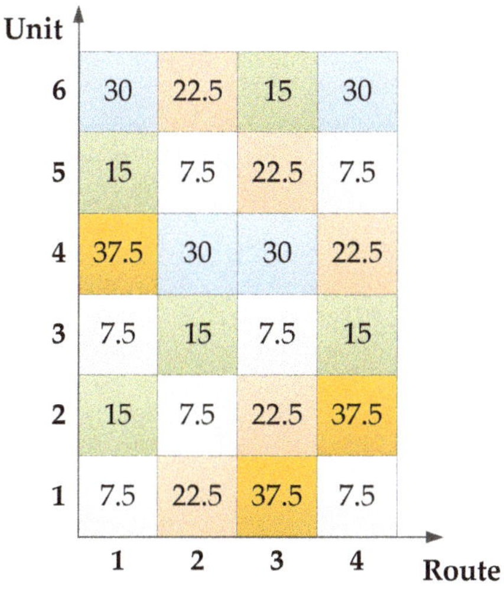


**Figure 3.** Prescription value setting in each operation unit.

Before the experiment, the working area was located by LocaSpace Viewer software (Beijing 3D Vision Technology Co., Ltd.). Then, the geographic position information of the working field was obtained, and the longitude and latitude were input into Microsoft Excel (Microsoft Corporation, USA) file. As a result, the ArcMap software was imported to generate the line map layer grid [12–14]. Eventually, the prescription map was created. By adding element classes to generate fishing nets, the grid can be edited and read. Finally, the longitude and latitude information of each unit and the dosage information were inserted by the linear difference method to generate the prescription map of the work area.

The prescription diagram mainly contains three layers of information, as shown in Figure 4. The first layer is raster information, which is a rectangular raster of equal size according to the effective spray amplitude and flight speed of UAV. The second layer is prescription value information layer, which is based on the grid dosage obtained by the expert system of pesticide application, and different colors represent different prescription values. The third level is the geographic information layer, which is mainly the latitude and longitude information of raster rows and columns.

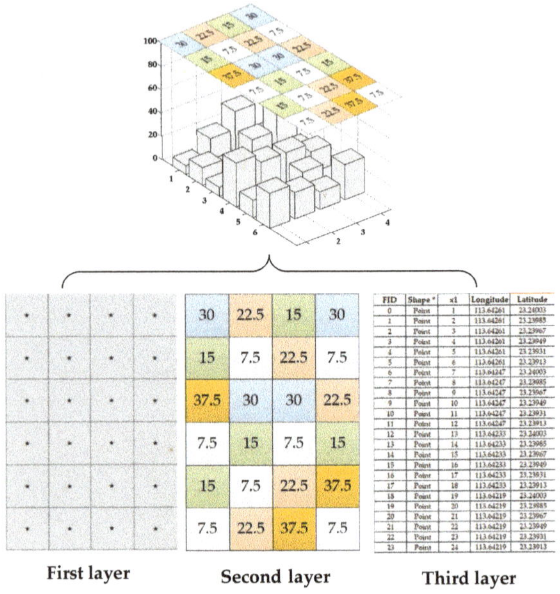


**Figure 4.** Structure chart of prescription map.

The prescription map generated by the ArcMap software simulation mainly includes information such as the latitude and longitude and the amount of spraying of each unit, and programs the information of each unit to be stored in the STM32 controller. During the spraying operation, the on-board high-precision GPS transmits the position information of the plant protection drone to the STM32 controller in real time, and matches the position information contained in the work prescription map to determine the unit area where the UAV is currently located. When the position information of UAV is successfully matched, the prescription value of the current location is obtained. The obtained spray quantity information is transmitted to the variable spray controller to perform spray operation. The schematic diagram of hardware representing the prescription graphic translation system is shown in Figure 5.

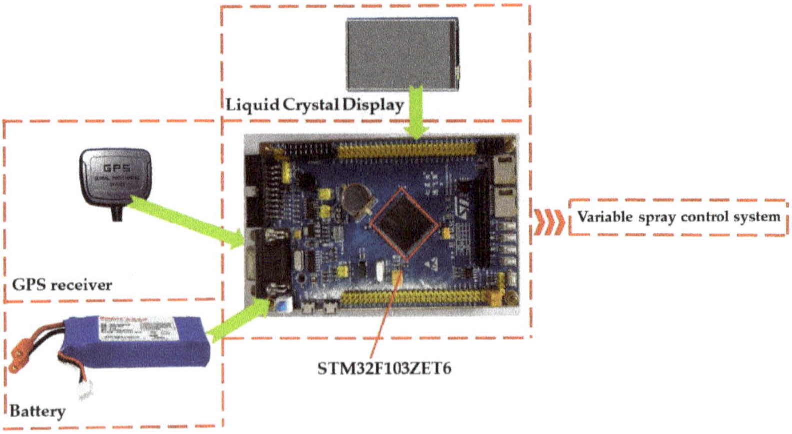


**Figure 5.** Schematic diagram of hardware representing a prescription graphic translation system.

## 3.2. Variable Spray System Design

The overall structure of the plant protection drone variable spray system is illustrated in Figure 6. The self-developed subsystems of prescription graphic translation and spray controller were installed in the M23 UAV (Shenzhen Hi-tech New Agriculture Technologies Co, Ltd., Shenzhen, China). Along with the medicine box, a miniature diaphragm pump (PLD-2201, Shijiazhuang Prandi Co, Ltd., Shijiazhuang, China) and a pressure nozzle (110-015 types, LECHLER Company, Germany) were used. Among these, the STM32 chip (STM32F103ZET6, Langyi Electronic Technology Co, Ltd., Shanghai, China) was the spray controller core. The Hall flow sensor (MJ-HZ06K, Mocho Technology Co, Ltd., Shenzhen, China) was used to measure the flow of the system, and the measuring range of the hall flow sensor was 0.1–1 L/min.

**Figure 6.** Physical diagram of the variable spraying system. **Note:** 1. Prescription map interpretation system and spray controller; 2. medicine case; 3. Hall flow sensor; 4. miniature diaphragm pump; 5. pressure nozzle.

The micro-diaphragm pump was controlled by the signal sent by the interpretation system. The liquid in the medicine box was transported to the nozzles and fractured into tiny droplets under pressure. The Hall flow sensor was used to measure the flow inside the system, and the flow rate information was fed back to the spray controller. Subsequently, the PWM was adjusted according to the deviation between the actual flow and the target flow. The schematic diagram of hardware representing the variable spray controller is shown in Figure 7.

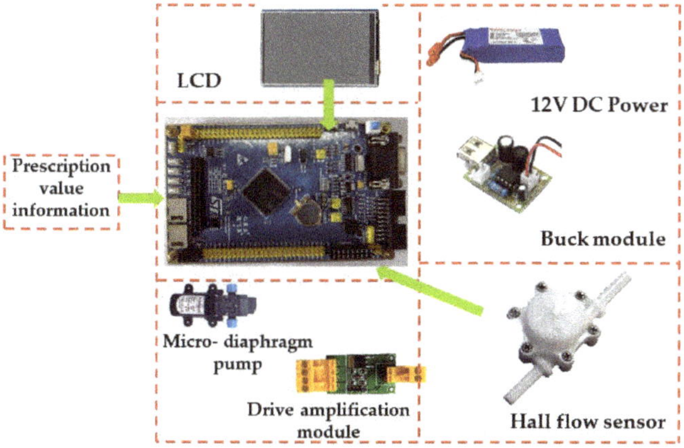


**Figure 7.** Schematic of hardware representing the variable spray controller.

### 3.3. Micro-Diaphragm Pump Drive

The micro-diaphragm pump was controlled by the PWM technology. When the spray controller received the prescription value information, the high-level time was used as the percentage of the whole cycle of change, thus obtaining the PWM square wave signals with different duty cycles. Since the PWM signal generated by the STM32 series microcomputer (MCU) was 5 V and the rated voltage of the micro-diaphragm pump was 12 V, the PWM square wave output from the I/O port of the MCU cannot directly drive the micro-diaphragm pump. As a result, the driving amplification effect of the metal–oxide–semiconductor (MOS) transistor was used. Moreover, the rotation of the micro-diaphragm pump was drive by the PWM square wave signal.

When the UAV was in flight operation, the variable spray system adjusted the rotational speed of the diaphragm pump by changing the duty ratio of the PWM square wave signal according to the prescription value. As a result, the flow rate of the system was adjusted. The period of the PWM square wave signal was set to 200 ms. When the duty ratio was lower than 40%, the micro-diaphragm pump could not be started. To ensure normal operation of micro-diaphragm pump, the duty ratio changed within the range of 40% to 100% during the test. In order to measure the change of flow clearly, the duty cycle of the PWM square wave signal was increased by 5% each time.

The Hall flow sensor was used to measure the single nozzle flow rate of the micro-diaphragm pump under different duty ratios. Additionally, the relationship between the flow rate of the micro-diaphragm pump and the duty ratio of PWM square wave signal was obtained. The result is shown in Figure 8. By using the cubic polynomial to fit the actual flow rate and duty cycle curve [15,16], the relationship between the flow rate of the variable system and the duty cycle of the PWM square-wave signal is:

$$duty = (20.776v^3 - 21.452v^2 + 8.242v - 0.431) \times 100\% \quad (1)$$

where $v$ is the flow rate of a nozzle in L/min, and *duty* is the duty cycle of the PWM square wave signal in %.

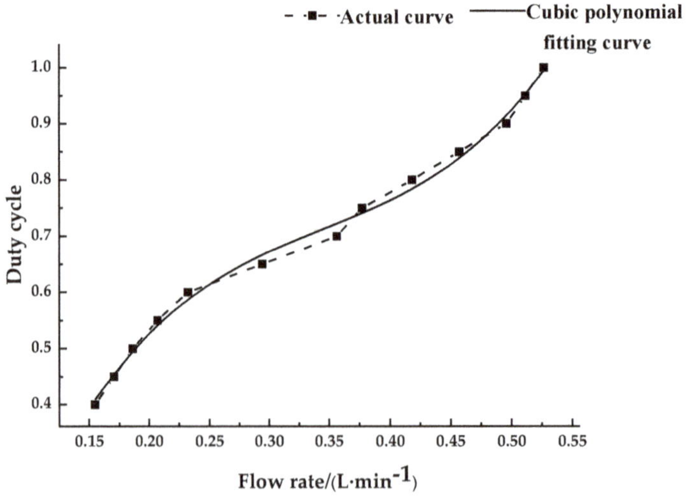


**Figure 8.** Velocity curve of the miniature diaphragm pump.

### 3.4. Nozzle Installation of Variable Spray System

The spray amplitude of the plant protection UAV is related to the installation distance of the nozzle. In order to get the theoretical spray size of the variable spray system designed by the project group, four standard angle sector nozzles (110-015 types, LECHLER Company, Germany) were installed side by side. The spray was gradually sharpened at the edge, and the theoretical spray angle of the nozzle was 110 degrees. The plant protection UAV spray system was dismantled and carried out by spraying with self-designed sprays. According to the basic requirement of the spray nozzle, the installation space of the adjacent sprinkler must be more than 50 cm, and thus a better spraying effect can be obtained when spraying [17,18]. Therefore, the installation space of the nozzle of the project group was 50 cm. A schematic diagram of the nozzle installation is shown in Figure 9.

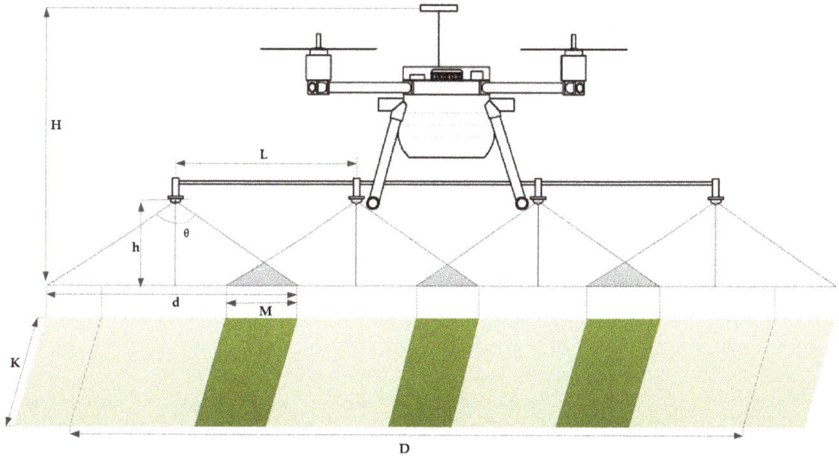


**Figure 9.** Schematic diagram of nozzle installation. **Note:** $H$ is the height of the GPS from the top of the crop in me; $h$ is the height of the nozzle from the top of the crop in m; $L$ is the installation distance of the adjacent nozzle in m; $d$ is the spray nozzle of the corresponding height in m; $M$ is the width of the overlap area of the adjacent nozzles in m; $K$ is a certain time, with the distance of the drone flight in m; $D$ is the effective spray width of the drone in m.

It can be seen from Figure 9 that the equation to calculate the spray width of a single nozzle is:

$$d = 2h \tan \frac{\theta}{2} \quad (2)$$

where $d$ is the nozzle spray of the corresponding height in m; $h$ is the height of the nozzle at the tip of the crop in m; and $\theta$ is the number of spray angles in °.

In order to achieve a uniform spraying effect, the complementary spraying range of the nozzle should be 25–30% of the single spraying range [19,20]. As shown in Figure 9, when the adjacent two nozzles are located in the complementary region, the effective spraying range of the four nozzles is 25% of the single spraying range. The effective pray width of the four nozzles can be expressed as:

$$D = \frac{11}{2} h \tan \theta \quad (3)$$

The height of the plant protection UAV is the height of the GPS from the ground. The height of rice, wheat and other crops is about 50 cm. The spray bar designed by the research team was installed on the frame of the drone. The height of the nozzle is about 70 cm away from the GPS. Therefore, when the flying operation height of the drone is about 2 m, regardless of the influence of other external environmental factors, it can be discerned from Equation (3) that the effective spray width of the four nozzles is about 6 m.

### 3.5. Control Program Design of Variable Spray System

The program was utilized to control the variable spray system and process various signals. It was written in Keil Software (ARM Germany GmbH, USA), which was mainly composed of the PWM duty cycle adjustment module, serial communication module, LCD module and PID control module. The program control flow chart is shown in Figure 10. When the system works, the GPS communication protocol (NMEA0183) is used to analyze the information acquired by the GPS in real time [21], so the position information of the current plant protection drone is obtained. Then, the prescription value is extracted through the prescription graphic translation system which processes the information. The prescription values are sent to the variable spray system. When the variable spray system receives the prescription value information, the timing and counter are turned on. The number of pulses of the flow sensor is fed back every 50 ms, and the current system instantaneous flow is obtained through calculation. The deviation between the instantaneous flow rate and the target flow of the system is used as the input of the PID controller. After the PID control algorithm is operated, a duty value is output, and the variable spray controller generates a PWM square wave with the corresponding duty ratio.

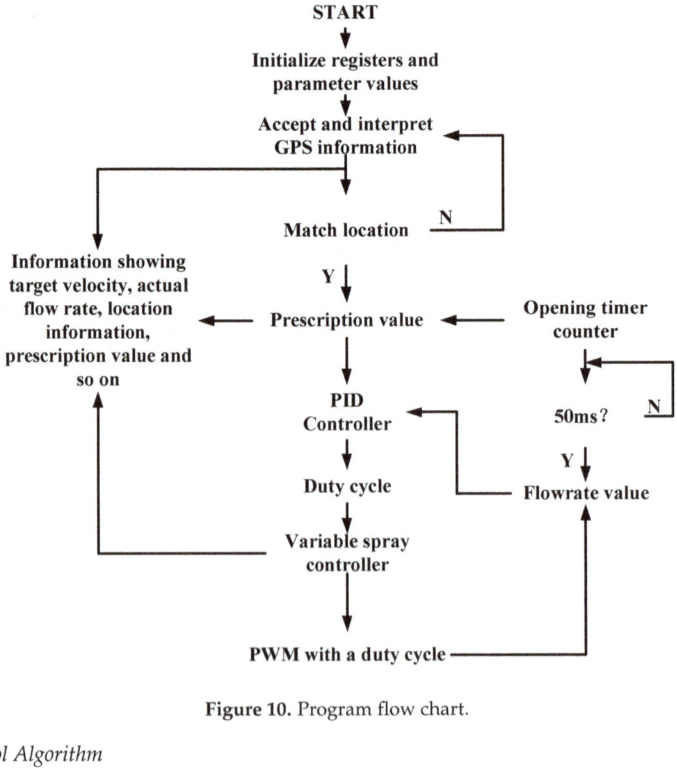


**Figure 10.** Program flow chart.

### 3.6. PID Control Algorithm

Due to the fact that the actual spray operation is affected by many factors, the system has a certain delay from receiving the flow change signal to adjusting the flow to the target value. Therefore, the actual flow will fluctuate around the target flow. The PID control algorithm is based on a control law of the system error. The algorithm is an optimal control adjusted by proportion (P), integration (I), and differentiation (D). It has the characteristics of a simple principle, high control precision, easy implementation and strong practicability [22]. The PID control algorithm has obvious effects on the process of the dynamic system calibration of continuous systems. In order to accurately control the flow of the system and ensure the accuracy and stability of the variable spray operation, the PID control algorithm is used to achieve the closed-loop control of the system [23]. The equation of PID control is as follows:

$$\begin{aligned} u(k) &= K_P^{e(k)} + \tfrac{TK_P}{T_I}\sum_{i=0}^{k} e(i) + K_P T_D \tfrac{e(k)-e(k-1)}{T} \\ &= K_P^{e(k)} + K_I \sum_{i=0}^{k} e(i) + K_D^{[e(k)-e(k-1)]} \end{aligned} \tag{4}$$

where $K_P$ is the proportional gain; $K_I = TK_P/T_I$ is the integral time constant; $K_D = K_P T_D$ is the differential time constant; $u(k)$ is the output of the control system at sampling time $k$; $e(k)$ is the system output deviation from the input quantity at sampling time $k$, and is described as $e(k) = y(k) - r(k)$, $y(k)$ is the output feedback value, and $r(k)$ is the reference input value.

The flow control process of the miniature diaphragm pump has the characteristics of large inertia lag and time variance. The process of adjusting the variable spray of the micro-diaphragm pump can be described as a second-order pure lag system. The transfer function is as follows:

$$G(s) = \frac{C(s)}{R(s)} = \frac{Ke^{-\tau s}}{(T_1 s + 1)(T_2 s + 1)} \tag{5}$$

where $K$ is the amplification factor; $\tau$ is the pure lag time in second; $T_1$ and $T_2$ are time coefficients.

In the process of adjusting the flow rate of the diaphragm pump, $K$, $\tau$, $T_1$ and $T_2$ are closely related to the performance of the system components and environmental factors [24]. The numerical value of the parameters was obtained through simulation. Therefore, the value of the amplification factor $K$ is 1, the value of the pure lag time $\tau$ is 0.25, and the values of the time constants $T_1$ and $T_2$ are 0.05 and 1.849, respectively. The variable spray system PID control model was created by the simulink component of MATLAB software (MathWorks, USA), as shown in Figure 11. According to the simulation model, when the given signal is a unit step signal, the PID simulation coefficient is debugged by the optimized parameter design method to obtain the optimal control effect. In actual use, the algorithm needs to be run through the STM32 chip. After repeated simulation and experiments, the values of $K_P$, $K_I$, and $K_D$ are set to 6, 0.8, and 0.5, respectively. When the input is the unit "1", the response curve obtained from the step response output of the system is shown in Figure 12.

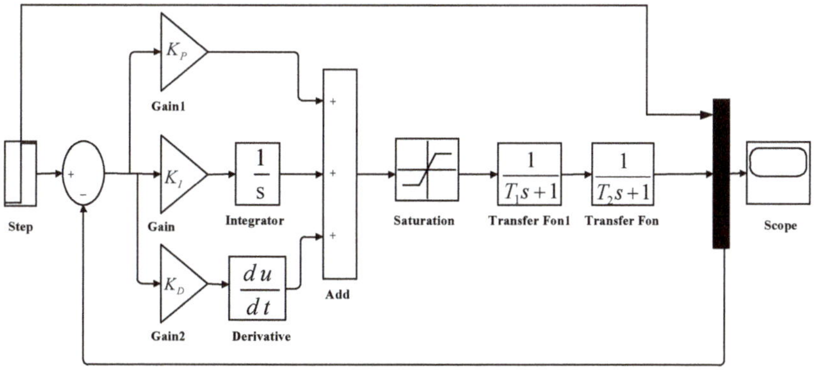


**Figure 11.** PID regulation simulation model.

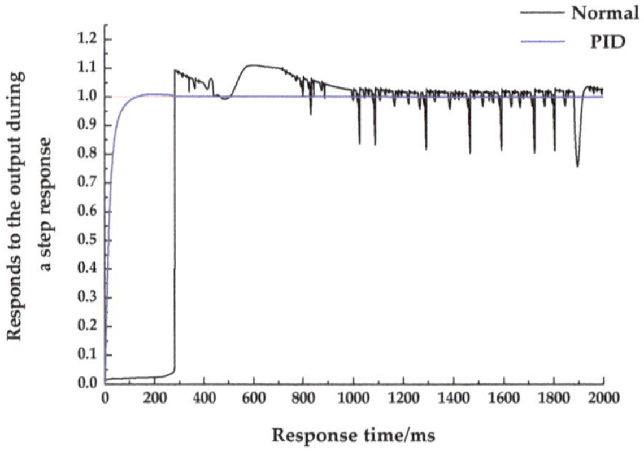


**Figure 12.** PID control and unregulated step response curve.

According to the step response curve of the system in Figure 12, the target flow is set to "1". When the PID algorithm is not added, the system receives the change signal of the flow and has a certain delay in response. About 300 ms of time is needed to change the flow to the target value; moreover, after reaching the target value, the actual flow value fluctuates around the target value, which has a certain deviation from the target flow. However, after the PID control, the response time is shortened to about 15 ms. At the same time, the actual flow reaches the target flow value and then stabilizes in the vicinity of the target value. To a certain extent, the system response speed and stability are improved.

## 4. Experiment

### *4.1. Effect of Duty Ratio on Droplet Size and Spray Angle*

The spray droplet volume and spray angle are important parameters for evaluating atomization effect [25]. Since the particle size and spray angle of the pressure nozzle change with the change of the fluid pressure, a laser particle size analyzer (DP-02, Zhuhai Omega Instrument Co., Ltd.) was used to measure droplet volume. Additionally, a single-lens reflex camera (ILCE-5100, Sony Corporation of Japan) was used to photograph the spray angle. The average particle size was tested at different duty cycles, and the results are shown in Figure 13.

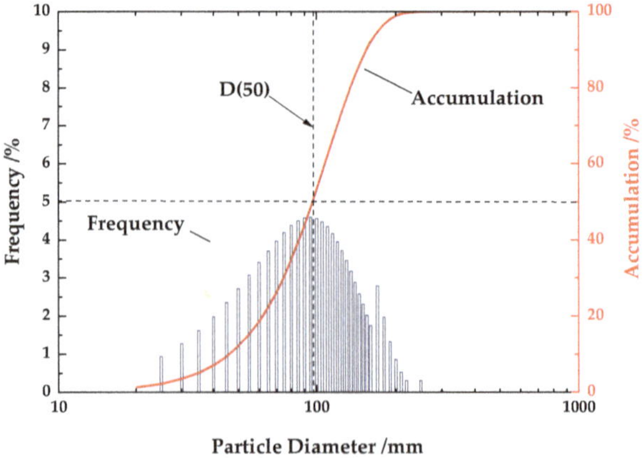


**(a)**

**Figure 13.** *Cont.*

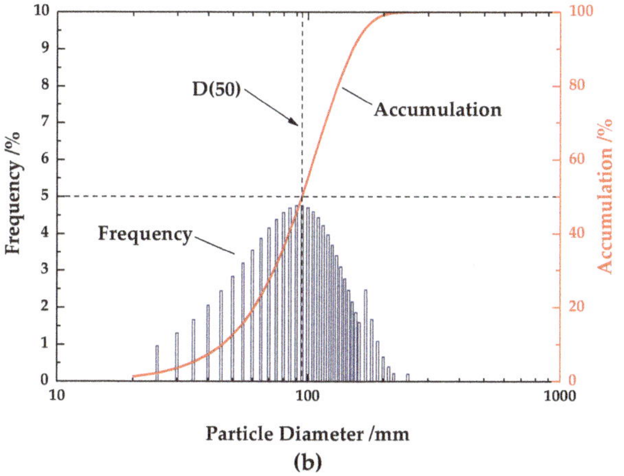


**Figure 13.** Droplet spectra of different PWM square wave duty ratios: (**a**) droplet spectrum with a duty ratio of 40%; (**b**) droplet spectrum with a duty ratio of 100%.

In Figure 13, frequency denotes the percentage of droplets with a certain size in the whole droplet group, and accumulation denotes the percentage of droplets arranged from small to large, to a certain size of droplets accumulating in the whole droplet group. The droplet size at 50% position is denoted as droplet volume diameter D (50). It can be seen from Figure 13 that the droplets produced by different nozzles are more uniform and the droplet spectrum is normal. The droplet size percentage is accumulated according to the order of droplet size from small to large. When the PWM duty ratio is 40% and 100%, the droplet volume diameter values are 99.81 and 96.26 μm, respectively.

The spray angles of PWM duty ratios between 40% and 100% are shown in Figure 14, and the particle size and spray angle of the other duty ratios are shown in Table 1. The experimental data in Table 1 demonstrate that the volume of the spray droplets varies between 94 and 100 μm, when the duty ratio is varied from 40% to 100%. Moreover, the volume of the volume gradually decreases as the duty ratio increases. The coefficient of variation is 5.52%, the spray angle varies between 97° and 105°, and the coefficient of variation is 2.77%. As the duty cycle increases, the flow rate of the system increases, and the pressure at the nozzle also increases, increasing atomization energy and promoting droplet breakage. The experimental results of droplet size show that different flow rates can be achieved by changing the duty cycle of PWM square wave signal to regulate the flow rate.

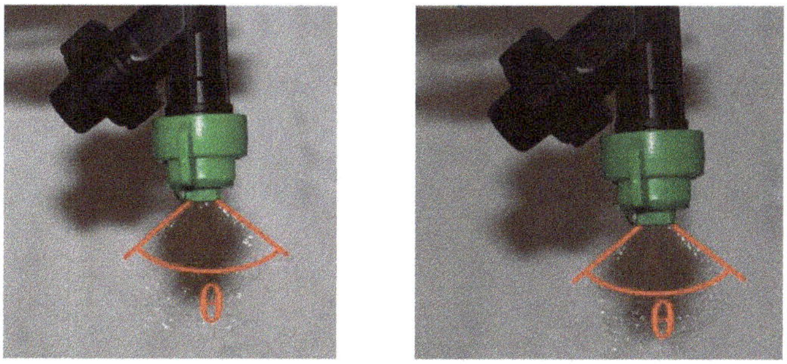

(**a**) (**b**)

**Figure 14.** Spray angle of different PWM square wave duty ratios with a duty cycle of 40% (**a**) and a duty cycle of 100% (**b**). **Note:** θ is the spray angle.

**Table 1.** Diameter of droplet volume and spray angle at different duty ratios of PWM square waves.

| Duty Cycle (%) | Volume Diameter (μm) | Spray Angle (°) |
|---|---|---|
| 40 | 99.81 | 98 |
| 45 | 98.72 | 99 |
| 50 | 97.85 | 97 |
| 55 | 97.46 | 99 |
| 60 | 97.32 | 98 |
| 65 | 97.07 | 99 |
| 70 | 96.73 | 100 |
| 75 | 96.77 | 98 |
| 80 | 96.49 | 101 |
| 85 | 96.26 | 102 |
| 90 | 95.79 | 104 |
| 95 | 95.65 | 105 |
| 100 | 94.59 | 105 |

### *4.2. Analysis of Actual Flow and Theoretical Flow Error*

A laboratory experiment was used to observe the real-time flow of the system conveniently. The flow information measured by the Hall flow sensor was displayed through the LCD screen in real time. The target flow rates under different duty cycles can be obtained by using the relationship between the duty cycle of the PWM square wave signal and nozzle flow rate. Table 2 is a comparison of the actual flow and the target flow of a single nozzle under different duty cycles.

**Table 2.** System flow deviation.

| Duty Cycle (%) | Target Flow Rate ($L \cdot min^{-1}$) | Actual Flow Rate ($L \cdot min^{-1}$) | Deviation (%) |
|---|---|---|---|
| 40 | 0.1550 | 0.1508 | 2.71 |
| 45 | 0.1705 | 0.1682 | 1.35 |
| 50 | 0.1860 | 0.1805 | 1.34 |
| 55 | 0.2072 | 0.1950 | 5.89 |
| 60 | 0.2324 | 0.2290 | 1.46 |
| 65 | 0.2945 | 0.2894 | 1.73 |
| 70 | 0.3564 | 0.3487 | 2.16 |
| 75 | 0.3774 | 0.3683 | 2.41 |
| 80 | 0.4184 | 0.4097 | 2.08 |
| 85 | 0.4572 | 0.4493 | 1.73 |
| 90 | 0.4959 | 0.4850 | 2.20 |
| 95 | 0.5116 | 0.5064 | 1.02 |
| 100 | 0.5268 | 0.5159 | 2.07 |

The actual flow measurement data measured by the Hall flow sensor were fed back to the controller chip. The PID algorithm adjusts the system flow according to the error between the target flow and the actual flow. It can be seen from Table 2 that the system flow is regulated by the duty cycle of PWM square wave signal. The actual flow changes with the change of duty cycle and is stable near the target flow value. Additionally, the average deviation of flow regulation is 2.16%, which indicates that the system can adjust the flow well.

### *4.3. Experiments Outdoors*

#### 4.3.1. Experimental Scheme

The stability and sensitivity of variable spray system were tested by outdoor spray deposition experiment, the experiment was carried out in a paddy field of the Zengcheng Research and Teaching Base (113°38′15″ E, 23°14′37″ N) of South China Agricultural University, Guangzhou. The experimental site is the same as the prescription map. The outdoor experiment site photo of the plant protection drone variable spray system is shown in Figure 15.

**Figure 15.** Spray test site.

According to the prescription diagram, the prescription values of adjacent units are different. In order to explore the uniformity of droplet deposition in each unit and the droplet deposition at the boundary of adjacent units, the sampling bands such as S1, S2 were shown in Figure 16, in which the operational units each were set to 10 m × 10 m. Figure 16 shows only the layout of the sampling points in the adjacent operational units, and the other units were arranged in the same way. In UAV flying one sortie, the total number of sampling bands was 124. A sampling band was set up in the center of each operation unit, such as S4 and S10. In order to study the change of spray volume at the boundary of the operation unit, the sensitivity of the variable spray system was evaluated. The sampling bands were set on the boundary of the adjacent operation unit and to be 1 m and 2 m away from the boundary line on its both sides, as shown in Figure 16. S1, S7 and S13 were the sampling bands on the boundary line; S5, S6, S8 and S9 were the sampling bands on the left and right sides of the dividing line, respectively. The amount of droplet deposition, the sediment density and the sedimentation uniformity of aviation plant protection operations are important parameters reflecting the quality of spraying [26]. In order to study the law of droplet deposition between various regions, a Rhodamine B (soluble fluorescent tracer) solution with a concentration of 5 g/L was used instead of the pesticide solution, and a Mylar card with a size of 50 mm × 80 mm and water-sensitive paper with a size of 28 mm × 75 mm were used to collect the sprayed droplets to analyze the deposition density and deposition uniformity. Nine sampling points were evenly arranged at an interval of 1 m on each collection belt, and both the Mylar card and water-sensitive paper were fixed at a height of about 50 cm from each sampling point. The layout of Mylar card and water-sensitive paper is shown in Figure 17.

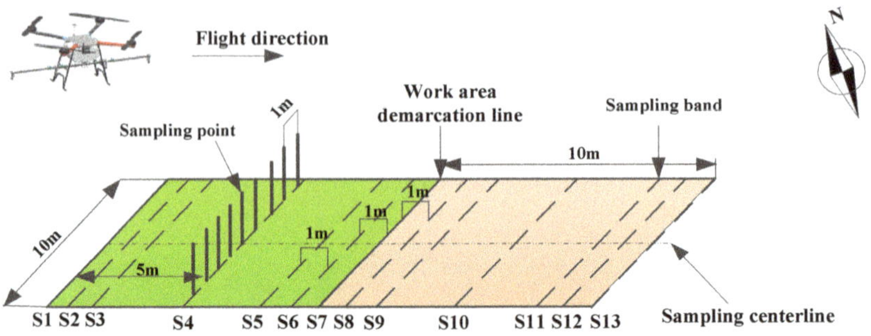


**Figure 16.** Inner sampling bands of adjacent operation units. **Note:** S1, S2, S3, S4, S5, S6, S7, S8, S9, S10, S11, S12, and S13 are the numbers of the collection bands.

**Figure 17.** Layout of the Mylar card and water-sensitive paper.

The experiment was conducted on 5 November 2018, and the mature rice was selected as the tested crop. A total of four sorties were tested, the flight speed of the plant protection drone was stable at 5 m/s and the flight altitude was 2 m. At the same time, a portable ultrasonic micro-automatic weather station (Hberw6-3, Shenzhen Hongyuan Technology Co, Ltd., China) was used to measure the environmental information, and the height of the weather station was set at 2 m above the ground. The test was carried out four times in total. The environment parameters are shown in Table 3.

**Table 3.** Test environment parameter list.

| Sorties | Temperature (°C) | Humidity (%) | Wind Speed and Direction ($m \cdot s^{-1}$) |
|---|---|---|---|
| 1 | 19.5 | 54.3 | 0.54/SW |
| 2 | 20.1 | 54.2 | 0.78/SW |
| 3 | 21.3 | 53.9 | 0.84/SW |
| 4 | 22.6 | 53.7 | 0.47/SW |

4.3.2. Test Data Processing

The amount of droplet deposition is a significant parameter reflecting the quality of droplet deposition per unit area [27]. The Mylar card that is a resin card collected by the test is eluted with 20 mL of distilled water, and six Rhodamine B solutions with different concentrations were set for calibration in the range of absorbance of the fluorescence spectrophotometer (F-380, Tianjin Gangdong Technology Development Co., Ltd., China). Concentrations of Rhodamine B solutions were 0.002 μg/mL, 0.005 μg/mL, 0.01 μg/mL, 0.02 μg/mL, 0.05 μg/mL, and 0.1 μg/mL. The specific method was as follows: three parts of each standard solution were prepared, and each repeated measurement was performed twice to monitor the influence of the cuvette on the measurement result. Therefore, the actual number of repeated measurements for each standard solution amount to six times, and the Rhodamine B solution concentration absorbance curve is shown in Figure 18.

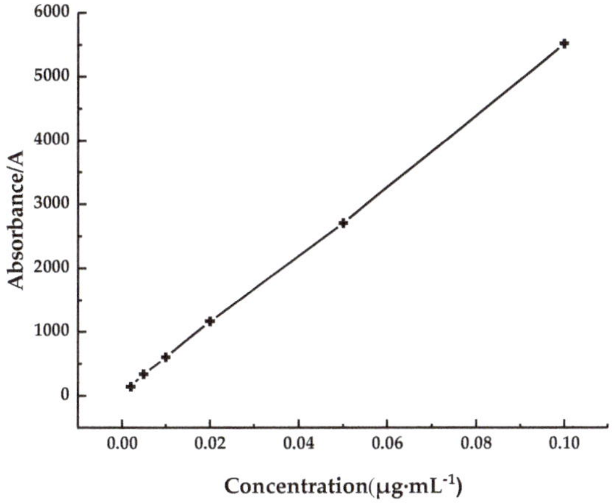


**Figure 18.** Rhodamine B solution standard concentration absorbance curve.

By linear regression fitting, the coefficient of determination $R^2$ is 0.9997. The standard curve gives the concentration of the sample solution, thereby calculating the deposition amount per unit area [28]. The calculation equation is as follows:

$$\beta_{dep} = \frac{(\rho_{sampl} - \rho_{blk})F_{cal}V_{dil}}{\rho_{spray}A_{col}} \tag{6}$$

where $\beta_{dep}$ is the amount of droplet deposition in g·cm$^{-2}$; $\rho_{sampl}$ is the reading of the sample solution fluorescence meter; $\rho_{blk}$ is the reading of the fluorescence meter of the eluent (distilled water for this test); $F_{cal}$ is the calibration coefficient in g·L$^{-1}$; $V_{dil}$ is the volume of the solution used to elute the collected sample in L; $\rho_{spray}$ is the concentration of the fluorescent tracer in the spray solution in %; $A_{col}$ is the area of the collected card in cm$^2$.

The water-sensitive paper collected by the test was scanned and analyzed by DepositScan software (USDA-ARS Application Technology Research Unit, Wooster, OH, USA), and the droplet deposition rate and deposition density under different prescription values were obtained [29]. The coefficient of variation is usually used to represent the uniformity of droplet deposition between different collection points for the same collection [30]. The calculation equation is:

$$CV = \frac{S}{\overline{X}} \times 100\% \tag{7}$$

$$S = \sqrt{\sum_{i=1}^{n} (X_i - \overline{X})^2 / (n-1)} \tag{8}$$

where $S$ is the standard deviation of the sampled specimens in the same collection zone; $X_i$ is the deposition amount of each collection point in μL·cm$^{-2}$; $\overline{X}$ is the average value of the deposition amount of the sampling points of the same collection zone in μL·cm$^{-2}$; and $n$ is the group sampling with the number of collection points.

## *4.4. Analysis of Experiments Results*

### 4.4.1. Droplet Deposition Density Analysis

During the experiment, the number of droplets per square centimeter of the water-sensitive paper in each sampling point was collected. The distribution uniformity of the droplet coverage tester was calculated by Equation (7). The grayscale image the water-sensitive paper after scanning is shown in Figure 19. The central sampling band of 6 working orders in the next route of the one or two sorties was selected, as shown in Table 4. When the water droplet coverage density is less than 15 per $cm^2$, it is regarded as an invalid sampling point.

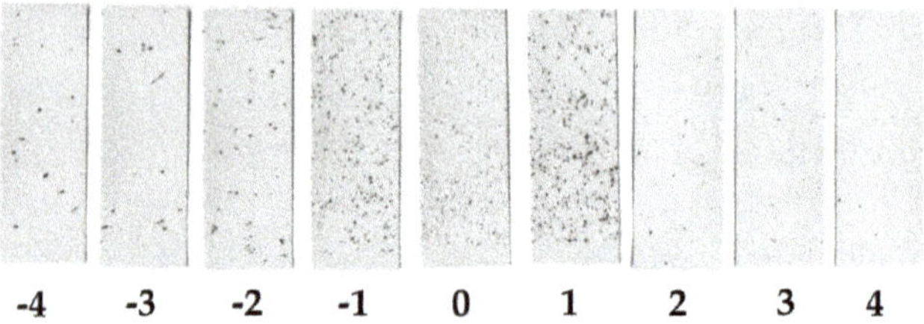


**Figure 19.** Grayscale image of water-sensitive paper.

**Table 4.** Statistics of droplet coverage density.

| Sortie | Unit | Sampling Point | | | | | | | | | Average | Coefficient of Variation |
|---|---|---|---|---|---|---|---|---|---|---|---|---|
| | | −4 | −3 | −2 | −1 | 0 | 1 | 2 | 3 | 4 | | |
| First | 1 | 6 | 16 | 19 | 29 | 44 | 31 | 23 | 14 | 9 | 25.14 | 38.07% |
| | 2 | 2 | 17 | 22 | 37 | 81 | 54 | 22 | 10 | 8 | 34.71 | 32.5% |
| | 3 | 0 | 15 | 19 | 26 | 47 | 34 | 17 | 6 | 0 | 23.43 | 237.77% |
| | 4 | 13 | 26 | 57 | 123 | 169 | 115 | 41 | 23 | 12 | 79.14 | |
| | 5 | 1 | 9 | 16 | 39 | 73 | 27 | 18 | 7 | 0 | 27 | 56.88% |
| | 6 | 0 | 14 | 23 | 45 | 137 | 67 | 24 | 16 | 3 | 46.57 | 72.48% |
| Second | 1 | 0 | 12 | 19 | 24 | 41 | 31 | 20 | 16 | 4 | 23.29 | 59.47% |
| | 2 | 9 | 16 | 22 | 37 | 93 | 49 | 29 | 14 | 0 | 37.14 | 28.08% |
| | 3 | 0 | 15 | 23 | 36 | 55 | 29 | 16 | 13 | 1 | 26.71 | 73.83% |
| | 4 | 4 | 21 | 33 | 59 | 143 | 37 | 20 | 12 | 0 | 46.43 | 47.9% |
| | 5 | 1 | 6 | 15 | 23 | 69 | 31 | 18 | 7 | 0 | 24.15 | 113.54% |
| | 6 | 3 | 14 | 29 | 67 | 115 | 89 | 32 | 15 | 3 | 51.57 | |

According to the data in Table 4, when the prescription values are different, the distribution of droplet deposition density is more intense in the middle and less intense on both sides. Because of the small natural wind speed during the experiment, the droplet deposition does not have obvious migration, and the peak value of deposition concentration is gathered near the sampling center line. The prescription values of operational units 1–6 are 7.5 $L/hm^2$, 15 $L/hm^2$, 7.5 $L/hm^2$, 37.5 $L/hm^2$, 15 $L/hm^2$, and 30 $L/hm^2$, and the normalized ratio of prescription value is 1:2:1:5:2:4.

In the first sortie, the normalized ratio of effective droplet deposition density of these six operation units is 1:1.38:0.93:3.15:1.07:1.85. In the second sortie, the normalized ratio of their effective droplet deposition density is 1:1.59:1.15:1.99:1.04:2.21. The actual normalized ratio of fog droplet deposition density of these six operation units is smaller than its theoretical ratio. The main reason is that the droplet does not fall due to the drift of droplets in the process of operation. The data in the table show that the effective sampling point number is from −3 to 3. According to the Civil Aviation Industry Standards of the People's Republic of China (MH/T1002.1-2016) with regard to the technical specifications for the quality of agricultural aviation ultra-low-capacity pesticide-spraying operations [31], it can be estimated that the actual spraying swath is about 5 m.

#### 4.4.2. Droplet Deposition Analysis

A total of four sorties were tested, with four routes under each sortie. The spraying prescription values of each unit were the same as those of the prescription map. The droplet depositions of four sorties were collected by the Mylar card. The average droplet deposition in various operational units was calculated by elution analysis of the Maylar card. The droplet deposition in the sampling zone at the center of six operational units on the third and fourth sorties was analyzed. The collection of droplet deposition data is shown in Figure 20.

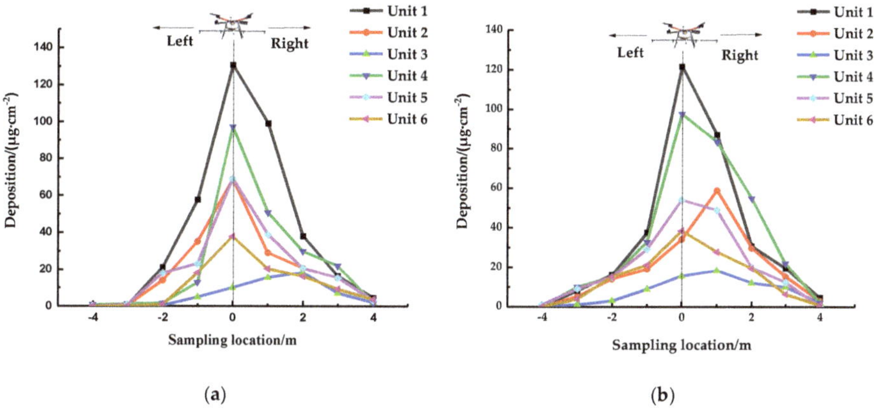


**Figure 20.** Droplet deposition in center line of operational unit: (**a**) third sortie; (**b**) fourth sortie.

The prescription values of the operational units 1–6 were set by the prescription map to be 37.5 L/hm$^2$, 225 L/hm$^2$, 7.5 L/hm$^2$, 30 L/hm$^2$, 22.5 L/hm$^2$, and 15 L/hm$^2$, respectively. It can be observed in Figure 20 that the distribution of droplet deposition in each area is basically the same. The amount of droplet deposition is related to the prescription value of each area. The peak value of spray deposition appears below the fuselage. The main reason is the drift of droplets on both sides of the fuselage due to the influence of the rotor wind field. The droplet distribution appears to be near the sampling center line. The droplets on each collection belt are mainly distributed at the sampling points of −2#, −1#, 0#, 1#, 2#, 3#. It can be estimated that the actual spraying swath is about 5 m. This is similar to the result of Table 4.

In order to visualize the droplet deposition changes in different operational units under four sorties, the average droplet deposition in each operation unit was calculated, which is shown in Figure 21.

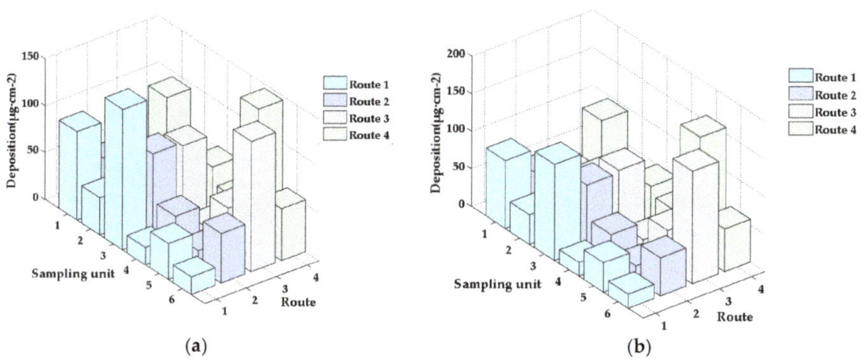


**Figure 21.** *Cont.*

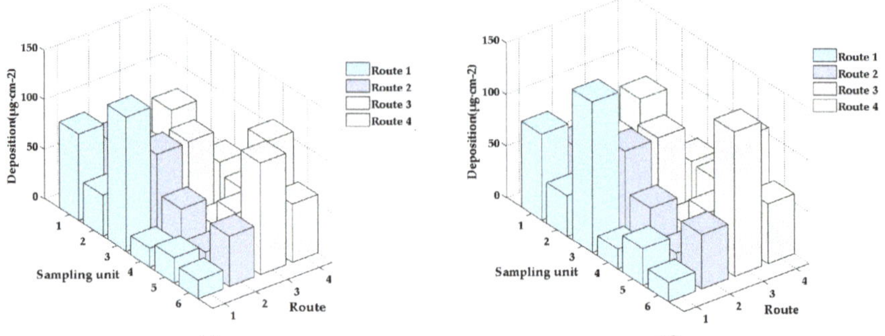


(c) (d)

**Figure 21.** Droplet depositions in various sorties: (**a**) first sortie; (**b**) second sortie; (**c**) third sortie; (**d**) fourth sortie.

The average deposition amounts of fog droplets in each operational unit of four sorties are basically the same. The variation trends of the deposition amount of fog droplets between different operational units on each route are the same, and the average deposition amounts of fog droplets in each operation unit are different when placed in different places. It can be observed that the variable spray system designed by the research group can accomplish variable spray operation according to the prescription value set in the prescription map.

#### 4.4.3. Droplet Deposition at the Boundary of Operation Units

In order to verify the sensitivity of the designed variable spray system, the sampling bands are set up to collect droplet deposition. The sampling bands are located 1 m and 2 m away from the operation unit boundary line. The droplet deposition amount of the units 1–4 of the first sortie was selected for analysis. And the test data are shown in Table 5. In Table 5, S4, S10 and S16 are the acquisition bands at the demarcation line of the operation unit. S3, S5, S9, S11, S15 and S17 are the numbers of acquisition band numbers which are 1 m away from the demarcation line on both sides of the demarcation line. S2, S6, S8, S12, S14 and S18 are the numbers of acquisition bands which are about 2 m away from the demarcation line on both sides of the demarcation line. S7, S13 and S19 are the numbers of acquisition bands located in the central position of the operation unit respectively. The layout of the sampling zone from S2 to S19 is in accordance with the method shown in Figure 16.

**Table 5.** Droplet deposition at the boundary of operational unit.

| Unit | Prescription Value ($L \cdot hm^{-2}$) | Sampling Band | Deposition ($\mu g \cdot cm^{-2}$) |
|---|---|---|---|
| 1 | 7.5 | S2 | 22.58 |
| | | S3 | 21.89 |
| Boundary | | S4 | 20.63 |
| 2 | 15 | S5 | 35.68 |
| | | S6 | 39.43 |
| | | S7 | 39.21 |
| | | S8 | 40.28 |
| | | S9 | 38.64 |
| Boundary | | S10 | 39.46 |
| 3 | 7.5 | S11 | 27.64 |
| | | S12 | 23.64 |
| | | S13 | 24.06 |
| | | S14 | 21.96 |
| | | S15 | 22.68 |

**Table 5.** *Cont.*

| Unit | Prescription Value (L·hm$^{-2}$) | Sampling Band | Deposition (µg·cm$^{-2}$) |
|---|---|---|---|
| Boundary | | S16 | 23.07 |
| 4 | 37.5 | S17 | 69.76 |
| | | S18 | 129.87 |
| | | S19 | 134.26 |

The diagram of droplet deposition in operation units 1–4 is shown in Figure 22.

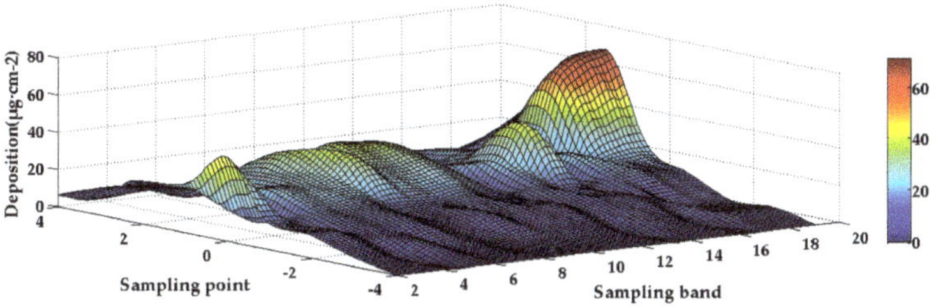


**Figure 22.** Droplet depositions in operation units.

It is known from Table 5 and Figure 22, when the UAV flies over the boundary band, the prescription interpretation system is interpreted to get the prescription value of the next operation unit and sent to the variable spray control system. The variable spray control system regulates the flow rate of the spray system according to the spraying prescription value. From Table 5, the droplet deposition at the boundary is similar to that of the working unit. The droplet deposition at the 1 m position behind the boundary is between the values at the center of the two working units. For example, the variation rates of the droplet deposition at S5, S11 and S17 are 63%, 29.95% and 217.67% respectively. On the other hand, the deposition of droplets on the acquisition belt, which are 2 m away from the demarcation line, is similar to that on the central line of the operation unit. Since the plant protection drone flies at 4 m/s during the test, the new prescription value is received. When the target value of the spray volume change value is about 2 m, the operation time of the system is 0.4s from receiving to square value to arriving at the target value of flow. Which reflects the sensitivity of the system. From Table 5 and Figure 22, it can be seen that the droplet depositions in the four operation units 1–4 vary with the prescription value. The normalized ratio of the droplet deposition in the center line of operation units 1–4 is 1:1.79:1.07:5.93, and the ratio of prescription value for these four operation units is 1:5:1:7. Due to the drift and adherence of droplets to fuselage, the actual deposition value is smaller than the prescription value, but the ratio is consistent, which reflects the effectiveness of variable spray system. The above analysis shows that the PWM-PID variable spray system designed based on the prescription can quickly adjust the flow according to the prescription map. The system has certain sensitivity and stability.

## 5. Conclusions

The variable spray system based on PWM-PID control can achieve rapid and accurate change of flow according to prescription information, effectively reducing herbicide use and enhance chemical effect.

(1) Using serial communication technology to receive the prescription value information after the prescription translation, the PWM technology was used to adjust the rotation speed of the micro-diaphragm pump to realize the variable spray, and the spray effect of the spray system was tested. The results show that the variable spray system designed by the research group ensures that the atomization effect is stable under the duty cycle of different PWM square wave signals,

and the coefficient of variation of the system flow rate with the duty cycle of the PWM square wave signal is 39.21%, which can satisfy various kinds of different spray requirements;

(2) The PID algorithm was used to control the flow adjustment process to reduce the steady-state time of the system, so that the deviation between the actual flow and the target flow is stable at 2.16%, indicating that the system can adjust the flow well;

(3) The outdoor sedimentation test shows that the variable spray system can quickly change the spray flow according to the prescription value of the working plot. Variable pulse spraying can be realized by PWM technology.

(4) Based on the data of experimental deposition and deposition density, the variable spray system can be stabilized within 0.4 s from receiving the prescription value to adjusting the flow rate to a predetermined value, and the effective injection rate of actual operation is about 5 m.

**Author Contributions:** Conceptualization, S.W., Q.Z., J.D. and Y.L.; methodology, W.S., Q.Z. and J.S.; software, S.W., Q.Y., and J.S.; validation, Q.Z., X.Y. and J.S.; formal analysis, S.W. and Y.L.; investigation, W.S.; resources, J.Z. and Y.L.; data curation, Q.Z. and J.S.; writing of the original draft preparation, Q.Y.; writing of review and editing, S.W.; visualization, S.W. and J.Z.; supervision, S.W.; project administration, S.W.; funding acquisition, S.W. and J.Z.

**Funding:** This research was funded by the Science and Technology Program of Guangzhou, China (Grant No. 201707010047), Science and Technology Program of Guangdong, China (Grant No.: 2016A020210100, and 2017A020208046), National Key Technologies Research and Development Program (Grant No.: 2016YFD0200700), and National Natural Science Foundation of Guangdong, China (Grant No.: 2017A030310383).

**Acknowledgments:** Thanks to Yilong Zhan, et al. from the National Center for International Collaboration Research on Precision Agricultural Aviation Pesticides Spraying Technology for helping the authors to complete the outdoor experiment.

**Conflicts of Interest:** The authors declare no conflicts of interest.

## References

1. Guo, Y.W.; Yuan, H.Z.; He, X.K.; Shao, Z.R. Analysis on the development and prospect of agricultural aviation protection in China. *Chin. J. Plant Prot.* **2014**, *10*, 78–82. (In Chinese)
2. Zhou, Z.Y.; Ming, R.; Zang, Y.; He, X.G.; Luo, X.W.; Lan, Y.B. Development status and countermeasures of agricultural aviation in China. *Trans. Chin. Soc. Agric. Eng.* **2017**, *33*, 1–13. (In Chinese)
3. Song, Y.; Sun, H.; Li, M.; Zhang, Q. Technology Application of Smart Spray in Agriculture: A Review. *Intell. Autom. Soft Comput.* **2015**, *21*, 319–333. [CrossRef]
4. Mogili, U.R.; Deepak, B.B.V.L. Review on application of drone systems in precision agriculture. *Procedia Comput. Sci.* **2018**, *133*, 502–509. [CrossRef]
5. Xue, X.Y. Develop an unmanned aerial vehicle based automatic aerial spraying system. *Comput. Electron. Agric.* **2016**, *128*, 58–66. [CrossRef]
6. He, X.K.; Bonds, J.; Herbst, A.; Langenakens, J. Recent development of unmanned aerial vehicle for plant protection in East Asia. *Int. J. Agric. Biol. Eng.* **2017**, *10*, 18–30.
7. Chen, Z.G.; Chen, M.X.; Wei, X.H.; Li, J.Y.; Li, L. Variable prescription pesticide spraying system for farmland based on the Beidou Navigation Satellite system. *J. Drain. Irrig. Mach. Eng.* **2015**, *33*, 965–970.
8. Perez-Ruiz, M.; Aguera, J.; Gil, A.; Slaughter, D.C. Optimization of agrochemical application in olive groves based on positioning sensor. *Precis. Agric.* **2011**, *12*, 564–575. [CrossRef]
9. Qiu, B.J.; Li, K.; Shen, C.J.; Xu, X.C.; Mao, H.P. Experiment on response characteristics of variable-rate continuous spraying system. *Trans. Chin. Soc. Agric. Mach.* **2010**, *41*, 32–35. (In Chinese)
10. Gonzalez, R.; Pawlowski, A.; Rodriguez, C.; Guzman, J.L.; Sanchez-Hermosilla, J. Design and implementation of an automatic pressure-control system for a mobile sprayer for greenhouse applications. *Span. J. Agric. Res.* **2012**, *10*, 939–949. [CrossRef]
11. Shahemabadi, A.R.; Moayed, M.J. An algorithm for pulsed activation of solenoid valves for variable rate application of agricultural chemical. *IEEE Int. Symp. Inf. Technol.* **2008**, *4*, 1–3.
12. Reyes, J.F.; Esquivel, W.; Cifuentes, D.; Ortega, R. Field testing of an automatic control system for variable rate fertilizer application. *Comput. Electron. Agric.* **2015**, *113*, 260–265. [CrossRef]

13. Farooque, A.A.; Chang, Y.K.; Zaman, Q.U.; Groulx, D.; Schumann, A.W.; Esau, T.J. Performance evaluation of multiple ground based sensors mounted on a commercial wild blueberry harvester to sense plant height, fruit yield and topographic features in real-time. *Comput. Electron. Agric.* **2013**, *91*, 135–144. [CrossRef]
14. Yalew, S.G.; Griensven, V.A.; Zaag, V.P. AgriSuit: A web-based GIS-MCDA framework for agricultural land suitability assessment. *Comput. Electron. Agric.* **2016**, *128*, 1–8. [CrossRef]
15. Jiang, H.; Zhang, L.; Shi, W. Effects of Operating Parameters for Dynamic PWM Variable Spray System on Spray Distribution Uniformity. *IFAC-PapersOnLine* **2016**, *49*, 216–220. [CrossRef]
16. Wang, D.S.; Zhang, J.X.; Li, W.; Xiong, B.; Zhang, S.L. Design and test of the dynamic variable spraying system of plant protection UAV. *Trans. Chin. Soc. Agric. Mach.* **2017**, *48*, 86–93. (In Chinese)
17. Chen, S.D.; Lan, Y.B.; Li, J.Y.; Zhou, Z.Y.; Liu, A.M.; Mao, Y.D. Effect of wind field below unmanned helicopter on droplet deposition distribution of aerial spraying. *Int. J. Agric. Biol. Eng.* **2017**, *10*, 67–77.
18. Gonzalez-de-Soto, M.; Emmi, L.; Perez-Ruiz, M.; Aguera, J.; Gonzalez-de-Santos, P. Autonomous systems for precise spraying–Evaluation of a robotised patch sprayer. *Biosyst. Eng.* **2016**, *146*, 165–182. [CrossRef]
19. El Aissaoui, A. *A Feasibility Study of Direct Injection Spraying Technology for Small Scale Farms: Modeling and Design of a Process Control System*; Universite de Liege: Liege, Belgique, 2015.
20. Yin, D.F.; Chen, S.R.; Pei, W.C.; Shen, B.G. Design of map-based indoor variable weed spraying system. *Trans. Chin. Soc. Agric. Eng.* **2011**, *27*, 131–135. (In Chinese)
21. Llorens, J.; Gil, E.; Llop, J.; Escola, A. Ultrasonic and LIDAR sensors for electronic canopy characterization in vineyards: Advances to improve pesticide application methods. *Sensors* **2011**, *11*, 2177–2194. [CrossRef]
22. Cai, X.; Walgenbach, M.; Doerpmond, M.; Schulze, L.P.; Sun, Y. Closed-Loop Control of Chemical Injection Rate for a Direct Nozzle Injection System. *Sensors* **2016**, *16*, 127. [CrossRef] [PubMed]
23. Deng, W.; Zhao, C.; Chen, L.; Wang, X. Constant pressure control for variable-rate spray using closed-loop proportion integration differentiation regulation. *J. Agric. Eng.* **2016**, *47*, 148–156. [CrossRef]
24. Qin, Y.; Sun, L.; Hua, Q.; Liu, P. A Fuzzy Adaptive PID Controller Design for Fuel Cell Power Plant. *Sustainability* **2018**, *10*, 2438. [CrossRef]
25. Xue, X.Y.; Tu, K.; Qin, W.C.; Lan, Y.B.; Zhang, H.H. Drift and deposition of ultra-low altitude and low volume application in paddy field. *Int. J. Agric. Biol. Eng.* **2014**, *7*, 23–28.
26. Salyani, M.; Fox, R.D. Performance of Image Analysis for Assessment of Simulated Spray Droplet Distribution. *Trans. ASAE* **1994**, *37*, 1083–1089. [CrossRef]
27. Kesterson, M.A.; Luck, J.D.; Sama, M.P. Development and preliminary evaluation of a spray deposition sensing system for improving pesticide application. *Sensors* **2015**, *15*, 31965–31972. [CrossRef]
28. Gil, E.; Llorens, J.; Llop, J.; Fabregas, X.; Gallart, M. Use of a terrestrial LIDAR sensor for drift detection in vineyard spraying. *Sensors* **2013**, *13*, 516–534. [CrossRef]
29. Zhu, H.; Salyani, M.; Fox, R.D. A portable scanning system for evaluation of spray deposit distribution. *Comput. Electron. Agric.* **2011**, *76*, 38–43. [CrossRef]
30. Yao, W.X.; Lan, Y.B.; Wang, J.; Wen, S.; Wang, G.B. Droplet drift characteristics of aerial spraying of AS350B3e helicopter. *Trans. Chin. Soc. Agric. Eng.* **2017**, *33*, 75–83. (In Chinese)
31. Zhai, C.Y.; Wang, X.; Liu, D.Y.; Ma, W.; Mao, Y.J. Nozzle flow model of high pressure variable-rate spraying based on PWM technology. *Adv. Mater. Res.* **2011**, *422*, 208–217. [CrossRef]

*applied sciences*

*Article*

# Sliding Mode Thau Observer for Actuator Fault Diagnosis of Quadcopter UAVs

**Ngoc Phi Nguyen and Sung Kyung Hong ***

Faculty of Mechanical and Aerospace Engineering, Sejong University, Seoul 05006, Korea; phinguyen.183@gmail.com
* Correspondence: skhong@sejong.ac.kr; Tel.: +82-02-3408-3772

Received: 6 September 2018; Accepted: 2 October 2018; Published: 11 October 2018

**Featured Application: This work addresses issues related to fault-tolerant control of quadcopter UAVs.**

**Abstract:** Fault diagnosis (FD) is one of the main roles of fault-tolerant control (FTC) systems. An FD should not only identify the presence of a fault, but also quantify its magnitude and location. In this work, we present a robust fault diagnosis method for quadcopter unmanned aerial vehicle (UAV) actuator faults. The state equation of the quadcopter UAV is examined as a nonlinear system. An adaptive sliding mode Thau observer (ASMTO) method is proposed to estimate the fault magnitude through an adaptive algorithm. We then obtain the design matrices and parameters using the linear matrix inequalities (LMI) technique. Finally, experimental results are presented to show the advantages of the proposed algorithm. Unlike previous research on quadcopter UAV FD systems, our study is based on ASMTO and can, therefore, determine the time variability of a fault in the presence of external disturbances.

**Keywords:** fault diagnosis; quadcopter UAV; fault-tolerant control; sliding mode observer; Thau observer

## 1. Introduction

Quadcopter unmanned aerial vehicles (UAVs) have been used in a variety of applications, due to their numerous advantages, such as small size, agility, low cost, mechanical simplicity, and indoor and outdoor operability, which have led to their increased popularity compared to other UAV systems. As a result, they have been investigated and tested in a range of environments and applications which include target tracking [1,2], fault detection and fault-tolerant control [3,4], and formation flight [5,6].

Particularly the topic of fault-tolerant control (FTC) has received a large amount of attention in the community, which led to quadcopter UAVs that are less error-prone and, thus, more reliable during flight. In general, there are two types of FTC: passive and active. Several studies investigated passive FTCs [7,8], which have the advantage that they do not require any fault diagnosis scheme, but the resulting disadvantage is that they have a lower fault tolerance [9]. To overcome this limitation, active FTCs have been introduced to improve said fault tolerance. Fault diagnosis (FD) is the essential requirement for active FTCs to determine the location and magnitude of faults. Through FD, active FTCs can be designed to compensate the effect of faults and, thus, improve flight control and stability, which makes FD the main task of active FTCs.

The FD approach has been studied by numerous authors. Freddi et al. [10,11] investigated a model-based fault diagnosis which can be used to monitor sensor faults and detect actuator faults. In this method, residuals are used to distinguish between system and observer outputs, but these methods are inaccurate and unsuitable for quantifying the magnitude of a fault. Ma and Zhang [12,13] proposed a method for fault estimation based on a Kalman filter, but their approach is of insufficient robustness with regard to disturbances if the transfer matrices are inaccurate. Several effective approaches, such as sliding mode observer [14,15], neural network [16,17], and adaptive observer [18,19], have been investigated, but none of these approaches focused on a real quadcopter UAV. Moreover, recent studies [20,21] used fuzzy methods for fault diagnosis problems, but these approaches do not focus on real quadcopters, and may be overly complex to implement in a flight controller. Only few studies focused on the problem of fault diagnosis in a real quadcopter UAV, verified through real flight data. While [4] used an actuator fault estimation with an adaptive observer based on $H_\infty$, and demonstrated the effectiveness of the proposed scheme, this method may not be sufficiently robust to external disturbances because the underlying mathematical model neglects both nonlinear terms and external disturbances. The most recent application of an adaptive Thau observer (ATO) for actuator fault diagnosis was proposed in [22]. While this approach is capable of handling model uncertainties in the nonlinear quadcopter model, it is rather complex and time-consuming because it uses system identification to find the drag terms, and the filter to eliminate sensor noise.

In the present study, we try to overcome these limitations by combining a sliding mode observer based on Walcott–Zak observer design [23], with ATO to handle the actuator fault diagnosis. This method is capable of accounting for time-varying actuator faults. We then derive the Lyapunov stability and other conditions to obtain the desired matrices and associated parameters. Finally, a straightforward method based on linear matrix inequalities (LMI) is proposed to allow relaxing the derived conditions, which is a useful feature in flight controllers. Unlike previous methods, our approach is simple and not overly time-consuming, which makes it amenable for use in real quadcopters. Moreover, our method can handle uncertainties of magnitudes that are unknown, a priori, through an adaptive law approach. By comparing our approach to [22], we found that the adaptive algorithm is capable of compensating for the drag terms leading to clear improvements in the results.

## 2. System Description

While the right and left (3 and 4) motors of the quadcopter rotate in the clockwise direction, the other motors rotate in counterclockwise direction (Figure 1). Each motor is located at a distance $L$ from the center of mass $o$.

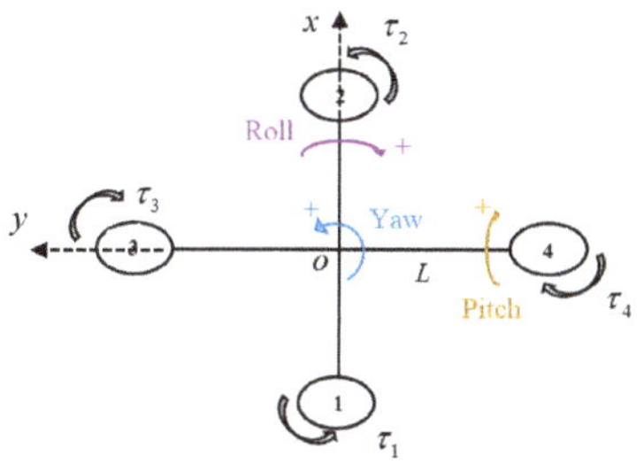


**Figure 1.** Schematic of the geometric configuration of the quadcopter unmanned aerial vehicle (UAV).

Assuming the control variables can be described as

$$\begin{cases} U_1 = T_1 + T_2 + T_3 + T_4 \\ U_2 = (T_3 - T_4)L \\ U_3 = (T_1 - T_2)L \\ U_4 = \tau_1 + \tau_2 - \tau_3 - \tau_4 \end{cases}, \tag{1}$$

where $\tau_i$ and $T_i$ represent the torque and thrust force produced by the $i$th motor, respectively; $U_1$ is the total thrust; $U_2$, $U_3$, $U_4$ are the torques in $\varphi, \theta, \psi$ directions, which correspond to roll, pitch, and yaw Euler angles, respectively (Figure 1).

Thrust force and torque are related to the rotational speed as follows:

$$T_i = b\Omega_i^2, \tag{2}$$

$$\tau_i = d\Omega_i^2, \tag{3}$$

where $b$, $d$ represent the thrust and drag coefficients, and $\Omega_i$ represents the rotational speed of the $i$th motor.

Inserting Equations (2) and (3) into (1) yields

$$\begin{cases} U_1 = b(\Omega_1^2 + \Omega_2^2 + \Omega_3^2 + \Omega_4^2) \\ U_2 = b(\Omega_3^2 - \Omega_4^2) \\ U_3 = b(\Omega_1^2 - \Omega_2^2) \\ U_4 = d(\Omega_1^2 + \Omega_2^2 - \Omega_3^2 - \Omega_4^2) \end{cases}. \tag{4}$$

The quadcopter dynamic model has previously been formulated as follows [22,24]:

$$\begin{cases} I_x\ddot{\varphi} = U_2 + (I_y - I_z)\dot{\theta}\dot{\psi} - J_T\dot{\theta}\Omega - K_\varphi\dot{\varphi} \\ I_y\ddot{\theta} = U_3 + (I_z - I_x)\dot{\varphi}\dot{\psi} - J_T\dot{\varphi}\Omega - K_\theta\dot{\theta} \\ I_z\ddot{\psi} = U_4 + (I_x - I_y)\dot{\varphi}\dot{\theta} - K_\psi\dot{\psi} \end{cases}, \tag{5}$$

where $I_x$, $I_y$, $I_z$ represent the moments of inertia along the $x$, $y$, $z$ directions, respectively; $K_\varphi$, $K_\theta$, $K_\psi$ are drag coefficients; $J_T$ is the moment of inertia of each motor, and $\Omega = \Omega_3 + \Omega_4 - \Omega_1 - \Omega_2$.

We consider drag terms as disturbances, and they can be compensated by adaptive law, which is discussed in the "nonlinear observer for fault diagnosis" section. By defining the state vector $x^T = \begin{bmatrix} \varphi & \theta & \psi & \dot{\varphi} & \dot{\theta} & \dot{\psi} \end{bmatrix}$, control input vector $u^T = \begin{bmatrix} U_2 & U_3 & U_4 \end{bmatrix}$, and output vector $y^T = [\begin{matrix} \varphi & \theta & \psi & \dot{\varphi} & \dot{\theta} & \dot{\psi} \end{matrix}]$, Equation (5) can be described in the state equation as

$$\begin{cases} \dot{x}(t) = Ax(t) + p(x,u) + Bu(t) + E_d d(t) \\ y = Cx(t) \end{cases}, \tag{6}$$

where $E_d$ is disturbance matrix, $d(t) \in R^s$ is disturbance vector, $A = \begin{bmatrix} 0 & 0 & 0 & 1 & 0 & 0 \\ 0 & 0 & 0 & 0 & 1 & 0 \\ 0 & 0 & 0 & 0 & 0 & 1 \\ 0 & 0 & 0 & 0 & 0 & 0 \\ 0 & 0 & 0 & 0 & 0 & 0 \\ 0 & 0 & 0 & 0 & 0 & 0 \end{bmatrix}$, $B = \begin{bmatrix} 0 & 0 & 0 \\ 0 & 0 & 0 \\ 0 & 0 & 0 \\ 1/I_x & 0 & 0 \\ 0 & 1/I_y & 0 \\ 0 & 0 & 1/I_z \end{bmatrix}$, $C = I_{6\times 6}$, and $p(x,u) = \begin{bmatrix} 0 \\ 0 \\ 0 \\ (\dot{\theta}\dot{\psi}(I_y - I_z) - J_T\dot{\theta}\Omega)/I_x \\ (\dot{\varphi}\dot{\psi}(I_z - I_x) - J_T\dot{\varphi}\Omega)/I_y \\ \dot{\varphi}\dot{\theta}(I_x - I_y)/I_z \end{bmatrix}$. When an actuator fault occurs, Equation (6) can be described as

$$\begin{cases} \dot{x}(t) = Ax(t) + p(x,u) + Bu(t) + Ff(t) + E_d d(t) \\ y = Cx(t) \end{cases}, \tag{7}$$

where $F$ is the fault matrix, and $f(t) \in R^l$ is an actuator fault vector.

## 3. Nonlinear Observer for Fault Diagnosis

### 3.1. Standard Thau Observer for Fault Detection

According to the state Equation (7), the two following conditions must be met by the Thau observer design:

**C1** the pair $(C,\ A)$ is observable.

**C2** the nonlinear term $p(x,u)$ is continuously differentiable and assumed to be Lipschitz, with a constant $\gamma$, i.e., $\|p(x_1(t),u(t)) - p(x_2(t),u(t))\| \le \gamma\|x_1 - x_2\|$.

From the above conditions, the state Equation (7), based on Thau observer, can be constructed as [20]:

$$\begin{cases} \dot{\hat{x}}(t) = A\hat{x}(t) + p(\hat{x},u) + Bu(t) + K(\hat{y}(t) - y(t)) \\ \hat{y} = C\hat{x}(t) \end{cases}, \tag{8}$$

where $K$ is the observer gain matrix which is determined by

**Lemma 1.** [11]: *If the given observer gain matrix in Equation (8) satisfies*

$$K = P_\varepsilon^{-1}C^T, \tag{9}$$

*then matrix $P_\varepsilon$ can be obtained from the Lyapunov equation*

$$A^TP_\varepsilon + P_\varepsilon A - C^TC + \varepsilon C^TP_\varepsilon = 0, \tag{10}$$

*where $\varepsilon$ is a positive constant such that $P_\varepsilon \ge 0$, and the state space model Equation (6) is an asymptotic estimation with $\lim\limits_{t\to\infty} e(t) = \lim\limits_{t\to\infty}(\hat{x}(t) - x(t)) = 0$.*

### 3.2. Adaptive Sliding Mode Thau Observer for Fault Diagnosis

The following conditions and lemmas are given for the ASMTO design:

**C3** $f(t)$ and $\dot{f}(t)$ are norm-bounded, i.e., $\|f(t)\| \le f_1$, $\|\dot{f}(t)\| \le f_2$, with $f_1, f_2 > 0$.

**C4** There exists an unknown constant that satisfies $\|d(t)\| \le N$.

**Lemma 2.** *For a given symmetric matrix $P \ge 0$ and scalar $\mu > 0$, the following inequality must be satisfied:*

$$2x^Ty \le \frac{1}{\mu}x^TPx + \mu y^TP^{-1}y. \tag{11}$$

**Lemma 3.** *If C2 holds, there exists a matrix $P \ge 0$ such that*

$$2e^TP(p(x_1,u) - p(x_2,u)) \le \gamma^2e^TPPe + e^Te. \tag{12}$$

If all the above conditions and lemmas hold, then the ASMTO has a form

$$\begin{cases} \dot{\hat{x}}(t) = A\hat{x}(t) + p(\hat{x},u) + Bu(t) + E_dv(t) + F\hat{f}(t) + K(\hat{y}(t) - y(t)) \\ \hat{y} = C\hat{x}(t) \end{cases}, \tag{13}$$

where $\hat{x}(t) \in R^n$, $\hat{f}(t) \in R^l$, $\hat{y}(t) \in R^q$ are the observer state vector, fault estimation of $f(t)$, and observer output vector, respectively. $K$ is the Thau observer gain matrix and $v(t)$ is given by the following algorithm:

$$\begin{aligned} \dot{n}(t) &= \alpha\|F_1e_y(t)\| \\ v(t) &= -n(t)\tfrac{F_1e_y(t)}{\|F_1e_y(t)\|} \end{aligned}, \tag{14}$$

where $\alpha$ is a constant and $F_1$ is discussed in the "stability analysis" section.

### 3.3. Stability Analysis

Denote

$$\begin{array}{c} e_x = \hat{x}(t) - x(t) \\ \tilde{n}(t) = n(t) - N \\ e_y = \hat{y}(t) - y(t) \\ e_f = \hat{f}(t) - f(t) \end{array}. \quad (15)$$

Then, the error dynamics can be obtained from (7), (13), and (15) as

$$\begin{array}{c} \dot{e}_x(t) = (A - KC)e_x + p(\hat{x}, u) - p(x, u) \\ +Fe_f + E_d(v(t) - d(t)) \end{array}. \quad (16)$$

**Theorem 1.** *For a given observer gain, K , if there exist matrices* $P = P^T > 0$, $G = G^T > 0$, $F_1$, *and* $F_2$ *such that*

$$\begin{bmatrix} P(A - KC) + (A - KC)^T P + \gamma^2 PP + I & 0 \\ 0 & \frac{\sigma+1}{\sigma} G \end{bmatrix} < 0, \quad (17)$$

$$E_d^T P = \frac{1}{\sigma} F_1 C, \quad (18)$$

$$F^T P = \frac{1}{\sigma} F_2 C, \quad (19)$$

*where* $\sigma$ *is positive constant, then, the fault estimation algorithm can be described as*

$$\dot{\hat{f}}(t) = -\Gamma F_2 e_y + \sigma \Gamma \hat{f}(t), \quad (20)$$

*where* $\Gamma$ *is the learning rate matrix,* $\Gamma = \Gamma^T > 0$.

**Remark 1.** *The adaptive law in Equation (20) uses both error dynamics and fault vector information. While the proportional term can lead to a rapid improvement in system response, the fault vector can eliminate the error of estimation.*

**Proof.** Considering the following Lyapunov function.

$$V(t) = e_x^T P e_x + \frac{1}{\sigma} e_f^T \Gamma^{-1} e_f + \frac{1}{\sigma} \tilde{n}^T \alpha^{-1} \tilde{n} \quad (21)$$

Then, its time derivative $\dot{V}(t)$ is

$$\begin{aligned} \dot{V}(t) &= \dot{e}_x^T(t) P e_x(t) + e_x^T(t) P \dot{e}_x(t) \\ &\quad + \tfrac{2}{\sigma} e_f^T(t) \Gamma^{-1} \dot{e}_f(t) + \tfrac{2}{\sigma} \dot{n}(t) \alpha^{-1} \tilde{n}(t) \\ &= e_x^T(t) \left[ P(A - KC) + (A - KC)^T P \right] e_x(t) \\ &\quad + 2 e_x^T(t) P E_d (v(t) - d(t)) \\ &\quad + 2 e_x^T(t) P F e_f(t) + \tfrac{2}{\sigma} \| F_1 e_y(t) \| (n(t) - N) \\ &\quad + 2 e_x^T(t) P (p(\hat{x}, u) - p(x, u)) \\ &\quad + \tfrac{2}{\sigma} e_f^T \Gamma^{-1} \dot{\hat{f}}(t) - \tfrac{2}{\sigma} e_f^T \Gamma^{-1} \dot{f}(t) \end{aligned}. \quad (22)$$

Using Theorem 1, Lemma 2, and Lemma 3, one can see that

$$\begin{aligned} & 2e_x^T(t)PFe_f(t) + \tfrac{2}{\sigma}e_f^T\Gamma^{-1}\dot{\hat{f}}(t) \\ & = 2e_x^T(t)PFe_f(t) + \tfrac{2}{\sigma}e_f^T\Gamma^{-1}\left(-\Gamma F_2 e_y + \sigma\Gamma\hat{f}(t)\right) \\ & = 2e_f^T\hat{f}(t) \\ & \leq e_f^T G e_f + \hat{f}^T(t)G^{-1}\hat{f}(t) \\ & \leq e_f^T G e_f + f_1^2\lambda_{\max}(G^{-1}) \end{aligned}, \tag{23}$$

$$\begin{aligned} & 2e_x^T(t)PE_d(v(t) - d(t)) \\ & = \tfrac{2}{\sigma}(F_1 e_y(t))^T\left(-n(t)\tfrac{F_1 e_y(t)}{\|F_1 e_y(t)\|} - d(t)\right) \\ & < -\tfrac{2}{\sigma}\|F_1 e_y(t)\|(n(t) - N) \end{aligned}, \tag{24}$$

where $\lambda_{\max}$ is the maximum eigenvalue of the associated matrix.

From Lemma 2, one can see that

$$\begin{aligned} -\tfrac{2}{\sigma}e_f^T(t)\Gamma^{-1}\dot{f}(t) & = \tfrac{2}{\sigma}\left(-e_f^T(t)\right)\left(\Gamma^{-1}\dot{f}(t)\right) \\ & \leq \tfrac{1}{\sigma}(e_f^T(t)Ge_f(t) \\ & \quad +\dot{f}^T(t)\Gamma^{-1}G^{-1}\Gamma^{-1}\dot{f}(t)) \\ & \leq \tfrac{1}{\sigma}(e_f^T(t)Ge_f(t) \\ & \quad +f_2^2\lambda_{\max}(\Gamma^{-1}G^{-1}\Gamma^{-1})) \end{aligned}. \tag{25}$$

According to Lemma 3, we obtain

$$\begin{aligned} & e_x^T(t)[P(A-KC) + (A-KC)^T P]e_x(t) \\ & +2e_x^T(t)P(p(\hat{x},u) - p(x,u)) \\ & \leq e_x^T(t)[P(A-KC) + (A-KC)^T P \\ & \quad +\gamma^2 PP + I]e_x(t) \end{aligned}. \tag{26}$$

With (23), (24), (25), and (26), Equation (22) becomes

$$\begin{aligned} \dot{V}(t) & = e_x^T(t)[P(A-KC) + (A-KC)^T P \\ & \quad +\gamma^2 PP + I]e_x(t) \\ & \quad +e_f^T G e_f + f_1^2\lambda_{\max}(G^{-1}) \\ & \quad +\tfrac{1}{\sigma}(e_f^T(t)Ge_f(t) + f_2^2\lambda_{\max}(\Gamma^{-1}G^{-1}\Gamma^{-1})) \\ & = e_x^T(t)[P(A-KC) + (A-KC)^T P \\ & \quad + \gamma^2 PP + I]e_x(t) \\ & \quad \tfrac{\sigma+1}{\sigma}e_f^T(G)e_f + \eta \\ & \leq \xi^T(t)\Theta\xi(t) + \eta \end{aligned}, \tag{27}$$

where $\eta = f_1^2\lambda_{\max}(G^{-1}) + \frac{1}{\sigma}f_2^2\lambda_{\max}(\Gamma^{-1}G^{-1}\Gamma^{-1}))$, $\xi(t) = \begin{bmatrix} e_x^T(t) & e_f^T(t) \end{bmatrix}$, and $\Theta = \begin{bmatrix} P(A-KC) + (A-KC)^T P + \gamma^2 PP + I & 0 \\ 0 & \frac{\sigma+1}{\sigma}G \end{bmatrix}$.

If $\Theta < 0$, then $\dot{V}(t) < 0$ for $\sigma\|\xi(t)\|^2 > \eta$, where $\sigma = \lambda_{\min}(-\Theta)$. This means that $(e_x(t), e_y(t))$ converges to a small set, according to Lyapunov stability theory [25]. □

**Remark 2.** *It is difficult to solve Equations (17)–(19) simultaneously, and this problem can be addressed using the LMI technique. Therefore, we modify Equations (18) and (19) to [25]*

$$\begin{bmatrix} \eta_1 I & E_d^T P - F_1 C \\ (E_d^T P - F_1 C)^T & \eta_1 I \end{bmatrix} > 0, \tag{28}$$

$$\begin{bmatrix} \eta_2 I & F^T P - \frac{1}{\sigma} F_2 C \\ (F^T P - \frac{1}{\sigma} F_2 C)^T & \eta_2 I \end{bmatrix} > 0. \tag{29}$$

## 4. Experimental Results

### *4.1. Experimental Setup and Parameters*

For safety purposes, the quadcopter test bed was developed in the guidance, navigation, and control (GNC) lab (Figure 2). The fault diagnosis algorithm from Section 3 was tested on a DJI F450 quadcopter. The algorithm was implemented on a Pixhawk2 flight controller using C++ program from Eclipse software [26]. The flight controller used firmware version 3.5. In the experimental setup, a remote control was used to inject faults by limiting the pulse width modulation (PWM) of the motors, which allowed us to switch between stabilized and fault modes. During testing, the Mission Planner (MP), a commercially available software, was used to monitor flight data through Xbee (Telemetry) communication [27]. Since the MP has some limitations with regard to parameter monitoring, the fault estimation data had to be obtained through a C++ program that writes them to a log file. The experimental procedure is summarized in Figure 3.

**Figure 2.** DJI F450 quadcopter.

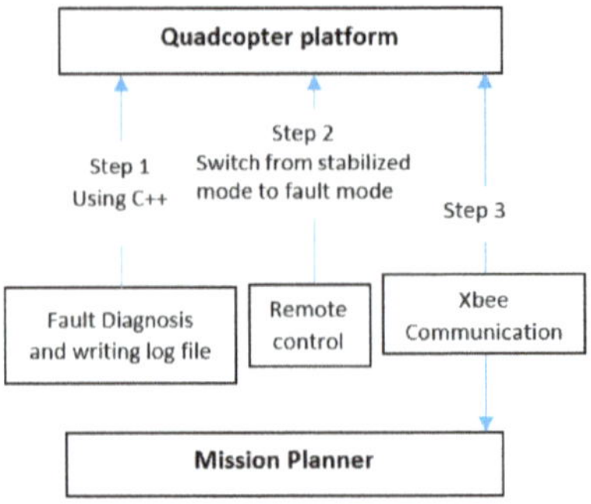


**Figure 3.** Experimental procedure.

The DJI F450 parameter values are shown in Table 1. For the experiment, the matrices were chosen as follows: the fault matrix $F = B$, and the disturbance matrix $E_d = \begin{bmatrix} 1 & 1 & 1 & 1 & 1 & 1 \end{bmatrix}^T$. The pair $(A, C)$ is observable and condition C1 is satisfied, $p(x, u)$ is continuously differentiable, and satisfies condition C2 as it only contains multiplications and divisions. Thus, all conditions are met, and the proposed scheme is applicable.

**Table 1.** DJI F450 quadcopter parameters.

| Parameter | Description | Value |
|---|---|---|
| $L$ | Arm length | 0.225 m |
| $b$ | Thrust coefficient | $9.8 \times 10^{-6}$ N/m$^2$ |
| $d$ | Drag coefficient | $1.6 \times 10^{-7}$ |
| m | Mass | 2 kg |
| $I_x$; $I_y$; $I_z$ | Moments of inertia | 0.0035; 0.0035; 0.005 kg·m$^2$ |
| $J_T$ | Rotor inertia | $2.8 \times 10^{-6}$ kg·m$^2$ |

We used the following learning rate $\Gamma = diag(0.005, 0.005, 0.005)$ and sampling time $T = 0.0025$ s for the experimental test bed. The matrices obtained from the ASMTO are $F_1 = \begin{bmatrix} 101.77 & 101.77 & 101.77 & 101.77 & 101.77 & 101.77 \end{bmatrix}$, $F_2 = \begin{bmatrix} -26 & 5 & 5 & 13,946 & 5 & 5 \\ 5 & -24 & 5 & 5 & 12,887 & 5 \\ 3 & 3 & -16 & 3 & 3 & 8701 \end{bmatrix}$ $G = 100 \times I_{6\times6}$, $K = \begin{bmatrix} 100 & 0 & 0 & 1 & 0 & 0 \\ 0 & 100 & 0 & 0 & 1 & 0 \\ 0 & 0 & 100 & 0 & 0 & 1 \\ 1 & 0 & 0 & 100 & 0 & 0 \\ 0 & 1 & 0 & 0 & 100 & 0 \\ 0 & 0 & 1 & 0 & 0 & 100 \end{bmatrix}$, $P = \begin{bmatrix} 101.8 & 0.04 & 0.04 & -0.19 & 0.04 & 0.04 \\ 0.04 & 101.81 & 0.04 & 0.04 & -0.19 & 0.04 \\ 0.04 & 0.04 & 101.8 & 0.04 & 0.04 & -0.19 \\ -0.19 & 0.04 & 0.04 & 101.8 & 0.04 & 0.04 \\ 0.04 & -0.19 & 0.04 & 0.04 & 101.8 & 0.04 \\ 0.04 & 0.04 & -0.19 & 0.04 & 0.04 & 101.8 \end{bmatrix}$.

The 30% partial loss fault is injected artificially into motor 1 by limiting the PWM of the motor at time $t$ = 7 s. This is achieved by changing from stabilized mode to fault mode using the remote control. The fault percentage is user-controllable, and can be set in the C++ program. The moments of motors M2, M3, and M4 remained zero while the moment of motor M1 decreases because of the actuator fault (Figure 4).

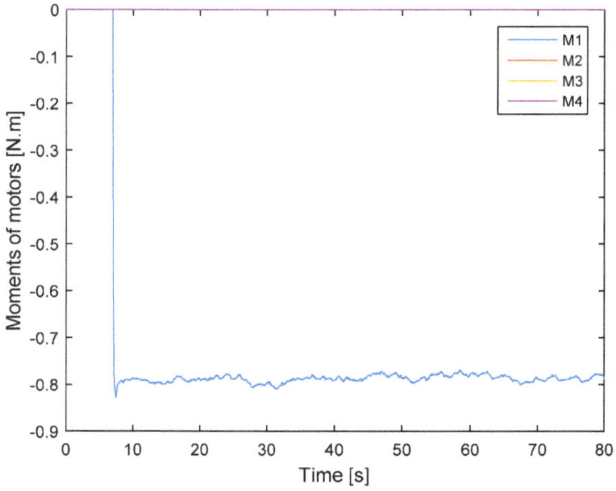


**Figure 4.** Motor offsets caused by the fault.

### *4.2. Robust Fault Diagnosis Result*

The in-flight attitude response is shown in Figure 5. While Figure 6 shows the fault offset estimations M1 to M4, the real value and fault offset estimation for M1 are shown in Figure 7. It can be seen from Figure 6 that the estimation values of M2, M3, and M4 are affected by that of M1 from 7 to 18 s and, then, they converge to zero. Moreover, from the Figure 7, we see that the fault offset estimation value using ASMTO converges to the desired value with high accuracy. Figure 8 compares the real controller output offset and its estimation. From this Figure, we see that the estimation value can smoothly track the real one. Although ASMTO does not use noise filtering and identification technique for drag terms, which is presented in [20], the estimation values still obtain the high accuracy and smooth tracking.

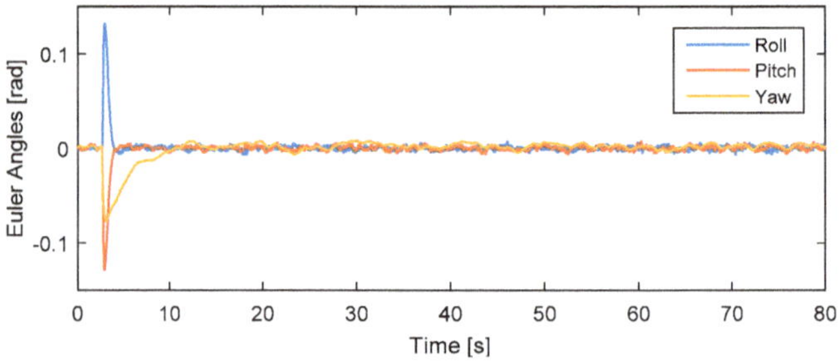


**Figure 5.** The attitude angles.

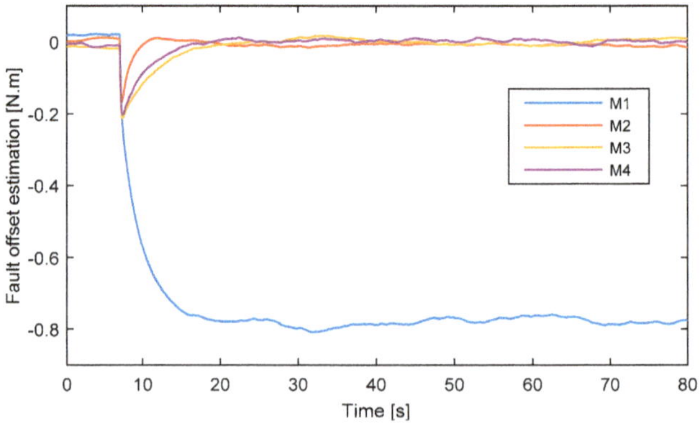


**Figure 6.** Fault estimation.

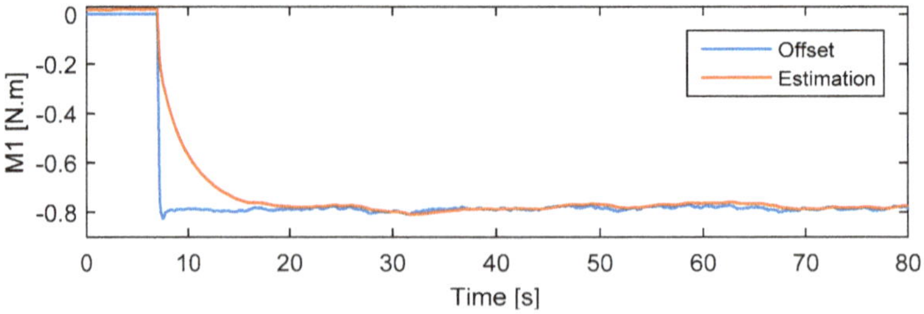


**Figure 7.** M1 offset by fault and its estimation.

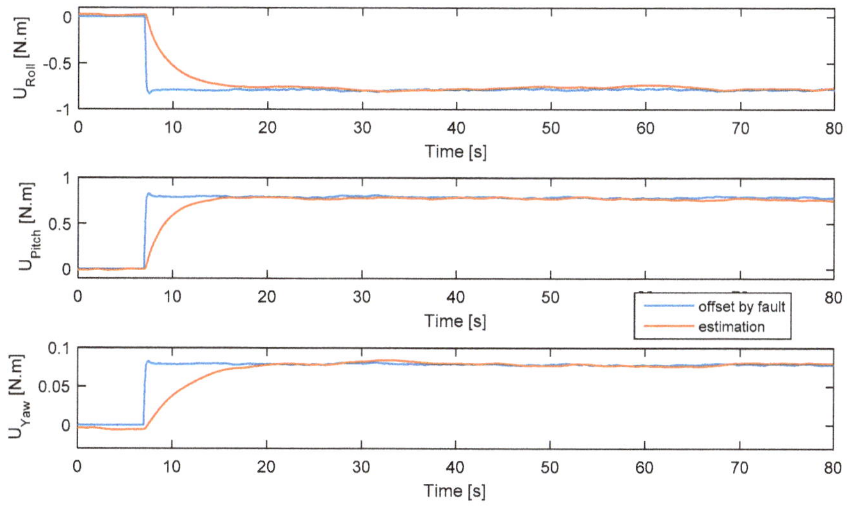


**Figure 8.** Controller output offset by fault and its estimation.

**Remark 3.** *From Equations (15) and (20), it is easy to show that*

$$\dot{e}_f = -\Gamma F_2 e_y + \sigma\Gamma e_f + \sigma\Gamma f(t) - \dot{f}(t). \tag{30}$$

The speed with which the estimation converges depends on the fault characteristics and the ASMTO design parameters. From Equation (30), we can see that $\Gamma$ and $\sigma$ need to be tuned in order to obtain better estimations that are adapted to the fault characteristics. Normally, the value of $\sigma$ would be fixed in this algorithm.

**Remark 4.** *Since the moments of roll and pitch are at least one order of magnitude larger than the moment of yaw, the latter has a larger error in its fault estimation. We resolved this problem by using an amplification and reduction technique [22].*

**Remark 5.** *The fault-tolerant controller was not the main focus of this paper, and we only used an attitude controller for the actuator fault diagnosis. Cases with more than one actuator fault were considered to be beyond the scope of this work, and were excluded mainly due to safety concerns. Moreover, the partial loss fault should be smaller than the nominal thrust of the quadcopter.*

## 5. Conclusions

In this paper, a robust fault diagnosis method based on a Thau observer has been investigated for use on a quadcopter UAV under actuator fault, using a nonlinear modelling approach. Contrary to previous studies, the proposed scheme not only detects time-varying faults, but also works with an unknown upper bound of the associated disturbances. The stability of the error system could be demonstrated under the presence of an actuator fault. The experimental results could prove the effectiveness of this new method. In our future work, we will attempt to relax Equations (17)–(19) and implement an FTC for an attitude and position controller that will be based on the actuator fault estimation information.

**Author Contributions:** Conceptualization, N.P.N. and S.K.H.; Methodology, N.P.N. and S.K.H.; Software, N.P.N.; Validation, N.P.N.; Formal Analysis, N.P.N.; Investigation, N.P.N.; Resources, N.P.N.; Data Curation, N.P.N.; Writing-Original Draft Preparation, N.P.N.; Writing-Review & Editing, S.K.H.; Visualization, N.P.N.; Supervision, S.K.H.; Project Administration, S.K.H.; Funding Acquisition, S.K.H.

**Funding:** This research was funded and conducted under the Competency Development Program for Industry Specialists of the Korean Ministry of Trade, Industry and Energy (MOTIE), operated by Korea Institute for Advancement of Technology (KIAT), grant number N0002431.

**Acknowledgments:** This research was supported by Guidance, Navigation, and Control Lab at Sejong University.

**Conflicts of Interest:** The authors declare no conflict of interest.

## References

1. Ren, W.; Beard, R.W. Trajectory tracking for unmanned air vehicles with velocity and heading rate constraints. *IEEE Trans. Control Syst. Technol.* **2004**, *12*, 706–716. [CrossRef]
2. Bonna, R.; Camino, J.F. Trajectory Tracking Control of a Quadcopter Using Feedback Linearization. In Proceedings of the XVII International Symposium on Dynamic Problems of Mechanics, Natal-Rio Grande Do Norte, Brazil, 22–27 February 2015.
3. Amoozgar, M.H.; Chamseddine, A.; Zhang, Y. Experimental test of a two-stage Kalman filter for actuator fault detection and diagnosis of an unmanned quadcopter helicopter. *J. Intell. Robot. Syst.* **2013**, *70*, 107–117. [CrossRef]
4. Chen, F.; Lei, W.; Tao, G.; Jiang, B. Actuator Fault Estimation and Reconfiguration Control for Quad-rotor Helicopter. *Int. J. Adv. Robot. Syst.* **2016**, *13*. [CrossRef]
5. Zhao, W.; Go, T.H. Quadcopter formation flight control combining MPC and robust feedback linearization. *J. Frankl. Inst.* **2014**, *351*, 1335–1355. [CrossRef]
6. Mahmood, A.; Kim, Y. Decentralized formation flight control of quadcopters using robust feedback linearization. *J. Frankl. Inst.* **2017**, *354*, 852–871. [CrossRef]
7. Zhao, Q.; Jiang, J. Reliable state feedback control system design against actuator failures. *Automatica* **1998**, *34*, 1267–1272. [CrossRef]
8. Tao, G.; Chen, S.; Joshi, S.M. An adaptive actuator failure compensation controller using output feedback. *IEEE Trans. Autom. Control* **2002**, *47*, 506–511. [CrossRef]
9. Zhang, Y.; Jiang, J. Bibliographical review on reconfigurable fault-tolerant control systems. *Annu. Rev. Control* **2008**, *32*, 229–252. [CrossRef]
10. Freddi, A.; Longhi, S.; Monteriù, A. A diagnostic thau observer for a class of unmanned vehicles. *J. Intell. Robot. Syst.* **2012**, *67*, 61–73. [CrossRef]
11. Freddi, A.; Longhi, S.; Monteriù, A. A model-based fault diagnosis system for a mini-quadrotor. In Proceedings of the 2009 7th Workshop on Advanced Control and Diagnosis, Zielona Gora, Poland, 19–20 November 2009; pp. 19–20.
12. Ma, L. Development of Fault Detection and Diagnosis Techniques with Applications to Fixed-Wing and Rotarywing UAVs. Master's Thesis, Concordia University, Montréal, QC, Canada, 2011.
13. Ma, L.; Zhang, Y.M. Fault detection and diagnosis for GTM UAV with dual unscented Kalman filter. In Proceedings of the AIAA Guidance, Navigation, and Control Conference, Toronto, ON, Canada, 2–5 August 2010; p. 7884.
14. Veluvolu, K.C.; Defoort, M.; Soh, Y.C. High-gain observer with sliding mode for nonlinear state estimation and fault reconstruction. *J. Frankl. Inst.* **2014**, *351*, 1995–2014. [CrossRef]
15. Chen, F.; Zhang, K.; Jiang, B.; Wen, C. Adaptive sliding mode observer-based robust fault reconstruction for a helicopter with actuator fault. *Asian J. Control* **2016**, *18*, 1558–1565. [CrossRef]
16. Fekih, A.; Xu, H.; Chowdhury, F.N. Neural networks based system identification techniques for model based fault detection of nonlinear systems. *Int. J. Innov. Comput. Inf. Control* **2007**, *3*, 1073–1085.
17. Rajakarunakaran, S.; Venkumar, P.; Devaraj, D.; Surya Prakasa Rao, K. Artificial neural network approach for fault detection in rotary system. *J. Appl. Soft Comput.* **2008**, *8*, 740–748. [CrossRef]
18. Zhang, K.; Jiang, B.; Shi, P. Adaptive Observer-Based Fault Diagnosis with application to satellite attitude control system. In Proceedings of the Second International Conference on Innovative Computing, Information and Control, Kumamoto, Japan, 5–7 September 2007; p. 508.
19. Wang, H.; Daley, S. Actuator fault diagnosis: An adaptive observer-based technique. *IEEE Trans. Autom. Control* **1996**, *41*, 1073–1078. [CrossRef]
20. Li, L.; Chadli, M.; Ding, S.X.; Qiu, J.; Yang, Y. Diagnostic Observer Design for T-S Fuzzy Systems: Application to Real-Time Weighted Fault Detection Approach. *IEEE Trans. Fuzzy Syst.* **2018**, *26*, 805–816. [CrossRef]

21. Youssef, T.; Chadli, M.; Karimi, H.R.; Wang, R. Actuator and sensor faults estimation based on proportional integral observer for TS fuzzy model. *J. Frankl. Inst.* **2017**, *354*, 2524–2542. [CrossRef]
22. Cen, Z.; Noura, H.; Susilo, T.B.; Youmes, Y.A. Robust Fault Diagnosis for Quadrotor UAVs Using Adaptive Thau Observer. *J. Intell. Robot. Syst.* **2014**, *73*, 573–588. [CrossRef]
23. Walcott, B.L.; Corless, M.J.; Zak, S.H. Comparative study of nonlinear state-observation techniques. *Int. J. Control* **1987**, *45*, 2109–2132. [CrossRef]
24. Zhang, Y.; Chamseddine, A. Fault tolerant flight control techniques with application to a quadrotor UAV testbed. In *Automatic Flight Control Systems—Latest Developments*; Lombaerts, T., Ed.; InTech: Vienna, Austria, 2012; pp. 119–150.
25. Zhang, K.; Jiang, B.; Cocquempot, V. Adaptive observer-based fast fault estimation. *Int. J. Control Autom. Syst.* **2008**, *6*, 320–326.
26. Editing/Building with Eclipse on Windows. Available online: http://ardupilot.org/dev/docs/editing-the-code-with-eclipse.html (accessed on 31 August 2018).
27. Telemetry. Available online: http://ardupilot.org/copter/docs/common-telemetry-landingpage.html (accessed on 31 August 2018).

*applied sciences*

MDPI

*Article*

# Design of Wing Root Rotation Mechanism for Dragonfly-Inspired Micro Air Vehicle

**Jae Hyung Jang and Gi-Hun Yang ***

Robotics Group, Korea Institute of Industrial Technology, Ansan-si 15588, Gyeonggi-do, Korea; jhjang7@kitech.re.kr

* Correspondence: yanggh@kitech.re.kr; Tel.: +82-31-8040-6389

Received: 7 August 2018; Accepted: 29 September 2018; Published: 10 October 2018

**Abstract:** This paper proposes a wing root control mechanism inspired by the drag-based system of a dragonfly. The previous mechanisms for generating wing rotations have high controllability of the angle of attack, but the structures are either too complex or too simple, and the control of the angle of attack is insufficient. In order to overcome these disadvantages, a wing root control mechanism was designed to improve the control of the angle of attack by controlling the mean angle of attack in a passive rotation mechanism implemented in a simple structure. Links between the proposed mechanism and a spatial four-bar link-based flapping mechanism were optimized for the design, and a prototype was produced by a 3D printer. The kinematics and aerodynamics were measured using the prototype, a high-speed camera, and an F/T sensor. In the measured kinematics, the flapping amplitude was found to be similar to the design value, and the mean angle of attack increased by approximately 30° at a wing root angle of 0°. In the aerodynamic analysis, the drag-based system implemented using the wing root control mechanism reduced the amplitude of the force in the horizontal direction to approximately 0.15 N and 0.1 N in the downstroke and upstroke, respectively, compared with the lift-based system. In addition, at an inclined stroke angle, the force in the horizontal direction increased greatly when the wing root angle was 0° at the inclined stroke angle, while the force in the vertical direction increased greatly at a wing root angle of 30°. This means that the flight mode can be controlled by controlling the wing root angle. As a result, it is shown that the wing root control mechanism can be applied to the MAV (micro air vehicle) to stabilize hovering better than the MAV using a lift-based system and can control the flight mode without changing the posture.

**Keywords:** biomimetic robot; micro air vehicle; flapping; drag-based system; dragonfly

## 1. Introduction

Small birds and insects are good objects to mimic for developing a micro air vehicle (MAV) for stable flight at low Reynolds fluid [1–7]. The RoboBee developed by researchers at Harvard [1], the robot hummingbird of DARPA [2], the beetle robot of Konkuk University [3], and the tailless aerial robot inspired by the flies of Delft [7] are representative robots developed by mimicking the flight of small birds and insects. Among the insects mimicked, the dragonfly has the most stable hovering ability, the ability to switch flight modes without changing posture, and ability to fly backwards [8,9]. In order to achieve this high maneuverability, the dragonfly uses characteristics such as the phase difference between the forewing and hind wing [8,10–12], independent control of each wing [13], and a drag-based system in hovering flight [14,15].

Among these characteristics, we focus on the drag-based system. Most insects, except the dragonfly, use a lift-based system in which they stroke their wings in the horizontal direction of the body during hovering. On the other hand, dragonflies use a drag-based system in which they stroke their wings in an inclined direction to the body. This drag-based system not only provides more stable flight but also improves maneuverability by maintaining posture when changing flight mode [14,15]. Dragonflies that use a drag-based system rely on drag force for 76% of the force required for hovering, and most vertical forces for hovering flight are obtained on the downstroke [15]. For this reason, in order to maximize their vertical force, dragonflies increase the drag force through a large angle of attack during the downstroke, while minimizing the drag force through a small angle of attack in the upstroke [16,17].

The aforementioned angle of attack is an important factor that enables most insects, including dragonflies, to control the magnitude and direction of the force generated when flapping their wings [18]. Most small birds and insects produce an angle of attack by rotating their wings in the longitudinal direction when the direction of the flapping stroke is reversed, such as from the upstroke to the downstroke, or from the downstroke to the upstroke [19,20]. These wing rotations are divided into passive rotation using aerodynamic force and inertial force and active rotation that is directly controlled through the muscles. It is not known exactly what type of wing rotation insects use, but it has been proven that passive rotation explicitly occurs during flapping [21]. A variety of mechanisms for generating wing rotation for the FW-MAV (flapping wing micro air vehicle) have been proposed in studies mimicking the flight characteristics of insects.

These mechanisms are divided into passive [22] and active rotation mechanisms [23,24] like insect wing rotation. A passive rotation mechanism has been produced by creating a wing frame using a flexible material or loose membrane. For the active rotation mechanism, a method of artificially generating rotation using a spring, and a method of simultaneously implementing flapping and rotation with a single actuator using a spatial four-bar link mechanism have been presented. Various attempts have been made to develop a wing rotation mechanism, but there is a clear limit to mimicking the drag-based system of dragonflies. In the case of the active rotation mechanism, since wing rotation is directly generated using the power of an actuator, it is suitable to implement the desired movement through mechanical design. Moreover, the flapping pattern is easily switched from an asymmetric flapping pattern to a symmetric flapping pattern or vice versa by adding degrees of freedom to the MAV. However, this mechanism is difficult to apply in practical MAV development. The reason is because not only is the structure complex, such that the weight of the airframe increases, and friction and torsion occur to a large extent, but additional energy is also required to generate wing rotation. On the other hand, the passive rotation mechanism has a relatively simple structure and is widely used in the development of MAV [1–5,25,26]. A passive rotation mechanism has also been developed to control the magnitude of the angle of attack by adding degrees of freedom to the wing roots in a passive rotation mechanism in which the wing's membrane is loosely constructed. The passive rotation mechanism, which depends on the proposed external force, is suitable for developing an MAV with the same angle of attack in the up-down stroke, such as a symmetric flapping pattern. However, it is not suitable for developing MAVs with asymmetric flapping patterns that require a different angle of attack in up-down strokes. Thus, both proposed wing rotation mechanisms have limitations in terms of mimicking the drag-based system of a dragonfly.

The study proposes a mechanism that mimics the flight characteristics of dragonflies, using a drag-based system for hovering flight and a lift-based system for forward flight. The mechanism can be applied to an MAV using a passive rotation mechanism and can implement an asymmetric flapping pattern by changing the mean angle of attack. The proposed mechanism consists of a slide-crank mechanism designed to operate independently of flapping. The flapping motion was implemented using a spatial four-bar link mechanism, while the passive rotation mechanism was fabricated using the design parameters of the wing, optimized for the symmetric flapping pattern from previous research [27]. A high-speed camera was used for kinematic analysis of the prototype, and force measurement tests were conducted to determine whether the horizontal force amplitude of the drag-based system implemented by the proposed mechanism was reduced compared with that of the lift-based system, as well as the force direction variations with respect to changes in the wing root angle.

## 2. Flight Characteristics of Dragonflies

### *2.1. Kinematics of Dragonfly Flight*

The kinematics [28,29] and aerodynamics [14–16,20,30,31] of dragonflies have been studied extensively in the past. Wang analyzed the stroke amplitude and wing angle of attack placed at 70% from the wing root in forward flight using an ultrahigh-speed camera [28]. In the same way, Azuma measured the stroke amplitude and angle of attack in hovering flight of the dragonfly [29]. However, the maximum and minimum stroke angle and mean angle of attack must be clearly defined in order to proceed with the aerodynamic experiment on the MAV. In this study, these parameters were defined as listed in Table 1, and the coordinate system of the kinematics is shown in Figure 1. The figure shows the flapping motion in the relative coordinate system $x'y'z'$ rotated by $\alpha$ with respect to the absolute coordinate system xyz, where $\alpha$ is the stroke angle, $\beta$ is the stroke amplitude, and $\delta$ is the angle of attack.

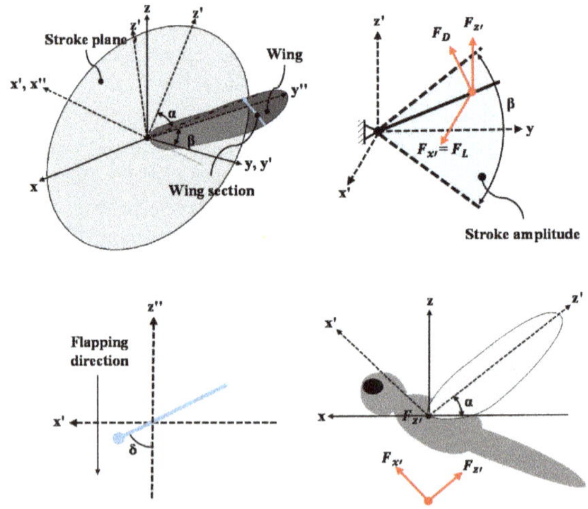


**Figure 1.** Coordinate system of flapping motion.

**Table 1.** Kinematics of dragonfly for experiment.

| | | Angle of Attack | | Stroke Amplitude | Stroke Angle |
|---|---|---|---|---|---|
| | | Hover | Forward | | |
| Angle | Down stroke | 75° | 45° | −40°~40° | 60° |
| | Up stroke | 15° | 45° | | |
| | Mean angle of attack | 30° | 0° | | |

### *2.2. Aerodynamics of Dragonfly Flight*

The systems used by organisms for swimming or flying are divided into drag-based systems and lift-based systems, depending on the type of force [14,15]. The lift-based system uses lift force in the manner of most fish or flying insects. Lift forces are obtained through up-down strokes, whereas drag force is generated in the opposite direction with the each up-down stroke, and the net force generated for one cycle is close to zero, as shown in Figure 2a. The drag-based system is based on the drag force, used in rowing locomotion or jellyfish swimming. In order to generate enough drag force to fly or swim, dragonflies flap their wings, and with each down-stroke maintain a large angle of attack to increase the drag force. In contrast, the upstroke minimizes the drag force by flapping with a small angle of attack, as shown in Figure 2b. This method allows the creature to generate force in the direction of movement through the drag force.

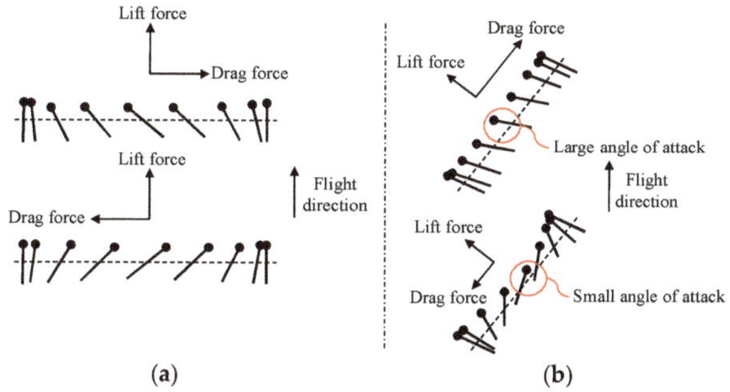


**Figure 2.** Stroke motion of a (**a**) lift-based system and (**b**) drag-based system.

Wang [14] mentioned that the drag-based system is more stable than the lift-based system in stationary flight, while the lift-based system has superior performance in forward flight. In addition, Vogel [15] experimentally proved that the drag-based system is more stable in stationary flight. Unlike most insects that use a lift-based system, dragonflies use both a drag-based system in hovering flight and lift-based system in forward flight. This system conversion is achieved by changing their angle of attack, which gives them more stable and superior flight abilities.

Dickinson experimentally defined lift and drag coefficients for when the thin wing moves in low Reynolds fluid as the following equations.

$$C_L = 0.225 + 1.58\sin(2.13\alpha - 7.20), \quad (1)$$

$$C_D = 1.92 - 1.55\sin(2.04\alpha - 9.82). \quad (2)$$

where $C_L$ is the lift coefficient, $C_D$ is the drag coefficient, and $\alpha$ is the angle of attack. The equation shows that the drag coefficient is a function of the angle of attack. In the relationship between the angle of attack and drag force, in order to use the drag-based system at hovering, the dragonfly increases the drag force using a large angle of attack at the downstroke and minimizes the drag force using a small angle of attack at the upstroke. On the other hand, in forward flight, the dragonfly flies by generating lift force using the same angle of attack in the up-down stroke like other insects. Thus, the rapid conversion from a drag-based system to a lift-based system without changing the stroke angle is achieved by changing the mean angle of attack.

## 3. Development of a Wing Root Rotation Mechanism

### 3.1. Passive Wing Rotation Mechanism

In the flight of insects, wing rotation is an essential factor for a positive angle of attack at flapping. To generate the wing rotation passively, this study used a wing rotation mechanism using a loose wing membrane, as shown in Figure 3 [27]. The proposed wing rotation mechanism was fabricated with a wing membrane larger than the designed wing membrane to generate wing rotation, and this changes the angle of attack by rotating the wing root. This wing rotation made by the loose wing membrane not only generates more lift force by the camber wing structure in flight but also increases the efficiency while gliding [32–35].

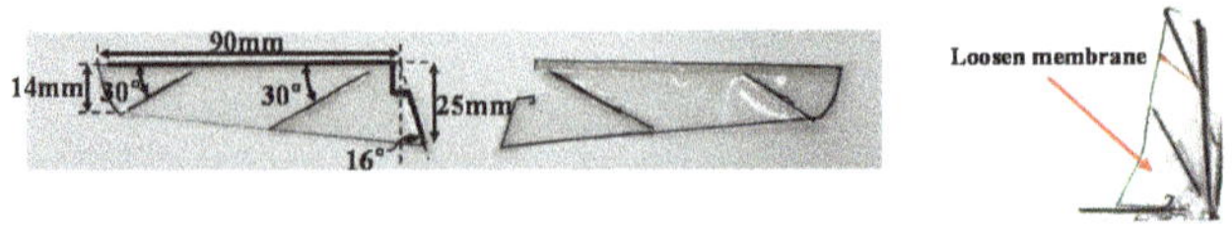


**Figure 3.** (**a**) Passive wing rotation mechanism using a loose wing membrane and (**b**) loosen membrane in moving wing.

The frame of the wing is made of carbon rods to reduce the weight, and the wing membrane is made of 25 μm thick polypropylene film. For the passive rotation of the wing, a polyolefin tube attached at the leading edge and root of the wings reduce the friction at wing rotation. In addition, two 0.6 mm carbon rods attached to the wing membrane maintain the rigidity of the wing after rotation, as shown Figure 3a.

### 3.2. Concept of Wing Root Rotation Mechanism

The purpose of this study is to develop a passive wing rotation mechanism that mimics the flight characteristics of dragonflies and can change the mean angle of attack by rotating the wing root. The mean angle of attack changes by wing root rotation with respect to the horizontal direction of the wing when the wing has the same angle of attack at each up-down stroke. Figure 4 shows the variation of the wing angle of attack range with this passive wing rotation mechanism. We developed the mechanism for changing the mean angle of attack based on a slide-crank link mechanism, where the crank is the wing root and rotates on the y″-axis.

To obtain a lift-based system with the same angle of attack at each up-down stroke, the wing root direction should be perpendicular to the stroke direction. However, the lift-based system of the wing root rotation mechanism has a singularity position problem when designed using a general slide-crank link mechanism. In order to solve this problem, we designed a slide-crank link mechanism with a three-bar link, as shown Figure 5. The wing root of the crank link with the three-bar link is located below the general slide-crank link mechanism's wing root. Therefore, the crank link is located above the singularity position when the wing root direction is perpendicular to the stroke direction. As a result, the wing root can rotate without singularity.

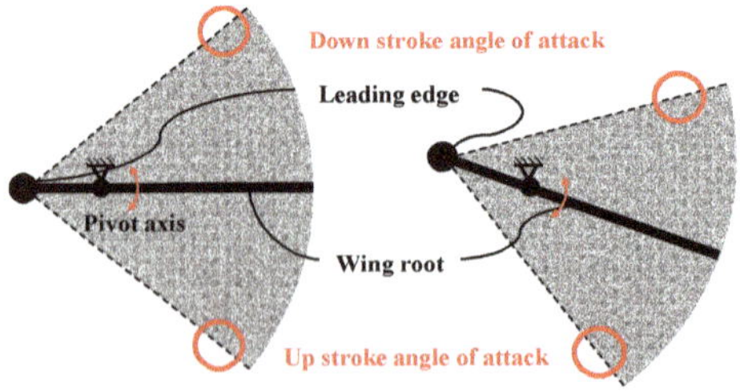


**Figure 4.** Range of angle of attack with respect to the wing root angle.

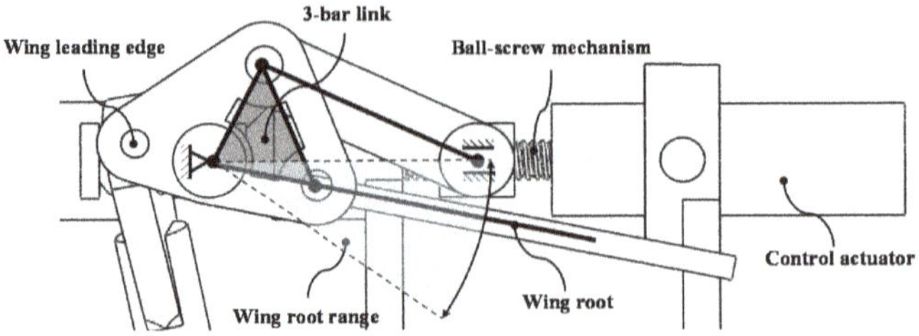


**Figure 5.** Slide-crank mechanism with three-bar link.

### 3.3. Kinematics of Wing Root Rotation Mechanism

Figure 6 shows a schematic of the proposed slide-crank mechanism with the three-bar link to define the relationship among variables, and the kinematics of this mechanism was analyzed by vector analysis.

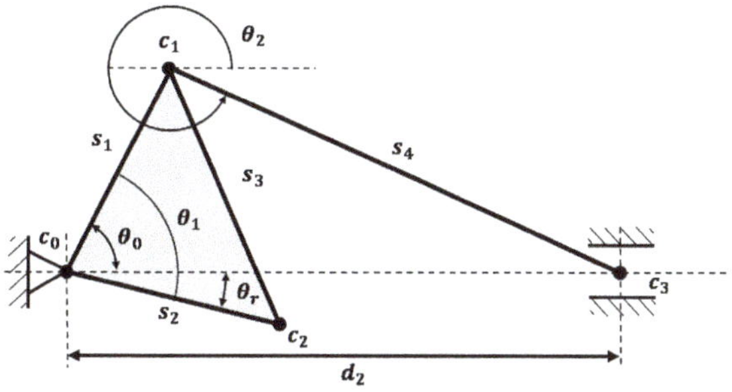


**Figure 6.** Schematic kinematics of wing root rotation mechanism.

The wing root angle $\theta_r$ is defined by $\theta_0$ and $\theta_1$, as in Equation (3).

$$\theta_r = \theta_1 - \theta_0 \tag{3}$$

where $\theta_1$ is a constant value, and $\theta_0$ is a dependent variable described by $d_2$, which is an independent variable.

$$\theta_0 = \cos^{-1}\left(\frac{d_2^2 + s_1^2 - s_4^2}{2s_1 d_2}\right) \tag{4}$$

As a result, $\theta_r$ can be obtained as Equation (5) by substituting (4) into (3).

$$\theta_r = \theta_1 - \cos^{-1}\left(\frac{d_2^2 + s_1^2 - s_4^2}{2s_1 d_2}\right) \tag{5}$$

After the kinematic analysis, each variable was defined for the design of the wing root control mechanism in Table 2. $s_1$, minimum value of $d_2$, and $\theta_2 - (\theta_0 + 180^\circ)$ are the values to avoid physical collision. These design parameters were obtained by trial and error in 3D modeling. $s_4$ was obtained by applying the proposed values to (6).

$$s_4 = \sqrt{d_2^2 + s_1^2 \cos^2(\theta_2 - (\theta_0 + 180^\circ)) - s_1^2} + s_1 \cos(\theta_2 - (\theta_1 + 180^\circ)) \tag{6}$$

Lastly, $\theta_1$, which makes $\theta_r$ equal to zero when $d_2$ is the minimum, is defined by (7).

$$\theta_1 = \cos^{-1}\left(\frac{s_1^2 + d_2^2 - s_2^2}{2s_2 d_2}\right) \tag{7}$$

Figure 7 shows the results of the kinematics calculated by the defined design parameters. Figure 7a shows the simulation output motion ($\theta_r$) with respect to input motion ($d_2$) and describes that the output motion is normally driven by input motion. The blue line is the $s_4$ link, the red line is the $s_4$ link, the blue dot is the input motion of linear motion, and the red dot is output motion of the wing root.

To implement the lift-based system and drag-based system of the dragonfly, each mean angle of attack should be 0° and 30°. Therefore, it is necessary to define the input variable for each system. The wing root angle for the lift-based system was defined as 0 at the above link optimization, but $d_2$ for the drag-based system, which is −30°, was not defined. Figure 7b shows relationship between $d_2$ and $\theta_r$, and the desired value of $d_2$ when $\theta_r$ is −30°. As a result, we confirmed that the MAV implements a drag-based system when $d_2$ is 16.25 mm. Table 2 lists the design parameters of the wing root rotation mechanism.

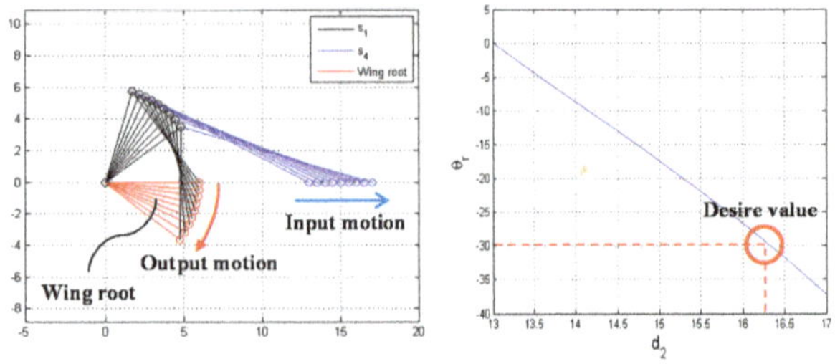


(a) (b)

**Figure 7.** (**a**) Kinematics of input motion (slide motion) and output motion (wing root rotation). (**b**) Graph of relationship between $\theta_r$ and $d_2$ to define $d_2$ with respect to desired $\theta_r$ (30°).

**Table 2.** Design parameters of wing root rotation mechanism.

| Design Parameters | $s_1$ | $s_2$ | $s_4$ | $\theta_1$ | $d_{2,\min}$ |
|---|---|---|---|---|---|
| Value | 6 mm | 6 mm | 12.62 mm | 72.97° | 13 mm |

## 4. Design of Flapping Mechanism

### *4.1. Analysis of Kinematics*

In this work, we used the spatial four-bar-based flapping mechanism to convert from rotational motion to a couple of flapping motions using a rotary actuator. The spatial four-bar-based flapping mechanism consists of two spherical joints, a crank link, and a flapping link. In this mechanism, the factor considered for the design is the limited range of spherical joints. Figure 8 shows the angle limit of the spherical joint; when the $y$-axis is the joint axis, it has an angle limit of $-30°$ to $30°$ in the $x$-axis direction and $z$-axis direction.

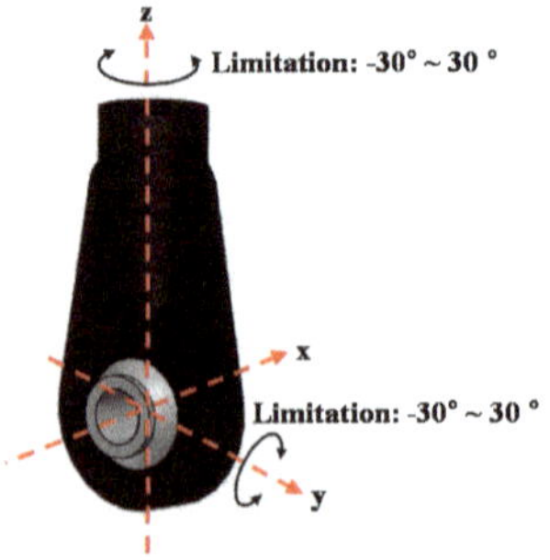


**Figure 8.** Limitation of spherical joint.

The schematic of the flapping mechanism is shown in Figure 9, where $O_1$, $O_2$ are the three-axis spherical joints, $O_0$, $O_3$ are one-axis joints; $\theta_1$ is the crank angle; $\theta_3$ is the flapping angle; and $\theta_{2,x}$, $\theta_{2,z}$ are the angles of the spherical joint. The kinematics of the flapping mechanism was analyzed by vector analysis; the input variable is $\theta_1$, which is the crank angle. The flapping angle ($\theta_3$) is defined by the crank angle and expressed as (8).

$$\theta_3 = \sin^{-1}\left(\frac{l_2^2 - (k_x^2 + k_y^2 + k_z^2 + l_3^2)}{2l_3\sqrt{k_x^2 + k_z^2}}\right) - \sin^{-1}\left(\frac{k_x}{\sqrt{k_x^2 + k_z^2}}\right) \tag{8}$$

In the equation, $k_x = l_{4,x}$, $k_y = l_{4,y} - l_1 \cos(\theta_1)$, and $k_z = l_{4,z} - l_1 \sin(\theta_1)$. Then, using the defined $\theta_3$, the spherical joint angles $\theta_{2,x}$, and $\theta_{2,z}$ are described by (9) and (10), respectively.

$$\theta_{2,x} = \cos^{-1}\left(\frac{k_x + l_3 \cos(\theta_3)}{l_2}\right) \tag{9}$$

$$\theta_{2,z} = \cos^{-1}\left(\frac{k_y}{l_2 \sin(\theta_{2,z})}\right) \tag{10}$$

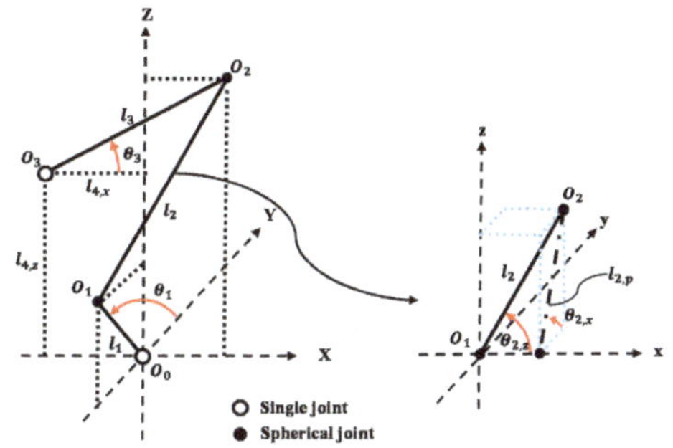


**Figure 9.** Schematic kinematics of spatial four-bar link-based flapping mechanism.

## 4.2. *Define Link Length of Flapping Mechanism*

We defined the link length of the flapping mechanism so that the stroke amplitude is within a given range, and $\theta_{2,z}$, $\theta_{2,x}$ are within the angle limits of the spherical joint. The mechanism was determined to have a stroke amplitude of $-40°$ to $40°$, an angle of the $x$-axis of the spherical joint within 60–120°, and an angle of the $z$-axis of the spherical joint within 60–90°. In the processing, the limit of the $z$-axis angle of the spherical joint was set as 90° rather than 120° to avoid physical collision at 90°. The link length was defined in three steps: (1) the determination of $l_{2,p}$ with respect to the $x$-axis limit of the spherical joint; (2) the calculation of $l_3$, $l_{4,z}$ with respect to the stroke amplitude; and (3) the determination of $l_{4,x}$ with respect to the $z$-axis limit of the spherical joint.

To determine the link length with respect to the $\theta_{2,x}$ range, $l_{2,p}$, which is projected on the $y$-$z$ plane of $l_2$, as shown in Figure 10a, is defined as (11).

$$l_{2,p} = \frac{l_1}{\cos(\theta_{2,x})} \tag{11}$$

The maximum and minimum values of $\theta_2$ for calculating $l_{2,p}$ are the same because the central axis of the crank link and flapping link are located on the same plane, as shown in Figure 10b. This means that if the link value is determined for one of the maximum and minimum values, then the other value is also defined. Therefore, $l_{2,p}$ was obtained by applying $\theta_{2,min}$ to (11), after $\theta_{2,min}$ was defined as 77°.

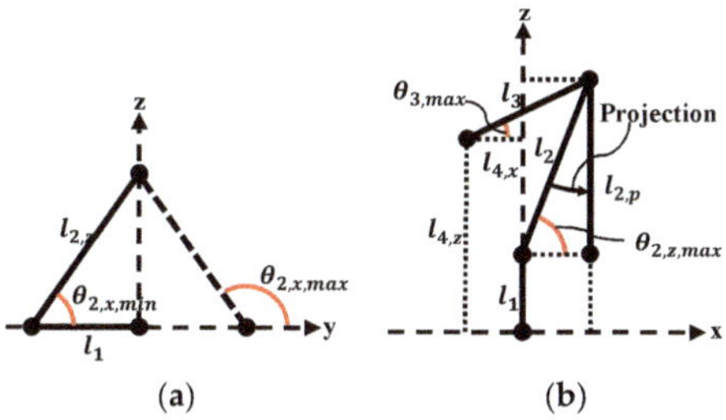


**Figure 10.** (**a**) Projected on the $y$-$z$ plane; (**b**) projected on the $x$-$z$ plane of flapping mechanism.

Second, in order to calculate $l_3$ and $l_{4,z}$, and having the stroke amplitude defined in Table 1, the relation between $l_3$ and $l_{4,z}$ is expressed as (12) and (13) when $\theta_3$ is the maximum and minimum, respectively.

$$l_{3,\max} = \frac{(l_{2,p} + l_1 - l_{4,z})}{\sin(\theta_{3,\max})} \tag{12}$$

$$l_{3,\min} = \frac{(l_{2,p} - l_1 - l_{4,z})}{\sin(\theta_{3,\min})} \tag{13}$$

In addition, to determine $l_3$ and $l_{4,z}$ for the desired stroke amplitude, the object function was defined as (14), which is the absolute value of the difference between (12) and (13).

$$O_{(l_3,l_4)} = |l_{3,\max} - l_{3,\min}| \tag{14}$$

Afterwards, $l_3$ and $l_{4,z}$ were determined by the defined object function and line search method. $l_{2,p}$ is different from the previously defined value, but the defined $l_{2,p}$ was used because the difference in values is within the tolerance range. Finally, we defined the link length for the angle limit of $\theta_{2,z}$, as shown in Figure 10b; $\theta_{2,z}$ has the maximum value when $\theta_3$ is the maximum. The maximum value of $\theta_{2,z}$ was defined as 80°, and $l_{4,x}$ was obtained by applying the variables to (15).

$$l_{4,x} = l_3 \cos(\theta_3) - \frac{l_{2,p}}{\tan(\theta_{2,z,\max})} \tag{15}$$

The values of $\theta_3$, $\theta_{2,x}$, $\theta_{2,z}$, and the link lengths obtained from the kinematic equations are summarized in Table 3. Figure 11 shows each angular value according to the crank angle change. In the graph, we confirmed that $\theta_{2,x}$ and $\theta_{2,z}$ are within the angle limit of the spherical joint, and $\theta_3$ is within the desired stroke amplitude.

**Table 3.** Value of design parameters of the flapping mechanism.

| $l_1$ | $l_2$ | $l_3$ | $l_{4,x}$ | $l_{4,z}$ | $\theta_{2,x}$ | $\theta_{2,z}$ | $\theta_3$ |
|---|---|---|---|---|---|---|---|
| 6 mm | 26.8 mm | 9.331 mm | −4.43 mm | 26.67 mm | 76.85°~103.1° | 79.46°~84.19° | −40°~39.91° |

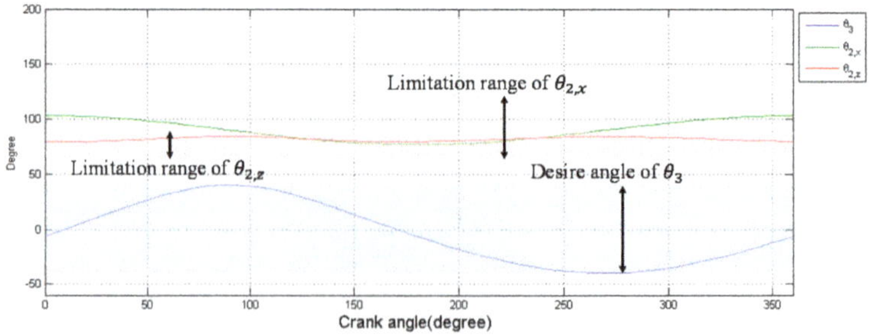


**Figure 11.** Values of $\theta_{2,x}$, $\theta_{2,z}$, and $\theta_3$ with respect to the crank angle.

## 5. Design and Fabrication of Prototype

In order to combine the wing mechanism using a rotary actuator and the wing root rotation mechanism using a linear actuator, both mechanisms must be designed to move independently. Figure 12a shows the side view of the prototype design for the rotation mechanism, and Figure 12b shows the front view of the prototype design for the flapping mechanism. Figure 12a shows the main flapping link, slide flapping link, and wing root rotation link for independently driving two actuators. The wing root rotation link rotates with two flapping links as pivot axes and always moves in the same plane as the flapping link. The slide flapping link rotates with the spindle of the ball screw as the pivot axis and performs linear motion as the nut of the ball screw. Because the slide flapping link can simultaneously implement rotary motion and linear motion as the nut of the ball screw, the designed prototype has both independent flapping motion and wing root rotation motion.

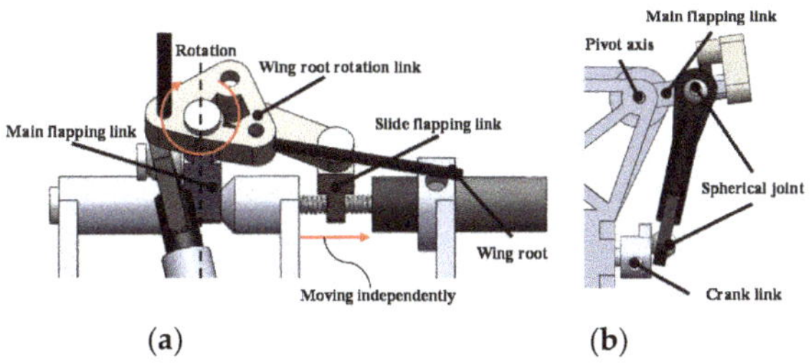


**Figure 12.** (**a**) Side view of the prototype design for the rotation mechanism; (**b**) front view of prototype design for the flapping mechanism.

Figure 13a,b shows the designed prototype and fabricated prototype, respectively. The prototype was designed using 3D CAD (Solidworks 2013, Dassault Systemes, France) software and design parameters determined for link optimization. The parts of the prototype, except a spherical joint link, were manufactured by an SLA (stereo lithography apparatus)-type 3D printer (ProJet 5000, resolution: 100 µm, 3D Systems, Rock Hill, SC, USA) using PC (polycarbonate). The joints were fixed with a 2 mm rivet for low friction and to affix without additional processing. The actuator for flapping is an EC6 (φ6, brushless, 2-watt, Hall sensors, maxon motor, Suisse, www.maxon-motor.com), the controller is a DEC 24/2, and a 37.5:1 gearbox ratio was used for the flapping mechanism. The LCP06-A03V-0026(D&J WITH Co., Geumcheon-Gu, Seoul, Korea) was used as the actuator of the wing root rotation mechanism, and 2 mm bolt was used for the spindle.

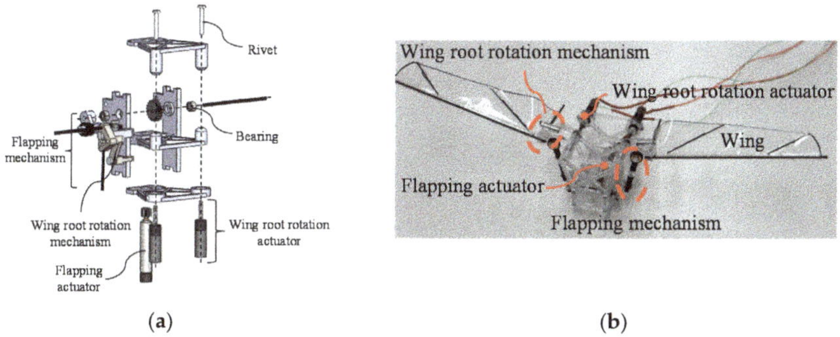


**Figure 13.** (**a**) Designed prototype and (**b**) front view of fabricated prototype.

## 6. Experiment and Analysis

### 6.1. Experimental Setup

This study proposes a mechanism for using the drag-based system in a hovering and lift-based system in forward flight by varying the mean angle of attack. We measured two factors to verify whether the drag-based and lift-based systems were implemented according to wing root rotation changes in the proposed mechanism. First, we took the front and side views of the MAV using a high-speed camera and measured the kinematics using the photographs. To obtain 30 images per cycle of flapping, the MAV was taken at a rate of 450 FPS (frames per second) at flapping frequency of 15 Hz. The angle of attack at flapping was obtained by measuring the length of the wing membrane at 70% from of wing root of the image.

Second, the aerodynamic force was measured by a force sensor. Figure 14 shows the force sensor, test stand, and signal line. High sampling rates and high resolution were required for force measurements during flapping. Therefore, the Nano 17 F/T sensor (resolution: 1/80 N, ATI Industrial Automation, Apex, NC, USA) was used for force measurement. The $x$-axis of the force sensor was installed at the front of the MAV and the $z$-axis was installed along the vertical direction of the MAV to measure data from each lift and drag. The tests proceeded at a flapping frequency of 10–15 Hz, and to obtain 500 data points per flapping cycle, the data was measured at 5000 RPS (rate per second) at 10 Hz and 7500 RPS (rate per second) at 15 Hz. Figure 15 shows the entire flowchart of the force and image data.

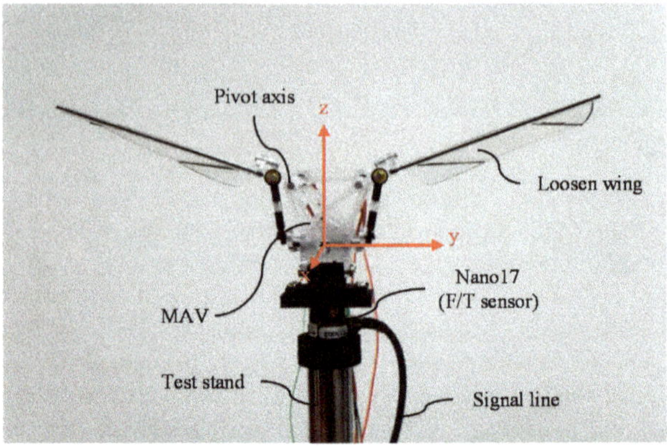


**Figure 14.** Experimental setup.

According to Caetano et al. [36], it has been determined that the frequency of the aerodynamic force is 2 times higher than flapping frequency, and the inertial force frequency is 3 times higher than the flapping frequency. Therefore, in this study, the measured data were filtered by a lowpass filter with a cutoff frequency 2 times larger than the flapping frequency to estimate the aerodynamic force.

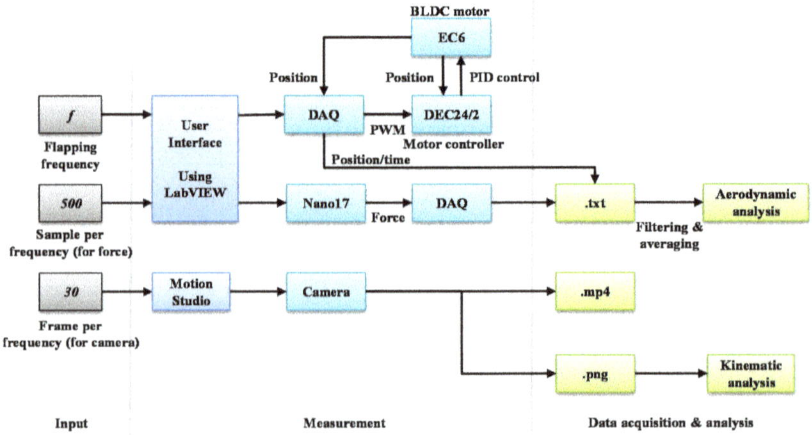


**Figure 15.** Flowchart of force and image data.

## 6.2. *Analysis of Kinematics*

### 6.2.1. Stroke Amplitude

Figure 16a shows the MAV at the minimum and maximum amplitudes taken with a high-speed camera. Figure 16b shows the measured amplitude during one cycle of the flapping link and wing by analyzing the image taken. In order to extract the kinematics from the image, the pivot, spherical joint tip, and wing tip were marked. Then, each mark was connected by a line, as shown in the left image in Figure 16a. The kinematics was extracted by measuring the angle between the connected lines. The amplitude of the flapping link has an error of approximately 4° with the designed amplitude by the joint's tolerance. On the other hand, the wing amplitude has an error of approximately 20° and is caused by the flexion of the wing frame.

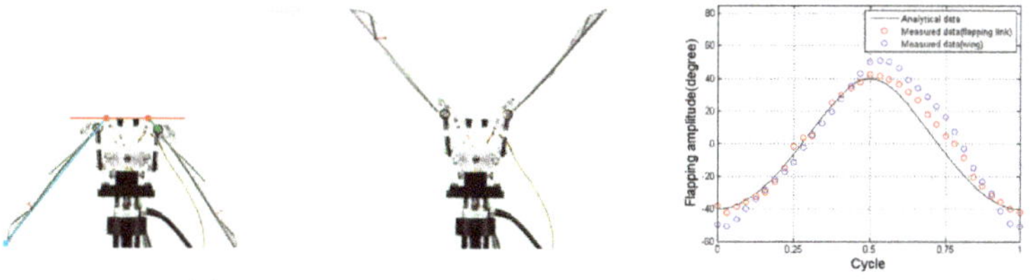


(a) (b)

**Figure 16.** (**a**) Flapping amplitude image of prototype; (**b**) flapping amplitude comparison between designed and measured value.

### 6.2.2. The Wing Angle of Attack

Figure 17 shows each angle of attack at upstroke and downstroke when the wing root angle is 0° and 30°.

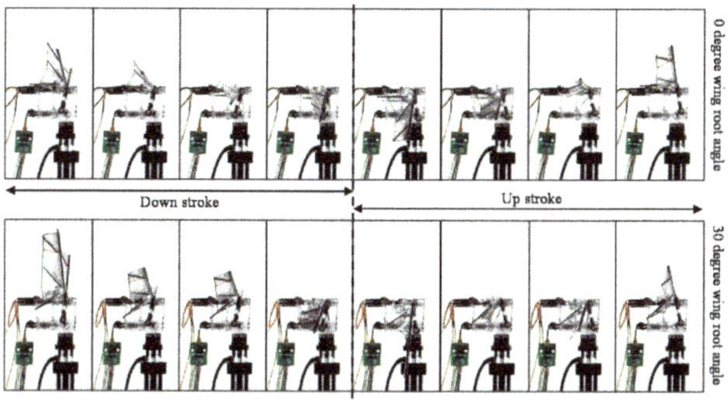


**Figure 17.** Angle of attack at each upstroke and downstroke at wing root angles of 0° and 30°.

Figure 18 shows the angle of attack of each wing at wing root angles of 0° and 30°. The red straight line is the wing root angle of 0°, the blue straight line is the wing root angle of 30°, and the dotted line is the mean of each angle of attack. Since the angle of attack is defined based on the wing root, it has a negative value in the upstroke and a positive value in the downstroke. The angle of attack at a wing root angle of 0° was symmetric with respect to the angle of attack of 0°, and the mean angle of attack was measured close to 0°. On the other hand, the angle of attack at the wing root angle of 30° had a large angle of attack in the downstroke and a small angle of attack in the upstroke, with mean angle of attack of 30°. These results indicate that the wing root angle and mean angle of attack have a linear relationship and that the angle of attack at the flapping of the MAV is similar to that of a real dragonfly.

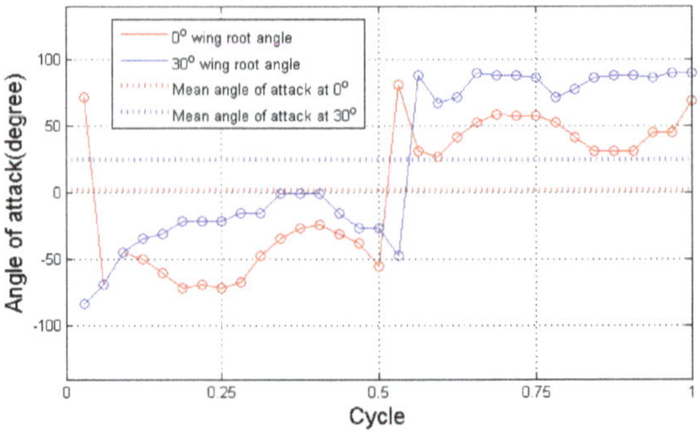


**Figure 18.** The measured angles of attack.

## 6.3. Analysis of Aerodynamics

The lift and drag forces acting on the object are given by (16) and (17), respectively, in the quasi-steady model.

$$F_L = \frac{1}{2}\rho C_L S U^2 \tag{16}$$

$$F_D = \frac{1}{2}\rho C_D S U^2 \tag{17}$$

where $F_L$ and $F_D$ are the lift force and drag force, respectively, $\rho$ is the density of air, $S$ is the area of the object, $U$ is the velocity of the object, and $C_L$ and $C_D$ are the lift and drag coefficients, respectively. Lift force and drag force were defined with respect to wing, and the wing generates a drag force in the opposite direction of the moving direction and a lift force in the vertical direction of the moving direction. In the equation, the lift and drag are proportional to the velocity squared of the object because all parameters except the velocity are constant. $F_{x'}$ and $F_{z'}$, which were measured in the experiment, have the relationship indicated in (18) and (19) with $F_L$ and $F_D$ as shown Figure 1b. Therefore, $F_{x'}$ and $F_{z'}$ should increase with the flapping frequency squared of the object.

$$F_{x'} = F_L \tag{18}$$

$$F_{z'} = F_D \sin(\theta_s) \tag{19}$$

Figure 19a shows the relationship between each $F_{x'}$, $F_{z'}$ and flapping frequency of the MAV for wing root angles of 0° and 30° during flapping. The red dotted line ($F_t$) denotes the estimated value of $F_{x'}$ and $F_{z'}$ of the wing over the flapping frequency square through the characteristics of (16) and (17). The blue straight line is the measured lift force, and the black straight line is $F_{z'}$. Figure 19b shows the difference between the estimated and measured values. The difference between the measured force and estimated value is not over 0.006 N regardless of the flapping frequency, as shown in the figure. Therefore, $F_{x'}$ and $F_{z'}$ in the proposed MAV wing are proportional to the flapping frequency squared, and it is possible to estimate the force generated by increasing the flapping frequency.

In the left graph in Figure 19a, $F_{x'}$ at a 0° wing root angle increases in the positive direction with flapping frequency, while $F_z$ increases in the negative direction or appears near 0. On the other hand, $F_{x'}$ at a wing root angle of 30°, shown in the right graph, is lower than that of the wing root angle of 0° but has a similar tendency, and $F_{z'}$ increases with increasing flapping frequency. A comparison of the two results shows that the magnitude of $F_{x'}$ and $F_{z'}$ that occurs in the wing changes when the wing root angle changes.

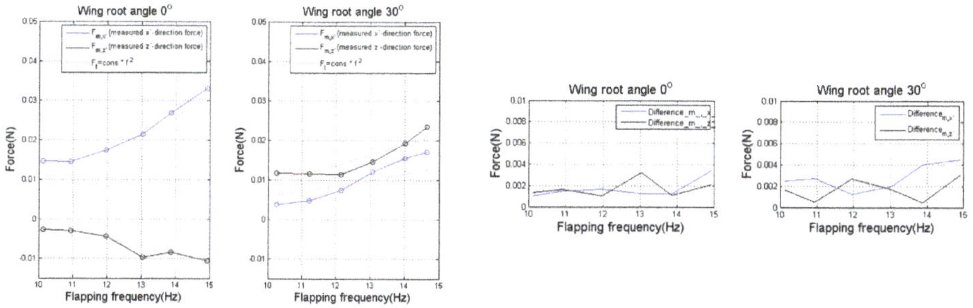


(a) (b)

**Figure 19.** (a) Measured $F_{x'}$ and $F_{z'}$ and (b) difference with respect to flapping frequency.

The top graph in Figure 20 shows $F_{z'}$ at a 0° wing root angle when the MAV flaps at a frequency of 15 Hz, while the bottom graph shows $F_{z'}$ at a 30° wing root angle. The top and bottom image in Figure 20 show the angle of attack when the wing is horizontal to the ground. The top two images indicate the angle of attack at a 0° wing root angle, while the bottom two images indicate the angle of attack at a 30° wing root angle. The force generated is an aerodynamic force, which is $F_{z'}$ of the MAV wing. The aerodynamic force at a 0° wing root angle does not show a significant difference when up-down strokes are compared. In contrast, the peak value after vibration due to the inertial force in the downstroke at a 30° the wing root angle means that the wing receives a large average $F_{z'}$. The decreasing tendency after the vibration due to the inertial force in the upstroke means that the wing receives a small average $F_{z'}$. The experimental results demonstrate that flapping at a wing root angle of 30° produces a reasonable average $F_{z'}$ in the up-down stroke.

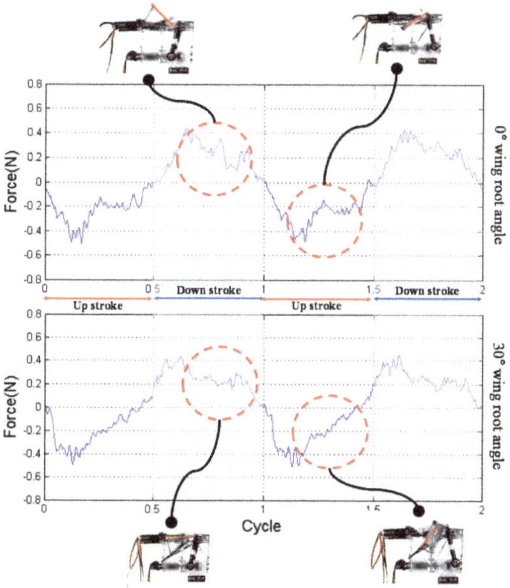


**Figure 20.** Drag force comparison between 0° and 30° wing root angle at 15 Hz flapping frequency.

#### 6.3.1. Comparison between Horizontal Force Amplitude

The drag-based system used in hovering by a real dragonfly flaps in the inclined stroke angle direction of 60°, as shown in Figure 21. On the other hand, the direction of $F_{x'}$ measured in the study was generated toward the vertical direction of the stroke angle, and the direction of the drag force was generated toward the horizontal direction of the stroke angle. To measure the horizontal force generated by the wing root rotation mechanism at the same stroke angle as the drag-based system of a real dragonfly, the horizontal force was calculated by applying $F_{x'}$ and $F_{z'}$ measured at the wing root angle of 30° to (21).

$$F_V = F_{x'} \cos(\theta_s) + F_{z'} \sin(\theta_s) \tag{20}$$

$$F_H = F_{x'} \sin(\theta_s) - F_{z'} \cos(\theta_s) \tag{21}$$

where $F_V$ is the vertical force, $F_H$ is the horizontal force, and $\theta_s$ is the stroke angle. Since the lift-based system used for hovering has a horizontal directional stroke angle, the horizontal force generated during flapping of the MAV was defined as the $F_{z'}$ of the 0° wing root angle.

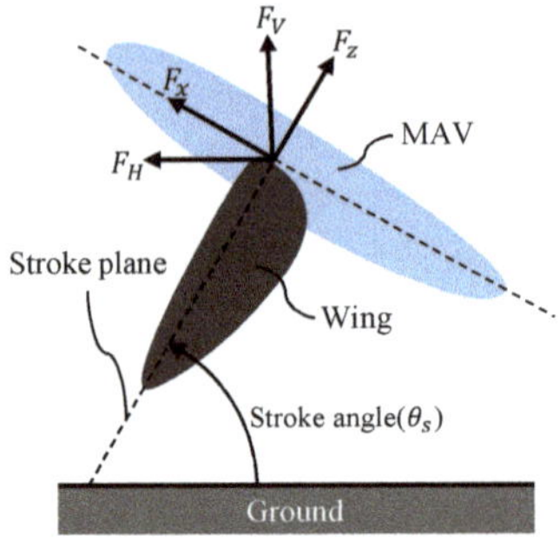


**Figure 21.** Force direction at an inclined stroke angle.

Figure 22 shows the horizontal forces of the two systems defined by the measured force. The black line is the horizontal force of the lift-based system implemented at a 0° the wing root angle, and the blue line is the horizontal force of the drag-based system implemented at a 30° wing root angle. The dotted line is the raw data and the straight line is the graph after applying the moving average filter, which averages a window of 30 data points. In drag-based systems, the MAV generated a large horizontal force in the upstroke because the lift force has more effect on horizontal force, and lift force was generated more at the upstroke. In contrast, the lift-based system has the same horizontal force amplitude in both the upstroke and downstroke. Comparison of the amplitudes of the two forces indicates that the horizontal force amplitude of the lift-based system is approximately 1.8 times larger than the horizontal force amplitude of the drag-based system. The difference between the two amplitudes is 0.15 N and 0.1 N in each downstroke and upstroke, respectively.

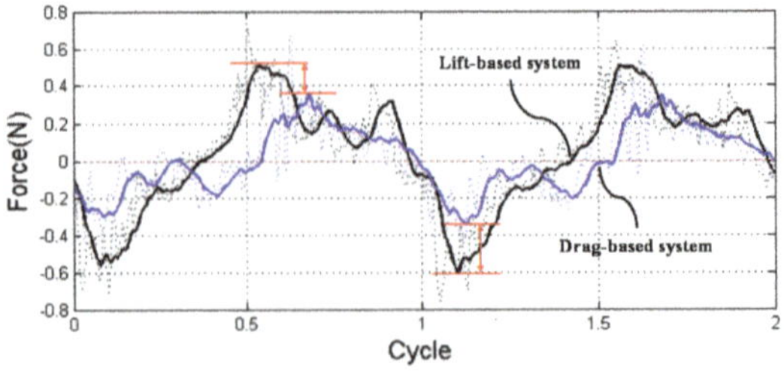


**Figure 22.** Horizontal force comparison between drag-based system and lift-based system at hovering.

6.3.2. Force Direction Change with Respect to Wing Root Angle

Figure 23 shows the vertical and horizontal forces generated in the wing during flapping at a stroke angle of 60°. The vertical and horizontal forces were calculated by Equations (17) and (18), respectively. The graph on the left shows the force generated at a wing root angle of 0° and the graph on the right shows the force generated at a wing root angle of 30°. At a wing root angle of 0°, the horizontal force increased with increasing flapping frequency, and the vertical force is 0.005 N, which is lower than the horizontal force. This means that the MAV generates force in the appropriate direction for forward flight when it has a wing root angle of 0° at flapping. Comparison with a 0° wing root angle shows that the vertical force at a wing root angle of 30° increases with increasing flapping frequency, and the horizontal force has a value close to 0 N. Therefore, when MAV flaps with a wing root angle of 30°, the wing generates a force in the direction for hovering. Thus, the wing root angle is an important factor in determining the flight mode of the MAV, and the flight mode can be changed by changing the wing root angle.

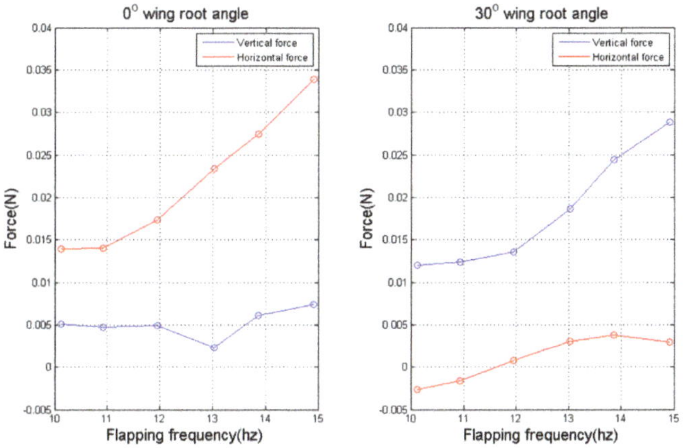


**Figure 23.** Vertical force and horizontal force with respect to flapping frequency at an inclined stroke angle.

## 7. Conclusions

In this study, we propose a mechanism to change the mean angle of attack of MAVs by rotating the wing roots. The links used in the flapping mechanism and the wing root rotation mechanism were optimized, and the graphs show the motion of the optimized mechanism. To confirm whether the proposed mechanism has stable hovering ability and can change flight mode without a posture change, a prototype incorporating the flapping mechanism and wing root rotation mechanism was fabricated, and the force generated by the MAV was measured. The prototype was designed using a ball-screw mechanism to independently control the wing root rotation mechanism and flapping mechanism. The measured force showed that the MAV with the wing root rotation mechanism can fly more stably when hovering and can change the flight mode by changing only the wing root angle. In the future, the wing root rotation mechanism is expected to fly more stably than the MAV with a lift-based system, and it will be possible to develop an MAV that mimics the interaction between the forewing and hindwing of an actual dragonfly.

**Author Contributions:** Supervision and Funding Acquisition, G.-H.Y.; Writing-Review & Editing, G.-H.Y.; Conceptualization, J.J.; Writing-Original Draft Preparation, J.J.; Design and Fabrication J.J.; Data acquisition and analysis, J.J.

**Funding:** This work was supported in part by the Korea Institute of Industrial Technology as under "Development of Soft Robotics Technology for Human-Robot Coexistence Care Robots (KITECH EO180026)."

**Conflicts of Interest:** The authors declare no conflict of interest.

## References


1. Wood, R.J. Design, fabrication, and analysis of a 3DOF, 3 cm flapping wing mav. In Proceedings of the 2007 International Conference on Intelligent Robots and Systems, San Diego, CA, USA, 29 October–2 November 2007; pp. 1576–1581.
2. Keennon, M.; Klingebiel, K.; Won, H.; Andriukov, A. Development of the Nano Hummingbird: A tailless flapping wing micro air vehicle. In Proceedings of the 50th AIAA Aerospace Sciences Meeting, Nashville, TN, USA, 9–12 January 2012; p. 588.
3. Phan, H.V.; Park, H.C. Generation of Control Moments in an Insect-link Tailless Flapping-wing Micro Air Vehicle by Changing the Stroke-plane Angle. *J. Bionic Eng.* **2016**, *13*, 449–457. [CrossRef]
4. Rosen, M.H.; Pivain, G.L.; Sahai, R.; Jafferis, N.T.; Wood, R.J. Development of a 3.2 g Untethered Flapping-Wing Platform for Flight Energetics and Control Experiments. In Proceedings of the 2016 IEEE International Conference on Robotics and Automation (ICRA), Stockholm, Sweden, 16–21 May 2016; pp. 3227–3233.
5. Ramezani, A.; Shi, X.; Chung, S.J.; Hutchinson, S. Bat Bot (B2), A Biologically Inspired Flying Machine. In Proceedings of the 2016 IEEE International Conference on Robotics and Automation (ICRA), Stockholm, Sweden, 16–21 May 2016; pp. 3219–3226.
6. Hsiao, F.Y.; Yang, L.J.; Lin, S.H.; Lin, S.H.; Chen, C.L.; Shen, J.F. Autopilots for Ultra Lightweight Robotic Birds: Automatic Altitude Control and System integration of a Sub-10g Weight Flapping-Wing Micro Air Vehicle. *IEEE Control Syst. Mag.* **2012**, *14*, 35–48.
7. Karásek, M.; Muijres, F.T.; Wagter, C.D.; Remes, B.D.W.; de Croon, G.C.H.E. A tailless aerial robotic flapper reveals that flies use torque coupling in rapid banked turns. *Science* **2018**, *361*, 1089–1094. [CrossRef] [PubMed]
8. Alexander, D. Unusual phase relationships between the forewing and hindwings in flying dragonflies. *J. Exp. Biol.* **1984**, *109*, 379–383.
9. Norberg, R. Hovering flight of the dragonfly: *Aeschna juncea* L. Kinematics and aerodynamics. *Swim. Fly. Nat.* **1975**, *2*, 763–780.
10. Wang, J.K.; Sun, M. A computational study of the aerodynamics and forewing-hindwing interaction of a model dragonfly in forward flight. *J. Exp. Biol.* **2005**, *208*, 3785–3804. [CrossRef] [PubMed]
11. Hu, Z.; Deng, X.Y. Aerodynamic interaction between forewing and hindwing of a hovering dragonfly. *Acta Mech. Sin.* **2014**, *30*, 787–799. [CrossRef]
12. Usherwood, J.R.; Lehmann, F.O. Phasing of dragonfly wings can Improver aerodynamic efficiency by removing swirl. *J. R. Soc. Interface* **2008**, *5*, 1303–1307. [CrossRef] [PubMed]
13. Wang, Z.J. Dissecting insect flight. *Annu. Rev. Fluid Mech.* **2005**, *37*, 183–210. [CrossRef]
14. Wang, Z.J. The role of drag in insect hovering. *J. Exp. Biol.* **2004**, *207*, 4147–4155. [CrossRef] [PubMed]
15. Vogel, S. *Lift in Moving Fluids*; Princeton University: Princeton, NJ, USA, 1996.
16. Ellington, C.P. The aerodynamics of Hovering Insect Flight. III. Kinematics. *Philos. Trans. R. Soc. Lond. B* **1984**, *305*, 41–78. [CrossRef]
17. Ellington, C.P. The Novel Aerodynamics of Insect Flight: Applications to Micro-air Vehicles. *J. Exp. Biol.* **1999**, *202*, 3439–3448. [PubMed]
18. San, S.P.; Dickinson, M.H. The Control of Flight Force by a Flapping Wing: Lift and Drag Production. *J. Exp. Biol.* **2001**, *204*, 2607–2626.
19. Dickinson, M.H. The Effects of Wing Rotation on Unsteady Aerodynamic Performance at Low Reynolds Numbers. *J. Exp. Biol.* **1994**, *192*, 179–206. [PubMed]
20. Dickinson, M.H. Wing rotation and the Aerodynamic Basis of Insect Flight. *Science* **1999**, *284*, 1954–1960. [CrossRef] [PubMed]
21. Bergou, A.J.; Xu, S.; Wang, J. Passive wing pitch reversal in insect flight. *J. Fluid Mech.* **2007**, *591*, 321–337. [CrossRef]
22. Whitney, J.P.; Wood, R.J. Aerodynamics of passive rotation in flapping flight. *J. Fluid Mech.* **2010**, *60*, 197–220. [CrossRef]
23. McIntosh, S.H.; Agrawal, S.K.; Khan, Z. Design of a Mechanism for Biaxial Rotation of a Wing for a Hovering Vehicle. *IEEE/ASME Trans. Mechatron.* **2006**, *11*, 145–153. [CrossRef]

24. Fenelon, M.A.A.; Furukawa, T. Design of an active flapping wing mechanism and a micro aerial vehicle using a rotary actuator. *Mech. Mach. Theory* **2010**, *45*, 137–146. [CrossRef]
25. Karásek, M.; Hua, A.; Nan, Y.; Lalami, M.; Preumont, A. Pitch and Roll Control Mechanism for a Hovering Flapping Wing MAV. *Int. J. Micro Air Veh.* **2014**, *6*, 253–264. [CrossRef]
26. Nguyen, Q.V.; Park, H.C.; Goo, N.S.; Byun, D. Characteristics of a Beetle's Free Flight and a Flapping-Wing system that Mimics Beetle Flight. *J. Bionic Eng.* **2010**, *7*, 77086. [CrossRef]
27. Karásek, M. Robotic Hummingbird: Design of a Control Mechanism for a Hovering Flapping Wing Micro Air Vehicle. Ph.D. Thesis, Department of Mechanical, Engineering and Robotics, Delft University, Delft, The Netherlands, 2014.
28. Wang, H.; Zeng, L.; Liu, H.; Yin, C. Measuring wing kinematics, flight trajectory and body attitude during forward flight and turning maneuvers in dragonflies. *J. Exp. Biol.* **2003**, *206*, 745–757. [CrossRef] [PubMed]
29. Azuma, A.; Azuma, S.; Watanabe, I.; Furuta, T. Flight Mechanics of a Dragonfly. *J. Exp. Biol.* **1985**, *116*, 79–107.
30. Okamoto, M.; Yasuda, K.; Azuma, A. Aerodynamic Characteristics of the Wings and Body of a Dragonfly. *J. Exp. Biol.* **1996**, *199*, 281–294. [PubMed]
31. Hu, Z.; McCauley, R.; Schaeffer, S.; Deng, X. Aerodynamics of Dragonfly Flight and Robotic Design. In Proceedings of the 2009 IEEE International Conference on Robotics and Automation (ICRA), Kobe, Japan, May 12–17 May 2009; pp. 3061–3066.
32. Koehler, C.; Lisang, Z.; Gaston, Z.; Wan, H.; Dong, H. 3D reconstruction and analysis of wing deformation in free-flying dragonflies. *J. Exp. Biol.* **2012**, *215*, 3018–3027. [CrossRef] [PubMed]
33. Agrawal, A.; Agrawal, S.K. Design of Bio-inspired Flexible Wings for Flapping-Wing Micro-sized Air Vehicle Applications. *Adv. Robot.* **2009**, *23*, 979–1002. [CrossRef]
34. Du, G.; Sun, M. Effects of wing deformation on aerodynamic forces in hovering hoverflies. *J. Exp. Biol.* **2010**, *213*, 2273–2283. [CrossRef] [PubMed]
35. Pelletier, A.; Mueller, T.J. Low Reynolds Number Aerodynamics of Low-Aspect_Ratio, Thin/Flat/Cambered-Plat Wings. *J. Aircr.* **2000**, *37*, 825–832. [CrossRef]
36. Caetano, J.V.; Percin, M.; van Oudheusden, B.W.; Remes, B.; de Wagter, C.; de Croon, G.C.H.E.; de Visser, C.C. Error Analysis and Assessment of Unsteady Forces Acting on a Flapping Wing Micro Air Vehicle: Free-Flight versus Wind Tunnel Experimental Methods. *Bioinspir. Biomim.* **2015**, *10*, 056004. [CrossRef] [PubMed]

applied sciences

MDPI

*Article*

# Robust Adaptive Path Following Control of an Unmanned Surface Vessel Subject to Input Saturation and Uncertainties

**Yunsheng Fan *, Hongyun Huang and Yuanyuan Tan**

School of Marine Electrical Engineering, Dalian Maritime University, Dalian 116026, China; iamhongyun@gmail.com (H.H.); tanyuanyuan@dlmu.edu.cn (Y.T.)

* Correspondence: yunsheng@dlmu.edu.cn

Received: 17 February 2019; Accepted: 25 April 2019; Published: 1 May 2019

**Abstract:** This paper investigates the path following control problem of an unmanned surface vessel (USV) subject to input saturation and uncertainties including model parameters uncertainties and unknown time-varying external disturbances. A nonlinear robust adaptive control scheme is proposed to address the issue, more specifically, steering a USV to follow the desired path at a certain velocity assignment despite the involved disturbances, by utilizing the finite-time currents observer based line-of-sight (LOS) guidance and radial basis function neural networks (RBFNN). Backstepping and Lyapunov's direct method are the main design frameworks. Based on the finite-time currents observer and adaptive control technique, an improved LOS guidance law is proposed to obtain the desired approaching angle to the desired path, making compensations for the effects of unknown time-varying ocean currents. Then, a kinetic controller with the capability of uncertainties estimation and disturbances rejection is proposed based on the RBFNNs, where the adaptive laws including leakage terms estimate the approximation error and the unknown time-varying disturbances. Subsequently, sophisticated auxiliary control systems are employed to handle input saturation constraints of actuators. All error signals of the closed-loop system are proved to be locally uniformly ultimately bounded (UUB). Numerical simulations demonstrated the effectiveness and robustness of the proposed path following control method.

**Keywords:** unmanned surface vessel; path following; integral line-of-sight; finite-time currents observer; radial basis function neural networks; input saturation

## 1. Introduction

Unmanned surface vessel (USV) as an intelligent and autonomous marine equipment has received more and more attention from the control community, for broad application in the cluttered ocean environment, especially in cases where human intervention is not possible [1]. Generally speaking, three different types of control technologies play a crucial role in the development of USVs: path following control, trajectory tracking control and set-point control [2]. Many researchers propose lots of relevant control strategies and address the issues to a various extent. This study continues along the works and contributions of the predecessors. This paper aims at the path following control, mainly discussing the guidance and control of a USV. Path following is usually defined as steering a vessel to follow the desired path at a certain speed, which is not specified with temporal constraint [3]. Although there are considerable theoretical studies regarding the path following and practical engineering achievements, practical studies of the path following control for USVs have progressed haltingly amid great difficulties. It is essential to develop a highly accurate and robust path following controller for a USV when executing various vital missions. Therefore, under the circumstance of severe sea state, the safe operation and mission execution can be guaranteed.

Two aspects play crucial roles in the path following control scheme: guidance and control [4]. Guidance can refer to the popular and effective line-of-sight (LOS) guidance, refining missile guidance approach or marine guidance [5,6]. This special guidance law exploits the geometry relationships to generate a yaw angle known as the approaching angle, which is fed into the control system. In other words, the control system tracks the reference yaw angle signal together with the specified velocity tracking. Hence, the performance of the path following heavily depends on the guidance system. It turns out that classic LOS guidance is simple and effective [7], albeit with limitations in the case of being exposed to the complex ocean environment induced by waves, wind, and ocean currents. Moreover, the traditional LOS guidance will cause large cross-tracking error when the marine surface vehicles are in steady state, which strictly depends on the path curvature and the magnitude of the drift force. Therefore, the traditional LOS was extended to various forms such as integral LOS (ILOS) guidance and adaptive LOS (ALOS) guidance by Fossen. In [8], an ILOS guidance with integral action is proposed to handle the constant and irrotational ocean currents and other environmental disturbances including wind or waves. In [9], a new ILOS with time-varying lookahead distance is presented with the capability of canceling the effects of constant environmental disturbances, like constant ocean currents. In [10], the direct and indirect adaptive ILOS based path following controller is proposed to deal with the ocean currents. Another way to improve the LOS guidance is to estimate the sideslip angle caused by the external disturbance as an observer-based strategy. In [11], a novel adaptive LOS (ALOS) guidance with small computation footprint is proposed where the adaptive laws dominate the sideslip angle compensation rate. The sideslip angle is treated as an unknown constant, which significantly limits the application of ALOS. In [12], a reduced-order extended state observer (ESO) based LOS guidance is proposed to deal with the time-varying sideslip angle, which is appropriate for straight-line and curved path following. In [13], the magnitude and convergence speed of the sideslip angle is considered, which can be identified by the constructed finite-time sideslip angle observer. However, on the one hand, it should be noted that the aforementioned literature (e.g., [8–10]) merely solve the problem in the kinematic level, i.e., ignoring the along-tracking error. On the other hand, the method in [10] only adds the ocean currents in the kinematic model in terms of the relative surge and sway velocities, whereas the along-tracking error is omitted.

The dynamics control system as the execution system of a USV is another crucial constituent in the path following scheme. In general, USVs do not have an independent actuator in their sway direction. This nonholonomic constraint and underactuated nature makes the control system design much more challenging [14]. To satisfy the successful execution of missions and achieve expected performance and robustness in the full range of work space from calm ocean environment to severe ocean environment along with inaccuracy system parameters, various nonlinear control methods have been proposed [15–22]. In [15], external disturbances rejection method based on the reduced-order linear ESO is proposed. In [16], Do proposed a global path-tracking controller for underactuated ships under deterministic and stochastic loads where weak nonlinearly and strong Lyapunov design method is introduced and the estimations of disturbances are updated by projection algorithms. In [17,18], nonlinear disturbance observers are utilized to estimate the ubiquitous external disturbances in the dynamics model of marine vehicles, where the disturbance estimation errors are proved to be UUB. In [19], a nonlinear adaptive PI sliding mode tracking controller is proposed to solve the environmental disturbance problems by relaxing the assumption of knowing the upper bound of the disturbance. In [20,21], a novel adaptive switching-gain-based control method is proposed for a general uncertain Euler–Lagrange system, which is not only insensitive to the nature of uncertainties but can also alleviate the overestimation–underestimation problem. Moreover, intelligent control such as fuzzy and neural control has been applied to deal with the uncertainties and disturbances of underactuated marine surface ships. In [22], Wang proposed an adaptive online constructive fuzzy controller where the fuzzy approximator is used to approximate the unknown disturbances. Besides, robust radial basis function neural network control laws and iterative neural network control laws are proposed in [23,24], respectively, which are devoted to identifying and compensating the dynamical uncertainties and external disturbances. In addition, it should be noted that few of the aforementioned path following control schemes take into account input saturation. In fact, the practical constraint of actuators determined by the maximum forces and moments would degrade the performance of the control system or even make it unstable. Therefore, it is essential to implement the emulate of the constraint on the control laws design process for reliability and robustness.

Motivated by the observations and considerations mentioned above, a finite-time currents observer based ILOS guidance is proposed to deal with unknown time-varying ocean currents, which is suitable for any desired parametric path with high accuracy control performance. Subsequently, adaptive control laws based on the RBFNN are designed, which solve input saturation with sophisticated auxiliary systems simultaneously. A robust adaptive controller is developed to address the path following problem, which is verified to be effective via numerical simulations. The main contributions of the paper are summarized as follows.

(1) A finite-time currents observer based LOS guidance is presented to obtain the desired yaw angle and estimate the unknown time-varying ocean currents precisely, which significantly influences the performance of the control subsystem.
(2) The RBF neural networks are incorporated into the kinetic controller to solve the uncertainties, which does not require any prior knowledge of the dynamics of the USV and disturbances, and the adaptive laws are designed to estimate the compound bounds of approximating errors and external time-varying disturbances.
(3) The input constraint effect is analyzed with auxiliary systems and the states of auxiliary systems are utilized to make compensations for input saturation, which attenuates the challenge of the actuators.

The rest sections are organized as follows. Section 2 presents the preliminaries and problem formulation. Section 3 provides the design details of the guidance, the kinetic controller, and the stability analysis. Simulation results are presented and analyzed in Section 4. Finally, Section 5 concludes the paper.

## 2. Preliminaries and Problem Formulation

### 2.1. RBFNN Approximation

Consider an unknown smooth nonlinear function $f(\mathbf{x}) : \mathbb{R}^m \to \mathbb{R}$ can be approximated on a compact set $\boldsymbol{\Omega} \in \mathbb{R}^m$ by the following RBFNN:

$$f(\mathbf{x}) = \mathbf{W}^{*T}\boldsymbol{\varphi}(\mathbf{x}) + \varepsilon \tag{1}$$

where $\mathbf{x} \in \boldsymbol{\Omega}$ is the input vector, $\varepsilon$ is the approximation error and satisfies $|\varepsilon| \leqslant \bar{\varepsilon}$ and $\bar{\varepsilon}$ is a constant, and the node number of the NN is $l > 1$. $\mathbf{W}^* \in \mathbb{R}^l$ represents the optimal weight vector, which is defined by

$$\mathbf{W}^* = \arg \min_{\hat{\mathbf{W}} \in R^l} \left\{ \sup_{\mathbf{x} \in \boldsymbol{\Omega}} \left| f(\mathbf{x}) - \hat{\mathbf{W}}^T \boldsymbol{\varphi}(\mathbf{x}) \right| \right\} \tag{2}$$

where $\hat{\mathbf{W}}$ is the estimation of $\mathbf{W}^*$. $\boldsymbol{\varphi}(\mathbf{x}) = [\varphi_1, \varphi_2, \ldots, \varphi_l]^T : \boldsymbol{\Omega} \to \mathbb{R}^l$ represents the radial function vector, the element of which is chosen as the Gaussian function:

$$\varphi_i(\mathbf{x}) = \exp\left( -\frac{(x - b_i)^T (x - b_i)}{c_i^2} \right) (i = 1, 2, \ldots, l) \tag{3}$$

where $\mathbf{b} = [b_1, b_2, \ldots, b_l]^T$ and $\mathbf{c} = [c_1, c_2, \ldots, c_l]^T$ are the centers of receptive field and spread of the Gaussian function, respectively.

*2.2. USV Model*

As illustrated in Figure 1, the position and orientation describe the horizontal plane motion of a USV neglecting roll, pitch, and heave, where $\{i\}$ and $\{b\}$ represent the inertial frame and body fixed frame, respectively. The desired continuous path $P_d(\theta) = [x_d(\theta), y_d(\theta)]^T$ is parameterized by a scalar variable $\theta$ and $P = [x, y]^T$ is the position coordinate. The kinematic equations of a USV can be described by relative velocities as follows [23]

$$\begin{cases} \dot{x} = u_r \cos\psi - v_r \sin\psi + V_x \\ \dot{y} = u_r \sin\psi + v_r \cos\psi + V_y \\ \dot{\psi} = r \end{cases} \tag{4}$$

where $u_r$ and $v_r$ are the relative surge and sway velocities; $x$, $y$, and $\psi$ express the position and orientation in $\{i\}$; and $V_x$ and $V_y$ are the ocean currents represented in $\{b\}$. The dynamics of a USV is expressed as follows [25]

$$\begin{cases} \dot{u}_r = \underbrace{\frac{m_{22}}{m_{11}} v_r r - \frac{d_{11}}{m_{11}} u_r - \sum_{i=2}^{3} \frac{d_{ui}}{m_{11}} |u_r|^{i-1} u_r}_{f_u} + \frac{1}{m_{11}} \tau_u + \frac{1}{m_{11}} \tau_{wu} \\ \dot{v}_r = \underbrace{-\frac{m_{11}}{m_{22}} u_r r - \frac{d_{22}}{m_{22}} v_r - \sum_{i=2}^{3} \frac{d_{vi}}{m_{11}} |v_r|^{i-1} v_r}_{f_v} + \frac{1}{m_{22}} \tau_{wv} \\ \dot{r} = \underbrace{\frac{(m_{11} - m_{22})}{m_{33}} u_r v_r - \frac{d_{33}}{m_{33}} r - \sum_{i=2}^{3} \frac{d_{ri}}{m_{11}} |r|^{i-1} r}_{f_r} + \frac{1}{m_{33}} \tau_r + \frac{1}{m_{33}} \tau_{wr} \end{cases} \tag{5}$$

where the positive constant parameters $m_{jj}(j = 1,2,3)$ are the inertia including added mass; $d_{ii}, d_{ui}, d_{vi}$, and $d_{ri}(i = 2,3)$ are the linear and quadratic hydrodynamic damping in surge, sway, and yaw; $\tau_{wu}, \tau_{wv}$, and $\tau_{wr}$ are the unknown time-varying environmental disturbances; and $\tau_u$ and $\tau_r$ are the available control inputs of the surge force and the yaw moment thereby viewing it as the underactuated control problem. Since the model parameters are directly related to the operation conditions [26], the parameters of the model is uncertain. Actually, due to the physical constraint, the control inputs are subject to nonlinear saturations, which is shown as follows

$$\tau_i = \begin{cases} \tau_{i\max}, & \tau_{i0} > \tau_{i\max} \\ \tau_{i0}, & \tau_{i\min} \leqslant \tau_{i0} \leqslant \tau_{i\max} \\ \tau_{i\min}, & \tau_{i0} < \tau_{i\min} \end{cases} \tag{6}$$

where $\tau_{i\min}$ and $\tau_{i\max}$ $(i = u, r)$ are the minimum and maximum control inputs produced by the actuators, referring to actual constraints of the motor's rotational speed and rudder deflection; and $\tau_{i0}$ is the command control input of the path following controller.

**Assumption 1.** *Assume that all states of a USV are measurable.*

**Assumption 2.** *The time-varying ocean currents $\mathbf{v} = [V_x, V_y]^T$ are assumed to be irrotational and bounded, and there exist a positive constant M, such that $\|\dot{\mathbf{v}}\| \leqslant M, M > 0$. The disturbances $\tau_{wi}(i = u, v, r)$ are unknown time-varying and bounded, and the first derivative of them are also bounded such that $|\dot{\tau}_{wi}| \leqslant \bar{\tau}_w$, where $\bar{\tau}_w$ is unknown constant.*

**Remark 1.** *Note that in (5) the off-diagonal terms of the inertia and damping are ignored. No matter a large scale surface vessel or a highly maneuverable unmanned surface vessel, these terms are relatively small than the main diagonal terms. Therefore, it is reasonable to omit these terms.*

**Remark 2.** *The external disturbances $\tau_{wu}, \tau_{wv}$, and $\tau_{wr}$ in Equation (5) represent the compound disturbances of the wind and wave disturbances. The ocean currents as the form of hydrodynamic terms with relative velocities are represented in Equation (4).*

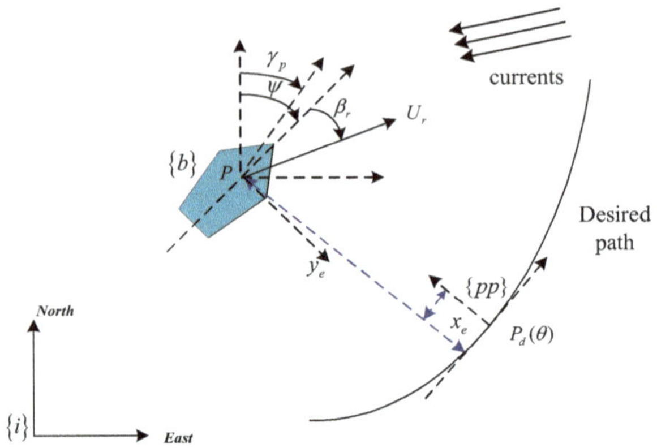


**Figure 1.** USV path following guidance information illustration.

### 2.3. Control Objective

In Figure 1, a local path parallel reference frame is denoted as {pp}. To arrive at{pp}, {i} should be rotated an angle $\gamma_p(\theta) = \text{atan2}\,(y'_d(\theta), x'_d(\theta))$, where the notation $y'_d(\theta) = dy_d(\theta)/d\theta$ is used. Therefore, the position errors can be given as follows

$$\begin{bmatrix} x_e \\ y_e \end{bmatrix} = \underbrace{\begin{bmatrix} \cos\gamma_p & -\sin\gamma_p \\ \sin\gamma_p & \cos\gamma_p \end{bmatrix}^T}_{\mathbf{R}^{\mathbf{T}}(\gamma_p)} (P - P_d(\theta)) \tag{7}$$

where $x_e$ and $y_e$ are the along-tracking error and cross-tracking error, respectively, and $\mathbf{R}(\gamma_p)$ is the rotation matrix. Meanwhile, we also have

$$\dot{P}_d(\theta) = \mathbf{R}(\gamma_p)\left[U_{pp}, 0\right]^T \tag{8}$$

where $\left[U_{pp}, 0\right]^T$ is the velocity of {pp} with respect to {i}, represented in {pp}.

The path following control problem is concerned with designing control laws to reach and then keep following the desired path. Once the path is reached, the vessel can maintain a desired surge velocity assignment of $u_d$. It is worth noting that a constant speed profile is frequently chosen in many cases. Therefore, the control objective is as follows:

$$\sup_{t>0}|x_e| \leqslant \delta_1, \sup_{t>0}|y_e| \leqslant \delta_2, \sup_{t>0}|u_r - u_d| \leqslant \delta_3 \tag{9}$$

where $\delta_1$, $\delta_2$, and $\delta_3$ are small positive constants. Meanwhile, it is guaranteed that all error signals of the closed-loop system are locally UUB.

**Assumption 3.** *The desired path should be sufficiently smooth such that its first derivative $\dot{P}_d$ is bounded. In addition, the desired speed assignment $u_d$ and its first derivative are bounded.*

## 3. Main Results

This section presents the design details of the path following controller to satisfy the control objective that concluded in Section 2. The whole controller consists of two parts: the modified ILOS guidance module and the kinetic controller, which is shown in Figure 2. The ILOS guidance is utilized to calculate the desired yaw angle, where the constructed finite-time currents observer can provide the fast and precise estimation of the unknown time-varying ocean currents. By incorporating the RBFNNs into the backstepping design method, the kinetic controller are developed. Finally, the stability analysis is presented to validate the feasibility of the proposed control approach.

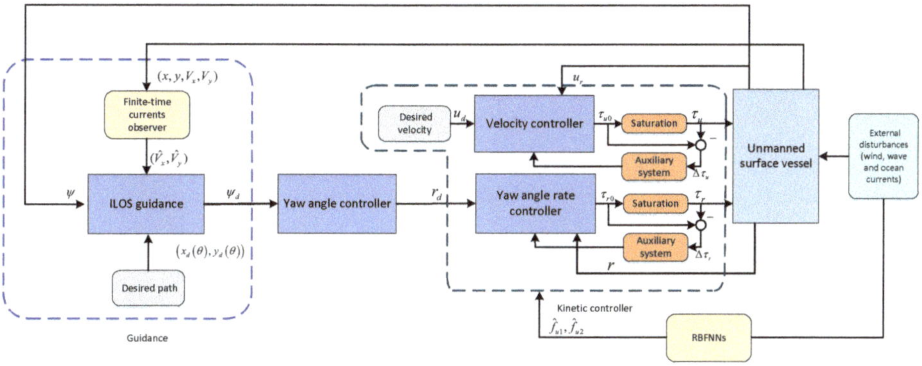


**Figure 2.** The structure diagram of the path following controller.

### *3.1. Guidance*

This section is devoted to designing the ILOS guidance to calculate the approaching angle with respect to the desired path. Then, the time derivative of Equation (7) is taken to obtain the position errors dynamics

$$\begin{cases} \dot{x}_e = u_r \cos(\psi - \gamma_p) - v_r \sin(\psi - \gamma_p) + y_e \dot{\gamma}_p + \theta_x - U_{pp} \\ \dot{y}_e = u_r \sin(\psi - \gamma_p) + v_r \cos(\psi - \gamma_p) - x_e \dot{\gamma}_p + \theta_y \end{cases} \tag{10}$$

where $\theta_x = V_c \cos(\beta_c - \gamma_p)$ and $\theta_y = V_c \sin(\beta_c - \gamma_p)$; $V_c = \sqrt{V_x^2 + V_y^2} \leqslant V_{c\max}$ and $\beta_c = \text{atan2}(V_y, V_x)$. By using the kinematic relationship given by Equation (8), we have $\dot{\theta} = U_{pp}/\sqrt{\dot{x}_d^2 + \dot{y}_d^2}$. In addition, the virtual target velocity $U_{pp}$ can be regarded as the extra degree-of-freedom to stabilize the cross-tracking error [5]. To prescribe the desired yaw angle $\psi_d$ for $\psi$, the following form of the guidance law is taken

$$\psi_d = \gamma_p - \beta_r + \text{atan2}\left(-\left(y_e + \alpha_y\right), \Delta\right) \tag{11}$$

where $\beta_r = \text{atan2}\left(v_r, u_r\right)$; $\Delta$ is the specific look-ahead distance; and $\alpha_y$ is the virtual control input. It should be noted that $\alpha_y$ is introduced to shape the dynamics of the ILOS guidance by inherently adding an integral action.

#### 3.1.1. Estimations of Ocean Currents

Before designing the kinematic controller, the unknown ocean currents need be identified precisely. The finite-time currents observer for $V_x$ and $V_y$ are designed as follows:

$$\begin{cases} \dot{\hat{\boldsymbol{\varepsilon}}}_e = \boldsymbol{\Lambda} + \mathbf{g} \\ \boldsymbol{\Lambda} = -\rho_1 \mathbf{L}^{1/2} sig^{1/2}(\hat{\boldsymbol{\varepsilon}}_e - \boldsymbol{\varepsilon}_e) + \hat{\boldsymbol{v}} \\ \dot{\hat{\boldsymbol{v}}} = -\rho_2 \mathbf{L} sig(\hat{\boldsymbol{v}} - \boldsymbol{\Lambda}) \end{cases} \tag{12}$$

where $sig^{\alpha}(\bullet) = |\bullet|^{\alpha}\text{sign}(\bullet)$; $\hat{\boldsymbol{\varepsilon}}_e = [\hat{x}, \hat{y}]^T$; $\mathbf{g} = [U_r \cos(\psi + \beta_r), U_r \sin(\psi + \beta_r)]^T$, $U_r = \sqrt{u_r^2 + v_r^2}$; $\hat{\boldsymbol{v}} = [\hat{V}_x, \hat{V}_y]^T$; $\mathbf{L} = diag(l_1, l_2) > 0$; $\rho_1 > 0, \rho_2 > 0$.

**Theorem 1.** *Considering the proposed finite-time currents observer in Equation (12), the unknown time-varying ocean currents $\boldsymbol{v}$ can be precisely estimated within a finite time.*

**Proof.** According to Equation (4), we have

$$\dot{\boldsymbol{\varepsilon}}_e = \mathbf{g} + \boldsymbol{v} \tag{13}$$

Together with the finite-time currents observer, we have

$$\begin{cases} \dot{\tilde{\boldsymbol{\varepsilon}}}_e = -\rho_1 \mathbf{L}^{1/2} sig^{1/2}(\hat{\boldsymbol{\varepsilon}}_e - \boldsymbol{\varepsilon}_e) + \tilde{\boldsymbol{v}} \\ \dot{\tilde{\boldsymbol{v}}} = -\rho_2 \mathbf{L} sig(\hat{\boldsymbol{v}} - \boldsymbol{\Lambda}) - \dot{\boldsymbol{v}} \\ \quad \in -\rho_2 \mathbf{L} sig(\hat{\boldsymbol{v}} - \boldsymbol{\Lambda}) + [-M, M] \end{cases} \tag{14}$$

where $\tilde{\boldsymbol{\varepsilon}}_e = \hat{\boldsymbol{\varepsilon}}_e - \boldsymbol{\varepsilon}_e$ and $\tilde{\boldsymbol{v}} = \hat{\boldsymbol{v}} - \boldsymbol{v}$.

In light of Lemma 2 in [27], one can immediately have observer errors $\tilde{\boldsymbol{\varepsilon}}_e$ and $\tilde{\boldsymbol{v}}$ converge to zero within a finite time. It implies that there exists a finite time $T_0$ and finite constants $\varepsilon_b$ and $v_b$ such that

$$\begin{cases} \|\tilde{\boldsymbol{\varepsilon}}_e\| < \varepsilon_b, \|\tilde{\boldsymbol{v}}\| < v_b, \ \forall t < T_0 \\ \tilde{\boldsymbol{\varepsilon}}_e \equiv 0, \tilde{\boldsymbol{v}} \equiv 0, \ \forall t \geqslant T_0 \end{cases} \tag{15}$$

This concludes the proof. □

Therefore, the parametric currents estimation errors $\tilde{\theta}_x = \theta_x - \hat{\theta}_x$ and $\tilde{\theta}_y = \theta_y - \hat{\theta}_y$ satisfy the following inequality

$$\begin{cases} \tilde{\theta}_k < \sqrt{2(V_{c\max}^2 + v_b^2)} := \bar{\theta}, \forall t < T_0 \\ \tilde{\theta}_k \equiv 0, \forall t \geqslant T_0 \end{cases} \quad (k = x, y) \tag{16}$$

**Remark 3.** *With the availability of $\hat{V}_x$ and $\hat{V}_y$ of the finite-time currents observer, the estimations of parametric currents $\hat{\theta}_x$ and $\hat{\theta}_y$ can be reliably obtained.*

### 3.1.2. Design of Kinematic Controller

In this section, the kinematic control laws including the virtual control law and path variable update law are presented.

Substituting Equation (11) into Equation (10) results in

$$\begin{cases} \dot{x}_e = U_r \cos\left(\psi - \gamma_p + \beta_r\right) + \theta_x - U_{pp} + y_e \dot{\gamma}_p \\ \dot{y}_e = -\dfrac{U_r \left(y_e + \alpha_y\right)}{\sqrt{\left(y_e + \alpha_y\right)^2 + \Delta^2}} + U_r \phi\left(y_e, \tilde{\psi}\right) \tilde{\psi} + \theta_y - x_e \dot{\gamma}_p \end{cases} \tag{17}$$

where $\tilde{\psi} = \psi - \psi_d$ and $\phi\left(y_e, \tilde{\psi}\right) = \frac{\sin\tilde{\psi}}{\tilde{\psi}} \frac{\Delta}{\sqrt{\left(y_e+\alpha_y\right)^2+\Delta^2}} - \frac{\cos\tilde{\psi}-1}{\tilde{\psi}} \frac{\left(y_e+\alpha_y\right)}{\sqrt{\left(y_e+\alpha_y\right)^2+\Delta^2}}$. Note that the upper bound of $\phi\left(y_e, \tilde{\psi}\right)$ is 1.73.

The virtual control $\alpha_y$ is designed to cancel $\theta_y$ asymptotically [10]

$$\frac{U_r \alpha_y}{\sqrt{\left(y_e + \alpha_y\right)^2 + \Delta^2}} = \hat{\theta}_y \tag{18}$$

Solving for $\alpha_y$ provides a credible solution (the negative root) as follows

$$\alpha_y = \frac{y_e \left(\hat{\theta}_y / U_r\right)^2 - \hat{\theta}_y / U_r \sqrt{\Delta^2 \left(1 - \left(\hat{\theta}_y / U_r\right)^2\right) + y_e^2}}{1 - \left(\hat{\theta}_y / U_r\right)^2} \tag{19}$$

In this case, the boundedness of $\hat{\theta}_y / U_r$ must be ensured such that $\left|\hat{\theta}_y / U_r\right| < 1$. Thus, the boundedness of the parametric ocean currents is defined such that $\theta_y < M_\theta < U_r$.

The desire target velocity $U_{pp}$ is chosen to stabilize $x_e$ as follows

$$U_{pp} = U_r \cos\left(\psi - \gamma_p + \beta_r\right) + k_1 x_e + \hat{\theta}_x \tag{20}$$

where $k_1$ is a positive design parameter. Thus, the path variable update law can be obtained according to Equation (8):

$$\dot{\theta} = \frac{U_{pp}}{\sqrt{\dot{x}_d^2 + \dot{y}_d^2}} \tag{21}$$

Therefore, substituting Equations (18) and (19) into Equation (17), we have

$$\begin{cases} \dot{x}_e = -k_1 x_e + \tilde{\theta}_x + y_e \dot{\gamma}_p \\ \dot{y}_e = -\dfrac{U_r y_e}{\sqrt{\left(y_e + \alpha_y\right)^2 + \Delta^2}} + U_r \phi\left(y_e, \tilde{\psi}\right) \tilde{\psi} + \tilde{\theta}_y - x_e \dot{\gamma}_p \end{cases} \tag{22}$$

### 3.2. Design of Kinetic Controller

In this section, a kinetic controller is designed to track the desired approaching angle and the desired surge velocity. Let $\tilde{r} = r - r_d$ and $\tilde{u}_r = u_r - u_d$ be the attitude tracking error and surge velocity tracking error, respectively, where $r_d$ is the virtual control law and $u_d$ is the desired surge velocity.

According to Equation (4), the time derivative of $\tilde{\psi}$ is given by

$$\dot{\tilde{\psi}} = r - \dot{\psi}_d \tag{23}$$

To stabilize Equation (23), the desired intermediate control law for $r_d$ is chosen as"

$$r_d = -k_2 \tilde{\psi} + \dot{\psi}_d - U_r \phi y_e \tag{24}$$

where $k_2$ is a positive design parameter.

From Equation (5), the dynamics of $\tilde{r}$ and $\tilde{u}_r$ are given as follows

$$\begin{cases} m_{33} \dot{\tilde{r}} = m_{33} f_r + \tau_r + \tau_{wr} - m_{33} \dot{r}_d \\ m_{11} \dot{\tilde{u}}_r = m_{11} f_u + \tau_u + \tau_{wu} - m_{11} \dot{u}_d \end{cases} \tag{25}$$

Since the inertia and damping parameters are unknown, RBFNNs are employed to handle the unknown parts

$$\begin{cases} m_{33} f_r - m_{33} \dot{r}_d = \mathbf{W}_1^{*T} \boldsymbol{\varphi}(\mathbf{z}) + \varepsilon_1 \\ m_{11} f_u - m_{11} \dot{u}_d = \mathbf{W}_2^{*T} \boldsymbol{\varphi}(\mathbf{z}) + \varepsilon_2 \end{cases} \tag{26}$$

where $\mathbf{W}_i^*(i=1,2)$ is the ideal constant weight matrix satisfying $\|\mathbf{W}_i{}^*\| \leqslant W_{iM}$; $\boldsymbol{\varphi}(\mathbf{z})$ is the radial basis function with $\mathbf{z} = [u_r, v_r, r]^T$ being the input vector to the RBFNNs; and $\varepsilon_i(i=1,2)$ is the approximation error with unknown constant upper bound such that $|\varepsilon_i| \leqslant \bar{\varepsilon}_i$. Denote the unknown parts as $m_{33}f_r - m_{33}\dot{r}_d = f_{u1}$ and $m_{11}f_u - m_{11}\dot{u}_d = f_{u2}$. Furthermore, there exists bounded functions $\delta_1$ and $\delta_2$ such that $|\varepsilon_1| + |\tau_{wr}| \leqslant \delta_1$, and $|\varepsilon_2| + |\tau_{wu}| \leqslant \delta_2$.

Therefore, the nominal control inputs are chosen as follows by considering the input saturation

$$\begin{cases} \tau_{r0} = -k_3\tilde{r} - \tilde{\psi} - \hat{\mathbf{W}}_1\boldsymbol{\varphi}(\mathbf{z}) - \hat{\delta}_1 h(\tilde{r}) + k_{\zeta 1}\sigma_1 \\ \tau_{u0} = -k_4\tilde{u}_r - \hat{\mathbf{W}}_2\boldsymbol{\varphi}(\mathbf{z}) - \hat{\delta}_2 h(\tilde{u}_r) + k_{\zeta 2}\sigma_2 \end{cases} \tag{27}$$

with update laws

$$\begin{cases} \dot{\hat{\mathbf{W}}}_1 = \Gamma_1\left(\boldsymbol{\varphi}(\mathbf{z})\tilde{r} - \iota_1\hat{\mathbf{W}}_1\right) \\ \dot{\hat{\delta}}_1 = \xi_1\left(\tilde{r}h(\tilde{r}) - \lambda_1(\hat{\delta}_1 - \delta_1^0)\right) \end{cases} \tag{28}$$

$$\begin{cases} \dot{\hat{\mathbf{W}}}_2 = \Gamma_2\left(\boldsymbol{\varphi}(\mathbf{z})\tilde{u}_r - \iota_2\hat{\mathbf{W}}_2\right) \\ \dot{\hat{\delta}}_2 = \xi_2\left(\tilde{u}_r h(\tilde{u}_r) - \lambda_2(\hat{\delta}_2 - \delta_2^0)\right) \end{cases} \tag{29}$$

where $k_3$ and $k_4$ are positive design parameters; $k_{\zeta 1}$ and $k_{\zeta 2}$ are positive design parameters; $\hat{\delta}_i(i=1,2)$ is the estimation of $\delta_i$; $h(\tilde{r}) = \tanh(\tilde{r}/\chi_r)$, $h(\tilde{u}_r) = \tanh(\tilde{u}_r/\chi_u)$, $\chi_j(j=r,u)$ is a positive constant; $\boldsymbol{\Gamma}_i(i=1,2)$ is a positive define design matrix; $\lambda_i(i=1,2)$ is a small design parameter; $\delta_j^0(j=r,u)$ is the prior estimation of $\delta_j$; and $\sigma_i(i=1,2)$ is the state of the auxiliary system.

To compensate for the constraint effects of input saturation, auxiliary dynamic systems [28] are given as follows

$$\dot{\sigma}_1 = \begin{cases} -k_{\sigma_1}\sigma_1 - \dfrac{|\tilde{r}\Delta\tau_r| + 0.5\Delta\tau_r{}^2}{\sigma_1^2}\sigma_1 + \Delta\tau_r, & |\sigma_1| \geqslant \mu_1 \\ 0, & |\sigma_1| < \mu_1 \end{cases} \tag{30}$$

and

$$\dot{\sigma}_2 = \begin{cases} -k_{\sigma 2}\sigma_2 - \dfrac{|\tilde{u}_r\Delta\tau_u| + 0.5\Delta\tau_u^2}{\sigma_2^2}\sigma_2 + \Delta\tau_u, & |\sigma_2| \geqslant \mu_2 \\ 0, & |\sigma_2| < \mu_2 \end{cases} \tag{31}$$

where $k_{\sigma_i}(i=1,2)$ are positive design parameters; $\mu_i(i=1,2)$ are small positive constants; $\Delta\tau_r = \tau_r - \tau_{r0}$; $\Delta\tau_u = \tau_u - \tau_{u0}$.

Therefore, the closed-loop attitude and surge velocities tracking errors dynamics become by virtue of Equations (25) to (27)

$$\begin{cases} m_{33}\dot{\tilde{r}} = -k_3\tilde{r} - \tilde{\psi} - \tilde{\mathbf{W}}_1\boldsymbol{\varphi}(\mathbf{z}) - \tilde{\delta}_1 h(\tilde{r}) + k_{\varsigma 1}\sigma_1 + \varepsilon_1 + \tau_{wr} + \Delta\tau_r. \\ m_{11}\dot{u}_e = -k_4 u_e - \tilde{\mathbf{W}}_2\boldsymbol{\varphi}(\mathbf{z}) - \tilde{\delta}_2 h(u_e) + k_{\varsigma 2}\sigma_2 + \varepsilon_2 + \tau_{wu} + \Delta\tau_u. \end{cases} \tag{32}$$

where $\tilde{\mathbf{W}}_i = \hat{\mathbf{W}}_i - \mathbf{W}_i^*(i=1,2)$ are weight matrix estimation errors; and $\tilde{\delta}_i = \hat{\delta}_i - \delta_i(i=1,2)$ are the adaptive terms estimation errors.

### 3.3. Stability Analysis

In this section, the main theorem of the path following controller is presented.

**Theorem 2.** *Consider the USV model in Equations (4) and (5) in the presence of uncertainties and unknown external time-varying disturbances under input saturation, and suppose that Assumptions 1 and 2 are satisfied, under the guidance law in Equation (11) along with the finite-time currents observer in Equation (12). The given path is parameterized by $\theta$ with the update laws in Equation (21), and the desired velocity $u_d$ is given as well. The control laws in Equation (27) together with the adaptive laws in Equations (28) and (29) and the auxiliary* *systems in Equations (30) and (31) are incorporated to assist in handling input saturation, guaranteeing that all tracking error signals are locally UUB.*

**Proof.** Consider the following Lyapunov function

$$V = \frac{1}{2}x_e^2 + \frac{1}{2}y_e^2 + \frac{1}{2}\tilde{\psi}^2 + \frac{1}{2}m_{33}\tilde{r}^2 + \frac{1}{2}m_{11}\tilde{u}_r^2 + \frac{1}{2}\sum_{i=1}^{2}\sigma_i^2 + \frac{1}{2}\sum_{i=1}^{2}\tilde{\mathbf{W}}_i^T\mathbf{\Gamma}_i^{-1}\tilde{\mathbf{W}}_i + \frac{1}{2\xi_i}\sum_{i=1}^{2}\tilde{\delta}_i^2 \tag{33}$$

Taking the time derivative of Equation (33) along with Equations (22)–(24) and (32) yields

$$\begin{aligned}\dot{V} \leqslant & -k_1x_e^2 - \frac{U_r}{\sqrt{(y_e+\alpha_y)^2+\Delta^2}}y_e^2 - k_2\tilde{\psi}^2 - k_3\tilde{r}^2 - \tilde{\mathbf{W}}_1\boldsymbol{\varphi}(\mathbf{z})\tilde{r} - \hat{\delta}_1 h(\tilde{r})\tilde{r} \\ & + k_{\zeta 1}\sigma_1\tilde{r} + \delta_1\tilde{r} + \Delta\tau_r\tilde{r} - k_4\tilde{u}_r^2 - \tilde{\mathbf{W}}_2\boldsymbol{\varphi}(\mathbf{z})\tilde{u}_r - \hat{\delta}_2 h(\tilde{u}_r)\tilde{u}_r + k_{\zeta 2}\sigma_2\tilde{u}_r + \delta_2\tilde{u}_r + \Delta\tau_u\tilde{u}_r \\ & + \tilde{\theta}_x x_e + \tilde{\theta}_y y_e + \sum_{i=1}^{2}\sigma_i\dot{\sigma}_i + \sum_{i=1}^{2}\tilde{\mathbf{W}}_i^T\mathbf{\Gamma}_i^{-1}\dot{\hat{\mathbf{W}}}_i + \frac{1}{\xi_i}\sum_{i=1}^{2}\tilde{\delta}_i\dot{\hat{\delta}}_i\end{aligned} \tag{34}$$

(1) When $|\sigma_i| \geqslant \mu_i (i = 1,2)$, according to Equations (27) to (30) and Young's equalities $k_{\zeta 1}\sigma_1\tilde{r} \leqslant \frac{1}{2}\tilde{r}^2 + \frac{1}{2}k_{\zeta 1}^2\sigma_1^2$, $\sigma_1\Delta\tau_r \leqslant \frac{1}{2}\sigma_1^2 + \frac{1}{2}\Delta\tau_r^2$, $k_{\zeta 2}\sigma_2\tilde{u}_r \leqslant \frac{1}{2}\tilde{u}_r^2 + \frac{1}{2}k_{\zeta 2}^2\sigma_2^2$, $\sigma_2\Delta\tau_u \leqslant \frac{1}{2}\sigma_2^2 + \frac{1}{2}\Delta\tau_u^2$, $\tilde{\theta}_x x_e \leqslant \frac{1}{2}\tilde{\theta}^2 + \frac{1}{2}x_e^2$, and $\tilde{\theta}_y y_e \leqslant \frac{k_1}{2}\tilde{\theta}^2 + \frac{1}{2k_1}y_e^2$, we have

$$\begin{aligned}\dot{V} \leqslant & -\left(k_1 - \frac{1}{2}\right)x_e^2 - \left(\frac{U_r}{\sqrt{y_e^2+\Delta^2}} - \frac{1}{2k_1}\right)y_e^2 - k_2\tilde{\psi}^2 - \left(k_{\sigma 1} - \frac{1}{2} - \frac{1}{2}k_{\zeta 1}^2\right)\sigma_1^2 - \left(k_3 - \frac{1}{2}\right)\tilde{r}^2 \\ & - \left(k_{\sigma 2} - \frac{1}{2} - \frac{1}{2}k_{\zeta 2}^2\right)\sigma_2^2 - \left(k_4 - \frac{1}{2}\right)\tilde{u}_r^2 - \sum_{i=1}^{2}\iota_i\tilde{\mathbf{W}}_i^T\hat{\mathbf{W}}_i - \sum_{i=1}^{2}\lambda_i\tilde{\delta}_i\left(\hat{\delta}_i - \delta_i^0\right) \\ & + \tilde{r}\left(\delta_1 - \hat{\delta}_1 h(\tilde{r})\right) + \tilde{\delta}_1 h(\tilde{r})\tilde{r} + \tilde{u}_r\left(\delta_2 - \hat{\delta}_2 h(\tilde{u}_r)\right) + \tilde{\delta}_2 h(\tilde{u}_r)\tilde{u}_r + \frac{1+k_1}{2}\tilde{\theta}^2\end{aligned} \tag{35}$$

Consider the following inequality of the hyperbolic tangent function holds for any $\chi > 0$ and for any $\varpi \in \mathbb{R}$ [29]

$$0 \leqslant |\varpi| - \varpi\tan\left(\frac{\varpi}{\chi}\right) \leqslant \kappa_\chi\chi \tag{36}$$

where $\kappa_\chi$ is a constant that satisfies $\kappa_\chi = e^{-(\kappa_\chi+1)}$, i.e., $\kappa_\chi = 0.2785$.

It is worth noting that the following equalities hold

$$-\sum_{i=1}^{2}\iota_i\tilde{\mathbf{W}}_i^T\hat{\mathbf{W}}_i \leqslant -\sum_{i=1}^{2}\frac{\iota_i}{2}\tilde{\mathbf{W}}_i^T\tilde{\mathbf{W}}_\mathbf{i} + \sum_{i=1}^{2}\frac{\iota_i}{2}\|\mathbf{W}_i^*\|^2 \tag{37}$$

$$-\sum_{i=1}^{2}\lambda_i\tilde{\delta}_i\left(\hat{\delta}_i - \delta_i^0\right) \leqslant -\sum_{i=1}^{2}\frac{\lambda_i}{2}\tilde{\delta}_i^2 + \sum_{i=1}^{2}\frac{\lambda_i}{2}\left(\delta_i - \delta_i^0\right)^2 \tag{38}$$

$$\tilde{r}\left(\delta_1 - \hat{\delta}_1\tan\left(\frac{\tilde{r}}{\chi_r}\right)\right) + \tilde{\delta}_1\tan\left(\frac{\tilde{r}}{\chi_r}\right)\tilde{r} \leqslant \delta_1\left(|\tilde{r}| - \tilde{r}\tan\left(\frac{\tilde{r}}{\chi_r}\right)\right) \leqslant 0.2785\chi_r\delta_1 \tag{39}$$

$$\tilde{u}_r\left(\delta_2 - \hat{\delta}_2\tan\left(\frac{\tilde{u}_r}{\chi_u}\right)\right) + \tilde{\delta}_2\tan\left(\frac{\tilde{u}_r}{\chi_u}\right)\tilde{u}_r \leqslant \delta_2\left(|\tilde{u}_r| - \tilde{u}_r\tan\left(\frac{\tilde{u}_r}{\chi_u}\right)\right) \leqslant 0.2785\chi_u\delta_2 \tag{40}$$

Substituting Equations (36)–(39) into Equation (35), we have

$$\begin{aligned}\dot{V} \leqslant & -\left(k_1-\frac{1}{2}\right)x_e^2-\left(\frac{U_r}{\sqrt{y_e^2+\Delta^2}}-\frac{1}{2k_1}\right)y_e^2-k_2\tilde{\psi}^2-\left(k_3-\frac{1}{2}\right)\tilde{r}^2-\left(k_4-\frac{1}{2}\right)\tilde{u}_r^2 \\ & -\sum_{i=1}^{2}\left(k_{\sigma i}-\frac{1}{2}-\frac{1}{2}k_{\zeta i}^2\right)\sigma_i^2-\sum_{i=1}^{2}\frac{\iota_i}{2}\tilde{\mathbf{W}}_i^T\tilde{\mathbf{W}}_{\mathrm{i}}-\sum_{i=1}^{2}\frac{\lambda_i}{2}\tilde{\delta}_i^2+\sum_{i=1}^{2}\frac{1}{2}\left(\delta_i-\delta_i^0\right)^2 \\ & +\sum_{i=1}^{2}\frac{\iota_i}{2}W_{iM}{}^2+0.2785\left(\chi_r\delta_1+\chi_u\delta_2\right)+\frac{1+k_1}{2}\bar{\theta}^2 \\ \leqslant & -\kappa_1 V+\vartheta_1\end{aligned} \tag{41}$$

where $\kappa_1=\min\left\{2k_1-1,\frac{2U_r}{\sqrt{y_e^2+\Delta^2}}-\frac{1}{k_1},2k_2,\frac{2k_3-1}{m_{33}},\frac{2k_4-1}{m_{11}},\min\limits_{i=1,2}\left(2k_{\sigma i}-1-k_{\zeta i}^2\right),\min\limits_{i=1,2}\left(\iota_i\lambda_{\min}(\mathbf{\Gamma}_i)\right),\min\limits_{i=1,2}\left(\lambda_i\xi_i\right)\right\}$, $k_1=\max\left\{\frac{1}{2},\frac{\sqrt{y_e^2+\Delta^2}}{2U_r}\right\}$, $\lambda_{\min}(\bullet)$ denotes the minimum eigenvalue of a matrix, and $\vartheta_1=\sum_{i=1}^{2}\frac{1}{2}\left(\delta_i-\delta_i^0\right)^2+\sum_{i=1}^{2}\frac{\iota_i}{2}W_{iM}{}^2+0.2785\left(\chi_r\delta_1+\chi_u\delta_2\right)+\frac{1+k_1}{2}\bar{\theta}^2$.

(2) When $|\sigma_i|<\mu_i(i=1,2)$, we have $\sum_{i=1}^{2}\sigma_i\dot{\sigma}_i=0$. According to Equations (30) and (31) and the inequalities $\sum_{i=1}^{2}\frac{1}{2}k_{\zeta i}^2\sigma_i^2\leqslant-\sum_{i=1}^{2}\frac{1}{2}k_{\zeta i}^2\sigma_i^2+\sum_{i=1}^{2}k_{\zeta i}^2\mu_i^2$, $\tilde{r}\Delta\tau_r\leqslant\frac{1}{2}\tilde{r}^2+\frac{1}{2}\Delta\tau_r^2$, and $\tilde{u}_r\Delta\tau_u\leqslant\frac{1}{2}\tilde{u}_r^2+\frac{1}{2}\Delta\tau_u^2$, Equation (34) becomes

$$\begin{aligned}\dot{V} \leqslant & -\left(k_1-\frac{1}{2}\right)x_e^2-\left(\frac{U_r}{\sqrt{y_e^2+\Delta^2}}-\frac{1}{2k_1}\right)y_e^2-k_2\tilde{\psi}^2-\left(k_3-1\right)\tilde{r}^2-\left(k_4-1\right)\tilde{u}_r^2 \\ & -\sum_{i=1}^{2}\frac{k_{\zeta i}^2}{2}\sigma_i^2-\sum_{i=1}^{2}\frac{\iota_i}{2}\tilde{\mathbf{W}}_i^T\tilde{\mathbf{W}}_i-\sum_{i=1}^{2}\frac{\lambda_i}{2}\tilde{\delta}_i^2+\sum_{i=1}^{2}\frac{1}{2}\left(\delta_i-\delta_i^0\right)^2+\sum_{i=1}^{2}\frac{\iota_i}{2}W_{iM}{}^2 \\ & +0.2785\left(\chi_r\delta_1+\chi_u\delta_2\right)+\sum_{i=1}^{2}k_{\zeta i}^2\mu_i^2+\frac{1}{2}\Delta\tau_r^2+\frac{1}{2}\Delta\tau_u^2+\frac{1+k_1}{2}\bar{\theta}^2 \\ \leqslant & -\kappa_2 V+\vartheta_2\end{aligned} \tag{42}$$

where $\kappa_2=\min\left\{2k_1-1,\frac{2U_r}{\sqrt{y_e^2+\Delta^2}}-\frac{1}{k_1},2k_2,\frac{2k_3-2}{m_{33}},\frac{2k_4-2}{m_{11}},\min\limits_{i=1,2}\left(k_{\zeta i}^2\right),\min\limits_{i=1,2}\left(\iota_i\lambda_{\min}(\mathbf{\Gamma}_i)\right),\min\limits_{i=1,2}\left(\lambda_i\xi_i\right)\right\}$, $k_1=\max\left\{\frac{1}{2},\frac{\sqrt{y_e^2+\Delta^2}}{2U_r}\right\}$; $\vartheta_2=\sum_{i=1}^{2}\frac{1}{2}\left(\delta_i-\delta_i^0\right)^2+\sum_{i=1}^{2}\frac{\iota_i}{2}W_{iM}{}^2+0.2785\left(\chi_r\delta_1+\chi_u\delta_2\right)+\sum_{i=1}^{2}k_{\zeta i}^2\mu_i^2+\frac{1}{2}\Delta\tau_r^2+\frac{1}{2}\Delta\tau_u^2+\frac{1+k_1}{2}\bar{\theta}^2$.

Synthesizing Equations (41) and (42), we have

$$\dot{V}\leqslant-\kappa V+\vartheta \tag{43}$$

where $\kappa=\min\{\kappa_1,\kappa_2\}$ and $\vartheta=\max\{\vartheta_1,\vartheta_2\}$ with the design parameters satisfying the conditions: $k_1>\max\left(\frac{1}{2},\frac{\sqrt{y_e^2+\Delta^2}}{2U_r}\right),k_2>0,k_3>1,k_4>1,k_{\zeta 1}>0,k_{\zeta 2}>0,k_{\sigma 1}>\frac{1}{2}k_{\zeta 1}^2+\frac{1}{2},k_{\sigma 2}>\frac{1}{2}k_{\zeta 2}^2+\frac{1}{2}$. Then, the following inequality can be obtained

$$0\leqslant V\leqslant\left(V(0)-\frac{\vartheta}{\kappa}\right)e^{-\kappa t}+\frac{\vartheta}{\kappa} \tag{44}$$

In conclusion, it follows the definition of $V$ that the tracking error signals $\eta_e=[x_e,y_e,\tilde{\psi},\tilde{r},\tilde{u}_r]^T$ are locally UUB, which ultimately converges to the compact sets $\mathbf{\Omega}_1=\left\{\eta_e\in\mathbb{R}^5|\,\|\eta_e\|\leqslant\sqrt{2\left(V(0)-\frac{\vartheta}{\kappa}\right)e^{-\kappa t}+\frac{2\vartheta}{\kappa}}\right\}$. The ultimate compact set can be easily tuned by adjusting the design parameters. Meanwhile, the parameters estimation errors $\tilde{\mathbf{W}}_1, \tilde{\mathbf{W}}_2, \tilde{\delta}_1$, and $\tilde{\delta}_2$ are locally UUB. Theorem 2 is thus proved. □

### 3.4. Sway Dynamics

For the sway velocity dynamics, considering the Lyapunov function $V_v = \frac{1}{2}v_r^2$ and taking the time derivative of it based on Equation (5) yields

$$\begin{aligned} \dot{V}_v &= v_r \dot{v}_r \\ &= -\frac{m_{11}}{m_{22}} u_r r v_r - \frac{d_{22}}{m_{22}} v_r^2 - \frac{d_{v2}}{m_{11}} |v_r| v_r^2 - \frac{d_{v3}}{m_{11}} v_r^4 + \frac{1}{m_{22}} \tau_{wv} \\ &\leqslant -\chi v_r + \frac{1}{m_{22}} \bar{\tau}_w \end{aligned} \tag{45}$$

where $\chi = \frac{m_{11}}{m_{22}} u_r r$. Based on the above analyses, the boundedness of $\tilde{u}_r$ and $\tilde{r}$ is guaranteed, thus $\chi$ is bounded. It should be noted that $d_{vi}$ and $m_{ii} (i = 2, 3)$ are positive constants. According to Krstic, M.; Kanellakopoulos, I.; Kokotovic, P. V. [30], we have

$$\begin{aligned} V_v &\leqslant V_v(t_0) e^{-\chi(t-t_0)} + \bar{\tau}_w \frac{1 - e^{-\chi(t-t_0)}}{m_{22}\chi} \\ &\leqslant V_v(t_0) + \frac{\bar{\tau}_w}{m_{22}\chi} \end{aligned} \tag{46}$$

Therefore, the boundedness of the sway velocity $v_r$ is guaranteed.

## 4. Simulations

The effectiveness and robustness of the proposed path following control method were evaluated based on the platform of MATLAB. The Cybership II [31] was taken as the control object whose parameters were as follows: $m_{11} = 25.8, m_{22} = 33.8, m_{33} = 2.76, d_{11} = 0.9257, d_{22} = 2.8909, d_{33} = 0.5$. For simplicity, we ignored the off-diagonal terms of the inertia and damping. The maximum actuated force and moment were 2 N and 1.5 Nm. From the beginning of the simulations, the ocean currents with time-varying speed were given as $V_x = 0.08\sin(0.1t)$ m/s and $V_y = 0.04\sin(0.1t)$ m/s. The time-varying external disturbances were generated with the first-order Markov process $\dot{\tau}_{wu} + \varsigma_1 \tau_{wu} = w_1, \dot{\tau}_{wv} + \varsigma_2 \tau_{wv} = w_2, \dot{\tau}_{wr} + \varsigma_3 \tau_{wr} = w_3$, where $w_i$ and $\varsigma_i (i = 1, 2, 3)$ are zero-mean Gaussian white noise and constants, respectively [32]. The parameters of the controller are listed in Table 1. The node numbers and widths of RBFNNs were chosen as: node number $l = 21$ and the widths $b_i = 3 (i = 1, 2, \ldots, l)$. The neural active region was chosen as $[|u|, |v|, |r|] \in [[0, 2], [0, 1.5], [0, 1.5]]^T$. Performance comparisons between the proposed finite-time currents observer based ILOS guidance with adaptive RBFNN (FCONN) controller and the indirect adaptive observer based [10] ILOS guidance with adaptive RBFNN (IAONN) controller are presented in the following two control scenarios.

**Table 1.** Parameters of the path following controller.

| Notation | Value | Natation | Value | Natation | Value | Natation | Value | Natation | Value |
|---|---|---|---|---|---|---|---|---|---|
| $k_1$ | 1 | $l_2$ | 4 | $\Gamma_1 ii (i = 21)$ | 500 | $\xi_2$ | 20 | $\chi_r$ | 0.01 |
| $k_2$ | 2 | $\rho_1$ | 0.01 | $\Gamma_2 ii (i = 21)$ | 50 | $\lambda_1$ | 0.01 | $\chi_u$ | 1 |
| $k_3$ | 4 | $\rho_2$ | 0.03 | $\iota_1$ | 0.05 | $\lambda_2$ | 0.01 | $\Delta$ | 2.51 |
| $k_4$ | 5 | $k_{\zeta_1}$ | 1.2 | $\iota_2$ | 0.05 | $\delta_1^0$ | 0.1 | | |
| $l_1$ | 100 | $k_{\zeta_2}$ | 1.2 | $\xi_1$ | 50 | $\delta_1^0$ | 0.1 | | |

Case 1: The desired path and speed assignment were chosen as $\boldsymbol{P}_d = [\theta, \theta]^T$ and $u_d = 0.5$ m/s. The initial states were given as $[x(0), y(0), u_r(0), v_r(0), r]^T = [0\ \text{m}, 2\ \text{m}, 0.01\ \text{m/s}, 0\ \text{m/s}, 0\ \text{rad/s}]^T$.

The simulation results are shown in Figure 3a–g. Table 2 summarizes the performance indices based on the integrated absolute error (IAE) and the time integrated absolute error (ITAE), which were used to evaluate the transient performance and steady-state performance. The proposed FCONN control method could drive the USV following the desired path with a high-precision and better transient process (Figure 3a). Their detailed distinction is more clearly shown in Figure 3b,c, specifically the smaller along- and cross-tracking errors and the smaller heading and surge velocity tracking errors. Meanwhile, the lower IAE and ITAE metrical values of the cross-tracking error revealed the better transient and steady-state performance. Figure 3d shows that the proposed finite-time currents observer could identify the time-varying currents accurately, whereas the IAONN control scheme had an obvious oscillation during the transient process, as well as lower accuracy in the steady state. Figure 3d demonstrates that the RBFNNs could capture the unknown dynamical uncertainties precisely and Figure 3e presents the compound bounds and their estimation, which played decisive roles in driving the dynamics state $r$ and $u_r$ to their real value. Figure 3f depicts the profile of the control inputs where the input saturation (IS) problem Was effectively compensated by the auxiliary system. The control inputs of the proposed method were in the specified region. Hence, in the case, the proposed FCONN path following control method was more effective and robust according to these simulation results.

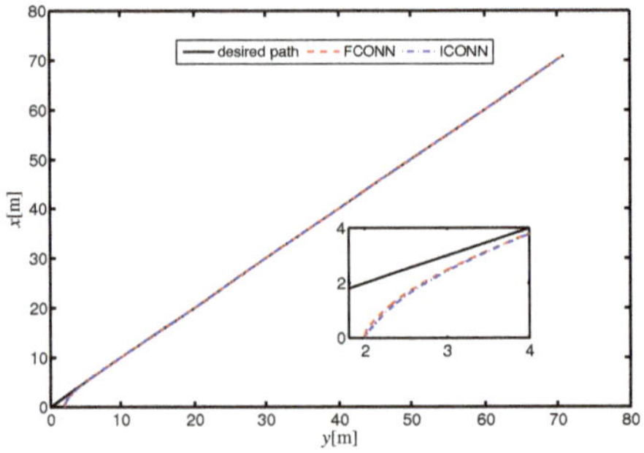


(a) line path following.

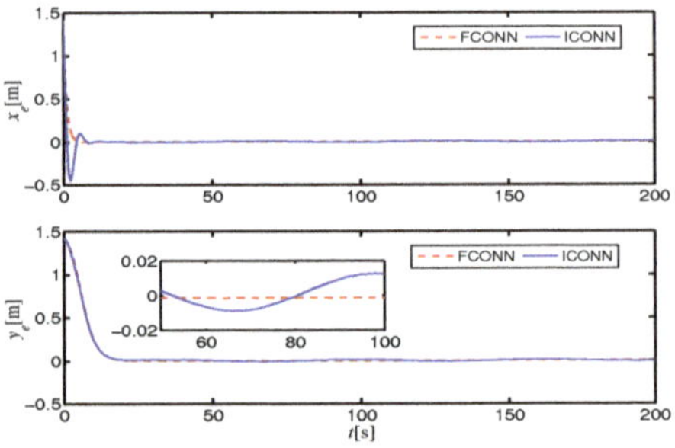


(b) Along- and cross-tracking error in Case 1.

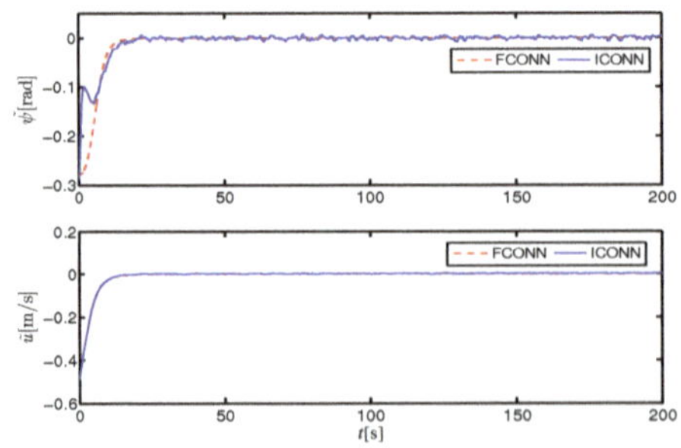


(c) Profile of the heading and surge velocity tracking in Case 1.

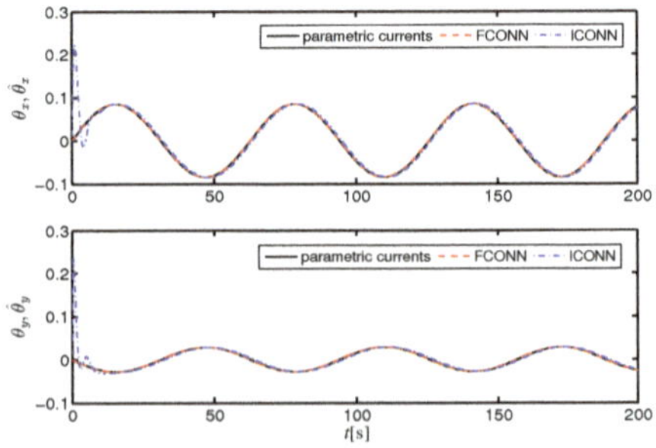


(d) Estimations of parametric currents in Case 1.

**Figure 3.** *Cont.*

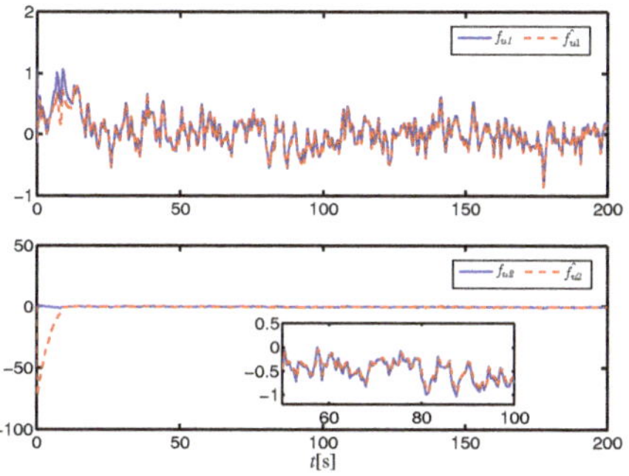


(e) Estimations of dynamical uncertainties using RBFNN in Case 1.

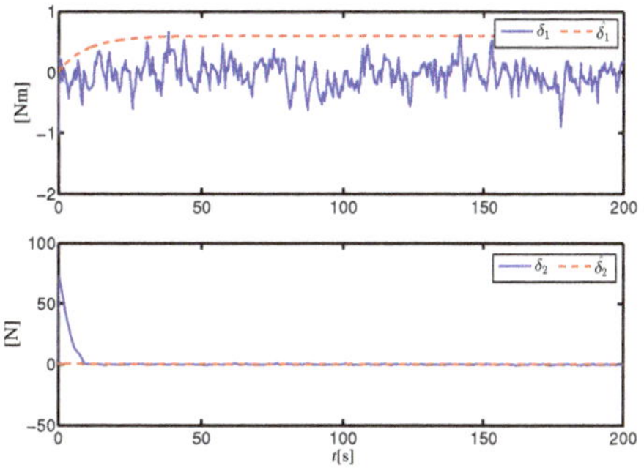


(f) Estimations of compound bound in Case 1.

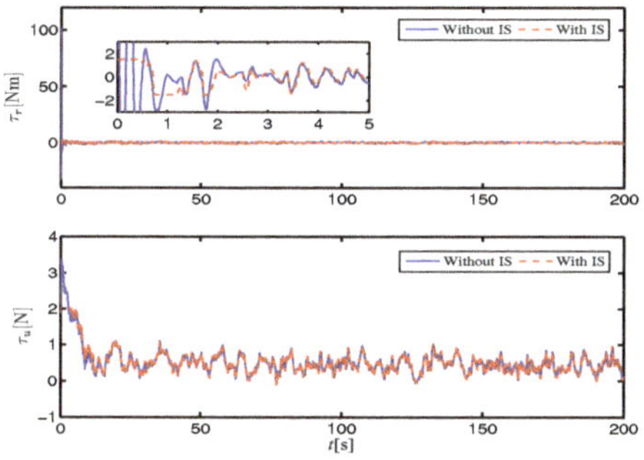


(g) Profile of the control inputs in Case 1.

**Figure 3.** Line path following results.

**Table 2.** Performance indices in these two path following scenarios.

| Control Law | Line Path Following | | Curvilinear Path Following | |
|---|---|---|---|---|
| | IAE($\cdot 10^3$) $\int_0^t \lvert y_e \rvert d\tau$ | ITAE($\cdot 10^3$) $\int_0^t t \lvert y_e \rvert d\tau$ | IAE($\cdot 10^4$) $\int_0^t \lvert y_e \rvert d\tau$ | ITAE($\cdot 10^4$) $\int_0^t t \lvert y_e \rvert d\tau$ |
| FCONN | 1.92 | 5.78 | 2.58 | 9.01 |
| IAONN | 4.18 | 13.12 | 7.32 | 20.32 |

Case 2: Similar to Case 1, another comparison is presented to verify the performance in the case of following a curvilinear path with the same design parameters. The desired path and speed assignment were chosen as $\boldsymbol{P}_d = \theta, [10sin(0.1\theta)]^T$ and $u_d = 0.5$ m/s. The initial states were given as $[x(0), y(0), u_r(0), v_r(0), r]^T = [0\text{ m}, 2\text{ m}, 0.01\text{ m/s}, 0\text{ m/s}, 0\text{ rad/s}]^T$. The simulations results are shown in Figure 4a–g and the performance quantification indices are summarized in Table 2. As illustrated in Figure 4a, although following the curvilinear path, the proposed FCONN method behaved almost the same in both control scenarios. As shown in Figure 4b, the along- and cross-tracking errors oscillated to varying degrees for the poor performance of IAONN, whereas the position errors of FCONN could smoothly and steadily converge to a small neighborhood around zero within a short time. Moreover, the smaller IAE and ITAE metrical values verified it. Figure 4c shows the slight oscillation of the heading and surge velocity tracking error of IAONN. Moreover, in Figure 4d, the poor performance of estimating ocean currents of IAONN undoubtedly degraded the tracking performance. Figure 4e,f shows the exceptional performance of the dynamical uncertainties estimation and disturbance rejection

of the FCONN method. In addition, the control inputs were in the specified region for introducing the auxiliary system, as depicted in Figure 4g. Overall, the proposed control method achieved satisfactory performance and robustness in both cases with fast and accurate estimations of the unknown time-varying ocean currents and dynamical uncertainties, and satisfactory rejection of external disturbances.

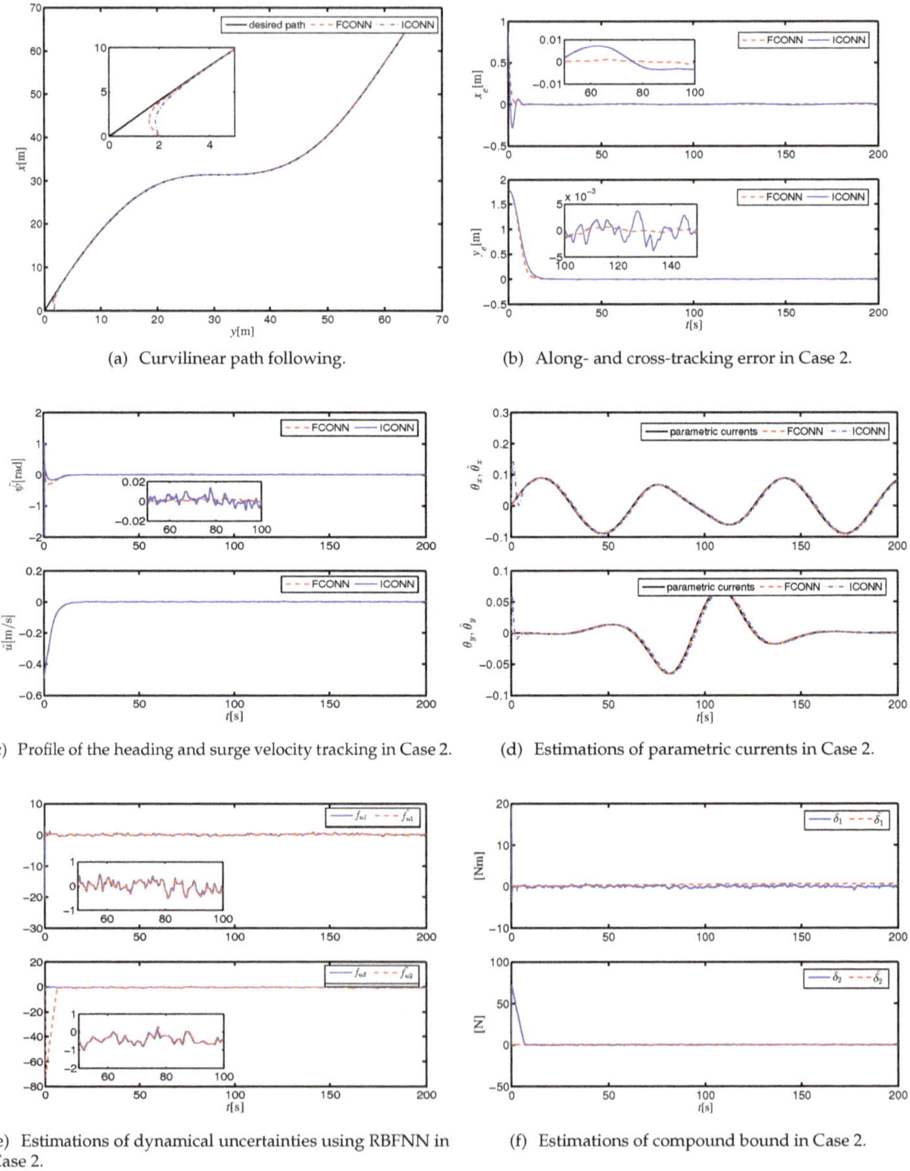


(a) Curvilinear path following.

(b) Along- and cross-tracking error in Case 2.

(c) Profile of the heading and surge velocity tracking in Case 2.

(d) Estimations of parametric currents in Case 2.

(e) Estimations of dynamical uncertainties using RBFNN in Case 2.

(f) Estimations of compound bound in Case 2.

**Figure 4.** *Cont.*

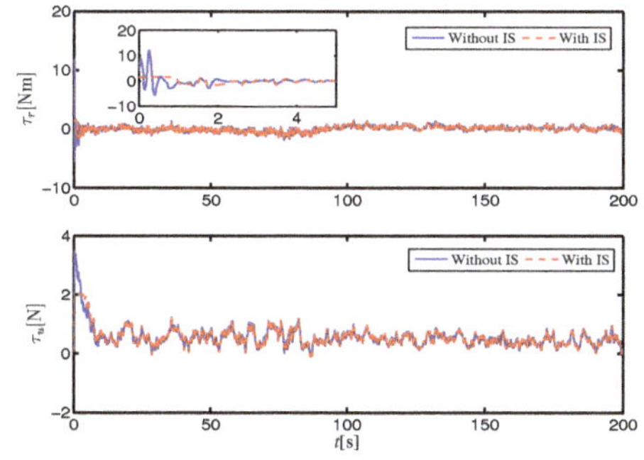


(g) Profile of the control inputs in Case 2.

**Figure 4.** Curvilinear path following results.

## 5. Conclusions

In this paper, a path following control scheme for a USV subject to input saturation and uncertainties has been proposed by resorting to the finite-time currents observer based ILOS guidance, the adaptive RBFNN, and the auxiliary dynamic system. The finite-time currents observer based ILOS guidance is applied to obtain the desired yaw angle, where the incorporated finite time currents observer can provide the precise estimations of the unknown time-varying ocean currents. Simultaneously, the RBF neural networks and the adaptive laws with leakages terms can provide the precise estimations of dynamical uncertainties and the compound bounds of the approximation errors and external disturbances, without knowing any prior knowledge of the time-varying disturbance. The auxiliary control system is introduced to handle input saturation of the actuators. It has been proved that all error signals of the closed-loop system are locally UUB. Finally, both linear and curved path following are presented and compared with the preceding control method. Simulations results have verified that the proposed control method can achieve satisfactory performance and robustness. Future work will cover the aspect of the position error constraint to ensure that the USV can work in these situations including the narrow passage and the channel between obstacles, as well as the precise estimation of the sideslip angle.

**Author Contributions:** Y.F., H.H. and Y.T. conceived the framework and wrote the paper. Y.F. and H.H. designed the controller and H.H. and Y.T. were responsible for the formula analyses. H.H. made the simulations and Y.T. analyzed the data. Y.F., H.H. and Y.T. discussed the results and contributed to the whole manuscript.

**Funding:** This research was funded by National Nature Science Foundation under grant number 51609033, in part by Natural Science Foundation of Liaoning Province under Grant number 201801732, and in part by Fundamental Research Funds for the Central Universities under Grant number 3132016312.

**Conflicts of Interest:** The authors declare no conflict of interest.

## References

1. Liu, Z.; Zhang, Y.; Yu, X.; Yuan, C. Unmanned surface vehicles: An overview of developments and challenges. *Annu. Rev. Control.* **2016**, *41*, 71–93. [CrossRef]
2. Ashrafiuon, H.; Muske, K.R.; Mcninch, L.C. Review of nonlinear tracking and setpoint control approaches for autonomous underactuated marine vehicles. In Proceedings of the 2010 American Control Conference, Baltimore, MD, USA, 30 June–2 July 2010; pp. 5203–5211.
3. Do, K.D.; Pan, J. Control of ships and underwater vehicles. In *Advances in Industrial Control*; Springer: London, UK, 2009; pp. 40–42, ISBN 978-1-84882-729-5.
4. Caccia, M.; Bibuli, M.; Bono, R.; Bruzzone, G. Basic navigation, guidance and control of an unmanned surface vehicle. *Auton. Robot.* **2008**, *25*, 349–365. [CrossRef]

5. Breivik, M.; Fossen, T.I. Path following for marine surface vessels. In Proceedings of the Oceans '04 MTS/IEEE Techno-Ocean '04, Kobe, Japan, 9–12 November 2004; pp. 2282–2289.
6. Fossen, T.I.; Breivik, M.; Skjetne, R. Line-of-sight path following of underactuated marine craft. In Proceedings of the IFAC Manoeuvring and control of Marine Craft, Girona, Spain, 17–19 September 2003; pp. 211–216.
7. Fossen, T.I.; Pettersen, K.Y. On uniform semiglobal exponential stability (usges) of proportional line-of-sight guidance laws. *Automatica* **2014**, *50*, 2912–2917. [CrossRef]
8. Borhaug, E.; Pavlov, A.; Pettersen, K.Y. Integral LOS control for path following of underactuated marine surface vessels in the presence of constant ocean currents. In Proceedings of the 47th IEEE conference on Decision and Control, Cancun, Mexico, 9–11 September 2008; pp. 4984–4991.
9. Lekkas, A.M.; Fossen, T.I. Integral los path following for curved paths based on a monotone cubic hermite spline parametrization. *IEEE Trans. Control. Syst. Technol.* **2014**, *22*, 2287–2301. [CrossRef]
10. Fossen, T.I.; Lekkas, A.M. Direct and indirect adaptive integral line-of-sight path-following controllers for marine craft exposed to ocean currents. *Int. J. Adapt. Control. Signal Process.* **2017**, *31*, 445–463. [CrossRef]
11. Fossen, T.I.; Pettersen, K.Y.; Galeazzi, R. Line-of-sight path following for dubins paths with adaptive sideslip compensation of drift forces. *IEEE Trans. Control. Syst. Technol.* **2015**, *23*, 820–827. [CrossRef]
12. Liu, L.; Wang, D.; Peng, Z. Eso-based line-of-sight guidance law for path following of underactuated marine surface vehicles with exact sideslip compensation. *IEEE J. Ocean. Eng.* **2017**, *42*, 477–487. [CrossRef]
13. Wang, N.; Sun, Z.; Yin, J.; Su, S.F.; Sharma, S. Finite-time observer based guidance and control of underactuated surface vehicles with unknown sideslip angles and disturbances. *IEEE Access* **2018**, *6*, 14059–14070. [CrossRef]
14. Wang, W.; Huang, J.; Wen, C.; Fan, H. Distributed adaptive control for consensus tracking with application to formation control of nonholonomic mobile robots. *Automatica* **2014**, *50*, 1254–1263. [CrossRef]
15. Miao, J.; Wang, S.; Tomovic, M.M.; Zhao, Z. Compound line-of-sight nonlinear path following control of underactuated marine vehicles exposed to wind, waves, and ocean currents. *Nonlinear Dyn.* **2017**, *89*, 1–19. [CrossRef]
16. Do, K.D. Global robust adaptive path-tracking control of underactuated ships under stochastic disturbances. *Ocean. Eng.* **2016**, *111*, 267–278. [CrossRef]
17. Liu, S.; Liu, Y.; Wang, N. Nonlinear disturbance observer-based backstepping finite-time sliding mode tracking control of underwater vehicles with system uncertainties and external disturbances. *Nonlinear Dyn.* **2017**, *88*, 1–12. [CrossRef]
18. Du, J.; Hu, X.; Krstić, M.; Sun, Y. Robust dynamic positioning of ships with disturbances under input saturation. *Automatica* **2016**, *73*, 207–214. [CrossRef]
19. Sun, Z.; Zhang, G.; Yi, B.; Zhang, W. Practical proportional integral sliding mode control for underactuated surface ships in the fields of marine practice. *Ocean. Eng.* **2017**, *142*, 217–223. [CrossRef]
20. Roy, S.; Roy, S.B.; Kar, I.N. Adaptive-robust control of euler-lagrange systems with linearly parametrizable uncertainty bound. *IEEE Trans. Control. Syst. Technol.* **2017**, *26*, 1842–1850. [CrossRef]
21. Roy, S.; Roy, S.B.; Kar, I.N. A New Design Methodology of Adaptive Sliding Mode Control for a Class of Nonlinear Systems with State Dependent Uncertainty Bound. In Proceedings of the 15th International Workshop on Variable Structure Systems, Graz, Austria, 9–11 September 2018; pp. 414–419.
22. Wang, N.; Er, M.J.; Sun, J.C.; Liu, Y.C. Adaptive robust online constructive fuzzy control of a complex surface vehicle system. *IEEE Trans. Cybern.* **2017**, *46*, 1511–1523. [CrossRef]
23. Zheng, Z.; Sun, L. Path following control for marine surface vessel with uncertainties and input saturation. *Neurocomputing* **2016**, *177*, 158–167. [CrossRef]
24. Liu, L.; Wang, D.; Peng, Z. Path following of marine surface vehicles with dynamical uncertainty and time-varying ocean disturbances. *Neurocomputing* **2016**, *173*, 799–808. [CrossRef]
25. Zheng, Z.; Feroskhan, M. Path following of a surface vessel with prescribed performance in the presence of input saturation and external disturbances. *IEEE/ASME Trans. Mechatronics* **2017**, *22*, 2564–2575. [CrossRef]
26. Roy, S.; Shome, S.N.; Nandy, S.; Ray, R.; Kumar, V. Trajectory following control of auv: A robust approach. *J. Inst. Eng. India Series C* **2013**, *94*, 253–265. [CrossRef]
27. Shtessel, Y.B.; Shkolnikov, I.A.; Levant, A. Smooth second-order sliding modes: Missile guidance application. *Automatica* **2007**, *43*, 1470–1476. [CrossRef]

28. Chen, M.; Ge, S.S.; Ren, B. Adaptive tracking control of uncertain mimo nonlinear systems with input constraints. *Automatica* **2011**, *47*, 452–465. [CrossRef]
29. Polycarpou, M.M.; Ioannou, P.A. A robust adaptive nonlinear control design. *Automatica* **1996**, *32*, 423–427. [CrossRef]
30. Krstic, M.; Kanellakopoulos, I.; Kokotovic, P.V. Nonlinear and adaptive control. In *Lecture Notes in Control and Information*; Sciences: New York, NY, USA, 1995; pp. 511–514, ISBN 9780471127321.
31. Fredriksen, E.; Pettersen, K.Y. Global k-exponential way-point maneuvering of ships: Theory and experiments. *Automatica* **2009**, *42*, 677–687. [CrossRef]
32. Fossen, T.I. How to incorporate wind, waves and ocean currents in the marine craft equations of motion. In Proceedings of the 9th IFAC Conference on Manoeuvring and Control of Marine Craft, Arenzano, Italy, 19–21 September 2012; pp. 126–131.

MDPI
St. Alban-Anlage 66
4052 Basel
Switzerland
Tel. +41 61 683 77 34
Fax +41 61 302 89 18
www.mdpi.com

*Applied Sciences* Editorial Office
E-mail: applsci@mdpi.com
www.mdpi.com/journal/applsci

www.ingramcontent.com/pod-product-compliance
Lightning Source LLC
LaVergne TN
LVHW071934080526
838202LV00064B/6607
* 9 7 8 3 0 3 9 2 1 9 1 6 2 *